MW00837925

The Way of Analysis

Jones and Bartlett Books in Mathematics

Advanced Calculus, Revised Edition
Lynn H. Loomis and Shlomo Sternberg

Calculus with Analytic Geometry, Fourth Edition
Murray H. Protter and Philip E. Protter

College Geometry
Howard Eves

Differential Equations: Theory and Applications
Ray Redheffer

Discrete Structures, Logic, and Computability
James L. Hein

Fundamentals of Modern Elementary Geometry
Howard Eves

Introduction to Differential Equations
Ray Redheffer

Introduction to Fractals and Chaos
Richard M. Crownover

Introduction to Numerical Analysis
John Gregory and Don Redmond

Lebesgue Integration on Euclidean Space
Frank Jones

Logic, Sets, and Recursion
Robert L. Causey

Mathematical Models in the Social and Biological Sciences
Edward Beltrami

The Poincaré Half-Plane: A Gateway to Modern Geometry
Saul Stahl

The Theory of Numbers: A Text and Source Book of Problems
Andrew Adler and John E. Coury

Wavelets and Their Applications
Mary Beth Ruskai, et al.

The Way of Analysis
Robert S. Strichartz

The Way of Analysis

Robert S. Strichartz

Mathematics Department
Cornell University
Ithaca, New York

Jones and Bartlett Publishers

Boston London

Editorial, Sales, and Customer Service Offices
Jones and Bartlett Publishers
One Exeter Plaza
Boston, MA 02116
1-800-832-0034
617-859-3900

Jones and Bartlett Publishers International
7 Melrose Terrace
London W6 7RL
England

Library of Congress Cataloging-in-Publication Data

Strichartz, Robert S.
The way of analysis/ Robert S. Strichartz.
p. cm.
Includes Index.
ISBN 0-86720-471-0
1. Mathematical analysis. I. Title.
QA300.S888 1995
515–dc20 95-1913

CIP

The quotations in problem 1 in problem set 1.1.3 (page 7) are of proverbial or biblical origin, except for a. from "Home on the Range," by Brewster M. Higley, and g. from *Macbeth*, by William Shakespeare.

Printed in the United States of America
99 98 97 96 95 10 9 8 7 6 5 4 3 2 1

Contents

Preface

> Do not ask permission to understand.
> Do not wait for the word of authority.
> Seize reason in your own hand.
> With your own teeth savor the fruit.

Mathematics is more than a collection of theorems, definitions, problems and techniques; it is a way of thought. The same can be said about an individual branch of mathematics, such as analysis. Analysis has its roots in the work of Archimedes and other ancient Greek geometers, who developed techniques to find areas, volumes, centers of gravity, arc lengths, and tangents to curves. In the seventeenth century these techniques were further developed, culminating in the invention of the calculus of Newton and Leibniz. During the eighteenth century the calculus was fashioned into a tool of bold computational power and applied to diverse problems of practical and theoretical interest. At the same time the foundation of analysis—the logical justification for the success of the methods—was left in limbo. This had practical consequences: for example, Euler—the leading mathematician of the eighteenth century—developed all the techniques needed for the study of Fourier series, but he never carried out the project. On the contrary, he argued in print *against* the possibility of representing functions as Fourier series, when this proposal was put forth by Daniel Bernoulli, and his argument was based on fundamental misconceptions concerning the nature of functions and infinite series.

In the nineteenth century, the problem of the foundation of analysis was faced squarely and resolved. The theory that was developed forms most of the content of this book. We will describe it in its logical order, starting from the most basic concepts such as sets and numbers and building up to the more involved concepts of limits, continuity, derivative, and integral. The actual historical order of discovery was almost the reverse; much like peeling a cabbage, mathematicians began with the outermost layers and worked their way inward. Cauchy and Bolzano began the process in the 1820s by developing the theory of functions without defining the real numbers. The first rigorous definition of the real number system came in the work of Dedekind, Weierstrass, and Heine in the 1860s. Set theory came later in the work of Cantor, Peano, and Frege.

The consequences of the nineteenth century foundational work were enormous and are still being felt today. Perhaps the least important consequence was the establishment of a logically valid explanation of the calculus. More important, with the clearing away of the conceptual murk, new problems emerged with clarity and were developed into important theories. We will give some illustrations of these new nineteenth century discoveries in our discussions of differential equations, Fourier series, higher dimensional calculus, and manifolds. Most important of all, however, the nineteenth century foundational work paved the way for the work of the twentieth century. Analysis today is a subject of vast scope and beauty, ranging from the abstract to the concrete, characterized both by the bold computational power of the eighteenth century and the logical subtlety of the nineteenth century. Most of these developments are beyond the scope of this book or at best merely hinted at. Still, it is my hope that the reader, after having entered so deeply along the way of analysis, will be encouraged to continue the study.

My goal in writing this book is to communicate the mathematical ideas of the subject to the reader. I have tried to be generous with explanations. Perhaps there will be places where I belabor the obvious, nevertheless, I think there is enough truly challenging material here to inspire even the strongest students. On the other hand, there will inevitably be places where each reader will find difficulties in following the arguments. When this happens, I suggest that you write your questions in the margins. Later, when you go over the material, you may find that you can answer the question. If not, be sure to ask your instructor or another student; often, it is a minor misunderstanding that causes confusion and can easily be cleared up. Sometimes, the inherent difficulty of the material will demand considerable effort on your part to attain understanding. I hope you will not become frustrated in the process; it is something which all students of mathematics must confront. I believe that what you learn through a process of struggle is more likely to stick with you than what you learn without effort.

Understanding mathematics is a complex process. It involves not only following the details of an argument and verifying its correctness, but seeing the overall strategy of the argument, the role played by every hypothesis, and understanding how different theorems and definitions fit together to create the whole. It is a long-term process; in a sense, you cannot appreciate the significance of the first theorem until you have learned the last theorem. So please be sure to review old material; you may find the chapter summaries useful for this purpose. The mathematical ideas presented in this book are of fundamental importance, and you are sure to encounter them again in further studies in both pure and applied mathematics. Learn them well and they will serve you well in the future. It may not be an easy task, but it is a worthy one.

To the Instructor

This book is designed so that it may be used in several ways, including

1. a one-semester introductory real analysis course,
2. a two-semester real analysis course not including Lebesgue integration,
3. a two-semester real analysis course including an introduction to Lebesgue integration.

There are many optional sections, marked with an asterisk ($*$), that can be covered or omitted at your discretion. There is some flexibility in the ordering of the later chapters. Thus you can design a course in accordance with your interests and requirements. There are three chapters on applications (Chapter 11, Ordinary Differential Equations; Chapter 12, Fourier Series; and Chapter 13, Implicit Functions, Curves and Surfaces). These topics are often omitted, or treated very briefly, in a real analysis course because they are covered in other courses. However, they serve an important purpose in illustrating how the abstract theory may be applied to more concrete situations. I would urge you to try to fit as much of this material as time allows into your course.

The chapters may be divided into four groupings:

1. functions of one variable: 1, 2, 3, 4, 5, 6, 7, 8;

2. functions of several variables: 9, 10, 15;

3. applications: 11, 12, 13;

4. Lebesgue integration: 14, 15.

Note that Chapter 15, Multiple Integrals, may be used either with or without the Lebesgue integral.

The first 10 chapters are designed to be used in the given order (sections marked with an asterisk may be omitted or postponed). If you are not covering the Lebesgue integral, then selections from Chapter 15 (15.1.1, 15.2.1, and 15.2.3) can be covered any time after Chapter 10. The applications, Chapters 11, 12, and 13, can be done in any order. It is advisable to do at least some of Chapter 12, Fourier series, before doing the Lebesgue Integral, Chapter 14. In Chapter 6 (section 6.1.3) I have included a preview of some results in integration theory that are covered in detail later in the book—this is the only place I have violated the principle of presenting full proofs of all results in the order they are discussed. I think this is a reasonable compromise, in view of the facts that (a) the students will want to use these results in doing exercises, and (b) to present proofs at this point in the text would require long detours.

Here are some concrete suggestions for using this book.

1. *One-semester course*: do Chapters 1–8 in order, omitting all sections marked with an asterisk. This will cover the one variable theory. If time remains at the end, return to omitted sections.

2. *Two-semester course* (*without Lebesgue integrals*)

   First semester: do chapters 1–7 in order, including most sections marked with an asterisk.

Second semester: do Chapters 8–10; then 15.1.1, 15.2.1, 15.2.3, then Chapters 11–13, including some sections marked with an asterisk.

3. *Two-semester course (with Lebesgue integrals)*

   First semester: do Chapters 1–7 in order, including most sections marked with an asterisk, but omit 6.2.3.

   Second semester: do Chapters 8–10; then selections from chapters 11–13; then chapters 14 and 15.

This book contains a generous selection of exercises, ranging in difficulty from straightforward to challenging. The most difficult ones are marked with an asterisk.

All the main results are presented with complete proofs; indeed the emphasis is on a careful explanation of the ideas behind the proofs. One important goal is to develop the reader's mathematical maturity. For many students, a course in real analysis may be their first encounter with rigorous mathematical reasoning. This can be a daunting experience but also an inspiring one. I have tried to supply the students with the support they will need to meet the challenge.

My recommendation is that students be required to read the material before it is discussed in class. (This may be difficult to enforce in practice, but here is one suggestion: have students submit brief written answers to a question based on the reading and also a question they would like to have answered in class.) The ability to read and learn from a mathematical text is a valuable skill for students to develop. This book was written to be read—not deciphered. If I have perhaps coddled the students too much, I'm sure they won't complain about that!

The presentation of the material in this book is often informal. A lot of space is given to motivation and a discussion of proof strategies. Not every result is labeled as a theorem, and sometimes the precise statement of the result does not emerge until after the proof has been given. Formulas are not numbered, and theorems are referred to by name and not number. To compensate for the informality of the body of the text, I have included summaries at the end of each chapter (except the first) of all the main results, in standard dry mathematical format. The students should find these chapter summaries handy both for review purposes and for references.

I have tried to give some historical perspective on the material presented, but the basic organization follows logical rather than historical order. I use conventional names for theorems, even if this perpetuates injustices and errors (for example, I believe it is more important to know what the Cauchy-Schwartz inequality is than to decide whether or not Bunyakowsky deserves some/most/all of the credit for it). One important lesson from the historical record is that abstract theorems did not grow up in a vacuum: they were motivated by concrete problems and proved their worth through a variety of applications. This text gives students ample opportunity to see this interplay in action, especially in Chapters 11, 12, and 13.

In order to give the material unity, I have emphasized themes that recur. Also, many results are presented twice, first in a more concrete setting. For example, I develop the topology of the real line first, postponing the general theory of metric spaces to Chapter 9. This is perhaps not the most efficient route, but I think it makes it easier for the students. Whenever possible I give the most algorithmic proof, even if it is sometimes harder (for example, I construct a Fourier series of a continuous function that diverges at a point). I have tried to emphasize techniques that can be used again in other contexts. I construct the real number system by Cauchy completion of the rationals, since Cauchy completion is an important technique. The derivative in one variable is defined by best affine approximation, since the same definition can be used in $\mathbb{R}^n$. Chapter 7, Sequences and Series of Functions, is presented entirely in the context of functions of one variable, even though most of the results extend easily to the multivariable setting, or more generally to functions on metric spaces.

A good text should make the job of teaching easier. I hope I have succeeded in providing you with a text that you can easily teach from. I would appreciate receiving any comments or suggestions for improvements from you.

Acknowledgments

I would like to thank the students at Cornell who learned this material from me as I developed the preliminary versions of the text over the past dozen years. I am grateful for your criticism as well as your encouragement. I needed both, even if I was more gracious about the encouragement (special thanks to Chris Wittemann). I am also grateful to Graeme Bailey, Dan Barbasch, Eugene Dynkin, Archil Gulisashvili, and Oscar Rothaus, who also taught using this material.

I am especially grateful to Mark Barsamian, who went over the text line-by-line from the point of view of the student, and made me change just about everything. He helped me to improve the organization and style of presentation and pointed out possible sources of ambiguity and confusion. I know that the text is much stronger as a result of his criticism, and I feel more confident that I have fulfilled my promise to write a book that students can understand.

I am grateful to Carl Hesler, my editor at Jones and Bartlett, who has encouraged me throughout the long process of turning my rough lecture notes into a polished book, and I especially appreciate his confidence in the value of this work for the mathematical community.

I would like to thank June Meyermann for her outstanding job preparing the manuscript in LaTeX and David Larkin who produced many of the figures using Mathematica.

Chapter 1

Preliminaries

1.1 The Logic of Quantifiers

1.1.1 Rules of Quantifiers

Logic plays a central role in mathematics. While other considerations—such as intuition, agreement with empirical evidence, taste, esthetics, wishful thinking, personal ambition—may influence the way mathematicians think and act, there is always a central core of mathematical reasoning that is supposed to be logically sound. But is it? There are really two questions here. The first concerns the structure of the logic itself; the second concerns how it is used. Mathematicians have made a careful study of logic, and the results are rather impressive. The main result, called the *completeness theorem for first-order predicate calculus*, shows that the logical reasoning we are going to use in this book, if used correctly, is both sound and incapable of being improved (in other words, additional new forms of logical reasoning would not enable us to do more than we already can). The interested reader is referred to any text on mathematical logic for a full discussion, which is beyond the scope of this work.

But what about the second question. Do mathematicians make mistakes in logic? Of course they do! Carelessness and wishful thinking are the main culprits, and there is little that can be done other than constant vigilance and rechecking of work. At the very least, however, every mathematician and student of mathematics should clearly understand the rules of logic so that deliberate mistakes are not made.

Ignorance of logic is no excuse! Some of the theorems we are going to study in this book were originally stated incorrectly or given incorrect proofs because of misunderstandings concerning the logic of quantifiers. So our first responsibility is to make perfectly clear the meaning and use of "there exists" and "for all". Of course preliminary to this we need the logic of connectives. Fortunately, this is a straightforward matter, summarized by the following truth table.

TRUTH TABLE

A	B	not A	A and B	A or B	A implies B	A if and only if B
T	T	F	T	T	T	T
T	F	F	F	T	F	F
F	T	T	F	T	T	F
F	F	T	F	F	T	T

We will not employ special symbols such as $\wedge$ for "and", $\rightarrow$ for "implies", but we will use the usual circumlocutions such as "if A then B" for "A implies B". Notice that the use of connectives is purely finitistic; although the individual statements may refer to the infinite, the use of connectives does not introduce any new infinite reference. The statement "A or B" means A is true or B is true (or both), whatever A or B may be. Of course a proof of "A or B" may not tell us which of the two is true. Notice that the double negative "not not A" is identical to the statement "A". The statement "A implies B" is logically equivalent to its *contrapositive* "not B implies not A" but is distinct from its *converse* "B implies A". The logic of connectives, also called the *propositional calculus*, will be assumed as part of the mathematical background of the reader.

Quantifiers introduce the infinite, potential or actual, into our finite language. We can write "$2+1 = 1+2$, $2+2 = 2+2$, $2+3 = 3+2$, $2+4 = 4+2, \ldots$" and assume that the three dots are self-explanatory, or we can write "for all natural numbers $x, x + 2 = 2 + x$". In either case, the finite sentence we have written is supposed to convey an infinite amount of information. The infinite enters because the set over which x varies—the natural numbers—is infinite. In this context we can, if we wish, regard this infinite as only potential; any individual will experience only a finite number of instances of $x+2=2+x$. However, in order to understand the way of analysis, we will have to accept the completed infinite. This may create problems, both mathematical and psychological, but it is inescapable.

In this book, the quantifiers used will always refer to variables whose domains are clearly specified sets. Thus we may say "for all sets of sets of integers", but we will not say "for all sets". The "for all" is the *universal* quantifier, and the "there exists" is the *existential* quantifier. The universal quantifier is a grand "and", while the existential quantifier is a grand "or". The statement "for all x in $U, A(x)$" where U is a prescribed set and A is a statement in which the variable x appears has the meaning that $A(x)$ is true for every value of x in U. If the elements in U could be enumerated $u_1, u_2, u_3, \ldots$, then "for all x in $U, A(x)$" would mean "$A(u_1)$ and $A(u_2)$ and $A(u_3)$ and ...". If the set U is finite, then "for all x in $U, A(x)$" is exactly the finite conjunction. Similarly "there exists x in U such that $A(x)$" means $A(x)$ is true for at least one x in U or "$A(u_1)$ or $A(u_2)$ or ..." if $u_1, u_2, \ldots$ enumerates U.

To deny a universal statement we need find only one counterexample, so the negation of "for all x in $U, A(x)$" is the existential statement "there exists x in U such that not $A(x)$". But to negate an existential statement we must show that every possible instance is false, so the negation of "there exists x in U such that $A(x)$" is the universal "for all x in U, not $A(x)$". This is to be expected if we recall that the negation of a conjunction (and) is the disjunction (or) of the negations, and the negation of a disjunction is the conjunction of the negations (universal and existential quantifiers being grand conjunctions and disjunctions, respectively). To paraphrase: *commuting a quantifier with a negation changes the type of quantifier.*

The real fun comes when there is more than one quantifier in a statement. For example, the commutative law of addition of integers can be stated as follows: for all integers x, for all integers $y, x+y=y+x$. Clearly the order of the quantifiers is immaterial, and in virtue of this fact we will abbreviate the statement as follows: for all integers x and y, $x+y=y+x$. Similarly, the order of two or more consecutive existential quantifiers is immaterial, so we say "there exist integers x and y such that $x+y=2$ and $x+2y=3$". To paraphrase, *quantifiers of like type commute.* But the situation is different for multiple quantifiers of differing type. Consider the statement: everyone has a mother. We can express this via quantifiers as follows: for every person x there exists a person y such that y is the mother of x. If we reverse the order of quantifiers the statement becomes: there exists a person y such that for every person x, y is the mother of x. Clearly this is not the same statement as before; rather it says "y is the mother of us all". This is a stronger statement, less likely to be true; in fact, it is false.

Since this is a crucial point, it is worth looking into more closely. The statement "there exists y in U such that for every x in $V, A(x, y)$" asserts that one y will make $A(x, y)$ true no matter what x is. The statement "for every x in V there exists y in U such that $A(x, y)$" asserts only that $A(x, y)$ can be made true by choosing y depending on x. It really asserts the existence of a function $y = f(x)$ (in the example above we can call it the mother function) such that $A(x, f(x))$ is true. For the existential-universal form the function must be constant. To paraphrase: *the existential-universal implies the universal-existential but not vice versa*; *the universal-existential is equivalent to asserting the existence of a function from the domain of the universally quantified variable to the domain of the existentially quantified variable.*

1.1.2 Examples

To understand the way of analysis we have to deal with sentences that contain many quantified variables; we have to understand what these sentences mean and how to form their negation. This involves only applying the above principles several times; but it can get confusing, especially when a lot of intricate mathematical ideas are involved at the same time. To gain some confidence with the process we will examine some examples where the mathematics is simple.

Goldbach's conjecture states: every even natural number greater than 2 is the sum of two primes. This can be written: for every x in the set E of even natural numbers greater than 2, there exists p in the set P of prime natural numbers and there exists q in P such that $x = p + q$. Since the last two consecutive quantifiers are both existential, we can combine them: for every x in E there exists p and q in P such that $x = p + q$. To make the dependence of p and q on x clearer we could write them as functions $p(x)$ and $q(x)$ from E to P, and Goldbach's conjecture becomes: there exist functions $p(x)$ and

$q(x)$ from E to P such that $x = p(x) + q(x)$.

What is the negation of Goldbach's conjecture? We can calculate it in stages as follows: (not) (for every x in E) (there exist p and q in P) $(x = p + q)$ is equivalent to (there exists x in E) (not) (there exists p and q in P) $(x = p+q)$ is equivalent to (there exists x in E) (for all p and q in P) (not) $(x = p + q)$. In other words, there exists x in E such that for all p and q in P, $x \neq p+q$. The negation of Goldbach's conjecture thus asserts the existence of a counterexample: an even number x (greater than 2) that is not the sum of two primes. Looking back at our computation of this rather simple fact, we see that it was a bit long-winded. Clearly to form the negation of a statement with a string of quantifiers we simply change the type of each quantifier, preserving the order, and negate whatever follows the string of quantifiers. From now on we will do this without further ado.

Goldbach's conjecture being rather difficult—no one has succeeded in proving or disproving it—mathematicians have looked at some weaker statements (also without success) in the same vein. For example, "Goldbach's conjecture has at most a finite number of counterexamples". This can be written several ways. We can say, there exists a natural number n such that for all x in E there exist p and q in P such that $x \leq n$ or $x = p + q$. This is perhaps a somewhat artificial form in that the p and q whose existence are asserted for $x \leq n$ are completely arbitrary and irrelevant. However, it has the advantage of placing all the quantifiers first. A second version is: there exists a natural number n such that for all x in E, $x > n$ implies there exist p and q in P such that $x = p + q$. A third version is: there exists a natural number n such that for all x in E_n, the set of even numbers greater than n, there exist p and q in P such that $x = p + q$.

What is the negation of the sentence "Goldbach's conjecture has at most a finite number of counterexamples"? From the first version we find: for all natural numbers n there exists x in E such that for all p and q in $P, x > n$ and $x \neq p + q$—in other words, the existence of counterexamples greater than any prescribed n. (Here we have used the propositional logic equivalence of not (A or B) and not A and not B.)

Another weakening of Goldbach's conjecture can be formulated as "every even number is the sum of at most a fixed number of primes". We can write this as follows: there exists a natural number k such that for every x in E there exist $p_1, p_2, \ldots, p_j$ in P such that $j \leq k$ and $x = p_1 + p_2 + \cdots + p_j$. The negation of this statement is: for every natural number k there exists x in E such that for all $j \leq k$ and $p_1, \ldots, p_j$ in P, $x \neq p_1 + p_2 + \cdots + p_j$.

Here is a statement with four consecutive quantifiers: for every natural number n there exists a natural number m such that for every natural number x there exist non-negative integers $a_1, a_2, \ldots, a_m$ such that $x = a_1^n + a_2^n + \cdots + a_m^n$. What does it mean? It says that given n, there is a number m depending on n—we might write $m(n)$—such that every number is the sum of $m(n)$ n-th powers. This statement is known as Waring's problem, and there are some rather difficult proofs of it. Its negation would say: there exists a natural number n such that for every natural number m there exists a natural number x such that for all non-negative numbers $a_1, a_2, \ldots, a_m$, $x \neq a_1^n + a_2^n + \cdots + a_m^n$.

In deciphering complicated strings of quantifiers it is amusing to imagine a game played with the devil. Every time an existential quantifier appears it is your move, and every time a universal quantifier appears it is the devil's move. You make a choice of the quantified variable from the specified set that makes things as good as possible, while the devil does his best to mess you up. If you have a strategy to beat the devil every time, then the statement is true; otherwise, it is false.

A *direct proof* of a quantified statement is one that gives the strategy for beating the devil explicity. However, we will also allow *indirect proofs*, in which we only prove (by contradiction) that such a strategy must exist. Clearly a direct proof is preferable, if it can be found. There is a minority school of thought, called *Intuitionism* or *Constructivism*, that holds that we should not accept indirect proofs. The majority of mathematicians reject this as counterproductive, for the following reason: by allowing indirect proofs, we create a much richer mathematical world, that contains more direct proofs than would have been discovered (presumably) had indirect proofs been rejected outright.

In the best of all possible worlds (from the point of view of students), mathematicians would be required to write all quantified sentences with the quantifiers at the beginning of the sentence (in the correct order). Similarly, all impliction sentences (or statements of theorems) would have the hypotheses first and the conclusions second. In this world, they don't do it that way! The English language allows many different forms of expression for the same ideas, and mathematicians help themselves freely. This is bound to cause confusion. As you do the exercises and look back at the examples already discussed, try to develop the insight to recognize hidden quantifiers or transposed word orders. It is a skill that will serve you well in what follows.

1.1.3 Exercises

1. For each of the following famous sayings, rewrite the statement making all the quantifiers explicit. Then form the negation of the statement. Finally, recast the negation in a form similar to the original saying.

   a. The skies are not cloudy all day.

   b. Man cannot live by bread alone.

   c. The sun never sets on the British Empire.

   d. To every thing there is a season, and a time for every purpose under heaven.

   e. The devil makes work for idle hands.

   f. Sufficient unto the day is the evil thereof.

   g. All our yesterdays have lighted fools the way to dusty death.

2. For each of the following mathematical statements, rewrite the statement making all the quantifiers explicit. Then form the negation of the statement. Finally recast the negation in a form similar to the original statement.

   a. Every positive integer has a unique prime factorization.

   b. The only even prime is 2.

   c. Multiplication of integers is associative.

   d. Two points in the plane determine a line.

   e. The altitudes of a triangle intersect at a point.

   f. Given a line in the plane and a point not on it, there exists a unique line passing through the given point parallel to the given line.

g. Any partitioning of the integers into a finite number of disjoint subsets has the property that one of the subsets contains arbitrarily long arithmetic progressions.

h. If there are more letters than mailboxes, at least one mailbox must get more than one letter.

3. Each of the following true statements is in universal-existential form. Write the corresponding statement with the order of quantifiers reversed, and show why it is false.

   a. Every line segment has a midpoint.

   b. Every non-zero rational number has a rational reciprocal.

   c. Every non-empty subset of the positive integers has a smallest element.

   d. There is no largest prime.

1.2 Infinite Sets

1.2.1 Countable Sets

Set theory, like logic, plays a central role in mathematics today. In contrast to the settled state of logic, however, we are in a state of grievous ignorance about some of the basic properties of sets and are likely to remain so. The reason for this is that our finite minds have difficulty penetrating to the core of the infinite; at best we can play with pebbles on the shore of a vast ocean. Fortunately we possess some mighty pebbles that will enable us to enter the way of analysis. Some of the material in the section may be familiar to you, but in any case it is worth reviewing because of its importance.

The basis for our intuition of the infinite is the set of natural numbers $1, 2, 3, \ldots$. As Galileo observed, this set can be put in one-to-one correspondence with a proper subset, say the even numbers—an observation that convinced Galileo that he shouldn't toy with the infinite. Cantor was braver, adopting the definition that two sets have the same *cardinality* if their elements can be put in one-to-one correspondence and facing squarely the fact that infinite sets can have the same cardinality as proper subsets. For finite sets this cannot happen, of course.

This leads immediately to the question: do all infinite sets have the same cardinality? A set with the same cardinality as the natural numbers is called *countable*; the one-to-one correspondence with $1, 2, 3, \ldots$ amounts to a counting or enumeration of the set. The elements of a countable set can be listed $u_1, u_2, u_3, \ldots,$ although the order of the elements may have nothing to do with any relationships between the elements. The listing is merely a convenient way of displaying the one-to-one correspondence with the natural numbers. In order to construct a set that is not countable (*uncountable*) we attempt to build larger sets by the natural processes of set theory.

Suppose we have two countable sets, $A = (a_1, a_2, \ldots)$ and $B = (b_1, b_2, \ldots)$. Is their union countable? Clearly we can splice the two listings $a_1, b_1, a_2, b_2, \ldots$ to obtain a listing of $A \cup B$. This listing may involve duplication, since we have not assumed the sets A and B are disjoint, so it may not be a one-to-one correspondence. However, the problem is easily fixed by tossing out the duplications as they arise. In other words we use the simple lemma: *if there is a mapping of the natural numbers onto a set U (not necessarily one-to-one), then U is either finite or countable.* In this case $A \cup B$ cannot be finite since it contains the infinte subset A.

A similar argument shows that the union of any finite number of countable sets is countable. What about the union of a countable number of countable sets? Suppose $A_1, A_2, A_3, \ldots$ are sets and each one is countable. Let us denote the elements of A_1 by $a_{11}, a_{12}, \ldots,$ the elements of A_2 by $a_{21}, a_{22}, \ldots,$ and in general the elements of A_k by $a_{k1}, a_{k2}, \ldots$. We can then write all the elements of the union $\bigcup_{k=1}^{\infty} A_k$ in an infinite matrix as

$$\begin{matrix} a_{11} & a_{12} & a_{13} & \cdots \\ a_{21} & a_{22} & a_{23} & \cdots \\ a_{31} & a_{32} & a_{33} & \cdots \\ \cdots & \cdots & \cdots & \cdots \end{matrix}$$

The k-th row of this matrix enumerates the set A_k. Of course the matrix may contain duplications, but this can be handled as before. The question of the countability of the union set then boils down to the question: are the elements of the above infinite matrix countable? The answer is obtained by turning your head 45° counterclockwise (or equivalently turning the paper 45° clockwise). The matrix then looks like

$$\begin{array}{c} a_{11} \to \\ \to a_{21} \to a_{12} \to \\ \to a_{31} \to a_{22} \to a_{13} \to \\ \vdots \quad \vdots \end{array}$$

an infinite triangle and can be counted by following the arrows. Thus we cannot construct an uncountable set by the process of unions. Incidentally, the same argument shows that the Cartesian product $A \times B$ of two countable sets is also countable (recall that the Cartesian product $A \times B$ is the set of all ordered pairs (a, b) where a varies over A and b varies over B). One can similarly show that any finite Cartesian product $A_1 \times A_2 \times \cdots \times A_k$ of countable sets is countable, but the same is not true of the Cartesian product of a countable number of countable sets (see exercise set 1.2.3, number 5).

1.2.2 Uncountable Sets

To escape beyond the countable we need a more powerful set operation, called the *power set.* If A is any set, then the *power set*, 2^A, is the set of all subsets of A. That is, the elements x of 2^A are the subsets of A (in other words, x is a set, all of whose elements are elements of A). The reason behind the notation 2^A is that we can "parametrize" the subsets of A by attaching a two-element set, say $(\text{Yes}_a, \text{No}_a)$, to each element a of A. A particular subset x of A is then uniquely determined by a choice of one of the two, Yes_a if a is in x or No_a if a is not in x, for each a in A. Alternatively we can think of the choice Yes_a or No_a as giving a function from A to the two-element set (Yes, No). It is important to understand that the choice Yes_a or No_a is completely arbitrary—we do not assume that it is given by a particular rule that we could describe, even potentially, in our finite language. This is essential because the number of subsets that are potentially describable in any finite language are at most countable and rather limited in many ways (when the set A is the natural numbers then these describable sets are called *recursive*). To consider 2^A for any infinite set A is thus to admit into mathematics an object well beyond the scope of our full comprehension. This is an order of magnitude greater than the admission of the completed infinite of the natural numbers, because every natural number has a potential name in our finite language.

Cantor's proof that 2^N is uncountable, where N is the set of natural numbers, goes as follows: Suppose 2^N is countable; then we will obtain a contradiction. Let $u_1, u_2, u_3, \ldots$ be the supposed enumeration of 2^N. This means that each u_k is a set of natural numbers, and every set of natural numbers must appear on the list. The contradiction arises from the fact that one can now describe a set v that is not on the list. The set v is the set that contains the number k if and only if u_k does not contain k. Symbolically, $v = \{k : k \text{ is not in } u_k\}$. The set v is unambiguously defined, in fact by a statement in our finite language, but the definition of v depends on the particular enumeration $u_1, u_2, u_3, \ldots$. Now the very construction of v guarantees that it does not appear on the list, because it differs from u_k in the matter of the number k; one contains k and the other doesn't. This contradicts the fact that $u_1, u_2, u_3, \ldots$ was supposed to list every subset of N, and thus the impossibility of such an enumeration is demonstrated.

Despite the simplicity of Cantor's proof, it contains some subtle points that deserve to be explicated. For example, how come we can't repair the error of having left out the set v simply by inserting it somewhere on the list, say first: $v, u_1, u_2, \ldots$? Now we have an enumeration that contains v. And yet, doesn't the same argument show that v is not on the list? The key to overcoming these sophistries is to remember that the definition of v depended on the particular enumeration. Thus the set constructed by the argument from the new enumeration $v, u_1, u_2, \ldots$ is a different set; call it v'. We could of course add v' to the enumeration, but then the argument would construct still another set. Thus in fact what the proof shows is that given any countable set $u_1, u_2, u_3, \ldots$ of subsets of N, it is possible to construct a subset of N different from all of them. This construction is often called a *diagonalization* argument, since if we represent the enumeration $u_1, u_2, u_3, \ldots$ by an infinite matrix of Yes's and No's, with a Yes in the $j - k$ place meaning that j is in u_k, then v is obtained by changing all the diagonal entries. This kind of argument is used quite frequently in mathematics. We will also encounter later a different kind of diagonalization argument that shows how we can pass to a subsequence a countable number of times.

The conclusion that 2^N is of greater cardinality than N (we say that a set A has greater cardinality than a set B if A cannot be put in one-to-one correspondence with B, but a proper subset of A can be put in one-to-one correspondence with B) raises the interesting question of whether there are sets of intermediate cardinality. In other words, does there exist a subset A of 2^N that is of greater cardinality than N but of lesser cardinality than 2^N; or on the contrary, must every infinite subset A of 2^N be capable of being put in one-to-one correspondence with either N or 2^N, as Cantor's famous Continuum Hypothesis asserts? We now have very convincing evidence, in the work of Kurt Gödel and Paul Cohen, that this question will never be answered. Gödel showed that the Continuum Hypothesis cannot be disproved, and Cohen showed that it cannot be proved. They constructed different "models" of set theory in which all the usual axioms of set theory are valid and the Continuum Hypothesis is in one case true and in another case not true. The usual interpretation of these results is that we do not have enough axioms for set theory to distinguish between these wildly different "set theories". But we don't have a clue where to search for new axioms for set theory nor any legitimate way to decide whether or not to accept new axioms.

It is possible that some intuitively appealing set-theoretic axiom may yet be discovered that will settle the Continuum Hypothesis, but at present this is just idle speculation. At any rate, we have no right to expect that our finite reasoning can fully illuminate the uncountable infinite.

Fortunately, for the analysis we are going to do in this book, these set-theoretic questions will not enter. In fact one might say that the way of analysis consists of using the finite to get at the countably infinite (by constructing arbitrarily large finite segments of countable sets) and then using the countably infinite to get at the uncountably infinite by an appropriate approximation process. The uncountably infinite will remain largely potential, just as in number theory the infinite is largely potential. Also, we will be dealing only with sets that are built up from the natural numbers in a finite number of steps; that is, we may deal with sets of sets of numbers or sets of sets of sets of numbers, and the like, although we will introduce abbreviations that will hide this fact. We will certainly never deal with such linguistic contortions as "the set of all sets", that leads to the famous paradoxes of naive set theory such as "the set of all sets not containing themselves". All the set theory we will use is capable of being axiomatized, although we shall not do so.

1.2.3 Exercises

1. Prove that every subset of N is either finite or countable. (Hint: use the ordering of N.) Conclude from this that there is no infinite set with cardinality less than that of N.

2. Is the set of all finite subsets of N countable or uncountable? Give a proof of your assertion.

3. Prove that the rational numbers are countable. (**Hint**: they can be written as the union over $k \in N$ of the sets $Q_k = \{\pm j/k : j \in N\}$.)

4. Show that if a countable subset is removed from an uncountable set, the remainder is still uncountable.

5. Let $A_1, A_2, A_3, \ldots$ be countable sets, and let their Cartesian product $A_1 \times A_2 \times A_3 \times \cdots$ be defined to be the set of all sequences $(a_1, a_2, \ldots)$ where a_k is an element of A_k. Prove that the Cartesian product is uncountable. Show that the same conclusion holds if each of the sets $A_1, A_2, \ldots$ has at least two elements.

6. Let A be a set for which there exists a function f from A to N with the property that for every natural number k, the subset of A given by the solutions to $f(a) = k$ is finite. Show that A is finite or countable.

7. Generalize the diagonalization argument to show that 2^A has greater cardinality than A for every infinite set A.

1.3 Proofs

1.3.1 How to Discover Proofs

This book is about proofs. Not only will you read the proofs presented in the text, but in the exercises you will be asked to prove things. *What is a proof? How do you go about finding one? How should you write a proof?* Presumably, these are questions you have thought about before in the course of your mathematical education. These are not easy questions; and although there is a reasonable consensus in the mathematical community on the answers, there is still some disagreement on where to put the emphasis.

In principle, a proof is a sequence of logical deductions from the hypotheses to the conclusion of the statement being proved. The statement is typically in the form of an implication, say "P and Q implies R". In this case there are two hypotheses, P and Q, and one conclusion, R. All terms involved in the expressions P, Q, and R should be previously defined. The first step in reading (or finding) a proof is to *make sure you understand all the hypotheses and the conslusions and remember the definitions of all the terms used.* Sometimes, important hypotheses are hidden in the fine print. *Read the fine print.*

The reasoning allowed in a proof involves the accepted principles of logical reasoning and the application of axioms or previously established theorems. A common source of error in proofs involves applying a theorem without verifying all the hypotheses of that theorem. This may result in an incorrect proof (if the theorem does not apply) or an incomplete proof (if the theorem does apply but some of the key ideas of the proof are involved in verifying this). Another common source of error involves making extra assumptions, beyond what is given in the statement. Also, if the conclusion of the statement says "for all $x \ldots$", it is not correct to prove the statement just for $x = 2$.

When faced with the task of finding a proof for a given statement, it is a good idea to make a list of *what you know* (the hypotheses, and sometimes also the definitions of key terms) and *what you want to show* (the conclusion, sometimes spelled out by supplying definitions of terms). Then search for theorems that connect the two. It is advisable to work simultaneously backward and forward. If your hypotheses contain the hypotheses of a theorem that you know, you may deduce the conclusion of that theorem. This is working forward. Or, you may know a theorem whose conclusion is the conclusion you are trying to show. Then the hypotheses of this theorem become the new target (what you have to show). This is working backward. Often, the fit between the theorem you want to use and the way you want to use it is not perfect. For example, in working forward, the hypotheses of the theorem you want to use may not all be exactly on your list of what you know, but you may be able to deduce them (with a little work) from what you know. In this way you can build a chain of deductions from your hypotheses and a chain of deductions to your conclusion.

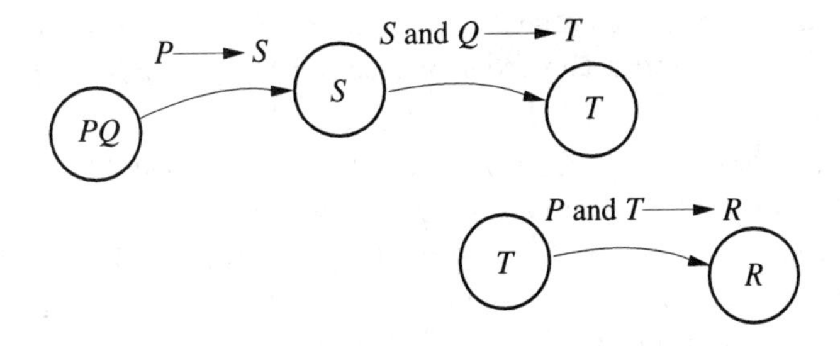

Figure 1.3.1:

If you are lucky, at some point these two chains will meet, as in Figure 1.3.1. Then you can join them to form a single forward chain of deduction from hypotheses to conclusion, the proof you sought. Such a simple strategy may be called "follow your nose". It should work for the simpler problems. Here is a good tip if you get stuck. A good problem will require that *all* the hypotheses be used in the proof. So if you are wondering what to do next, go down the list of what you know and see if anything hasn't been used yet. Then figure out a way to use it—that is, look for theorems that have it (or a consequence of it) as one hypothesis.

If "follow your nose" doesn't lead to a proof, you may have to resort to a "construction". This means that you introduce a new object that is not mentioned in either the hypotheses or conclusions of the statement. For example, if the statement concerns two numbers x and y, you might need to consider the average $z = 1/2(x + y)$. This is a simple example, but constructions can be much more elaborate; their discovery requires ingenuity and creativity. There is no set formula for how to go about the process, but you will find that experience is helpful. The goal of the construction is to make more theorems applicable to the problem. You may get a hint that a construction is needed if there is a theorem that seems likely to be relevant but doesn't quite apply to the objects mentioned in the problem.

After you have discovered a proof, and checked that it is a valid proof, you will want to write it up. There are two approaches you can take. You can write it in the order in which you discovered it, or you can try to rearrange the order of reasoning to "polish" the argument. For example, in Figure 1.3.1, you might first have observed that "P

and T imply R", then that "P implies S", and finally that "S and Q imply T". A proof in the order of discovery might read as follows:

Proof: We need to show R. But the Purple Peril Theorem says P and T imply R; and since we already know P, it suffices to show T to complete the proof. But observe that P implies S by the definition of grunginess, and Q and S imply T by the previous lemma. This completes the proof. QED

Or, you could rearrange the deductions to run forward. The result might read like this:

Proof: We are assuming P and Q. Then P implies S by the definition of grunginess, and Q and S imply T by the previous lemma. Thus we have both P and T, the hypotheses of the Purple Peril Theorem, whose conclusion is T. QED

The argument for preferring the rearranged proof is that it is easier to follow. On the other hand, the first proof indicates the reasoning used in the process of discovery and so may make more sense to the reader. A little common sense may help one decide between the two options (or a compromise, that retains some of the original order of discovery but does some pruning and straightening in between).

One important issue with which you will have to grapple is the amount of detail you include in the proof. Often there are very simple arguments that can either be omitted altogether or dismissed with a phrase like "it is obvious that". If all these simple arguments were included, the proof might become so long that the reader would "lose the forest for the trees". On the other hand, you don't want to leave out any non-trivial parts of the argument or make a mistake in reasoning. The ultimate test of a written proof is whether or not it convinces the reader. This, of course, depends on the reader! For a written assignment that is being graded, you had better include enough detail to convince the grader that you know how to supply the missing arguments. For the proofs presented in the text, I have tried to supply enough detail so that you can fill in whatever is omitted.

A word about *proof by contradiction* (also known as *indirect proof, reductio ad absurdum*, or *proof by the contrapositive*). It is an accepted logical principle that "not Q implies not P" is the same as "P implies Q". So to prove the statement "P implies Q", it is legitimate to start by assuming Q is false and reach a contradiction. Along the way you may use both the hypothesis P and the assumed not Q in the argument. Often this allows you to successfully find a proof that otherwise might be elusive. However, it is my experience that students tend to abuse proof by contradiction in the following way. The student will assume Q is false, then give an argument that produces Q as a consequence, and find the contradiction between Q and not Q. But in fact, buried in the middle of the student's proof will be a direct proof that P implies Q. Generally speaking, this might be the case with your proof if the contradiction involves Q and not Q. If so, check to see whether you really need not Q in your argument. If you didn't use not Q, then it is preferable to present the proof as a direct proof.

Why is the direct proof preferred over the indirect proof? Usually a direct proof is more constructive. If the conclusion is an existential statement, "there exists x such that ...", the direct proof may include an algorithm for finding x. This gives more information than simply showing that the non-existence of such an x always leads to a contradiction.

1.3.2 How to Understand Proofs

One of the goals of this book is to get you to understand the proofs of the theorems presented. There are many levels of understanding, and they will involve considerable effort on your part. On the first level of understanding, you should be able to read through the proof and be convinced that the reasoning is correct. There must be no "proof by intimidation". You have to see all the individual steps of the argument; and this means going through the proof line by line, checking all the details that are presented, and filling in the details that are left out. I have tried to present the proofs clearly, but it will probably happen that some statements are not clear to you on first reading. Put a mark in the text when this happens and be sure to return to these points and clear them up, either by yourself or with the help of someone else.

The second level of understanding involves grasping the structure of the proof. What are the main theorems that are used? What constructions are involved? What are the main difficulties, and how are they overcome? You should have in your mind an outline of the proof, with some hints as to how to fill in the details. Without actually having to memorize anything, you should retain enough to be able to reconstruct the proof in your own words without referring to the book.

The third level of understanding involves seeing how every hypothesis enters into the proof. You should be able to point out where each hypothesis is used; better still, you should be able to give a counterexample to the theorem without that hypothesis. You should explore the possibilities of weakening the hypotheses or strengthening the conclusion. Is the result still true, or can you find a counterexample? If the proof involves a construction, try to understand why that particular construction was chosen. In short, this level of understanding requires that you take the proof apart and put it back together to see how it ticks (this cliché refers to the distant past when watches actually ticked, and you could actually take them apart). When you have achieved this level of understanding, you should have no difficulty remembering the exact statement of the theorem.

There is another level of understanding, although this refers to the theorem as a whole, not just the proof. This level refers to the way the theorem is used and how it can be generalized. Obviously, you cannot achieve this level right away. Working the problems will help give you a feel for what is involved in using a theorem. By the time you reach the end of this book you will have seen how many of the theorems in the beginning of the book are used. As you continue your mathematical education you will see how the themes developed in this book recur and how theorems are used and generalized. By the same token, you should find as you study this book that your understanding of previously learned mathematics is broadened and deepened. In mathematics, truly, nothing should be forgotten.

1.4 The Rational Number System

We take the rational number system as our starting point in the construction of the real number system. We could, of course, give a detailed construction of the rational numbers in terms of more primitive notions. However, every mathematical work must start somewhere, with some common notions that are accepted without formal development. It is appropriate, since we are entering upon the way of analysis, that is characterized by infinite processes, that we begin with a system that is essentially finite and yet is as close to our goal as possible—so that we do not excessively weary ourselves with preliminaries. The rational number system serves this purpose admirably. (There are an infinite number of rationals, but we do not require infinite notions to define the algebraic and order structures of the rationals.)

It is assumed that the reader has had ample experience in dealing with the rational number system and feels comfortable with its properties. We give here a summary of the properties we will be using.

A rational number is a number of the form p/q, where p and q are integers and q is not zero. The expressions p/q and p'/q' denote the same rational number if and only if $pq' = p'q$. Every rational number has a unique irreducible expression p/q where q is positive and as small as possible.

Arithmetic is defined for rational numbers by

$$\frac{p}{q} + \frac{p'}{q'} = \frac{pq' + p'q}{qq'}$$

and

$$\frac{p}{q} \cdot \frac{p'}{q'} = \frac{p \cdot p'}{q \cdot q'}.$$

These operations are well defined in that if we replace one of the expressions in the sum or product by a different expression for the same rational number, then the result will be a different expression for the same rational number (so $1/2 + 1/3 = 5/6$ and $2/4 + 1/3 = 10/12$ but $5/6$ and $10/12$ are expressions for the same rational number). This may seem like a trivial point, but it is essential that we observe it. Later, when we define real numbers, we will encounter a similar situation in that there will be many different expressions for the same real number, and when we define operations like addition on real numbers we will also be obliged to verify that the result is independent of that particular expression is chosen.

The rational numbers form a *field* under the operations of addition and multiplication defined above. This means

1. addition and multiplication are each commutative and associative, $a+b = b+a$, $(a+b)+c = a+(b+c)$, $a \cdot b = b \cdot a$, $(a \cdot b) \cdot c = a \cdot (b \cdot c)$;

2. multiplication distributes over addition, $a \cdot (b + c) = a \cdot b + a \cdot c$;

3. 0 is the additive identity, $0 + a = a$, and 1 is the multiplicative identity, $1 \cdot a = a$;

4. every rational has a negative, $a + (-a) = 0$, and every non-zero rational has a reciprocal, $a \cdot 1/a = 1$.

The field axioms imply all the usual laws of arithmetic and allow the definition of subtraction and division by a non-zero rational number.

The rational numbers also possess an order. A rational is positive if it has the expression p/q with p and q positive. It is negative if it has the expression p/q with p negative and q positive. Every rational is either positive, negative, or zero. In terms of this we define $a < b$ if $b - a$ is positive and $a \leq b$ if $b - a$ is positive or zero (non-negative).

The order and the arithmetic are connected. The sum and product of positive rational numbers are positive. These properties express the fact that the rational numbers form an *ordered field.* All the usual properties relating order and arithmetic (such as $a > b$ and $c \geq d$ imply $a + c > b + d$) are easily deducible from them.

For example, we can define the absolute value $|a| = a$ if $a \geq 0$ and $-a$ if $a < 0$ and prove the *triangle inequality* $|a+b| \leq |a|+|b|$ for rational numbers. This property will be used frequently and generalized broadly (the terminology "triangle inequality" comes from a generalization to vectors in the plane, where $|a + b|$ is interpreted as the length of the third side of a triangle whose two other sides have length $|a|$ and $|b|$). The triangle inequality is also frequently used in transposed form

$$|a - b| \geq |a| - |b|.$$

In addition to the connections with arithmetic, there are some other properties of the ordering of the rational numbers that are noteworthy. One is called the *Axiom of Archimedes*: *for every positive rational number* $a > 0$ *there exists an integer* n *such that* $a > 1/n$. (Note that the term "axiom" is used here for historic reasons only; it is not taken as an axiom for the rational numbers but rather is a theorem that can be proven for the rational number system). The reciprocal version of this is that every positive rational number is less than some integer. This will turn out to be a very crucial property—one that is also possessed by the real number system.

Another interesting property of the ordering of rational numbers is that between any two distinct rationals there is an infinite number of other rationals. Thus there is no next largest rational. If you have ever studied the concept of *well-ordering* you will recognize the fact that the ordering of the rationals is *not* one.

The rational number system, with its arithmetic and ordering, forms such a simple and elegant mathematical model that it is tempting to want to stay within its comfortable domain. What more could one demand of a number system? Why not do analysis here? To find out, turn to the next chapter.

1.5 The Axiom of Choice*

In thinking about infinite sets, we are inclined to adopt forms of reasoning that arise from our intuitive ideas about finite sets. This transference of ideas from the finite to the infinite is by no means routine and often has consequences that are unforseen. Such an innocent principle as the law of the excluded middle—that a statement must either be true or false—results in the non-constructive nature of mathematics. We can prove "there exists x such that blah" by showing that "for all x not blah" leads to a contradiction, without offering a clue as to how to find the x whose existence is asserted. With the exception of the Intuitionist and Constructivist schools of thought, most mathematicians accept this sort of non-constructivity routinely, with the feeling that the problem of actually finding the x is a legitimate, but different, mathematical problem.

The Axiom of Choice is another principle—obvious for finite sets and transferred to infinite sets by analogy—that leads to non-constructive mathematics. In many ways the use of this axiom leads to a higher degree of non-constructiveness than the use of the law of the excluded middle. (Here I am referring to an intuitive conception of the degree of non-constructiveness rather than a formal mathematical theory.) For this reason, the axiom of choice has received careful scrutiny and perhaps a bit of notoriety as well. Since we will be using it as a valid method of reasoning, we will take the time here at the beginning to discuss it in detail.

If A and B are two sets that are non-empty, we define the Cartesian product $A \times B$ to be the set of ordered pairs (a, b) with a in A and b in B. For finite sets A and B this is a completely straightforward definition, and even for infinite sets A and B it causes little problem. If A is non-empty it must contain at least one element a_1 and if B is non-empty it must contain at least one element b_1, so (a_1, b_1) is in $A \times B$. *The Cartesian product of two non-empty sets is non-empty.* Of course we may not have a constructive procedure for obtaining the element (a_1, b_1) of $A \times B$, if we lack a constructive method for getting a_1 or b_1. But the process of pairing does not add to the non-constructivity. If we can "construct" a_1 and b_1, then we can "construct" (a_1, b_1).

The same ideas can be used to create the Cartesian product $A_1 \times A_2 \times \cdots \times A_n$ of n sets for any finite n. This is the set of ordered n-tuples $(a_1, a_2, \ldots, a_n)$ where each a_j is an element of A_j. Once again, the Cartesian product of non-empty sets is non-empty.

We encounter the axiom of choice when we try to extend these ideas to an infinite collection of sets. Suppose $A_1, A_2, \ldots$ is a countable collection of sets. The Cartesian product $A_1 \times A_2 \times \cdots$ is defined to be the set of sequences $(a_1, a_2, \ldots)$ where each a_n belongs to A_n. The *countable axiom of choice* asserts that if the sets A_n are all non-empty, then the Cartesian product is also non-empty. The term "choice" refers to the fact that any particular element $(a_1, a_2, \ldots)$ of the Cartesian product arises from the "choice" of one a_n from each set A_n.

There are two important points that need to be emphasized here. The first is that it is usually not necessary to invoke this axiom in order to show that the Cartesian product is non-empty. In most particular cases we know enough about the sets A_n to produce a sequence $(a_1, a_2, \ldots)$ by other methods of reasoning. For example, if the A_n are lines in the plane (being considered as sets of the points on the lines), we can take a fixed origin in the plane and define a_n to be the point on the line closest to the origin. We thus use a theorem of Euclidean geometry to produce the element $(a_1, a_2, \ldots)$. From the point of view of "choice", we have replaced the infinite simultaneous *unspecified* choice of the axiom with an infinite simultaneous specified choice. When a specified choice is available, the axiom of choice is unnecessary.

The second point is that it is the infinite number of sets involved and not the infinity of cardinality of the sets A_n that requires the axiom. We might even need to use the axiom if each of the sets A_n contains only two elements. It might be that each A_n contains two elements, and we might even possess procedures for choosing one of the two elements of A_n, but the procedures might be so unrelated to each other that we cannot specify a general procedure. Thus the countable axiom of choice, applied to sets of cardinality two, leads to non-constructive existence.

Normally the countable axiom of choice is used to justify the inclusion in an argument of the making of a countable number of simultaneous choices where we cannot (or are too lazy to) make the choices in a specific way. In such applications we can avoid all mention of the Cartesian product, since it is the individual sequence $(a_1, a_2, \ldots)$ that we need. Such uses are relatively uncontroversial and do not lead to any worse non-contructiveness than the use of the law of the excluded middle.

The general axiom of choice refers to an arbitrary—perhaps uncountable—collection $\mathcal{A}$ of non-empty sets A and asserts the possibility of making a choice of one element from each of the sets A (formally, a function f, called a *choice function*, whose domain is $\mathcal{A}$, and such that $f(A)$ is a point in A for each set A in $\mathcal{A}$). Although the general concept of set as an arbitrary collection of elements would naturally lead us to accept this axiom—if each of the sets A is non-empty, why shouldn't there be a choice function?—it does lead to a level of non-constructivity that is mind-boggling.

Here is an example. The general axiom of choice can be used to show that there exists an *ultrafilter* on the set of natural numbers. Intuitively, an ultrafilter is a collection of the "big" sets of numbers. It must possess the property that if A contains B and B is big, then A is big and also that the intersection of any finite number of big sets is also big. These two consistency conditions define the notion of *filter*. There are many filters; the simplest example is to define the complements of finite sets to be big. The *ultra* denotes the additional property that every set of numbers must either be big or else its complement must be big. This means for instance that either the even numbers or the odd numbers must be big (not both, for their intersection is empty). Thus the ultra-filter makes an arbitrary choice—in a consistent manner—between each set and its complement. Clearly there is no specific way to make such a choice, and it is beyond the imagination how such a choice could be made. The use of the words "there exists" in the phrase "there exists an ultrafilter on the natural numbers" thus involves a further step away from the constructive. For this reason, any use of the general axiom of choice deserves special mention and comment. Fortunately we will not need to use this axiom in this work.

Chapter 2

Construction of the Real Number System

2.1 Cauchy Sequences

2.1.1 Motivation

We have an intuitive concept of the real number system; it is the number system that should be used for measurements of space and time as well as for other quantities such as mass, temperature, and pressure that are thought of as varying continuously rather than discretely. This intuitive real number system has been used by mathematicians since at least the period of ancient Greek mathematics, but it was not until the second half of the nineteenth century that a satisfactory formal mathematical system was constructed that could serve in its place. In fact, the whole history of the discovery of the foundations of analysis reads backward—much as one peels a cabbage starting from the outermost leaves and working inward—so mathematicians started by giving precise definitions for the most advanced concepts such as derivative and integral (Cauchy and Bolzano, 1820s) in terms of an intuitive real number system, then worked inward to construct the real number system (Weierstrass, Dedekind, Meray, Heine, Cantor, 1860s and 1870s) in terms of an intuitive set theory, and then worked to the core of axiomatic set theory (Peano, Frege, Zermelo, Russell, Whitehead, Frankel, starting in the 1890s and extending well into the twentieth century). Whether or not this is really the core or merely another inner leaf surrounding a more elementary core is left for the future to answer. The order of eating the cabbage does not have to correspond to the order of peeling, so in this book we will discard the core and start with the inner leaves.

We formulate the problem as follows: *construct a mathematical system, by means of precise, unambiguous definitions and theorems proved by purely logical reasoning, accepting as given the logic, set theory, and rational number system discussed in the previous chapter, which has as much as possible in common with the intuitive real number system.* This problem is clearly not as well formulated as we might like. It is by no means clear that a solution is possible or that there are not many different solutions; we do not even have a strict standard by which to judge what constitutes a solution. Nevertheless it is an important problem, and we shall study in detail one solution, called the *Cauchy completion of the rationals.* This solution is equivalent to several others, which we will discuss later. Together we can refer to these as the *classical real number system.* Since there is no completely objective method to evaluate how closely this system conforms to the intuitive concept of real numbers, we must rely on the consensus of the majority of working mathematicians and users of mathematics to ratify the choice of the classical real number system as a worthy and successful solution. As a student of mathematics, you are invited to study this system and become part of the consensus or, if you wish, to oppose the consensus. (There are two other seriously competing number systems, called the *constructive real number system* (E. Bishop, 1970) and the *non-standard real number system* (A. Robinson, 1960). Each has a following of mathematicians who believe that the alternative system is in fact a better solution to the problem as formulated above than the classical real number system. These systems are discussed briefly at the end of the chapter. Neither would make a good basis for a book on this level because both require a thorough understanding of the classical real number system.)

Before discussing the construction of the classical real number system, we should at least attempt to spell out some of the properties of the intuitive real number system that we expect the formal mathematical system to possess. It should certainly contain the rational numbers, and it should have an arithmetic with similar properties. In addition to the arithmetic operations, the rational numbers possess a compatible notion of order that is of great importance, so we want the real number system to have a similar order. We can summarize the above by saying we want the real number system to be an *ordered field*, that is, a set $\mathbb{R}$ with two operations, addition $(x+y)$ and multiplication $(x \cdot y)$, an additive unit 0, a multiplicative unit 1, and an order relation $(x < y)$, such that the ordered field axioms described in Section 1.4 are verified. (We will recall these axioms when we prove that the system we construct is in fact an ordered field.)

So far we have not discussed any requirements for the real number system that are not already possessed by the rational number system. Nevertheless we know that the rational number system is not large enough for even simply geometry, let alone analysis. There are "numbers", such as $\sqrt{2}$, for which we have intuitive evidence favoring inclusion in the real number system, which are not included in the rational number system. For $\sqrt{2}$ the evidence is especially striking, consisting of drawing the diagonal of a square whose sides have length one and quoting the Pythagorean theorem. While this "construction" of $\sqrt{2}$ is extremely picturesque, it is somewhat misleading in that it only involves a finite number of steps; if we were to insist that all "constructions" involve only a finite number of steps we would never be able to follow the way of analysis. Of course there are many other examples of numbers such as π, e, $\sqrt{2}^{\sqrt{2}}$ for which no finite construction exists; still it is particularly simple to discuss $\sqrt{2}$ and to learn an important point from the discussion.

What do we know about $\sqrt{2}$? By definition it is the positive solution to the equation $x^2 = 2$. (You probably remember the proof that no rational number can satisfy this equation: if $x = p/q$ is a rational number factor out all powers of two, so $x = 2^k p_1/q_1$ with p_1 and q_1 odd and k an arbitrary integer. Then $x^2 = 2^{2k}p_1^2/q_1^2$ and $x^2 = 2$ lead to $2^{2k-1}p_1^2 = q_1^2$, an even $=$ odd contradiction for k positive or $2^{1-2k}q_1^2 = p_1^2$ if k is not positive.) Perhaps you remember an algorithm for computing the decimal expansion of $\sqrt{2}$. It is somewhat cumbersome, and in fact there is a simpler and more efficient (you get more accuracy for the same amount of labor) method, which is a lot more fun. Choose a first guess x_1, and take for the second guess the average of the first guess and two divided by the first guess:

$$x_2 = \frac{1}{2}\left(x_1 + \frac{2}{x_1}\right).$$

If the first guess were exactly $\sqrt{2}$, then the second guess would also be $\sqrt{2}$; but if you guessed too low, then $2/x_1$ would be greater then $\sqrt{2}$ and the average would be closer to $\sqrt{2}$, and similarly if you guessed too high. The process can then be iterated, producing a third guess $x_3 = 1/2(x_2 + 2/x_2)$ and so on. In this way we get a sequence $x_1, x_2, x_3, \ldots$ of better and better approximations to $\sqrt{2}$, and with a little calculus one can show that convergence is quite rapid (we will discuss this example later). It is quite likely that your pocket calculator uses this procedure, or something similar, when it tells you $\sqrt{2} = 1.414\ldots$ (the calculator stops iterating when the iteration produces no change in the number of decimals retained).

The key point of this discussion is that when we calculate $\sqrt{2}$ numerically, what we actually obtain is a sequence of approximations to $\sqrt{2}$, whether it be the successive partial decimals $1, 1.4, 1.41, 1.414$ or the successive guesses $x_1, x_2, x_3, \ldots$. In any particular computation we obtain only a finite number of approximations, but in principle the approximation could continue indefinitely. *The evidence for the existence of $\sqrt{2}$ as a number is then that we can approximate it by other numbers whose existence we already know* (the partial decimals are all rational numbers, and the same is true of the sequence $x_1, x_2, \ldots$ above, provided the first guess x_1 is a rational number). The same can be said, for example, for $\pi, e, \sqrt{2}^{\sqrt{2}}$. Thus we want the real numbers to possess a property not shared by the rational numbers, which we will call *completeness.* For now we can describe this intuitively by the condition that anything that can be approximated arbitrarily closely by real numbers must also be a real number; later we will give the formal counterpart of this statement and prove that the real number system we construct does have this property. In the meantime we will use this intuitive description as a clue to how to proceed.

As a counterweight to the intuitive notion of completeness, we want a principle that will keep the real number system from being too large (for example, to exclude imaginary numbers like $\sqrt{-1}$). The simplest such principle, called the *density of the rational numbers*, is that there are rational numbers arbitrarily close to any real number. This means that if we represent the real numbers and rational numbers graphically by points along a line, we will not see any difference. The "holes" in the rational number system due to the absence of irrational numbers like $\sqrt{2}$ are not visible, because of the rational numbers nearby.

Having rejected the rational number system because it is too small, let us pause to consider another possible easy way out: infinite decimal expansions. All the numbers we have been talking about have infinite decimal expansions, and we could just as easily think of the computations above of $\sqrt{2}$ as merely producing more and more digits of this expansion. On a certain level we would not be far from wrong to define the real number system as the set of all infinite decimals $\pm N.a_1, a_2, a_3, \ldots$, where N is a nonnegative integer, and each a_j is a digit from 0 to 9. Of course there is nothing special about the base 10; we could use any other base just as well. Because of the familiarity of infinite decimals, this proposal is quite appealing. However, it has two technical drawbacks. The first is that the decimal expansion is not unique: .999... and 1.000... are the same number. This is usually met by the ad hoc requirement that the decimal cannot end in an infinite string of zeroes. The second drawback is that it is somewhat awkward to define addition and multiplication, because long carries could change earlier digits. For example, what is the first digit in the sum

$$\begin{array}{rr} & .199999999\ldots \\ + & \underline{.100000000\ldots} \end{array}$$

where the ... here means that we don't know what comes next? Is it a 2 or a 3? We can't say in advance how many more digits we need to compute before we know, despite the fact that we already know that the sum is extremely close to .3.

But aside from these technical problems, which can be overcome, there is a more important reason why mathematicians prefer not to describe the real number system in terms of infinite decimal expansions. This reason is that the infinite decimal is only one way of describing real numbers and, although it has its uses, is a somewhat peculiar one (we mentioned two such peculiarities above). It would be pedagogically and psychologically unsound to devote minute attention to the peculiarities of this system of representing numbers, since these peculiarities do not shed any light on the path we intend to follow. Instead we will look for a deeper method, one that will cast a shadow forward as well as backward. In the end we will show that all the numbers in our system have infinite decimal expansions, so that we have an equivalent system.

2.1.2 The Definition

Let us examine more closely the idea of approximating exotic numbers, like $\sqrt{2}$, by sequences of more prosaic numbers. What is it about the sequence $1, 1.4, 1.41, 1.414, \ldots$ that gives us confidence there is some number being approximated? We might say that there is a "coming together of terms", unlike the sequences $1, 2, 3, \ldots$ or $1, 2, 1, 2, 1, 2, \ldots$, which do not appear to approximate anything. The key question, which can be proposed first in the intuitive real number system and was first solved by Cauchy in that context, is the following: *what condition on a sequence of numbers is necessary and sufficient for the sequence to converge to a limit but does not explicitly involve the limit*? Of course if we knew in advance the number x to which the sequence $x_1, x_2, \ldots$ of numbers is supposed to converge, we could express the convergence in the typical way: *for all natural numbers n, there exists a natural number m (depending on n), such that $|x - x_k| < 1/n$ for all $k \geq m$.* In other words, given any prescribed error $1/n$, if we go far enough out in the sequence (beyond m) the terms all differ from x by at most $1/n$. This is the standard definition of limit. Commonly the error $1/n$ is denoted ϵ and is allowed to be any positive quantity. But this is merely an equivalent variant since we can always find $1/n$ smaller than ϵ. In this book we will use $1/n$ rather than ϵ because it simplifies matters. Regardless of which variant we use, it should be recognized that this definition is only precise if we know what "number" means.

Cauchy's problem was how to get the limit x out of the definition of limit! His solution was to observe on an informal level that *if the numbers x_k are getting close to x, they must be getting close to each other.* To translate this into a precise statement, however, requires some care. Suppose we try the most obvious condition: that consecutive terms get close together. While this is true of convergent sequences, it is also true of the sequence $1, 1\frac{1}{2}, 2, 2\frac{1}{3}, 2\frac{2}{3}, 3, 3\frac{1}{4}, 3\frac{1}{2}, 3\frac{3}{4}, 4, 4\frac{1}{5}, \ldots$, which does not converge. It is not sufficient that consecutive terms be close; we need *all terms* beyond a certain point to be close. This is called the *Cauchy criterion*: *for all natural numbers n there exists a natural number m*

(depending on n) such that for all $j \geq m$ and $k \geq m$, $|x_j - x_k| \leq 1/n$. In other words, beyond the m-th term in the sequence, all terms differ from one another by at most $1/n$. A sequence that satisfies the Cauchy criterion is called a *Cauchy sequence.* Clearly any convergent sequence is a Cauchy sequence, because if we go far enough out in the sequence that x_j and x_k differ from the limit x by at most $1/2n$, then they will differ from each other by at most $1/2n + 1/2n = 1/n$. But the definition of Cauchy sequence does not involve the limit, as we wanted, so it is not immediately clear that every Cauchy sequence has a limit. Cauchy claimed to have proved this, but on a rigorous level his proof had to be bogus since he never defined "number". Nevertheless, we should consider carefully an informal proof that the statement "every Cauchy sequence converges to a real number" accords with our intuitive concepts, especially the idea of completeness.

Informal Proof: Let $x_1, x_2, x_3, \ldots$ be a Cauchy sequence of real numbers. We want there to be a real number x that is the limit. What should x be? Suppose we want to determine x to an accuracy of $1/n$. Then by the Cauchy criterion there exists m (depending on n) such that all terms beyond the m-th differ from each other by at most $1/n$. If we plot all the numbers x_k for $k \geq m(n)$ on a line, they will lie in a segment of width at most $1/n$ and the limit presumably must also lie in that segment, as shown in Figure 2.1.1.

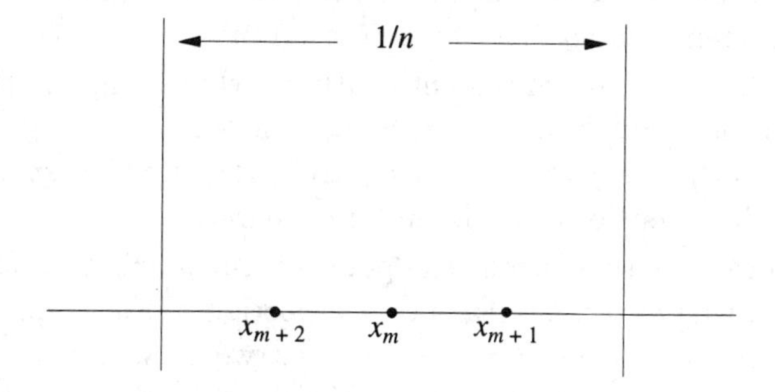

Figure 2.1.1:

Now we examine what happens when we increase n, say to n'. Then there exists m', beyond which the terms all differ from one another by at most $1/n'$; in other words, by going farther in the sequence, the segment that contains all the terms, and presumably the limit, narrows, as shown in Figure 2.1.2.

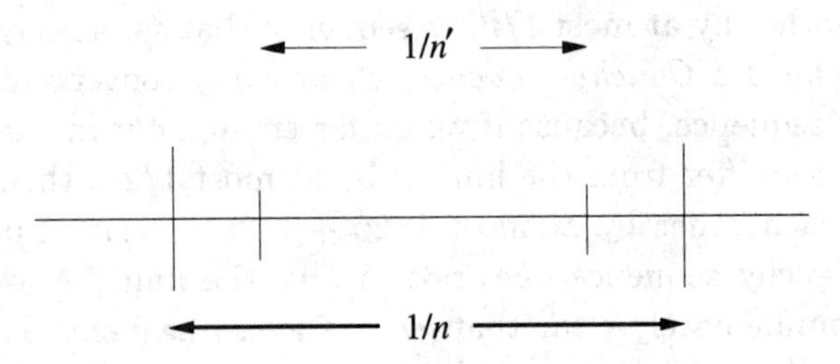

Figure 2.1.2:

Considering the situation for all values of n, we come to the conclusion that there exists a nested sequence of segments of length $1/n$ in which the limit presumably lies. This suggests that the limit x is exactly that number that is in all those segments. If there were no such number, this would suggest a "hole" in our number system, something that the idea of completeness is supposed to prevent. In any event we can "compute" this number x to any desired accuracy, say $1/n$, by taking any value in the n-th segment (say x_m for m large enough that x_m lies in the n-th segment). Finally the sequence $x_1, x_2, x_3, \ldots$ must converge to this limit since for any error $1/n$, there exists m (depending on n), such that all the terms beyond the m-th and the number x all lie in the m-th segment and so differ by at most $1/n$.

We can illustrate this argument neatly in what I will call the "two-dimensional picture". Draw a graph that plots x_n on the y-axis over the point $1/n$ on the x-axis, as in Figure 2.1.3, drawing straight-line segments in between to make the picture clearer.

The limit of the sequence is the point on the y-axis that the graph hits. The Cauchy criterion means that a portion of the graph is boxed in by a sequence of concentric rectangles (two of which are drawn), whose x-coordinates go from 0 to $1/m$ (the condition $k \geq m$) and whose y-coordinates lie in a segment of length $1/n$ (these are the segments discussed above). The two-dimensional picture lets you "see" how the Cauchy criterion forces the graph to hit the y-axis at a precise spot.

At present we have no way of making this informal proof precise. Nevertheless it is important because it will motivate our construction

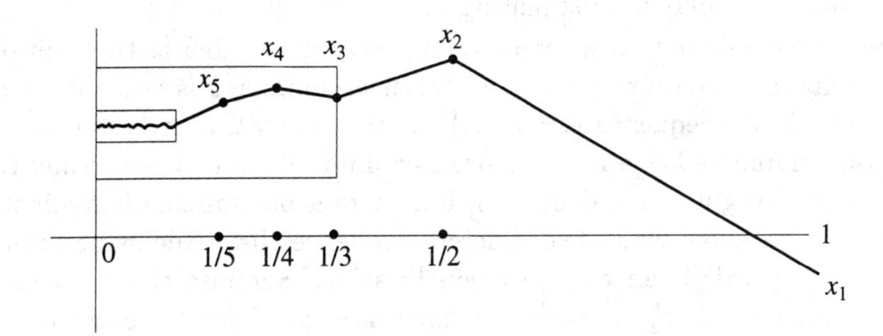

Figure 2.1.3:

of the real number system. Since we already have the rational numbers and we believe that every Cauchy sequence should have a real number as limit, we should certainly add to the rational numbers a real number to be the limit of each Cauchy sequence of rational numbers (a Cauchy sequence of rational numbers does not necessarily have a rational number as limit, as the sequence $1, 1.4, 1.41, 1.414, \ldots$ approximating the irrational number $\sqrt{2}$ shows). Suppose we had a number system that contained rational numbers and limits of Cauchy sequences of rational numbers; would that be enough? One might first guess "No", since there would then be Cauchy sequences of irrational numbers with which to contend. They would have to have limits, and wouldn't these constitute a new variety of real numbers? Perhaps so, but perhaps not, because more than one Cauchy sequence can have the same limit (compare $1, 1/2, 1/3, 1/4, \ldots$ with $0, 0, 0, \ldots$). It might be that every Cauchy sequence of real numbers $x_1, x_2, x_3, \ldots$ has the same limit as a Cauchy sequence of rational numbers $y_1, y_2, y_3, \ldots$. In fact the following informal proof appears convincing: just choose for y_n any rational number that differs from x_n by at most $1/n$ (here we use the intuitive property that there are rational numbers arbitrarily close to any real number). The error involved in changing from x_n to y_n gets smaller as we go out in the sequence and so shouldn't change the limit. Another informal demonstration that we should get every real number as a limit of Cauchy sequences of rational numbers is the infinite decimal expansion; if a real number x has an infinite decimal representation $\pm N.\ a_1 a_2 a_3 \ldots$, then $\pm N, \pm N.a_1, \pm N.a_1 a_2, \ldots$ is a Cauchy sequence of rational numbers approximating x.

We now believe that we can obtain every real number as the limit of a Cauchy sequence of rational numbers (if the number x is rational, then we can take the sequence $x, x, x, \ldots$) and that every Cauchy sequence of rational numbers has a real number as a limit. We are almost ready to make this insight into a definition, but there is one missing ingredient. We need to know when two different sequences have the same limit. Fortunately this is an easy problem to solve. Suppose the sequences $x_1, x_2, \ldots$ and $x'_1, x'_2, \ldots$ have the same limit x. Then for every error $1/n$ there is a position m in the sequence $x_1, x_2, \ldots$ beyond which all the x_k differ from x by at most $1/2$, and similarly there exists a position m' in the sequence $x'_1, x'_2, \ldots$ beyond which all the x'_k differ from x by at most $1/n$. If we go beyond the larger of the two, m and m', then x_k will differ from x'_k by at most $1/n + 1/n = 2/n$. Clearly it is just a change of notation to arrive at the formulation *for every* n *there exists* m *(depending on* n*) such that for all* $k \geq m, |x_k - x'_k| \leq 1/n$. Two Cauchy sequences with this property will be called *equivalent* (at present this definition can only be formally stated for Cauchy sequences of rational numbers, since we have not defined the meaning of $|x_k - x'_k| \leq 1/n$ for real numbers x_k and x'_k; however, once we have done so, the definition will apply verbatim to sequences of real numbers). We have thus shown, informally, that if two Cauchy sequences have the same limit, then they are equivalent. What about the converse? If two Cauchy sequences $x_1, x_2, \ldots$ and $x'_1, x'_2, \ldots$ are equivalent, by how much can their limits x and x' differ? Certainly not by more than $1/n + 1/n$ for any n. This implies that the two Cauchy sequences have the same limit, at least if we believe the axiom of Archimedes (every positive number is greater than $1/n$ for some n). We have seen that the axiom of Archimedes is true for rational numbers, and it is one of the properties we want to keep when we pass from the rationals to the reals.

Let us summarize the informal conclusions we have obtained so far: *if the real number system is to be an ordered field containing the rational numbers for which completeness, density of the rationals, and the axiom of Archimedes hold, then it should consist of limits of Cauchy sequences of rationals with two such limits being considered equal if and only if the Cauchy sequences are equivalent.*

The use of the term "equivalent" in mathematics requires that the properties of reflexivity (A is equivalent to A), symmetry (A is equivalent to B implies B is equivalent to A), and transitivity (A is equivalent to B and B is equivalent to C implies A is equivalent to C) be satisfied. If these properties hold, then we have a true equivalence relation, and we can divide the set on which the relation is defined into disjoint subsets, called *equivalence classes*, such that A and B belong to the same equivalence class if and only if A is equivalent to B. In the present case the first two properties are obvious. The transitivity is not hard to show.

Lemma 2.1.1 *The equivalence of Cauchy sequences of rationals is transitive.*

Proof: Let $x_1, x_2, \ldots, y_1, y_2, \ldots,$ and $z_1, z_2, \ldots$ be Cauchy sequences of rationals such that the $x-y$ and $y-z$ pairs are equivalent. To show the $x-z$ pair is equivalent we must show that given any error $1/n$, we can find an m (depending on n) such that for all $k \geq m$, $|x_k - z_k| \leq 1/n$. But by the $x-y$ equivalence there exists m_1 (depending on n) such that for $k \geq m_1$, $|x_k - y_k| \leq 1/2n$. Similarly by the $y-z$ equivalence there exists m_2 (depending on n) such that for all $k \geq m_2$, $|y_k - z_k| \leq 1/2n$. Thus by taking m to be the larger of m_1 and m_2, for all $k \geq m$ we have

$$\begin{aligned} |x_k - z_k| &= |(x_k - y_k) + (y_k - z_k)| \\ &\leq |x_k - y_k| + |y_k - z_k| \leq \frac{1}{2n} + \frac{1}{2n} = \frac{1}{n} \end{aligned}$$

by the triangle inequality for rationals. QED

Notice that this formal proof is a little artificial in the choice of $1/2n$ as the error in the first two estimates, which leads to the final estimate with an error of $1/n$. Alternatively, we could have started with the estimates $|x_k - y_k| \leq 1/n$ and $|y_k - z_k| \leq 1/n$ and come up with the conclusion that $|x_k - z_k| \leq 2/n$ for all $k \geq m$ and then remarked that we can obtain the desired conclusion by a change of variable. Either form of proof is acceptable. Notice that the idea behind this simple proof is that if x is close to y and y is close to z, then x is close to z. The triangle inequality gives this idea quantitative form, and then it is inserted into the universal-existential-universal form of sentence that defines equivalence. This proof is a simple prototype of many of the proofs we will be seeing and so I have somewhat belabored the point.

Eventually the steps involved in constructing the formal proof from the idea of the proof, or conversely in extracting the idea of the proof from the formal proof, should become routine (except that ideas of proofs are often quite subtle and difficult).

We can now proclaim the formal definition of the real number system:

Definition 2.1.1 *Let C denote the set of all Cauchy sequences $x_1, x_2, \ldots$ of rational numbers, and let $\mathbb{R}$ denote the set of equivalence classes of elements of C. We call $\mathbb{R}$ the real number system, or the reals, for short, and the elements x of $\mathbb{R}$ (which are equivalence classes of Cauchy sequences of rationals) are called real numbers. We will also say that a particular Cauchy sequence in the equivalence class x converges to x or has x as a limit.*

One should not get too hung up on the formal nature of this definition. Although we have defined the real number x to *be* the equivalence class of Cauchy sequences, this is largely a linguistic convention. We could equally well think of the particular Cauchy sequences as specifying the real number, or labelling it, with the different elements of an equivalence class providing different labels for the same number. We could also try to pick out a particular Cauchy sequence from each equivalence class, and in fact we will do this when we discuss the infinite decimal expansions. From the point of view of the mathematics that results, all these approaches are equivalent. Quite frankly, no mature mathematician thinks of an equivalence class of Cauchy sequences of rationals every time the word "real number" appears. This definition, or any equivalent one, is merely a device to begin studying the properties of the real number system; the properties of the system eventually lead to the individual mathematician's mental conception of the real numbers. Since we have as yet no properties in sight, we had best not linger on the linguistic conventions of the definition.

We want to think of the reals as an enlargement of the rationals, so we want the rationals to form a subset of the reals. This is not strictly speaking the case in our formal definition, but we have the informal idea that the rational number r should be the limit of the Cauchy sequence $r, r, \ldots$, and so we should attempt to *identify* the rational number r with the real number which is the equivalence class of $r, r, \ldots$. To make this identification at this stage requires only that we verify that distinct rationals $r \neq s$ are identified with distinct reals ($r, r, \ldots$ not equivalent to $s, s, \ldots$). This verification is simple, since $r \neq s$ implies there exists N with $|r-s| \geq 1/N$ by the axiom of Archimedes for rationals, and this prevents $r, r, \ldots$ from being equivalent to $s, s, \ldots$. The identification also brings with it a future committment to consistency: *whenever a property or operation is defined for the reals that is already defined for the rationals, the two definitions must coincide on the rational subset of the reals.* Fortunately, this will never be difficult to verify.

2.1.3 Exercises

1. Show that there is an uncountable number of Cauchy sequences of rational numbers equivalent to any given Cauchy sequence of rational numbers.

2. Show that every real number can be given by a Cauchy sequence of rationals $r_1, r_2, \ldots$, where none of the rational numbers $r_1, r_2, \ldots$ is an integer.

3. What kinds of real numbers are representable by Cauchy sequences of integers?

4. Suppose $x_1, x_2, \ldots$ and $y_1, y_2, \ldots$ are two sequences of rational numbers. Define the shuffled sequence to be $x_1, y_1, x_2, y_2, \ldots$. Prove that the shuffled sequence is a Cauchy sequence if and only if $x_1, x_2, \ldots$ and $y_1, y_2, \ldots$ are equivalent Cauchy sequences.

5. Prove that if a Cauchy sequence $x_1, x_2, \ldots$ of rationals is modified by changing a finite number of terms, the result is an equivalent Cauchy sequence.

6. Give a proof that any infinite decimal expansion $\pm N.a_1, a_2, a_3, \ldots$ gives a Cauchy sequence $\pm N, \pm N.a., \pm N.a_1, a_2, \ldots$.

7. Show that the Cauchy sequence $.9, .99, .999, \ldots$ is equivalent to $1, 1, 1, \ldots$.

8. Can a Cauchy sequence of positive rational numbers be equivalent to a Cauchy sequence of negative rational numbers?

9. Show that if $x_1, x_2, \ldots$ is a Cauchy sequence of rational numbers there exists a positive integer N such that $x_j \leq N$ for all j.

2.2 The Reals as an Ordered Field

2.2.1 Defining Arithmetic

To begin the justification of the definition of the real number system that we have chosen, we want to transfer the basic properties of order and arithmetic from the rationals to the reals. The basic idea for accomplishing this is to work term-by-term on the Cauchy sequences of rationals. This requires a certain amount of detailed checking that everything makes sense and is well defined. Let us consider addition first. What should $x + y$ mean if x and y are real numbers? Suppose $x_1, x_2, \ldots$ is a particular Cauchy sequence of rational numbers in the equivalence class that defines x, and similarly let $y_1, y_2, \ldots$ be a Cauchy sequence of rationals that represents y. Recall our intuition that the terms x_k in the first sequence are approximating x. Shouldn't $x_k + y_k$ then approximate $x + y$? Surely this is what working term-by-term suggests. But is this a reasonable definition? The sequence $x_1 + y_1, x_2 + y_2, \ldots$ is clearly a sequence of rationals, but is it a Cauchy sequence? Presumably it is, but this is the first thing we have to check to determine if the definition makes sense. The second thing we have to check is more subtle but equally important. We chose particular Cauchy sequences $x_1, x_2, \ldots$ and $y_1, y_2, \ldots$ out of the equivalence classes defining x and y. Suppose we chose different ones, i.e., $x'_1, x'_2, \ldots,$ equivalent to $x_1, x_2, \ldots$ and $y'_1, y'_2, \ldots$ equivalent to $y_1, y_2, \ldots$; would $x'_1 + y'_1, x'_2 + y'_2, \ldots$ then be a Cauchy sequence equivalent to $x_1 + y_1, x_2 + y_2, \ldots$? This is certainly necessary if the sum $x + y$ is to be well defined as the equivalance class of $x_1 + y_1, x_2 + y_2, \ldots$. To summarize: *when defining an operation on real numbers, we need to verify first that the operation preserves Cauchy sequences of rationals and then to verify that it respects equivalence classes.*

Lemma 2.2.1

a. *Let $x_1, x_2, \ldots$ and $y_1, y_2, \ldots$ be Cauchy sequences of rationals. Then $x_1 + y_1, x_2 + y_2, \ldots$ is also a Cauchy sequence of rationals.*

b. *In addition, let $x'_1, x'_2, \ldots$ be a Cauchy sequence of rationals equivalent to $x_1, x_2, \ldots$, and let $y'_1, y'_2, \ldots$ be a Cauchy sequence of rationals equivalent to $y_1, y_2, \ldots$. Then $x'_1 + y'_1, x'_2 + y'_2, \ldots$ is equivalent to $x_1 + y_1, x_2 + y_2, \ldots$.*

Proof: The argument is very similar to the proof of the transitivity of equivalence given in the last section. For part a, given any error $1/n$, there exists m_1 such that $|x_j - x_k| \leq 1/2n$ for j, $k \geq m_1$ and there exists m_2 such that $|y_j - y_k| \leq 1/2n$ for j, $k \geq m_2$, because $x_1, x_2, \ldots$ and $y_1, y_2, \ldots$ are Cauchy sequences. Then by taking m to be the larger of m_1 and m_2, we have

$$\begin{aligned} |(x_j + y_j) - (x_k + y_k)| &= |(x_j - x_k) + (y_j - y_k)| \\ &\leq |x_j - x_k| + |y_j - y_k| \leq \frac{1}{2n} + \frac{1}{2n} = \frac{1}{n} \end{aligned}$$

for $j, k \geq m$.

For part b, given any error $1/n$, there exists m_1 such that $|x_k - x'_k| \leq 1/2n$ for $k \geq m_1$ and there exists m_2 such that $|y_k - y'_k| \leq 1/2n$ for $k \geq m_2$, because of the equivalence of $x_1, x_2, \ldots$ and $x'_1, x'_2, \ldots$ and the equivalence of $y_1, y_2, \ldots$ and $y'_1, y'_2, \ldots$. If we take m to be the larger of m_1 and m_2, then we have

$$\begin{aligned} |(x_k + y_k) - (x'_k + y'_k)| &= |(x_k - x'_k) + (y_k - y'_k)| \\ &\leq |x_k - x'_k| + |y_k - y'_k| \leq \frac{1}{2n} + \frac{1}{2n} = \frac{1}{n} \end{aligned}$$

for $k \geq m$. QED

Definition 2.2.1 *The real number $x + y$ is the equivalence class of the Cauchy sequence $x_1 + y_1, x_2 + y_2, \ldots$, where $x_1, x_2, \ldots$ represents x and $y_1, y_2, \ldots$ represents y.*

By the lemma, $x_1 + y_1, x_2 + y_2, \ldots$ is a Cauchy sequence and the equivalence class depends only on x and y and not on the particular Cauchy sequences $x_1, x_2, \ldots$ and $y_1, y_2, \ldots$ chosen.

The story for multiplication is similar. We want to define $x \cdot y$ as the equivalence class of $x_1 y_1, x_2, y_2, \ldots$. Why is this a Cauchy sequence? Look at a typical difference $x_j y_j - x_k y_k$. We have information about $x_j - x_k$ and $y_j - y_k$, so we write

$$x_j y_j - x_k y_k = y_j(x_j - x_k) + x_k(y_j - y_k)$$

(this is also the idea behind the formula for the derivative of a product). We can make $x_j - x_k$ and $y_k - y_k$ small by taking j and k large, but we still have to control the factors y_j and x_k that multiply them. Note that we don't have to make both factors in a product small in order to make the product small; it is enough to make one factor small and the other bounded. Thus we need an upper bound for y_j and x_k, independent of j and k.

Lemma 2.2.2 *Every Cauchy sequence of rationals is bounded. That is, there exists a natural number N(depending of the sequence $x_1, x_2, \ldots$) such that for all k, $|x_k| \le N$.*

Proof: By the Cauchy criterion there exists m such that $|x_j - x_k| \le 1$ for $j, k \ge m$. If we choose N larger than $|x_m| + 1$, then

$$\begin{aligned} |x_j| &= |(x_j - x_m) + x_m| \le |x_j - x_m| + |x_m| \\ &\le |x_m| + 1 \le N \end{aligned}$$

for all $j \ge m$. If we also take N greater than $|x_1|, \ldots, |x_{m-1}|$, then we will trivially have $|x_j| \le N$ for $j < m$. This imposes only a finite number of conditions on N, so such a number can be found. QED

Lemma 2.2.3

a. *Let $x_1, x_2, \ldots$ and $y_1, y_2, \ldots$ be Cauchy sequences of rationals. Then $x_1y_1, x_2y_2, \ldots$ is also a Cauchy sequence of rationals.*

b. *In addition, let $x'_1, x'_2, \ldots$ be a Cauchy sequence of rationals equivalent to $x_1, x_2, \ldots$, and let $y'_1, y'_2, \ldots$ be a Cauchy sequence of rationals equivalent to $y_1, y_2, \ldots$. Then $x'_1y'_1, x'_2y'_2, \ldots$ is equivalent to $x_1y_1, x_2y_2, \ldots$.*

Proof:

a. Given any error $1/n$, there exists (as before) m such that $|x_j - x_k| \le 1/n$ and $|y_j - y_k| \le 1/n$ for all $j, k \ge m$. Also by the previous lemma there exists N such that $|x_j| \le N$ and $|y_j| \le N$ for j. Note that N is fixed once and for all and does not depend on n. This is crucial, for we have

$$\begin{aligned} |x_jy_j - x_ky_k| &= |y_j(x_j - x_k) + x_k(y_j - y_k)| \\ &\le |y_j||x_j - x_k| + |x_k||y_j - y_k| \\ &\le N \cdot \frac{1}{n} + N \cdot \frac{1}{n} = \frac{2N}{n} \end{aligned}$$

for all $j, k \geq m$. By a change of variable (replacing m by m' determined by $2Nn$ rather than n) we have $|x_j y_j - x_k y_k| \leq 1/n$ for all $j, k \geq m'$, which proves that $x_1 y_1, x_2 y_2, \ldots$ is a Cauchy sequence.

b. Let N be an upper bound for all four Cauchy sequences. Given any error $1/n$, we use the equivalence of $x_1, x_2, \ldots$ and $x'_1, x'_2, \ldots$ to find m_1 such that $|x_j - x'_j| \leq 1/2Nn$ (guess why this choice!) if $j \geq m_1$ and similarly find m_2 such that $|y_j - y'_j| \leq 1/2Nn$ if $j \geq m_2$. Taking for m the larger of m_1 or m_2, we have

$$\begin{aligned} |x_j y_j - x'_j y'_j| &= |y_j(x_j - x'_j) + x'_j(y_j - y'_j)| \\ &\leq |y_j||x_j - x'_j| + |x'_j||y_j - y'_j| \\ &\leq N \cdot \frac{1}{2Nn} + N\frac{1}{2Nn} = \frac{1}{n} \end{aligned}$$

for all $j \geq m$, proving the equivalence of $x_1 y_1, x_2 y_2, \ldots$ and $x'_1 y'_1, x'_2 y'_2, \ldots$. QED

Definition 2.2.2 *The real number $x \cdot y$ is the equivalence class of the Cauchy sequence $x_1 y_1, x_2 y_2, \ldots$, where $x_1, x_2, \ldots$ and $y_1, y_2, \ldots$ respectively represent x and y.*

2.2.2 The Field Axioms

We have now defined an arithmetic of real numbers. The rational numbers being identified with a subset of the real numbers, we should check that the arithmetics are consistent; that is, if we add the rationals r and s as rationals, $r+s$, we get the same real number $(r+s, r+s, \ldots)$ as we get by adding the real numbers $r, r, \ldots$ and $s, s, \ldots$. Clearly this is trivially true. Less trivially, we want the arithmetic of the reals to have all the properties of arithmetic of rationals. These properties are exactly the field axioms (and their consequences). Recall that a set F with two operations $x + y$ and $x \cdot y$ defined is called a *field* if the following axioms hold:

i. addition is commutative ($x + y = y + x$) and associative ($(x + y) + z = x + (y + z)$).

ii. there exists an additive identity 0, such that $0 + x = x$.

iii. every element x has a negative $-x$, such that $-x + x = 0$ (this implies that subtraction is always possible, $x - y$ being $x + (-y)$).

iv. multiplication is commutative and associative.

v. there exists a multiplicative unit 1 (distinct from 0), such that $1 \cdot x = x$.

vi. every element x except 0 has a reciprocal x^{-1}, such that $x^{-1} \cdot x = 1$ (this implies that division by non-zero numbers is always possible).

vii. multiplication distributes over addition, $x \cdot (y + z) = x \cdot y + x \cdot z$.

Theorem 2.2.1 *The real numbers form a field.*

Proof: Clearly the zero is the equivalence class of $0, 0, \ldots$, and the one is the equivalence class of $1, 1, \ldots$. Almost all the field axioms are trivial to establish and entail little more than quoting the analogous axiom for the rationals. The exception is axiom vi, the existence of reciprocals, so let us go into the details. Intuitively it is clear that we want to define x^{-1} by $x_1^{-1}, x_2^{-1}, \ldots$ if x is defined by $x_1, x_2, \ldots$. We meet here a preliminary obstacle in that x_j^{-1} is not defined if $x_j = 0$. Even supposing this is never the case, we will need to verify that $x_1^{-1}, x_2^{-1}, \ldots$ is a Cauchy sequence. For this we will need to estimate the difference $x_j^{-1} - x_k^{-1}$, knowing that $x_j - x_k$ is small. Now we have $x_j^{-1} - x_k^{-1} = x_j^{-1} x_k^{-1} (x_k - x_j)$, so again we are in the situation of having a product in which one factor is small. We need to show that the other factors are bounded. Thus we need an upper bound for x_j^{-1} or equivalently a positive lower bound for x_j. The piece of information that we have at our disposal, and which we have not yet used, is that x is not zero, which means $x_1, x_2, \ldots$ is not in the equivalence class of zero. This property, which we will be able to identify later as the axiom of Archimedes, we formulate as a separate lemma.

Lemma 2.2.4 *Let x be any real number different from zero. Then there exists a natural number N such that for every Cauchy sequence $x_1, x_2, \ldots$ in the equivalence class of x, there exists m such that $|x_j| \geq 1/N$ for all $j \geq m$. The number m of exceptions will depend on the particular Cauchy sequence, but the lower bound $1/N$ will not.*

Proof: What does $x \neq 0$ mean in terms of a particular Cauchy sequence $x_1, x_2, \ldots$ in the equivalence class x? We need to form the negation of the statement that $x_1, x_2, \ldots$ is equivalent to $0, 0, \ldots$; in other words, the negation of the statement: for all n there exists m such that $j \geq m$ implies $|x_j| \leq 1/n$. The negation is: there exists n such that for all m there exists $j \geq m$ such that $|x_j| > 1/n$. This is close to what we want but not quite right. We want the estimate $|x_j| \geq 1/N$ to hold for all $j \geq m$, not just for an infinite set of j's ("for all m there exists $j \geq m$" is equivalent to "for an infinite set of j's"). To get this added information we need to use the fact that $x_1, x_2, \ldots$ is a Cauchy sequence, a fact we have not yet used and without which the conclusion is false (as in the sequence $0, 1, 0, 1, 0, 1, \ldots$). The Cauchy criterion makes the terms eventually all become close together, so we can't have an infinite number satisfy $|x_j| > 1/n$ without all but a finite number satisfying a slightly weaker estimate, say $|x_j| \geq 1/2n$. To make this precise, we use the Cauchy criterion with error $1/2n$: there exists m such that for all $j, k \geq m, |x_j - x_k| \leq 1/2n$. By then choosing one particular value of $j \geq m$ such that $|x_j| > 1/n$, which we know exists by the first step of the proof, we have

$$\begin{aligned} |x_k| &= |(x_k - x_j) + x_j| \geq |x_k - x_j| - |x_j| \\ &\geq \frac{1}{n} - \frac{1}{2n} = \frac{1}{2n} \end{aligned}$$

for all $k \geq m$. (Note the use of the transposed-triangle inequality $|a - b| \geq |a| - |b|$.)

This establishes the desired estimate with $N = 2n$ for each particular Cauchy sequence in the equivalence class. However, it is not yet clear that we can find a single value for N that will work for all Cauchy sequences in the equivalence class. To see this we need one more observation: if there exists m such that $|x_j| \geq 1/N$ for all $j \geq m$ and if $x'_1, x'_2, \ldots$ is equivalent to $x_1, x_2, \ldots$, then there exists m' such that $|x'_j| \geq 1/2N$ for all $j \geq m'$. This will complete the proof since once we have the lower bound $1/N$ for one representative Cauchy sequence, we obtain the lower bound $1/2N$ for all Cauchy sequences in the equivalence class. To establish the observation we need only choose m' greater than m such that $|x_j - x'_j| \leq 1/2N$ for all $j \geq m'$ (using the equivalence of the sequences), and then

$$\begin{aligned} |x'_j| &= |(x'_j - x_j) + x_j| \\ &\geq |x_j| - |x_j - x'_j| \\ &\geq \frac{1}{N} - \frac{1}{2N} = \frac{1}{2N} \end{aligned}$$

for all $j \geq m'$. QED

Having completed the proof of the lemma, we return to the proof of the theorem, namely the existence of x^{-1}. Let $x_1, x_2, \ldots$ be a Cauchy sequence representing x. By the lemma, all but a finite number of terms are non-zero. If we modify all the zero terms, say replace them by ones, we get an equivalent sequence. Assuming this done (without changing notation), we can form the sequence of rationals $x_1^{-1}, x_2^{-1}, \ldots$. As before we need to show two things: i) it is a Cauchy sequence, and ii) the equivalence class of $x_1^{-1}, x_2^{-1}, \ldots$ does not depend on the particular choice of $x_1, x_2, \ldots$.

To prove it is a Cauchy sequence we let N be the natural number given by the lemma. Then for any given error $1/n$, there exists m such $|x_j - x_k| \leq 1/N^2 n$ for all $j, k \geq m$, since $x_1, x_2, \ldots$ is a Cauchy sequence. If we also choose m large enough so that $|x_j| \geq 1/N$ for all $j \geq m$ as the lemma asserts we can, then for $j \geq m$ we have

$$\begin{aligned} |x_j^{-1} - x_k^{-1}| &= |x_j^{-1} x_k^{-1}(x_k - x_j)| \\ &\leq |x_j^{-1}||x_k^{-1}||x_k - x_j| \\ &\leq N \cdot N \cdot \frac{1}{N^2 n} = \frac{1}{n}, \end{aligned}$$

so $x_1^{-1}, x_2^{-1}, \ldots$ is a Cauchy sequence.

If $x_1', x_2', \ldots$ is any equivalent Cauchy sequence (again assuming that a finite number of terms are modified if necessary so that all x_j' are non-zero), we want to show that $x_1'^{-1}, x_2'^{-1}, \ldots$ is equivalent to $x_1^{-1}, x_2^{-1}, \ldots$. Again we choose N as in the lemma; then given $1/n$ we choose m so that $|x_j - x_j'| \leq 1/N^2 n$ for all $j \geq m$ (the equivalence of $x_1, x_2, \ldots$ and $x_1', x_2', \ldots$) and also so that $|x_j| \geq 1/N$ and $|x_j'| \geq 1/N$ for all $j \geq m$. Then for all $j \geq m$ we have

$$\begin{aligned} |x_j^{-1} - x_j'^{-1}| &= |x_j^{-1} x_j'^{-1}(x_j' - x_j)| \\ &\leq |x_j^{-1}||x_j'^{-1}||x_j' - x_j| \\ &\leq N \cdot N \cdot \frac{1}{N^2 n} = \frac{1}{n}. \end{aligned}$$

We have thus constructed a unique real number x^{-1} for every non-zero x, and it is clear that $x^{-1} \cdot x = 1$ by multiplying $x_1^{-1}, x_2^{-1}, \ldots$ and $x_1, x_2, \ldots$ directly. QED

2.2.3 Order

Having established the field axioms for the real number system, we have completed the transference of arithmetic from the rationals to the reals, since all the usual facts of arithmetic are consequences of the field axioms. We turn next to the concept of *order*. Every rational number is either positive, negative, or zero (only one of the above!); and for two rational numbers r and s, we say $r > s$, $r < s$, or $r = s$ according to if $r - s > 0$, $r - s < 0$, or $r - s = 0$. For the real numbers we want a similar ordering, and the first step is to decide for each real x whether $x > 0$, $x < 0$, or $x = 0$. Clearly it is tempting to say $x > 0$ if $x_j > 0$ for all j where $x_1, x_2, \ldots$ is a Cauchy sequence of rationals representing x. But there are some troubles with this idea, since it depends on the particular choice of Cauchy sequence—a modification of say the first term to $x_1 = -1$ will spoil things without changing the number x. We might then be tempted to try the condition: there exists m such that $x_j > 0$ for $j \geq m$. But there is still a problem here, since the sequence $1, 1/2, 1/3, \ldots$ represents zero but still satisfies the condition. Thus while the condition might well be necessary for positivity (in fact it is), it is not sufficient.

To understand the difficulty better, let's look at the two-dimensional picture, both for the sequence $1, 1/2, 1/3, \ldots$ and for a general sequence representing a point above the x-axis, as shown in Figure 2.2.1.

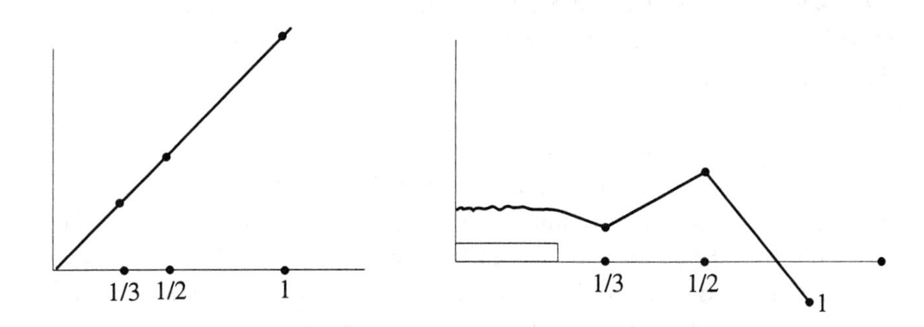

Figure 2.2.1:

Notice that in the second picture we can slip a rectange under the graph. If the height of this rectangle is $1/N$ and the base is the interval $[0, 1/m]$, then we are saying $x_j \geq 1/N$ for all $j \geq m$ (we could equally well say $x_j > 1/N$ by replacing N by $N+1$). You should think of this condition as showing that the tail of the sequence is *bounded away from zero*; this is a stronger condition than merely being positive since it asserts that the separation from zero remains greater than a fixed amount $1/N$ for all the terms beyond the m-th (of course for an individual number there will be no distinction between being positive and being bounded away from zero, as we will show in the proof of the axiom of Archimedes).

Definition 2.2.3 *A real number x is said to be positive if there exist natural numbers N and m such that $x_j \geq 1/N$ for all $j \geq m$, where $x_1, x_2, \ldots$ represents x. The number m depends on the particular Cauchy sequence, but the number N does not (the proof of the previous lemma shows that once we have verified the condition for one Cauchy sequence, it is valid for all equivalent ones). A real number x is called negative, $x < 0$, if $-x$ is positive. Note that if x is rational, these definitions are consistent with the definitions of positivity and negativity for rational numbers, since the rationals satisfy the axiom of Archimedes.*

Theorem 2.2.2 *Each real number is either positive, negative, or zero, but only one of the three. The sum and product of positive numbers are positive.*

Remarks A field with a notion of positivity with the properties given by this theorem is called an *ordered field.* Thus the reals are an ordered field.

Proof: Zero is not positive, as the Cauchy sequence $0, 0, \ldots$ clearly shows. Since $-0 = 0$, zero is not negative either. Let x be any non-zero number; we need to show that either x or $-x$ is positive, but not both. Let $x_1, x_2, \ldots$ be a Cauchy sequence representing x. By the previous lemma there exists N and m such that $|x_j| \geq 1/N$ for all $j \geq m$. But the signs of the rational numbers cannot keep changing, because each sign change produces a jump of at least $2/N$ between terms and this would violate the Cauchy criterion. Thus by increasing m if necessary, we have either $x_j \geq 1/N$ for all $j \geq m$ or $x_j \leq -1/N$ for all $j \geq m$. In the first case x is positive, in the second case $-x$ is positive, and both can't occur for the same x. Finally the verification that the sum or product of positive numbers is positive is essentially trivial, since the sum or product of the lower bounds will give lower bounds. QED

We can use the concept of positive number to define inequalities for real numbers. Thus $x > y$ means $x - y$ is positive, $x \geq y$ means $x > y$ or $x = y$, $|x| < y$ means $y - x$ and $y + x$ are positive, and so on. We define $|x|$ to be x if $x > 0, -x$ if $x < 0$, and $|0| = 0$. In verifying inequalities for real numbers it is convenient to be able to pass from inequalities involving the rational approximations; this requires that the inequalities be non-strict ($\leq$ or $\geq$) but not strict ($<$ or $>$). For example, $1/n > 0$, but the Cauchy sequence $1, 1/2, 1/3, \ldots$ represents the number 0, and we do not have $0 > 0$. The next lemma gives the positive result.

Lemma 2.2.5 *Let x and y be real numbers defined by Cauchy sequences $\{x_k\}$ and $\{y_k\}$ of rational numbers. If $x_k \leq y_k$ for all $k \geq m$, then $x \leq y$.*

Remarks The converse is not true (can you give a counterexample?).

Proof: If not, then $x > y$; so $x - y$ is positive. But then by the definition, $x_k - y_k > 1/n$ for k large and some $1/n$, contradicting $x_k \leq y_k$ for all $k \geq m$. QED

Basically all the properties of inequalities for rationals are true for reals. The next two are especially important.

Theorem 2.2.3 (*Triangle inequality*) *$|x + y| \leq |x| + |y|$ for real numbers x and y.*

Proof: Apply the previous lemma to the triangle inequality for rationals $|x_k + y_k| \leq |x_k| + |y_k|$ where $\{x_k\}$ and $\{y_k\}$ are Cauchy sequences of rationals defining x and y. QED

The triangle inequality is frequently used in the form $|x - z| \leq |x - y| + |y - z|$, in transposed form $|x - y| \geq |x| - |y|$, and for the sums of more than two numbers.

Theorem 2.2.4 (*Axiom of Archimedes*) *For any positive real number $x > 0$ there exists a natural number n such that $x \geq 1/n$.*

Proof: We have already shown that there exists n such that $x_j \geq 1/n$ for all $j \geq m$, where $\{x_j\}$ is a Cauchy sequence of rationals defining x. By the lemma this implies $x \geq 1/n$. QED

This theorem is frequently used in the following form: if $|x| \leq 1/n$ for every natural number n, then $x = 0$.

The fact that the real numbers form an ordered field means that all the familiar algebraic identities, such as $(x+y)^2 = x^2 + 2xy + y^2$, which only involve the operations of arithmetic, are valid for real numbers; and the same is true for inequalities such as $x^2 + y^2/x^2 \geq 1$. The reason for this is that such identities and inequalities are consequences of the ordered field axioms. We will use such "facts" freely from now on without special mention. In exercises 8 and 9 you will be asked to derive some of these facts from the axioms of an ordered field, and this will give you some confidence that the ordered field axioms are really sufficient to contain this aspect of elementary algebra.

We conclude this section with a precise formulation of the density of the rationals in the reals. This is essentially built into the definition, since the Cauchy sequence of rationals $\{x_k\}$ defining the real number x consists of rational numbers x_k that are approximating x, and the density of the rationals in the reals simply says that every real can be approximated arbitrarily closely by rationals.

Theorem 2.2.5 (*Density of Rationals*) *Given any real number x and error $1/n$, there exists a rational number y such that $|x - y| \leq 1/n$.*

Proof: Let $\{x_k\}$ be any Cauchy sequence of rational defining x. Given the error $1/n$, there exists m such that $|x_k - x_j| \leq 1/n$ if $j, k \geq m$. Choose $y = x_m$, so that $|x_k - y| \leq 1/n$ for every $k \geq m$. By the lemma we have $|x - y| \leq 1/n$. QED

2.2.4 Exercises

1. Write out a proof of the commutative and associative laws for addition of real numbers.

2. Show that the real number system is uncountable and, in fact, has the same cardinality as the set of all subsets of the integers.

3. If x is a real number, show that there exists a Cauchy sequence of rationals $x_1, x_2, \ldots$ representing x such that $x_n < x$ for all n.

4. Let x be a real number. Show that there exists a Cauchy sequence of rationals $x_1, x_2, \ldots$ representing x such that $x_n \leq x_{n+1}$ for every n.

5. Prove that there are an infinite number of rational numbers in between any two distinct real numbers.

6. Let x be a positive real number. Prove that there exists a Cauchy sequence of rationals of the special form p^2/q^2, p and q integers, representing x.

7. Prove $|x - y| \geq |x| - |y|$ for any real numbers x and y. (**Hint**: use the triangle inequality).

8. Prove the following identities from the field axioms:

   a. $(x + y)^2 = x^2 + 2xy + y^2$.
   b. $(x + a/x)^2 - 4a = (x - a/x)^2, \quad x \neq 0$.
   c. $ax^2 + bx + c = a(x - b/2a)^2 + c - b^2/4a$.
   d.
   $$(x + y)^n = \sum_{k=0}^{n} \binom{n}{k} x^k y^{n-k} \text{ where } \binom{n}{k} = \frac{n!}{k!(n-k)!}.$$

9. Prove the following inequalities from the ordered field axioms:

   a. $x^2 + y^2/x^2 \geq 1, \quad x \neq 0$.
   b. $2xy \leq x^2 + y^2$.
   c. $x/y > x$ if $x > 0$ and $0 < y < 1$.

10. Show that if a real number x can be represented by a Cauchy sequence of positive rationals, then $x \geq 0$. What does this tell you about real numbers that can be represented by two equivalent Cauchy sequences of rationals, one consisting of only positive rationals and the other consisting of only negative rationals.

11. Prove that no real number satisfies $x^2 = -1$.

12. Define $x^3 = x \cdot x^2$. Prove that if $x_1, x_2, \ldots$ represents x, then $x_1^3, x_2^3, \ldots$ represents x^3.

2.3 Limits and Completeness

2.3.1 Proof of Completeness

At this stage in our development of the real number system, we have succeeded—with a lot of hard work—in arriving at about where we started. We have shown that the real number system is an ordered field. But the rational number system was also an ordered field. Now we need to show that we have really plugged up all the holes (such as $\sqrt{2}$) in the rational number system. This property is called *completeness* and can be succinctly described by saying that if we repeated the process whereby the reals were constructed from the rationals, starting instead from the reals, then we would not end up with anything new.

Our construction of real numbers was based on sequences of rational numbers. We now want to consider sequences of real numbers $x_1, x_2, x_3, \ldots$. Here each x_j is a real number and so, strictly speaking, is a symbol that stands for an equivalence class of Cauchy sequences of rational numbers. Again I must emphasize the desirability of doing a bit of mental gymnastics: think of the real number x_j as a single entity (like a pebble), and yet be capable at times of recalling the definition as an equivalence class of Cauchy sequences of rationals (the pebble is actually an amalgam of molecules, each of which is composed of atoms, each of which is composed of ...). We can now apply the Cauchy criterion to this sequence of reals: *for every n there exists m such that $|x_j - x_k| \leq 1/n$ if $j, k \geq m$.* This is a meaningful statement since the inequality $|x_j - x_k| \leq 1/n$ is meaningful for real numbers. A sequence of real numbers that satisfies the Cauchy criterion is called a *Cauchy sequence.* The intuition involved is the same as for the definition of Cauchy sequences of rationals: the terms of the sequence get closer and closer together as you go out in the sequence.

Since not every Cauchy sequence of rational numbers had a limit that was a rational number, we were motivated to invent real numbers to be these limits. We do not have to invent any new numbers to be limits of Cauchy sequences of real numbers. To see this we need first to formalize the idea of limit. When is a real number x the limit of the sequence of real numbers $x_1, x_2, \ldots$? Clearly when the terms x_k in the sequence get closer and closer to x. We can use the order of the reals to define the inequality $|x_k - x| \leq 1/n$ and, hence, the *definition of limit:* $x = \lim_{k\to\infty} x_k$ *if for every natural number* n *there exists a natural number* m *such that* $k \geq m$ *implies* $|x_k - x| \leq 1/n$. Notice that while the limit of a sequence of real numbers need not always exist, if it exists it is unique. This is because if y were another limit, then we would have

$$\begin{aligned}|x-y| &= |(x-x_k)-(y-y_k)| \\ &\leq |x-x_k|+|y-y_k| \leq \frac{1}{n}+\frac{1}{n}\end{aligned}$$

if k is large enough; hence, $x - y = 0$ by the Axiom of Archimedes.

We should also verify, for consistency of notation, that if $\{x_k\}$ is a Cauchy sequence of rationals defining x, then $\lim_{k\to\infty} x_k = x$. Indeed, given the error $1/n$, we can find m such that $j, k \geq m$ implies $|x_j - x_k| \leq 1/n$, since $\{x_k\}$ is a Cauchy sequence. But then $j \geq m$ also implies $|x_j - x| \leq 1/n$, since this follows from $|x_j - x_k| \leq 1/n$ for all k large, which is what we have for $k \geq m$. Thus $\lim x_j = x$.

Theorem 2.3.1 (*Completeness of the Reals*) *A sequence* $x_1, x_2, \ldots$ *of real numbers has a limit if and only if it is a Cauchy sequence.*

Proof: The fact that the existence of the limit implies the Cauchy criterion is trivial: if m is such that $k \geq m$ implies $|x - x_k| \leq 1/n$, then $j, k \geq m$ implies $|x_j - x_k| \leq 2/n$ by the triangle inequality. The non-trivial part is the converse.

Suppose then that the sequence $x_1, x_2, \ldots$ satisfies the Cauchy criterion. We need to construct the limit as a real number. This means we have to find a Cauchy sequence of rationals $y_1, y_2, \ldots$ to define y and then prove $\lim_{k\to\infty} x_k = y$. The idea is that we want to take for y_k a rational number close to x_k, say so close that $|x_k - y_k| \leq 1/k$. This is possible by the density of rationals. Then it is a simple matter to show that $\{y_k\}$ is a Cauchy sequence. Given an error $1/n$, choose m so that $|x_j - x_k| \leq 1/2n$ for $j, k \geq m$ (this is possible because $\{x_j\}$ is a Cauchy sequence). Then

$$\begin{aligned}|y_j - y_k| &\leq |y_j - x_j| + |x_j - x_k| + |x_k - y_k| \\ &\leq \frac{1}{j} + \frac{1}{2n} + \frac{1}{k} \leq \frac{1}{2n} + \frac{2}{m},\end{aligned}$$

which can be made less than $1/n$ if m is chosen greater than $4n$ (there is no harm in increasing m). Thus $\{y_k\}$ is a Cauchy sequence of rational numbers and, hence, defines a real number y.

It remains to show $\lim_{k\to\infty} x_k = y$. But again this is easy, since

$$|y - x_k| \le |y - y_k| + |y_k - x_k| \le |y - y_k| + \frac{1}{k}.$$

Since $y_1, y_2, \dots$ represents y, we know that we can make $|y - y_k| \le 1/2n$ for $k \ge m$, and hence $|y - x_k| \le 1/2n + 1/k \le 1/n$ if $m \ge 2n$ also. Thus $\lim_{k\to\infty} x_k = y$. QED

Next we state a theorem that summarizes the basic properties of limits. Essentially it says that limits preserve the arithmetic and order properties of real numbers. We have actually seen all these statements before in terms of Cauchy sequences of rational numbers.

Theorem 2.3.2

a. *If* $\lim_{k\to\infty} x_k = x$ *and* $\lim_{k\to\infty} y_k = y$, *then*

$$\lim_{k\to\infty} (x_k + y_k) = x + y \quad \textit{and} \quad \lim_{k\to\infty} (x_k y_k) = xy.$$

If in addition $y \ne 0$, *then there exists* m *such that* $y_k \ne 0$ *for* $k \ge m$ *and* $\lim_{k\to\infty} x_k/y_k = x/y$.

b. *If* $x_k \ge y_k$ *for all* $k \ge m$, *then* $x \ge y$.

We will not go through the formal proofs of these statements, as they are merely repetitions of arguments already given. Notice that in the case of the quotient, some of the terms x_k/y_k may be undefined if $y_k = 0$, but since this cannot happen for $k \ge m$, it does not really matter. Also part b would not be valid with strict inequalities, since $1/k > 0$ but $\lim_{k\to\infty} 1/k = 0$.

2.3.2 Square Roots

We can illustrate the abstract ideas that we have been developing in a concrete example by discussing square roots. It was, after all, the square root of 2 that started the whole idea of irrational numbers. We can now show that within the real number system, all positive numbers have square roots.

Theorem 2.3.3 *Let x be any positive real number. Then there exists a unique positive real number y such that $y^2 = x$ (we then write $y = \sqrt{x}$).*

Proof: It is easy to show uniqueness, for if also $z^2 = x$, then $y^2 - z^2 = 0$. But $y^2 - z^2 = (y-z)(y+z)$; and since the reals form a field, we must either have $y - z = 0$ or $y + z = 0$. Since both y and z were assumed positive, we must have $y + z > 0$, so $y - z = 0$ and hence $y = z$.

To prove existence we use a method that we call *divide and conquer.* This is an idea we will use many times in the pages to come. We start by finding two numbers y_1 and z_1 such that $y = \sqrt{x}$ lies between them. This is easy. If $x > 1$, then $x^2 = x + x(x-1) > x$ since $x(x-1) > 0$. From $1 < x < x^2$ we would expect to have $1 < y < x$ if $y = \sqrt{x}$. Thus we set $y_1 = 1$ and $z_1 = x$. (Similarly, if $0 < x < 1$, we can take $y_1 = x$ and $z_1 = 1$, because $x^2 < x < 1$. If $x = 1$ we can take $y = 1$ and we are done.) Note that we are not claiming to have proved $y_1 \leq y \leq z_1$ (this does not make sense because we have not yet constructed y), but we have proved $y_1^2 \leq x \leq z_1^2$, which is intuitively an equivalent statement.

Now we divide and conquer. The interval y_1 to z_1 has midpoint $m_1 = (y_1 + z_1)/2$. We choose the next interval y_2 to z_2 to be either the left (y_1 to m_1) or the right interval (m_1 to z_1), depending on the relative size of m_1^2 and x. If $m_1^2 > x$ we take $y_2 = y_1$ and $z_2 = m_1$, while if $m_1^2 < x$ we take $y_2 = m_1$ and $z_2 = z_1$ (if $m_1^2 = x$, then $\sqrt{x} = m_1$ and we're done). The two possibilities are illustrated in Figure 2.3.1. The point of this choice is that we still have $y_2^2 \leq x \leq z_2^2$,

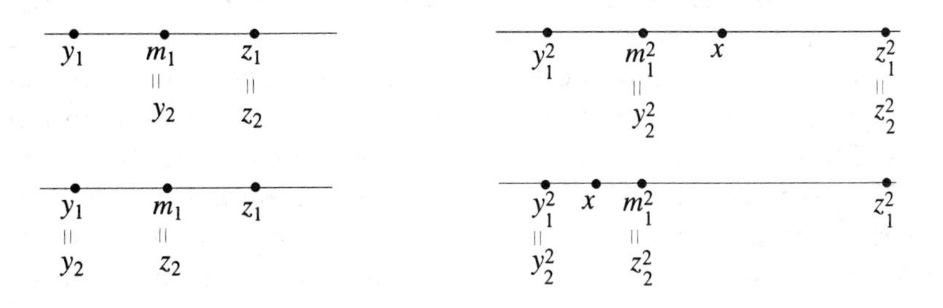

Figure 2.3.1:

but we have divided the distance between y_1 and z_1 in half when we pass to y_2 and z_2. Now we iterate; in other words, we repeat the process of going from y_1, z_1 to y_2, z_2 but starting with y_2, z_2 to obtain y_3, z_3 (so

$m_2 = (y_2 + z_2)/2$ and either $y_3 = m_2$ and $z_3 = z_2$ if $m_2^2 < x$ or $y_3 = y_2$ and $z_3 = m_2$ if $x < m_2^2$), and then y_4, z_4, and so on. We repeat the process of dividing infinitely often (unless we happen to hit $\sqrt{x}$ exactly at one of the midpoints). In this way we obtain two sequences $y_1, y_2, \ldots$ increasing and $z_1, z_2, \ldots$ decreasing such that

a. $y_k^2 \le x \le z_k^2$ for all k, and

b. $z_k - y_k = (z_1 - y_1)/2^{k-1}$.

Now we claim that $y_1, y_2, \ldots$ and $z_1, z_2, \ldots$ are Cauchy sequences and that they are equivalent. This is easy to see. Given an error $1/n$ we take m large enough that $(z_1 - y_1)/2^{m-1} < 1/n$. Then by condition b, $0 \le z_m - y_m < 1/n$, and all y_k and z_k for $k \ge m$ lie in the interval from y_m to z_m. Thus $|y_j - y_k| < 1/n, |z_j - z_k| < 1/n$, and $|y_j - z_k| < 1/n$ for $j, k \ge m$. This proves the claim.

Let y be the real number equal to the common limit of these two sequences. Passing to the limit in a yields $y^2 \le x \le y^2$, which implies $y^2 = x$. QED

The divide and conquer algorithm is not very efficient, since at each iteration we only cut the error in half. To reduce the error by 10^{-3} would require 10 iterations. In exercise 6 the reader is invited to give a different proof based on the more efficient algorithm described in section 2.1.1. However, the proof is much trickier.

Of course you still cannot take square roots of negative numbers within the real number system; to do this it is necessary to enlarge the system still further to the complex numbers. We will discuss this extension later. For now we will just point out that it is rather different from the extension rationals → reals in at least two important ways. The first way is that it is algebraic, not involving infinite processes. The second is that the complex number system has a completely separate interpretation and intuitive image, whereas both the real and rational number systems have a common intuitive basis in the idea of magnitude.

2.3.3 Exercises

1. Write out a proof that $\lim_{k\to\infty}(x_k + y_k) = x + y$ if $\lim_{k\to\infty} x_k = x$ and $\lim_{k\to\infty} y_k = y$ for sequences of real numbers.

2. Prove that every real number has a unique real cube root.

3. Let $x_1, x_2, \ldots$ be a sequence of real numbers such that $|x_n| \leq 1/2^n$, and set $y_n = x_1 + x_2 + \cdots + x_n$. Show that the sequence $y_1, y_2, \ldots$ converges.

4. Let $ax^2 + bx + c$ be a quadratic polynomial with real coefficients a, b, c and positive discriminant, $b^2 - 4ac > 0$. Prove that

$$\frac{-b \pm \sqrt{b^2 - 4ac}}{2a}$$

are the unique real roots.

5. For every sequence $k_0, k_1, k_2, \ldots$ of non-negative integers, let

$$x_n = k_0 + \cfrac{1}{k_1 + \cfrac{1}{k_2 + \cfrac{1}{\ddots \cfrac{}{k_{n+1} + \cfrac{1}{k_n}}}}}$$

be the associated sequence of continued fractions. Prove that $x_0, x_1, x_2, \ldots$ is a Cauchy sequence of rationals and that every positive real number arises as a limit.

6. *Suppose $x > 1$, and define the sequence $y_1, y_2, \ldots$ by $y_1 = x$ and $y_{k+1} = T(y_k)$ for $T(y) = (y + x/y)/2$.

   a. Show $y - T(y) = (y^2 - x)/2y$ and $T(y)^2 - x = (y^2 - x)^2/4y^2$.

   b. Show $0 \leq y - T(y) \leq (y^2 - x)/2$ and $0 \leq T(y)^2 - x \leq (y^2 - x)^2/4$ for $y \geq 1$ and $y^2 \geq x$.

   c. Show $0 \leq y_k - y_{k+1} \leq (y_k^2 - x)/2$ and $0 \leq y_{k+1}^2 - x \leq (y_k^2 - x)^2/4$.

   d. Show that $y_1, y_2, \ldots$ is a Cauchy sequence and if $y = \lim_{k\to\infty} y_k$, then $y^2 = x$.

   e. Show that if $|y_k^2 - x| \leq 10^{-3}$, then $|y_{k+1}^2 - x| \leq 10^{-6}/4$ and $|y_{k+2}^2 - x| \leq 10^{-13}$.

7. Prove that $a > b > 0$ implies $\sqrt{a} > \sqrt{b} > 0$.

8. Prove that if $\lim_{k\to\infty} x_k = x$ and $x_k \geq 0$ for all k, then $\lim_{k\to\infty} \sqrt{x_k} = \sqrt{x}$.

9. Prove the completeness of the integers (every Cauchy sequence of integers converges to an integer). Why is this result not very interesting?

10. Prove that the irrational numbers are dense in $\mathbb{R}$.

2.4 Other Versions and Visions

2.4.1 Infinite Decimal Expansions

We have now completed the basic task of establishing the real number system. Ahead of us lies the more challenging task of exploring the deeper properties of this system. But before plunging ahead, we should pause to consider some alternate ways we might have proceeded. First we will discuss other versions of the same system: infinite decimal expansions and Dedekind cuts. Then we will briefly discuss other visions, essentially different mathematical systems that offer a competing view of what the real number system should be.

Let's consider infinite decimal expansions. From the decimal expansion we obtain immediately a Cauchy sequence of rationals by truncating (if $\sqrt{2} = 1.414\ldots$, then $1, 1.4, 1.41, 1.414, \ldots$ is a Cauchy sequence of rationals defining $\sqrt{2}$). In fact we can say that an infinite decimal expansion is a special kind of Cauchy sequence of rationals, one for which $x_k = n + \sum_{j=1}^{k} a_j/10^j$ where n is an integer and $0 \leq a_j \leq 9$ (for negative numbers this must be slightly modified). In fact it is trivial to verify that this is a Cauchy sequence since

$$x_k - x_m = \sum_{j=m+1}^{k} \frac{a_j}{10^j} \leq 10^{-m} \quad \text{if } k \geq m.$$

A less trivial matter is the question whether every real number (defined by a Cauchy sequence of rationals) has an infinite decimal expansion. The naive approach to solving this problem would be to write out the infinite decimal expansion of each rational number x_n in the Cauchy sequence defining x and to hope that these expansions eventually settle down to the expansion for x. Of course the example $1.1, .99, 1.01, .999, 1.001, .999, \ldots$ shows that this procedure may not always work. However, the key observation is that it can only fail for numbers like $x = 1$, which have two distinct infinite decimal expansions. Such numbers are of the form $m/10^k$; and since we know how to write infinite decimal expansions for them, we can use the naive procedure on all the other numbers.

Thus assume x is a real number such that $x \neq m/10^k$ for any m or k. For simplicity let us assume $x > 0$. Let $x_1, x_2, x_3, \ldots$ be any Cauchy sequence of rational numbers defining x. Consider the infinite decimal expansions of the rational numbers $x_1, x_2, \ldots$. We want to show that eventually they all agree to any number of terms desired. For example, consider the first three terms $n \cdot a_1 a_2 a_3$. The rational numbers r whose infinite decimal expansions begin this way are those satisfying the inequalities $n + a_1/10 + a_2/100 + a_3/1000 < r < n + a_1/10 + a_2/100 + a_3/1000 + 1/1000$ (what happens at the endpoint is a matter of convention). We want to show that eventually all the rational numbers x_k satisfy one such inequality, namely the one satisfied by x. Indeed x must satisfy such an inequality because we have assumed it is not of the form $n + a_1/10 + a_2/100 + a_3/1000$ (in more detail, choose the largest number of the form $n.a_1 a_2 a_3$ such that $n.a_1 a_2 a_3 < x$; then we must have $x < n.a_1 a_2 a_3 + 1/1000$ because $x \neq n.a_1 a_2 a_3 + 1/1000$ and if we had $x > n.a_1 a_2 a_3 + 1/1000$, then $n.a_1 a_2 a_3$ would not be the largest).

We claim that the first three terms of the infinite decimal expansions of the rational numbers x_k are all $n.a_1 a_2 a_3$ for k sufficiently large. To do this we clearly need the following result.

Lemma 2.4.1 *If $y < x < z$ for any three real numbers and if $\{x_k\}$ is a Cauchy sequence of rationals defining x, then there exists m such that $y < x_k < z$ for all $k \geq m$.*

Proof: The idea of the proof is represented by Figure 2.4.1 where the brackets [ ] indicate an interval $[x - 1/n, x + 1/n]$ that lies within the interval (y, z). By the axiom of Archimedes we can always choose the error $1/n$ small enough to achieve this fit (since $x - y$ and $z - x$

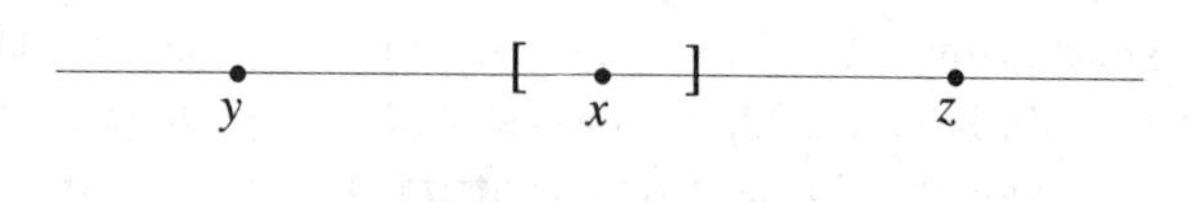

Figure 2.4.1:

are both positive by assumption, they must both be bounded below by $1/n$ for some n, and these inequalities translate into the above picture). Since $\lim_{k\to\infty} x_k = x$, there must be an m such that x_k lies within the interval in brackets for all $k \geq m$, and hence $y < x_k < z$. QED

Taking $y = n.a_1a_2a_3$ and $z = n.a_1a_2a_3+1/1000$ in the lemma shows that all the infinite decimal expansions of the x_k beyond a certain point begin with $n.a_1a_2a_3$. In this way we can find any number of terms in the decimal expansion of x, so the infinite decimal expansion of x is defined unambiguously for every real number not of the form $n+\sum_{j=1}^{m} a_j/10^j$. However, there is one severe shortcoming to this procedure: there is no a priori bound for how far out in the sequence $\{x_k\}$ you have to go before, say, a_1 is determined. For example, suppose you know that for all $j, k \geq m_0$, $|x_j - x_k| \leq 1/100$. Does this mean that the decimal expansions of all x_j for $j \geq m_0$ agree up to a_1? Well, it depends! If one such x_j is 1.427, then all the others must lie between 1.417 and 1.437 and so must begin 1.4. But if one such x_j is 1.4005, then all we can say is that all the others must lie between 1.3905 and 1.4105, so we don't know if the infinite decimal expansion begins 1.3 or 1.4. Futhermore, we cannot say in advance how accurately we need to control the variation in the x_j before we can decide between 1.3 and 1.4. All we know is that if we continue long enough, we will eventually get a decision (this is only true because we have assumed x is not of the form $n + \sum_{j=1}^{m} a_j/10^j$).

We have now shown that every real number has an infinite decimal expansion. To complete the identification of the real number system with the system of infinite decimal expansions, we would have to define the arithmetic (addition and multiplication) of infinite decimals and the order relation $x > y$ and show they agree with the arithmetic and order relation already defined for the real number system. This would actually be a rather noxious task, since we have to deal with the possibility of infinite carries. Thus we will have to be content with the plausibility of the outcome. We will have no further need for infinite decimal expansions in this work.

2.4.2 Dedekind Cuts*

The other version of the real number system that we should discuss is the Dedekind cut construction. The method of Dedekind can actually be traced back to the work of the Greek mathematician Eudoxes, which is included in Euclid's text. It is based entirely upon the ordering of the rational numbers. The idea is that a real number x creates a division of the rational numbers into two sets: those greater than x and those less than x. If x itself is rational, then we have also to do something with x; and by convention we can lump it with the big guys. Thus, assuming we have constructed a real number system, we want to associate to each real number x the set L_x of all *rational* numbers less than x. What kind of set is L_x? Clearly it contains some but not all rational numbers; and if r is in L_x, then every rational number less than r is also in L_x. The convention that x should not belong to L_x if x is rational means that L_x does not contain a largest rational number. It turns out that these properties characterize the sets of rational numbers that are of the form L_x. That is, any subset L of the rational numbers that satisfies:

1. L is not empty;
2. L is not all the rationals;
3. if r is in L and $q < r$, then q is in L;
4. if r is in L, there exists s in L with $s > r$

must be L_x for some real x. Since we have already constructed the real numbers, we can prove this statement as a theorem.

Dedekind's idea, however, was to use this proposition as a definition, and by means of it to construct the real number system. In other words, let us return to our initial position of knowing only the rational number system. It is certainly possible to think of sets of rational numbers L satisfying 1–4. Call such sets *Dedekind cuts*, and define the Dedekind real number system to be the set of all Dedekind cuts! We can identify the rational number q with the Dedekind cut $L_q = \{r$ rational: $r < q\}$ and so embed the rational number system. We can also define arithmetic and order on Dedekind cuts. For example, $L_1 + L_2$ is the cut L consisting of all rationals of the form $r_1 + r_2$ where r_1 is in L_1 and r_2 is in L_2 (multiplication, alas, is more tricky to define, because the product of two negative numbers is positive—it requires a segregation according to sign). Of course such a definition requires a proof that L is actually a cut and that $L_{q_1} + L_{q_2} = L_{(q_1+q_2)}$ if q_1 and q_2 are rationals so that the sum for rationals agrees with the sum for cuts. The order relation $L_1 < L_2$ is the same as containment, $L_1 \subset L_2$. One can then show, with some work, that the Dedekind real number system is a complete ordered field.

What we want to see now is that the Dedekind real number system is identical to the real number system that we have constructed. We have already seen how to associate a Dedekind cut L_x to a real number x. The fact that distinct real numbers $x \neq y$ give rise to distinct cuts $L_x \neq L_y$ is just the observation that there exist rational numbers in between x and y—a fact we observed as a consequence of the axiom of Archimedes. The fact that every Dedekind cut is of the form L_x, which we mentioned but have yet to prove, will show that the correspondence $x \to L_x$ is onto.

So let L be a Dedekind cut; that is, a subset of the rational numbers satisfying the properties 1–4. We want to find a real number x such that $L = L_x$. The idea of the proof is divide and conquer. We know from the first two properties of L that there is at least one rational, call it p_1, in L, and at least one rational q_1, not in L. Now $q_1 > p_1$ by property 3, and clearly x (if it exists) must lie in between, as shown in Figure 2.4.2.

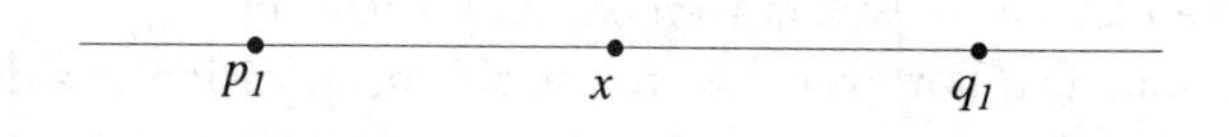

Figure 2.4.2:

Consider the midpoint $(p_1 + q_1)/2$. If it lies in L, then x (if it exists) must lie between it and q_1, as shown in Figure 2.4.3; whereas if it does not lie in L, then x (if it exists) must lie between p_1 and it, as in Figure 2.4.4.

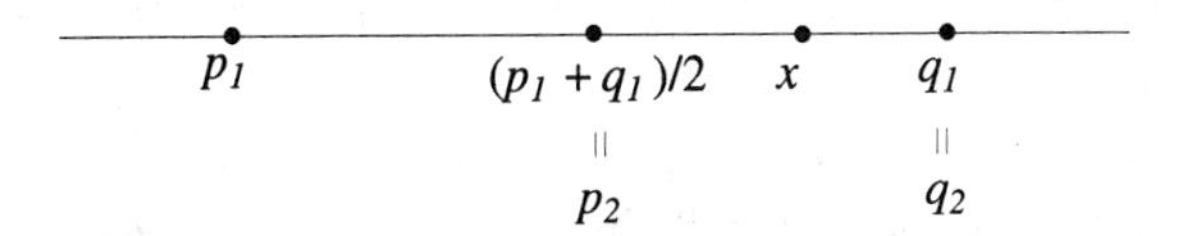

Figure 2.4.3:

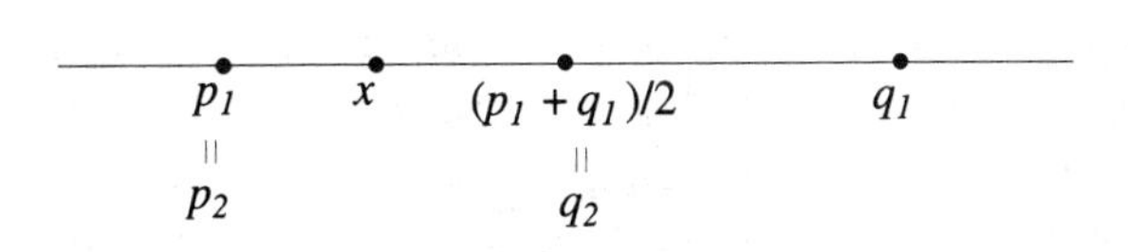

Figure 2.4.4:

In the first case let $p_2 = (p_1 + q_1)/2$ and $q_2 = q_1$, and in the second case let $p_2 = p_1$ and $q_2 = (p_1 + q_2)/2$. In other words we replace one of the two points p_1 or q_1 by the midpoint and leave the other alone. In this way we still have p_2 in L and q_2 not in L, but the gap between p_1 and q_1 has been halved. We then repeat the process, dividing the interval p_2, q_2 in half and choosing for p_3, q_3 the half such that p_3 is in L and q_3 is not in L. Repeating the process indefinitely we obtain sequences $p_1, p_2, \ldots$ and $q_1, q_2, \ldots$ of rationals. It is easy to see that these are Cauchy sequences; and they are equivalent, so define a real number x. It seems plausible that for this $x, L_x = L$. Let us in fact prove it. Remember that L_x was defined as the cut consisting of all rational numbers less than x. L was the cut with which we started. Suppose q is a rational number in L. Why is q in L_x? Here we have to use property 4. There exists r in L greater than q. Eventually the sequence $p_1, p_2, \ldots$ must exceed r (whatever the initial distance $q_1 - r$, the distance $q_k - p_k$ must eventually be smaller—since it is halved at each step—and so $p_k > r$, as shown in Figure 2.4.5).

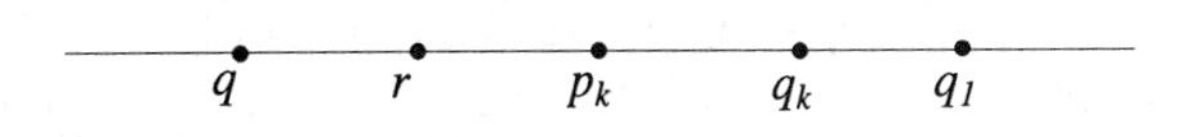

Figure 2.4.5:

From $p_k \geq r$ for all large k we obtain $x \geq r$ and so $x > q$. This says q is in L_x.

Conversely suppose q is in L_x; i.e., $q < x$. We want to show q is in L. We claim $q \leq p_k$ for some k, for if not, then $q \geq p_k$ for all k and hence $q \geq x$, contradicting $q < x$. Thus $q \leq p_k$ and so q is in L by property 3.

This shows $L = L_x$ and so establishes a one-to-one correspondence between the real numbers and the Dedekind cuts. Of course to show that the two number systems are the same we have to verify that the arithmetic operations and ordering relations agree. For example, is the cut $L_{(x+y)}$ associated with the real number $x + y$ the same as the sum $L_x + L_y$ of the cuts associated with the numbers x and y? Remember that $L_{(x+y)} = \{\text{rationals } q < x+y\}$ while $L_x+L_y = \{\text{rationals } q = q_1+q_2$ where $q_1 < x$ and $q_2 < y\}$. Clearly $L_x + L_y \subseteq L_{(x+y)}$ because if $q_1 < x$ and $q_2 < y$, then $q_1 + q_2 < x + y$. For the reverse inclusion we need to see that if q is a rational number less than $x + y$, then it can be written $q = q_1 + q_2$ where q_1 and q_2 are rational numbers with $q_1 < x$ and $q_2 < y$. But by the axiom of Archimedes $x + y - q > 1/n$ for some n, so if we take q_1 any rational such that $x - 1/2n < q_1 < x$, then $q_2 = q - q_1$ is less than the largest value for q, which is $x + y - 1/n$, minus the small value q_1, which is $x - 1/2n$, as shown in Figure 2.4.6. In other words, by making q_1 closer to x than q is to $x + y$, this forces

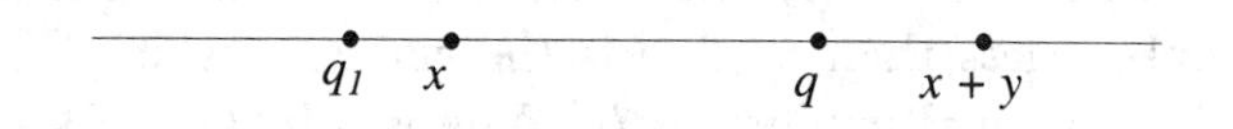

Figure 2.4.6:

$q_2 = q - q_1$ to be less than y.

The argument for products is similar, although more complicated, while the argument for the order relations is extremely simple (can you give it?).

Thus the method of Dedekind cuts gives a third version of the same real number system. Comparing it to the method of Cauchy sequences, we can see some advantages and disadvantages. The main advantage of the Dedekind cut method is that it involves less set theory: we only have to deal with a countable set of rationals and its power set—the set of subsets of the rationals. For the Cauchy sequence method we have to deal with equivalence classes of Cauchy sequences, involving sets of sets of rationals, just to get a single real number.

On the other hand, we are going to have to deal with the concept of Cauchy sequences eventually, whereas the concept of a Dedekind cut is not of much significance once the real number system is established. But the most compelling argument in favor of the Cauchy sequence method must look to the future. The construction of the reals from the rationals via Cauchy sequences is a proto-type of a general construction that is used frequently in mathematics. Thus there is an advantage to becoming familiar with this construction in its most concrete example. (It is also true that there are some constructions in mathematics that are generalizations of the method of Dedekind cuts, but these are less frequently encountered. Probably, a well-educated mathematician should be familiar with both approaches.)

2.4.3 Non-Standard Analysis*

We have constructed the real number system in which the Axiom of Archimedes is valid—there is no positive number less than $1/n$ for all n. Another way of saying this is that there are no infinitesmals. Nevertheless, there is a body of informal mathematics that is based on the concept of infinitesmals. We know that many of the mathematicians who contributed to the development of the calculus during the seventeenth and eighteenth centuries believed that the true foundations of the subject should be based on infinitesmals; and even after the general acceptance of the non-infinitesmal foundations in the nineteenth century, the use of infinitesmals in informal or heuristic arguments persisted. It would seem plausible, then, that there should exist a logically satisfactory foundation for analysis based on infinitesmals. Such a foundation was finally discovered around 1960 by Abraham Robinson, who called it *non-standard analysis.* The reason it took so long to discover is that it is a rather sophisticated mathematical system, requiring a strong background in mathematical logic and general topology in order to be understood. While I cannot begin to describe this system precisely here, I can give some indication of its general features.

We start with the real number system $\mathbb{R}$, and we enlarge it to the non-standard real number system $^*\mathbb{R}$, which contains, in addition to real numbers, infinitesmal numbers and infinite numbers (the reciprocals of the infinitesmals). The general finite number in $^*\mathbb{R}$ is the sum of a real number and an infinitesmal. The real number is uniquely determined by the finite non-standard number and is called its standard part. Thus we have the mental picture of the real line enlarged by surrounding each real point by a cloud, or galaxy, of infinitesmally close points. The cloud about the point 0 is the set of infinitesmals, and the reciprocals of this cloud form the infinite non-standard numbers. The set of infinitesmals is uncountable; in fact, it has cardinality equal to the set of subsets of real numbers, which is greater than the cardinality of the real numbers.

Every concept involving real numbers has an extension to the non-standard real numbers, and every true statement about the real numbers is also true about the non-standard real numbers, if properly interpreted. For example, the non-standard real numbers form a complete ordered field. However, the interpretation of statements in the non-standard real numbers requires a great deal of caution. For example, consider the Axiom of Archimedes. Since the non-standard real numbers contain infinitesmals, numbers $x > 0$ such that $x < 1, x < 1/2, x < 1/3, \ldots$, it is clear that the Axiom of Archimedes in its *usual* interpretation is false for $^*\mathbb{R}$. Nevertheless, since the Axiom of Archimedes is true for $\mathbb{R}$, it must be true for $^*\mathbb{R}$ in the *appropriate* interpretation. To find the appropriate interpretation we write out the statement formally: for every positive x in $\mathbb{R}$ there exists n in N (the natural numbers) such that $x > 1/n$. We then must substitute the non-standard version for everything—not just x in $\mathbb{R}$ but also n in N. There is a non-standard notion of natural number, a subset $^*N$ of $^*\mathbb{R}$ that is larger than N. It contains infinite integers as well as finite integers. Thus the true interpretation is: for every positive x in $^*\mathbb{R}$ there exists n in $^*N$ (the non-standard natural numbers) such that $x > 1/n$. If x is an infinitesmal, then $x > 1/n$ for some infinite integer n.

The real power of the non-standard real number system is that essentially all heuristic reasoning concerning infinitesmals can be made into valid logical proofs. For example, the derivative $f'(x)$ can be computed by forming the difference quotient $f(x + h) - f(x)/h$ when h is an infinitesmal and by taking the standard part of this non-standard real number (taking the standard part replaces the dubious practice of discarding infinitesmals in the final limit). Similarly, limits of infinite sequences $x_1, x_2, \ldots$ become the standard part of x_n for n infinite, and integrals can be interpreted as infinite sums of infinitesmals. Because the non-standard real number system encompasses valid methods of reasoning that were formerly thought of as merely plausible arguments, it has enabled mathematicians to prove theorems that they might otherwise have not discovered.

On the other hand, there are also several drawbacks to the non-standard real number system that must be pointed out:

1. It is not unique. There are many non-standard real number systems $^*\mathbb{R}$ that are not equivalent to each other but work just as well, so there is no reason to prefer one to the another. Intuitively one would expect the real number system to be unique—of course, one first has to deal with the more serious problem of the non-uniqueness of set theory, but once a version of set theory is chosen the real number system $\mathbb{R}$ is uniquely determined.

2. The "construction" if $^*\mathbb{R}$ (there are now several equivalent ones) involves the use of the axiom of choice in a more serious way than the "construction" of $\mathbb{R}$. This is not to say that our "construction" of $\mathbb{R}$ is truly "constructive". When we say there exists a Cauchy sequence of rationals ... we do not imply the existence of an algorithm for producing the rationals in the sequence. Our arguments also frequently use the countable axiom of choice—we assume that a countable number of unspecified choices can be made. However, the very definition of $^*\mathbb{R}$ requires the uncountable axiom of choice—simultaneously an uncountable number of unspecified choices must be made. This puts an additional level of abstraction between the intuition and the non-standard real number system.

3. Any theorem about the real number system that can be proved using the non-standard real number system can be proved without it. This is a meta-mathematical theorem; there is even a translation mechanism for taking a non-standard proof and replacing it with a standard one. In fact, the first well-known published theorem discovered using non-standard analysis was followed in the same journal by a "translation" so that mathematicians unfamiliar with non-standard analysis could follow the proof. From a practical standpoint, however, this is a rather trivial drawback,

since we are interested in theorems whose proofs we can discover rather than whose proofs exist. Since non-standard analysis is a proven aid to the discovery of proofs, it helps us learn more mathematics.

4. Non-standard analysis is more difficult than standard analysis to learn. We will take this drawback as the incontestable reason for ending this discussion of non-standard analysis.

2.4.4 Constructive Analysis*

At the opposite philosophic pole to non-standard analysis, we find Erret Bishop's theory, enunciated in his book *Foundations of Constructive Analysis* (Academic Press, 1967). The idea behind his work is that we have failed to make the distinction in meaning between "there exists blah" and "there exists an algorithm for constructing blah". This failure is embedded in our logic, which allows us to pass in an argument from a constructive existence to a non-constructive existence. A simple example of this is as follows: let $x_1, x_2, \ldots$ be the Cauchy sequence of rationals defined by $x_n = 0$ if every even number (≥ 4) less than n is the sum of two primes, while otherwise $x_n = 1/k$ where k is the first even number (≥ 4) less than n that is not the sum of two primes. Clearly there is an algorithm (involving finding all primes less than n, forming all sums of pairs, and comparing the list with all even numbers less than n) for computing x_n in a finite number of steps. The number this Cauchy sequence represents, call it x, is 0 if Goldbach's conjecture is true and $x > 0$ if Goldbach's conjecture is false. However, our proof that $x = 0$ or $x > 0$ does not provide us with an algorithm for deciding which case is true—for if it did it would provide us with an algorithm for settling Goldbach's conjecture. Such an algorithm might exist, but no one expects its discovery to be a routine consequence of real analysis.

Thus the theorems proved in the standard theory of the real number system are lacking in constructive content. We don't know if things asserted to exist can actually be found, even if every hypothesis of the theorem is true constructively. If we are interested in the constructive content of our theorems we must initiate a new mathematical development. Such a posteriori development is given by the theory of *recursive* functions; however, the school of Bishop rejects this development because it allows too broad a notion of algorithm (a *recursive* sequence x_n consists of a finite set of instructions that enables you to compute x_n in a finite length of time given n; the problem is that there may not be a proof that the length of the computation is finite even though this is in fact the case). Instead, Bishop proposes that we begin a priori with only algorithmic existence and use a system of logic that allows us to deduce only algorithmic existence. He develops a substitute for the real number system, called the *constructive real number system*, and he is able to prove substitute constructive versions of the theorems in this book.

The definition of the constructive real number system bears a superficial resemblance to the definition of the real number system. It is based on Cauchy sequences of rationals, but with the additional restriction that there must be an algorithm for demonstrating the rate of convergence (Bishop uses the technical trick of requiring $|x_n - x_m| < 1/n + 1/m$ for all his Cauchy sequences). However, the actual number system that results from this definition is radically different from the usual one. The most striking distinction is that the act of doing mathematics actually changes the constructive real number system. The reason for this is that there is a large "don't know" category of sequence $x_1, x_2, \ldots$ of rationals for which we neither have a proof of the Cauchy criterion nor a disproof. These sequences do not define constructive real numbers *now*. However, if tomorrow we discover an algorithmic proof that one of these does satisfy the Cauchy criterion, then that sequence defines a new constructive real number. Another way the system may change is that we may "learn" that two numbers x and y are really equal, if we discover an algorithmic proof that the Cauchy sequences defining them are equivalent.

It is easy to dismiss the work of Bishop and his disciples as that of a group of reactionary religious zealots intent on bullying the mathematical community into accepting its peculiar orthodox tenets. This is probably the attitude of most mainstream mathematicians, and it is aided and abetted by some overenthusiastic pronouncements from the constructivist school. However, I believe it is more valuable to look on this work as an interesting, and potentially fruitful, separate branch of mathematics. There is no doubt that the theorems of constructive analysis are legitimate mathematical discoveries, even when interpreted in conventional terms. They spell out in precisely what ways conventional statements can be made algorithmic. Of course it is the responsibility of constructive mathematicians to come up with mathematical discoveries that will be exciting and profound enough to capture the imagination of the larger mathematical community, if they want their work to be taken very seriously.

Constructive analysis should not be confused with applied mathematics, where the goal is not just algorithms but algorithms that can be carried out efficiently. Applied mathematics has not been notably influenced by the work of the constructivists. It has, on the contrary, benefitted greatly from mathematics developed in a completely non-constructive spirit, which leads to the paradox that constructive mathematics may be best advanced by allowing free use of non-constructive reasoning.

One can also argue that mathematics, as it develops, becomes progressively less constructive. Even the notion of "construction" has evolved from the purely finite notion in Greek mathematics (straightedge and compass) to the contemporary idea of an algorithm giving a finite set of instructions for obtaining successive approximations. Constructivism can then be seen as a "reaction" to this "progress".

In the remainder of this book, we will freely use non-constructive ideas, although whenever possible we will give constructive demonstrations because they usually convey more information. Thus when we assert that something exists, we do not necessarily imply that we have provided explicitly or implicitly a method for finding it. The reader who has misgivings about this may wish to study Bishop's work, but I recommend that this be done only after the non-constructive theory has been well digested.

2.4.5 Exercises

1. Let L be the set of all negative rational numbers and those positive rational numbers satisfying $r^2 < 2$. Show that L is a Dedekind cut that represents the number $\sqrt{2}$.

2. If A is a set of real numbers and L_x is the Dedekind cut associated with each x in A, show that the union of all the cuts L_x is itself either a Dedekind cut or the set of all rationals.

3. If x and y are positive reals, show that L_{xy} consists of all non-positive rationals and all positive rationals of the form rs where r and s are positive rationals in L_x and L_y.

4. Give a simple non-constructive proof that there exist positive irrational numbers a and b such that a^b is rational by considering the possibilities for $\sqrt{2}^{\sqrt{2}}$ and $(\sqrt{2}^{\sqrt{2}})^{\sqrt{2}}$. Does this proof allow you to compute a to within an error of $1/100$?

5. Let a_k, b_k, c_k, n_k for $k = 1, 2, \ldots$ be an enumeration of all quadruples of positive integers with $n_k \geq 3$. Let $x_j = 0$ if $a_k^{n_k} + b_k^{n_k} \neq c_k^{n_k}$ for all $k \leq j$ and otherwise $x_j = 1/k$ where k is the smallest integer $\leq j$ for which $a_k^{n_k} + b_k^{n_k} = c_k^{n_k}$. Prove that $x_1, x_2, \ldots$ is a Cauchy sequence. What is the relationship between the Axiom of Archimedes for the real number given by this Cauchy sequence and Fermat's Last Theorem?

2.5 Summary

2.1 Cauchy Sequences

Definition *A Cauchy sequence of rational numbers is a sequence $x_1, x_2, \ldots$ of rational numbers such that for every n there exists m such that $|x_j - x_k| \leq 1/n$ for all j and $k \geq m$.*

Definition *Two Cauchy sequences $x_1, x_2, \ldots$ and $y_1, y_2, \ldots$ are* **equivalent** *if for every n there exists m such that $|x_k - y_k| \leq 1/n$ for $k \geq m$.*

Lemma 2.1.1 *The equivalence of Cauchy sequences of rationals is an equivalence relation (it is symmetric, reflexive, and transitive).*

Definition 2.1.1 *A* **real number** *is an equivalence class of Cauchy sequences of rationals.*

2.2 The Reals as an Ordered Field

Definition 2.2.1 *The real number $x + y$ is the equivalence class of*

$x_1+y_1, x_2+y_2, \dots$ *where* $x_1, x_2, \dots$ *and* $y_1, y_2, \dots$ *are Cauchy sequences of rationals representing* x *and* y.

Lemma 2.2.2 *A Cauchy sequence is bounded.*

Definition 2.2.2 *The real number* $x \cdot y$ *is the equivalence class of* $x_1y_1, x_2y_2, \dots$ *where* $x_1, x_2, \dots$ *and* $y_1, y_2, \dots$ *are Cauchy sequences representing* x *and* y.

Definition *A field is a set with two operations, addition and multiplication, such that both operations are commutative and associative, multiplication distributes over addition, there exist distinct additive identity* 0 *and multiplicative identity* 1, *and every element* x *has an additive inverse* $-x$ *and a multiplicative inverse* x^{-1} *(if* $x \neq 0$*).*

Theorem 2.2.1 *The real numbers form a field.*

Lemma 2.2.4 *Every real number not equal to zero is bounded away from zero (there exists* N *such that for every Cauchy sequence* $x_1, x_2, \dots$ *representing* x *there exists* m *with* $|x_j| \geq 1/N$ *for all* $j \geq m$*).*

Definition 2.2.3 *A real number* x *is positive if there exists* N *and* m *and* $x_1, x_2, \dots$ *a Cauchy sequence representing* x *such that* $x_j \geq 1/N$ *for all* $j \geq m$. *A real number is negative if* $-x$ *is positive.*

Theorem 2.2.2 *The reals form an ordered field—every real number is either positive, negative, or zero, and sums and products of positive numbers are positive.*

Definition $x > y$ *means* $x - y$ *is positive,* $|x| = x$ *if* $x \geq 0$ *and* $-x$ *if* $x < 0$.

Theorem 2.2.3 (*Triangle Inequality*) $|x + y| \leq |x| + |y|$.

Theorem 2.2.4 (*Axiom of Archimedes*) *If* $x > 0$ *there exists* n *such that* $x \geq 1/n$.

Theorem 2.2.5 (*Density of Rationals*) *Given x a real number and n, there exists a rational number y such that $|x - y| \leq 1/n$.*

2.3 Limits and Completeness

Definition *A Cauchy sequence of real numbers is a sequence $x_1, x_2, \ldots$ of real numbers such that for every n there exists m such that $|x_j - x_k| \leq 1/n$ for all j and $k \geq m$.*

Definition *For x a real number and $x_1, x_2, \ldots$ a sequence of real numbers, $x = \lim_{k\to\infty} x_k$ if for every n there exists m such that $k \geq m$ implies $|x_k - x| \leq 1/n$.*

Theorem 2.3.1 (*Completeness of Reals*) *Every Cauchy sequence of real numbers has a real limit.*

Theorem 2.3.2

a. *Limits commute with addition, multiplication, and division (with non-zero denominator).*

b. *Limits preserve non-strict inequalities.*

Theorem 2.3.3 *Every positive real has a unique real positive square root.*

2.4 Other Versions and Visions

Theorem *Every real number has an infinite decimal expansion.*

Definition *A Dedekind cut is a set L of rational numbers satisfying*

1. *L is not empty;*
2. *L is not all the rationals;*
3. *if r is in L and $q < r$, then q is in L;*
4. *if r is in L there exists s in L with $s > r$.*

Theorem *There is a one-to-one correspondence between real numbers and Dedekind cuts.*

Chapter 3

Topology of the Real Line

3.1 The Theory of Limits

3.1.1 Limits, Sups, and Infs

In this chapter we delve deeper into the properties of real numbers, sequences of real numbers, and sets of real numbers. Many of the concepts introduced in this chapter will reappear in a broader context (metric spaces) in Chapter 9. By considering these concepts first in the concrete case of the real number line, you will have the opportunity to develop an intuition for them. This will make it easier to appreciate the generalizations that follow.

We can think of the real number system as representing a geometric line. We are interested in properties that have a qualitative geometric nature, which is the meaning of the word "topology". One of the fundamental concepts is that of limit, which we have already discussed but repeat its definition for emphasis. *If $x_1, x_2, x_3, \ldots$ is a sequence of real numbers and if x is a real number such that given any error $1/n$ there exists a place in the sequence m, such that $|x - x_j| < 1/n$ for all $j \geq m$, then we say x is the limit of the sequence, $x = \lim_{j\to\infty} x_j$.* A limit need not exist, but if it does it is unique. We have already motivated this definition by the idea that any number from the sequence beyond the mth place is very close to x. We can also think of this in a geometric way. For each $1/n$, the set of all real numbers y that satisfy the inequality $|x - y| < 1/n$ is the open interval $x - 1/n < y < x + 1/n$ of length $2/n$ centered at x. Think of this interval as comprising a neighborhood of x (we will give this a precise definition later). As we increase n these neighborhoods shrink, giving a nested picture, as shown in Figure 3.1.1,

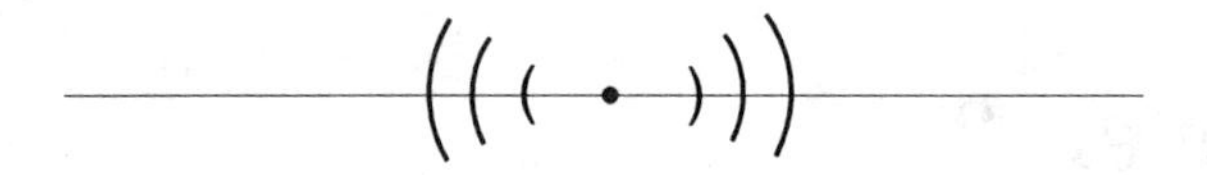

Figure 3.1.1:

more and more accurately pinpointing the location of x. The definition of the limit says that $\lim_{j\to\infty} x_j = x$ if the sequence eventually $(j \geq m)$ lies entirely in each neighborhood.

It is sometimes convenient to give a meaning to the expressions $\lim_{j\to\infty} x_j = +\infty$ and $\lim_{j\to\infty} x_j = -\infty$. If we think of the half-infinite intervals $\{x : x > n\}$ and $\{x : x < n\}$ as defining neighborhoods of $+\infty$ and $-\infty$, respectively (allowing n to be any integer, positive or negative), then the definition is again the same: $\lim_{j\to\infty} x_j = +\infty$ if the sequence eventually lies entirely in each neighborhood of $+\infty$. Sometimes it is convenient to think of the symbols $+\infty$ and $-\infty$ as standing for new numbers in what is called the *extended real number system*. The system consists of the real numbers together with $+\infty$ and $-\infty$. It has an obvious order ($+\infty$ is the biggest, $-\infty$ the smallest) and a limited arithmetic, with rules like $x + (+\infty) = +\infty$ for x real, but certain expressions, such as $+\infty + (-\infty)$, must remain undefined.

Of course not every sequence that fails to have a real number limit will have $+\infty$ or $-\infty$ as a limit. A sequence that jumps about, such as $0, 1, 0, 1, 0, 1, \ldots$, is an obvious counterexample. (Some mathematicians, such as Leibniz and Euler, felt that this sequence *should* have $1/2$ as limit; nevertheless *it does not* under the definition we have adapted, since the sequence never lies in the neighborhood $1/4 < x < 3/4$ about $1/2$.) However, we can consider weaker notions that provide some information about where the points of a sequence lie, and these notions turn out to have great importance. The simplest of these are the supremum and infimum, which are based on the order properties of the real numbers. These concepts can be defined not just for sequences but for any sets of real numbers. We will use the abbreviation sup and inf.

If E is any finite set of real numbers, we can define sup E and inf E to be the largest and smallest numbers in E, respectively. But if E is an infinite set of real numbers, there may not be a largest or smallest number in E. For example, the set of positive integers has no largest number; the set of positive reals has no smallest number. Nevertheless, if we imagine the set E as represented geometrically on the line, there should be a left-most and right-most point (possibly $-\infty$ or $+\infty$) indicating the range of the set; these endpoints might or might not belong to the set. We will call them inf E and sup E. How are we to define them? First let us develop the intuition that leads us to believe they exist; this will indicate the definition we want to adopt, and then we will have to give a proof that inf E and sup E exist. For simplicity we will deal only with sup E; the treatment of inf E is completely analogous, or one may simply say that inf E is minus the sup of $-E$. To avoid triviality we assume E is non-empty.

If sup E is supposed to indicate the top-most extent of E, then every point in E must be less than or equal to it. Thus, in searching for sup E, we can confine attention to numbers y with the property that every number x in E satisfies $x \leq y$. (The example of a finite set should convince you that we want $x \leq y$ and not $x < y$.) Such numbers are called *upper bounds* of E. Not every set possesses upper bounds, for example, the set of positive integers. We say E is *bounded from above* if it has any upper bounds or *unbounded from above* if it has none. By convention we set sup $E = +\infty$ if E is unbounded from above.

Among all upper bounds for E (if they exist), which shall we choose for sup E? Obviously the smallest one! But are we sure that a smallest one exists? We have already mentioned that some sets of numbers, such as the positive numbers, have no smallest element. Well, suppose we start with one upper bound, call it y_1. If it is not the smallest, then pick a smaller one, y_2.

Continuing in this way, we could pick a sequence $y_1, y_2, y_3, \ldots$ of upper bounds that get smaller and smaller and hope that $\lim_{j\to\infty} y_j = y$ would give us sup E. Unfortunately, we have to be a bit more careful to really go down far enough at each step—for it is clear from Figure 3.1.2 only that y is an upper bound for E, not that it is the smallest.

To bring the sequence $\{y_j\}$ down close to E we must consider some points in E as well as upper bounds. Along with y_1 choose a point

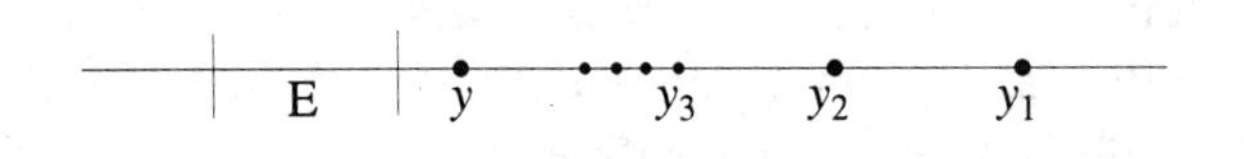

Figure 3.1.2:

x_1 in E (we assumed E is not empty). Since y_1 is an upper bound for $E, x_1 \leq y_1$, and clearly the sup must lie somewhere in between. Again we use a divide and conquer argument. Consider the midpoint, $(x_1 + y_1)/2$. If it is an upper bound for E, choose it for y_2, and set $x_2 = x_1$. If it is not an upper bound for E, then it fails because there is some point x_2 in E bigger than the midpoint; in this case we take $y_2 = y_1$. The two cases are illustrated in Figure 3.1.3.

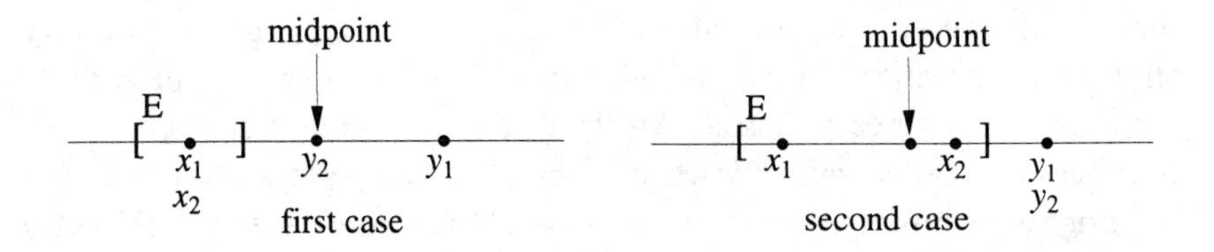

Figure 3.1.3:

In either case, we have replaced the original pair x_1, y_1 with a new pair x_2, y_2 with x_2 in E and y_2 an upper bound to E, with the distance apart $|x_2 - y_2|$ at most half the original distance $|x_1 - y_1|$ (in the first case $|x_2 - y_2| = |x_1 - y_1|/2$ since y_2 is the midpoint, while in the second case $|x_2 - y_2| < |x_1 - y_1|/2$ since x_2 is greater than the midpoint). By iterating this argument we obtain an increasing sequence $x_1, x_2, x_3, \ldots$ of points in E and a decreasing sequence $y_1, y_2, y_3, \ldots$ of upper bounds for E such that $|x_n - y_n| \leq |x_1 - y_1|/2^n$. It is an easy matter to conclude from this that these are equivalent Cauchy sequences and, hence, converge to the same limit; call it y. What can we say about this point y?

1. It is an upper bound for E. In fact this follows from the more general fact that *a limit of upper bounds for E is an upper bound*

for E. The condition of being an upper bound for E is given by non-strict inequalities $x \leq y$ for all x in E, and non-strict inequalities are preserved by limits (thus $x \leq y_j$ for all y_j implies $x \leq \lim_{j\to\infty} y_j$, and this reasoning applies at each point x of E).

2. It is the least upper bound of E: if y' is another upper bound for E, then $y \leq y'$. The reason for this is that $x_j \leq y'$ since x_j is in E and y' is an upper bound for E and so $y = \lim_{j\to\infty} x_j \leq y'$ since the non-strict inequality is preserved in the limit.

Clearly there is only one number with these two properties—and this deserves to be defined as sup E. The terminology *least upper bound*, abbreviated l.u.b., is used synonomously, while greatest lower bound, abbreviated g.l.b., is used for inf. We restate the important theorem that we have established.

Theorem 3.1.1 *For every non-empty set* E *of real numbers that is bounded above, there exists a unique real number* sup E *such that*

1. sup E *is an upper bound for* E*;*

2. *if* y *is any upper bound for* E*, then* $y \geq$ sup E.

A closely related theorem concerns sequences whose terms are increasing (there is an analogous result for decreasing sequences). A sequence $x_1, x_2, \ldots$ is called *monotone increasing* if $x_{j+1} \geq x_j$ for every j. Note that we do not demand strict inequality, which explains the use of the awkward adverb "monotone". If a monotone increasing sequence is unbounded, then it has limit $+\infty$ (once $x_j > n$ we have $x_k > n$ for all $k \geq j$). But suppose it is bounded; it would appear to rise to some finite limit, namely, the sup of the set of numbers $\{x_j\}$.

It is important to understand the distinction between the *sequence* $x_1, x_2, \ldots$ and the *set* $\{x_1, x_2, \ldots\}$. The sequence has the numbers in a specified order, and numbers may be repeated. In the set, elements are unordered, and repeated numbers are treated no differently from unrepeated numbers. Thus the set associated to the sequence $3, 2, 1, 2, 1, \ldots$ is just the three element set $\{1, 2, 3\}$. We will sometimes follow conventional usage and denote a sequence by $\{x_j\}$. It should be clear from context that we mean the sequence and not the set, even though the notation makes no distinction.

Now we define the sup of a sequence to be the sup of the associated set, which we can write either as sup $\{x_j\}$ or $\sup_j x_j$. Let us verify that indeed $\lim_{j\to\infty} x_j = \sup\{x_j\}$ if the sequence is monotone increasing. Let y denote the sup. We know $x_k \leq y$ for every k since y is an upper bound. Since y is the least upper bound, we know $y - 1/n$ is *not* an upper bound, for any choice of $1/n$. What does this mean? It means that $x_j \leq y - 1/n$ must fail for some x_j. This is the same as $x_j > y - 1/n$, and because the sequence is monotone increasing, $x_k > y - 1/n$ for every $k \geq j$, as indicated in Figure 3.1.4.

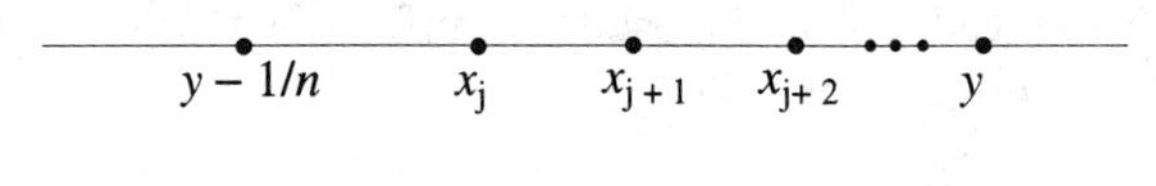

Figure 3.1.4:

The combined inequalities $x_k \leq y$ and $x_k > y - 1/n$ show $|x_k - y| \leq 1/n$ for every $k \geq j$, and this says exactly $\lim_{j\to\infty} x_j = y$. We have shown:

Theorem 3.1.2 *A monotone increasing sequence that is bounded from above has a finite limit, and the limit equals the* sup.

3.1.2 Limit Points

Now we turn to the case of a general sequence, which may not have a limit. We have defined the sup and inf of the sequence (the sup may be $+\infty$ and the inf $-\infty$). The interval between the inf and sup contains all the points in the sequence and is the smallest such interval. Nevertheless, it provides only a very crude indication of where the sequence really lies. For instance, the convergent sequence $0, 3, 1, 1, 1, \ldots$ has inf $= 0$ and sup $= 3$. What do 0 and 3 have to do with the limit? We need a more refined concept, one that is not influenced by only a finite number of terms.

Suppose we take two convergent sequences $x_1, x_2, \ldots$ and $y_1, y_2, \ldots$ that have different limits, x and y, and shuffle them to form the sequence $x_1, y_1, x_2, y_2, \ldots$. The shuffled sequence will not have a limit; nevertheless the two values x and y are connected to the sequence in some weaker but still important way. Suppose we look at a neighborhood of x, the interval from $x - 1/n$ to $x + 1/n$, where we choose $1/n$ so small that y and all points in a neighborhood of y do not lie in this neighborhood (we are assuming $x \neq y$, so this is possible by the axiom of Archimedes). The situation is illustrated in Figure 3.1.5. Will it be true that all the terms of the sequence beyond a certain place lie in the neighborhood? No, because the y_j's eventually all lie in the neighborhood of y.

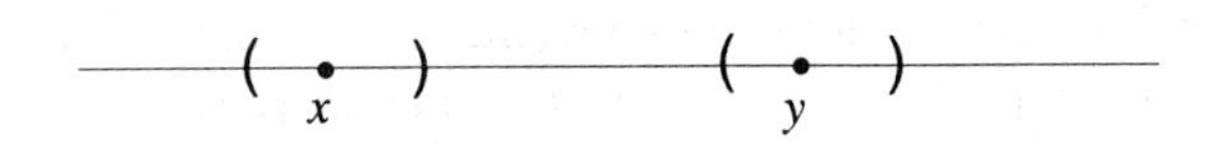

Figure 3.1.5:

However, all the x_j's eventually will lie in the neighborhood of x. We can state this as follows: *an infinite number of terms of the shuffled sequence lie in each neighborhood of x.* Comparing this with the definition of limit, we see that the strong statement, "all terms beyond the mth" has been replaced by a weaker statement, "an infinite number of terms". We use this weakening to define the concept of *limit-point* (the expressions *accumulation point* and *cluster point* are frequently used synonomously).

Definition 3.1.1 *If $\{x_j\}$ is a sequence of real numbers and x a real number, we say x is a limit-point of the sequence if for every error $1/n$, there are an infinite number of terms x_j satisfying $|x - x_j| < 1/n$.*

An equivalent way of formulating this is the following: given any n and m, there exists $j \geq m$ such that $|x_j - x| < 1/n$. You should be able to show that these are equivalent. By convention we say $+\infty$ is a limit-point of the sequence if for every n there are infinitely many terms satisfying $x_j > n$. You should be able to verify that this is equivalent to the condition that the sequence is unbounded from above.

In the example of the shuffled sequence $x_1, y_1, x_2, y_2, \ldots$, the two numbers x and y are both limit-points of the sequence, and there are no other limit-points (why?). In a way this is typical. In order to understand this we need to think about subsequences. The concept of subsequence is extremely simple to comprehend but remarkably difficult to notate. If $x_1, x_2, x_3, \ldots$ is a sequence, a subsequence is any other sequence obtained by crossing out some (possibly infinitely many) terms, keeping the same order for the remaining terms. For an exact definition we first need to specify the class of functions $m(n)$ from the non-negative integers to the non-negative integers that are increasing, $m(n+1) > m(n)$ for all n. Call such a function a *subsequence selection function* (it will pick out the position of the terms that remain after the crossing out). Then $\{y_j\}$ is a *subsequence* of $\{x_j\}$ if there exists a subsequence selection function such that $y_n = x_{m(n)}$. In place of the compound subscripts (it gets worse if you take a subsequence of a subsequence) we will follow the convention of using primes to denote subsequences, so $\{x'_j\}$ denotes a subsequence of $\{x_j\}$. Note that in the shuffled sequence $x_1, y_1, x_2, y_2, \ldots$ each of the original sequences is a subsequence. (What are the corresponding subsequence selection functions?) Notice the connection in this example between limit-points (x and y) and limits of subsequences. This is true in general.

Theorem 3.1.3 *Let $\{x_j\}$ be any sequence of real numbers. A real number (or even an extended real number) x is a limit-point of $\{x_j\}$ if and only if there exists a subsequence $\{x'_j\}$ such that $\lim_{j\to\infty} x'_j = x$.*

Proof: This theorem is almost completely trivial if you understand the definitions. We assume that x is a real number. The case $x = \pm\infty$ is treated similarly. First suppose there exists a subsequence with x as a limit. Given any error $1/n$, the subsequence approaches within $1/n$ of x beyond a certain place, and since all the terms of the subsequence belong to the sequence, there are an infinite number of terms of the sequence within $1/n$ of x. Thus x is a limit-point.

For the converse we have to do a little work; namely, we have to construct a subsequence with limit x, assuming x is a limit-point of the sequence $\{x_j\}$. To make life easy we will choose the subsequence $\{x'_j\}$ so that $|x'_n - x| < 1/n$; this clearly implies $\lim_{j\to\infty} x'_j = x$. How do we choose x'_n? By the definition of limit-point there are infinitely many x_j satisfying $|x_j - x| < 1/n$; we choose $x'_1, x'_2, \ldots$ in order such that, after choosing $x'_1, \ldots, x'_{n-1}$, we take for x'_n some x_j beyond $x'_1, \ldots, x'_{n-1}$ in the original sequence with $|x_j - x| < 1/n$. In this way $\{x'_j\}$ is a subsequence of $\{x_j\}$ with limit x. QED

Thus if the sequence $\{x_j\}$ has a limit-point x, we can think of it as a kind of shuffling of a sequence converging to x with another sequence but not necessarily an even shuffling. Of course the structure of the set of limit-points of a sequence can be quite complex, as the following example illustrates. Let $\{x_j\}$ be a sequence in which every rational number appears infinitely often. Such a sequence is easily constructed by applying the diagonalization argument to the rectangular array

$$\begin{array}{ccccccccc} & r_1 & \nearrow & r_2 & \nearrow & r_3 & \nearrow & \cdots \\ & r_1 & \nearrow & r_2 & \nearrow & r_3 & \nearrow & \cdots \\ \nearrow & r_1 & \nearrow & r_2 & \nearrow & r_3 & \nearrow & \cdots \\ \nearrow & \vdots & \nearrow & & \nearrow & & & \end{array}$$

where each row is the same enumeration of the rational numbers; thus the sequence $\{x_j\}$ is $r_1, r_1, r_2, r_1, r_2, r_3, r_1, r_2, r_3, r_4, \ldots$. It has the property that every real number is a limit-point! Indeed if x is a real number there is a Cauchy sequence of rationals converging to it, and *any* sequence of rationals is a subsequence of $\{x_j\}$ (why?).

A convergent sequence has only one limit-point, namely its limit, since every subsequence of a convergent sequence converges to the same limit. The converse is also true: if a sequence has only one limit-point (counting $+\infty$ and $-\infty$ as possible limit-points), then it is convergent. This is not so obvious, but it will emerge from further considerations.

Does every sequence have a limit-point? We will show that this is the case (allowing $+\infty$ and $-\infty$). In fact there are two special limit-points, the largest, called limsup, and the smallest, called liminf. We could simply define limsup to be the sup of the set of limit-points, but this begs the question of the existence of limit-points. Instead we will write down a formula for limsup. Note that limsup is quite different from sup; for the sequence $2, 1, 1, 1, \ldots$, the sup is 2, but the only limit-point is 1, and this is the limsup. Nevertheless, the sup is a good starting point for finding limsup. Let's assume the sequence is bounded above, for otherwise we take $+\infty$ by convention for both sup and limsup. The trouble with the sup, as the example shows, is that it might be x_1, which has nothing to do with the limiting behavior of the sequence (just as in the case of the limit, any finite number of terms of a sequence can be changed without changing the limit-points). We could try to fix things by throwing the rascal out—consider the sup over all

x_j in the sequence except x_1—write this $\sup_{j>1}\{x_j\}$. This is not much of an improvement, for now x_2 might be the culprit. If we continue to throw the rascals out, considering in turn $\sup_{j>2}\{x_j\}, \sup_{j>3}\{x_j\}, \ldots,$ we will never achieve our objective, but we may be approximating it. Since at each stage we are taking the sup over a smaller set (these sets are infinite, so smaller refers to containment, not cardinality), the sups are decreasing. That is, if we let $y_k = \sup_{j>k}\{x_j\}$, then $y_{k+1} \leq y_k$, so the sequence $\{y_k\}$ is monotone decreasing and so has a limit, its inf (possibly $-\infty$). If we write out the expression for this limit, $y = \lim_{k\to\infty} \sup_{j>k}\{x_j\}$, then we are sorely tempted, on purely linguistic terms, to define this to be $\limsup_{j\to\infty} x_j$.

To justify the definition we need to verify two facts:

1. limsup is a limit-point of the sequence.

2. limsup is the sup of the set of limit-points of the sequence.

Incidentally fact 2 does not imply fact 1 a priori, since there exist sets that do not contain their sups.

Proof of fact 1: Suppose first that y is finite. Since y is the limit of $\sup_{j>k} x_j$, given any $1/n$ we can find m such that $|y - \sup_{j>k} x_j| \leq 1/2n$ for all $k \geq m$. Since $\sup_{j>k} x_j$ is finite, we can find x_l for $l > k$ such that $|x_l - \sup_{j>k} x_j| \leq 1/2n$, so $|y - x_l| \leq 1/n$. In fact we can find an infinite number of x_l (since $l > k$ and we can take any $k \geq m$). This shows y is a limit-point of the sequence.

If $y = +\infty$, then $\{\sup_{j>k} x_j\}$ is unbounded above, hence $\{x_j\}$ is unbounded above; so $+\infty$ is a limit-point. Finally, if $y = -\infty$, then given any $-n$ there exists m such that $\sup_{j>k} x_j \leq -n$ for all $k \geq m$. Thus there are infinitely many x_j with $x_j \leq -n$, so $-\infty$ is a limit-point of the sequence.

Proof of fact 2: We have to show that $y =$ limsup is an upper bound for the set of limit-points—since we know by fact 1 that it is a limit-point, this will show it is the least upper bound. Thus what we need to show is that if x is any limit-point of the sequence, then $x \leq y$. If x is a limit-point let $\{x'_j\}$ be a subsequence converging to x. Let us compare x'_{k+1} with $y_k = \sup_{j>k} x_j$. Since $\{x'_j\}$ is a subsequence, x'_{k+1} is one of the x_j with $j > k$, so y_k is the sup of a set containing x'_{k+1}, hence $x'_{k+1} \leq y_k$. Since this holds for each k, we have $x \leq y$ since limits preserve non-strict inequality (you should understand why the $k+1$ in place of k makes no difference).

We can reformulate the above discussion by adopting the following definition.

Definition 3.1.2 *The* limsup *of a sequence is the extended real number* $\text{limsup}_{k\to\infty} x_k = \lim_{k\to\infty} \sup_{j>k} x_j$. *Similarly, the* liminf *is defined by* $\text{liminf}_{k\to\infty} x_k = \lim_{k\to\infty} \inf_{j>k} x_j$.

We have proved the following theorem.

Theorem 3.1.4 *The* limsup *of a sequence is a limit-point of the sequence and is the* sup *of the set of limit-points of the sequence.*

Because the proof is non-trivial, it would be a good exercise for you to try to write out the analogous proof that $\liminf_{k\to\infty} x_k = \lim_{k\to\infty} \inf_{j>k} x_k$ is a limit-point and the inf of the set of limit-points. Incidentally, since the sequence $\{\sup_{j>k} x_j\}$ is monotone decreasing and the sequence $\{\inf_{j>k} x_j\}$ is monotone increasing, we can also write $\text{limsup}_{k\to\infty} x_k = \inf_k \sup_{j>k} x_j$ and $\text{liminf}_{k\to\infty} x_k = \sup_k \inf_{j>k} x_j$.

The difference of the limsup and liminf, sometimes called the *oscillation* of the sequence, measures the spread of the set of limit-points. We are *not* asserting that every value in between is a limit-point; in the sequence $0, 1, 0, 1, \ldots$, the limsup is 1 and the liminf is 0, and there are no other limit-points. But, if the limsup and liminf are equal, we would expect the sequence to converge to their common value. To see why this is true, call the common value x. Suppose first that x is finite. We have two sequences $\{y_k\}$ and $\{z_k\}$ converging to x, where $y_k = \sup_{j>k} x_j$ and $z_k = \inf_{j>k} x_j$, by the definition of limsup and liminf. Note that $z_k \leq x_j \leq y_k$ if $j > k$. If we choose k large enough so that $|x - y_k| < 1/n$ and $|x - z_k| < 1/n$, which we can do because x is the limit of $\{y_k\}$ and $\{z_k\}$, then

$$x - \frac{1}{n} < z_k \leq x_j \leq y_k < x + \frac{1}{n}$$

for all $j > k$, so $|x - x_j| < 1/n$ for all $j > k$, proving x is the limit of $\{x_j\}$. Thus if limsup = liminf is finite, the common value is the limit. (Incidentally, it is not so easy to prove this using only the property of limsup being the sup of the limit-points, etc.) From this we deduce the immediate corollary: if a sequence $\{x_j\}$ has only one limit-point and the limit-point is finite, then the sequence is convergent. Notice that we have to know there is only one limit-point among the extended real numbers, for a sequence like $0, 1, 0, 2, 0, 3, 0, 4, \ldots$ has the two limit-points 0 and $+\infty$ but is not convergent. Another way to state this is to require that the sequence be *bounded* (bounded means bounded above and below; this is sometimes expressed concisely as $|x_j| \le M$ for all j). A bounded sequence has finite limsup and liminf, since limsup $\le$ sup and inf $\le$ liminf.

Theorem 3.1.5 *A bounded sequence is convergent if and only if the* limsup *equals the* liminf *or, equivalently, if and only if it has only one limit-point.*

Finally, we need to consider the case when limsup $=$ liminf $= +\infty$, say. The condition liminf $= +\infty$ means $z_k = \inf_{j>k} x_j$ can be made greater than any n by taking k large enough; but then $x_j \ge z_k > n$ for all $j > k$, and this means $\lim_{j\to\infty} x_j = +\infty$.

3.1.3 Exercises

1. Compute the sup, inf, limsup, liminf, and all the limit points of the following sequences $x_1, x_2, \ldots$ where

   a. $x_n = 1/n + (-1)^n$,

   b. $x_n = 1 + (-1)^n/n$,

   c. $x_n = (-1)^n + 1/n + 2\sin n\pi/2$.

2. If a bounded sequence is the sum of a monotone increasing and a monotone decreasing sequence ($x_n = y_n + z_n$ where $\{y_n\}$ is monotone increasing and $\{z_n\}$ is monotone decreasing) does it follow that the sequence converges? What if $\{y_n\}$ and $\{z_n\}$ are bounded?

3. If E is a set and y a point that is the limit of two sequences, $\{x_n\}$ and $\{y_n\}$ such that x_n is in E and y_n is an upper bound for E, prove that $y = \sup E$. Is the converse true?

4. Prove $\sup(A \cup B) \ge \sup A$ and $\sup(A \cap B) \le \sup A$.

5. Prove $\limsup\{x_n + y_n\} \le \limsup\{x_n\} + \limsup\{y_n\}$ if both lim sups are finite, and give an example where equality does not hold.

6. Is every subsequence of a subsequence of a sequence also a subsequence of the sequence?

7. Construct a sequence whose set of limit points is exactly the set of integers.

8. Write out the proof that $+\infty$ is a limit-point of $\{x_n\}$ if and only if there exists a subsequence whose limit is $+\infty$.

9. Can there exist a sequence whose set of limit points is exactly $1, 1/2, 1/3, \ldots$? (**Hint**: what is the liminf of the sequence?)

10. Prove that the set of limit-points of a shuffled sequence $x_1, y_1, x_2, y_2, \ldots$ is exactly the union of the set of limit-points of $\{x_j\}$ and the set of limit-points of $\{y_j\}$. Is the same true if the shuffling is not regular?

11. Consider a sequence obtained by diagonalizing a rectangular array.

$$\begin{array}{ccccccccc} & a_{11} & \nearrow & a_{12} & \nearrow & a_{13} & \nearrow & \cdots \\ & a_{21} & \nearrow & a_{22} & \nearrow & a_{23} & \nearrow & \cdots \\ \nearrow & a_{31} & \nearrow & a_{32} & \nearrow & a_{33} & \nearrow & \cdots \\ \nearrow & \vdots & \nearrow & & \nearrow & & & \end{array}$$

Prove that any limit-point of any row or column of the array is a limit-point of the sequence. Do you necessarily get all limit-points this way?

12. Say two sequences are *equivalent* if they differ in only a finite number of terms (there exists m such that $x_j = y_j$ for all $j \ge m$). Prove that this is an equivalence relation. Show that equivalent sequences have the same set of limit-points.

3.2 Open Sets and Closed Sets

3.2.1 Open Sets

We have already had many occasions to use inequalities and sets defined by inequalities. Perhaps you have noticed that on some occasions the distinction between strict and non-strict inequalities is not essential—for example, in the definition of limit, we could require either $|x_k - x| < 1/n$ or $|x_k - x| \leq 1/n$ and it would make no difference; however on other occasions the distinction is essential, for example, non-strict inequalities are preserved in the limit, but strict inequalities may not be. We are now going to delve further into the matter, from the point of view of the sets defined by the inequalities. The set determined by the strict inequalities $a < x < b$ (for $a < b$) we will call an *open interval*, written (a, b); while the set determined by the non-strict inequalities $a \leq x \leq b$ we will call a *closed interval*, written $[a, b]$. For the open interval we will also allow $a = -\infty$ or $b = +\infty$ or both. It may seem a trifling matter whether or not the endpoints are included in the interval, but it makes a significant difference for certain questions. In the open interval, every point is surrounded by a sea of other points. This is the qualitative feature we will want when we define the notion of open sets; it is certainly not true of the endpoints of the closed interval. On the other hand, an open interval (a, b) seems to be "missing" its endpoints. Although they are not points in the interval, they can be approached arbitrarily closely from within the interval. It is as if they had been unfairly omitted. The closed interval has all the points it should from this point of view, and this is the "closed" aspect that we will generalize when we define a closed set.

Let us begin with open sets. The idea of "open" suggests that one should always be able to go a little further, that one should never reach the end. Thus we will *define an open set A of real numbers to be a set with the property that every point x of A lies in an open interval (a, b) that is contained in A.* The open interval may vary with x, and it may be very small. If A itself is an open interval, then it is trivially an open set because A contains the open interval A that contains each point x in A (this is an example where the interval does not have to vary with x). A closed interval $[a, b]$ is *not* an open set because the point a in A does not lie in any open interval contained in A. The union of two open intervals, say $A = (0,1) \cup (1,2)$, is open, since any point in A lies in either the open interval $(0,1)$ or the open interval $(1,2)$. The fact that the point 1 does not lie in an interval contained in A is not relevant, because 1 does not belong to A. The empty set is an open set because it satisfies the definition trivially (since it contains no points, there is nothing to verify). In what follows we are mainly interested in non-empty open sets, but we phrase the results so that they remain true for the empty set as well.

The union of any number (finite and infinite) of open intervals is an open set, simply because any point in the union must belong to one of the open intervals. Conversely, every open set is a union of open intervals. In fact, if A is an open set, then each point x in A, according to the definition, lies in some open interval I_x contained in A. Then A is the union of all the intervals I_x, $A = \bigcup_{x \epsilon A} I_x$; for x is in I_x for each x in A, so $A \subseteq \bigcup_{x \epsilon A} I_x$; on the other hand each I_x is contained in A, so $\bigcup_{x \epsilon A} I_x \subseteq A$.

Actually we can say a little more precisely what every open set is like. For if two open intervals intersect at all, they must overlap, and so their union can be combined into a single open interval, as shown in Figure 3.2.1.

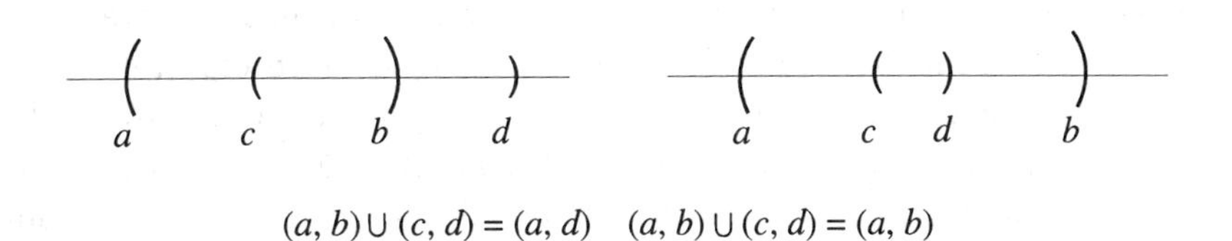

Figure 3.2.1:

Thus a union of open intervals can be simplified to a union of *disjoint* open intervals (disjoint means no two intersect). (Strictly speaking this requires a more elaborate proof, since there may be multiple overlaps. We leave the details to exercise set 3.2.3, number 15). How many open intervals? There could be any finite number or an infinite number. As an example of an infinite number, consider the set $(1/2,1) \cup (1/4,1/2) \cup (1/8,1/4) \cup \cdots$ shown in Figure 3.2.2.

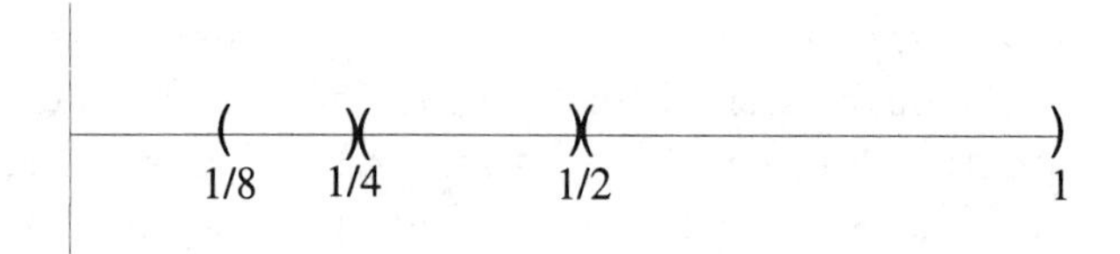

Figure 3.2.2:

This set can also be described as the interval $(0,1)$ with the points $1/2, 1/4, 1/8, \ldots$ deleted. From this second description it is not obvious that we are dealing with a disjoint union of intervals. Of course the general disjoint union of intervals can be much more complicated than this, with the sizes of the intervals distributed in incredibly complex ways. We can, however, assert that the cardinality of the collection of disjoint intervals is at most countable. That is, we cannot have an uncountable union of disjoint open intervals. To understand why this is so we have to reason about the length of the intervals. Let us call $\mathcal{A}$ the collection of disjoint intervals. Consider $\mathcal{A}_1$, the subset of $\mathcal{A}$ of those intervals of length greater than 1. The set $\mathcal{A}_1$ is at most countable because every interval of length greater than 1 must contain at least one integer, and the disjointness means that no two intervals in $\mathcal{A}_1$ can contain the same integer. Next consider $\mathcal{A}_2$, the subset of $\mathcal{A}$ of those intervals of length greater than $1/2$. Every interval in $\mathcal{A}_2$ contains at least one half-integer (a number $m/2$ where m is an integer), so again $\mathcal{A}_2$ is at most countable. Continue in this way to define $\mathcal{A}_n$ to be the subset of $\mathcal{A}$ of those intervals of length at most $1/n$. By the same reasoning all the sets $\mathcal{A}_n$ are at most countable. But $\mathcal{A} = \bigcup_n \mathcal{A}_n$, for every open interval has a length greater than $1/n$ for some n. Since $\mathcal{A}$ is a countable union of at most countable sets, it is countable.

Thus we have a structure theorem for open sets: *every open set of real numbers is a disjoint union of a finite or countable number of open intervals.* The intervals that comprise the union are uniquely determined; we will leave the proof to exercise set 3.2.3, number 16. This structure theorem has no analogue in higher dimensions and so is rarely emphasized. (For example, if you want to prove something about open sets of real numbers, it would be better not to use the structure theorem if you can avoid it, for then your proof would stand a better chance of generalizing.)

Next we study the closure properties of the class of open sets; in other words, what operations can you perform on open sets and still come out with an open set?

Theorem 3.2.1

1. *The union of any number of open sets is an open set.*

2. *The intersection of a finite number of open sets is an open set.*

Notice also that we have to restrict the intersections in property 2 to finitely many, while there is no such restriction for unions (the number of open sets in the union is even allowed to be uncountable). Here is an example that shows why we need this restriction. Take a closed interval, say $[0, 1]$. We have observed that this is not an open set. Yet we can easily get it as a countable intersection of, say, the open intervals $(-1/n, 1 + 1/n)$, as shown in Figure 3.2.3.

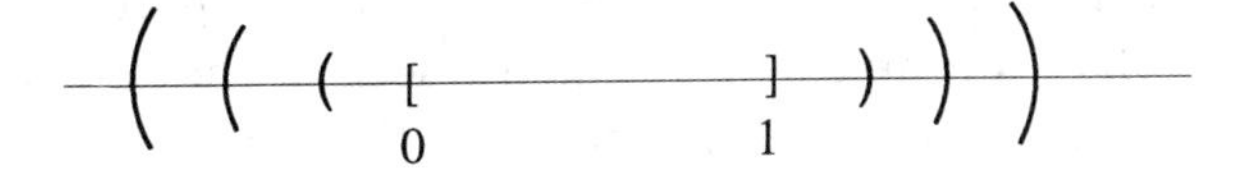

Figure 3.2.3:

Proof of Property 1: Let $\mathcal{A}$ denote any collection of open sets and $\bigcup_{\mathcal{A}} A$ their union. A point x in the union must belong to one particular open set A. Since A is open, x must lie in an open interval I contained in A and, hence, in the union. Thus the union is open.

Proof of Property 2: Let $A_1, \ldots, A_n$ be a finite number of open sets, and let $A = A_1 \cap \cdots \cap A_n$ be their intersection. Any point x in A lies in all the A_k. (If A is empty there is nothing more to do, since the empty set is open.) Since each A_k is open, x lies in an open interval (a_k, b_k) contained in A_k. So we have a picture like Figure 3.2.4 ($n = 3$).

We want an open interval containing x that lies entirely in A. We can't take any one of the intervals (a_k, b_k), because all we know about

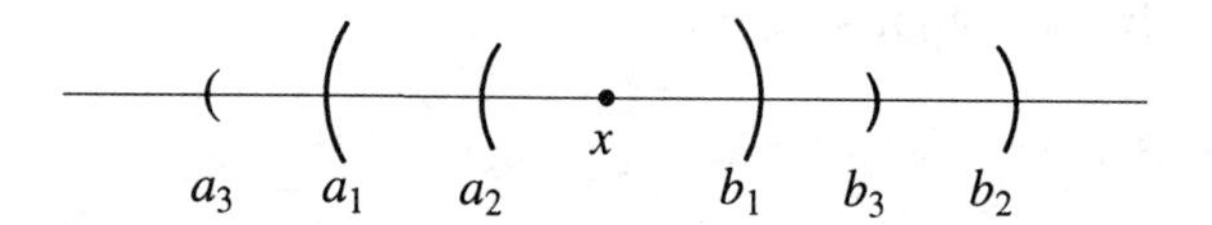

Figure 3.2.4:

it is that it lies in A_k. The only hope is to take the intersection of all the intervals, since that will lie in A. But is the intersection an open interval containing x? A glance at Figure 3.2.4 will convince you that the intersection is the open interval (a, b) where a is the largest of the a_k's and b is the smallest of the b_k's and of course it contains x since it is the intersection of intervals containing x. (Note that this is where the argument breaks down for infinite intersections: if there were an infinite number of intervals the intersection might not be an open interval.) QED

This rather innocent theorem on the closure properties of open sets turns out to have an unexpected significance. In the general theory of topology, properties 1 and 2 are chosen as the *axioms* for the abstract notion of "open set" (together with a trivial axiom that the empty set and the whole universe are open sets). We will not discuss the general theory in this work, but from time to time we will point out some of the ways in which abstract theories are forshadowed in concrete instances.

What is the significance of the open sets? To explain this let's consider a closely related concept. A *neighborhood* of a point x is defined to be any open set containing x (sometimes the word "neighborhood" is taken to mean any set containing an open set containing x, and the narrower meaning we have ascribed to the term is denoted by "open neighborhood"). A particularly simple neighborhood of x is an open interval containing x, or even an interval of the form $(x - 1/n, x + 1/n)$. By the definition of open set, every neighborhood of x contains an open interval containing x and, by the axiom of Archimedes, even one of the special form $(x - 1/n, x + 1/n)$. Thus, although there are an uncountable number of possible neighborhoods of x, if we are willing to shrink neighborhoods, we need only consider a countable number, the intervals $(x - 1/n, x + 1/n)$. In most instances when we use the term "neighborhood" it comes quantified, "for all neighborhoods of x blah-blah-blah", or "there exists a neighborhood of x such that phooey", and it usually does not change the meaning of the sentence to say "for every interval $(x - 1/n, x + 1/n)$ blah-blah-blah", or "there exists an interval $(x - 1/n, x + 1/n)$ such that phooey". This simple observation will play an important role in Chapter 14 as part of a general strategy to replace uncountable collections of sets by countable collections of sets.

An important and typical example of the use of the neighborhood concept is in the definition of limit. We have defined $\lim_{j\to\infty} x_j = x$ to mean that given any error $1/n$, we can make $|x_j - x|$ less than that error by taking j sufficiently large. But the condition $|x_j - x| < 1/n$ is exactly the statement that x_j lies in the neighborhood $(x - 1/n, x + 1/n)$ of x. Bearing in mind what we said about replacing general neighborhoods by special ones, we expect the following reformulation of the definition of limit to be equivalent to the old one: $\lim_{j\to\infty} x_j = x$ *if for every neighborhood of x, there exists m such that all x_j are in that neighborhood for $j \geq m$.* Indeed this says exactly the same as before for the special neighborhoods $(x - 1/n, x + 1/n)$, and since every neighborhood contains a special one, the statement is true for all neighborhoods if and only if it is true for the special ones.

What is the advantage of reformulating the definition of limit as above? It shows that the concept of open set alone suffices to define limit, without any reference to distance. We will find that this is true for many other concepts as well. As an exercise, try giving the definition of limit-point using neighborhoods. On the other hand, a concept like sup or inf, which involves the order properties of the real number system, cannot be defined by neighborhoods alone.

3.2.2 Closed Sets

We now turn to the closed intervals and, their generalization, the closed sets. We have said that the term "closed" is used to indicate that the set contains all the points that it "ought to" contain. Here we are thinking that a set "ought to" contain those numbers that can be approximated arbitrarily closely by numbers already in the set. In this sense the endpoints of an interval ought to be in the interval, and their absence in the open interval leads to the expectation that the open interval isn't closed.

To make these ideas precise we need to introduce the concept of *limit-point* for sets. This will be analogous to the concept of limit-point for sequences but with a subtle twist. Let A denote a set of real numbers, and let x denote a real number (perhaps in A, perhaps not). We want to say that x is a limit-point of A if there are points in A that approximate x arbitrarily closely. We can say this in terms of distance: given any error $1/n$, there exists a point y_n in A (depending on n) such that $|x - y_n| < 1/n$; or we can say it in terms of neighborhood: every neighborhood of x contains a point in A. It is easy to see that these statements are equivalent; however, they are not quite right. The reason is that they allow x to approximate x. If A contains x, then the above requirement is trivially satisfied by taking the point x in A, so we would end up having all the points of A automatically limit-points. While this makes perfect logical sense, it fails to capture the meaning we want, which is that of an infinite cluster of points in A around x (the synonym "cluster point" reinforces this point). Thus we must add to the definition a clause that eliminates x from consideration as an approximating point.

Definition 3.2.1 *x is a limit-point of A if given any error $1/n$, there exists a point y_n of A not equal to x satisfying $|y_n - x| < 1/n$ or, equivalently, if every neighborhood of x contains a point of A not equal to A.*

Note that this definition actually implies that every neighborhood of x contains an infinite number of points of A. For if on the contrary it only contained a finite number of points of A, say $a_1, \ldots, a_n$, we could find a smaller neighborhood of x that does not contain $a_1, \ldots, a_n$, as indicated in Figure 3.2.5.

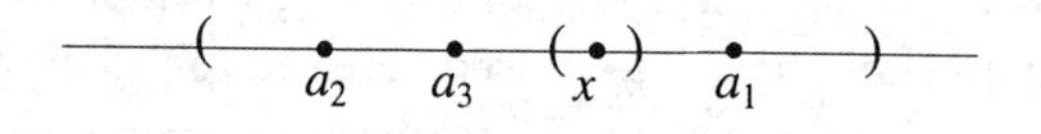

Figure 3.2.5:

If we reformulate the definition to include this, we can drop the clause excluding x: *x is a limit-point of A if every neighborhood of x contains infinitely many points of A.*

Note that if x is contained in an interval (closed or open) contained in A, then x is a limit-point of A. In particular, every point of an open set is a limit-point. But there are other ways of being a limit point. The set $\{1, 1/2, 1/4, \ldots\}$ has 0 as limit-point; in fact 0 is the only limit-point of this set. The set of integers has no limit-points, for no number has an infinite number of integers nearby.

What is the relationship between the concepts of limit-point for sequences and sets? If we denote the sequence by $x_1, x_2, \ldots$ and the set by $\{x_j\}$, are the limit-points of $x_1, x_2, \ldots$ the same as the limit points of $\{x_j\}$? Unfortunately, the answer is no. The simplest example is the sequence $5, 5, 5, \ldots$, which has the limit-point 5. The corresponding set contains just the one point 5, so it has no limit-points. Clearly the culprit here is repetition. If the sequence has no repetition, or only a finite number of repetitions, then the two concepts of limit-point coincide. In the general case we can only say that a limit-point of the set is a limit-point of the sequence. Incidentally, this confusion would *not* be cleared up if we adopted the convention of allowing x to approximate itself. See exercise set 3.2.3, number 3.

Now we can define a *closed set to be any set that contains all its limit-points.* Note that we do not require that all points of the set be limit-points (such sets are called *perfect sets*). A set with no limit-points, such as the empty set, or a finite set, is automatically closed. A closed interval $[a, b]$ (with $a \leq b$) is a closed set; in fact, it is a perfect set. A non-empty open interval (a, b) with a or b finite is not closed because the finite endpoint(s) are limit-points. However the whole line $(-\infty, \infty)$ *is* closed (note that we have *not* defined the possibility of $\pm\infty$ being limit-points of a set; although we could do so in the same way we did for sequences, it would confuse matters too much).

The whole story of closed sets is revealed in the following basic theorem:

Theorem 3.2.2 *A set is closed if and only if its complement is open.*

To simplify the proof we first prove a lemma. If B denotes any set in $\mathbb{R}$, write B' for its complement, $B' = \{x \text{ in } \mathbb{R} : x \text{ is not in } B)$.

Lemma 3.2.1 *A point x in B' is not a limit-point of B if and only if x is contained in an open interval contained in B'.*

Proof: x is not a limit-point of B means that there exists a neighborhood $(x - 1/n, x + 1/n)$ containing no points of B other than x. But x already is in B', so $(x - 1/n, x + 1/n)$ is the open interval (containing x) that is contained in B'. Conversely, if x is in (a, b) contained in B', then $(x - 1/n, x + 1/n)$ is contained in (a, b), hence B', for sufficiently small $1/n$. This implies x is not a limit-point of B. QED

Proof of Theorem 3.2.2: The definition that B be closed says "for all x, x is a limit-point of B implies x is in B". Replacing the implication by its contrapositive yields the equivalent statement "for all x, x is in B' implies x is not a limit-point of B". Since x is assumed to be in B', we can use the lemma to replace "x is not a limit-point of B" by the equivalent "x is contained in an open interval contained in B'". We now have the definition that B' be open: for all x in B', x is contained in an open interval contained in B'. QED

Note that this theorem does *not* say that if B is not closed, then B is open. Most sets are neither open nor closed. Can you give an example?

Using this theorem we can deduce many properties of closed sets from properties of open sets. For example, since union and intersection are interchanged by complementation, it follows that the closed sets are closed under finite unions and arbitrary intersections. Of course we can also prove this directly, and it is worth doing so.

Theorem 3.2.3

1. *The union of a finite number of closed sets is a closed set.*

2. *The intersection of any number of closed sets is a closed set.*

Proof:

a. Let $B_1, \ldots, B_n$ be closed sets, and let $B = B_1 \cup \cdots \cup B_n$. To show that B is closed, we have to show it contains all its limit-points. So let x be a limit-point of B. Does this mean x is a limit-point of B_1? Perhaps not, as Figure 3.2.6 suggests.

Of course x is a limit-point of B_2 in this case. This suggests that perhaps we must have x as a limit-point of one of the sets $B_1, \ldots, B_n$. If this is true it will certainly do the trick, for each of these sets is

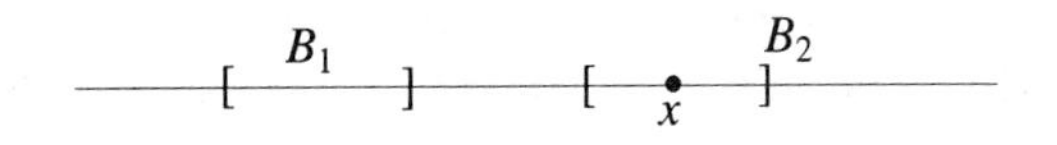

Figure 3.2.6:

closed, hence contains all its limit-points, so x would belong to one of these sets, hence to the union.

Let's try to prove it. What do we know? Since x is a limit-point of B, every neighborhood of x contains a point, not x, of B. Now B is the union of the B_j's, so to each neighborhood, say $(x - 1/k, x + 1/k)$ of x, there is a point y_k in the neighborhood in one of the B_j's. Now, we want to focus on each of the sets $B_1, B_2, \ldots, B_n$ in turn. The y_k's are distributed among them in some unspecified way. But since there are only a finite number of sets and an infinite number of points $y_1, y_2, \ldots$ to distribute among them, one of the sets must contain an infinite number of points. Call the set B_{29} and the points $y'_1, y'_2, y'_3, \ldots$. Since the points $y'_1, y'_2, y'_3, \ldots$ approximate x arbitrarily closely, x is a limit-point of B_{29}. (The fact that not *all* the points in the original sequence $y_1, y_2, \ldots$ belong to B_{29} is not important—the ones that do, being infinite in number, get into every neighborhood of x and so x is a limit-point.) Thus we have x in B_{29} because B_{29} is closed and, hence, x in B.

b. Let $\mathcal{B}$ be any collection of closed sets, and let $\bigcap_{B \in \mathcal{B}} B$ be the intersection. To show the intersection is closed we have to show it contains all its limit-points. So let x be a limit-point. It follows easily from the definition of limit-point that if you increase a set you do not lose any limit-points (you may gain some). Since each B in $\mathcal{B}$ contains the intersection, it follows that x is a limit-point of each B in $\mathcal{B}$. Since each B is closed, x belongs to each B and, hence, to the intersection. QED

Let's look at an example of a closed set that is perhaps a bit more complicated than you might expect. It is called *the Cantor set* and is the proto type of a large family of sets, which are called somewhat loosely *Cantor sets.* These sets are obtained from the closed interval $[0, 1]$ by removing a countable collection of open intervals, so they are the countable intersection of finite unions of closed intervals, hence closed sets. In this example we will successively remove the middle-third of every closed interval. In the first stage we remove the interval $(1/3, 2/3)$ and so are left with the two intervals $[0, 1/3] \cup [2/3, 1]$. In the second stage we remove the middle-third of each of these intervals, $(1/9, 2/9)$ and $(7/9, 8/9)$ respectively, leaving the four intervals $[0, 1/9] \cup [2/9, 1/3] \cup [2/3, 7/9] \cup [8/9, 1]$, as shown in Figure 3.2.7.

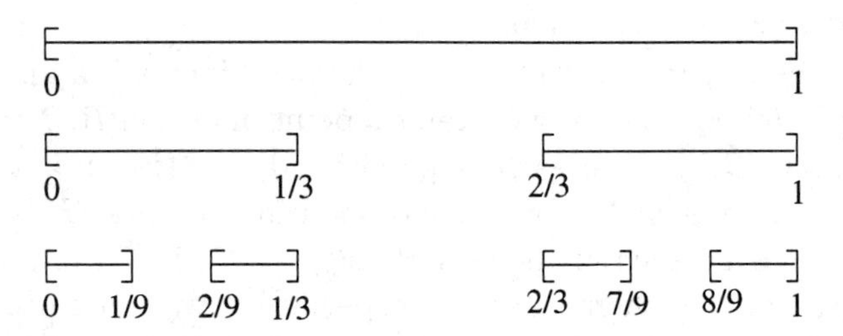

Figure 3.2.7:

Iterating this process infinitely often produces the Cantor set. There is another way to describe this set. Write the numbers between 0 and 1 in base-3 notation. Then the numbers in the middle third $(1/3, 2/3)$ are those that begin $.1\ldots$, those in the second set of middle thirds $(1/9, 2/9)$ and $(7/9, 8/9)$ are those that begin $.01\ldots$ and $.21\ldots$. Thus all the deleted numbers are those with a 1 in their base-3 expression, so the Cantor set consists of all numbers expressible with just 0's and 2's (note that numbers like $1/3$ and $2/3$ that have ambiguous expressions, $.1000\ldots$ or $.0222\ldots$ for $1/3$, $.2000\ldots$ or $.1111\ldots$ for $2/3$, are included in the Cantor set—as they should be—because they have one expression not involving 1's).

The Cantor set is a perfect set but it contains no intervals. It is uncountable (as all perfect sets must be, although this is more difficult to prove). We leave these facts as exercises.

We conclude this section with some concepts related to open and closed sets. A point x is said to be in the *interior* of a set A if A contains a neighborhood of x. Every point in an open set is in its interior. The concept of interior points is clearly a localized version of open set. Note that the interior of any set is automatically an open set.

In fact it is the largest open set contained in the set (see exercises).

If A is any set, the *closure* of A is the set consisting of all the points of A together with all the limit points of A. Thus a set is closed if and only if it is equal to its closure. The closure of A is always a closed set. This is not obvious from the definition but requires a proof. The issue here is that by adding the limit-points of A to A, we might conceivably produce new limit-points that were not there before. But we can rule out this possibility as follows: Suppose x is a limit-point of the closure of A. This means that every neighborhood of x, say $(x - 1/n, x + 1/n)$, contains points y_n not equal to x in the closure of A. If y_n belongs to the closure of A either y_n belongs to A or y_n is a limit-point of A. Now we want to show that x is a limit-point of A; so we need to show that $(x - 1/n, x + 1/n)$ contains a point of A. Now if y_n is in A we are done. If not, y_n is a limit-point of A, so the neighborhood $(y_n - 1/n, y_n + 1/n)$ contains points of A, say z_n (choose $z_n \neq x$, which is possible because we know there are infinitely many—hence at least two—points of A there). From $|z_n - y_n| < 1/n$ and $|x - y_n| < 1/n$ we obtain $|z_n - x| < 2/n$, so the neighborhood $(x - 2/n, x + 2/n)$ contains a point of A not equal to x. Clearly the factor of 2 is irrelevant, so x is a limit-point of A.

The closure of A is thus closed. Also it is clearly the smallest closed set containing A. One could very easily obtain the smallest closed set containing A by taking the intersection of all closed sets containing A—but this abstract construction is less informative.

If B is a subset of A such that A is contained in the closure of B (so $B \subseteq A \subseteq$ closure (B)), we say that B is *dense* in A. Put another way, B is dense in A (or we say B is a *dense subset* of A) if B is a subset of A, and every point in A is either a point of B or a limit-point of B. For example, the rational numbers are dense in the real numbers, since every real number is a limit of a sequence of rational numbers—that's how we constructed the real number system. The open interval (a, b) is dense in the closed interval $[a, b]$. Dense subsets are very convenient and are used in the following manner. If you want to prove that every point in A has a certain property that is preserved under limits, then it suffices to prove that every point in a dense subset B of A has that property. The dense subset might be simpler and smaller than A.

3.2.3 Exercises

1. Let A be an open set. Show that if a finite number of points are removed from A, the remaining set is still open. Is the same true if a countable number of points are removed?

2. Let $x_1, x_2, \ldots$ be a sequence, and let A be the set whose elements are $x_1, x_2, \ldots$. Show that a limit-point of A is a limit-point of the sequence. Show that if no point in A occurs more than a finite number of times in the sequence, then a limit-point of the sequence is a limit-point of the set.

3. Suppose that the definition of limit-point of a set is changed to the one first suggested (every neighborhood of x contains a point of the set—without requiring the point to be different from x). Give an example to show that it would still not be true that the limit-points of a sequence and the limit-points of the underlying set must be the same. Can you show that one contains the other?

4. Let A be a set and x a number. Show that x is a limit-point of A if and only if there exists a sequence $x_1, x_2, \ldots$ of distinct points in A that converges to x.

5. Let A be a closed set, x a point in A, and B be the set A with x removed. Under what conditions is B closed?

6. Prove that every infinite set has a countable dense subset. Give an example of a set A such that the intersection of A with the rational numbers is not dense in A.

7. Give an example of a set A that is not closed but such that every point of A is a limit-point.

8. Show that the set of limit-points of a sequence is a closed set.

9. Given a closed set A, construct a sequence whose set of limit-points is A. (**Hint**: use exercise 6.)

10. Show that the set of numbers of the form $k/5^n$, where k is an integer and n a positive integer, is dense in the line.

11. Show that the set of numbers in the interval $[0,1]$ having decimal expansions using only odd digits is closed. Describe this set by a Cantor-set type construction.

12. a) Show that the Cantor set is a perfect set that contains no open intervals. Show that it is uncountable. b) Are the same statements true of the set in exercise 11?

13. Define the *derived set* of a set A as the set of limit-points of A. Prove that the derived set is always closed. Give an example of a closed set A that is not equal to its derived set. Give an example of a set A such that the derived set of A is not equal to the derived set of the derived set of A. (Note: Cantor was originally led to study set theory in order to understand better the notion of derived set and to answer questions similar to the above.)

14. What sets are both open and closed?

15. Show that a union of open intervals can be written as a disjoint union of open intervals.

16. Show that an open set cannot be written in two different ways as a disjoint union of open intervals (except for a change in the order of the intervals).

3.3 Compact Sets

Infinite sets are more difficult to deal with than finite sets because of the large number of points they contain. Nevertheless, there is a class of infinite sets, called *compact sets*, that behave in certain limited ways very much like finite sets. The compact sets of real numbers turn out to be exactly the sets that are both closed and bounded, but this is a theorem and not the definition of compactness. The concept of compactness is not confined to sets of real numbers; we shall deal with it again later in other guises.

In what way can an infinite set behave like a finite set? Consider the infinite pigeon-hole principle: if an infinite number of letters arrive addressed to a finite number of people, then at least one person must receive an infinite number of letters. In more conventional terms, *if*

$x_1, x_2, \ldots$ is an infinite sequence of real numbers, and each x_j belongs to a finite set A then at least one element of A must be equal to x_j for an infinite number of j's. Now if A were an infinite set, this statement is obviously false. However, we could hope for a slightly weaker conclusion: that A contains a limit-point of the sequence. After all, a limit-point is one that is approximated arbitrarily closely infinitely often, and for most purposes such approximation is just as good as equality.

Let us take this property as the definition of compactness (there are several other equivalent conditions, and any of them could serve as well for the definition).

Definition 3.3.1 *A set A of real numbers is said to be compact if it has the property that every sequence $x_1, x_2, \ldots$ of real numbers that lies entirely in A has a (finite) limit-point in A (or, equivalently, has a subsequence that converges to a point in A).*

What kind of sets can be compact? Certainly only closed sets. For if A is not closed, it must have a limit-point y not in A; but then by the definition of limit-point we could construct a sequence $x_1, x_2, \ldots$ of points in A that converge to y. Thus y would be the sole limit-point of the sequence, and the defining condition of compactness would fail.

For similar reasons an unbounded set can never be compact. Indeed if A is unbounded we can find a sequence of points $x_1, x_2, \ldots$ in A such that $x_n > n$ or $x_n < -n$, and such a sequence clearly has no finite limit-point.

Thus only closed and bounded sets can be compact. We will show, conversely, that all closed and bounded sets are compact. The proof will be easy because we know that every bounded sequence has a finite limit-point.

Theorem 3.3.1 *A set of real numbers is compact if and only if it is closed and bounded.*

Proof: We have already seen the necessity of the conditions that the set be closed and bounded. Conversely, let A be closed and bounded, and let $x_1, x_2, \ldots$ be any sequence of points in A. Since A is bounded, the sequence is bounded; and we have proved that a bounded sequence possess a limit-point. But A is closed and contains all its limit-points, so we need only show that a limit-point of the sequence is a limit-point of the set. Actually this is a statement about limit-points that is *false* in general. Nevertheless we can still save the proof. Recall that y is a limit-point of the sequence if every neighborhood of y contains infinitely many points in the sequence. For y to be a limit-point of the set, every neighborhood of y must contain points of A *not equal to* y. Clearly the only way things could go wrong would be if *all* these points were equal to y. But if y ever appears in the sequence that means y is in A, which is what we are trying to prove. Thus to complete the proof we need only consider first the special case when the limit-point y appears in the sequence—in that case y is already in A; and then in the contrary case, when y never appears in the sequence, we can conclude that y is a limit-point of A and then that y is in A since A is closed. QED

We consider now another property of compact sets that resembles a property of finite sets. This may seem rather artificial at first, but it turns out to be extremely useful. We introduce the concept of a *cover* of a set A, which is any collection of sets, finite or infinite in number, whose union contains A. We have the obvious "picture" of the sets in the cover covering A. Now if A is a finite set, we clearly have no need for an infinite number of sets to cover it. Nevertheless, by very bad planning, we might find ourselves with an infinite collection $\mathcal{B}$ of sets that cover A. In that case we could certainly simplify things by throwing away all but a finite number of the sets in $\mathcal{B}$. We need only select one set containing each point of A and throw the rest away as redundant. We define a *subcover* of a *cover* to be a subset $\mathcal{B}_1$ of $\mathcal{B}$ that is still a cover (note that in $\mathcal{B}$ we are dealing with a set of sets; the "subset" in the definition refers to the big collection $\mathcal{B}$, not to the individual sets in $\mathcal{B}$ that are either accepted whole into $\mathcal{B}_1$ or discarded; similarly when we speak of *finite* covers we mean that the big set $\mathcal{B}$ should be finite, consisting say of $B_1, B_2, \ldots, B_n$, but the sets B_j themselves may be infinite). Then we can express this trivial property of finite sets as follows: *every cover contains a finite subcover.*

Again we have a statement that is obviously false for any infinite set; just consider a cover by sets with one element. The remarkable fact is that we obtain a true principle for compact sets by simply requiring the sets in the cover to be open sets! We call such a cover an *open cover.* Why should an open cover be better than an arbitrary cover?

Essentially because it is quite difficult to cover things with open sets. Two open intervals that intersect must overlap, and two open intervals that just nestle together fail to cover the point in between, as in Figure 3.3.1.

Figure 3.3.1:

Let us attempt to cover the closed interval $[0, 1]$ by open sets so that no finite subcover exists. We might try something like $(1/2, 2)$, $(1/3, 1)$, $(1/4, 1/2)$, $(1/5, 1/3)$, and so on, as in Figure 3.3.2.

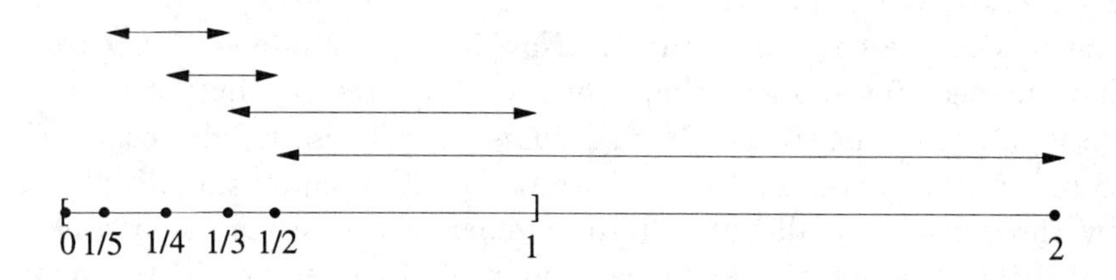

Figure 3.3.2:

Clearly we cannot remove any one of these sets without uncovering a point (if we remove $(1/(n+2), 1/n)$ we uncover the point $1/(n+1)$). Of course we have not quite covered the closed interval $[0, 1]$. We have missed the point 0. Well, why not just add one more set to the cover to cover 0. Aye, but there's the rub! The set must be open, and, therefore, it must contain a neighborhood of zero, say $(-1/n, 1/n)$. But then we can use this set and discard all but a finite number of the other sets; we need keep only $(1/2, 2)$ and $(1/(k+2), 1/k)$ for $k = 1, 2, \ldots, n-1$.

Well, perhaps we need to try a more ingenious method of covering. I will not pursue the matter but invite you to try your own ideas for covering the closed interval $[0, 1]$ by an infinite number of open sets so that no finite subcover exists. Remember that you must cover *every* point and that you must show that *no* finite subcover is possible, not merely that one particular attempt to find a finite subcover fails. I promise you it will be a frustrating experience.

Let's turn the problem around, then, and try to find sets that do not have the property that every open cover has a finite subcover. We have already found one such set, namely $(0, 1]$. Note that this set is not closed because it does not contain the limit-point 0. Can we show that every set that fails to be closed fails to have this property? Stated in contrapositive form, if every open cover of set A has a finite subcover, can we prove that A is closed? Suppose y is a limit point of A. We have to show y is in A. Suppose it were not. We want to construct an open cover of A with no finite subcover. Let's try the complements of closed intervals $[y - 1/n, y + 1/n]$—we take complements of closed intervals to get open sets. Clearly this covers the whole real line except for the point y and so covers A since y is not in A. Suppose it had a finite subcover. Since these are nested sets (they increase with n), this would mean that a single one contains A, as in Figure 3.3.3.

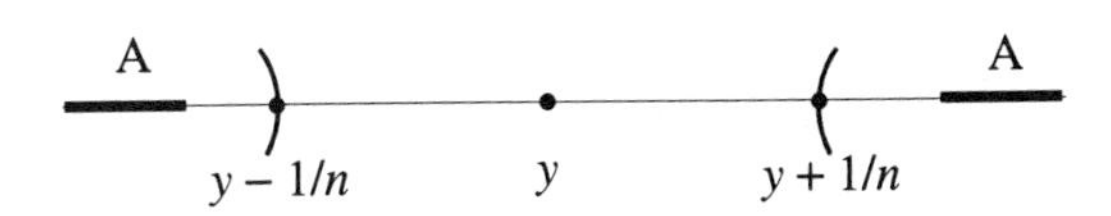

Figure 3.3.3:

But this means the neighborhood $(y - 1/n, y + 1/n)$ of y contains no points of A, contradicting the fact that y is a limit-point of A.

Thus only closed sets can have this property. We can also easily show that only bounded sets can have this property. For the open sets $(n, n+2)$, with n varying over all integers, cover the whole plane, hence any set A. If a finite number cover A, then A is bounded.

So far we have shown that *if a set has the property that every open cover has a finite subcover, then the set must be closed and bounded,* hence compact. The converse is also true and is called the *Heine-Borel Theorem.* You may very well feel at this point that such a theorem could not possibly be very interesting—I certainly felt that way when I first encountered this theorem. I hope that when you see the way this theorem is used, you will gain an appreciation for it.

Theorem 3.3.2 *Every open cover of a compact set has a finite subcover.*

Proof: Let A be a compact set and $\mathcal{B}$ an open cover. As a first step we show that $\mathcal{B}$ has a countable subcover. (This part of the argument does not use the fact that A is compact.) The idea is that, although there are an uncountable number of open sets, there exists a countable set of open sets—namely, the open intervals with rational endpoints—that suffices for most purposes. Clearly the open intervals with rational endpoints form a countable set because they are indexed by a pair of rational numbers—the endpoints. We will choose a subcover of $\mathcal{B}$ that is in one-to-one correspondence with a subset of the collection of open intervals with rational endpoints and, hence, is a finite or countable subcover. Here is how we do it: for each such interval I, we choose one set in $\mathcal{B}$ that contains I, if there are any. (If there are none, leave well enough alone.) We claim that we still have a cover of A. Indeed, start with any point x in A. We need to show it is covered. Since all of $\mathcal{B}$ is a cover, there must be a set B in $\mathcal{B}$ containing x.

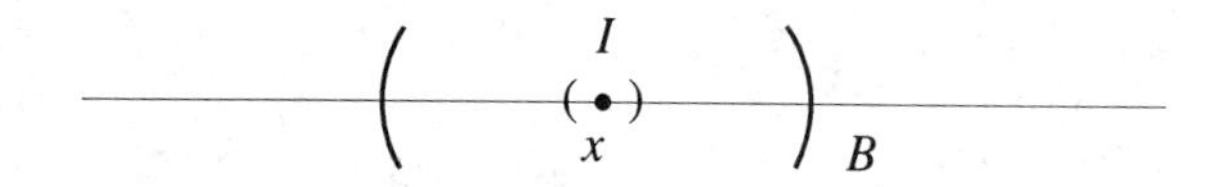

Figure 3.3.4:

Since B is open, it must contain an open interval I containing x, and by shrinking the interval I if need be, we can arrange for it to have rational endpoints, as shown in Figure 3.3.4. So now we have the existence of at least one set in $\mathcal{B}$ containing I; hence we must have selected a set in $\mathcal{B}$ containing I for our subcover, and since x is in I, it is covered.

Thus we have a subcover $B_1, B_2, \ldots$ that is at most countable (if it is finite then there is nothing more to prove). We now come to the heart of the proof: if we take n large enough, then $B_1, B_2, \ldots, B_n$ already covers A. Suppose not. Then for each n there is a point x_n of A that isn't covered by $B_1, B_2, \ldots, B_n$. Since A is compact, the sequence $x_1, x_2, \ldots$ has a limit-point x in A. Now, since the infinite collection $B_1, B_2, \ldots$ covers A, there must be a B_k that contains x. But $x_k, x_{k+1}, x_{k+2}, \ldots$ are not in B_k by the choice of the x_k. We now have a contradiction, because if x is a limit-point of the sequence $x_1, x_2, \ldots$, the neighborhood

B_k of x must contain infinitely many of them. The contradiction shows that $B_1, \ldots, B_n$ must cover A for some n. QED

The property that every open cover has a finite subcover is sometimes taken as the definition of compactness; in that case the Heine-Borel Theorem says that every closed, bounded set is compact. To bring these ideas full circle we should prove directly that if A is a set with the property that every open cover has a finite subcover, then every sequence of points in A has a limit-point in A. We already have a two-stage proof of this; the argument given before the Heine-Borel Theorem shows A is closed and bounded, and this implies A is compact. For a more direct proof, consider a sequence $x_1, x_2, \ldots$ in A with no limit-point in A. Then the closure of the set $\{x_1, x_2, \ldots\}$ is closed and contains no points of A other than $x_1, x_2, \ldots$. Now the complement of this set is open. Take it as the first set B_0 in an open cover. For the other sets $B_1, B_2, \ldots$ in the cover simply choose B_j to be an open interval containing x_j but no other x_k that is not equal to x_j. Then $B_0, B_1, B_2, \ldots$ is an open cover of A, since B_0 contains all points of A except $x_1, x_2, \ldots$. If this open cover is to have a finite subcover, then there can only be a finite number of distinct points in the sequence—for B_0 contains none of them, and each B_j for $j \geq 1$ contains exactly one of them. But if the sequence contains only a finite number of distinct points at least one of them must repeat infinitely often and then that point is a limit-point of the sequence in A.

To summarize, we have shown that the following three conditions on a set A are equivalent:

1. A is closed and bounded.

2. Every sequence of points in A has a limit-point.

3. Every open cover of A has a finite subcover.

There are other equivalent statements, but they are less important and are left as exercises. We conclude this section with an important property of compact sets. A sequence of sets $A_1, A_2, \ldots$ is called *nested* if A_n contains A_{n+1} for every n. For a nested sequence of sets, the intersection of the first n, $A_1 \cap A_2 \cap \cdots \cap A_n$, is equal to A_n, so we may think of the intersection of all of them, $\bigcap_{n=1}^{\infty} A_n$, as a kind of "limit" of the sequence. If the sets A_n are all non-empty we might expect this intersection to be non-empty also, but this need not be the case. If we take the sequence of open intervals $(0,1), (0,1/2), (0,1/3), (0,1/4), \ldots$, as shown in Figure 3.3.5,

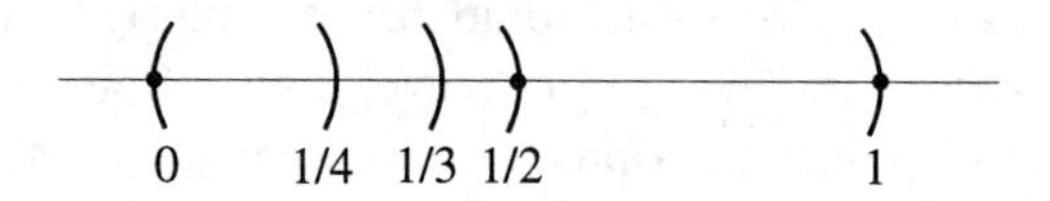

Figure 3.3.5:

there is no point in the intersection. The same is true if we take for A_n the closed set $\{x; x \geq n\}$. However, if the sets A_n are compact, this cannot happen.

Theorem 3.3.3 *A nested sequence of non-empty compact sets has a non-empty intersection.*

Proof: Let $A_1, A_2, \ldots$ denote the nested sequence. Choose points x_n in A_n. The sequence $x_1, x_2, \ldots$ lies in A_1 by the nesting and has a limit-point x in A_1 since A_1 is compact. But x is also a limit-point of $x_n, x_{n+1}, \ldots$ and since this sequence lies in A_n, x must lie in A_n (again we must argue separately that either x equals one of the points $x_n, x_{n+1}, \ldots$ or x is a limit-point of the set $\{x_n, x_{n+1}, \ldots\}$ and, hence, is in A_n since A_n is closed). Since x lies in A_n for every n, x is in the intersection, so the intersection is non-empty. QED

3.3.1 Exercises

1. Show that compact sets are closed under arbitrary intersections and finite unions.

2. Show that the following *finite intersection property* for a set A is equivalent to compactness: if $\mathcal{B}$ is any collection of closed sets such that the intersection of any finite number of them contains a point of A, then the intersection of all of them contains a point of A. (**Hint**: consider the complements of the sets of $\mathcal{B}$.)

3. If $B_1, \ldots, B_n$ is a finite open cover of a compact set A, can the union $B_1 \cup \cdots \cup B_n$ equal A exactly?

4. If $A \subseteq B_1 \cup B_2$ where B_1 and B_2 are disjoint open sets and A is compact, show that $A \cap B_1$ is compact. Is the same true if B_1 and B_2 are not disjoint?

5. For which compact sets can you set an upper bound on the number of sets in a subcover of an open cover?

6. For two non-empty sets of numbers A and B, define $A+B$ to be the set of all sums $a+b$ where a is in A and b is in B. Show that if A is open, then $A+B$ is open. Show that if A and B are compact, then $A+B$ is compact. Give an example where A and B are closed but $A+B$ is not.

7. Which of the analogous statements of exercise 6 are valid for the product set $A \cdot B$ (consisting of all products $a \cdot b$)? Can you modify the false ones slightly to make them true?

8. If A is compact, show that $\sup A$ and $\inf A$ belong to A. Give an example of a non-compact set A such that both $\sup A$ and $\inf A$ belong to A.

9. Show that every infinite compact set has a limit-point. Is the same true of closed sets? of open sets?

10. Find necessary and sufficient conditions for A to be the complement of a compact set.

3.4 Summary

3.1 The Theory of Limits

Definition $\lim_{j\to\infty} x_j = +\infty$ *if for every* n *there exists* m *such that* $x_j \geq n$ *for all* $j \geq m$.

Definition y *is an upper bound for a set* E *if* $x \leq y$ *for all* x *in* E.

Theorem 3.1.1 *For every non-empty set E of real numbers that is bounded above there exists a unique real number* $\sup E$ *such that*

1. $\sup E$ *is an upper bound for E.*

2. *if y is an upper bound for E then $y \geq \sup E$.*

Definition *A sequence $x_1, x_2, \ldots$ is said to be monotone increasing if $x_{j+1} \geq x_j$ for every j.*

Theorem 3.1.2 *A monotone increasing sequence that is bounded from above has a finite limit, equal to the* sup.

Definition 3.1.1 *A real number x is said to be a limit-point of a sequence $x_1, x_2, \ldots$ if for every n there exists an infinite number of terms x_j such that $|x - x_j| < 1/n$.*

Definition *A sequence $y_1, y_2, \ldots$ is said to be a subsequence of $x_1, x_2, \ldots$ if there is an increasing function $m(n)$ (meaning $m(n+1) > m(n)$ for all n) such that $y_n = x_{m(n)}$.*

Theorem 3.1.3 *x is a limit-point of $x_1, x_2, \ldots$ if and only if there exists a subsequence with limit x.*

Definition 3.1.2 $\text{limsup}_{k\to\infty} x_k = \lim_{k\to\infty} \sup_{j>k} x_j$ *and* $\text{liminf}_{k\to\infty} x_k = \lim_{k\to\infty} \inf_{j>k} x_j$.

Theorem 3.1.4 *The set of limit-points in the extended reals of a sequence is non-empty, containing* limsup, *which is its* sup, *and* liminf, *which is its* inf.

Theorem 3.1.5 *A bounded sequence converges if and only if it has only one limit-point (if and only if the* limsup *and* liminf *are equal).*

3.2 Open and Closed Sets

Definition *A set is open if every point of the set lies in an open interval entirely contained in the set.*

Theorem *A set of reals is open if and only if it is the disjoint union of at most countably many open intervals.*

Theorem 3.2.1 *Open sets are preserved under arbitrary unions and finite intersections.*

Definition *A neighborhood of a point is an open set containing the point.*

Theorem $x = \lim_{j\to\infty} x_j$ *if and only if every neighborhood of* x *contains all but a finite number of the* x_j.

Definition 3.2.1 x *is a limit-point of a set* A *if every neighborhood of* x *contains a point of* A *different from* x.

Definition *A set is said to be closed if it contains all its limit-points.*

Theorem 3.2.2 *A set is closed if and only if its complement is open.*

Theorem 3.2.3 *Closed sets are preserved under finite unions and arbitrary intersections.*

Definition *The Cantor set is the subset of* $[0, 1]$ *of all numbers expressible in base* 3 *with digits* 0 *and* 2.

Definition *The interior of a set* A *is the subset of all points which lie in an open interval entirely contained in* A.

Definition *The closure of a set is the union of the set and all its limit-points.*

Theorem *The closure of a set is closed.*

Definition *A subset* B *of* A *is said to be dense in* A *if the closure of* B *contains* A.

3.3 Compact Sets

Definition 3.3.1 *A set A of real numbers is said to be compact if every sequence of points in A has a limit-point in A.*

Theorem 3.3.1 *A set of real numbers is compact if and only if it is closed and bounded.*

Theorem 3.3.2 (*Heine-Borel*) *A set is compact if and only if it has the property that every open cover has a finite subcover.*

Theorem 3.3.3 *A nested sequence of non-empty compact sets has a non-empty intersection.*

Chapter 4

Continuous Functions

4.1 Concepts of Continuity

4.1.1 Definitions

In this section we introduce four important concepts: 1) functions, 2) continuity, 3) uniform continuity, and 4) limits of functions. There are some subtle distinctions to be made here, and it will be important to look at some examples and to pay attention to the motivation behind the definitions. It is more common to define limits of functions first and then base the definition of continuity on the notion of limits. However, it is easier to motivate the definition of continuity, so that is the order we follow.

The abstract notion of function is that of a correspondence between sets. Two sets, called the domain and the range, are given; and to each element x of the domain there is given an element $f(x)$ of the range. We will study functions whose domain and range are sets of real numbers, and unless explicitly stated otherwise, the term *function* will be reserved for this special case.

Definition 4.1.1 *A function consists of a domain D, a range R—both subsets of the real numbers, and a correspondence $x \to f(x)$ where x is a variable point in D and $f(x)$ is a point in R. We do not require that every point in R actually occurs as $f(x)$ for some x in D. We will call the image $f(D)$ the set of values $f(x)$ as x varies in D. The image is a subset of the range. We say the function is onto if the image equals the* *range. We say the function is one-to-one if* $f(x_1) \neq f(x_2)$ *if* $x_1 \neq x_2$ *for any* x_1 *and* x_2 *in the domain. We will usually take the range* R *to be the whole real line* $\mathbb{R}$*, and we will not distinguish between functions that are the same except for the range (for example,* $f(x) = x^2$ *with domain* $\mathbb{R}$ *and range* $\mathbb{R}$ *and* $f(x) = x^2$ *with domain* $\mathbb{R}$ *and range* $y \geq 0$*). However we must distinguish between functions with the same rule and different domains (*$f(x) = x^2$ *with domain* $\mathbb{R}$ *and* $f(x) = x^2$ *with domain* $[0,1]$ *are different functions). Concerning the correspondence* $x \to f(x)$*, we make no assumptions. It may be given by a recognizable rule, or by a bizarre rule, or by no describable rule at all. We only require that there is a unique value* $f(x)$ *that can or could be determined if the value of* x *is given. This might be described as an agnostic view of function. It is one of the broadest views possible—except that we are requiring that the function be single-valued.*

This is not the only concept of function that mathematicians have put forth. The concept of a function as a formula was very prevalent during the eighteenth century, but it led to great confusion since the concept of "formula" kept changing. Later, when we discuss Fourier series, we will discover that many of the functions in our agnostic sense do have formulas after all. In the twentieth century, various concepts of recursive function and constructible function in which the correspondence $x \to f(x)$ must be given in a manner that would in principle be computable, have been put forth. In my opinion it would be impossible to learn about these concepts without first mastering the agnostic theory presented in this work.

Basically there are three ways a function can be specified:

1. *Explicitly*, the rule $x \to f(x)$ is given by a formula, and the domain is specified.

2. *Implicitly*, the correspondence is given by the solution of an equation involving x and $f(x) = y$, and the domain consists of all x for which the equation has a solution. In this case you have to be careful about uniqueness of the solution, since we insist on single-valued functions. For example the equation $x^2 + y^2 = 1$ does not lead to a single-valued function, but together with the condition $y \leq 0$ we obtain the function $f(x) = -\sqrt{1-x^2}$ with domain $[-1,1]$.

3. By giving the *graph*, which is by definition the set of ordered pairs $(x, f(x))$ where x varies in the domain. We follow the usual Cartesian convention of picturing the ordered pairs of real numbers as points in the Euclidean plane.

The concept of function, even restricting the domain and range to sets of numbers, is in a way too general to be very interesting. Certainly in order to do any analysis we have to restrict further the kind of functions with which we deal. In this chapter we will discuss continuous functions. This is neither the largest nor the smallest class of functions that is convenient for analysis, but it is perhaps the most intuitive. Unfortunately there are many intuitive ideas behind the notion of continuity. Some of them are quite helpful and valid, while others may lead to confusion. Our goal is to provide a rigorous framework to which we can attach our intuitions and to make some subtle but important distinctions that would otherwise slip by.

One rather geometric intuitive idea of a continuous function is one whose graph consists of a single, connected piece. Now it is indeed possible to make a precise mathematical definition of *connected* sets in the plane so that the graph of a continuous function is connected. However, it turns out that there are functions that are not continuous whose graphs are also connected. Thus we will put aside the intuitive idea of continuous as "without a break" or "drawable without lifting the pen" and turn to other ideas.

Let us for the moment think of a function as a mathematical representation of a relationship between variables in the real world. Think of a measurement or experiment where the input is x and the output is $f(x)$. One of the things we require of experimental science is repeatability. If we put the input x in several times, we should always get $f(x)$ out. The requirement that a function be single-valued would seem to take care of this, but in fact it is not enough. The reason is that in any real situation we cannot control the input exactly, any more than we can measure the output exactly. So $f(x_1)$ and $f(x_2)$ might represent outputs of what we believe to be identical experiments if x_1 and x_2 are very close to each other. Therefore the requirement of repeatability is actually that very close values of the input should yield very close values of the output. This is exactly the intuitive idea behind continuity.

Let us look at the situation a little more closely. What do we mean by very close values of the input and output? Clearly we must be referring to some condition like $|x - x_0| < 1/n$ for the input and $|f(x) - f(x_0)| < 1/m$ for the output, where $1/n$ and $1/m$ are small errors. The statement "very close values of the input yield very close values of the output" then should translate into "$|x - x_0| < 1/n$ implies $|f(x) - f(x_0)| < 1/m$". But this is still vague, for we have not specified the relationships between the errors $1/n$ and $1/m$. We have the correct statement, but we have to decide on the quantifiers and their order. To help us decide, let's ask how small an error in the output we would accept. Would $1/10$ be good enough? Or $1/100$ or $1/10^{23}$? Isn't the tolerance for error in the output a relative judgment—one that should not be made once and for all? Today we might be happy with an error of $1/1,000$, but tomorrow we might want to do better. To build a mathematical theory on a fixed notion of acceptable error would be absurd. We must have the flexibility to make the error in the output as small as we like. Thus the first quantifier must be "for all errors $1/m$ in the output". To meet any given tolerance for error in the output, we may have to take drastic action to control the error in the input. Again we do not want to say in advance how small this error must be; only that some small error in the input will do the trick. Thus the second quantifier must be "there exists an error $1/n$ (depending on $1/m$) in the input". Altogether we now have: "for every $1/m$ there exists $1/n$ such that $|x - x_0| < 1/n$ implies $|f(x) - f(x_0)| < 1/m$". In the familiar $\epsilon - \delta$ formulation we would write ϵ for $1/n$ and δ for $1/m$.

We are not quite done, for we have not specified what x and x_0 are and how the errors $1/m$ and $1/n$ relate to them. This is by no means a trivial question. In fact there are two distinct concepts that await the resolution—in two different ways—of this question. Let us imagine that x_0 is the value—perhaps idealized—of the input variable in which we are interested. Then x represents a nearby value, which must be allowed to vary over the domain of the function with no control other than the error specification $|x - x_0| < 1/n$. This leads to the definition of *continuity at a point* x_0: *for every* $1/m$ *there exists* $1/n$ *such that for every* x *in the domain of the function with* $|x - x_0| < 1/n$ *we have* $|f(x) - f(x_0)| < 1/m$. The error $1/n$ depends on $1/m$ and on x_0; the dependence on x_0 need not be mentioned in this definition, since x_0 is fixed. However, when we define a continuous function to be a function that is continuous at every point of its domain, this dependence becomes important. The best way to see this is through an example.

Let's show that the function $f(x) = 1/x$ on the domain $x > 0$ is continuous. We fix a point x_0 in the domain. Given any error $1/m$, we need to find an error $1/n$ such that $|x - x_0| < 1/n$ and $x > 0$ (x in the domain) implies $|1/x - 1/x_0| < 1/m$. Now we compute $1/x - 1/x_0 = (x_0 - x)/xx_0$; and if we are to bound this from above, we need to bound xx_0 from below. Since x_0 does not vary, the problem is to keep x from getting close to zero. Thus we want to require something like $1/n < x_0/2$, for then $|x - x_0| < 1/n$ implies $x > x_0/2$, as in Figure 4.1.1.

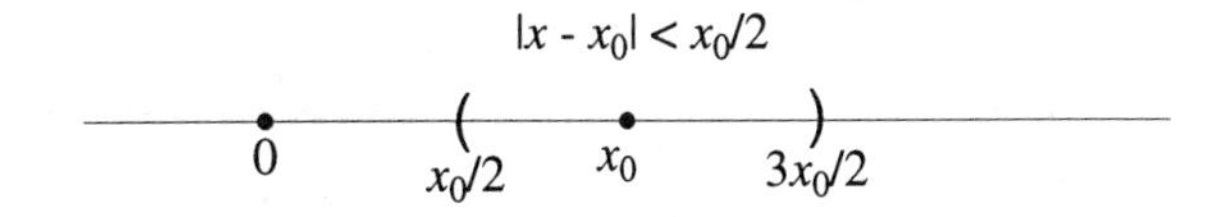

Figure 4.1.1:

Then

$$\begin{aligned}\left|\frac{1}{x} - \frac{1}{x_0}\right| &\leq \frac{|x - x_0|}{xx_0} \leq \frac{2}{x_0^2}|x - x_0| \\ &\leq \frac{2}{nx_0^2}\end{aligned}$$

and we need $2/nx_0^2 \leq 1/m$. Running the argument backward, if we choose $1/n$ to be less than $x_0/2$ and less than $x_0^2/2m$, we have $|x - x_0| < 1/n$ (the condition $x > 0$ is now redundant) implies $|1/x - 1/x_0| < 1/m$. This demonstrates the continuity of the function at each point of its domain. Note, however, that the error $1/n$ depends on the point x_0. As x_0 gets closer to zero the error $1/n$ must be made smaller in order to guarantee the same error $1/m$ in the output. This is clear from the graph of the function, as shown in Figure 4.1.2.

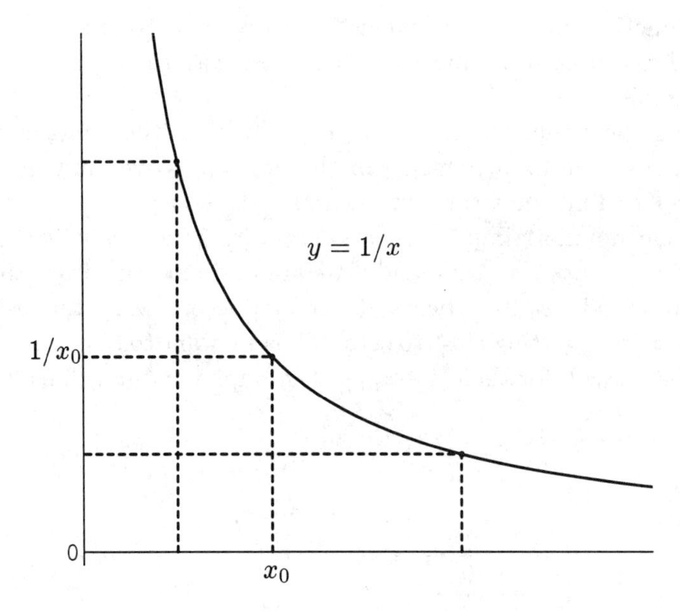

Figure 4.1.2:

Of course this function is not even bounded near $x = 0$, so we might expect trouble. The trouble with which we are dealing, however, can turn up even for bounded functions. For example, consider a function with domain $(0, 1]$ whose values zigzag between 0 and 1, hitting 0 at $1/n$ for n even and 1 at $1/n$ for n odd, as shown in Figure 4.1.3. If we want the error in the output less than $1/2$, we will have to restrict $|x - x_0|$ severely if x_0 is close to 0, and no condition $|x - x_0| < 1/n$ will work for all x_0 in the domain (for example, it fails for $x_0 = 1/n$).

In both examples the domain of the function isn't closed, and the trouble arises near a limit-point that is not in the domain. Later we will see that if the domain is compact this situation can't arise. If the error in the input $|x - x_0| < 1/n$ can be chosen so as to make the error in the output $|f(x) - f(x_0)| < 1/m$ for all points x and x_0 in the domain, then we have a stronger condition than continuity, which is called *uniform continuity*. We summarize the discussion in a formal definition.

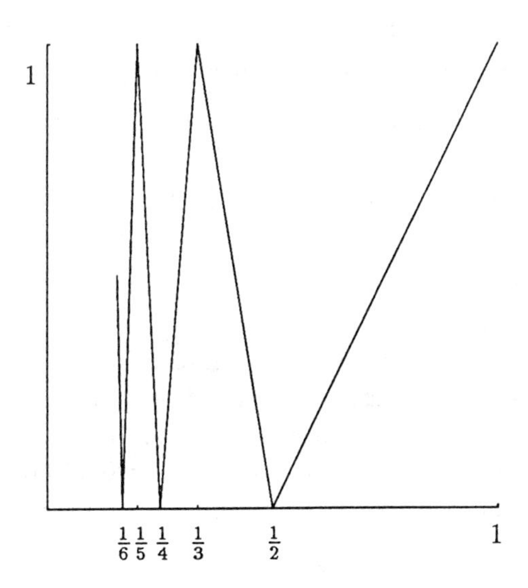

Figure 4.1.3:

Definition 4.1.2 *Let f be a function on a domain D. Let x_0 be a point of D. We say f is continuous at x_0 if for every $1/m$ there exists $1/n$ such that $|f(x) - f(x_0)| < 1/m$ for every x in D satisfying $|x - x_0| < 1/n$. We say f is continuous if for every x_0 in D and for every $1/m$ there exists $1/n$ (depending on x_0 and $1/m$) such that $|f(x) - f(x_0)| < 1/m$ for every x in D satisfying $|x - x_0| < 1/n$. We say f is uniformly continuous if for every $1/m$ there exists $1/n$ such that $|f(x) - f(x_0)| < 1/m$ for all x and x_0 in D satisfying $|x - x_0| < 1/n$.*

If f is continuous at x_0, then the values of $f(x)$ for x near x_0 are approximating the value $f(x_0)$. This suggests a close relationship between the concept of continuity of functions and limits of sequences. We can make the analogy stronger if we introduce a related concept, the *limit of a function.* Let f be a function defined on a domain D, and let x_0 be a limit-point of D. We do not require that x_0 be actually in D—frequently we will want to take D to be an open interval and x_0 an endpoint. The reason we require x_0 to be a limit-point of D is that we need values of x in D nearby. We want to define the *limit of $f(x)$ at x_0* to be the number, *if it exists*, that is approximated by $f(x)$ for x near x_0. For purposes of defining the limit we will ignore the value $f(x_0)$ if

x_0 happens to be in the domain—this being a convention chosen not for compelling reasons but for convenience in certain applications.

Definition 4.1.3 *Let f be a function defined on a domain D, and let x_0 be a limit-point of D. We say f has a limit at x_0 if there exists a number y, which we call the limit of f at x_0 and write $y = \lim_{x \to x_0} f(x)$, such that for every $1/m$ there exists $1/n$ such that $|f(x) - y| < 1/m$ for all x in D not equal to x_0 satisfying $|x - x_0| < 1/n$.*

The number y, if it exists, is unique for the same reason that the limit of a sequence, if it exists, is unique. There is no requirement that the limit have anything to do with $f(x_0)$, if x_0 happens to be in D, since we have made the convention of excluding the value $x = x_0$. Thus the function equal to 1 for all values of $x \neq 0$ and 0 for $x = 0$, as shown in Figure 4.1.4, has limit 1 at $x = 0$.

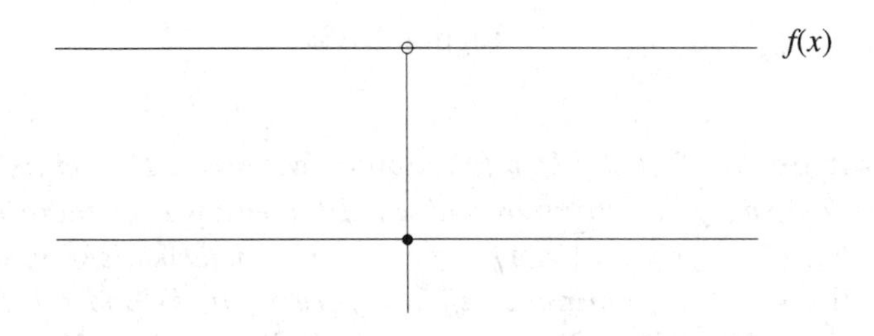

Figure 4.1.4:

Comparing the definition of limit of a function with limit of a sequence we see that the only difference is that the condition of going far out in the sequence is replaced by the condition of going near the point x_0.

The connection between the definitions of limit of a function and continuity is also clear. *A function is continuous at a point x_0 in its domain that is a limit-point of its domain if and only if f has a limit at x_0 and the value of $f(x_0)$ equals* $\lim_{x \to x_0} f(x)$. If x_0 is a point of the domain that is not a limit-point of the domain, then the definition of continuity is trivially valid—we take $1/n$ so that the only point of the domain satisfying $|x - x_0| < 1/n$ is $x = x_0$ itself. Thus every function is continuous at a non-limit-point (called an *isolated* point) of its domain, and we normally will not be interested in this trivial case.

4.1.2 Limits of Functions and Limits of Sequences

Now let us further examine the connection between the notion of limits of functions and sequences. Let x_0 be a limit-point of the domain D of a function, and let $x_1, x_2, \ldots$ be a sequence of points in D, none equal to x_0, that converges to x_0. Then we would expect the sequence $f(x_1), f(x_2), \ldots$ to converge to $\lim_{x \to x_0} f(x)$, if the limit exists, for by going far out in the sequence we make $|x_j - x_0| < 1/n$. Of course there are usually an uncountable number of points in the domain satisfying $|x - x_0| < 1/n$, so we cannot expect that the convergence of $f(x_1), f(x_2), \ldots$ for any one sequence will contain as much information as the existence of the limit of the function. In the example of the zigzag function in Figure 4.1.3, the function does not have a limit as $x \to 0$; but if we choose the sequence $1, 1/3, 1/5, 1/7, \ldots$ where the function takes on the value 1, then the sequence $f(1), f(1/3), \ldots$ is just $1, 1, 1, \ldots$ and so has the limit 1. Of course it was only by a careful choice of the sequence of points that we were able to come out with a limit; the sequence $1, 1/2, 1/3, 1/4, \ldots$ also converges to zero, but $f(1), f(1/2), (1/3), \ldots$ is the sequence $1, 0, 1, 0, \ldots$, which has no limit. This suggests that perhaps the convergence of $f(x_1), f(x_2), \ldots$ for *every* sequence $x_1, x_2, \ldots$ converging to x_0 is equivalent to the existence of $\lim_{x \to x_0} f(x)$. This is indeed the case.

Theorem 4.1.1 *Let x_0 be a limit-point of the domain D of a function f. Then $\lim_{x \to x_0} f(x)$ exists if and only if $f(x_1), f(x_2), \ldots$ converges for every sequence $x_1, x_2, \ldots$ of points of D, none equal to x_0, converging to x_0. It is not necessary to assume that the limit of all the sequences is the same, although this must be true, and the common limit is equal to the limit of the function.*

Proof: Suppose $\lim_{x \to x_0} f(x) = y$ exists, and let $x_1, x_2, \ldots$ converge to x_0. Given any error $1/m$, we want to make $|f(x_j) - y| < 1/m$ by going far out in the sequence. Now first by the existence of $\lim_{x \to x_0} f(x)$ we know we can make $|f(x) - y| < 1/m$ by taking $|x - x_0| < 1/n$, where $1/n$ depends on $1/m$ and x_0. We would like to apply this to $x = x_j$, so we must next use the convergence of the sequence $x_1, x_2, \ldots$ to x_0. Given $1/n$, there exists k such that $|x_j - x_0| < 1/n$ for all $j \geq k$, by this convergence. Also $x_j \neq x_0$ by assumption. Thus we can apply the implication $|x - x_0| < 1/n$ implies $|f(x) - y| < 1/m$ derived above to

$x = x_j$ if $j \geq k$, and we obtain $j \geq k$ implies $|f(x_j) - y| < 1/m$. Thus we have established the convergence of $f(x_1), f(x_2), \ldots$ to y.

Conversely, suppose the sequences $f(x_1), f(x_2), \ldots$ always converge if $x_1, x_2, \ldots$ converges to x_0, with the x_j in D and different from x_0. We want to show this implies the existence of the limit of the function $\lim_{x \to x_0} f(x)$.

First we claim all the sequences $f(x_1), f(x_2), \ldots$ have a common limit, for if not we could shuffle two sequences with different limits. In other words, if $x_1, x_2, \ldots$ and $y_1, y_2, \ldots$ both converge to x_0 and $f(x_1), f(x_2), \ldots$ converges to a and $f(y_1), f(y_2), \ldots$ converges to b with $a \neq b$, then the shuffled sequence $x_1, y_1, x_2, y_2, \ldots$ still has limit x_0 but the shuffled sequence $f(x_1), f(y_1), f(x_2), f(y_2), \ldots$ does not converge, contradicting the hypotheses. Thus there is a common limit, call it y, of all the sequences $f(x_1), f(x_2), \ldots$. This is the value that we will show is equal to $\lim_{x \to x_0} f(x)$.

Suppose $\lim_{x \to x_0} f(x) = y$ were false. Negating the definition leads to the statement: there exists $1/m$ such that for all $1/n$ there exists a point z_n in the domain, not equal to x_0, such that $|z_n - x_0| < 1/n$, and yet $|f(z_n) - y| \geq 1/m$. This is the statement that we must show leads to a contradiction. But if this statement were true, the sequence $\{z_n\}$ would converge to x_0 (since $|z_n - x_0| < 1/n$) and yet $f(z_1), f(z_2), \ldots$ would not converge to y, since $|f(z_n) - y| \geq 1/m$ for every n. This contradicts the fact that $f(z_1), f(z_2), \ldots$ converges to y that we just established. QED

This theorem has a very striking consequence regarding continuous functions. Suppose f is continuous, and let $x_1, x_2, \ldots$ be any sequence of points in the domain that converges to a point x_0 in the domain (with none of the x_j equal to x_0). Then x_0 is a limit-point of the domain; and since $\lim_{x \to x_0} f(x)$ exists and equals $f(x_0)$, we have that $f(x_1), f(x_2), \ldots$ converges to $f(x_0)$. We can paraphrase this by saying that *the image under f of a convergent sequence (converging to a point in the domain) is convergent.* We can easily remove the artificial requirement that none of the x_j equal x_0; however the condition that x_0 be in the domain is crucial, as the zigzag example shows. The converse is also true, so we can characterize continuous functions as those that preserve convergent sequences or, equivalently, as those that commute with sequential limits, $f(\lim_{j \to \infty} x_j) = \lim_{j \to \infty} f(x_j)$.

Theorem 4.1.2 *Let f be a function defined on a domain D. Then f is continuous if and only if for every sequence of points $x_1, x_2, \ldots$ that has a limit in D, the sequence $f(x_1), f(x_2), \ldots$ is convergent. It is not necessary to assume that the limit of the sequence $f(x_1), f(x_2), \ldots$ is equal to $f(\lim_{j\to\infty} x_j)$, but this follows from the hypotheses.*

Proof: The proof is quite similar to the proof of the previous theorem; we could in fact reduce it to the proof to the previous theorem, at the expense of a lot of special cases.

Suppose first f is continuous. Then if $x_1, x_2, \ldots$ converges to x_0 we can show $f(x_1), f(x_2), \ldots$ converges to $f(x_0)$ by the same argument as in the previous theorem: given $1/m$ we first find $1/n$ such that $|x - x_0| < 1/n$ implies $|f(x) - f(x_0)| < 1/m$ by continuity of f and then find k such that $j \geq k$ implies $|x_j - x_0| < 1/n$, by the convergence of $x_1, x_2, \ldots$ to x_0. Then $j \geq k$ implies $|f(x) - f(x_0)| < 1/m$.

For the converse we first use the shuffling argument to show the limit of the sequence $f(x_1), f(x_2), \ldots$ is the same for all sequences $x_1, x_2, \ldots$ converging to the point x_0; and since $x_0, x_0, \ldots$ is one such sequence and the limit of $f(x_0), f(x_0), \ldots$ is $f(x_0)$, it follows that the common limit of all these sequences is $f(x_0)$. Now if x_0 is not a limit-point of D there is nothing to prove, while if x_0 is a limit-point of D the previous theorem implies $\lim_{x\to x_0} f(x)$ exists and equals the common limit $f(x_0)$. Thus f is continuous at x_0. QED

4.1.3 Inverse Images of Open Sets

We now have two equivalent characterizations of continuity, and we will presently find a third. Recall that we were able to characterize convergent sequences entirely in terms of open sets: $x_1, x_2, \ldots$ converges to x_0 if for every neighborhood of x_0, all but a finite number of terms in the sequence lie in that neighborhood. Similarly we can rephrase the definition of continuity of f at x_0 as follows: *for every neighborhood of $f(x_0)$ there exists a neighborhood of x_0 that is mapped into the neighborhood of $f(x_0)$ by f.* If we denote by M the set $\{y : |y - f(x_0)| < 1/m\}$ and N the set $\{x : |x - x_0| < 1/n\}$, then the statement "f maps N into M" is the same as "$|x - x_0| < 1/n$ implies $|f(x) - f(x_0)| < 1/m$". Now if A is any subset of the range of f, we denote by $f^{-1}(A)$, the *inverse image* of A under f, the set of points x in the domain of f such that $f(x)$ is in A. The fact that f maps N into M is the same as saying N

is contained in $f^{-1}(M)$ and so continuity at x_0 becomes: *the inverse image of every neighborhood of* $f(x_0)$ *contains a neighborhood of* x_0. It is important to convince yourself that it is the inverse image that belongs in this statement. It is *not* true that the image of a neighborhood of x_0 under a continuous function f contains a neighborhood of $f(x_0)$; for example, a constant function is continuous, but its entire image is a single point.

The situation actually improves if we reformulate the definition of continuity on the whole domain rather than at a single point. For simplicity we assume that the domain is an open set. (We will return to this in Chapter 9.)

Theorem 4.1.3 *Let f be a function defined on an open domain. Then f is continuous if and only if the inverse image of every open set is an open set.*

Proof: First suppose f is continuous. Let A be an open set of real numbers. We want to show $f^{-1}(A)$ is open. To do this we need to show that every point in $f^{-1}(A)$ is contained in an open interval lying in $f^{-1}(A)$. Of course $f^{-1}(A)$ may be empty, but in that case there is nothing to prove since the empty set is open.

So suppose x_0 is in $f^{-1}(A)$. This means that x_0 is in the domain of f and $f(x_0)$ is in A. Since A is open, there is an open interval about $f(x_0)$, say $\{y : |y - f(x_0)| < 1/m\}$, contained in A. By the continuity of f, there is an open interval about x_0, $|x - x_0| < 1/n$, that is mapped into the interval about $f(x_0)$ (actually the definition says $|x-x_0| < 1/n$ and x in the domain implies $|f(x) - f(x_0)| < 1/m$, but since we have assumed the domain is open, we can arrange that $|x-x_0| < 1/n$ implies x is in the domain by taking $1/n$ small enough). But this implies that the interval $|x - x_0| < 1/n$ lies in $f^{-1}(A)$, which shows that $f^{-1}(A)$ is open.

Conversely, suppose $f^{-1}(A)$ is open for every open set A. We want to show that f is continuous at every point x_0 of the domain. To do this we have to show that given any $1/m$ we can find $1/n$ such that $|x - x_0| < 1/n$ implies $|f(x) - f(x_0)| < 1/m$. Let us choose for A the open set $\{y : |y - f(x_0)| < 1/m\}$. By hypothesis $f^{-1}(A)$ is open. Note that x_0 is in $f^{-1}(A)$, for $|f(x_0) - f(x_0)| = 0$. Since x_0 is a point of the open set $f^{-1}(A)$, there is an open interval about x_0, say $|x - x_0| < 1/n$, contained entirely in $f^{-1}(A)$. But this means that $|x - x_0| < 1/n$ implies $f(x)$ is in A, and by the choice of A, $|f(x) - f(x_0)| < 1/m$ as desired. QED

This last characterization of continuous function may be furthest from the intuition of continuity, but it is undeniably simple and elegant. It is extremely general—we will see that it is valid also for functions of several variables—and it is in fact taken to be the definition of continuity in general topology. There is a similar characterization in which open sets are replaced by closed sets, if the domain of f is closed. This is not at all surprising in view of the fact that complements of open sets are closed, but it is an important fact. We leave the details to exercise set 4.1.5, number 1. If the domain of f is the whole line, then both results apply. In particular, a set defined by an "open" condition like $f(x) > a$ is open and a set defined by a "closed" condition like $f(x) \geq a$ or $f(x) = a$ is closed. See exercise set 4.1.5, numbers 2 and 3 for related results.

4.1.4 Related Definitions

Continuity is a qualitative property, in that it concerns a relation between the error of the input and error of the output that is not specified. It is not surprising that there are many related quantitative properties—where the relation between the errors takes on a specific form. The simplest possible form is proportionality—the existence of a constant M such that $|x - x_0| < 1/Mm$ implies $|f(x) - f(x_0)| < 1/m$. It is simple to see that this is equivalent to the condition

$$|f(x) - f(x_0)| \leq M|x - x_0|.$$

If this holds with a fixed M for all x and x_0 in the domain we say f satisfies a *Lipschitz condition.* A typical function satisfying this condition is $f(x) = |x|$, where the constant M can be taken equal to 1. Other variants of this are the *Hölder conditions* of order α, $0 < \alpha \leq 1$ (sometimes referred to as Lipschitz conditions of order α):

$$|f(x) - f(x_0)| \leq M|x - x_0|^{\alpha}.$$

This implies continuity in that $|f(x) - f(x_0)| < 1/m$ if $|x - x_0| < 1/(mM)^{1/\alpha}$. Another way of thinking about these conditions is to define the *modulus of continuity* $\omega(x_0, \delta)$ of f to be the sup of the values $|f(x) - f(x_0)|$ as x varies over the interval $|x - x_0| < \delta$. The condition that f be continuous at x_0 is the statement that $\omega(x_0, \delta)$ as a function of δ has limit 0 as $\delta \to 0$, while the quantitative conditions describe the rate of convergence $\omega(x_0, \delta) \le M\delta$ for the Lipschitz condition.

Generally speaking, for every theorem about continuous functions, there is an analogous quantitative version for Lipschitz functions, which can be proved by similar methods.

There are many circumstances when we will want to consider limits and continuity that are one-sided. For example, consider the signum function

$$\operatorname{sgn} x = \begin{cases} +1 & \text{if } x > 0, \\ -1 & \text{if } x < 0. \end{cases}$$

The domain of this function is all $x \neq 0$, although sometimes one takes the convention $\operatorname{sgn} 0 = 0$ to have the function defined on the whole line. The graph is shown in Figure 4.1.5.

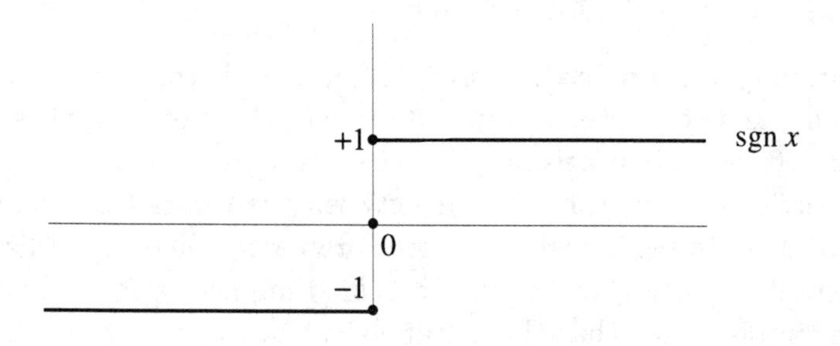

Figure 4.1.5:

This function does not have a limit at $x = 0$, because if you approach 0 from positive numbers the value is 1 while if you approach from negative numbers the value is -1. We would then like to say $\lim_{x\to 0^+} \operatorname{sgn} x = +1$ and $\lim_{x\to 0^-} \operatorname{sgn} x = -1$, and it is not hard to define such *limits from above and below* (or *right* and *left*) to make this so. By $\lim_{x\to x_0^+} f(x) = y$ (where x_0 is a limit-point of the domain) we will mean for every $1/n$ there exists $1/m$ such that $x_0 < x < x_0 + 1/m$, and x in the domain implies $|f(x) - y| < 1/n$, while by $\lim_{x\to x_0^-} f(x) = y$ we will mean the same condition with now $x_0 - 1/m < x < x_0$. Similarly we will say that f is *continuous from the right* at x_0 if $f(x_0) = \lim_{x \to x_0^+} f(x)$ and that f is *continuous from the left* at x_0 if $f(x_0) = \lim_{x \to x_0^-} f(x)$.

Clearly continuity is the same as continuity from both the left and right. The choice $\operatorname{sgn} 0 = 0$ makes the signum continuous from neither side at 0, but a different convention would allow us to have one or the other but not both.

The kind of discontinuity that signum has is called a *jump discontinuity* or *discontinuity of the first kind.* The definition is that both $\lim_{x \to x_0^+} f(x)$ and $\lim_{x \to x_0^-} f(x)$ should exist and be different. Any worse discontinuity, where one or another one-sided limit does not exist, is called a *discontinuity of the second kind.* If we adjoin 0 to the domain of the zigzag function (Figure 4.1.3) it will have such a discontinuity there. There is another technical kind of discontinuity, in which the limit exists at x_0 but is different from $f(x_0)$. This is called a *removable discontinuity,* because if we simply redefine the function at x_0 to equal its limit then we will have a continuous function there. An example is the function

$$f(x) = \begin{cases} 1 & \text{if} \quad x \neq 0, \\ 0 & \text{if} \quad x = 0. \end{cases}$$

It is not unreasonable to think of removable discontinuities as simply mistakes that can be, and should be, corrected.

The one-sided limits we have discussed involve *restricting the domain* of the function to one side of the point. There are also concepts of *upper semi-continuity* and *lower semi-continuity* that involve restricting the *range* of the function. These are less frequently used and will not be needed in this book.

Finally we note that it is often convenient to allow the extended real numbers $\pm\infty$ to appear in limits, in either the domain or range. We leave it to the reader to supply the obvious meaning for statement like

$$\lim_{x \to +\infty} f(x) = y \quad \text{and} \quad \lim_{x \to x_0} f(x) = +\infty.$$

4.1.5 Exercises

1. Let f be a function defined on a closed domain. Show that f is continuous if and only if the inverse image of every closed set is a closed set.

2. Let A be the set defined by the equations $f_1(x) = 0, f_2(x) = 0, \ldots, f_n(x) = 0$, where $f_1, \ldots, f_n$ are continuous functions defined on the whole line. Show that A is closed. Must A be compact?

3. Let A be the set defined by the equations $f_1(x) \geq 0, f_2(x) \geq 0, \ldots, f_n(x) \geq 0$ where $f_1, \ldots, f_n$ are continuous functions defined on the whole line. Show that A is closed. Show that the set defined by $f_1(x) > 0, \ldots, f_n(x) > 0$ is open.

4. Give a definition of $\lim_{x\to\infty} f(x) = y$. Show that this is true if and only if for every sequence $x_1, x_2, \ldots$ of points in the domain of f such that $\lim_{x\to\infty} x_n = +\infty$, we have $\lim_{n\to\infty} f(x_n) = y$.

5. Show that the function $f(x) = x^\beta$ on $[0,1]$ for $0 < \beta \leq 1$ satisfies a Hölder condition of order α for $0 < \alpha \leq \beta$ but not for $\alpha > \beta$.

6. Let f have a jump discontinuity at x_0. Show that if $x_1, x_2, \ldots$ is any sequence of points in the domain of f converging to x_0, with no x_j equal to x_0, then the sequence $f(x_1), f(x_2), \ldots$ has at most two limit-points.

7. Give an example of a continuous function with domain $\mathbb{R}$ such that the inverse image of a compact set is not compact.

8. Give an example of a continuous function with domain $\mathbb{R}$ such that the image of a closed set is not closed.

9. Show that the function $f(x) = x^2$ with domain $0 \leq x < \infty$ is one-to-one but the function $f(x) = x^2$ with domain $\mathbb{R}$ is not. What is the image of these functions? Are they uniformly continuous?

10. Show that a function that satisfies a Lipschitz condition is uniformly continuous.

11. If f is continuous on $\mathbb{R}$, is it necessarily true that $f(\text{limsup}_{n\to\infty} x_n) = \text{limsup}_{n\to\infty} f(x_n)$?

12. If f is a continuous function on $\mathbb{R}$, is it true that x is a limit-point of $x_1, x_2, \ldots$ implies $f(x)$ is a limit-point of $f(x_1), f(x_2), \ldots$?

13. Is the inverse image of a convergent sequence under a continuous function necessarily a convergent sequence?

14. Show that $f^{-1}(A \cup B) = f^{-1}(A) \cup f^{-1}(B)$ and $f^{-1}(A \cap B) = f^{-1}(A) \cap f^{-1}(B)$ for any function f. Is the same true of images (as opposed to inverse images)?

15. If f is defined on a finite open interval (a, b) and uniformly continuous, show that the limit of f exists at the endpoints and f can be extended to a uniformly continuous function on the closed interval.

4.2 Properties of Continuous Functions

4.2.1 Basic Properties

We begin with some simple observations on the preservation of continuity under arithmetic operations. Suppose f and g are defined on the same domain D. Then by $f + g$ we mean the function with domain D that takes the value $f(x) + g(x)$ at x. Even if f and g have different domains, we can define $f + g$ on the intersection of their domains. If f and g are both continuous, then $f + g$ will also be continuous. This follows immediately from the characterization of continuity in terms of taking convergent sequences to convergent sequences and from the fact that the sum of convergent sequences is convergent. This property is expressed by saying *the continuous functions are preserved by addition* or *the sum of continuous functions is continuous.* This is true whether we consider continuity at a point or continuity on the whole domain. (For uniform continuity and Lipschitz conditions, see the exercises.) Clearly the same is true for scalar multiples af, differences $f - g$, and products $f \cdot g$. For quotients f/g we must avoid dividing by zero. If we have f and g defined and continuous on D, then f/g will be defined and continuous on the subset of D of points where g is not zero. It may happen that f/g can be further defined and continuous at points where both f and g are zero, but this has to be determined on a case-by-case basis. We will return to this when we discuss l'Hôpital's Rule in Section 5.4.3.

Since the constant functions and the identity function $f(x) = x$ are easily seen to be continuous, if follows that all rational functions $p(x)/q(x)$ where p and q are polynomials are continuous on the domain $\{x : q(x) \neq 0\}$. This gives us a large collection of continuous functions.

Most of the special functions, such as $\sin x, \cos x, e^x, \log x$, are also continuous. We will prove this when we give the precise construction of these functions.

Continuous functions can also be created by gluing together continuous pieces. If f is defined and continuous on $[a, b]$ and g is defined and continuous on $[b, c]$ with $f(b) = g(b)$, then

$$h(x) = \begin{cases} f(x), & a \leq x \leq b, \\ g(x), & b \leq x \leq c, \end{cases}$$

is continuous on $[a, c]$. The proof is left to the exercises.

The maximum and minimum of two continuous functions are also continuous. If f and g are defined on the same domain D, let $\max(f, g)$ denote the function on D that at x takes the value $f(x)$ if $f(x) \geq g(x)$ or $g(x)$ if $g(x) \geq f(x)$; define $\min(f, g)$ similarly. (Note that the graphs of f and g may cross each other infinitely often, so we cannot reduce this to a gluing argument.)

Theorem 4.2.1 *If f and g are continuous, then* $\max(f, g)$ *and* $\min(f, g)$ *are continuous.*

Proof: Fix a point x_0 in the domain; and suppose $f(x_0) \geq g(x_0)$, so $\max(f, g)(x_0) = f(x_0)$. To show $\max(f, g)$ is continuous at x_0 we have to show that for each $1/m$ there exists $1/n$ such that $|x - x_0| < 1/n$ implies $|\max(f, g)(x) - \max(f, g)(x_0)| < 1/m$. We know that f and g are continuous at x_0, so given $1/m$ we can find $1/n$ (take the smaller of the two values for f and g) such that $|x - x_0| < 1/n$ implies $|f(x) - f(x_0)| < 1/m$ and $|g(x) - g(x_0)| < 1/m$. This $1/n$ will do the job for $\max(f, g)$. To see this we have to examine the two possibilities, $\max(f, g)(x) = f(x)$ or $g(x)$. In the first case $f(x) - f(x_0) = \max(f, g)(x) - \max(f, g)(x_0)$, so $|f(x) - f(x_0)| < 1/m$ gives the result we want. In the second case we have $g(x) \geq f(x)$ and $\max(f, g)(x) - \max(f, g)(x_0) = g(x) - f(x_0)$.

Now $g(x) - f(x_0) \geq f(x) - f(x_0)$ because $g(x) \geq f(x)$, while $g(x) - f(x_0) \leq g(x) - g(x_0)$ since $f(x_0) \geq g(x_0)$. So $|g(x) - f(x_0)| \leq \max(|f(x) - f(x_0)|, |g(x) - g(x_0)|) < 1/m$. QED

Using these kinds of ideas it is generally speaking possible to construct continuous functions that will do whatever you like. For example, suppose we want to construct a continuous function $f(x)$ on the line that is equal to zero exactly on a set A. We know (exercise set 4.1.5, number 1) that the set of solutions to $f(x) = 0$ is closed, so A will have to be closed, but no other restrictions on A are needed. To construct the function f we look at the complement of A, which is open. By our structure theorem for open sets, the complement of A is an at most countable union of disjoint open intervals so that the whole line is a disjoint union of A and some open intervals. The function f will be zero on A, and on each open interval we construct a tent (see Figure 4.2.1),

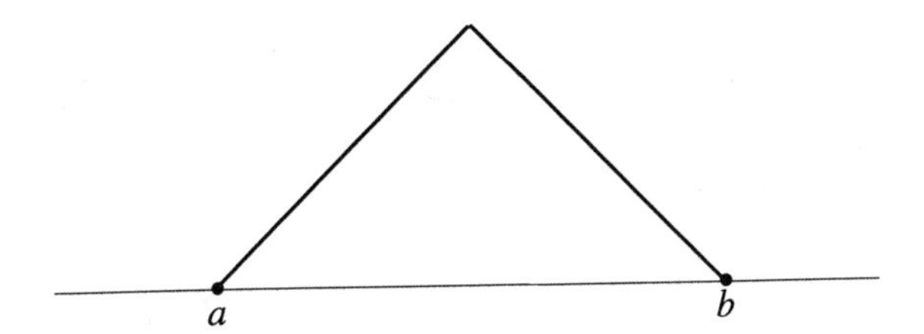

Figure 4.2.1:

which rises with slope 1 to the midpoint and falls with slope -1 to the end. For the infinite intervals (there are at most two of them, (a, ∞) and $(-\infty, b)$) we keep the slope $+1$ or -1 throughout. The total picture might resemble that in Figure 4.2.2. From the construction it is clear that $f = 0$ exactly on the set A.

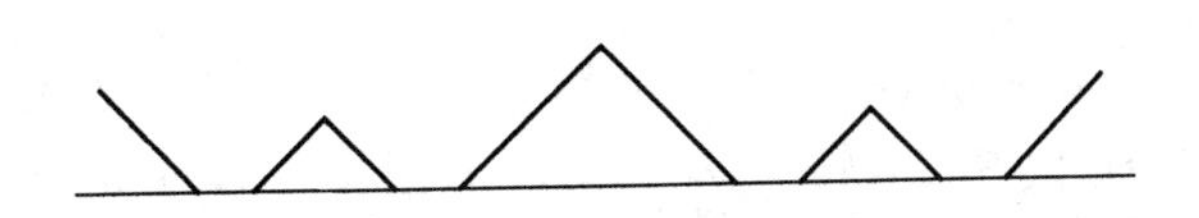

Figure 4.2.2:

It remains to see that f is continuous. But in fact f satisfies the Lipschitz condition $|f(x) - f(y)| \leq |x - y|$, which implies continuity. We leave the verification of the Lipschitz condition to the exercises.

We now turn to the properties of continuous functions. One property that seems obvious is that a continuous function must pass through all intermediate values on its way from one value to another. If $f(a) = y_1$ and $f(b) = y_2$, with f continuous on $[a, b]$ (or some larger domain) with say $y_1 < y_2$, then for any value of y in between, $y_1 < y < y_2$, there must exist at least one solution of $f(x) = y$ in (a, b). This is certainly evident graphically, as shown in Figure 4.2.3, if we imagine a moving horizontal line going from height y_1 to height y_2, the intersections of this line with the graph of f gives the solutions to $y = f(x)$.

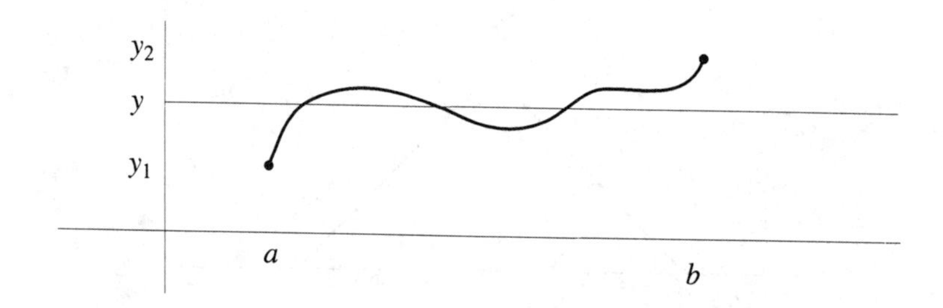

Figure 4.2.3:

Theorem 4.2.2 (*Intermediate Value Theorem*) *Let f be a continuous function on a domain containing $[a, b]$, with say $f(a) < f(b)$. Then for any y in between, $f(a) < y < f(b)$, there exists x in (a, b) with $f(x) = y$.*

Proof: We use the divide and conquer method. We repeatedly bisect the interval, retaining the half that might contain a solution. In other words, fixing a value of y, we consider a sequence of intervals $[a_1, b_1], [a_2, b_2], \ldots$ where $[a_1, b_1]$ is the original interval $[a, b]$ and $[a_2, b_2]$ is one of the halves $[a, a + b/2]$ or $[a + b/2, b]$—chosen so that $f(a_2) < y < f(b_2)$. Of course if $f(a + b/2) = y$, then you are done; otherwise either $f(a + b/2) < y$ in which case you take $[a + b/2, b]$ or $f(a + b/2) > y$ in which case you take $[a, a + b/2]$. Iterating this process we obtain either the solution we seek or a pair of sequences $a_1, a_2, \ldots$ and $b_1, b_2, \ldots$ such that $f(a_k) < y < f(b_k)$, where $b_k - a_k = 2^{1-k}(b - a)$. This condition shows that both sequences are Cauchy sequences and converge to a common limit x.

Since f is continuous, $f(x) = \lim_{k\to\infty} f(a_k) = \lim_{k\to\infty} f(b_k)$ while from $f(a_k) < y < f(b_k)$ we obtain $f(x) \le y \le f(x)$ in the limit, hence $f(x) = y$. QED

A frequently used special case of this theorem is the following: if a continuous function changes sign on an interval, then it has a zero in the interval. This is a valuable method for locating zeros of a function. We can use it to give a quick proof that a polynomial of odd degree (with real coefficients) has a real root. If the polynomial is $a_n x^n + a_{n-1}x^{n-1} + \cdots + a_0$ with say $a_n > 0$ and n odd, then it is not hard to see that the sign is determined by the leading term if $|x|$ is large—positive for $x > 0$ and negative for $x < 0$ because n is odd. Indeed

$$\begin{aligned} |a_{n-1}x^{n-1} + \cdots + a_0| &\le |a_{n-1}||x|^{n-1} + \cdots + |a_0| \\ &\le |a_n x^n| \left(\left| \frac{a_{n-1}}{a_n} x^{-1} \right| + \left| \frac{a_{n-2}}{a_n} x^{-2} \right| + \cdots + \left| \frac{a_0}{a_n} x^{-n} \right| \right) \end{aligned}$$

and the expression in parenthesis can be made less than one by taking $|x|$ large enough, so that $a_{n-1}x^{n-1} + \cdots + a_n$ is too small to change the sign of $a_n x^n$. Thus since the polynomial of odd degree assumes positive and negative values, it must have a root (we can even give an upper bound for the absolute value of the root in terms of the coefficients, namely the smallest value of x that makes the expression in parenthesis equal one). This argument does not work for polynomials of even degree, for the sign of the leading term is then always positive—this is as it should be, since such polynomials may $(x^2 - 1)$ or may not $(x^2 + 1)$ have real roots.

Another application of the intermediate value theorem is that the image of a function defined on an interval is also an interval—the end-points being the sup and inf of the image, which may or may not belong to the image. We leave the proof for an exercise.

4.2.2 Continuous Functions on Compact Domains

Next we consider some theorems concerning continuous functions on a compact domain. Since we have said that compact sets have properties analogous to finite sets, we would expect properties of *all* functions on a finite set to be valid for *continuous* functions on a compact set. Two very obvious properties of all functions on a finite set are that they are bounded and attain their maximum and minimum values.

Theorem 4.2.3 *Let f be a continuous function with domain D that is compact. Then f is bounded and there exist points y and z in D (not necessarily unique) such that $f(y) = \sup\{f(x) : x \text{ in } D\}$ and $f(z) = \inf\{f(x) : x \text{ in } D\}$.*

Proof: If f were not bounded above, there would exist a sequence of points $x_1, x_2, \ldots$ in D such that $f(x_j) \geq j$. Since D is compact, we could find a subsequence converging to a point x_0 in D. Since f is continuous at x_0, it would have to take this convergent subsequence to a convergent sequence, which contradicts $f(x_j) \geq j$. Similarly f is bounded below.

Since the set $\{f(x) : x \text{ in } D\}$, the image of f, is bounded above, it has a finite sup, and there must exist a sequence of values $\{f(x_j)\}$ converging to this sup. Again by the compactness of D, there must exist a convergent subsequence of $x_1, x_2, \ldots$; say $x'_1, x'_2, \ldots$ converges to y in D. Then $\{f(x'_j)\}$ is a subsequence of $\{f(x_j)\}$ and so also converges to the sup. By the continuity of f, $f(y) = f(\lim_{j\to\infty} x'_j) = \lim_{j\to\infty} f(x'_j)$, which equals the sup. QED

A function defined on a finite set takes on only a finite number of values; in other words, its image is also a finite set. To obtain the correct analogy we must replace *both* finite sets by compact sets.

Theorem 4.2.4 *If f is a continuous function on a compact domain, then the image of f is compact.*

Proof: Let $f(D)$ denote the image. To show it is compact we will show that every sequence in $f(D)$ has a limit-point in $f(D)$. Now the points in $f(D)$ are the values $f(x)$, so a sequence in $f(D)$ has the form $f(x_1), f(x_2), \ldots$ for points $x_1, x_2, \ldots$ in D. If the value $f(x_1)$ is assumed more than once then we could change x_1 without changing $f(x_1)$, so the sequence $x_1, x_2, \ldots$ is not uniquely determined by the sequence of values $f(x_1), f(x_2), \ldots$. But this turns out not to matter. The important thing is that we can pass from a sequence of values in the image to some sequence of points in the domain. Since the domain is compact, there is a limit-point x of the sequence $x_1, x_2, \ldots$

in D or, equivalently, there is a subsequence $x_1', x_2', \ldots$ that converges to x. Then $f(x_1'), f(x_2'), \ldots$ is a subsequence of the given sequence of values in the image, and by the continuity of f it converges to $f(x)$. Thus $f(x)$ is a point in the image that is a limit-point of the sequence $f(x_1), f(x_2), \ldots$. This shows the image is compact. QED

Note that the image of a closed set under a continuous function is not necessarily closed. As an example take $f(x) = 1/(1+x^2)$ defined on the whole line, shown in Figure 4.2.4. Then 0 is a limit-point of the image but is not in the image, which is $(0,1]$. This gives an example of a continuous function on a closed set that does not attain its inf.

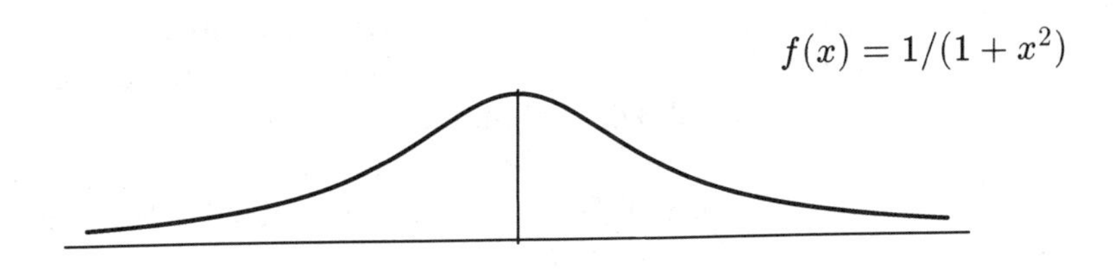

Figure 4.2.4:

Finally we have a result without any analogue for functions on finite sets: the uniform continuity of continuous functions on compact sets. If you look back at the examples of functions that are not uniformly continuous (Figure 4.1.3, or exercise set 4.1.5, number 9) you will notice that the domains are not compact.

Theorem 4.2.5 (*Uniform Continuity Theorem*) *Let f be a function on a compact domain D that is continuous. Then it is uniformly continuous.*

Proof: First let us recall what the issue is here. For f to be continuous means that given the error $1/m$ and the point x_0, we can find an error $1/n$ such that $|x - x_0| < 1/n$ and x in the domain implies $|f(x) - f(x_0)| < 1/m$. We do not know how the error $1/n$ varies with the point x_0. Uniform continuity means that we can find a value of $1/n$ that will work for all points x_0 in the domain. This is what we are going to prove must be true if the domain is compact.

Let us consider what would have to be true for uniform continuity to fail. To negate a statement that begins "for all $1/m$ there exists $1/n$", we have to begin with "there exists $1/m$ such that for all $1/n$". The negation of "$|x-y| < 1/n$ implies $|f(x) - f(y)| < 1/m$" would be an example of two points x, y in the domain such that $|x - y| < 1/n$ and $|f(x) - f(y)| \geq 1/m$. Since there must be one such example for each $1/n$, we should label the points x_n, y_n. We now have the full statement that f is not uniformly continuous: there exists $1/n$ such that for all $1/n$ there exist two points x_n, y_n in the domain such that $|x_n - y_n| < 1/n$ but $|f(x_n) - f(y_n)| \geq 1/m$. We have to show that this leads to a contradiction and so is impossible.

Since D is assumed compact, the obvious first step is to replace the sequences $x_1, x_2, \ldots$ and $y_1, y_2, \ldots$ by convergent subsequences. The condition $|x_n - y_n| \leq 1/n$ implies that both subsequences converge to the same limit, call it x_0. Calling the subsequences $x'_1, x'_2, \ldots$ and $y'_1, y'_2, \ldots$ we have $|f(x'_n) - f(y'_n)| \geq 1/m$ and yet both $\lim_{n\to\infty} f(x'_n) = f(x_0)$ and $\lim_{n\to\infty} f(y'_n) = f(x_0)$ by the continuity of f at the point x_0. This is a contradiction; we cannot have $\lim_{n\to\infty}(f(x'_n) - f(y'_n)) = f(x_0) - f(x_0) = 0$ and $|f(x'_n) - f(y'_n)| \geq 1/m$ for all n. QED

These theorems concerning continuous functions on compact sets can be used in a relative way for continuous functions on non-compact domains. Suppose for example that f is a continuous function on an open interval (a, b). Then the restriction of f to any compact subinterval $[c, d]$ is a continuous function on a compact set. Thus f is bounded, uniformly continuous, and attains its sup and inf on the set $[c, d]$. However, all these statements must be interpreted relative to the domain $[c, d]$ and say something different from what they would say for the domain (a, b)—where they may be false. For example, $f(x) = 1/x$ on the domain $(0, \infty)$ is continuous but is unbounded and not uniformly continuous. On the domain $[c, d]$, however, for $0 < c < d < \infty$ the function $f(x) = 1/x$ shown in Figure 4.2.5 is bounded and uniformly continuous, attaining its sup at $x = c$ and its inf at $x = d$.

4.2.3 Monotone Functions

We conclude this section with a discussion of *monotone functions.* We have seen that bounded monotone sequences have limits, so we would

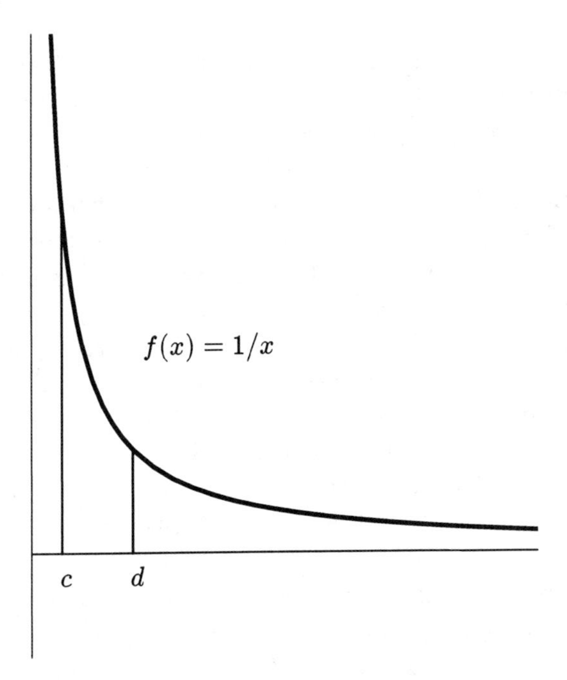

Figure 4.2.5:

expect some analogous result for monotone functions. We say f is *monotone increasing* on a domain D if $x < y$ for two points in the domain implies $f(x) \leq f(y)$ and *monotone decreasing* if $x < y$ for two points in the domain implies $f(x) \geq f(y)$. For simplicity we will restrict our discussion to the case when the domain is an interval. Now a monotone function can have jumps (the signum, for example), so we cannot prove that a monotone function is continuous. The analogue of the convergence theorem for monotone sequences is that one-sided limits always exist, so a bounded monotone function has at worst jump discontinuities.

Theorem 4.2.6 (*Monotone Function Theorem*) *Let f be a monotone function defined on an interval. Then the one-sided limits $\lim_{x \to x_0^+} f(x)$ and $\lim_{x \to x_0^-} f(x)$ both exist (allowing $+\infty$ or $-\infty$) at all points x_0 in the domain. These limits are finite except perhaps at the endpoints.*

Proof: Let f be monotone increasing, and consider first a point x_0 in the interior of the domain. Then the sequence $f(x_0-1)$, $f(x_0-1/2), f(x_0-1/3), \dots$ is monotone increasing (the first few terms may be undefined, but eventually $x_0 - 1/n$ is in the domain since x_0 is an interior point) and so has a limit y that must be finite since $f(x_0 - 1/n) \leq f(x_0)$. We claim this value y must be $\lim_{x \to x_0^-} f(x)$. The reason for this is simply that any point x less than x_0 but near it must be squeezed between some x_n and x_{n+1}, and by the monotonicity $f(x_n) \leq f(x) \leq f(x_{n+1})$, which forces $f(x)$ to be close to y. To make this more precise, suppose we are given an error $1/m$. By the convergence of the sequence $\{f(x_0 - 1/n)\}$ there must exist k such that $y - 1/m \leq f(x_0 - 1/n) \leq y$ for all $n \geq k$. Then if $x_0 - 1/k < x < x_0$ we have $x_0 - 1/n \leq x \leq x_0 - 1/n + 1$ for some $n \geq k$, so $f(x_0 - 1/n) \leq f(x) \leq f(x_0 - 1/n + 1)$, which implies $y - 1/m \leq f(x) \leq y$. Thus $\lim_{x \to x_0^-} f(x) = y$.

A similar argument shows the existence of $\lim_{x \to x_0^+} f(x)$, where this time we use the convergence of the monotone decreasing sequence $f(x_0+1), f(x_0+1/2), f(x_0+1/3), \dots$. At the endpoints of the interval we can also show the existence of the one-sided limit, allowing the possibility of $+\infty$ and $-\infty$. We leave the details as an exercise. QED

Corollary 4.2.1 *A monotone function on an open interval is continuous at all points except at an at most countable number of points where it has a jump discontinuity.*

Proof: We have to show that there are at most a countable number of points of discontinuity. Let us define the *jump* at a jump discontinuity x_0 to be $\lim_{x \to x_0^+} f(x) - \lim_{x \to x_0^-} f(x)$. If the function is monotone increasing, then the jumps are all positive and it would seem plausible that the sum of all the jumps between a and b should be at most $f(b) - f(a)$. We will use this idea cautiously in fashioning the proof.

Let $[c, d]$ be any compact interval contained in the domain, and consider the set of jump discontinuities in $[c, d]$ for which the jump exceeds $1/m$. We claim this set is finite—if we can prove it we are done because the set of all discontinuities is a countable union of such sets, where we vary $1/m$ and $[c, d]$ over countable sets.

In fact we can show that the number of jump discontinuities in (c, d) with jump exceeding $1/m$ is bounded by $m(f(d) - f(c))$. Suppose $x_1, \dots, x_n$ are distinct jump discontinuities in (c, d). Then we can shuffle them inside a sequence $c, y_1, \ldots, y_{n-1}, d$ so that $c < x_1 < y_1 < x_2 < y_2 < \cdots < y_{n-1} < x_n < d$. Because the function is monotone increasing, the jump at x_1 is bounded above by $f(y_1) - f(c)$ (see Figure 4.2.6),

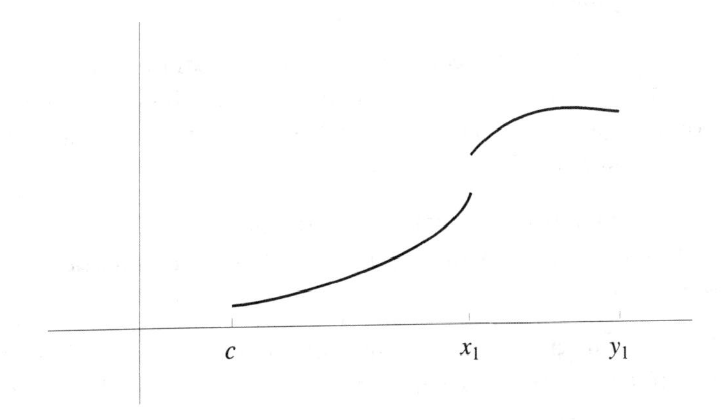

Figure 4.2.6:

the jump at x_2 is bounded above by $f(y_2) - f(y_1)$, and so on. We leave the proof of this as an exercise for the reader. Adding up, the sum of the jumps at the points $x_1, \ldots, x_n$ is at most $f(d) - f(c)$. Thus if each jump is at most $1/m$, there are at most $m(f(d) - f(c))$ points. QED

Despite the apparent simplicity of this result, you should not be lulled into thinking that the general monotone function is anything like the simple pictures you can draw. For example, if $r_1, r_2, \ldots$ is an enumeration of the rational numbers and if $f_k(x)$ is the function shown in Figure 4.2.7,

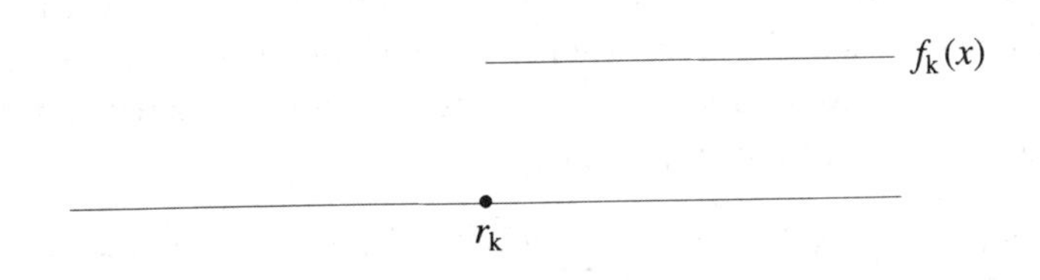

Figure 4.2.7:

which is one for $x \geq r_k$ and zero for $x < r_k$, then the function $f(x) = \sum_{k=1}^{\infty} 2^{-k} f_k(x)$ is a monotone increasing function with a jump discontinuity at every rational number. We postpone the discussion of this example to the chapter on infinite series.

4.2.4 Exercises

1. If f is monotone increasing on an interval and has a jump discontinuity at x_0 in the interior of the domain, show that the jump is bounded above by $f(x_2) - f(x_1)$ for any two points x_1, x_2 of the domain surrounding x_0, $x_1 < x_0 < x_2$.

2. If f is monotone increasing on an interval (a, b), write out the complete proof that $\lim_{x \to b} f(x)$ exists either as a real number or $+\infty$.

3. If the domain of a continuous function is an interval, show that the image is an interval. Give examples where the image is an open interval.

4. If a continuous function on an interval takes only a finite set of values, show that the function is constant.

5. Suppose f and g both satisfy a Lipschitz condition on an interval ($|f(x) - f(y)| \leq M|x - y|$ for all x and y in the interval). Show that $f + g$ also satisfies a Lipschitz condition.

6. Show that if f and g are bounded and satisfy a Lipschitz condition on an interval, then $f \cdot g$ satisfies a Lipschitz condition. Give a counterexample to show that it is necessary to assume boundedness.

7. Let f be a monotone function on an interval. Show that if the image of f is an interval, then f is continuous. Give an example of a non-monotone function on an interval whose image is an interval but that is not continuous.

8. Let $f = p + g$ where p is a polynomial of odd degree and g is a bounded continuous function on the line. Show that there is at least one solution of $f(x) = 0$.

9. If f and g are uniformly continuous, show that $f + g$ is uniformly continuous.

10. If f and g are uniformly continuous and bounded, show that $f \cdot g$ is uniformly continuous. Give a counterexample to show that it is necessary to assume boundedness.

11. If f is a continuous function on a compact set, show that either f has a zero or f is bounded away from zero ($|f(x)| > 1/n$ for all x in the domain, for some $1/n$).

12. If f and g are continuous functions and the domain of g contains the image of f, show that the composition $g \circ f$ defined by $g \circ f(x) = g(f(x))$ is continuous. If f and g are uniformly continuous, is $g \circ f$ uniformly continuous? What about Lipschitz conditions?

13. If f is continuous on $[a, b]$ and g is continuous on $[b, c]$, show that

$$h(x) = \begin{cases} f(x), & a \leq x \leq b, \\ g(x), & b \leq x \leq c, \end{cases}$$

is continuous if and only if $f(b) = g(b)$.

14. Show that the function constructed in Section 4.2.1 to vanish exactly on the closed set A satisfies a Lipschitz condition.

15. Give an example of a function on $\mathbb{R}$ that assumes its sup and inf on every compact interval and yet is not continuous.

16. Let f be a function defined on the extended reals $\mathbb{R} \cup \{\pm\infty\}$ but whose range is $\mathbb{R}$, which is continuous in the usual sense for points in $\mathbb{R}$, and $\lim_{x \to \pm\infty} f(x) = f(\pm\infty)$. Prove that f is bounded and attains its sup and inf (possibly at the points $\pm\infty$). Prove that f is uniformly continuous when restricted to $\mathbb{R}$.

17. Give an example of a function on $\mathbb{R}$ that has the intermediate value property for every interval (it takes on all values between $f(a)$ and $f(b)$ on $a \leq x \leq b$) but fails to be continuous at a point. Can such a function have jump discontinuities?

4.3 Summary

4.1 Concepts of Continuity

Definition 4.1.1 *A function consists of a domain D, a range R, and a correspondence $x \to f(x)$ assigning a point $f(x)$ of R to each point x of D. The image $f(D)$ is the set of all values $f(x)$. The function is onto if the image equals the range and is one-to-one if $x \neq y$ implies $f(x) \neq f(y)$.*

Definition 4.1.2 *A function f is said to be continuous at a point x_0 of its domain D if for every m there exists n such that $|x - x_0| < 1/n$ and x_0 in D implies $|f(x) - f(x_0)| < 1/m$. We say f is continuous if it is continuous at every point of D. We say f is uniformly continuous if for every m there exists n such that $|x - y| < 1/n$ and x and y in D implies $|f(x) - f(y)| < 1/m$.*

Definition 4.1.3 $y = \lim_{x \to x_0} f(x)$ *for x_0 a limit-point of the domain D means for every m there exists n such that $|x - x_0| < 1/n$ and x in D implies $|f(x) - y| < 1/m$.*

Theorem 4.1.1 $\lim_{x \to x_0} f(x)$ *exists if and only if $f(x_1), f(x_2), \ldots$ converges for every sequence $x_1, x_2, \ldots$ of points of D not equal to x_0 but converging to x_0.*

Theorem 4.1.2 *A function is continuous if and only if it takes convergent sequences to convergent sequences.*

Theorem 4.1.3 *A function on an open domain is continuous if and only if the inverse image of every open set is open.*

Definition *A function satisfies a Lipschitz condition if $|f(x) - f(x_0)| \leq M|x - x_0|$ for some M and all x and x_0 in the domain.*

Definition *A function f is said to have a limit from the right at x_0 equal to y, written $\lim_{x \to x_0^+} f(x) = y$ if for every $1/n$ there exists $1/m$ such that $x_0 < x < x_0 + 1/m$ implies $|f(x) - y| \leq 1/n$. Similarly we* *define limits from the left, written* $\lim_{x \to x_0^-} f(x)$. *We say* f *has a jump discontinuity at* x_0 *if it has limits from both sides at* x_0 *and they are different.*

4.2 Properties of Continuous Functions

Theorem *Continuity is preserved under addition, multiplication, and division (if the denominator never vanishes).*

Theorem 4.2.1 *If* f *and* g *are continuous, so is* $\max(f, g)$ *or* $\min(f, g)$.

Theorem 4.2.2 (*Intermediate Value Theorem*) *A continuous function* f *on a closed interval* $[a, b]$ *assumes all values between* $f(a)$ *and* $f(b)$.

Theorem *A polynomial of odd degree has a real zero.*

Theorem 4.2.3 *A continuous function on a compact set is bounded and attains its* sup *and* inf.

Theorem 4.2.4 *The image of a continuous function on a compact set is compact.*

Theorem 4.2.5 *A continuous function on a compact set is uniformly continuous.*

Theorem 4.2.6 *A monotone function on an interval has one-sided limits at all points of the domain, finite except perhaps at the endpoints.*

Corollary 4.2.1 *A monotone function on an interval has at most a countable number of discontinuities, all of which are jump discontinuities.*

Chapter 5

Differential Calculus

5.1 Concepts of the Derivative

5.1.1 Equivalent Definitions

In this chapter we are going to review the highlights of the differential calculus, supplying precise definitions and proofs for all results. From a computational point of view you will not learn very much new—you will still compute the derivative of $(\sin(\sin 1/(1+x^2)))^3$ in the same way you always have. From a conceptual point of view, however, we will pursue two goals. First, we want to provide a sound logical foundation for this enormously successful branch of mathematics. (This is not to say that this is the only possible foundation or even that it is the foundation that Newton and Leibniz had in mind but were not able to formulate clearly. Another possible foundation can be based on Abraham Robinson's Non-Standard Analysis, but this is considerably more difficult to describe.) Our second goal is to clarify certain concepts in the rather simple one-dimensional case so that we will be better prepared to deal with the many beautiful and more complicated generalizations to higher dimensions and beyond.

The idea of the derivative comes from the intuitive concepts of rate of change, velocity, and slope of a curve, which are thought of as instantaneous or infinitesimal versions of the basic difference quotient $(f(x)-f(x_0))/(x-x_0)$ where f is a function defined on a neighborhood of x_0. This is often written $(f(x_0+h)-f(x_0))/h$, which is clearly equivalent if we set $x-x_0=h$. The difference quotient has the immediate interpretation as the ratio of changes in the variables x and $f(x)$ over the interval from x_0 to x, as shown in Figure 5.1.1.

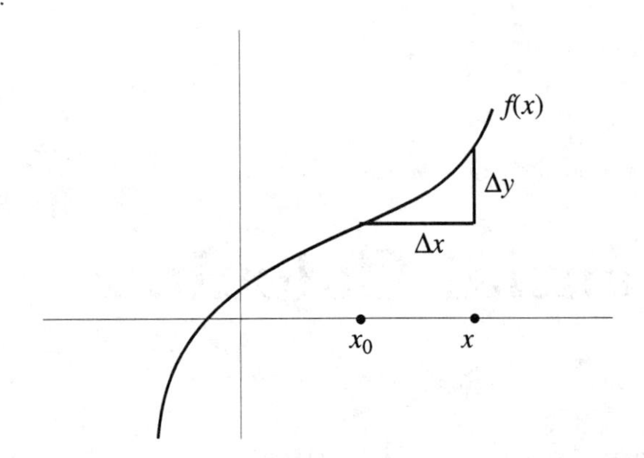

Figure 5.1.1:

The intuitive idea of letting x get closer and closer to x_0 (without actually reaching it) is captured in the definition of the limit of a function: form $q(x) = (f(x) - f(x_0))/(x - x_0)$ as a function of x, and take its limit as $x \to x_0$. We call this limit, if it exists, the derivative $f'(x_0)$ at x_0, and we say the function is differentiable at x_0. We only allow real number limits and explicitly exclude the possibility that $f'(x_0)$ might equal $+\infty$ or $-\infty$ when we say f is differentiable at x_0. Note that our explicit refusal to consider the value at $x = x_0$ in the definition of the limit pays off here, since the difference quotient is undefined, $(f(x_0) - f(x_0))/(x_0 - x_0)$, at $x = x_0$. Thus x_0 is not in the domain of the difference quotient $q(x)$, which consists of the domain of f with x_0 removed.

This is the definition of derivative that is usually given in modern calculus books, without a complete explanation of the limit concept. Let us write out the complete statement defining $f'(x_0)$ with the meaning of the limit made explicit:

Definition 5.1.1 *Let f be a function defined in a neighborhood of x_0. Then f is said to be differentiable at x_0 with derivative equal to the real number $f'(x_0)$ if for every error $1/m$ there exists an error $1/n$ such that*

$|x - x_0| < 1/n$ *and* $x \neq x_0$ *implies*

$$\left| \frac{f(x) - f(x_0)}{x - x_0} - f'(x_0) \right| \leq \frac{1}{m}.$$

Now observe that since $x - x_0 \neq 0$, we can multiply the inequality above by $|x - x_0|$ to obtain the equivalent inequality

$$|f(x) - f(x_0) - f'(x_0)(x - x_0)| \leq \frac{1}{m}|x - x_0|.$$

This new inequality admits an interesting interpretation. We think of $f(x) - f(x_0) - f'(x_0)(x - x_0)$ as the difference of two functions—the original function $f(x)$ and the function $f(x_0) + f'(x_0)(x - x_0)$. Here x_0 is thought of as a constant, as are $f(x_0)$ and $f'(x_0)$, so this second function is simply an *affine function* $ax + b$, where $a = f'(x_0)$ and $b = f(x_0) - x_0 f'(x_0)$. We use the term *affine* rather than *linear* because we want to reserve the term linear for the special case $b = 0$. An affine function is one whose graph is a straight line, and any non-vertical straight line is the graph of an affine function (for the function to be linear the line must pass through the origin). Clearly the affine functions are extremely simple to understand—almost any question about an affine function that one can imagine asking can be answered by a simple computation.

Now the existence of the derivative of f at x_0 is a statement about the difference between the original function and the affine function

$$g(x) = f(x_0) + f'(x_0)(x - x_0),$$

which we will think of as a statement about how well g approximates f. The problem is that there are many different notions of approximation, and we must pick out the one that is pertinent here and distinguish it from others. To do this let us rewrite the definition in terms of the functions f and g: *for every* $1/m$ *there exists* $1/n$ *such that* $|x - x_0| < 1/n$ *implies* $|f(x) - g(x)| \leq |x - x_0|/m$. (According to the definition we should require $x \neq x_0$, but it is trivially true that $|f(x_0) - g(x_0)| \leq 0$ because $f(x_0) = g(x_0)$ by the form of g.) Notice that this is a statement about what happens for x near x_0. It is a *local* approximation property. How close does x have to be to x_0? That will depend on the choice of $1/m$, but it will never hurt us to make it closer, and we may be forced to make it very close indeed. For a particular value of x, not equal to x_0, the statement may say nothing at all, since this value of x may not satisfy any of the conditions $|x - x_0| < 1/n$. It is only as x is varied closer to x_0 that the statement implies that $g(x)$ approximates $f(x)$ well.

Now let us look at the graphs of $f(x)$ and the various affine functions that might be $g(x)$ (see Figure 5.1.2). Since $g(x)$ is supposed to be approximating x at x_0 we may as well have the graphs cross at x_0; in other words take $f(x_0) = g(x_0)$. However, there are many affine functions whose graph crosses the graph of f at the point $(x_0, f(x_0))$. These functions have the form $f(x_0) + a(x - x_0)$ for any real constant a. What distinguishes the unique correct choice? It is what we visually identify as tangency—the extremely close touching of the graphs for x near x_0.

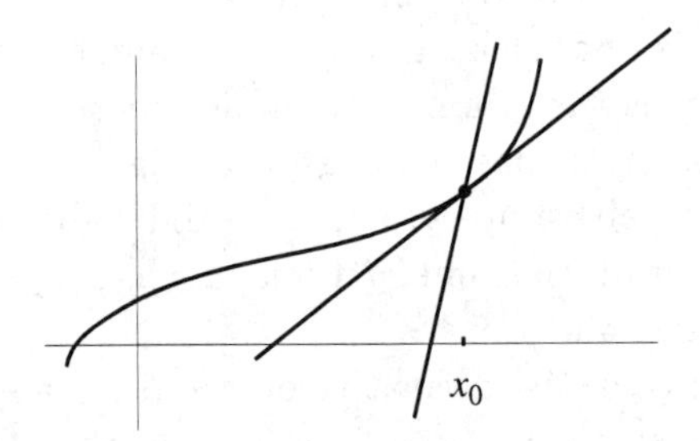

Figure 5.1.2:

It is not just that $f(x) - g(x)$ tends to 0 as x tends to x_0—this would happen with any choice of a—but it goes to zero *faster* than for any other choice of a. In fact the condition "$|x - x_0| < 1/n$ implies $|f(x) - g(x)| \leq |x - x_0|/m$" means exactly that it goes to zero *faster* than $|x - x_0|$.

If we compare any two distinct affine functions $g_1(x) = f(x_0) + a_1(x - x_0)$ and $g_2(x) = f(x_0) + a_2(x - x_0)$ passing through $(x_0, f(x_0))$ we find $|g_1(x) - g_2(x)| = |a_1 - a_2|\,|x - x_0|$, which goes to zero at a rate proportional to $|x - x_0|$. This is fundamentally different from the rate of vanishing of the difference $f(x) - g(x)$. In fact it shows the uniqueness of the affine function $g(x) = f(x_0) + f'(x_0)(x - x_0)$ corresponding to the choice $a = f'(x_0)$ and, hence, the uniqueness of the derivative. If a_1 is different from $f'(x_0)$, then the difference $f(x) - g_1(x)$ can be estimated from below by the triangle inequality,

$$\begin{aligned} |f(x) - g_1(x)| &= |(f(x) - g(x)) + (g(x) - g_1(x))| \\ &\geq |g(x) - g_1(x)| - |f(x) - g(x)| \\ &\geq |a_1 - f'(x_0)|\,|x - x_0| - \frac{1}{m}|x - x_0| \end{aligned}$$

if $|x - x_0| < 1/n$. By taking $1/m$ less than $|a_1 - f'(x_0)|/2$ we have

$$|f(x) - g_1(x)| \geq |a_1 - f'(x_0)|\,|x - x_0|/2$$

for all x satisfying $|x - x_0| < 1/n$ for some value of $1/n$. This means we can never make $f(x) - g_1(x)$ go to zero faster than $|x - x_0|$. We can see this clearly in Figure 5.1.2 by the way the graphs of $f(x)$ and $g_1(x)$ cross cleanly (the technical term *transversal* is sometimes used to describe this).

We can formulate the condition that $f(x) - g(x)$ vanishes at $x = x_0$ at a faster rate than $|x - x_0|$ in a convenient manner by introducing "big Oh" and "little oh" notation.

Definition 5.1.2 *Let f and g denote arbitrary functions defined near $x = x_0$. We say $f(x) = O(g(x))$ as $x \to x_0$ (read f is "big Oh" of g) if there exists $1/n$ and a positive constant c such that $|x - x_0| < 1/n$ implies $|f(x)| \leq c|g(x)|$ (or equivalently, the ratio f/g remains bounded for $|x - x_0| < 1/n$). We say $f(x) = o(g(x))$ as $x \to x_0$ (read f is "little oh" of g) if for every $1/m$ there exists $1/n$ such that $|x - x_0| < 1/n$ implies $|f(x)| < |g(x)|/m$ (or equivalently, $\lim_{x\to x_0} f(x)/g(x) = 0$). Note that $o(g(x))$ is a stronger statement than $O(g(x))$; $o(g(x))$ implies $O(g(x))$ but not conversely.*

Usually the function $g(x)$ in the definition is taken to be something relatively simple, such as a power of $|x - x_0|$, so that the condition gives a comparative statement concerning the size of f near x_0 and a standard of decay or growth. The simplest choice is $g \equiv 1$. Then $f(x) = O(1)$ as $x \to x_0$ means f is bounded near $x = x_0$, while $f(x) = o(1)$ as $x \to x_0$ means $\lim_{x\to x_0} f(x) = 0$. Continuity at x_0 can be expressed by $f(x) - f(x_0) = o(1)$ as $x \to x_0$. The choice $g(x) = |x - x_0|$ enables us to express differentiability as $f(x)-g(x) = o(|x-x_0|)$ as $x \to x_0$ where g is the affine function $f(x_0)+f'(x_0)(x-x_0)$. For any other affine functions $g_1(x) = f(x_0) + a(x - x_0)$ we have merely $f(x) - g(x) = O(|x - x_0|)$.

Let us define a *best affine approximation* to f at x_0 to be an affine function $g(x)$ such that $f(x) - g(x) = o(|x - x_0|)$ as $x \to x_0$. We have seen that f is differentiable at x_0 if and only if it has a best affine approximation at x_0, in which case the best affine approximation is unique and equals $f(x_0) + f'(x_0)(x - x_0)$. Thus the derivative here appears as the slope of the best affine approximation, and the graph of the best affine approximation is the tangent line to the graph of $f(x)$ at the point $(x_0, f(x_0))$.

Having defined differentiability at a point, we define differentiability on an open set A simply to mean differentiability at every point of A. (One might think, in analogy with continuity, that one would also want to consider a stronger condition of uniform differentiability, where the relationship between the errors is specified independently of the point x_0. In this regard see exercise set 5.2.4, numbers 8 and 9.) Thus *f is differentiable on A if for every x_0 in A there exists a constant $f'(x_0)$ such that for every $1/m$ there exists $1/n$ (depending on $1/m$ and x_0) such that $|x-x_0| < 1/n$ implies $|f(x)-(f(x_0)+f'(x_0)(x-x_0))| \leq 1/m$.* Note that if f is defined and differentiable on A, then the derivative $x_0 \to f'(x_0)$ can also be viewed as a function defined on A. This is the point of view that we will adopt. Note that it requires a non-trivial change of perspective, since the definition of $f'(x_0)$ involves holding x_0 fixed.

5.1.2 Continuity and Continuous Differentiability

Differentiability of f at x_0 implies that f is continuous at x_0. Indeed, continuity at x_0 would require that we can make $f(x) - f(x_0)$ small, whereas differentiability at x_0 means we can make $f(x)-f(x_0)-f'(x_0)(x - x_0)$ small. Since $f'(x_0)$ is fixed, we can also make $f'(x_0)(x-x_0)$ small and, hence, $f(x)-f(x_0)$ small. More precisely, we choose $1/n$ so that $|x-x_0| < 1/n$ implies $|f(x)-f(x_0)-f'(x_0)(x-x_0)| \leq |x - x_0|$. Then by the triangle inequality we have

$$|f(x) - f(x_0)| \leq |f'(x_0)(x - x_0)| + |x - x_0| \leq (1 + |f'(x_0)|)|x - x_0|.$$

(In fact, this argument only uses the condition $f(x) - f(x_0) - f'(x_0)(x - x_0) = O(|x - x_0|)$, which is weaker than differentiability.) This is a kind of local Lipschitz condition that implies continuity as follows: given $1/m$, choose $1/k$ such that both $k \geq n$ and $k > (1 + |f'(x_0)|)m$. Then $|x - x_0| < 1/k$ implies $|f(x) - f(x_0)| \leq (1 + |f'(x_0)|)|x - x_0| < 1/m$.

Since differentiability at a point implies continuity at a point, it follows that differentiability on an open set implies continuity on that set. Later we will show that if the derivative is also bounded, then the function is uniformly continuous; in fact it will satisfy a Lipschitz condition.

Differentiability of a function implies continuity of the function, but it does *not* imply continuity of the derivative. This is a rather subtle point, since the obvious attempt to create a counterexample doesn't work. Since the simplest function that fails to be continuous is one with a jump discontinuity, such as the signum, it would seem plausible that to create a function with a discontinuous derivative one would take a function like $|x|$ (shown in Figure 5.1.3) whose derivative is sgn x. The problem with this example is that the function is not differentiable at $x = 0$, the very point where the discontinuity in the derivative occurs. Indeed the difference quotient at $x = 0$ is $+1$ for positive values and -1 for negative values, so it can't have a limit (it has two distinct one-sided limits).

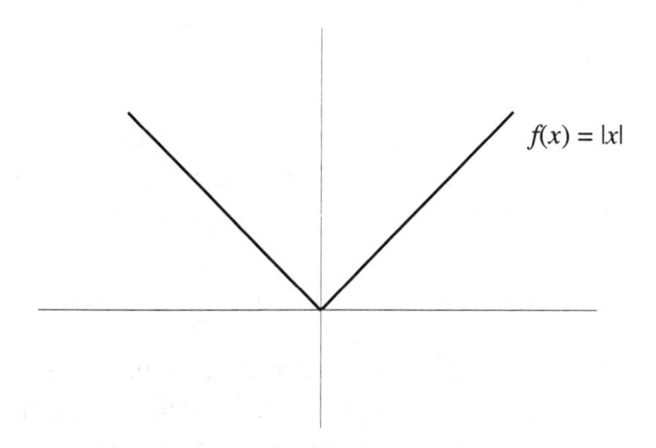

Figure 5.1.3:

No amount of patching will make this an example of a function differentiable at *every* point of an open set with a derivative that is discontinuous. In fact later we will prove that jump discontinuities never occur in a derivative that exists at every point. Thus we have to look for discontinuities of the second kind. The picture we have in mind is a function that oscillates more and more rapidly as x approaches x_0, but the size of the oscillations also decreases, tending to zero. We must then control the relative heights and widths of the oscillation rather carefully. The idea is that by making the heights decrease rapidly enough we can make the derivative exist and equal zero at the point $x = x_0$ and then by making the width decrease even more rapidly we can make the derivative discontinuous (even unbounded).

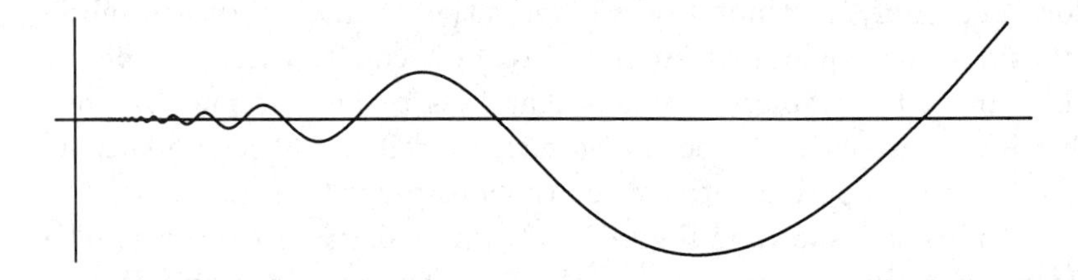

Figure 5.1.4:

An explicit example is the function

$$f(x) = \begin{cases} x^2 \sin(1/x^2), & x \neq 0, \\ 0, & x = 0, \end{cases}$$

shown in Figure 5.1.4. We may compute the derivative in the usual fashion for $x \neq 0$ (we are assuming here that the usual laws of differential calculus are valid—facts that we will eventually prove). We find

$$f'(x) = 2x \sin \frac{1}{x^2} - \frac{2}{x} \cos \frac{1}{x^2} \quad \text{if } x \neq 0;$$

and this function is clearly unbounded as $x \to 0$, so no way of defining $f'(0)$ could possibly make it continuous. Notice here that the speed of oscillation of the factor $\sin(1/x^2)$ overcomes the decay of x^2 to produce the unbounded derivative.

The usual procedures of the differential calculus do not provide a computation of $f'(0)$, let alone a guarantee that the derivative exists at

$x = 0$. Indeed it would be natural to guess that the derivative could not possibly exist at $x = 0$ because of all the oscillations nearby. However, it turns out that the decay of the factor x^2 is enough to overwhelm the oscillations and produce a zero derivative at $x = 0$. In fact the difference quotient at $x = 0$ is

$$\frac{f(x) - f(0)}{x - 0} = \frac{x^2 \sin(1/x^2) - 0}{x - 0} = x \sin \frac{1}{x^2},$$

which clearly has limit equal to 0 since $|\sin(1/x^2)| \leq 1$. Thus f is differentiable at $x = 0$, and $f'(0) = 0$. This is an everywhere differentiable function with a derivative that is discontinuous at $x = 0$. (In fact, by modifying this example it is possible to produce an everywhere differentiable function whose derivative is nowhere continuous.)

A function whose derivative exists and is continuous is called *continuously differentiable* or of *class* C^1. We will see that many of the theorems of differential calculus do not require this hypothesis; nevertheless, it is a very frequently encountered condition, and one could make a case for the viewpoint that there is very little importance attached to the game of trying to eliminate this hypothesis from theorems. In deference to established traditions we will play the game for a while.

Before passing to the study of properties of differentiable functions, we will discuss briefly the intuitive notion of smoothness of the graph and its relation to differentiability. A rule of thumb that is frequently expounded in calculus courses is that a continuous function is one whose graph can be drawn without a break (without lifting pen from paper or chalk from blackboard), while a differentiable function is one whose graph is sufficiently smooth so that you can run your finger along it without getting cut. There is a good deal of truth to this maxim, although the example of $x^2 \sin(1/x^2)$ should convince you of the superficial nature of the assumption that you can always "draw" the graph of a function. Nevertheless, if you can draw the graph, then sharp corners do indicate points where the derivative fails to exist. However, there is another reason the derivative can fail to exist even when the graph is smooth—namely, the tangent can become vertical. An example is the function $f(x) = \sqrt[3]{x}$ defined for all real x (the cube root of a negative number is negative), whose graph is the graph of $x = y^3$, as shown in Figure 5.1.5. At $x = 0$ the tangent is vertical and the derivative fails to exist

$$\left(\frac{f(x)-f(0)}{x-0}\right)=\frac{\sqrt[3]{x}}{x}=\frac{1}{(\sqrt[3]{x})^2} \text{ tends to } +\infty \text{ as } x\to 0),$$

even though the graph is perfectly smooth.

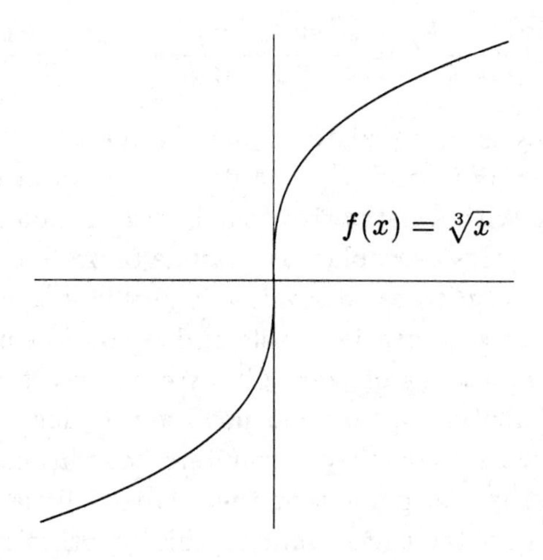

Figure 5.1.5:

5.1.3 Exercises

1. Show that $f(x)=O(|x-x_0|^2)$ as $x\to x_0$ implies $f(x)=o(|x-x_0|)$ as $x\to x_0$, but give an example to show that the converse is not true.

2. Show that $f(x)=O(|x-x_0|^k)$ and $g(x)=O(|x-x_0|^k)$ imply $(f+g)(x)=O(|x-x_0|^k)$. Is the same true of "little oh"?

3. Show that $f(x)=O(|x-x_0|^k)$ and $g(x)=o(|x-x_0|^j)$ imply $f\cdot g(x)=o(|x-x_0|^{k+j})$.

4. Show that $f(x)=O(|x-x_0|^k)$ implies $f(x)/x-x_0=O(|x-x_0|^{k-1})$ if $k\geq 1$.

5. Show that if $f(x)=o(|x-x_0|)$ as $x\to x_0$, then $f'(x_0)$ exists. What is $f'(x_0)$? What does this tell you about $x^2\sin(1/x^{1{,}000})$?

6. Show that $x \sin(1/x)$ fails to have a derivative at $x = 0$ and even the one-sided limits $(\lim_{x\to 0^+}(f(x) - f(0))/(x - 0)$, etc.) fail to exist.

7. Give an example of a differentiable function whose tangent line at a point fails to stay on one side of the graph (above or below) even locally (when restricted to any neighborhood of the point).

8. Show that if f is an affine function, it is equal to its own best affine approximation at every point. What does this tell you about the derivative of f?

9. A "zoom" on the graph of $y = f(x)$ near (x_0, y_0) (with $y_0 = f(x_0)$) with magnification factor M (the same in both x and y directions) is the graph of the function defined by $f(x_0 + x/M) = y_0 + y/M$. Prove that if f is differentiable at x_0, then the zoom converges to the straight line through the origin with slope $f'(x_0)$, as $M \to \infty$. What happens to the zoom of $|x|$ near the origin?

5.2 Properties of the Derivative

5.2.1 Local Properties

The basic idea of the differential calculus is to relate properties of the function to properties of its best affine approximation at a point. The simplest such properties involve questions of increase, decrease, and maxima and minima. We study first these properties at a point.

Definition 5.2.1 *Let f be a function defined in a neighborhood of a point x_0. We say f is monotone increasing at x_0 if there exists a (perhaps smaller) neighborhood of x_0 such that $f(x_1) \le f(x_0) \le f(x_2)$ for all points x_1 and x_2 in the neighborhood satisfying $x_1 < x_0 < x_2$. We say f is strictly increasing at x_0 if there exists a neighborhood of x_0 such that $f(x_1) < f(x_0) < f(x_2)$ for all points x_1 and x_2 in the neighborhood satisfying $x_1 < x_0 < x_2$. We say that f is monotone increasing on an interval if $f(x_1) \le f(x_2)$ for all points in the interval satisfying $x_1 < x_2$; f strictly increasing on an interval is defined in the same way with strict inequalites $f(x_1) < f(x_2)$. We define monotone* *and strictly decreasing similarly, reversing the inequalities for* $f(x)$*. We say that* f *has a local maximum at* x_0 *if there exists a neighborhood of* x_0 *such that* $f(x) \leq f(x_0)$ *for all* x *in the neigborhood. We say that* f *has a strict local maximum at* x_0 *if there exists a neighborhood of* x_0 *such that* $f(x) < f(x_0)$ *for all* x *not equal to* x_0 *in the neighborhood. We define local minimum and strict local minimum similarly, reversing the inequalities.*

These definitions are fairly obvious. One way to think of them is to draw the horizontal and vertical lines through the graph of f at the given point, dividing the plane into four quadrants, as in Figure 5.2.1.

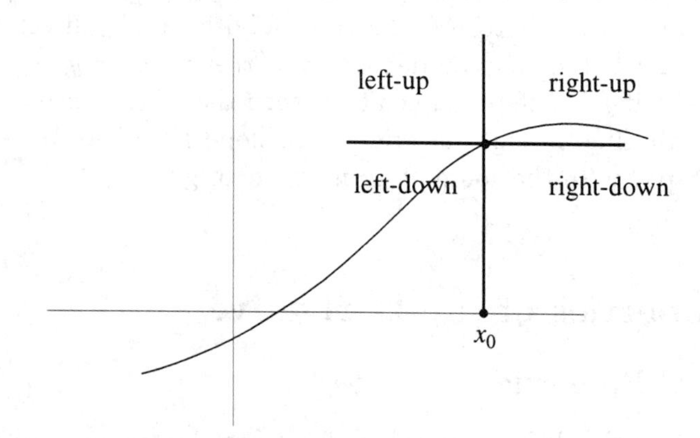

Figure 5.2.1:

The function is increasing if the graph passes from the left-down to the right-up quadrant, decreasing if the graph passes from left-up to right-down, and so on. For strict behavior we exclude the boundary of the quadrants. The condition "there exists a neighborhood of x_0" in all the definitions means we only look at a piece of the graph near the point.

One possible source of confusion is that the fact that f is monotone increasing at x_0 is not the same as saying f is monotone increasing on a neighborhood of x_0. To say f is monotone increasing in a neighborhood of x_0 we would want to know $x_1 < x_2$ implies $f(x_1) \leq f(x_2)$ for every x_1 and x_2 in the neighborhood, which is a stronger condition. For example, the zig-zag function illustrated in Figure 5.2.2 is monotone increasing at $x = 0$, but it is not monotone increasing in any neighborhood of 0. The same remark applies to strict increasing.

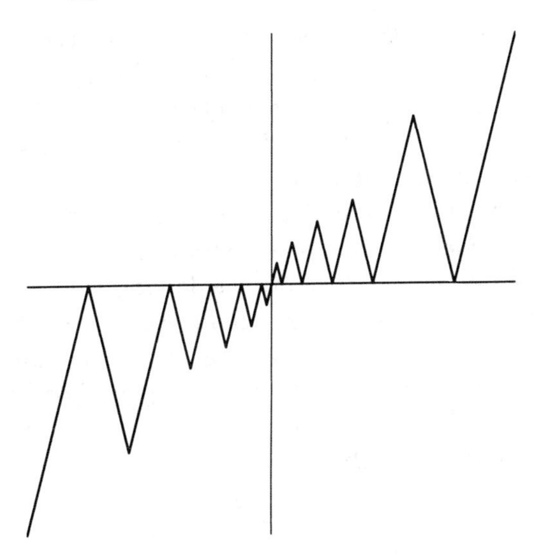

Figure 5.2.2:

On an intuitive level we expect these local properties of a function to correspond to the same properties for the best affine approximation at the point in question, and the properties of the best affine approximation hinge on the sign of the derivative. Since the derivative is the limit of the difference quotient $(f(x) - f(x_0))/(x - x_0)$ and the sign of the difference quotient is related to the relative values of $f(x)$ and $f(x_0)$, it is not difficult to establish the basic facts. Since there is a loss of information in passing to the limit, in that strict inequalities may not be preserved, we can't get a perfect match-up of conditions. Let us start with the true implications.

Theorem 5.2.1 *Let f be defined in a neighborhood of x_0, and let f be differentiable at x_0.*

a. *If $f'(x_0) > 0$, then f is strictly increasing at x_0. Similarly, if $f'(x_0) < 0$, then f is strictly decreasing at x_0.*

b. *If f is monotone increasing at x_0, then $f'(x_0) \geq 0$. Similarly, if f is monotone decreasing at x_0, then $f'(x_0) \leq 0$.*

c. *If f has a local maximum or minimum at x_0, then $f'(x_0) = 0$.*

Proof:

a. Since the limit of the difference quotient is strictly positive, there must be a neighborhood of x_0 in which the difference quotient is strictly positive. For x in this neighborhood $(f(x) - f(x_0))/(x - x_0) > 0$, so $f(x) - f(x_0) > 0$ if $x > x_0$ while $f(x) - f(x_0) < 0$ if $x < x_0$, showing that f is strictly increasing at x_0.

b. If f is monotone increasing at x_0, then there exists a neighborhood of x_0 for which the difference quotient is ≥ 0. Since non-strict inequality is preserved in the limit, the derivative at x_0 is also ≥ 0.

c. Suppose f has a local maximum at x_0. Then the difference quotient formed for $x < x_0$ will be ≥ 0, while the difference quotient formed for $x > x_0$ will be ≤ 0, for x in a neighborhood of x_0. Since we can take the limit from either side, the derivative at x_0 must be both ≥ 0 and ≤ 0, hence zero. QED

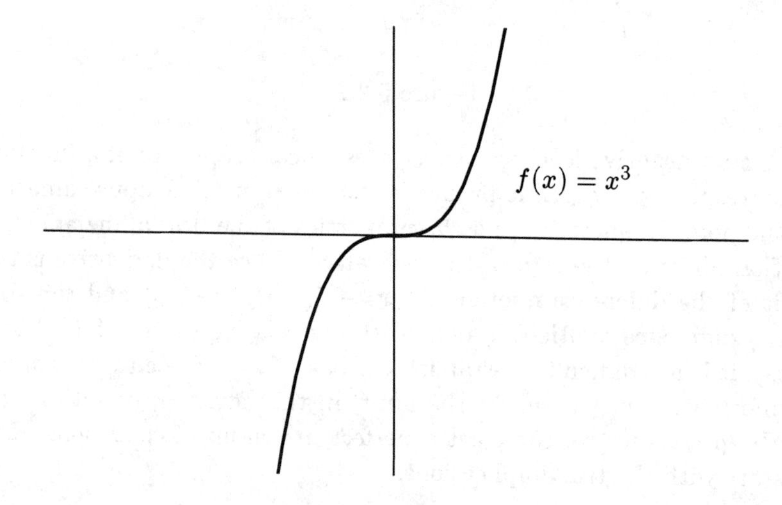

Figure 5.2.3:

Almost as important as what the theorem says is what it does not say. If the function is strictly increasing we cannot conclude that the derivative is positive, as the function $f(x) = x^3$ at $x = 0$ shows (see Figure 5.2.3). Similarly if the derivative is zero at a point we cannot draw any conclusions. Also it is necessary to assume the function is differentiable at the point in question; $f(x) = |x|$ has a local minimum at $x = 0$ but is not differentiable there.

Part c of the theorem will be extremely useful to us and forms the basis of many familiar applications of calculus. It is usually attributed to Fermat, but in fact Fermat developed a somewhat different method for locating maxima and minima. Fermat observed that in a neighborhood of a strict local maximum (say $f(x_0) = M$), the function assumes smaller values exactly twice but the maximum value only once, as in Figure 5.2.4.

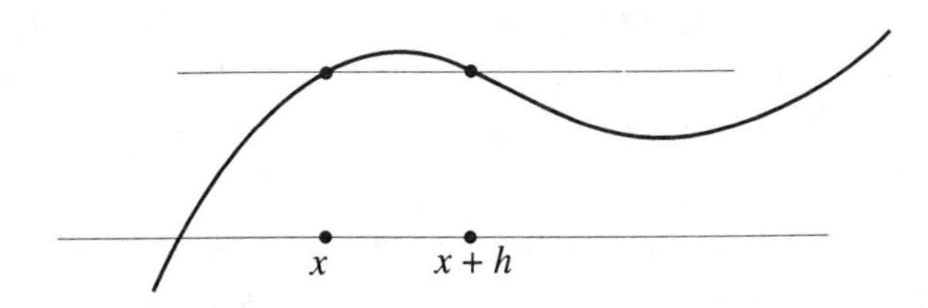

Figure 5.2.4:

Thus Fermat looks at the equation $f(x+h) = f(x)$, finds the non-zero solution h ($h = 0$ is always a solution), expresses x as a function of h, and then sets $h = 0$ (strictly speaking one must take the limit as $h \to 0$). Thus for $f(x) = x^2$ his method sets $(x+h)^2 = x^2$. The two solutions are $h = 0$ and $h = -2x$. For the non-zero solution we find $x = -1/2h$; and when $h \to 0$ we find the minimum at $x = 0$. For $f(x) = x^3$ the equation $(x+h)^3 = x^3$ has only $h = 0$ as a real root, so Fermat's method does not get fooled by the critical point $x = 0$, which is neither a maximum nor a minimum. Despite the fact that Fermat's method has this advantage over the usual method, it has been largely forgotten because it requires the explicit solution of the equation $f(x+h) = f(x)$, which is often intractible.

5.2.2 Intermediate Value and Mean Value Theorems

We turn now to properties of functions that are differentiable on an open interval. The two main theorems are the intermediate value theorem (not to be confused with the theorem of the same name in Section

4.2.1) and the mean value theorem. The proofs are quite similar, based on the observation that to get a solution of $f'(x) = 0$ we can take a local maximum or minimum and then by subtracting an appropriate affine function we can get $f'(x)$ to take on other values. The intermediate value theorem is something of a curiosity, since its conclusion is a consequence of the other intermediate value theorem if we assume the derivative is continuous, which we will frequently do for other reasons. The mean value therorem, on the other hand, is one of the most useful theorems in analysis. It turns out that its proof is not made any simpler by assuming the derivative is continuous.

Intermediate Value Theorem *Let $f(x)$ be differentiable on an open interval (a, b). Then its derivative has the intermediate value property: if $x_1 < x_2$ are any two points in the interval, then $f'(x)$ assumes all values between $f'(x_1)$ and $f'(x_2)$ on the interval (x_1, x_2).*

Proof: First let us prove the theorem in the case when the value we want $f'(x)$ to assume is zero. This means zero must lie between $f'(x_1)$ and $f'(x_2)$, so one must be positive and one must be negative, say $f'(x_1) < 0 < f'(x_2)$. This is shown in Figure 5.2.5.

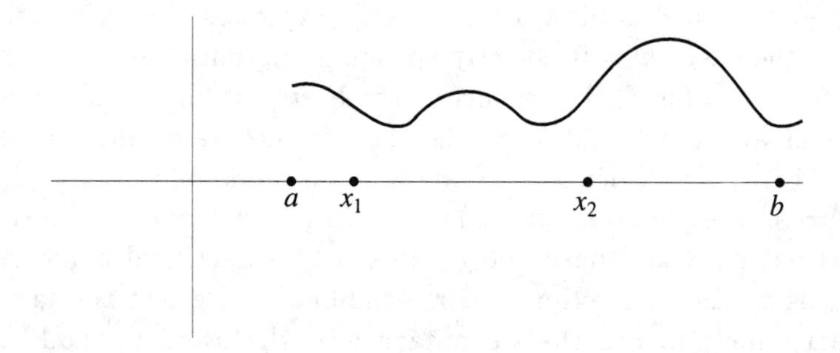

Figure 5.2.5:

To find a solution to $f'(x) = 0$ we need only show that f has a local maximum or minimum in the open interval (x_1, x_2). In the case we are considering we can show it must have at least one local minimum—it may, of course, have more than one, and it may have local maxima as well. To find a local minimum we look for a point where f attains its inf on the closed interval $[x_1, x_2]$. Here we use the fact that differentiability implies continuity of f, and then the fact that continuous functions on compact intervals attain their inf. If the inf is attained at an interior point x_0 of the interval, then x_0 is a local minimum (because (x_1, x_2) is a neighborhood of x_0 such that $f(x) \geq f(x_0)$ for all points in the neighborhood.) However, if the inf were attained at an endpoint we could not assert that we have a local minimum, as in the example $f(x) = x$ (Figure 5.2.6).

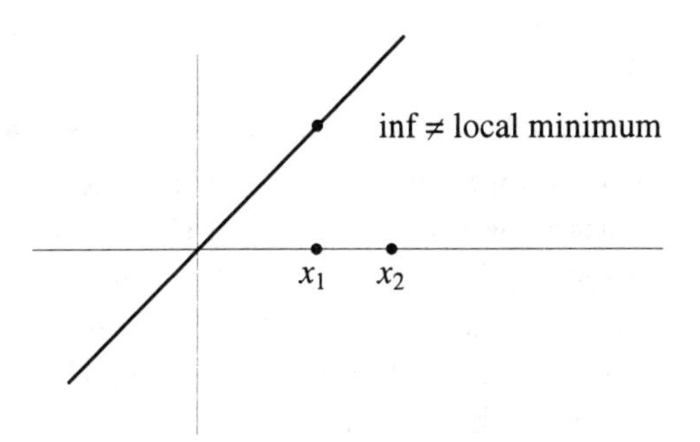

Figure 5.2.6:

Thus we have to use the hypotheses $f'(x_1) < 0 < f'(x_2)$ to show that the inf cannot occur at either endpoint. It can't occur at the left endpoint x_1, because there the function is strictly decreasing, so $f(x) < f(x_1)$ for x in the interval near x_1, and it can't occur at the right endpoint x_2, because there the function is strictly increasing, so $f(x) < f(x_2)$ for x in the interval near x_2. (If we had assumed the reverse inequalities $f'(x_2) < 0 < f'(x_1)$, then a similar argument would show that f can't attain its sup at either endpoint.)

We have thus proved that $f'(x)$ attains an intermediate value if that value happens to be zero. Now we need to see how the general case can be reduced to this special case. Suppose we want to show that $f'(x)$ can attain the value y_0, where say $f'(x_1) < y_0 < f'(x_2)$. The linear function $g(x) = y_0 x$ has derivative everywhere equal to y_0, so we need to solve $F'(x) = 0$ where $F = f - g$ (we are using here the elementary property $F' = f' - g'$, whose proof we will be given in Section 5.3.1). But $F'(x_1) = f'(x_1) - y_0 < 0$ and $F'(x_2) = f'(x_2) - y_0 > 0$, so we can apply the previous argument to F. QED

The intermediate value theorem explains why it is so tricky to get a function whose derivative is discontinuous—the discontinuity can't be a jump because then the intermediate value property would be violated. The proof of the intermediate value theorem is a good warm-up exercise for the proof of the mean value theorem.

Mean Value Theorem *Let f be a continuous function on a compact interval $[a, b]$ that is differentiable at every point in the interior. Then there exists a point x_0 in the interior where $f'(x_0) = (f(b) - f(a))/(b - a)$.*

The great significance of the mean value theorem is that it enables us to obtain information about the derivative from the computation of a single difference quotient, without passing to the limit. Of course we lose the precision of knowing the point x_0 exactly. The intuitive content of the theorem is that the difference quotient is the slope of the secant line joining the two points $(a, f(a))$ and $(b, f(b))$ of the graph, while the derivative is the slope of the tangent line. If we translate the secant line parallel to itself up or down, then at the moment it loses contact with the graph it should be tangent to it, as shown in Figure 5.2.7.

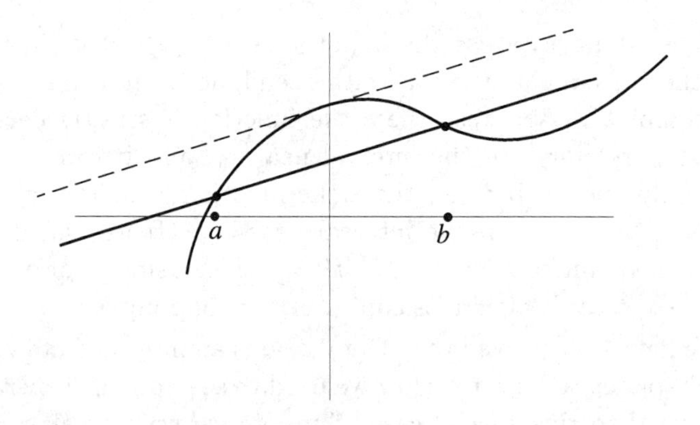

Figure 5.2.7:

This intuition can be made to form the basis of the proof if we first reduce to the case where the lines are all horizontal and then notice that a horizontal line loses contact with the graph at exactly the local maximum and minimum points.

Proof: We begin with the special case when $f(a) = f(b)$, so the difference quotient is zero (this case is known as *Rolle's Theorem*). Then we are looking for a point where $f'(x) = 0$, so it will suffice to find a local maximum or minimum. We have assumed that f is continuous on the compact interval $[a, b]$, so we know that f attains its sup or inf. If either happens at an interior point we have our local maximum or minimum. Thus we need only consider the special case where f attains both its sup and inf at the endpoints. This is not an impossible occurrence (for example, $f(x) = x$); but because we have assumed $f(b) = f(a)$, this implies f must be constant on $[a, b]$, so $f'(x) = 0$ at any point in the interval.

Next we reduce the general case to the special case. Let $g(x)$ be the affine function passing through the two points $(a, f(a))$ and $(b, f(b))$, as shown in Figure 5.2.8. The slope of the graph is obviously the difference quotient, so $g(x)$ has derivative equal to $(f(b) - f(a))/(b - a)$ at every point. By subtracting it off, $F = f - g$, we reduce the problem of solving $f'(x_0) = (f(b) - f(a))/(b - a)$ to that of solving $F'(x_0) = 0$.

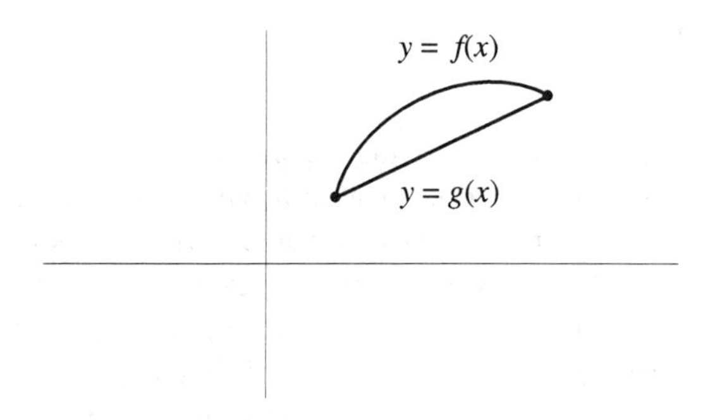

Figure 5.2.8:

But clearly F vanishes at both endpoints because f and g attain the same values at the endpoints. Thus $F(b) = F(a)$, so the previous argument applies.

You can easily verify that

$$g(x) = \left(\frac{f(b) - f(a)}{b - a} \right) (x - a) + f(a)$$

is the formula for the affine function, and one can then verify algebraically that $g'(x) = (f(b) - f(a))/(b - a)$ while $f = g$ at $x = a$ and $x = b$—facts that are evident from the graph. QED

5.2.3 Global Properties

Using the mean value theorem we can obtain relations between the derivative on an interval and the behavior of the function on the interval.

Theorem 5.2.2 *Let f be a differentiable function on the interval (a, b).*

a. *f is monotone increasing on (a, b) if and only if $f'(x) \geq 0$ for every x in the interval. Similarly f is monotone decreasing on (a, b) if and only if $f'(x) \leq 0$ on the interval.*

b. *If $f'(x) > 0$ for every x in the interval, then f is strictly increasing on the interval. Similarly $f'(x) < 0$ implies that f is strictly decreasing.*

c. *If $f'(x) = 0$ for every x in the interval, then f is constant on the interval.*

Proof

a. If f is monotone increasing on the interval, then the same argument as in the pointwise case (the difference quotients are all ≥ 0 and non-strict inequality is preserved in the limit) shows that $f'(x_0) \geq 0$ at any particular point x_0 in the interval. Conversely, suppose $f'(x) \geq 0$ on the interval. Let x_1 and x_2 be two points in the interval with $x_1 < x_2$. Then apply the mean value theorem to f on the closed interval $[x_1, x_2]$. Notice that the continuity hypothesis of the mean value theorem is satisfied since differentiability implies continuity. The mean value theorem gives us the identity $f'(x_0) = (f(x_2) - f(x_1))/(x_2 - x_1)$ for some point x_0. Since $f'(x_0) \geq 0$, we obtain $f(x_2) - f(x_1) \geq 0$, proving that f is monotone increasing.

b. We argue exactly as in part a to get $f'(x_0) = (f(x_2) - f(x_1))/(x_2 - x_1)$. Since now we are assuming $f'(x_0) > 0$, we obtain $f(x_2) - f(x_1) > 0$, proving that f is strictly increasing.

c. This time from $f'(x_0) = (f(x_2) - f(x_1))/(x_2 - x_1)$ and $f'(x_0) = 0$ we obtain $f(x_2) - f(x_1) = 0$. Since this is true for any two points in the interval, f is constant. QED

Notice that we have done somewhat better in part a than in the pointwise case (from $f'(x_0) \geq 0$ at one point we can deduce nothing)—we have an if and only if statement. The converse of part b is of course false, as the familiar example $f(x) = x^3$ shows. One can in fact show that $f'(x) \geq 0$ on an interval implies that f is strictly increasing unless $f'(x) = 0$ on an open subinterval. We leave this as an exercise. Part c can be used to prove the uniqueness of the indefinite integral.

Another easy application of the mean-value theorem shows that a differentiable function with bounded derivative satisfies a Lipschitz condition. Indeed, suppose f is continuous on $[a, b]$ and differentiable in the interior and $|f'(x)| \leq M$ for all x in (a, b). Then for any x_1, x_2 in the interval, $f(x_2) - f(x_1) = f'(x_0)(x_2 - x_1)$ for some x_0, so $|f(x_2) - f(x_1)| \leq M|x_2 - x_1|$, which is the desired Lipschitz condition. The converse is not true, for $f(x) = |x|$ satisfies a Lipschitz condition but is not differentiable at $x = 0$. Lipschitz functions have slightly rougher graphs than differentiable functions—as a rule of thumb they can have corners but not cusps (see Figure 5.2.9).

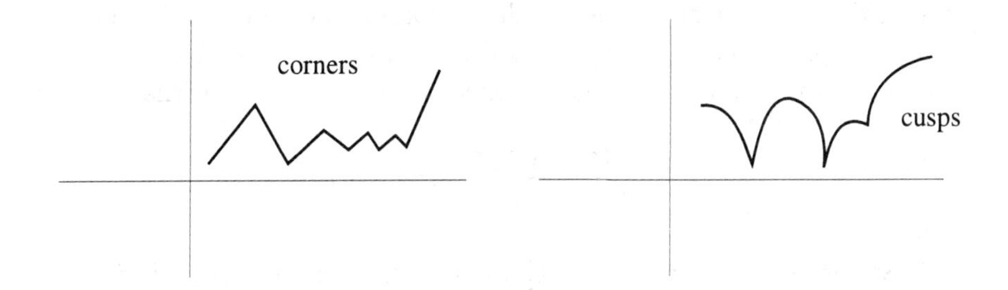

Figure 5.2.9:

Because of the greater flexibility that corners allow, there are some applications where it is important to allow Lipschitz functions. (A very profound theorem of Lebesgue—well beyond the scope of this work—says that Lipschitz functions must be differentiable at some points and, in fact, at most points in a certain well-defined sense.)

5.2.4 Exercises

1. Let f and g be continuous functions on $[a, b]$ and differentiable at every point in the interior, with $g(a) \neq g(b)$. Prove that there

exists a point x_0 in (a,b) such that

$$\frac{f(b)-f(a)}{g(b)-g(a)} = \frac{f'(x_0)}{g'(x_0)}.$$

(**Hint:** apply the mean value theorem to the function

$$(f(b)-f(a))g(x)-(g(b)-g(a))f(x).)$$

This is sometimes called the *second mean value theorem.*

2. If f is a function satisfying

$$|f(x)-f(y)| \leq M|x-y|^{\alpha}$$

for all x and y and some fixed M and $\alpha > 1$, prove that f is constant. (**Hint:** what is f'?) It is rumored that a graduate student once wrote a whole thesis on the class of functions satisfying this condition!

3. Is the converse of the mean value theorem true, in the sense that if f is continuous on $[a,b]$ and differentiable on (a,b), given a point x_0 in (a,b) must there exist points x_1, x_2 in $[a,b]$ such that

$$\frac{f(x_2)-f(x_1)}{x_2-x_1} = f'(x_0)?$$

4. Suppose f is defined on $[a,b]$ and g is defined on $[b,c]$ with $f(b) = g(b)$. Then define

$$h(x) = \begin{cases} f(x) & \text{if } a \leq x \leq b, \\ g(x) & \text{if } b \leq x \leq c. \end{cases}$$

Give an example where f and g are differentiable but h is not. Give a definition of one-sided derivatives $f'(b)$ and $g'(b)$ and show that the equality of these is a necessary and sufficient condition for h to be differentiable, given that f and g are differentiable.

5. Draw a picture of the graph of a function that is strictly increasing at a point but is not even monotone increasing in a neighborhood of that point.

6. Show that if f is differentiable and $f'(x) \geq 0$ on (a, b), then f is strictly increasing provided there is no subinterval (c, d) with $c < d$ on which f' is identically zero.

7. Draw the graph of a function that has a local maximum that is not a strict local maximum but is not constant on an interval.

8. Suppose f is continuously differentiable on an interval (a, b). Prove that on any closed subinterval $[c, d]$ the function is uniformly differentiable in the sense that given any $1/n$ there exists $1/m$ (independent of x_0) such that $|f(x) - f(x_0) - f'(x_0)(x - x_0)| \leq |x - x_0|/n$ whenever $|x - x_0| < 1/m$. (**Hint:** use the mean value theorem and the uniform continuity of f' on $[c, d]$.)

9. *Show that the converse to problem 8 is also valid: if f is uniformly differentiable on an interval, then f is continuously differentiable. (**Hint:** for two nearby points x_1 and x_2, consider the difference quotient $(f(x_2) - f(x_1))/(x_2 - x_1)$ as an approximation to both $f'(x_1)$ and $f'(x_2)$.)

10. If f assumes a local maximum or minimum at an endpoint of its domain $[a, b]$, what can you say about the one-sided derivative (assuming it exists)? **Warning:** The answer depends on which endpoint it is.

11. Prove that if f' is constant, then f is an affine function.

12. Give an example of a function that is differentiable on (a, b) but cannot be made continuous on $[a, b]$ by any definition of $f(a)$ or $f(b)$. Can you give an example where f is bounded?

13. If f is a differentiable function, prove that between any two zeroes of f there must be a zero of f'.

5.3 The Calculus of Derivatives

5.3.1 Product and Quotient Rules

In this section we derive the familiar rules of the differential calculus. These rules are all analogous to—and derived from—rules of the calculus of differences. One might find it surprising that the calculus of derivatives is somewhat simpler than the calculus of differences—but then the whole success of calculus is due to similar surprises.

If f is a function, we write $\Delta_h f(x) = f(x+h) - f(x)$ whenever it is defined. Think of h as a fixed number and Δ_h as an "operator", something that takes the function f as input and produces the function $f(x+h) - f(x)$ of x as output. The concept of operator is a distinctly twentieth century idea. We will use it only in an informal way in this work, but you will encounter it frequently if you continue your studies. We can also think of the derivative as an operator, taking the function f as input and producing the function f' as output. We are claiming here that the properties of the derivative operator are related to properties of the difference operator. Since $f'(x) = \lim_{h\to 0} \Delta_h f(x)/h$ by the definition of the derivative, this is not at all surprising.

Suppose we apply arithmetic operations on functions to obtain new functions. What happens to the differences? This is a matter of simple computation:

$$\begin{aligned} \Delta_h(f \pm g)(x) &= (f(x+h) \pm g(x+h)) - (f(x) \pm g(x)) \\ &= \Delta_h f(x) \pm \Delta_h g(x); \end{aligned}$$

$$\begin{aligned} \Delta_h(f \cdot g)(x) &= f(x+h)g(x+h) - f(x)g(x) \\ &= f(x+h)(g(x+h) - g(x)) + g(x)(f(x+h) - f(x)) \\ &= f(x+h)\Delta_h g(x) + g(x)\Delta_h f(x); \end{aligned}$$

$$\begin{aligned} \Delta_h(f \cdot g)(x) &= f(x)(g(x+h) - g(x)) + g(x+h)(f(x+h) - f(x)) \\ &= f(x)\Delta_h g(x) + g(x+h)\Delta_h f(x); \end{aligned}$$

$$\begin{aligned} \Delta_h\left(\frac{f}{g}\right)(x) &= \frac{f(x+h)}{g(x+h)} - \frac{f(x)}{g(x)} \\ &= \frac{f(x+h)g(x) - f(x)g(x+h)}{g(x+h)g(x)} \\ &= \frac{g(x)(f(x+h) - f(x)) - f(x)(g(x+h) - g(x))}{g(x+h)g(x)} \\ &= \frac{g(x)\Delta_h f(x) - f(x)\Delta_h g(x)}{g(x+h)g(x)}. \end{aligned}$$

Now to obtain the analogous rules for derivatives, we simply divide these identities by h and take the limit as $h \to 0$. For sums and products there is no difficulty, since the limits interchange with these operations. For the quotient we need to assume that g is not zero at the point in question.

Theorem 5.3.1 *If f and g are differentiable at x_0, then $f \pm g$ and $f \cdot g$ are also differentiable at x_0, and $(f \pm g)'(x_0) = f'(x_0) \pm g'(x_0)$, $(f \cdot g)'(x_0) = f(x_0)g'(x_0) + f'(x_0)g(x_0)$. In addition, if $g(x_0) \neq 0$, then f/g is differentiable at x_0, and*

$$\left(\frac{f}{g}\right)'(x_0) = \frac{g(x_0)f'(x_0) - f(x_0)g'(x_0)}{g(x_0)^2}.$$

Proof: We write out the complete proof for the quotient. The assumptions about differentiability mean $\lim_{h\to 0} \Delta_h f(x_0)/h = f'(x_0)$ and $\lim_{h\to 0} \Delta_h g(x_0)/h = g'(x_0)$. The fact that g is differentiable at x_0 implies that it is continuous at x_0, so $\lim_{h\to 0} g(x_0 + h) = g(x_0)$. Since we are assuming $g(x_0) \neq 0$, we have also $\lim_{h\to 0} g(x + h) \neq 0$. Altogether

$$\begin{aligned}
&\lim_{h\to 0} \frac{\Delta_h (f/g)(x_0)}{h} \\
&\quad = \lim_{h\to 0} \frac{g(x_0)\Delta_h f(x_0) - f(x_0)\Delta_h(x_0)}{hg(x_0 + h)g(x_0)} \\
&\quad = \frac{g(x_0)\lim_{h\to 0} \Delta_h f(x_0)/h - f(x_0)\lim_{h\to 0} \Delta_h g(x_0)/h}{g(x_0)\lim_{h\to 0} g(x_0 + h)},
\end{aligned}$$

which shows the limit exists (meaning f/g is differentiable at x_0) and equals

$$\frac{g(x_0)f'(x_0) - f(x_0)g'(x_0)}{g(x_0)^2}.$$

The proof for the sum and product is similar. For the product we can use either identity for the difference of a product—in the limit they yield the same formula.

We obtain immediately similar results about differentiability on an interval and even continuous differentiability since the formulas for derivatives preserve continuity (the condition $g(x) \neq 0$ for all x in the interval is required for the quotient, of course). What happens to the derivative of a quotient when both f and g are zero is a very interesting question—one we will be able to answer after we discuss Taylor's theorem.

5.3.2 The Chain Rule

The next basic calculus formula is the chain rule for the derivative of the composition $g \circ f(x) = g(f(x))$.

We begin with a direct approach to the computation in terms of d-ifference quotients; this approach encounters some obstacles, which will lead us to rethink the whole problem. We compute first the difference of the composition:

$$\begin{aligned} \Delta_h g \circ f(x) &= g(f(x+h)) - g(f(x)) \\ &= g(f(x) + (f(x+h) - f(x))) - g(f(x)) \\ &= (\Delta_{\Delta_h f(x)} g)(f(x)). \end{aligned}$$

This rather cumbersome notation means we take the difference of the function g at the point $f(x)$ with increment $\Delta_h f(x)$. Of course the domain of g must include $f(x)$ and $f(x+h)$ for this to be meaningful. To compute the derivative we want to divide by h and let $h \to 0$. To simplify the notation we let $z = \Delta_h f(x)$. Then

$$\begin{aligned} \frac{\Delta_h g \circ f(x)}{h} &= \frac{(\Delta_z g)(f(x))}{h} \\ &= \frac{\Delta_z g(f(x))}{z} \cdot \frac{z}{h} \\ &= \frac{\Delta_z g(f(x))}{z} \cdot \frac{\Delta_h f(x)}{h} \end{aligned}$$

and so the limit will exist (and equal the product) if the limits of both $\Delta_z g(f(x))/z$ and $\Delta_h f(x)/h$ exist. Now if we fix $x = x_0$ and assume that f is differentiable at x_0, then we have $\lim_{h\to 0} \Delta_h f(x_0)/h = f'(x_0)$. For the other factor, if we assume g is differentiable at the point $f(x_0)$, we have $\lim_{z\to 0} \Delta_z g(f(x_0))/z = g'(f(x_0))$. This is not quite what is called for, since we are taking the limit as h goes to zero, and z is defined in terms of h, namely $z = \Delta_h f(x_0)$. Now we do know that f is continuous at x_0, because it is differentiable, so $\lim_{h\to 0} z = 0$, which seems to imply $\lim_{h\to 0} \Delta_z g(f(x_0))/z = g'(f(x_0))$, leading to the familiar chain rule $f \circ g'(x_0) = g'(f(x_0))f'(x_0)$. However, if you try to make this into a precise proof you will come upon one very sticky point: if $z = 0$ nothing is defined. Of course nothing is defined if $h = 0$, either, but this is excluded in the definition of the derivative, while there is nothing to exclude $z = 0$. In fact $z = 0$ whenever $f(x_0 + h) = f(x_0)$, which may well happen for values of h arbitrarily close to zero. Thus we have only established the chain rule if there exists a neighborhood of zero such that $\Delta_h f(x_0) \neq 0$ for non-zero h in the neighborhood.

We can make a separate proof in the contrary case, showing that then both $f'(x_0) = 0$ and $g \circ f'(x_0) = 0$. However this results in a rather awkward proof, so we will try another tack, leaving the completion of the first approach for the exercises.

Let us think about the chain rule in terms of best affine approximations. If f is differentiable at x_0, then

$$f(x) = f(x_0) + f'(x_0)(x - x_0) + o(x - x_0).$$

[Strictly speaking, we should write $f(x) = f(x_0) + f'(x_0)(x - x_0) + R(x)$ with $R(x) = o(x - x_0)$ as $x \to x_0$, and this is the meaning of our abbreviation. This is a convenient and standard notational short-hand. You should keep in mind, however, that different occurrences of the same symbol $o(x - x_0)$ might refer to different functions.] If g is differentiable at $y_0 = f(x_0)$, then $g(y) = g(y_0) + g'(y_0)(y - y_0) + o(y - y_0)$. Substituting $y = f(x)$ we obtain

$$\begin{aligned} g(f(x)) &= g(f(x_0)) + g'(f(x_0))(f(x) - f(x_0)) + o(y - y_0) \\ &= g(f(x_0)) + g'(f(x_0))(f'(x_0)(x - x_0) \\ &\quad + o(x - x_0)) + o(y - y_0) \\ &= g(f(x_0)) + g'(f(x_0))f'(x_0)(x - x_0) + \text{remainder}, \end{aligned}$$

where the remainder term is the sum of $g'(f(x_0))o(x - x_0)$ and $o(y - y_0)$. To complete the proof we have to show this remainder is $o(x - x_0)$. Before doing this we should provide some interpretation for these computations. The formula $f(x) = f(x_0) + f'(x_0)(x - x_0) + o(x - x_0)$ is the same as $f(x) - f(x_0) = f'(x_0)(x - x_0) + o(x - x_0)$ or even $\Delta_h f(x_0) = f'(x_0)h + o(h)$ if we set $x = x_0 + h$. We interpret this to say that for values of x near x_0, changes in the x variable are multiplied by the magnification factor $f'(x_0)$—aside from a small remainder term—in order to obtain changes in $y = f(x)$. Similarly $g(y) - g(y_0) = g'(y_0)(y - y_0) + o(y - y_0)$ means changes in y, for y near y_0, get multiplied by the factor $g'(y_0)$—again aside from a small error—in passing to changes in $g(y)$. Thus in the composition $g \circ f$ the change in

x first gets multiplied by $f'(x_0)$ and then by $g'(y_0) = g'(f(x_0))$, hence altogether by $g'(f(x_0))f'(x_0)$, before producing a change in $g \circ f$. Notice that the error in the first stage also gets multiplied by $g'(f(x_0))$ and is then added to the error in the second stage. We can now put this all together into a complete proof.

Theorem 5.3.2 (*Chain Rule*) *Let f be defined in a neighborhood of x_0 and differentiable at x_0, and let g be defined in a neighborhood of $f(x_0)$ and differentiable at $f(x_0)$. Then $g \circ f$ is differentiable at x_0 and $(g \circ f)'(x_0) = g'(f(x_0))f'(x_0)$.*

Proof: We need to show that given any $1/m$ we can make

$$|g(f(x)) - g(f(x_0)) - g'(f(x_0))f'(x_0)(x - x_0)| \le \frac{|x - x_0|}{m}$$

by taking x close enough to x_0. From our discussion we know that we want to break this into two parts as follows:

$$\begin{aligned} & g(f(x)) - g(f(x_0)) - g'(f(x_0))f'(x_0)(x - x_0) \\ &= [g(f(x)) - g(f(x_0)) - g'(f(x_0))(f(x) - f(x_0))] \\ &\quad + [g'(f(x_0))(f(x) - f(x_0) - f'(x_0)(x - x_0))]. \end{aligned}$$

If we can show that each of the terms in brackets is at most $|x-x_0|/2m$, we will be done. For the second term this is just the differentiability of f at x_0, which enables us to get

$$|f(x) - f(x_0) - f'(x_0)(x - x_0)| < \frac{|x - x_0|}{2m|g'(f(x_0))|}$$

by taking x close to x_0 (if $g'(f(x_0)) = 0$ the whole term is zero and there is nothing to prove). For the first term we have to work a little harder. By the differentiability of g at $f(x_0) = y_0$ we can make

$$|g(y) - g(y_0) - g'(y_0)(y - y_0)| \le \frac{|y - y_0|}{km}$$

if $|y - y_0| < 1/n$ by choosing n to depend on km (the value of k will be chosen later). Substituting $y = f(x)$ and $y_0 = f(x_0)$ we obtain that $|f(x) - f(x_0)| < 1/n$ implies

$$\begin{aligned} |g(f(x)) - g(f(x_0)) &- g'(f(x_0))(f(x) - f(x_0))| \\ &\le \frac{|f(x) - f(x_0)|}{km}. \end{aligned}$$

Finally, by the differentiability of f at x_0, we can make both $|f(x) - f(x_0)| < 1/n$, so the above applies, and $|f(x) - f(x_0)| \leq M|x - x_0|$ where $M = 1 + |f'(x_0)|$ (recall the proof that differentiability implies Lipschitz continuity) by taking x close enough to x_0. This gives the estimate M/km for the first term in brackets, and we need only take $k = 2M$ to complete the proof. QED

When we come to discuss the differential calculus in several variables we will also have a chain rule, and we will be able to adapt the above proof to that context.

5.3.3 Inverse Function Theorem

The last of the important formulas in the calculus of derivatives is the rule for differentiating functions given implicitly. We are going to have to postpone a full discussion until a later chapter, because it requires the differential calculus in two variables. Here we will discuss a special case, that of inverse functions. In the abstract definition of function, if f is one-to-one from its domain D to its range R and is onto R (the image $f(D)$ is all of R), then the inverse function f^{-1} with domain R and range D is defined by $f^{-1}(y) = x$ if and only if $f(x) = y$. If I_D and I_R denote the identity functions on D and R respectively, then $f^{-1} \circ f = I_D$ and $f \circ f^{-1} = I_R$.

To begin the discussion we will assume that f is a numerical function and f and f^{-1} are both differentiable. This is a big assumption, and we will have to return to the point later. What we can see easily is that the chain rule establishes an identity involving the derivatives of f and f^{-1}. Note that $I(x) = x$ is the identity function on $\mathbb{R}$, so $f^{-1}(f(x)) = x$. Since this is an equality between functions, we may differentiate both sides of the equation. (Unfortunately, mathematical notation is sometimes ambiguous about whether an equation $f(x) = g(x)$ is meant to hold for all x or just for some particular value x. From the equality $f(x) = g(x)$ at one point x we can conclude nothing about the derivatives of f and g at that point; but if $f(x) = g(x)$ for all x in the domains, then f and g are the same function, so $f'(x) = g'(x)$ for all x because f' and g' are the same function.) From the chain rule we obtain $(f^{-1})'(f(x))f'(x) = 1$, hence

$$\begin{aligned}(f^{-1})'(f(x)) &= \frac{1}{f'(x)}, \quad \text{or} \\ (f^{-1})'(y) &= \frac{1}{f'(x)} \quad \text{if } y = f(x), \quad \text{or} \\ (f^{-1})'(y) &= \frac{1}{f'(f^{-1}(y))}.\end{aligned}$$

The last form is what we obtain by differentiating $f(f^{-1}(y)) = y$ and solving for the derivative of f^{-1}.

We might attempt to paraphrase this relation as saying the derivative of the inverse function is the reciprocal (the multiplicative inverse) of the derivative of the function. However, this is only part of the story, because it doesn't say where the derivatives are evaluated. It is not true that $(f^{-1})'(x)$ has any relation to $1/f'(x)$, for example. A good way to think about the situation is via the graphs of f and f^{-1}. Since $y = f(x)$ if and only if $x = f^{-1}(y)$, the graph of f^{-1} is obtained from the graph of f by interchanging the axes. This is the same as reflecting the graph in the diagonal line $y = x$, as shown in Figure 5.3.1.

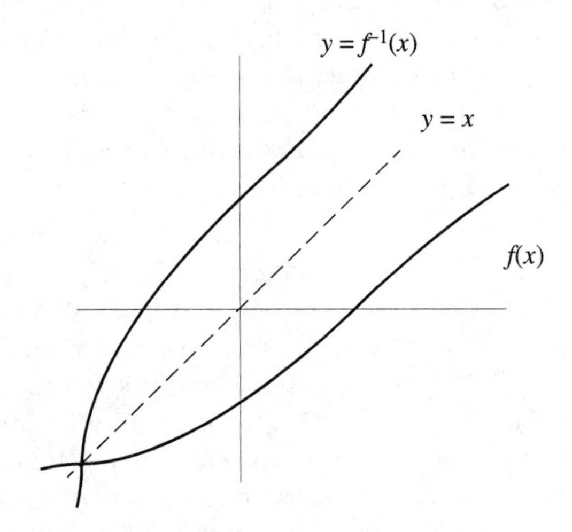

Figure 5.3.1:

Now the tangent line to the graph of f at the point (x_0, y_0) is also reflected in $y = x$ into the tangent line to the graph of f^{-1} at the point

(y_0, x_0), as shown in Figure 5.3.2.

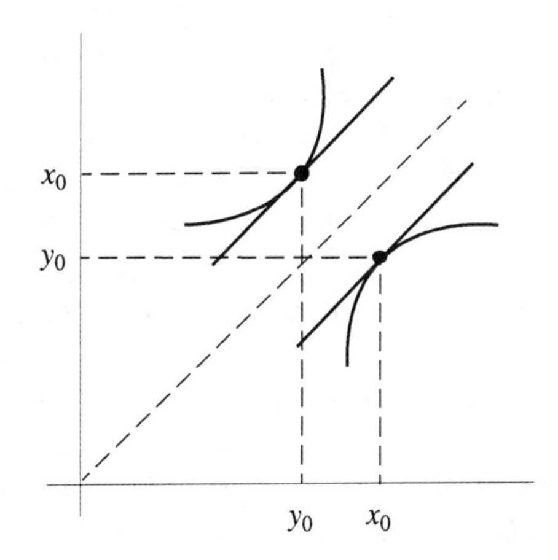

Figure 5.3.2:

The interchange of axes clearly changes the slope of the tangent line $\Delta y/\Delta x$ into its reciprocal $\Delta x/\Delta y$, while the first coordinate of the point of tangency is also clearly changed from x_0 to y_0.

The formal computation of $(f^{-1})'$, made under the assumption that f^{-1} exists and is differentiable, has given us a good intuitive grasp of the situation. Now we have to come to grips with three important problems:

1. How do we know if f^{-1} exists?

2. How do we know if f^{-1} is differentiable?

3. What happens when $f'(x) = 0$?

It turns out that all these problems are interconnected. We will start with a differentiable function f, and to avoid splitting hairs let us assume that f is continuously differentiable on an interval (a, b). How can f be one-to-one? If f is strictly increasing or strictly decreasing, then f cannot assume one value twice. Actually the converse is true; we will leave the proof as an exercise, since we will not need this result. How can we assure that f is strictly increasing or decreasing? By making $f'(x) > 0$ or $f'(x) < 0$ on the interval. If we make this assumption, then we can define f^{-1} with domain equal to the image of f, which is also an open interval. We also avoid the problem of what happens at a point where f' is zero. In fact it is clear from the reflection picture that if $f'(x_0)$ is zero, hence the tangent is horizontal, and if f^{-1} exists, then the tangent to the graph of f^{-1} must be vertical at the corresponding point $y_0 = f(x_0)$, so f^{-1} is not differentiable at y_0. The simplest example of this is $f(x) = x^3$ at $x = 0$. The inverse function $f^{-1}(y) = \sqrt[3]{y}$ is not differentiable at $y = 0$. Therefore we are not missing any positive results if we assume either $f'(x) > 0$ or $f'(x) < 0$ on the interval.

Inverse Function Theorem *Let f be a continuously differentiable function on an open interval (a, b), with image (c, d); and suppose either $f'(x) > 0$ or $f'(x) < 0$ on (a, b). Then f^{-1} with domain (c, d) and image (a, b) is continuously differentiable, and $(f^{-1})'(y) = 1/f'(x)$ if $y = f(x)$.*

Proof: Let x_0, y_0 denote fixed points such that $f(x_0) = y_0$, and let x, y denote variable points such that $f(x) = y$. Then

$$\lim_{x \to x_0} \frac{y - y_0}{x - x_0} = f'(x_0)$$

exists and is non-zero, so $\lim_{x \to x_0} (x - x_0)/(y - y_0) = 1/f'(x_0)$. Now $(x - x_0)/(y - y_0)$ is the difference quotient whose limit is the derivative of f^{-1} at y_0. The only problem is that in the definition of $(f^{-1})'$ we are supposed to take the limit as $y \to y_0$. Is this the same thing? To find out, let's write what we know and what we need to show. We know $\lim_{x \to x_0} (x - x_0)/(y - y_0) = 1/f'(x_0)$, so given any error $1/m$ there exists $1/n$ such that

$$\left| \frac{x - x_0}{y - y_0} - \frac{1}{f'(x_0)} \right| < \frac{1}{m}$$

provided $|x - x_0| < 1/n$ and $x \neq x_0$. What we need to show is $\lim_{y \to y_0} (x - x_0)/(y - y_0) = 1/f'(x_0)$, in other words given any error $1/m$ there exists $1/k$ such that

$$\left| \frac{x - x_0}{y - y_0} - \frac{1}{f'(x_0)} \right| < \frac{1}{m}$$

provided $|y - y_0| < 1/k$ and $y \neq y_0$. Thus to bridge the gap between these statements we need to show: *for any error* $1/n$ *there exists* $1/k$ *such that* $|x - x_0| < 1/n$ *and* $x \neq x_0$ *provided* $|y - y_0| < 1/k$ *and* $y \neq y_0$. Note that $y \neq y_0$ implies $x \neq x_0$ because f is one-to-one, so that part is easy. The rest of the statement is exactly the definition of the continuity of f^{-1} at y_0. This is the non-trivial part.

Why should f^{-1} be continuous? The idea is that since f' is never zero, it must be bounded away from zero in a neighborhood of x_0, so that changes in x must result in substantial changes in y—we get estimates that go in the opposite direction of the continuity estimates for f, so that when we switch over to f^{-1} the estimates will go in the correct direction. More precisely, first choose a neighborhood of x_0 such that $|f'(x)| \geq 1/N$ for some fixed N and all x in the neighborhood (here we use the continuity of the derivative). Then the mean value theorem $(f(x) - f(x_0))/(x - x_0) = f'(x_1)$ gives the estimate $|f(x) - f(x_0)| \geq |x - x_0|/N$ for all x in the neighborhood. Note that this is the reverse of the usual Lipschitz condition. But now $y = f(x)$ and $y_0 = f(x_0)$, so $|y - y_0| < 1/Nn$ implies $|x - x_0| \leq N|y - y_0| < 1/n$ as desired. This proves the continuity of f^{-1} and so completes the proof of the differentiability of f^{-1} with the correct derivative. Finally $(f^{-1})(y) = 1/f'(f^{-1}(y))$ is continuous because f' and f^{-1} are continuous and f' is never zero. QED

The inverse function theorem as stated is unique to one dimension, because only in one dimension do we have a criterion like strictly increasing or decreasing for the function to be one-to-one. There is a local version of the theorem, however, that can be generalized to higher dimensions, and for that reason we state it here.

Local Inverse Function Theorem *Let* f *be defined and continuously differentiable in a neighborhood of* x_0*, and suppose* $f'(x_0) \neq 0$*. Then there exists a neighborhood* (a, b) *of* x_0 *such that the restriction of* f *to* (a, b) *has a continuously differentiable inverse on the image* $(c, d) = f((a, b))$.

Proof: Since f' is assumed continuous and $f'(x_0) \neq 0$, we can find a neighborhood (a, b) of x_0 such that either $f'(x) > 0$ or $f'(x) < 0$ there, and then we apply the global theorem. QED

The inverse function theorem is an extremely useful and powerful theorem. Many important functions such as exp, sine, and cosine are best defined as inverses of functions given by explicit integrals. We will develop these ideas in a later chapter. The inverse function theorem in higher dimensions, which is a generalization of the local version only, is used to prove the differentiability (and existence) of functions defined implicitly.

5.3.4 Exercises

1. Define
$$x_+ = \begin{cases} x & \text{if } x \geq 0, \\ 0 & \text{if } x < 0. \end{cases}$$
Prove that $f(x) = x_+^k$ is continuously differentiable if k is an integer greater than one.

2. Show that $(x-a)_+^2(b-x)_+^2$ is a continuously differentiable function that is non-zero exactly on the interval (a, b).

3. Given a closed set A, construct a continuously differentiable function that has A as its set of zeroes.

4. Prove that if f is any one-to-one function on an interval (a, b), then either f is strictly increasing or strictly decreasing.

5. If f is differentiable on (a, b) and $f'(x) \neq 0$ for all x in the interval, prove that either $f'(x) > 0$ or $f'(x) < 0$ on the entire interval.

6. Give a proof of the chain rule arguing separately in the case when every neighborhood of zero contains a value of h for which $\Delta_h f(x_0) = 0$.

7. Prove that $f(x) = x^{1/k}$ can be defined on $[0, \infty)$ by the requirement that it be the inverse function of $g(x) = x^k$ on $[0, \infty)$, where k is any positive integer. Use the inverse function theorem to derive the usual formula for f'.

8. For any rational number r give a definition of $f(x) = x^r$ for $x > 0$ and show $f'(x) = rx^{r-1}$.

9. Show that x_+^r is continuously differentiable for any rational $r > 1$.

10. Show that a polynomial of even order ($\neq 0$) has either a global maximum or a global minimum but not both.

11. Show that the class of rational functions (polynomial divided by polynomial) is closed under the operation of differentiation.

12. *Show that no rational function has derivative equal to $1/x$.

13. Let f_n denote the nth iterate of $f, f_1 = f, f_2(x) = f(f_1(x)), \ldots$ $f_n(x) = f(f_{n-1}(x))$. Express f_n' in terms of f'. Show that if $a \leq |f'(x)| \leq b$ for all x, then $a^n \leq |f_n'(x)| \leq b^n$.

14. If f is a polynomial, show that f_n is a polynomial. What is the degree of f_n if f has degree N? Similarly, show that if f is a rational function, then f_n is a rational function.

15. A function is called *algebraic* if it satisfies a polynomial identity $\sum a_{jk}x^j f(x)^k = 0$ (finite sum, not all coefficients a_{jk} zero). Assuming $f(x)$ is differentiable, find a formula for f' in terms of f.

5.4 Higher Derivatives and Taylor's Theorem

5.4.1 Interpretations of the Second Derivative

If f is a differentiable function on an interval, then f' is also a function on that interval, which may or may not be differentiable. If f' is differentiable at x_0 we call its derivative $f''(x_0)$ the *second derivative* of f at x_0. If f'' exists for every point in the interval and is continuous we say f is *twice continuously differentiable* or f is C^2.

You are no doubt familiar with the interpretation of acceleration as a second derivative. The significance of acceleration in Newton's theory of mechanics guarantees that the notion of second derivative is of great importance. Another familiar application of second derivatives involves concavity of the graph and the problem of distinguishing local maxima and minima. Roughly speaking, *in an interval where f'' is positive, the graph lies above tangent lines and below secant lines, and only local minima can occur, while the reverse is true on an interval where f'' is negative.* This is illustrated in Figure 5.4.1.

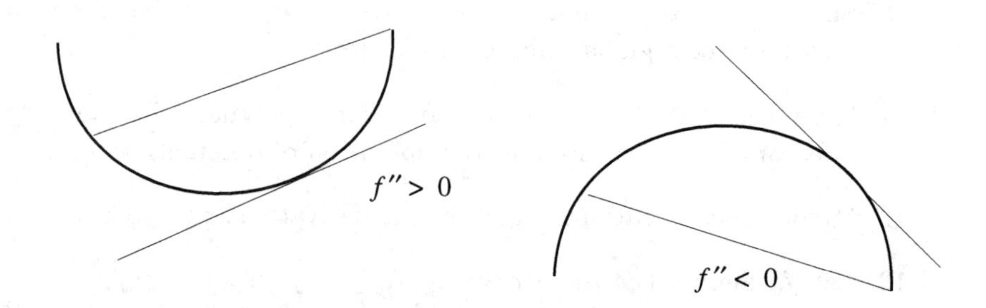

Figure 5.4.1:

We will establish these relationships in two theorems. The first deals with properties of the second derivative at a single point, and the second with properties of a continuous second derivative on an interval. The relationship involving the secant line only shows up in the second theorem.

Theorem 5.4.1 *Suppose f is differentiable in a neighborhood of x_0 and $f''(x_0)$ exists. Let $g(x) = f(x_0) + f'(x_0)(x - x_0)$ denote the best affine approximation to f at x_0.*

a. *If $f''(x_0) > 0$, then there exists a neighborhood of x_0 where $f(x) \geq g(x)$ (even $f(x) > g(x)$ for $x \neq x_0$), while if $f''(x_0) < 0$ there exists a neighborhood of x_0 where $f(x) \leq g(x)$.*

b. *If $f(x) \geq g(x)$ in a neighborhood of x_0, then $f''(x_0) \geq 0$; while if $f(x) \leq g(x)$ in a neighborhood of x_0, then $f''(x_0) \leq 0$.*

c. *Suppose also $f'(x_0) = 0$. If $f''(x_0) > 0$, then x_0 is a strict local minimum; while if $f''(x_0) < 0$, then x_0 is a strict local maximum.*

d. *If x_0 is a local minimum, then $f''(x_0) \geq 0$; while if x_0 is a local maximum, then $f''(x_0) \leq 0$.*

Proof: We begin with part c. If $f''(x_0) > 0$ this means the derivative of f' is positive at x_0, hence $f'(x)$ is strictly increasing at x_0. Since $f'(x_0) = 0$, this means that $f'(x) < 0$ if $x < x_0$ and $f'(x) > 0$ if $x > x_0$ for x near x_0, say $|x - x_0| < 1/n$. But this implies f is strictly decreasing on $(x_0 - 1/n, x_0)$ and strictly increasing on $(x_0, x_0 + 1/n)$, so x_0 is a strict local minimum. Notice that the proof used twice the relations between sign of the first derivative and behavior of the function: first in going from the sign of the derivative of f'' at a point to behavior of f' near the point and then in going from the sign of f' on an interval to behavior of f on the interval. The proof that $f''(x_0) < 0$ implies x_0 is a strict local maximum is analogous.

Next part d is essentially the contrapositive of part c, for a local minimum cannot be also a strict local maximum, so we cannot have $f''(x_0) < 0$ by part c.

Finally we can derive part a from part c and part b from part d. The function $f(x) - g(x)$ vanishes at x_0, and its derivative is also zero at x_0 while its second derivative is merely $f''(x_0)$. If $f''(x_0) > 0$, then part c implies that $f(x) - g(x)$ has a strict local minimum at x_0, hence $f(x) - g(x) \geq f(x_0) - g(x_0) = 0$ in a neighborhood of x_0, with strict inequality for $x \neq x_0$. Similarly, we can derive part b from part d. QED

The function $f(x) = x^4$ has a strict minimum at $x = 0$, but its second derivative is zero there, showing that we cannot improve part d to have strict inequality.

Theorem 5.4.2 *Let f be a C^2 function on an interval (a, b). Let g denote any affine function whose graph intersects the graph of f at two points (x_1, y_1) and (x_2, y_2) with $x_1 < x_2$. If $f''(x) > 0$ for all x in (x_1, x_2), then $f(x) < g(x)$ for all x in (x_1, x_2); while if $f''(x_0) < 0$ for all x in (x_1, x_2), then $f(x) > g(x)$ for all x in (x_1, x_2).*

Proof: Suppose $f'' > 0$ in the interval (x_1, x_2). Then the same is true for $h = f - g$ because $g'' = 0$. Note that h vanishes at the endpoints of the interval. We want to prove that h is negative on (x_1, x_2). If not, it would achieve a local maximum on (x_1, x_2), say at x_0, and by part d of the previous theorem this would imply $h''(x_0) \leq 0$, a contradiction. QED

The same proof also applies to non-strict inequalities. There is a converse to this theorem, which we leave to the exercises.

A different interpretation of the second derivative involves the notion of the second difference. This idea is implicit in the notation

d^2y/dx^2 and is useful in numerical solutions to differential equations. Recall that we defined the difference operator Δ_h with increment h as $\Delta_h f(x) = f(x+h) - f(x)$ and $f'(x)$ as the limit of $\Delta_h f(x)/h$ as $h \to 0$. The second difference operator Δ_h^2 is simply Δ_h applied twice:

$$\begin{aligned}
\Delta_h(\Delta_h f)(x) &= \Delta_h f(x+h) - \Delta_h f(x) \\
&= [f(x+h+h) - f(x+h)] - [f(x+h) - f(x)] \\
&= f(x+2h) - 2f(x+h) + f(x).
\end{aligned}$$

The second derivative should then be the limit of $\Delta_h^2 f(x)/h^2$ as $h \to 0$. The reason for this is that for small h

$$\frac{f(x+2h) - f(x+h)}{h} \approx f'(x+h)$$

and

$$\frac{f(x+h) - f(x)}{h} \approx f'(x),$$

so

$$\frac{\Delta_h^2 f(x)}{h^2} \approx \frac{f'(x+h) - f'(x)}{h} \approx f''(x).$$

To make this argument precise we will need to use the mean value theorem. A direct approach would be to replace the difference quotients $(f(x+2h) - f(x+h))/h$ and $(f(x+h) - f(x))/h$ by derivatives. However this would lead to the awkward situation of having $\Delta_h^2 f(x)/h^2 = (f'(x_2) - f'(x_1))/h$ where x_2 is in $(x+h, x+2h)$ and x_1 is in $(x, x+h)$, and we would have no control over $x_2 - x_1$. We need a more clever idea.

Theorem 5.4.3 *If f is C^2 on an interval, then $\lim_{h\to 0} \Delta_h^2 f(x)/h^2$ exists and equals $f''(x)$ for any x in the interval.*

Proof: We apply the mean value theorem to the function $g(x) = f(x+h) - f(x)$ on the interval $[x, x+h]$. Since $g'(x) = f'(x+h) - f'(x)$, we have the required differentiability and continuity of g if h is small enough. The mean value theorem gives $(g(x+h) - g(x))/h = g'(x_0)$ for x_0 in $(x, x+h)$, and this is exactly

$$\frac{\Delta^2 f(x)}{h} = f'(x_0 + h) - f'(x_0);$$

hence,

$$\frac{\Delta^2 f(x)}{h^2} = \frac{f'(x_0+h) - f'(x_0)}{h}.$$

This is similar to what we had before, with $x_1 = x_0$ and $x_2 = x_0 + h$, but now we have the difference $x_2 - x_1 = h$ under control. Applying the mean value theorem again to this difference quotient we obtain $\Delta_h^2 f(x)/h^2 = f''(x_1)$ for some point x_1 in $(x_0, x_0 + h)$. Since x_1 must lie between x and $x + 2h$ and f'' is assumed continuous, we may take the limit as $h \to 0$ and get $f''(x)$. QED

There is also a converse to this theorem, but it is much more difficult to prove.

5.4.2 Taylor's Theorem

We come now to the last, and in many ways the most important, interpretation of the second derivative. We have seen that the first derivative can be thought of as one of the constants in the best affine approximation to f at the point. Now an affine function is a polynomial of degree one, and the derivative shows up as the coefficient of the leading (highest degree) term. We can think of the best affine approximation as an improvement over approximation by the constant function $y = f(x_0)$, which is the best polynomial of degree-zero approximation to f at x_0. The affine approximation is an improvement because we only have $f(x) - f(x_0) = o(1) = o(|x - x_0|^0)$ while $f(x) - (f(x_0) + f'(x_0)(x - x_0)) = o(|x - x_0|)$ as $x \to x_0$. From this we would guess that by allowing polynomials of degree 2 we should be able to improve the approximation to $o(|x - x_0|^2)$. Now if f itself is a polynomial of degree at most 2, then

$$f(x) = f(x_0) + f'(x_0)(x - x_0) + \frac{1}{2}f''(x_0)(x - x_0)^2,$$

as can be verified by a simple computation. For more general f, we would expect

$$g_2(x) = f(x_0) + f'(x_0)(x - x_0) + \frac{1}{2}f''(x_0)(x - x_0)^2$$

to be the best approximation to f at x_0 by polynomials of degree at most 2. The second derivative thus appears as the coefficient of the leading term, except for the factor of $1/2$.

Now the expected improvement in using g_2 rather than the affine approximation concerns the rate of convergence as $x \to x_0$. For any particular value of x we do not know whether g_2 is a better approximation—it might very well be worse. However, the error being $o(|x-x_0|^2)$ means that by taking x close enough to x_0 we have a very good approximation, since $|x-x_0|^2$ is an order of magnitude smaller than $|x-x_0|$. The graphs of f and g_2 in Figure 5.4.2 show this in the order of contact at the point.

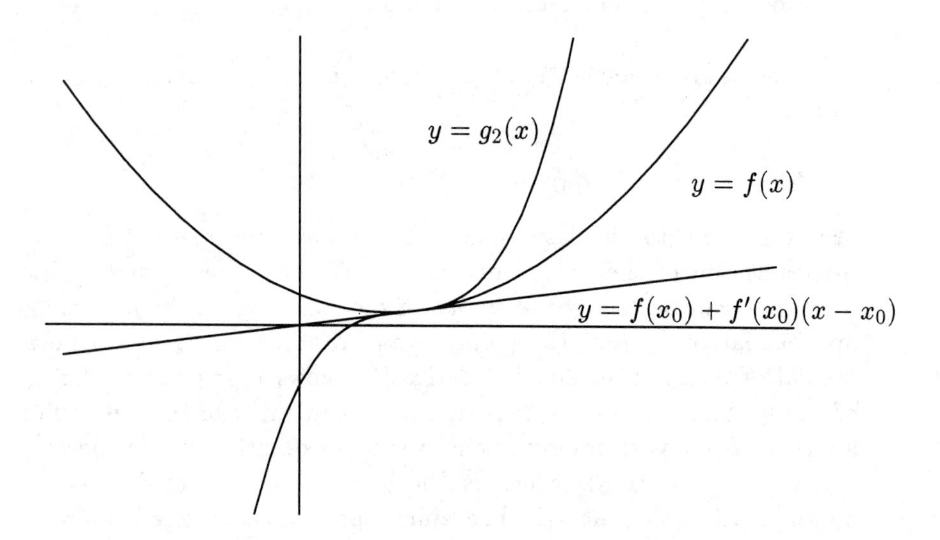

Figure 5.4.2:

Since the quadratic term $f''(x_0)(x-x_0)^2/2$ is $O(|x-x_0|^2)$ as $x \to x_0$, we can expect this order of improvement.

Theorem 5.4.4 *Let f be a C^2 function defined in a neighborhood of x_0, and let*

$$g_2(x) = f(x_0) + f'(x_0)(x - x_0) + \frac{1}{2}f''(x - x_0)^2.$$

Then $f - g_2 = o(|x - x_0|^2)$ as $x \to x_0$; in other words, given any error $1/m$, there exists $1/n$ such that $|x - x_0| < 1/n$ implies

$$|f(x) - g_2(x)| \leq \frac{1}{m}|x - x_0|^2.$$

Proof: The function g_2 was chosen so that $f(x_0) = g_2(x_0), f'(x_0) = g_2'(x_0)$, and $f''(x_0) = g_2''(x_0)$. Thus if we let $F = f - g_2$ we have $F(x_0) = 0, F'(x_0) = 0$, and $F''(x_0) = 0$, and F is C^2, being the difference of f and a quadratic polynomial. We need to show that we can make $|F(x)| \leq |x - x_0|^2/m$ by taking x near x_0, and we can do this by applying the mean value theorem twice to F. First, since $F(x_0) = 0$, we have $F(x) = F'(x_1)(x - x_0)$ for some x_1 between x_0 and x. Similarly, since $F'(x_0) = 0, F'(x_1) = F''(x_2)(x_1 - x_0)$ for some x_2 between x_0 and x_1. The fact that F is C^2 guarantees that the continuity and differentiability conditions of the mean value theorem are satisfied by F and F'. Putting together these two equations gives us $F(x) = F''(x_2)(x - x_0)(x_1 - x_0)$. Hence $|F(x)| \leq |F''(x_2)|\,|x - x_0|^2$ since $|x_1 - x_0| \leq |x - x_0|$ because x_1 lies between x_0 and x. Now $F''(x_0) = 0$ and F'' is continuous, so given $1/m$ there exists $1/n$ such that $|x - x_0| < 1/n$ implies $|F''(x)| < 1/m$. The point x_2 must lie in the neighborhood $|x - x_0| < 1/n$ also, since it lies between x_0 and x, so $|F''(x_2)| < 1/m$ and we have

$$|F(x)| \leq |F''(x_2)|\,|x - x_0|^2 \leq |x - x_0|^2/m$$

as required. QED

We will not discuss the problem of obtaining a converse kind of statement, deducing the existence of the second derivative from the existence of quadratic polynomial approximations, because such theorems are extremely difficult to prove and have few applications.

The theorem we have just established is clearly part two of a more general theorem, which is known as *Taylor's Theorem.* Let us write

$$\begin{aligned} T_n(x_0, x) &= f(x_0) + f'(x_0)(x - x_0) \\ &\quad + \frac{1}{2} f''(x_0)(x - x_0)^2 + \cdots + \frac{1}{n!} f^{(n)}(x_0)(x - x_0)^n, \end{aligned}$$

called the *Taylor expansion* of f at x_0. Here $f^{(n)}$ denotes the nth derivative of f, defined by induction $f^{(n)} = f^{(n-1)\prime}$. *We say f is C^n if all derivatives up to order n exist and are continuous.* We think of x_0 as a fixed point and $T_n(x_0, x)$ as a function of x. It is a polynomial of degree n and is uniquely determined by the requirement that at the point x_0 it agrees with f up to the nth derivative. If we need to discuss more than one function at a time we will write $T_n(f, x_0, x)$ instead of $T_n(x_0, x)$.

Theorem 5.4.5 (*Taylor's Theorem*) *Let f be C^n in a neighborhood of x_0. Then $f - T_n = o(|x - x_0|^n)$ as $x \to x_0$.*

Proof: Setting $F = f - T_n$, we have $F^{(k)}(x_0) = 0$ for $k = 0, 1, \ldots, n$. We need to show $F(x) = o(|x - x_0|^n)$ as $x \to x_0$. We apply the mean value theorem n times to the functions $F, F', \ldots, F^{(n-1)}$ to obtain points $x_1, x_2, \ldots, x_n$ all between x_0 and x_1, such that

$$\begin{aligned} F(x) &= F'(x_1)(x - x_0), \\ F'(x_1) &= F''(x_2)(x_1 - x_0), \ldots \\ F^{(n-1)}(x_{n-1}) &= F^{(n)}(x_n)(x_{n-1} - x_0), \end{aligned}$$

each time using $F^{(k)}(x_0) = 0$, and the hypothesis that f is C^n (hence F is C^n). Altogether we obtain

$$F(x) = F^{(n)}(x_n)(x - x_0)(x_1 - x_0) \cdots (x_{n-1} - x_0),$$

hence $|F(x)| \leq |F^{(n)}(x_n)|\,|x - x_0|^n$ and we complete the proof as before using $F^{(n)}(x_0) = 0$ and the continuity of $F^{(n)}$. QED

Taylor's theorem is extremely useful for understanding the behavior of a function near a point. It is important to realize, however, that the theorem does not assert anything about the behavior of $T_n(x_0, x)$ at a point x as n varies. The parameter n must be fixed in any application of the theorem, and x must vary close to x_0. We will return to this later when we discuss two seemingly related but in fact quite different topics: power series and the Weierstrass approximation theorem.

Taylor's theorem gives us a formula for the Taylor expansion $T_n(x_0, x)$, but it is not always necessary—or advisable—to use this formula to compute the Taylor expansion. We know $T_n(x_0, x)$ is uniquely determined among polynomials of degree n by the condition $f(x) - T_n(x_0, x) = o(|x - x_0|^n)$ as $x \to x_0$, for if we also had $f(x) - g(x) = o(|x - x_0|^n)$ as $x \to x_0$ for g a polynomial of degree n, then

$$\frac{g(x) - T_n(x_0, x)}{(x - x_0)^n} \to 0$$

as $x \to x_0$ and by writing $g(x) - T_n(x_0, x) = \sum_{k=0}^{n} a_k(x - x_0)^k$ we can show first $a_0 = 0$, then $a_1 = 0$, etc., since the lowest order non-zero term of $\sum_{k=0}^{n} a_k(x - x_0)^k$ dominates all the others as $x \to x_0$. This means that if we can obtain, by hook or crook, a polynomial g of degree n such that $f(x) - g(x) = o(|x - x_0|^n)$ as $x \to x_0$, then $g(x) = T_n(x_0, x)$. For example, if f is the product of two functions $f = f_1 \cdot f_2$ and we know the Taylor expansions of order n of f_1 and f_2, then we can obtain the Taylor expansion of order n of f by multiplying the Taylor expansions of f and g and retaining only the terms of degree up to n. Indeed the powers $(x-x_0)^k$ for $k > n$ are all $o(|x-x_0|^n)$ and so may be discarded. We will not state this—or other related results—as a formal theorem but rather enunciate a useful informal principle: *you can operate with Taylor expansions in the same way you can operate with functions, discarding higher order terms.* We will leave as exercises various special cases of this principle.

5.4.3 L'Hôpital's Rule*

As an application, let's look at the notorious L'Hôpital's rule for evaluating limits of quotients, $\lim_{x \to x_0} f(x)/g(x)$ when both limits $\lim_{x \to x_0} f(x)$ and $\lim_{x \to x_0} g(x)$ vanish. We will see that Taylor's Theorem leads to a conceptually clear proof of L'Hôpital's rule and also allows us to answer some related questions, such as: what is the derivative of a quotient f/g at a common zero of f and g?

Suppose f and g are C^1 and $g'(x_0) \neq 0$. We write

$$\begin{aligned} \frac{f(x)}{g(x)} &= \frac{f(x_0) + f'(x_0)(x - x_0) + o(|x - x_0|)}{g(x_0) + g'(x_0)(x - x_0) + o(|x - x_0|)} \\ &= \frac{f'(x_0)(x - x_0) + o(|x - x_0|)}{g'(x_0)(x - x_0) + o(|x - x_0|)} \end{aligned}$$

since we are assuming $f(x_0) = 0$ and $g(x_0) = 0$. Now we would like to cancel the common factor $x - x_0$, which is non-zero if $x \neq x_0$. This will change the $o(|x - x_0|)$ terms, which stand for functions with limit zero at x_0 after dividing by $x - x_0$, to $o(1)$ terms, which stand for functions with limit zero at x_0. Thus

$$\frac{f(x)}{g(x)} = \frac{f'(x_0) + o(1)}{g'(x_0) + o(1)}$$

and if $g'(x_0) \neq 0$ there is no difficulty in taking the limit as $x \to x_0$ to get

$$\lim_{x\to x_0} \frac{f(x)}{g(x)} = \frac{\lim_{x\to x_0}(f'(x_0)+o(1))}{\lim_{x\to x_0}(g'(x_0)+o(1))} = \frac{f'(x_0)}{g'(x_0)}.$$

Now let's ask the question: what is the derivative of $f(x)/g(x)$ at x_0? We assume, of course, that we define the value of $f(x)/g(x)$ at x_0 to be $f'(x)/g'(x)$ so as to have a continuous function. It is not obvious that $f(x)/g(x)$ is differentiable at x_0, and even if it is, the usual quotient formula for the derivative will not be very helpful. Instead we want to look at the difference quotient and use Taylor's theorem to find its limit. The difference quotient at x_0 is

$$\frac{1}{x-x_0}\left[\frac{f(x)}{g(x)} - \frac{f'(x_0)}{g'(x_0)}\right]$$

because $f'(x_0)/g'(x_0)$ is the value of the function at x_0. Let us assume that f and g are C^2 so that we can take the Taylor expansions to order 2. We obtain

$$\frac{1}{x-x_0}\left[\frac{f(x)}{g(x)} - \frac{f'(x_0)}{g'(x_0)}\right] = \frac{f(x)g'(x_0) - f'(x_0)g(x)}{(x-x_0)g(x)g'(x_0)} = \frac{A(x)}{B(x)}$$

where

$$\begin{aligned} A(x) &= g'(x_0)\left[f'(x_0)(x-x_0) + \frac{1}{2}f''(x_0)(x-x_0)^2 + o(|x-x_0|^2)\right] \\ &\quad -f'(x_0)\left[g'(x_0)(x-x_0) + \frac{1}{2}g''(x_0)(x-x_0)^2 + o(|x-x_0|^2)\right] \end{aligned}$$

and

$$B(x) = (x-x_0)g'(x_0)\left[g'(x_0)(x-x_0) + \frac{1}{2}g''(x_0)(x-x_0)^2 + o(|x-x_0|^2)\right]$$

since $f(x_0) = g(x_0) = 0$. Notice that the terms $f'(x_0)g'(x_0)(x-x_0)$ in the numerator cancel, and then we may factor out $(x-x_0)^2$ in the numerator and denominator to obtain simply

$$\frac{g'(x_0)\left[\frac{1}{2}f''(x_0)+o(1)\right] - f'(x_0)\left[\frac{1}{2}g''(x_0)+o(1)\right]}{g'(x_0)\left[g'(x_0) + \frac{1}{2}g''(x_0)(x-x_0) + o(|x-x_0|)\right]}.$$

It is clear that this has a limit as $x \to x_0$ equal to

$$\frac{g'(x_0)f''(x_0) - f'(x_0)g''(x_0)}{2g'(x_0)^2}$$

by the quotient formula for limits, so f/g is differentiable at x_0 and this is its derivative. We can also compute higher derivatives of f/g at x_0 by similar arguments; to get the nth derivative we need to assume that f and g are C^{n+1}.

So far we have been dealing only with the case $g'(x_0) \neq 0$. If $g'(x_0) = 0$ but $f'(x_0) \neq 0$ we can easily show that f/g does not have a finite limit as $x \to x_0$. If both $f'(x_0)$ and $g'(x_0)$ are zero, then we can go to higher order Taylor expansions (assuming more derivatives of f and g exist) to compute the limit of f/g. If f and g are C^n and $f^{(k)}(x_0) = 0$ and $g^{(k)}(x_0) = 0$ for $k = 0, 1, \ldots, n-1$ but $g^{(n)}(x_0) \neq 0$, then

$$\frac{f(x)}{g(x)} = \frac{\frac{1}{n!}f^{(n)}(x_0) + o(|x - x_0|^n)}{\frac{1}{n!}f^{(n)}(x_0) + o(|x - x_0|^n)} = \frac{f^{(n)}(x_0) + o(1)}{g^{(n)}(x_0) + o(1)},$$

which clearly has limit $f^{(n)}(x_0)/g^{(n)}(x_0)$ as $x \to x_0$. Using this generalized L'Hôpital's rule with $n = 2$ on the quotient

$$\frac{f'(x)g(x) - f(x)g'(x)}{g(x)^2},$$

which is the derivative of f/g for $x \neq x_0$, we can show, under the assumptions that f and g are C^3 and $f(x_0) = g(x_0) = 0$ but $g'(x_0) \neq 0$, that

$$\lim_{x \to x_0} \left(\frac{f}{g}\right)' = \frac{g'(x_0)f''(x_0) - f'(x_0)g''(x_0)}{2g'(x_0)^2}.$$

Since this is the value we computed for $(f/g)'$ at x_0, we have the continuity of $(f/g)'$ at x_0. We leave the details as an exercise.

The point of the above applications (and a number of exercises to follow) is that Taylor's theorem reduces certain kinds of problems to rather straightforward computations. Whenever the issue is the local behavior of a function near a point, it is the first technique you should try. Incidentally, I have not given the best possible results for L'Hôpital's rule. There are slightly weaker hypotheses that will also do.

5.4.4 Lagrange Remainder Formula*

For many applications of Taylor's theorem one needs a more precise form for the remainder, or error, term $o(|x-x_0|^n)$. We give now one such expression, the Lagrange remainder formula. We will give an integral remainder formula in the next chapter. The Lagrange remainder formula is really a generalization of the mean value theorem since it involves the value of a higher derivative of the function at an unspecified point. If we write the mean value theorem as $f(x) = f(x_0)+f'(x_1)(x-x_0)$ for some x_1 between x_0 and x, we can interpret this as a zero-order Taylor theorem with remainder $f'(x_1)(x-x_0)$. Note that the remainder looks exactly like the *next* term in the Taylor expansion except for the one change that the point x_1 appears instead of x_0 when we evaluate f'. This suggests the generalization

$$f(x) = T_n(x_0, x) + \frac{1}{(n+1)!} f^{(n+1)}(x_1)(x-x_0)^{n+1}$$

for some point x_1 between x_0 and x. Note that this would give the error as $O(|x-x_0|^{n+1})$, which is somewhat stronger than $o(|x-x_0|^n)$ (if $f(x) - T_n(x_0, x) = O(|x-x_0|^{n+1})$ then

$$\frac{f(x) - T_n(x_0, x)}{|x-x_0|^n} = O(|x-x_0|),$$

so it goes to zero as $x \to x_0$) but would also require assuming f is C^{n+1} since the derivative of order $n+1$ is involved (as in the case of the mean value theorem we could get away without assuming the continuity of $f^{(n+1)}$).

Lagrange Remainder Theorem *Suppose f is C^{n+1} in a neighborhood of x_0. Then for every x in the neighborhood there exists x_1 between x_0 and x such that*

$$f(x) = T_n(x_0, x) + \frac{1}{(n+1)!} f^{(n+1)}(x_1)(x-x_0)^{n+1}.$$

Proof: It would appear that we should attempt to give a proof by induction, based on the mean value theorem. However, if you look back at the proof that we gave for Taylor's theorem, you will see that we do not quite get the desired form for the remainder. Therefore we will take a different approach. We ask what could make the difference $f(x) - T_n(x_0, x)$ as bad as possible? If $f^{(n+1)}(x)$ were identically zero, then f would be a polynomial of degree n and so f would equal $T_n(x_0, x)$ exactly. It would appear then, that to make $f(x) - T_n(x_0, x)$ big we need to have $f^{(n+1)}(x)$ big, and it seems reasonable that taking $f^{(n+1)}(x)$ equal to a constant M would do the most damage among all possibilities with $f^{(n+1)}(x)$ bounded by $|M|$. Of course if $f^{(n+1)}(x) = M$ for all x, then f is a polynomial of degree $n + 1$ and so

$$f(x) = T_{n+1}(x_0, x) = T_n(x_0, x) + \frac{1}{(n+1)!} f^{(n+1)}(x_1)(x - x_0)^{n+1}$$

for any x_1. More generally, if M_+ and M_- denote the sup and inf of $f^{(n+1)}$ on the interval between x_0 and x, then it is reasonable to expect that $f(x)$ should lie between the extremes

$$T_n(x_0, x) + \frac{1}{(n+1)!} M_+ (x - x_0)^{n+1}$$

and

$$T_n(x_0, x) + \frac{1}{(n+1)!} M_- (x - x_0)^{n+1}.$$

This will allow us to complete the proof since $f^{(n+1)}$ assumes all values between M_- and M_+ on the interval between x_0 and x.

Suppose for simplicity that $x > x_0$. We need to show that

$$\frac{1}{(n+1)!} M_- (x - x_0)^{n+1} \leq f(x) - T_n(x_0, x) \leq \frac{1}{(n+1)!} M_+ (x - x_0)^{n+1}$$

where $M_- \leq f^{(n+1)} \leq M_+$ on $[x_0, x]$. Let us write

$$g(x) = f(x) - T_n(x_0, x) - \frac{1}{(n+1)!} M_- (x - x_0)^{n+1}.$$

We need to show $g(x) \geq 0$; in fact we will show g is non-negative on the interval $[x_0, x]$. Now the way we have constructed g—subtracting off the nth order Taylor expansion from f and then subtracting $M_-(x - x_0)^{n+1}/(n + 1)!$ guarantees that g and all its derivatives up to order n vanish at x_0 (all these derivatives of $(x - x_0)^{n+1}$ vanish at

x_0), and furthermore $g^{(n+1)}$ is non-negative on the interval $[x_0, x]$ because $g^{(n+1)} = f^{(n+1)} - M_-$ and M_- is the inf of $f^{(n+1)}$ on the interval (notice that here the factor $1/(n+1)!$ cancels the $(n+1)$-derivative of $(x - x_0)^{n+1}$).

Next we use reverse induction to show $g^{(n)}, g^{(n-1)}, \ldots, g', g$ are all non-negative on the interval. First we note that $g^{(n)}(x_0) = 0$ and $g^{(n)}$ is monotone increasing because its derivative $g^{(n+1)}$ is non-negative, so $g^{(n)}$ is non-negative. Once we know $g^{(n)}$ is non-negative we can apply the same reasoning to $g^{(n-1)}$ and so on. Finally when we have $g(x) \geq 0$ we have established $M_-(x - x_0)^{n+1}/(n+1)! \leq f(x) - T_n(x_0, x)$. The other inequality $f(x) - T_n(x_0, x) \leq M_+(x - x_0)^{n+1}/(n+1)!$ is established by analogous reasoning.

The case $x < x_0$ is actually a little more complicated, because we have the same estimates when $n + 1$ is even but the reverse estimates

$$\begin{aligned} \frac{1}{(n+1)!} M_+(x - x_0)^{n+1} &\leq f(x) - T_n(x_1, x_0) \\ &\leq \frac{1}{(n+1)!} M_-(x - x_0)^{n+1} \end{aligned}$$

when $n + 1$ is odd (when $(x - x_0)^{n+1}$ is negative). We will leave the details as an exercise. Notice that the reversal of the estimates does not in any way destroy the final step in the argument, that we must have $f(x) - T_n(x_0, x) = f^{(n+1)}(x_1)/(n+1)!$ for some x_1 because $f^{(n+1)}$ assumes all values between M_- and M_+. QED

5.4.5 Orders of Zeros*

Taylor's theorem allows us to generalize many concepts and theorems about polynomials to more general functions that are sufficiently differentiable. Here we discuss the notion of the *order* of a zero. If g is a polynomial and $g(x_0) = 0$, the order of the zero at x_0 is defined to be the highest integer k such that $(x - x_0)^k$ divides $g(x)$; in other words $g(x) = g_1(x)(x - x_0)^k$ for some polynomial $g_1(x)$ and $g_1(x_0) \neq 0$. Now it is simple algebra to show that the order k is characterized by the fact that $g(x_0) = 0, g'(x_0) = 0, \ldots, g^{(k-1)}(x_0) = 0$ but $g^{(k)}(x_0) \neq 0$. If the order is one we say x_0 is a *simple zero*. The order gives the number of times $(x - x_0)$ divides the polynomial. It also gives the rate at which $g(x)$ tends to zero as $x \to x_0$; namely, $g(x) = O(|x - x_0|^k)$.

Now let f be an arbitrary function that vanishes at x_0 and is of class C^k. We can say that f has a *zero of order* k at x_0 if $f(x_0) = 0$, $f'(x_0) = 0, \ldots, f^{(k-1)}(x_0) = 0$ but $f^{(k)}(x_0) \neq 0$; in other words if the polynomial $T_k(x_0, x)$ has a zero of order k at x_0. Zeroes of order 1, 2, 3 are shown in Figure 5.4.3. Taylor's theorem then says $f(x) = a_k(x - x_0)^k + o(|x - x_0|^k)$ where $a_k = f^{(k)}(x_0)/k!$ is non-zero, so $f(x) = O(|x - x_0|^k)$ as $x \to x_0$. Thus the rate at which f tends to zero is the same as for a polynomial with the same order zero. We can also deduce that if the zero is of odd order the functions must change sign near x_0, while if the zero is of even order it does not change sign near x_0.

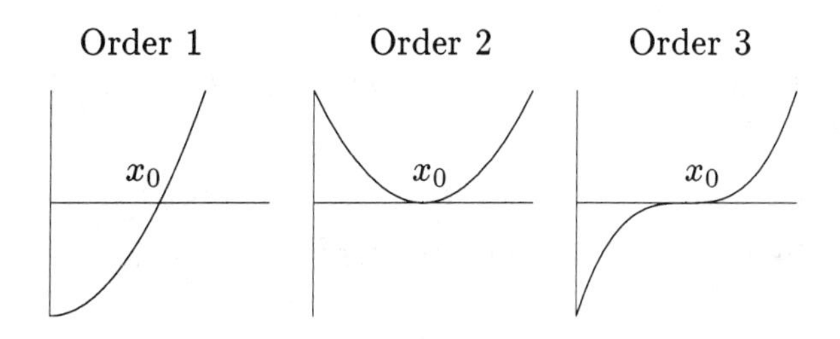

Figure 5.4.3:

Finally we can also prove an analogue of factorization: $f(x) = f_1(x)(x - x_0)^k$ where $f_1(x)$ is a continuous function if f has zero of order k at x_0. This is in fact a variation on l'Hôpital's rule, which we leave as an exercise.

We can paraphrase this discussion by saying *the zeroes of order* k *of an arbitrary function are qualitatively like* $(x - x_0)^k$. A natural question that arises is: are there any other kinds of zeroes? If we call the zeroes we have been discussing *zeroes of finite order*, it turns out that there are also zeroes of infinite order. For this we must assume the function is C^∞, which means that derivatives of all orders exist (hence must be continuous). A C^∞ function is said to have a zero of infinite order at x_0 if $f^{(n)}(x_0) = 0$ for all n—so that all the Taylor expansions $T_n(x_0, x)$ vanish identically. It is not obvious that such zeros exist (except for $f \equiv 0$), but the function e^{-1/x^2} at $x = 0$ gives one example. We will discuss this further in the chapter on transcendental functions.

5.4.6 Exercises

1. Suppose f is a C^2 function on an interval (a, b) and the graph of f lies above every secant line. Prove that $f''(x) \le 0$ on the interval.

2. Suppose $f'(x_0) = 0, f''(x_0) = 0, \ldots, f^{(n-1)}(x_0) = 0$ and $f^{(n)}(x_0) > 0$, for a C^n function f. Prove that f has a local minimum at x_0 if n is even and that x_0 is neither a local maximum nor a local minimum if n is odd.

3. If f is C^2 on an interval prove that

$$\lim_{h \to 0} \frac{f(x+h) - 2f(x) + f(x-h)}{h^2} = f''(x).$$

The expression $f(x+h) - 2f(x) + f(x-h)$ is called the *symmetric second difference.*

4. Let f and g be C^3 functions with $f(x_0) = g(x_0) = 0$ but $g'(x_0) \neq 0$. Show that the derivative of f/g is continuous at x_0.

5. Under the same hypotheses as exercise 4 show that $(f/g)''$ exists at x_0 and compute it.

6. If f and g are C^{n+1} and $f(x_0) = g(x_0) = 0$ but $g'(x_0) \neq 0$ show that f/g is C^n near x_0 and find a formula for $(f/g)^{(n)}(x_0)$.

7. If f and g are C^3 functions and $f(x_0) = f'(x_0) = g(x_0) = g'(x_0) = 0$ but $g''(x_0) \neq 0$ show that $(f/g)'$ exists at x_0 and compute it.

8. For each of the following functions defined for $x \neq 0$ find $\lim_{x \to 0} f(x)$ and $f'(0)$ if the function is appropriately defined at $x = 0$. You may use the familiar formulas for derivatives of sine and cosine:

   a. $f(x) = \sin x/x$,
   b. $f(x) = (1 - \cos x)/x^2$,
   c. $f(x) = (x^2 - x)/\sin x$,
   d. $f(x) = x/(1 - \cos x - \sin x)$.

9. Suppose f is a C^n function on an interval and $T_n(x_0, x)$ is the same function of x for all x_0 in the interval. What can you say about f?

10. Complete the proof of the Lagrange remainder formula for $x < x_0$.

11. Suppose f and g are C^n functions with Taylor expansions denoted $T_n(f, x_0, x)$ and $T_n(g, x_0, x)$. Prove that $T_n(f, x_0, x) + T_n(g, x_0, x)$ is the Taylor expansion of $f + g$ at x_0.

12. Under the same hypotheses as exercise 11, show that the Taylor expansion of $f \cdot g$ at x_0 is obtained by taking $T_n(f, x_0, x)T_n(g, x_0, x)$ and retaining only the powers of $(x - x_0)$ up to n.

13. Suppose f is a C^n function in a neighborhood of x_0 and g is C^n in a neighborhood of $f(x_0)$. Let $\sum_{k=0}^n a_k(x - x_0)^k$ be $T_n(f, x_0, x)$, and let $\sum_{j=0}^n b_j(y - y_0)^j$ be $T_n(g, y_0, y)$ where $y_0 = f(x_0)$. Show that $T_n(g \circ f, x_0, x)$ is obtained from

$$b_0 + \sum_{j=1}^n b_j \left(\sum_{k=1}^n a_k (x - x_0)^k \right)^j$$

by retaining only the powers of $(x - x_0)$ up to n.

14. If $f(x) = 1/(1 + x)$ show that $T_n(f, 0, x) = \sum_{k=0}^n (-1)^k x^k$.

15. Suppose f is a C^n function in a neighborhood of x_0 and suppose $f(x_0) \neq 0$. Let $T_n(f, x_0, x) = \sum_{k=0}^n a_k(x - x_0)^k$. Show that $T_n(1/f, x_0, x)$ is obtained from

$$\frac{1}{a_0} \left(1 + \sum_{j=1}^n (-1)^j \left(\sum_{k=1}^n \frac{a_k}{a_0} (x - x_0)^k \right)^j \right)$$

by retaining only the powers of $(x - x_0)$ up to n. (**Hint**: use exercises 13 and 14.)

16. Compute the Taylor expansions to order 3 for each of the following functions at the points $x_0 = 0$ and $x_0 = 1$:

    a. $f(x) = (x^2 + 1)^{25}$,
    b. $f(x) = x/(x^2 + 1)$,
    c. $f(x) = (1 + x + 2x^2)\sin^2 \pi x$,
    d. $f(x) = \cos(1 + x^2)$.

17. Suppose f is C^1 on an interval and f' satisfies the Hölder condition of order α, $|f'(x)-f'(y)| \leq M|x-y|^\alpha$ for all x and y in the interval, where α is a fixed value, $0 < \alpha \leq 1$. Show that $|\Delta_h^2 f(x)| \leq c|h|^{1+\alpha}$. How does the constant c relate to the constant M?

18. Let f be a C^n function. Show that the derivative of $T_n(f, x_0, x)$ is equal to $T_{n-1}(f', x_0, x)$.

19. Suppose f has a zero of order j at x_0 and g has a zero of order k at x_0. What can you say about the order of zero of the function $f+g, f \cdot g, f/g$ at x_0?

20. Use the second-order Taylor theorem with Lagrange remainder to estimate $\sqrt{101}$.

21. Apply Taylor's theorem with Lagrange remainder to $(x+y)^a$ for a rational to obtain a form of the binomial theorem.

22. *a. Let f be a C^2 function on $[a, b]$ and let $x_1, \ldots, x_n$ be points in $[a, b]$. Show that $f''(x) \geq 0$ on $[a, b]$ implies

$$f\left(\frac{1}{n}(x_1 + \cdots + x_n)\right) \leq \frac{1}{n}(f(x_1) + \cdots + f(x_n))$$

while $f''(x) \leq 0$ implies the reverse inequality.

b. More generally, let $p_1, \ldots, p_n$ be positive and satisfy the condition $p_1 + \cdots + p_n = 1$. Show $f''(x) \geq 0$ on $[a, b]$ implies $f(p_1x_1 + \cdots + p_nx_n) \leq p_1f(x_1) + \cdots + p_nf(x_n)$, while $f''(x) \leq 0$ implies the reverse inequality.

23. a. If f is C^n on an interval and has $n+1$ distinct zeroes, prove that $f^{(n)}$ has at least one zero on the interval.

b. If f is C^n on an interval and $f^{(n)}$ never vanishes, then f has at most n zeroes on the interval.

c. A polynomial of degree n has at most n real zeros.

d. *If f is C^2 on an interval and x_1, x_2, x_3 are three distinct points on the interval, then there exists y in the interval with

$$\begin{aligned}&(x_1 - x_2)f(x_3) + (x_2 - x_3)f(x_1) + (x_3 - x_1)f(x_2)\\ &= -\frac{1}{2}f''(y)(x_1 - x_2)(x_2 - x_3)(x_3 - x_1).\end{aligned}$$

(**Hint:** subtract a quadratic polynomial to reduce to part a).

24. If f is C^2, prove that f cannot have a local maximum or minimum at an inflection point (note that an inflection point is defined as a point where f'' changes sign; it is not enough that f'' vanish at the point).

5.5 Summary

5.1 Concepts of the Derivative

Definition 5.1.1 *A function f defined in a neighborhood of x_0 is said to be differentiable at x_0 with derivative $f'(x_0)$ if*

$$\lim_{x \to x_0} \frac{f(x) - f(x_0)}{x - x_0} = f'(x_0)$$

or, equivalently, if for every m there exists n such that $|x - x_0| < 1/n$ implies $|f(x) - g(x)| \leq |x - x_0|/m$ where $g(x) = f(x_0) + f'(x_0)(x - x_0)$, called the best affine approximation to f at x_0.

Definition 5.1.2 *For functions f and g defined in a neighborhood of x_0, we say $f(x) = O(g(x))$ as $x \to x_0$ if $|f(x)| \leq c|g(x)|$ for some constant c in a neighborhood of x_0. We say $f(x) = o(g(x))$ as $x \to x_0$ if $\lim_{x \to x_0} f(x)/g(x) = 0$.*

Theorem *If f is differentiable at x_0, then f is continuous at x_0.*

Definition *A function is said to be differentiable on an open set if it is differentiable at each point of the set. It is said to be continuously differentiable (C^1) if the derivative is a continuous function on the set.*

Example

$$f(x) = \begin{cases} x^2 \sin(1/x^2), & x \neq 0, \\ 0, & x = 0, \end{cases}$$

is differentiable but not C^1.

5.2 Properties of the Derivative

Definition 5.2.1 *A function f defined in a neighborhood of x_0 is said to be monotone (resp. strictly) increasing at x_0 if there exists a neighborhood of x_0 on which $x_1 < x_0 < x_2$ implies $f(x_1) \leq f(x_0) \leq f(x_2)$ (resp. $f(x_1) < f(x_0) < f(x_2)$). It is said to have a local maximum (resp. strict local maximum) if there exists a neighborhood of x_0 on which $f(x) \leq f(x_0)$ (resp. $f(x) < f(x_0)$ for $x \neq x_0$). A function defined on an interval is said to be monotone (resp. strictly) increasing on the interval if $x < y$ implies $f(x) \leq f(y)$ (resp. $f(x) < f(y)$) for all x and y in the interval. Similar definitions apply to monotone and strict decreasing and local minimum and strict local minimum by reversing the inequalities.*

Theorem 5.2.1 *$f'(x_0) > 0$ implies f is strictly increasing at x_0. If f is monotone increasing at x_0, then $f'(x_0) \geq 0$. If f has a local maximum at x_0, then $f'(x_0) = 0$.*

Intermediate Value Theorem *If f is differentiable on (a, b), then $f'(x)$ assumes all values between $f'(x_1)$ and $f'(x_2)$ on the interval (x_1, x_2).*

Mean Value Theorem *If f is continuous on $[a, b]$ and differentiable on (a, b), then $f'(x_0) = (f(b) - f(a))/(b - a)$ for some x_0 in (a, b).*

Theorem 5.2.2 *Let f be differentiable on (a, b).*

1. f is monotone increasing on (a, b) if and only if $f'(x) \geq 0$ on (a, b).
2. *$f'(x) > 0$ on (a, b) implies f is strictly increasing on (a, b).*
3. *$f'(x) = 0$ on (a, b) implies f is constant on (a, b).*

Theorem *If f is differentiable on (a, b) and f' is bounded, then f satisfies a Lipschitz condition uniformly on (a, b).*

5.3 The Calculus of Derivatives

Theorem 5.3.1 *If f and g are differentiable at x_0, then so are $f \pm g$, $f \cdot g$, and f/g (if $g(x_0) \neq 0$) with the familiar formulas for the derivatives.*

Theorem 5.3.2 (*Chain Rule*) *If f is differentiable at x_0 and g is differentiable at $f(x_0)$, then $g \circ f$ is differentiable at x_0 and $(g \circ f)'(x_0) = g'(f(x_0))f'(x_0)$.*

Inverse Function Theorem *Let f be C^1 on (a,b) with image (c,d), and suppose $f'(x) > 0$ on (a,b) (or $f'(x) < 0$ on (a,b)). Then f^{-1} exists on (c,d) and is C^1 with $(f^{-1})'(y) = 1/f'(x)$ if $y = f(x)$.*

Local Inverse Function Theorem *Let f be C^1 in a neighborhood of x_0 with $f'(x_0) \neq 0$. Then there exists a neighborhood (a,b) of x_0 such that f restricted to (a,b) has a C^1 inverse on $(c,d) = f((a,b))$.*

5.4 Higher Derivatives and Taylor's Theorem

Definition *If f' is defined in a neighborhood of x_0 and differentiable at x_0, then f is said to be twice differentiable at x_0 with second derivative $f''(x_0)$ equal to $(f')'(x_0)$. If $f''(x)$ exists and is continuous on an interval we say f is twice continuously differentiable (C^2).*

Theorem 5.4.1 *Suppose $f''(x_0)$ exists, and let $g(x) = f(x_0) + f'(x_0)(x - x_0)$ be the best affine approximation to f at x_0.*

1. *If $f''(x_0) > 0$, then $f(x) > g(x)$ on a neighborhood of x_0, for $x \neq x_0$ (the graph of f lies above the tangent line).*

2. *If $f(x) \geq g(x)$ in a neighborhood of x_0, then $f''(x_0) \geq 0$.*

3. *If $f'(x_0) = 0$ and $f''(x_0) > 0$, then x_0 is a strict local minimum.*

4. *If x_0 is a local minimum, then $f''(x_0) \geq 0$.*

Theorem 5.4.2 *If f is C^2 on (a,b) and $f''(x) > 0$ on (a,b), then the graph of f lies below any secant line.*

Theorem 5.4.3 *If f is C^2 on an interval, then*

$$\lim_{h\to 0}\frac{\Delta_h^2(x)}{h^2} = f''(x)$$

on the interval.

Theorem 5.4.4 *If f is C^2 in a neighborhood of x_0, then $f(x)-g_2(x) = o(|x-x_0|^2)$ as $x \to x_0$ where*

$$g_2(x) = f(x_0) + f'(x_0)(x-x_0) + \frac{1}{2}f''(x_0)(x-x_0)^2.$$

Definition *$f^{(n)}(x_0) = (f^{(n-1)})'(x_0)$ by induction, if it exists. We say f is C^n if $f^{(k)}$ exists and is continuous for all $k \le n$. The Taylor expansion of order n at x_0 for a C^n function is defined by*

$$T_n(x_0, x) = \sum_{k=0}^{n}\frac{1}{k!}f^{(k)}(x_0)(x-x_0)^k.$$

Theorem 5.4.5 (*Taylor's Theorem*) *If f is C^n, then $f(x)-T_n(x_0,x) = o(|x-x_0|^n)$ as $x \to x_0$.*

L'Hôpital Rule *If f and g are C^1 with $f(x_0) = g(x_0) = 0$ but $g'(x_0) \neq 0$, then*

$$\lim_{x\to x_0}\frac{f(x)}{g(x)} = \frac{f'(x_0)}{g'(x_0)}.$$

If f and g are C^2, then f/g is differentiable at x_0 with

$$\left(\frac{f}{g}\right)'(x_0) = \frac{g'(x_0)f''(x_0) - f'(x_0)g''(x_0)}{2g'(x_0)^2}.$$

If f and g are C^3, then f/g is C^1.

Lagrange Remainder Theorem *If f is C^{n+1}, then*

$$f(x) = T_n(x_0, x) + \frac{1}{(n+1)!}f^{(n+1)}(x_1)(x-x_0)^{n+1}$$

for some x_1 between x_0 and x.

Definition *If f is C^k we say f has a zero of order k at z_0 if $f(x_0) = f'(x_0) = \cdots = f^{(k-1)}(x_0) = 0$ but $f^{(k)}(x_0) \neq 0$.*

Theorem *If f has a zero of order k at x_0, then $f(x) = O(|x - x_0|^k)$ as $x \to x_0$.*

Chapter 6

Integral Calculus

6.1 Integrals of Continuous Functions

6.1.1 Existence of the Integral

In this section we will prove the existence of the definite integral $\int_a^b f(x)\,dx$ of a continuous function f on a compact interval $[a,b]$, following the usual approach. That is, we partition the interval into n subintervals $[x_{k-1}, x_k]$ where $a = x_0 < x_1 < \cdots < x_n = b$ and form the approximating sum $\sum_{k=1}^{n} f(x_k)(x_k - x_{k-1})$. The value $f(x_k)(x_k - x_{k-1})$ is the area of the rectangle shown in Figure 6.1.1 with base $[x_{k-1}, x_k]$ and height $f(x_k)$, which should be approximately the area under the graph of f over the interval $[x_{k-1}, x_k]$ if f does not vary much. Thus, the approximating sum should be close to the area under the graph of f (with the convention that regions below the x-axis are counted with a minus sign). The integral should then be obtained as a limit of these approximating sums as the size of the subintervals decreases to zero, and the number n of subintervals increases without bound. The notion of "limit" we are using here is somewhat different from the notion of a limit of a function or limit of a sequence, but it is very much in the same spirit. We will give the precise definition below, but before doing so we need to discuss some of the intuitive ideas that will be needed in the proof and that will also play a role in our later extension of the notion of integral to more general (not necessarily continuous) functions.

The choice of the point x_k at which we evaluate f is somewhat arbitrary. We could just as well choose any other point a_k in the interval

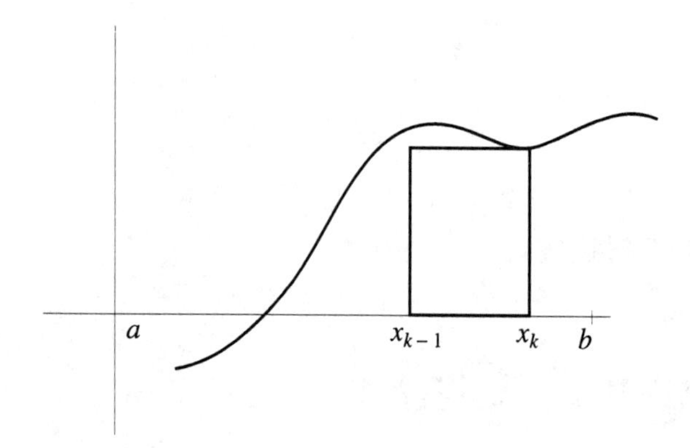

Figure 6.1.1:

$[x_{k-1}, x_k]$. In fact let us write $S(f, P) = \sum_{k=1}^{n} f(a_k)(x_k - x_{k-1})$ where P denotes the partition and a_k is any point in $[x_{k-1}, x_k]$ and call this a *Cauchy sum.* Two important special cases are when we choose $f(a_k)$ as large (or as small) as possible. If M_k denotes the sup of f on $[x_{k-1}, x_k]$ and m_k the inf, then $\sum_{k=1}^{n} M_k(x_k - x_{k-1})$ and $\sum_{k=1}^{n} m_k(x_k - x_{k-1})$ are the largest and smallest Cauchy sums we can form (since f is assumed continuous, it attains its sup and inf on the closed subintervals). We call these the *upper and lower Riemann sums*, written

$$S^+(f, P) = \sum_{k=1}^{n} M_k(x_k - x_{k-1}),$$

$$S^-(f, P) = \sum_{k=1}^{n} m_k(x_k - x_{k-1}).$$

(See Figure 6.1.2). Clearly any Cauchy sum $S(f, P)$ must lie between these. Also, from the intuitive properties of area, it is clear that the area under the graph of f must lie somewhere in between the upper and lower Riemann sums.

Definition 6.1.1 *For any partition P, the maximum interval length is the maximum length of the subintervals of the partition. We say that the limit of $S(f, P)$ exists and equals the number $\int_a^b f(x)\,dx$ if for every error $1/N$ there exists $1/m$ such that $|S(f, P) - \int_a^b f(x)\,dx| \le 1/N$ for any partition P with maximum interval length $\le 1/m$.*

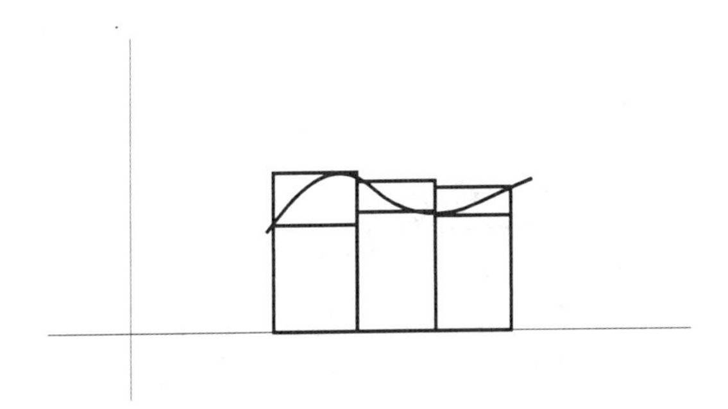

Figure 6.1.2:

Since the "area under the graph of f" is only an intuitive concept, we cannot use it to prove the existence of the integral; rather we want to use the integral to give a precise mathematical counterpart to the intuitive concept of area. Our strategy will be to show that the upper and lower Riemann sums converge to a common limit. Then the intuitive "area" should also be equal to this limit, because it is squeezed in between. The same argument will show that any Cauchy sums $S(f,P)$ will converge to the limit. Thus we need to show that $S^+(f,P)$ and $S^-(f,P)$ converge to a common limit as the size of the intervals in P tends to zero.

Why is this true? Let's first look at a somewhat simpler question that contains the crux of the matter: what will make $S^+(f,P) - S^-(f,P)$ small? The difference of the upper and lower Riemann sums is the area of the little rectangles in Figure 6.1.3, or $\sum_{k=1}^{n}(M_k - m_k)(x_k - x_{k-1})$. The sum of the lengths of the bases of these rectangles is clearly the length of the full interval $b - a = \sum_{k=1}^{n}(x_k - x_{k-1})$, which does not depend on the particular partition. The heights of the rectangles, $M_k - m_k$, are the variations of the function f over the subintervals $[x_{k-1}, x_k]$, and these can be made small by taking the subintervals sufficiently small, by the continuity of f. Since we need to make them all small simultaneously, it is clear that we should use uniform continuity. Because f is assumed continuous on the compact interval $[a,b]$, we know that f is uniformly continuous: given any error $1/N$ there exists $1/m$ such that $|x - y| < 1/m$ implies $|f(x) - f(y)| < 1/N$.

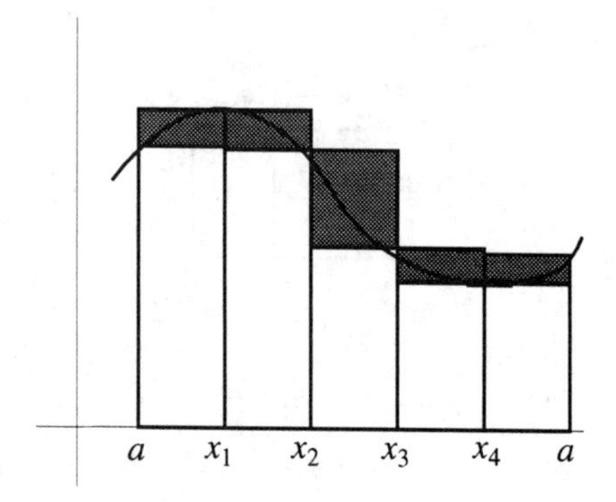

Figure 6.1.3:

Thus if we choose the partition P such that $x_k - x_{k-1} < 1/m$ for all k, then $M_k - m_k < 1/N$ since M_k and m_k are values assumed by f on $[x_{k-1}, x_k]$. Thus

$$\begin{aligned} S^+(f,P) - S^-(f,P) &= \sum_{k=1}^{n}(M_k - m_k)(x_k - x_{k-1}) \\ &\leq 1/N \sum_{k=1}^{n}(x_k - x_{k-1}) = 1/N(b-a), \end{aligned}$$

which can be made as small as desired since $b - a$ is fixed.

Being able to make the upper and lower Riemann sums close to each other for a fixed partition does not in itself imply the existence of the limit defining the integral. It is still conceivable that the values of these Riemann sums vary a lot as we vary the partition. This in fact does not happen, but to see why we need a new idea: the upper Riemann sums decrease and the lower Riemann sums increase when we add points to the partitions. Let us say that a partition P_1 is a *refinement* of P if P_1 contains P (we think of the partitions as consisting of the endpoints of the subintervals); in other words P_1 is obtained by further partitioning the subintervals of P. Since the sup of f over the smaller subintervals of P_1 will be smaller than the sup of f over the containing subintervals of P, we will have $S^+(f, P_1) \leq S^+(f, P)$. Similarly $S^-(f, P_1) \geq S^-(f, P)$. Thus if we consider a sequence of partitions $P_1, P_2, \ldots$, each of which is a refinement of the previous one, the sequence $\{S^+(f, P_k)\}$ will be monotone decreasing and $\{S^-(f, P_k)\}$ will be monotone increasing, so each will have a limit (the limits are finite since all Cauchy sums are bounded above and below by $b - a$ times the sup and inf of f on the interval). If the maximum interval length P_k tends to zero as $k \to \infty$, then the argument above shows $S^+(f, P_k) - S^-(f, P_k) \to 0$, so the two limits must be equal. We can take the limiting value as the definition of $\int_a^b f(x)\,dx$. Of course we then need to verify that the limit does not depend on the particular sequence of partitions $P_1, P_2, \ldots$. This is not hard to accomplish, using the ideas we have already encountered.

Theorem 6.1.1 (*Existence of the Integral*) *Let f be a continuous function on $[a, b]$. Then the limit of the Cauchy sums $S(f, P)$ exists. The integral $\int_a^b f(x)\,dx$ is also equal to the* inf *of the upper Riemann sums $S^+(f, P)$ and the* sup *of the lower Riemann sums $S^-(f, P)$ as P varies over all partitions.*

Proof: Let U denote the set of values of upper Riemann sums $S^+(f, P)$ and L denote the set of values of lower Riemann sums $S^-(f, P)$. We claim that $\inf U = \sup L$. Indeed $\inf U \geq \sup L$ because every element in U is greater than every element of L (if P_1 and P_2 are two partitions and P_3 is the union of the points in P_1 and P_2, so that P_3 is a refinement of both P_1 and P_2, then $S^+(f, P_1) \geq S^+(f, P_3) \geq S^-(f, P_3) \geq S^-(f, P_2)$). If P_k is any sequence of partitions such that P_{k+1} is a refinement of P_k and the maximum interval length of P_k tends to zero, then $\lim_{k\to\infty} S^+(f, P_k) = \lim_{k\to\infty} S^-(f, P_k)$, so we cannot have $\inf U > \sup L$.

Let $\int_a^b f(x)\,dx$ equal the common value of $\inf U$ and $\sup L$. Given the error $1/N$, choose $1/m$ so that $|x-y| < 1/m$ implies $|f(x)-f(y)| \leq 1/N(b-a)$. If P is any partition with maximum interval length at most $1/m$, then

$$\begin{aligned} S^+(f, P) - S^-(f, P) &= \sum_{k=1}^{n} (M_k - m_k)(x_k - x_{k-1}) \\ &\leq 1/N(b-a) \sum_{k=1}^{n} (x_k - x_{k-1}) = 1/N \end{aligned}$$

and so any Cauchy sum $S(f, P)$ lies in the interval $[S^-(f, P), S^+(f, P)]$ of length at most $1/N$. But $\int_a^b f(x)\,dx$ also lies in this interval ($\inf U \leq$

$S^+(f,P)$ and $\sup L \geq S^-(f,P)$), so the two differ by at most $1/N$. QED

From the definition we obtain immediately the linearity properties of the integral:

$$\int_a^b (f(x)+g(x))\,dx = \int_a^b f(x)\,dx + \int_a^b g(x)\,dx$$
$$\int_a^b cf(x)\,dx = c\int_a^b f(x)\,dx, \quad c \text{ constant},$$

and the additivity:

$$\int_a^c f(x)\,dx = \int_a^b f(x)\,dx + \int_b^c f(x)\,dx$$

if $a < b < c$. By defining the integral for $a > b$ by $\int_a^b f(x)\,dx = -\int_b^a f(x)\,dx$ and for $a = b$ by $\int_a^a f(x)\,dx = 0$, we can easily verify that the additivity continues to hold. We also have the basic estimate

$$m(b-a) \leq \int_a^b f(x)\,dx \leq M(b-a)$$

for $a < b$ where M is the sup and m the inf of f on $[a,b]$. This follows from the fact that $M(b-a)$ and $m(b-a)$ are Riemann upper and lower sums for the partition consisting of the single interval $[a,b]$. The expression $\int_a^b f(x)\,dx/(b-a)$ may be interpreted as an "average" value of f on the interval $[a,b]$, so our basic estimate can be interpreted as saying that the average lies between the maximum and the minimum values. This is certainly a property that one would expect an average to have. Another property one would expect from an average is linearity, and this is an immediate consequence of the linearity of the integral. This integral average is an "unbiased" average in that subintervals of equal length contribute to the average equally. For some applications it is desirable to give different weights to different portions of the interval. This can be accomplished by using a continuous, positive weight function $w(x)$ and defining the weighted integral average to be $\int_a^b f(x)w(x)\,dx/\int_a^b w(x)\,dx$. We leave to the exercises the verification that this satisfies the above properties of a reasonable average.

6.1.2 Fundamental Theorems of Calculus

We will frequently want to consider the integral with one of the end-points variable, to obtain a function, say $F(x) = \int_a^x f(t)\,dt$ for $a \leq x \leq b$ if f is continuous on $[a, b]$. We will informally call F the "integral" of f. We make the obvious remark that it is easier for the integral of f to exist than for the derivative of f to exist, contrary to the impression given in calculus courses, where one is impressed by the fact that it is easier to compute a "formula" for the derivative than to compute a "formula" for the integral. Our perspective now is that $\int_a^x f(x)\,dx$ *is* a formula for the integral; this formula does involve an infinite process, but we can always obtain approximations to $\int_a^x f(t)\,dt$ with any desired error. The restriction that f be continuous can be relaxed, and we will discuss this in detail later.

We can now obtain easily the fundamental theorem of the calculus or rather the two fundamental theorems—differentiation of the integral and integration of the derivative.

Theorem 6.1.2 (*Differentiation of the Integral*) *Let f be a continuous function on $[a, b]$, and let $F(x) = \int_a^x f(t)\,dt$ for $a \leq x \leq b$. Then F is C^1 and $F' = f$.*

Proof: From the basic properties of the integral we find the difference quotient for F is

$$\frac{F(x_0 + h) - F(x_0)}{h} = \frac{1}{h} \int_{x_0}^{x_0+h} f(t)\,dt.$$

Now $(1/h) \int_{x_0}^{x_0+h} f(t)\,dt$ is a kind of average value of f on the interval $[x_0, x_0 + h]$, so we would expect it to converge to $f(x_0)$ since f is continuous. (In Figure 6.1.4, it is the shaded area divided by the length of the base.)

More precisely, if $h > 0$, then $\int_{x_0}^{x_0+h} f(t)\,dt$ lies between hM and hm where M and m are the sup and inf of f on $[x_0, x_0 + h]$. Thus $m \leq (1/h) \int_{x_0}^{x_0+h} f(t)\,dt \leq M$. If $h < 0$, then

$$\frac{1}{h} \int_{x_0}^{x_0+h} f(t)\,dt = -\frac{1}{h} \int_{x_0+h}^{x_0} f(t)\,dt$$

and

$$-mh \leq \int_{x_0+h}^{x_0} f(t)\,dt \leq -Mh$$

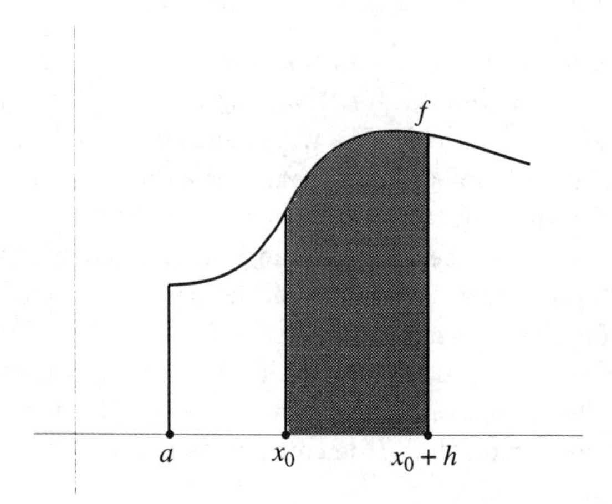

Figure 6.1.4:

where now M and m are the sup and inf of f on $[x_0 + h, x_0]$. Since $-h$ is positive, we obtain again $m \leq (1/h) \int_{x_0}^{x_0+h} f(t)\, dt \leq M$. Since f is continuous, both M and m tend to $f(x_0)$ as $h \to 0$, so

$$\lim_{h \to 0} \frac{1}{h} \int_{x_0}^{x_0+h} f(t)\, dt = f(x_0).$$

This shows $F' = f$; and since f was assumed continuous, we have F is C^1. QED

The antiderivative F is not the unique solution to the equation $g' = f$. However, any other solution must differ from F by a constant, for then $F - g$ would have derivative zero; but we have seen that a function with zero derivative on an interval must be constant.

Note that the previous theorem has a rather trivial analogue involving sums and differences: if $x_1, x_2, \ldots$ is any sequence of numbers and we form the sequence of sums $y_1, y_2, \ldots$ with $y_n = x_1 + x_2 + \cdots + x_n$, then $x_n = y_n - y_{n-1}$.

Similarly, the next theorem (integration of the derivative) is an "infinitesimal" version of the familiar fact that sums of differences "telescope": $\sum_{k=1}^{n} (x_k - x_{k-1}) = x_n - x_0$. It was first stated in the fourteenth century by Nicole Oresme in the form: the area under the graph of velocity is the distance traveled. He justified it by saying that the area was the sum of the areas of the vertical lines—thought of as rectangles with infinitesimal bases—and the areas of the vertical lines represent the distance traveled in an infinitesimal time interval, as indicated in Figure 6.1.5.

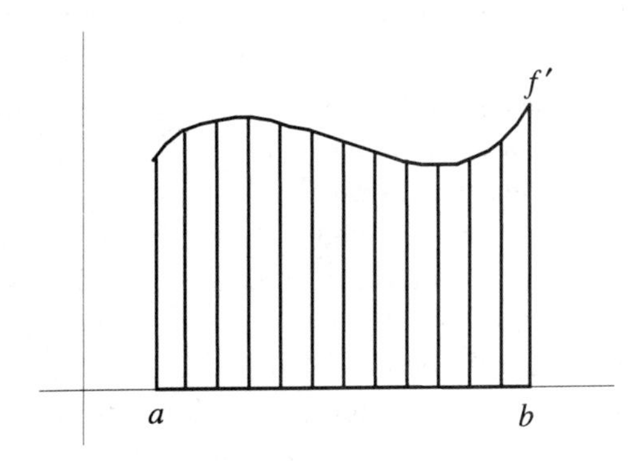

Figure 6.1.5:

Theorem 6.1.3 (*Integration of the Derivative*) *Let f be C^1 on $[a, b]$ (since the interval is closed, the derivative f' at the endpoints is a one-sided derivative). Then $\int_a^b f'(x)\,dx = f(b) - f(a)$.*

Proof: For any partition P of the interval, the Cauchy sum

$$\sum_{k=1}^{n} f'(a_k)(x_k - x_{k-1})$$

adds up terms $f'(a_k)(x_k - x_{k-1})$ that are approximately $f(x_k) - f(x_{k-1})$. In fact, by the mean value theorem, it is possible to choose a_k in the interval $[x_{k-1}, x_k]$ to make this exact equality: $f'(a_k)(x_k - x_{k-1}) = f(x_k) - f(x_{k-1})$. For this choice of a_k the Cauchy sum telescopes, so $S(f', P) = f(b) - f(a)$ exactly. Since we know that $\int_a^b f'(x)\,dx$ is the limit of *any* choice of Cauchy sum $S(f', P)$ as the maximum interval length tends to zero, it follows that it is the limit with this particular choice, hence $f(b) - f(a)$. QED

It is also possible to prove the integration of the derivative as a consequence of the differentiation of the integral, but this proof lacks intuitive appeal. We leave it as an exercise.

It is interesting to compare the integration of the derivative theorem with the mean value theorem. Both give the value of $(f(b) - f(a))/(b - a)$ in terms of the derivative f'. In the mean value theorem the derivative is evaluated at an unspecified point; in the integration of the derivative theorem this is replaced by the integral average $\int_a^b f'(t)\,dt/(b-a)$. In many applications you can use either result. (For example, as an exercise, try reproving the results in Section 5.2.3 by using the integration of the derivative theorem). Although the mean value theorem has the advantage that it holds with weaker hypotheses (the derivative does not have to be continuous), it also has the disadvantage of the unspecified nature of the point. A good rule of thumb is to try the mean value theorem first, but if you run into difficulties, to switch to the integration of the derivative theorem.

The two fundamental theorems enable us to define the *indefinite integral* or *primitive* of the continuous function f to be any C^1 function F such that $F' = f$. The differentiation of the integral theorem shows that an indefinite integral always exists, and we have seen that it is unique up to an additive constant. The integration of the derivative theorem shows us how to evaluate definite integrals using indefinite integrals. The two theorems can also be interpreted as saying the operations of differentiation and integration are inverse operations; however, since differentiation is *not* one-to-one, the inverse indefinite integration operator is only defined up to an additive constant.

As a consequence of the integration of the derivative and the product formula for derivatives we obtain the familiar integration by parts formula.

Theorem 6.1.4 (*Integration by Parts*) *Let f and g be C^1 on $[a, b]$. Then*

$$\int_a^b f(x)g'(x)\,dx = f(b)g(b) - f(a)g(a) - \int_a^b f'(x)g(x)\,dx.$$

Proof: Since $(f \cdot g)' = f'g + fg'$, we have $\int_a^b [f'(x)g(x) + f(x)g'(x)]\,dx = f(b)g(b) - f(a)g(a)$. QED

As an application we derive the integral remainder formula for Taylor's theorem. Assume f is C^{n+1}. We define the remainder $R_n(x_0 x)$ by the equation

$$f(x) = T_n(x_0, x) + R_n(x_0, x)$$

where $T_n(x_0, x)$ is the Taylor expansion of f to order n. We have already established Lagrange's remainder formula

$$R_n(x_0, x) = (x - x_0)^{n+1} f^{(n+1)}(x_1)/(n+1)!$$

where x_1 is some unspecified point between x_0 and x. For some applications, especially when one needs to vary x, the unspecific nature of x_1 causes difficulties (it gives no information about the derivative of the remainder, for example). The integral remainder formula is

$$R_n(x_0, x) = \frac{1}{n!} \int_{x_0}^{x} (x-t)^n f^{(n+1)}(t)\, dt.$$

Note that $\int_{x_0}^{x} (x-t)^n\, dt/n! = (x-x_0)^{n+1}/(n+1)!$, so both the integral remainder and Lagrange remainder are of the same order of magnitude, namely $O(|x-x_0|^{n+1})$. Notice that the Taylor expansion with integral remainder formula for $n = 0$ is $f(x) = f(x_0) + \int_{x_0}^{x} f'(t)\, dt$ just the integration of the derivative. To establish the integral remainder formula in general we simply apply integration by parts to

$$\frac{1}{n!} \int_{x_0}^{x} (x-t)^n f^{(n+1)}(t)\, dt$$

integrating $f^{(n+1)}$ and differentiating $(x-t)^n/n!$ to obtain

$$\begin{aligned} &\frac{1}{n!} \int_{x_0}^{x} (x-t)^n f^{(n+1)}(t)\, dt \\ &\quad = \frac{1}{(n-1)!} \int_{x_0}^{x} (x-t)^{n-1} f^{(n)}(t)\, dt - \frac{1}{n!}(x-x_0)^n f^{(n)}(x_0) \end{aligned}$$

and the proof can be completed by induction (see exercises).

Another important consequence of the fundamental theorems is the familiar change of variable formula.

Theorem 6.1.5 (*Change of Variable Formula*) *Let g be C^1 and increasing on $[a, b]$. Then for any continuous function f on $[g(a), g(b)]$, we have*

$$\int_{g(a)}^{g(b)} f(x)\, dx = \int_a^b f(g(x)) g'(x)\, dx.$$

Proof: Let F be an indefinite integral for f. Then $F(g(x))$ has derivative $f(g(x))g'(x)$ by the chain rule, so $\int_a^b f(g(x))g'(x)\,dx = F(g(b)) - F(g(a))$ by the integration of the derivative theorem. But this is the same value that the same theorem gives for $\int_{g(a)}^{g(b)} f(x)\,dx$ since $F' = f$. QED

As a special case we have the translation invariance of the integral.

Corollary 6.1.1 (*Translation Invariance*) *If f is continuous on $[a, b]$, then $\int_a^b f(x)\,dx = \int_{a-y}^{b-y} f(x+y)\,dx$ for any y.*

Proof: Use $g(x) = x + y$, and observe $g' \equiv 1$. QED

6.1.3 Useful Integration Formulas

In this section we discuss briefly three integration formulas that will be proved later in the text. From time to time we will need to use these formulas, and you may find them useful in doing some of the exercises. Strictly speaking, we should not be allowed to do this, but it would simply take us too far afield to define all the concepts and do the preparatory work needed to present the proofs here. The other extreme—avoiding all use of these theorems until after they are proved—would have the negative consequence that we would not be able to complete the discussion of other topics in the place where they naturally belong. So a healthy compromise seems in order. Of course you should check, when we eventually prove these theorems, that we have not used anything in the proof that was derived assuming the result.

The first result is the familiar arclength formula for the graph of a function.

Theorem 6.1.6 (*Arclength Formula*) *Let f be a C^1 function on $[a, b]$. Then the length of the curve given by the graph of the function f on $[a, b]$ is equal to $\int_a^b \sqrt{1 + f'(x)^2}\,dx$.*

Since we have not defined the length of a curve, we cannot begin to prove this result here. A thorough discussion will be given in Chapter 13, where we will give a more general result (for curves not given as graphs). We will need to use this result in Chapter 8 to motivate the definition of the trigonometric functions.

The next result is the general formula for differentiating a function defined by an integral. Here we allow the integrand to be a function of two variables $g(x,t)$, and let

$$f(x) = \int_{a(x)}^{b(x)} g(x,t)\,dt$$

where $a(x)$ and $b(x)$ are functions of x as well.

Theorem 6.1.7 *Let $a(x)$ and $b(x)$ be C^1 functions and $g(x,t)$ be a C^1 function of two variables. Then if $f(x) = \int_{a(x)}^{b(x)} g(x,t)\,dt$ we have*

$$f'(x) = b'(x)g(x,b(x)) - a'(x)g(x,a(x)) + \int_{a(x)}^{b(x)} \frac{\partial g}{\partial x}(x,t)\,dt.$$

Since we have not discussed functions of two variables yet and have not defined partial derivatives, we will have to postpone the complete proof until Chapter 10. However, we can indicate the main ideas of the proof rather easily. We write the difference quotient as a sum of three terms that isolate the three appearances of the x variable:

$$\begin{aligned} \frac{1}{h}(f(x+h) - f(x)) &= \frac{1}{h}\int_{b(x)}^{b(x+h)} g(x+h,t)\,dt \\ &\quad - \frac{1}{h}\int_{a(x)}^{a(x+h)} g(x+h,t)\,dt \\ &\quad + \int_{a(x)}^{b(x)} \frac{1}{h}(g(x+h,t) - g(x,t))\,dt. \end{aligned}$$

Each of the three terms will converge to the corresponding term in the formula. For the first two terms this is almost the differentiation of the integral theorem combined with the chain rule, except for the translation of x by h in $g(x+h,t)$. It is plausible that this does not affect the outcome since h is tending to zero. In the special case that g is only a function of t alone this problem does not arise (see exercises).

For the third term the issue is the interchange of the integral and the limit that defines $\partial g/\partial x$. We will discuss the question of interchanging limits and integrals in Chapter 7.

The last formula in this section is the interchange of integrals. Suppose $f(x, y)$ is a continuous function of two variables for x in $[a, b]$ and y in $[c, d]$. The exact definition is given in Chapter 9. We may consider the function $\int_c^d f(x, y)\,dy$ as a function of x. If it is continuous, which is in fact always true, we can take its integral

$$\int_a^b \left(\int_c^d f(x, y)\,dy \right) dx.$$

Similarly, we may perform the integrations in the reverse order.

Theorem 6.1.8 (*Interchange of Integrals*) *Let $f(x, y)$ be a continuous function for x in $[a, b]$ and y in $[c, d]$. Then $\int_c^d f(x, y)\,dy$ is a continuous function of x for x in $[a, b]$, $\int_a^b f(x, y)\,dx$ is a continuous function of y for y in $[c, d]$, and*

$$\int_a^b \left(\int_c^d f(x, y)\,dy \right) dx = \int_c^d \left(\int_a^b f(x, y)\,dx \right) dy.$$

We will prove this theorem in Chapter 15, where we show that both of the above iterated integrals are equal to the double integral $\iint f(x, y)\,dx\,dy$ defined by partitioning the rectangle $[a, b] \times [c, d]$ in the plane.

6.1.4 Numerical Integration

The existence of the integral as a limit of sums is a qualitative result in that we know the sums approximate the integral but don't know how fast the process converges. For practical purposes we would like to have a quantitative counterpart—an estimate for the difference of the sum and the integral. This is possible if we assume more about the function in terms of smoothness (this is quite reasonable because wiggling of the function is likely to cause greater errors). In this section we discuss briefly four methods of numerical integration and estimates for their errors.

Suppose f is differentiable and $|f'(x)| \leq M_1$ on $[a, b]$. Then if x and y_k are points in the interval $[x_{k-1}, x_k]$, we have $|f(x) - f(y_k)| \leq M_1|x - y_k|$ by the mean value theorem. Now the difference between $\int_{x_{k-1}}^{x_k} f(x)\,dx$ and $f(y_k)(x_k - x_{k-1})$ can be written $\int_{x_{k-1}}^{x_k} (f(x) - f(y_k))\,dx$

(think of the contribution $f(y_k)(x_k - x_{k-1})$ to the Cauchy sum as the integral of the constant function $f(y_k)$ over the interval $[x_{k-1}, x_k]$). The biggest this can be (in absolute value) is $\int_{x_{k-1}}^{x_k} M_1|x - y_k|\,dx$ because $f(x) - f(y_k)$ is at most $M_1|x - y_k|$ (in absolute value). This is an elementary integral that we can evaluate. Note that it is worst (largest) if y_k is one of the endpoints and best if y_k is the midpoint, as seen in Figure 6.1.6.

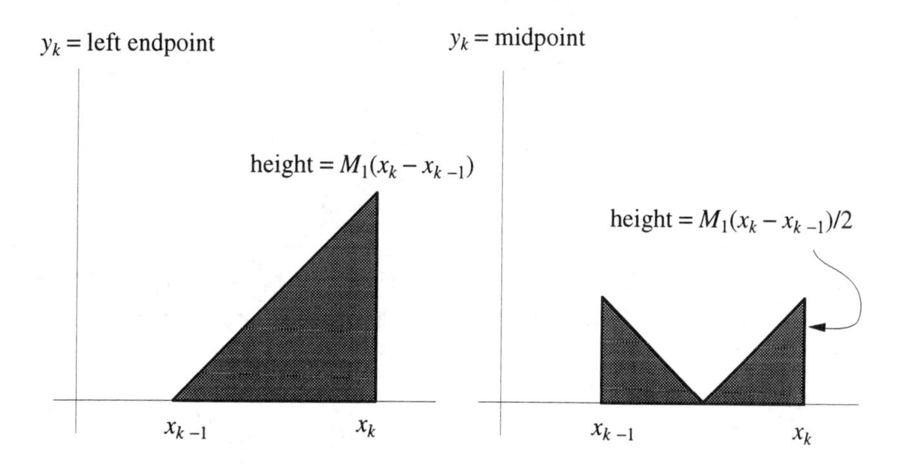

Figure 6.1.6:

At worst the error is $M_1(x_k - x_{k-1})^2/2$; and if we sum over all the intervals in the partition, the total error is bounded by

$$\sum_{k=1}^{n} M_1(x_k - x_{k-1})^2/2.$$

If we let δ denote the maximum interval length of the partition, the error bound is $\sum_{k=1}^{n} M_1\delta(x_k - x_{k-1})/2 = (b-a)M_1\delta/2$. This is called a *first-order* estimate because δ appears to the first power. Since δ is the only quantity we can control, by making the partition finer we are assured that we can make the error as small as we like but at a rather large cost in computation time and computation accuracy. For example, if $(b-a)M_1/2 \leq 1$ and we want an accuracy of 10^{-4}, we have to take $\delta = 10^{-4}$, which means about $10,000$ intervals in the partition.

We can do a lot better simply by taking the point y_k to be the midpoint of the interval, provided the function is smooth enough. Assume

f is C^2 and $|f''(x)| \le M_2$ on the interval. The point of using the *midpoint rule*, $y_k = (x_{k-1}+x_k)/2$, is that it gives the exact answer for any affine function on each subinterval. In other words, if $f(x) = ax+b$ on $[x_{k-1}, x_k]$, then $\int_{x_{k-1}}^{x_k} f(x)\,dx = f(y_k)(x_k - x_{k-1})$. (Geometrically, this just says that the area of a trapezoid is the product of the base times the midpoint altitude.) To exploit this fact we simply use the first-order Taylor expansion with Lagrange remainder about the point y_k, so $f(x) = f(y_k)+(x-y_k)f'(y_k)+R_1(x)$ with $R_1(x) = (x-y_k)^2 f''(z)/2$ for some point z in $[x_k, y_k]$ so that we have $|R_1(x)| \le M_2(x-y_k)^2/2$. Now we integrate the Taylor expansion:

$$\int_{x_{k-1}}^{x_k} f(x)\,dx = \int_{x_{k-1}}^{x_k} f(y_k)\,dx + f'(y_k)\int_{x_{k-1}}^{x_k} (x-y_k)\,dx + \int_{x_{k-1}}^{x_k} R_1(x)\,dx.$$

The first integral on the right is exactly $f(y_k)(x_k - x_{k-1})$, the midpoint rule; and the second integral is zero because $x - y_k$ is an odd function about y_k (this is where we require that y_k be the midpoint). If f were affine there would be no third term, so the midpoint rule would produce no error. But in any case, the error is at most

$$\int_{x_{k-1}}^{x_k} \frac{1}{2}M_2(x-y_k)^2\,dx = \frac{1}{6}M_2(x-y_k)^3\Big|_{x_{k-1}}^{x_k} = \frac{1}{24}M_2(x_k - x_{k-1})^3.$$

The total error using the midpoint rule is at most

$$\sum_{k=1}^{n} \frac{1}{24}M_2(x_k - x_{k-1})^3 \le \frac{1}{24}M_2\delta^2 \sum_{k=1}^{n}(x_k - x_{k-1}) = \frac{1}{24}M_2(b-a)\delta^2.$$

This is a *second-order* estimate, because of the factor δ^2. If $M_2(b-a)/24 \le 1$ and we want an error of at most 10^{-4} we only have to take $\delta = 10^{-2}$ or about 100 points in the partition.

A closely related method is the *trapezoidal rule*, which is obtained by replacing $f(y_k)$ by $\frac{1}{2}(f(x_{k-1}) + f(x_k))$, the average value at the endpoints, the point being that $\frac{1}{2}(f(x_{k-1})+f(x_k))(x_k - x_{k-1})$ is exactly the area of the trapezoid lying under the line segment joining the two points $(x_{k-1}, f(x_{k-1}))$ and $(x_k, f(x_k))$ on the graph of f, as shown in Figure 6.1.7. The trapezoidal rule also gives the exact integral for any affine function on each subinterval and also has a second-order error estimate. This is a little trickier to prove, and we leave it to the exercises.

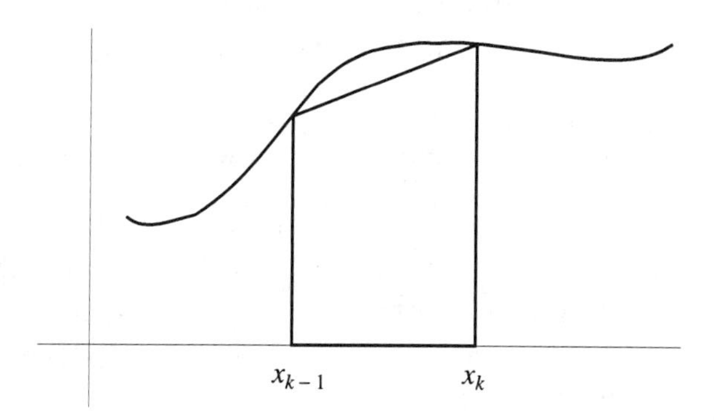

Figure 6.1.7:

Simpson's method uses

$$\sum_{k=1}^{n}(1/6)(f(x_{k-1}) + 4f(y_k) + f(x_k))(x_k - x_{k-1}).$$

It gives the exact integral for all cubic polynomials and has a fourth-order error estimate involving $M_4 = \sup|f''''(x)|$, assuming f is C^4. Notice that we have a trade-off: in order to obtain higher order estimates we must assume more smoothness for the function. For any fixed δ, it is not obvious which method will give the best approximation, since the constants multiplying the power of δ will vary. However, once we let δ get small, the higher order methods quickly win out. It usually pays to be clever.

6.1.5 Exercises

1. If $f(x) = \int_{a(x)}^{b(x)} g(t)\,dt$ where $a(x)$ and $b(x)$ are C^1 functions and g is continuous, prove that $f'(x) = b'(x)g(b(x)) - a'(x)g(a(x))$. (This is a special case of Theorem 6.1.7.)

2. Show that $|\int_a^b f(x)\,dx| \le \int_a^b |f(x)|\,dx$.

3. Derive the integration of the derivative theorem from the differentiation of the integral theorem. Can you prove the converse implication?

4. Prove the integral mean value theorem: if f is continous on $[a,b]$, then there exists y in (a,b) such that $\int_a^b f(x)\,dx = (b-a)f(y)$.

5. Let g be continuous on $[a,b]$, and let $f(x) = \int_a^x (x-t)g(t)\,dt$. Prove that f is a solution of the differential equation $f'' = g$ and the initial conditions $f(a) = f'(a) = 0$.

6. Suppose you want to compute a table of arctangents for values of x in $[0,1]$ using the formula $\arctan x = \int_0^x 1/(1+t^2)\,dt$. What size must you take for δ if you want the error to be at most $1/1,000$ using the midpoint rule?

7. Write out the complete proof of the integral remainder formula in Taylor's theorem.

8. Let f be a C^1 function on the line, and let $g(x) = \int_0^1 f(xy)y^2\,dy$. Prove that g is a C^1 function and establish a formula for $g'(x)$ in terms of f.

9. If f is continuous on $[ca, cb]$, show $\int_a^b f(cx)\,dx = (1/c)\int_{ca}^{cb} f(x)\,dx$ for $c > 0$. Is the same true for $c < 0$?

10. For a continuous, positive function $w(x)$ on $[a,b]$, define the weighted average operator A_w to be

$$A_w(f) = \int_a^b f(x)w(x)\,dx \Big/ \int_a^b w(x)\,dx$$

for continuous functions f. Prove that A_w is linear and lies between the maximum and minimum values of f.

11. Show that Simpson's rule gives the exact integral for any cubic polynomial.

12. Let $f(x)$ be a continuous function that is periodic of period a $(f(x+a) = f(x))$. Prove that $F(x) = \int_0^x f(t)\,dt$ is also periodic of period a if and only if $\int_a^b f(t)\,dt = 0$.

13. Let $F(x) = \int_0^x f(t)\,dt$ for continuous f. Show that F has strict local maxima and minima at points where f changes sign. Compare this to the $F'(x) = 0$ criterion.

14. *a. If f is C^2 on $[a, b]$ with $|f''(x)| \leq M_2$ and $f(a) = f(b) = 0$, prove that $|f(x)| \leq M_2|x - a|\,|x - b|$. (**Hint**: use the mean value theorem to write $f(x) = f'(y)(x - a)$, use Rolle's theorem to find a point z where $f'(z) = 0$, and use the mean value theorem to write $f'(y) = f''(w)(y - z)$.)

    b. Use part a on each subinterval to establish a second-order estimate for the trapezoidal rule.

6.2 The Riemann Integral

6.2.1 Definition of the Integral

Often we want to form the integral of a function that is not continuous. For example, it is convenient to think of the Cauchy sums as the integral of a function that is constant on each of the subintervals of the partition. This function is not continuous at the partition points, where it may have a jump discontinuity. It is not hard to show that the definition of integral we have given can be extended to functions that have a finite number of jump discontinuities. One can then ask if it can be extended further. The answer is yes, if the function has a finite number of any type of discontinuity and is bounded. Even if the number of discontinuities is countable. The trouble with each such theorem extending the definition of integration is that it raises still more questions. Are there other functions for which the definition of the integral also makes sense? No amount of piece-meal extension of the concept of the integral will put to rest the possibility that some important cases have been left out. However, there is a bold-stroke method for getting all possible extensions at once. This method was introduced by Bernard Riemann in the middle of the nineteenth century after many piece-meal methods had been discovered.

Here is Riemann's idea. We restrict attention to bounded functions on bounded intervals (so the graph of the function lies in a finite rectangle). This is an important restriction, because many new kinds of difficulties arise when the interval or the function is unbounded; later we will come back and deal with some of the more well-behaved unbounded cases when we discuss "improper integrals". The Riemann integral is thus a theory of "proper integrals". If we want the largest possible class of functions for which the definition of the integral makes sense, we should define the class of functions exactly in that way. Thus we will say that a bounded function on $[a, b]$ is *Riemann integrable* if the Cauchy sums $S(f, P)$ converge to some limit, called the *Riemann integral* of f and written $\int_a^b f(x)\,dx$ as before, as the maximum internal length of P tends to zero.

Notice that this definition is nothing but a linguistic trick—taking a theorem and making it a definition. Of course the theorem can be reinterpreted in light of the definition: every continuous function on $[a, b]$ is Riemann integrable. The important thing is that we have made a change in viewpoint. Instead of thinking of integration as a process attached to a specific class of functions (such as continuous functions), we think of it as a general process and search for functions to which it can be applied. Notice that this is the way we have dealt with the derivative from the start (you might argue that in a calculus course the derivative is presented piece-meal: first you learn to differentiate polynomials, then rational functions, and so forth).

Before we can make Riemann's idea precise we have to clarify one technical but important point. In establishing the existence of the integral of a continuous function, we saw that there were several e-quivalent ways of obtaining it. We could take the limit of $S(f, P_n)$ for a specific sequence of Cauchy sums and partitions, or the limit of $S(f, P)$ for all Cauchy sums and partitions, or the inf of upper Riemann sums $S^+(f, P)$ or the sup of lower Riemann sums $S^-(f, P)$. If we were to define the Riemann integrable functions by any one of these conditions, would we obtain the same class of functions? It turns out that the answer is yes, provided we avoid one rather silly choice. To understand why, we need to make a few simple observations about the sums $S(f, P)$. We are not assuming that f is continuous any more, merely that it is bounded. Thus the sup M_k and the inf m_k of f on the interval $[x_{k-1}, x_k]$ are still defined, but there do not have to exist points of the interval where f takes on these values. Thus the Riemann upper and lower sums $S^+(f, P) = \Sigma_{k=1}^n M_k(x_k - x_{k-1})$ and $S^-(f, P) = \Sigma_{k=1}^n m_k(x_k - x_{k-1})$ are not necessarily Cauchy sums $S(f, P) = \Sigma_{k=1}^n f(y_k)(x_k - x_{k-1})$. However the Cauchy sums lie in between, $S^-(f, P) \leq S(f, P) \leq S^+(f, P)$, and can be made as close as desired to the Riemann sums (this is with P fixed) simply by choosing y_k in $[x_{k-1}, x_k]$ so that $f(y_k)$ is as close as needed to the sup or inf.

Let us define the *oscillation* of f on P, $\mathrm{Osc}(f,P)$, to be the difference $S^+(f,P) - S^-(f,P)$. Since this also measures the spread of the different Cauchy sums $S(f,P)$ associated with P, it is clear that it is the important quantity in deciding whether f is Riemann integrable. If f is Riemann integrable we will have the integral $\int_a^b f(x)\,dx$ also lying in the interval between $S^-(f,P)$ and $S^+(f,P)$, so if $\mathrm{Osc}(f,P)$ is small we know any Cauchy sum $S(f,P)$ is close to the integral.

Now we need to relate the oscillation of f for different partitions. If P' is a refinement of P, then clearly $\mathrm{Osc}(f,P') \le \mathrm{Osc}(f,P)$, because the upper sums decrease and the lower sums increase as points are added to the partition. However we also need to compare the oscillation of f on partitions that are not necessarily refinements of P but are in some ways finer than P. Let δ be the *minimum* length of the subintervals in P. Then if P' is another partition with the *maximum* interval length less than δ, we claim $\mathrm{Osc}(f,P') \le 3\ \mathrm{Osc}(f,P)$. The factor 3 arises in this estimate because the subintervals of P' covering a fixed subinterval of P can be three times as long; it will turn out to be harmless when we try to make $\mathrm{Osc}(f,P')$ small.

Lemma 6.2.1 *If the maximum interval length of P' is less than the minimum length of subintervals of P, then* $\mathrm{Osc}(f,P') \le 3\ \mathrm{Osc}(f,P)$.

Proof: Look at one particular interval $[x'_{j-1}, x'_j]$ of P'. By the length assumptions, it overlaps at most two consecutive subintervals $[x_{k-1}, x_k]$ and $[x_k, x_{k+1}]$ of P. The oscillation of $M'_j - m'_j$ of f on $[x'_{j-1}, x'_j]$ is certainly less than the sum of the oscillations $M_k - m_k$ and $M_{k+1} - m_{k+1}$ over the intervals $[x_{k-1}, x_k]$ and $[x_k, x_{k+1}]$ that cover $[x'_{j-1}, x'_j]$. Thus we have

$$(M'_j - m'_j)(x'_j - x'_{j-1}) \le (M_k - m_k)(x'_j - x'_{j-1}) + (M_{k+1} - m_{k+1})(x'_j - x'_{j-1}).$$

We sum this inequality over all j, and on the left side we obtain exactly $\mathrm{Osc}(f,P')$. On the right side we obtain a rather complicated expression; however, if we sort the terms according to the index k of $(M_k - m_k)$, then we find $(M_k - m_k)$ multiplied by $(x'_j - x'_{j-1})$ for every subinterval of P' that intersects $[x_{k-1}, x_k]$. These subintervals of P' cover $[x_{k-1}, x_k]$ with perhaps some overlap at either end, as illustrated in Figure 6.2.1. But because the lengths of the subintervals of P' are less than $(x_k - x_{k-1})$, we certainly have the sum of the lengths of the covering intervals less than $3(x_k - x_{k-1})$. Thus $(M_k - m_k)$ is multiplied by at most $3(x_k - x_{k-1})$ on the right side, proving $\mathrm{Osc}(f, P') \leq 3\ \mathrm{Osc}(f, P)$. QED

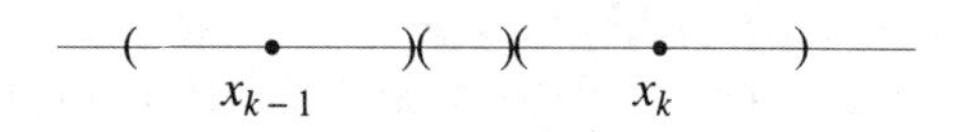

Figure 6.2.1:

We are now in a position to prove the equivalence of several possible ways of defining Riemann integrability and the Riemann integral.

Theorem 6.2.1 *Let f be a bounded function on $[a, b]$. Then the following are equivalent:*

a. *There exists a sequence P_j of partitions such that* $\mathrm{Osc}(f, P_j) \to 0$.

b. $\mathrm{Osc}(f, P) \to 0$ *as the maximum interval length of P tends to zero, in the sense that given any $1/n$ there exists $1/m$ such that* $\mathrm{Osc}(f, P) \leq 1/n$ *for any partition P with maximum interval length* $\leq 1/m$.

c. $\inf_P S^+(f, P) = \sup_P S^-(f, P)$.

d. *There exists a sequence of partitions P_j and a real number $\int_a^b f(x)\,dx$ such that $S(f, P_j) \to \int_a^b f(x)\,dx$ as $j \to \infty$ for every choice of Cauchy sums $S(f, P_j)$.*

e. *$S(f, P) \to \int_a^b f(x)\,dx$ as the maximum interval length of P tends to zero, in the sense that given any $1/n$ there exists $1/m$ such that $|S(f, P) - \int_a^b f(x)\,dx| \leq 1/n$ for any Cauchy sum $S(f, P)$ and any partition P with maximum interval length $\leq 1/m$.*

Proof: Clearly condition b is a stronger statement than condition a, but condition a implies condition b by the lemma, since once $\mathrm{Osc}(f, P_j) \leq 1/3n$, we have $\mathrm{Osc}(f, P) \leq 1/n$ if the maximum interval length of P is less than the minimum length of subintervals of P_j. Thus conditions a and b are equivalent. The same reasoning shows that conditions d and e are equivalent.

Now let's show that condition c is equivalent to condition a. Just as in the case of continuous functions, every upper sum $S^+(f,P)$ is greater than or equal to every lower sum $S^-(f,P')$, as follows by comparison with a partition P'', which is a common refinement of P and P'. Thus $\inf_P S^+(f,P) \geq \sup_P S^-(f,P)$ in any case. If we have equality, there must exist sequences P_j and P_j' of partitions such that $\lim_{j\to\infty} S^+(f,P_j) = \lim_{j\to\infty} S^-(f,P_j')$. By passing to common refinements P_j'' of P_j and P_j' we have $\lim_{j\to\infty} S^+(f,P_j'') = \lim_{j\to\infty} S^-(f,P_j'')$, which is the same thing as $\lim_{j\to\infty} \operatorname{Osc}(f,P_j'') = 0$. Thus condition c implies condition a. But conversely condition a implies $\lim_{j\to\infty} S^+(f,P_j) = \lim_{j\to\infty} S^-(f,P_j)$ for any sequence P_j, so $\inf_p S^+(f,P) = \sup_P S^-(f,P)$.

So far we have established the equivalences a $\leftrightarrow$ b $\leftrightarrow$ c and d $\leftrightarrow$ e. To complete the proof we will show condition d implies condition a and condition c implies condition d. Since the oscillation of f on P is the spread of the values of the Cauchy sums $S(f,P)$, the convergence of $S(f,P_j)$ implies $\operatorname{Osc}(f,P_j) \to 0$. So condition d implies condition a. But as before condition c implies $\lim_{j\to\infty} S^+(f,P_j) = \lim_{n\to\infty} S^-(f,P_j)$ for a sequence of partitions P_j and so $\lim_{j\to\infty} S(f,P_j)$ equals the common value $\inf_P S^+(f,P) = \sup_P S^-(f,P)$ for any Cauchy sums $S(f,P_j)$. This is condition d. QED

A function satisfying any one of the above equivalent conditions will be called *Riemann integrable.* Of course any continuous function is Riemann integrable, as was shown in Theorem 6.1.1. Condition a or b is useful because it does not involve explicitly the value of the integral. We note that for Riemann integrability it does *not* suffice to verify that $S(f,P_j)$ converges for one particular sequence of Cauchy sums. To see this we need to examine a famous example due to Dirichlet: the function f on $[0,1]$ equal to 1 if x is rational and 0 if x is irrational. This function is not continuous at any point, and there is no way to picture its graph. It is trivial to compute that $S^+(f,P) = 1$ and $S^-(f,P) = 0$ for this function and any partition because the sup and inf of f on any interval are 1 and 0, respectively. Thus f is not Riemann integrable. However one can easily choose Cauchy sums to converge to zero or one.

Dirichlet's function may strike you as rather pathological; indeed many of his contemporaries dismissed it as the work of a crackpot mentality. However, from the point of view of twentieth century mathematics, Dirichlet's function is rather tame in comparison with many functions that arise in solving very practical problems. Thus the inability of the Riemann theory of integration to deal with it is a sign of significant weakness. We shall also observe that there is rather compelling evidence for assigning the value zero to the integral of Dirichlet's function. We will return to this example when we discuss the Lebesgue theory of integration in a later chapter.

6.2.2 Elementary Properties of the Integral

It is a simple matter to show that the elementary properties of the integral of continuous functions are also true of the Riemann integral. Such properties now require an additional statement about Riemann integrability. Here is linearity: if f and g are Riemann integrable functions on $[a, b]$ and c is a real number, then $f + g$ and cf are also Riemann integrable functions on $[a, b]$ and

$$\begin{aligned}\int_a^b (f+g)(x)\,dx &= \int_a^b f(x)\,dx + \int_a^b g(x)\,dx \\ \int_a^b (cf)(x)\,dx &= c\int_a^b f(x)\,dx.\end{aligned}$$

Additivity says that if $a < b < c$ and f is defined on $[a, c]$ and the restrictions of f to $[a, b]$ and $[b, c]$ are Riemann integrable, then f is Riemann integrable on $[a, c]$ and

$$\int_a^c f(x)\,dx = \int_a^b f(x)\,dx + \int_b^c f(x)\,dx.$$

We leave the proofs as exercises. The basic estimate

$$m(b-a) \le \int_a^b f(x)\,dx \le M(b-a)$$

where M and m are the sup and inf of f on $[a, b]$ is also easily verified for Riemann integrable functions. This implies that the integral of a non-negative function is non-negative, and hence the integral preserves order: if $f(x) \ge g(x)$ for every x in $[a, b]$, then $\int_a^b f(x)dx \ge \int_a^b g(x)\,dx$.

It is also possible to show that the product of two Riemann integrable functions is Riemann integrable, although there is no formula for the integral of a product. The proof is based on the $\text{Osc}(f, P_j) \to 0$ criterion, as might be expected, and the estimate $\text{Osc}(fg, P) \leq M(\text{Osc}(f, P) + \text{Osc}(g, P))$ where M is the sup of $|f|$ and $|g|$ over the interval. We leave the details for the exercises, along with some related results on quotients and compositions.

It is important to be able to estimate the size of an integral; in other words to obtain a bound for $|\int_a^b f(x)\,dx|$. Now if g is any non-negative Riemann integrable function such that $|f(x)| \leq g(x)$ (we say *g dominates f*) for every x in $[a, b]$, then $-g(x) \leq f(x) \leq g(x)$, so

$$-\int_a^b g(x)\,dx \leq \int_a^b f(x)\,dx \leq \int_a^b g(x)\,dx$$

by the order preservation of the integral; hence, we obtain

$$\left|\int_a^b f(x)\,dx\right| \leq \int_a^b g(x)\,dx.$$

Often we will choose g in such a way that we can evaluate $\int_a^b g(x)\,dx$ explicitly.

Now $|f(x)|$ is the smallest non-negative function that dominates f. By the above argument we expect to have the estimate

$$\left|\int_a^b f(x)\,dx\right| \leq \int_a^b |f(x)|\,dx.$$

There is one hitch in the argument, however: how do we know $|f(x)|$ is Riemann integrable? It turns out to be true, but not obvious, that if f is Riemann integrable, then $|f|$ is also Riemann integrable. Before proving this let us observe that the converse is not true. We can have $|f|$ Riemann integrable without f being Riemann integrable; just consider the variant of Dirichlet's function that takes the value 1 for x rational and -1 for x irrational. This is still not Riemann integrable, but the absolute value of this function is identically one, hence continuous. This example helps to motivate the positive result in that it shows how taking the absolute value decreases the amount of oscillation in a function.

Theorem 6.2.2

a. *If f is Riemann integrable on $[a, b]$, then so is $|f|$.*

b. *If f and g are Riemann integrable on $[a, b]$, then so are* $\max(f, g)$ *and* $\min(f, g)$.

Proof:

a. We claim $\text{Osc}(|f|, P) \leq \text{Osc}(f, P)$. Indeed on any subinterval $[x_{k-1}, x_k]$ let M_k and m_k denote the sup and inf of f. If they both have the same sign, then the oscillation of $|f|$ on the subinterval is the same. If they have the opposite sign, then $M_k > 0$ and $m_k < 0$ and $M_k - m_k = M_k + (-m_k)$ is greater than either M_k or $-m_k$. But the sup of $|f|$ on the subinterval is the larger of M_k or $-m_k$, and the inf of $|f|$ is at least zero since $|f| \geq 0$. Thus the sup minus the inf for $|f|$ is less than $M_k - m_k$. Adding up over all the subintervals gives $\text{Osc}(|f|, P) \leq \text{Osc}(f, P)$. Since f is Riemann integrable, there exists a sequence P_j such that $\text{Osc}(f, P_j) \to 0$; hence, $\text{Osc}(|f|, P_j) \to 0$ and $|f|$ is Riemann integrable.

b. Let us consider

$$\max(f, g)(x) = \begin{cases} f(x) & \text{if } f(x) \geq g(x), \\ g(x) & \text{if } g(x) \geq f(x). \end{cases}$$

We claim $\text{Osc}(\max(f, g), P) \leq \text{Osc}(f, P) + \text{Osc}(g, P)$. This follows by considering the values that $\max(f, g)$ assumes on each subinterval—they can vary at most from the larger of the two sups to the smaller of the two infs. Then $\text{Osc}(\max(f, g), P) \to 0$ as the maximum length of the subintervals of P tends to zero since $\text{Osc}(f, P) \to 0$ and $\text{Osc}(g, P) \to 0$, so $\max(f, g)$ is Riemann integrable. QED

We now turn to the problem of showing that some discontinuous functions are Riemann integrable. We begin with a simple result.

Theorem 6.2.3 *Let f be a bounded function on $[a, b]$ that is continuous except at a finite number of points (we do not assume anything about the nature of the discontinuities at these points). Then f is Riemann integrable.*

Proof: Let $a_1, \ldots, a_N$ denote the points of discontinuity. Given any $1/n$, surround each of these points by an interval I_k of length at most

$1/n$. Then the function is continuous on the set consisting of $[a, b]$ with $\bigcup_{k=1}^{N} I_k$ removed. Let P denote any partition that contains all the intervals I_k. It will also contain some other interval J_k. In estimating $\text{Osc}(f, P)$ we will use a separate argument for contributions from the I and J type intervals. For the I type intervals we use the crude estimate that the length of each interval is at most $1/n$ and the oscillation of the function is at most $M - m$, where M and m are the sup and inf of f on $[a, b]$ (we assumed f was bounded, so M and m are finite). Thus each interval contributes at most $1/n(M - m)$, and there are N such intervals, so the total contribution is at most $N/n(M - m)$. For the contribution from the J type intervals we use the existence of the integral for continuous functions to conclude that it can be made as small as desired, say less than $1/n$, by taking the length of the subintervals J_k sufficiently small.

Thus we have shown that $\text{Osc}(f, P) \leq 1/n + N/n(M - m)$ if P is any partition containing the intervals I_k and the remaining subintervals of P are sufficiently small. Now the quantities N (number of discontinuities) and M and m (sup and inf of f on $[a, b]$) are fixed, so by taking n large we can make $\text{Osc}(f, P)$ as small as desired, proving f is Riemann integrable. QED

Notice that the idea of the proof is to cover the set of discontinuities of f by a union of a finite number of intervals whose lengths add up to a small number (N/n in the argument). Inside these intervals we have no control over f other than its boundedness, but this is enough in estimating $\text{Osc}(f, P)$ because the total lengths of the intervals is small. Outside these intervals the function f is continuous and we can control the oscillation as before.

6.2.3 Functions with a Countable Number of Discontinuities*

[Note: the arguments in this section use some elementary facts about infinite series that are discussed in detail in the next chapter.]

The arguments of the last section could be generalized to show the Riemann integrability of any function f that is continuous except for a set E with the property that given any $1/n$ there exists a finite covering of E by intervals whose lengths add up to at most $1/n$. Such a set is said to have *content zero.* (Sets of content zero will play a role in the theory of multiple integrals in Chapter 15. The general notion of *content* has been superseded by the notion of *measure*, which will be discussed in Chapter 14.) An example of a set of content zero is a countable set $a_1, a_2, \ldots$ that converges to the limit a_0, together with a_0; for then, given $1/n$, we cover a_0 with an interval of length $1/2n$, and this will cover all but a finite number of the a_j since $\lim_{j\to\infty} a_j = a_0$. The finite number of points left uncovered can then be covered by a finite number of intervals with lengths adding up to at most $1/2n$.

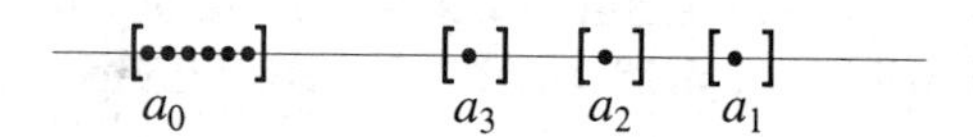

Figure 6.2.2:

However, not every countable set has content zero. The rational numbers in $[0, 1]$, for instance, can be covered by a finite number of intervals only if their lengths add up to one. Nevertheless, a bounded function with only a countable number of discontinuities is also Riemann integrable. This will require a new and more difficult kind of proof. Before embarking on it we will consider an example that shows there are functions that have discontinuities on the rational numbers alone (Dirichlet's function is not an example because it is discontinuous at every point). The idea of constructing the example is to place jumps at each rational points but to make the jumps very small so that the function will be continuous at all irrational points. The fact that we are forced to make the jumps small helps explain why such a function must be Riemann integrable.

Let $r_1, r_2, \ldots$ be any enumeration of the rational numbers. Then the function

$$q_k(x) = \begin{cases} 1 & \text{if} \quad x < r_k, \\ 0 & \text{if} \quad x \geq r_k \end{cases}$$

is continuous at every point except r_k. We will multiply q_k by a factor b_k and take the sum $f = \sum_{k=1}^{\infty} b_k q_k(x)$. For simplicity we take $b_k = 2^{-k}$, but any absolutely convergent series $\sum |b_k| < \infty$ will do as well. Then the function f can also be expressed $f(x) = \sum 2^{-k}$ where the sum extends over all k such that $x < r_k$. Since $\sum_{k=1}^{\infty} 2^{-k} = 1$, the function

f takes on values between 0 and 1. Now f is continuous at every irrational point x_0 because given any N we can find a neighborhood of x_0 that does not contain the first N rational numbers $r_1, \ldots, r_N$. The variation of f over this neighborhood is confined to the terms $2^{-k} r_k(x)$ for $k < N$, and the maximum these change is $\sum_{k=N+1}^{\infty} 2^{-k} = 2^{-N}$. Thus $|f(x) - f(x_0)| \leq 1/2^N$ in this neighborhood, so f is continuous at x_0. Roughly the same argument proves that f is discontinuous at each point r_k. Just write $f = 2^{-k} q_k + \sum_{j \neq k} 2^{-j} q_j$. Then $\sum_{j \neq k} 2^{-j} q_j$ is continuous at r_k by the argument just given, so f can't be continuous at r_k without implying $2^{-k} q_k$ is continuous there, which it clearly is not. Again it is not possible to draw a picture of the graph of this function.

Theorem 6.2.4 *Let f be a bounded function on $[a, b]$. If f is continuous except at a countable set of points, then f is Riemann integrable.*

Proof: The first step is to make precise the idea that f cannot jump around a lot at too many points. Define the oscillation of f at a point x_0, denoted $\mathrm{Osc}(f, x_0)$ to be the limit as $n \to \infty$ of the difference of the sup and inf of f on the interval $[x_0 - 1/n, x_0 + 1/n]$. Clearly if f is continuous at x_0, then $\mathrm{Osc}(f, x_0) = 0$, and conversely. Furthermore, the size of $\mathrm{Osc}(f, x_0)$ at a point of discontinuity gives a quantitative measurement of the amount of jumping around. If f has a jump discontinuity at x_0, then $\mathrm{Osc}(f, x_0)$ is the size of the jump (as long as $f(x_0)$ takes on some intermediate value between $\lim_{x \to x_0^+} f(x)$ and $\lim_{x \to x_0^-} f(x)$). Regardless of the type of discontinuity, $\mathrm{Osc}(f, x_0)$ is well defined because the difference of the sup and inf decreases with the size of the interval.

Now the set where $\mathrm{Osc}(f, x) < 1/n$ is open. To see this, notice that if $\mathrm{Osc}(f, x_0) < 1/n$, then $\mathrm{Osc}(f, x_0) \leq \delta$ for some $\delta < 1/n$. This means there is an interval $[x_0 - 1/k, x_0 + 1/k]$ on which the difference of the sup and inf of f is at most $1/n$, and hence the oscillation of f at every point of $(x_0 - 1/k, x_0 + 1/k)$ is less than $1/n$.

Since the set where $\mathrm{Osc}(f, x) < 1/n$ is open, its complement, where $\mathrm{Osc}(f, x) \geq 1/n$, is closed. Note that the set of discontinuities of f is the union over n of these closed sets. This fact, which is true of an arbitrary function, is sometimes stated as follows: the set of discontinuities of a function is an F_σ set (F_σ means a countable union of closed sets).

Now the hypothesis that the set of discontinuities is countable implies that the smaller set where $\text{Osc}(f, x) \geq 1/n$ is at most countable. The fact that it is also closed means that it is in some sense better behaved than the original set of discontinuities. In particular we want to show that it has content zero (it can be covered by a finite number of intervals whose lengths add up to as small a number as desired).

To do this we use the Heine-Borel theorem. Given $1/n$, we cover the closed countable set by a countable collection of open intervals whose lengths are

$$\frac{1}{2n}, \frac{1}{2^2 n}, \frac{1}{2^3 n}, \ldots.$$

Thus if $a_1, a_2, \ldots$ is the countable closed set (it must be compact because it is also bounded), we cover it by the intervals

$$\left(a_k - \frac{1}{2^{k+1}n}, \; a_k + \frac{1}{2^{k+1}n}\right).$$

By the Heine-Borel theorem a finite number of these intervals will cover, and the sum of lengths of these intervals is at most $\sum_{k=1}^{\infty} 1/2^k n = 1/n$. Thus $a_1, a_2, \ldots$ has zero content.

Now let us consider what we have accomplished. We have the interval $[a, b]$ divided into two sets. On one, where $\text{Osc}(f, x) < 1/n$, the function is fairly well behaved—it is not necessarily continuous, but at least whatever discontinuities exist are not too jumpy. The complementary set, where $\text{Osc}(f, x) \geq 1/n$, is small, in that we can cover it by a finite number of intervals of small length. We want to combine these two properties. However we still need to move cautiously and apply the Heine-Borel theorem one more time. First we cover the set where $\text{Osc}(f, x) \geq 1/n$ by a finite number of open intervals $I_1, \ldots, I_N$ such that the lengths add up to at most $1/n$ (the $1/n$ here is deliberately the same as the $1/n$ above). Now consider the complementary set, $[a, b]$ with $\bigcup_{j=1}^{N} I_j$ removed. This is again a compact set, and $\text{Osc}(f, x) < 1/n$ on this set. But $\text{Osc}(f, x) < 1/n$ means that x lies in an open interval in which f varies by at most $1/n$. Thus there exists a covering of the compact set $[a, b] - \bigcup_{j=1}^{N} I_j$ by such intervals and by the Heine-Borel theorem, a finite covering. By shrinking these intervals we can obtain a non-overlapping covering by closed intervals $J_1, \ldots, J_k$.

Altogether we have the interval $[a, b]$ partitioned into subintervals $I_1, \ldots, I_N$ and $J_1, \ldots, J_k$ with the properties:

1. The sum of the lengths of $I_1, \ldots, I_N$ is at most $1/n$.

2. The difference of the sup and inf of f on each J_j is at most $1/n$.

If P denotes the partition of $[a, b]$ consisting of these intervals we can estimate $\mathrm{Osc}(f, P)$ as follows: the contribution from the I type intervals is at most $(M - m)1/n$ where M and m are the sup and inf of f on $[a, b]$; while the contribution from the J type intervals is at most $1/n$ times the sum of the lengths of the J intervals, hence at most $(b - a)/n$. Thus $\mathrm{Osc}(f, P) \leq (M - m)/n + (b - a)/n$; and since $M - m$ and $b - a$ are fixed quantities, we can make $\mathrm{Osc}(f, P)$ as small as desired. Thus f is Riemann integrable. QED

6.2.4 Exercises

1. If $a < b < c$ and f is Riemann integrable on $[a, c]$, prove that f is Riemann integrable on $[a, b]$ (strictly speaking we should say that the restriction of f to $[a, b]$ is Riemann integrable).

2. Prove the linearity of the Riemann integral.

3. Prove the additivity of the Riemann integral.

4. Prove that if f and g are Riemann integrable on $[a, b]$, then $f \cdot g$ is Riemann integrable on $[a, b]$.

5. Prove that if f and g are Riemann integrable on $[a, b]$ and g is bounded away from zero (there exists $1/n$ such that $|f(x)| \geq 1/n$ for all x in $[a, b]$), then f/g is Riemann integrable.

6. Prove that if f is Riemann integrable on $[a, b]$ and $g(x) = f(x)$ for every x except for a finite number, then g is Riemann integrable.

7. *Let f be Riemann integrable on $[a, b]$ and let g be continuous on $[m, M]$, where M is the sup and m the inf of f on $[a, b]$. Prove that $g \circ f$ is Riemann integrable on $[a, b]$. (**Hint**: This is tricky. Not only do you need to use the uniform continuity of g, but you need to argue separately concerning the subintervals of a partition for which $\mathrm{Osc}(f, P)$ is small but $M_k - m_k$ is large.)

8. Let g be C^1 and increasing on $[a,b]$, and let f be Riemann integrable on $[g(a), g(b)]$. Prove that $f \circ g$ is Riemann integrable on $[a,b]$ and the change of variable formula holds.

9. a. If f is Riemann integrable on $[a,b]$, prove that $F(x) = \int_a^x f(t)\,dt$ is continuous.

   b. Prove it satisfies a Lipschitz condition.

10. If f is Riemann integrable on $[a,b]$ and continuous at x_0, prove that $F(x) = \int_a^x f(t)\,dt$ is differentiable at x_0 and $F'(x_0) = f(x_0)$. Show that if f has a jump discontinuity at x_0, then F is not differentiable at x_0.

11. If f is continuous on $[a,b]$ and differentiable on (a,b) and f' is Riemann integrable on $[a,b]$, show that $\int_a^b f'(x)\,dx = f(b) - f(a)$.

12. Prove that the complement of an F_σ set is a G_δ set (a countable intersection of open sets).

13. Give an example of an F_σ set that is not a G_δ set and a G_δ set that is not an F_σ set.

14. Prove that every open set is an F_σ set and every closed set is a G_δ set.

15. Prove that a G_δ set that is dense must be the whole line.

6.3 Improper Integrals*

6.3.1 Definitions and Examples

We frequently need to deal with integrals of functions that are unbounded and with integrals over unbounded intervals. In the Riemann theory of integration the only way to handle these is to take appropriate limits of integrals over smaller intervals. The term *improper integral* is used informally to denote any of a variety of such integrals. For example, consider the expression $\int_0^1 x^a\,dx$ where $a < 0$. (Strictly speaking, we have only defined x^a for a rational; the general case will be dealt with in Chapter 8. For now you can either assume that a is rational or else accept that the basic calculus formulas for x^a are also valid for all real a.) Since the function x^a is unbounded, we have not yet defined this integral. However, the function x^a is bounded and continuous on the interval $[\epsilon, 1]$ for any $\epsilon > 0$ and so $\int_\epsilon^1 x^a dx$ is defined. In fact we can compute it exactly as

$$\int_\epsilon^1 x^a\,dx = \begin{cases} \dfrac{1-\epsilon^{a+1}}{a+1}, & a \neq -1, \\ -\log\epsilon, & a = -1. \end{cases}$$

Thus we define the improper integral $\int_0^1 x^a\,dx$ to be $\lim_{\epsilon\to 0}\int_\epsilon^1 x^a\,dx$, and this is $1/(a+1)$ if $-1 < a < 0$; while the limit does not exist (or equals $+\infty$ in the extended real numbers) if $a \leq -1$.

Next let us look at an example where the interval of integration is infinite, $\int_1^\infty x^a dx$, again with $a < 0$. We define this to be $\lim_{y\to\infty}\int_1^y x^a\,dx$. Again we can compute explicitly

$$\int_1^y x^a\,dx = \begin{cases} \dfrac{y^{a+1}-1}{a+1}, & a \neq -1, \\ \log y, & a = -1, \end{cases}$$

and so the limit exists and equals $-1/(a+1)$ if $a < -1$ and fails to exist if $a \geq -1$. Notice that $a = -1$ is again the cut-off point between the existence and non-existence of the improper integral, but the inequality goes the other way. In particular, for no value of a does the improper integral $\int_0^\infty x^a\,dx$ exist. We can state these basic facts informally as follows: *the function x^a has an integrable singularity near $x = 0$ if and only if $a > -1$ and an integrable singularity near ∞ if and only if $a < -1$.*

So far we have looked at examples of improper integrals where the existence of the limit does not depend on cancellation of positive and negative values of the function. Suppose, to be specific, that f is a function defined on $[0, 1]$ that is Riemann integrable on $[\epsilon, 1]$ for every $\epsilon > 0$. We say that f has an *absolutely convergent improper integral* on $[0, 1]$ if $\lim_{\epsilon\to 0}\int_\epsilon^1 |f(x)|\,dx$ exists. This implies the existence of $\int_\epsilon^1 f(x)\,dx$ as well. To see this we use the Cauchy criterion. Given ϵ and δ with say $\epsilon < \delta$ then $\int_\epsilon^1 f(x)\,dx - \int_\delta^1 f(x)\,dx = \int_\epsilon^\delta f(x)\,dx$ by additivity and

$|\int_\epsilon^\delta f(x)\,dx| \le \int_\epsilon^\delta |f(x)|\,dx$ by our previous results. This means

$$\left|\int_\epsilon^1 f(x)\,dx - \int_\delta^1 f(x)\,dx\right| \le \int_\epsilon^1 |f(x)|\,dx - \int_\delta^1 |f(x)|\,dx,$$

so the Cauchy criterion for convergence of $\lim_{\epsilon\to 0}\int_\epsilon^1 |f(x)|\,dx$ implies the Cauchy criterion for convergence of $\lim_{\epsilon\to 0}\int_\epsilon^1 f(x)\,dx$. Similar reasoning shows the existence of $\lim_{y\to\infty}\int_1^y |f(x)|\,dx$ implies the existence of $\lim_{y\to\infty}\int_1^y f(x)\,dx$.

There are some important examples of improper integrals that are not absolutely convergent. One is the integral

$$\int_0^\infty \frac{\sin x}{x}\,dx = \lim_{y\to\infty}\int_0^y \frac{\sin x}{x}\,dx.$$

There is no difficulty near $x = 0$ because $\sin x/x$ is continuous there. However, $\int_0^y |\sin x/x|\,dx$ does not have a finite limit as $y \to \infty$, for much the same reason that $\sum_{n=1}^\infty 1/n$ diverges. Nevertheless, we can easily show that $\lim_{y\to\infty}\int_0^y (\sin x/x)\,dx$ exists by applying the Cauchy criterion. We have

$$\int_0^y \frac{\sin x}{x}\,dx - \int_0^z \frac{\sin x}{x}\,dx = \int_y^z \frac{\sin x}{x}\,dx = -\frac{\cos z}{z} + \frac{\cos y}{y} - \int_y^z \frac{\cos x}{x^2}\,dx$$

by integration by parts. Using the crude estimate $|\cos x| \le 1$ we find

$$\left|\int_0^y \frac{\sin x}{x}\,dx - \int_0^z \frac{\sin x}{x}\,dx\right| \le \frac{1}{z} + \frac{1}{y} + \int_y^z \frac{1}{x^2}\,dx = \frac{2}{y} \quad \text{if } z > y,$$

and this goes to zero as $y, z \to \infty$. It is possible to compute $\int_0^\infty (\sin x/x)dx$ exactly using methods of complex variables. You might be amused to look up the answer in a table of integrals.

Another important class of examples are the Cauchy principal value integrals. In these examples the singularity lies in the interior of the interval. Say f is defined on $[-1, 1]$ (possibly undefined at $x = 0$) and is bounded on $[-1, -\epsilon)$ and $(\epsilon, 1]$, for every $\epsilon > 0$, but unbounded on $(-\epsilon, \epsilon)$. We define the principal value integral P.V. $\int_{-1}^1 f(x)\,dx$ to be the limit $\lim_{\epsilon\to 0^+}\int_{-1}^{-\epsilon} + \int_\epsilon^1 f(x)\,dx$ if it exists. In other words, we cut away a symmetric neighborhood of the singularity and take the limit as the size of the deleted neighborhood goes to zero. For example, P.V. $\int_{-1}^1 (1/x)\,dx = 0$ although $f(x) = 1/x$ is not absolutely integrable. One can also show (see exercises) that P.V. $\int_{-1}^{1}(f(x)/x)\,dx$ always exists if f is C^1. The fact that the neighborhood is symmetric is important, since for example $\lim_{\epsilon\to 0+}\int_{-1}^{\epsilon}+\int_{2\epsilon}^{1}(1/x)\,dx = -\log 2$ while $\lim_{\epsilon\to 0+}\int_{-1}^{\epsilon}+\int_{\epsilon^2}^{1}(1/x)\,dx$ doesn't exist.

The question of the existence or non-existence of improper integrals boils down to the question of existence of limits, which we will take up in detail in the next chapter.

6.3.2 Exercises

1. For which values of a and b does the improper integral
$$\int_0^{1/2} x^a|\log x|^b dx$$
exist?

2. For which values of a and b does the improper integral
$$\int_2^{\infty} x^a|\log x|^b\,dx$$
exist?

3. If f is non-negative and $\int_1^{\infty} f(x)\,dx$ exists as an improper integral, must $\lim_{x\to\infty} f(x) = 0$? Must f be bounded? What can you say if $\lim_{x\to\infty} f(x)$ exists?

4. Let $p(x)$ be a polynomial with a simple zero at $x = 0$ but no other zeroes in $[-1, 1]$. Show that P.V. $\int_{-1}^{1} 1/p(x)\,dx$ exists. Also show that $\int_{-1}^{1}|p(x)|^{-a}dx$ exists for $0 < a < 1$.

5. Show that P.V. $\int_{-1}^{1}(f(x)/x)\,dx$ exists if f is C^1 on $[-1, 1]$.

6. Show that $\int_0^{\infty} x^{-a}\sin x\,dx$ exists for $0 < a < 2$.

7. If the improper integral $\int_{-\infty}^{\infty} f(x)\,dx$ exists, prove $\int_{-\infty}^{\infty} f(x+y)\,dx = \int_{-\infty}^{\infty} f(x)dx$ for all real y.

8. If f is positive and continuous on $(0, 1]$ and the improper integral $\int_0^1 f(x)\,dx$ exists, prove that the lower Riemann sums converge to the integral.

6.4 Summary

6.1 The Fundamental Theorem

Definition *Let P denote a partition of the interval $[a,b], a = x_0 < x_1 < \cdots < x_n = b$ and f a continuous function on $[a,b]$. A Cauchy sum $S(f,P)$ is any sum of the form $\sum_{k=1}^{n} f(a_k)(x_k - x_{k-1})$ where a_k is in $[x_{k-1}, x_k]$. The upper and lower Riemann sums are $S^+(f,P) = \sum_{k=1}^{n} M_k(x_k - x_{k-1})$ and $S^-(f,P) = \sum_{k=1}^{n} m_k(x_k - x_{k-1})$ where M_k and m_k denote the* sup *and* inf *of f on $[x_{k-1}, x_k]$. The maximum interval length of P is the maximum value of $x_k - x_{k-1}$.*

Theorem 6.1.1 (*Existence of the Integral*) *If f is continuous on $[a,b]$ there exists a real number $\int_a^b f(x)\,dx$ such that given N there exists m such that if the maximum interval length of P is less than $1/m$, then $|S(f,P) - \int_a^b f(x)\,dx| < 1/N$. Also $\int_a^b f(x)\,dx = \inf S^+(f,P) = \sup S^-(f,P)$ where P varies over all partitions.*

Theorem *The integral of continuous functions is linear and additive, and $m(b-a) \le \int_a^b f(x)\,dx \le M(b-a)$ where M and m denote the* sup *and* inf *of f over $[a,b]$.*

Theorem 6.1.2 (*Differentiation of the Integral*) *If f is continuous on $[a,b]$, then $F(x) = \int_a^x f(t)\,dt$ is C^1 and $F' = f$.*

Theorem 6.1.3 (*Integration of the Derivative*) *If f is C^1 on $[a,b]$, then $\int_a^b f'(x)\,dx = f(b) - f(a)$.*

Definition *F is called an indefinite integral or primitive of f if $F' = f$.*

Theorem 6.1.4 (*Integration by Parts*) *If f and g are C^1 on $[a,b]$ then*

$$\int_a^b f(x)g'(x)\,dx = f(b)g(b) - f(a)g(a) - \int_a^b f'(x)g(x)\,dx.$$

Theorem (*Integral Remainder Formula for Taylor's Theorem*) *If f is*

C^{n+1}, *then* $f(x) = T_n(x_0, x) + R_n(x_0, x)$ *where*

$$R_n(x_0, x) = \frac{1}{n!} \int_{x_0}^{x} (x-t)^n f^{(n+1)}(t)\, dt.$$

Theorem 6.1.5 (*Change of Variable*) *If* g *is* C^1 *and increasing on* $[a, b]$ *and* f *continuous on* $[g(a), g(b)]$,

$$\int_{g(a)}^{g(b)} f(x)\, dx = \int_a^b f(g(x))g'(x)\, dx.$$

Corollary 6.1.1 (*Translation Invariance*)

$$\int_a^b f(x)\, dx = \int_{a-y}^{b-y} f(x+y)\, dy.$$

Theorem 6.1.6 (*Arclength formula*) *The length of the graph of a* C^1 *function* f *on* $[a, b]$ *is* $\int_a^b \sqrt{1 + f'(x)^2}\, dx$.

Theorem 6.1.7 *Let* $f(x) = \int_{a(x)}^{b(x)} g(x, t)\, dt$ *for* $a(x), b(x)$, *and* $g(x, t)$ C^1 *functions. Then*

$$f'(x) = b'(x)g(x, b(x)) - a'(x)g(x, a(x)) + \int_{a(x)}^{b(x)} \partial g/\partial x(x, t)\, dt.$$

Theorem 6.1.8 (*Interchange of Integrals*) *Let* $f(x, y)$ *be continuous for* x *in* $[a, b]$ *and* y *in* $[c, d]$. *Then* $\int_c^d f(x, y)\, dy$ *and* $\int_a^b f(x, y)\, dx$ *are continuous and*

$$\int_a^b \left(\int_c^d f(x, y)\, dy \right) dx = \int_c^d \left(\int_a^b f(x, y)\, dx \right) dy.$$

Theorem *If f is C^1 on $[a,b]$ with $|f'(x)| \leq M_1$ and δ is the maximum interval length of P, then*

$$\left| \int_a^b f(x)\,dx - S(f,P) \right| \leq \frac{1}{2} M_1 (b-a)\delta.$$

Theorem (*Midpoint rule*) *If f is C^2 on $[a,b]$ with $|f''(x)| \leq M_2$ and $S(f,P)$ is formed by evaluating f at the midpoint of each subinterval, then*

$$\left| \int_a^b f(x)\,dx - S(f,P) \right| \leq \frac{1}{24} M_2 (b-a)\delta^2.$$

6.2 The Riemann Integral

Definition *A bounded function f on $[a,b]$ is said to be Riemann integrable with integral $\int_a^b f(x)\,dx$ if $S(f,P)$ converges to $\int_a^b f(x)\,dx$ as the maximum interval length of P tends to zero.*

Definition $\operatorname{Osc}(f,P) = S^+(f,P) - S^-(f,P)$.

Lemma 6.2.1 *If the maximum interval length of P' is less than the minimum length of subintervals of P, then*

$$\operatorname{Osc}(f,P') \leq 3 \operatorname{Osc}(f,P).$$

Theorem 6.2.1 *Let f be bounded on $[a,b]$. Then the following are equivalent:*

a. $\operatorname{Osc}(f,P_j) \to 0$ *for some sequence of partitions.*

b. $\operatorname{Osc}(f,P) \to 0$ *as the maximum interval length of P goes to zero.*

c. $\inf S^+(f,P) = \sup S^-(f,P)$.

d. $S(f,P_j)$ *converges for every choice of Cauchy sequence for some sequence of partitions.*

e. *f is Riemann integrable.*

Theorem *The Riemann integral is linear and additive, and $m(b-a) \leq \int_a^b f(x)\,dx \leq M(b-a)$ where M and m denote the* sup *and* inf *of f on $[a,b]$.*

Theorem 6.2.2 *If f and g are Riemann integrable on $[a,b]$, then so is $|f|$, $\max(f,g)$, and $\min(f,g)$.*

Theorem 6.2.3 *A bounded function with a finite set of discontinuities on $[a,b]$ is Riemann integrable.*

Example *There exists a function continuous at the irrational numbers but discontinuous at the rational numbers.*

Theorem 6.2.4 *A bounded function on $[a,b]$ continuous except at a countable set of points is Riemann integrable.*

6.3 Improper Integrals

Example *The function x^a has an integrable singularity near $x = 0$ if and only if $a > -1$ and an integrable singularity near ∞ if and only if $a < -1$.*

Definition *If f is defined on $[0,1]$ and Riemann integrable on $[\epsilon, 1]$ for every $\epsilon > 0$ we say f has an absolutely convergent improper integral on $[0,1]$ if $\lim_{\epsilon\to 0} \int_\epsilon^1 |f(x)|\,dx$ exists. This implies $\lim_{\epsilon\to 0}\int_\epsilon^1 f(x)\,dx$ exists.*

Example $\lim_{n\to\infty} \int_0^n (\sin x/x)\,dx$ *exists, but* $\lim_{n\to\infty}\int_0^n |\sin x/x|\,dx$ *does not.*

Definition *The Cauchy principal value integral P.V. $\int_{-1}^1 f(x)\,dx$ is defined to be $\lim_{\epsilon\to 0} \int_{-1}^{-\epsilon} + \int_\epsilon^1 f(x)\,dx$ if the limit exits.*

Chapter 7

Sequences and Series of Functions

7.1 Complex Numbers

7.1.1 Basic Properties of $\mathbb{C}$

So far we have been dealing with the real number system $\mathbb{R}$ and functions whose domain and range are subsets of $\mathbb{R}$. For many purposes it is important to consider also the complex number system $\mathbb{C}$. We can describe this system succinctly by defining a complex number to be a symbol $x+iy$, where x and y are real numbers and i is a formal symbol, which operationally is to be thought of as a solution of the equation $x^2 = -1$. We have already observed that there are no real solutions of this equation. The arithmetic of complex numbers is described by the formulas

$$(x + iy) + (x' + iy') = (x + x') + i(y + y'),$$

$$(x + iy) \cdot (x' + iy') = x \cdot x' - y \cdot y' + i(x \cdot y' + x' \cdot y),$$

which are obtained by adopting the usual rules of arithmetic together with the identity $i^2 = -1$. It is a familiar fact from algebra that the complex numbers satisfy the field axioms. Perhaps the only nontrivial one is the existence of multiplicative inverses, but we easily verify that

$$\frac{1}{x+iy} = \frac{x}{x^2+y^2} + i\left(-\frac{y}{x^2+y^2}\right)$$

does the trick. We leave the details as an exercise.

The construction of the complex numbers from the real numbers is purely algebraic (involving only finite operations). It is an example of a general procedure for enlarging any field by adjoining roots of an equation (in this case $x^2+1=0$). It is a remarkable fact that the complex numbers are *algebraically complete*, meaning that all polynomials with complex coefficients ($(\sqrt{5}+1)x^{27}+3x^6-\pi$, for example) have complex roots. Thus the complex numbers cannot be further enlarged algebraically. This fact is called the *fundamental theorem of algebra*. It will not be used in this book. For a proof, the reader can consult any text on complex variable theory.

The complex numbers do not possess an order that is mathematically relevant (one could impose an order, say lexicographic, but it would not have enough properties to make it worth studying). But there is a related concept of *absolute value* or *modulus* of a complex number, $|x+iy| = \sqrt{x^2+y^2}$. Note that x^2+y^2 is non negative, so the square root exists as a non negative real number. If we adopt the familiar convention of identifying the complex numbers with points in the plane ($x+iy$ corresponds to (x,y)), then $|x+iy|$ is the distance of the point to the origin by the Pythagorean theorem (see Figure 7.1.1).

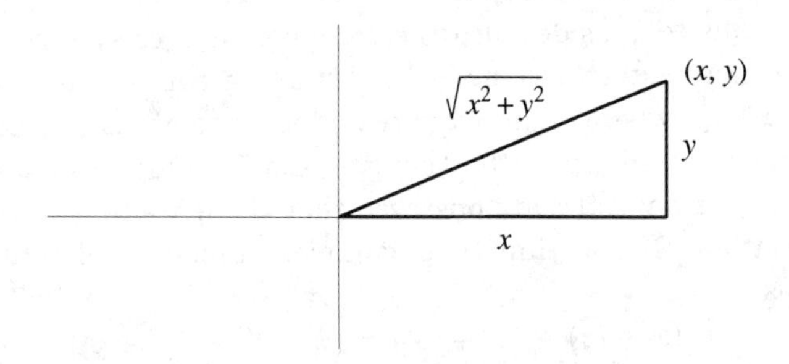

Figure 7.1.1:

The two most important properties of the absolute value are the multiplicativity, $|z\cdot z_1| = |z|\cdot|z_1|$, and the triangle inequality, $|z+z_1| \le |z|+|z_1|$, where z and z_1 are any complex numbers. The multiplicativity is established by a direct computation; if $z = x+iy$ and $z_1 = x_1+iy_1$, then

$$\begin{aligned}|z \cdot z_1|^2 &= |(xx_1 - yy_1) + i(xy_1 + x_1 y)|^2 \\ &= (xx_1 - yy_1)^2 + (xy_1 + x_1 y)^2 \\ &= x^2 x_1^2 + y^2 y_1^2 + x^2 y_1^2 + x_1^2 y^2\end{aligned}$$

because the cross terms cancel, and this is $(x^2 + y^2)(x_1^2 + y_1^2)$. Thus $|z \cdot z_1|^2 = |z|^2 |z_1|^2$, so the multiplicativity follows by taking the square-root.

There are many different proofs of the triangle inequality. First let's justify the name with a geometric interpretation. Consider the triangle in the plane with vertices at the origin, at z and at z_1. Then the lengths of the sides are $|z|, |z_1|$, and $|z_1 - z|$, as shown in Figure 7.1.2,

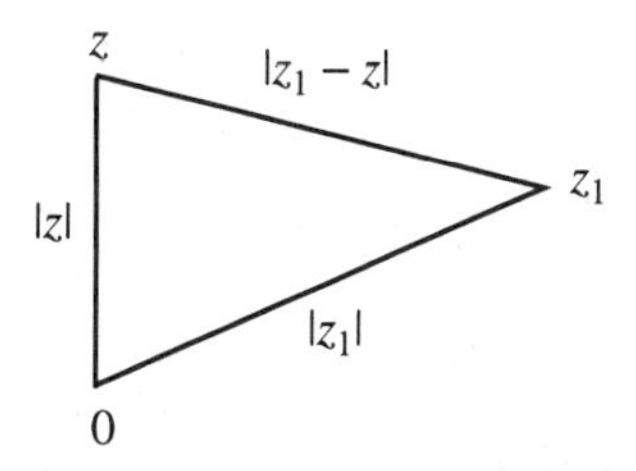

Figure 7.1.2:

and so $|z_1 - z| \leq |z| + |z_1|$ says the length of one side of the triangle is less than the sum of the lengths of the other two sides. This is equivalent to the inequality $|z_1 + z| \leq |z| + |z_1|$ since $|-z| = |z|$. It is then clear from the geometry that the inequality is an equality exactly when the triangle degenerates into a straight line with z and z_1 on the same side of the origin. Another way of saying this is $z_1 = rz$ where r is a non negative real number.

We can use this insight to fashion a proof. Let us hold z_1 fixed and vary z so that $|z| = c$ is also fixed and ask when $|z + z_1|$ is maximized.

The answer should be when $z = cz_1/|z_1|$, and if this is indeed the case we have $|z + z_1| = |z| + |z_1|$ for this particular choice and, hence, $|z + z_1| < |z| + |z_1|$ for every other choice (note that $|z| + |z_1|$ is not varying because z_1 and $|z| = c$ are fixed).

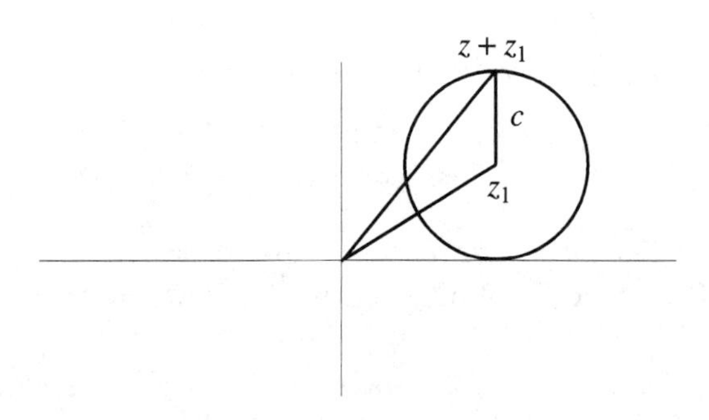

Figure 7.1.3:

Now we have the problem of maximizing $|z + z_1|$, or what is the same thing, maximizing $|z + z_1|^2 = (x + x_1)^2 + (y + y_1)^2$ given that $z_1 = x_1 + iy_1$ is fixed and $x^2 + y^2 = c^2$, as shown in Figure 7.1.3. But $(x + x_1)^2 + (y + y_1)^2 = x^2 + y^2 + x_1^2 + y_1^2 + 2xx_1 + 2yy_1$ and, since $x^2 + y^2 + x_1^2 + y_1^2 = c^2 + x_1^2 + y_1^2$ is fixed we need to maximize $2(xx_1+yy_1)$ given $x^2+y^2 = c^2$. This can be reduced to a simple calculus problem by solving $y = \pm\sqrt{c^2 - x^2}$ and finding the critical points of $f(x) = 2(xx_1 \pm y_1\sqrt{c^2 - x^2})$. Since $f'(x) = 2(x_1 \mp y_1 x/\sqrt{c^2 - x^2})$, we obtain $f'(x) = 0$ when $x_1 y = xy_1$. The two critical points occur when z and z_1 are colinear with the origin, but the maximum is clearly assumed when they both lie on the same side.

This is by no means the simplest proof of the triangle inequality, and we will give another after we discuss trigonometric functions.

A useful consequence of the triangle inequality is the inequality $||z| - |z_1|| \leq |z - z_1|$, which has the interpretation that the distance between points on concentric circles is at least the difference of the radii (see Figure 7.1.4). We leave the details to the exercises.

For our purposes, the main reason for introducing the absolute value is to use it to formulate topological properties of the complex numbers. For example, what do we mean by the limit of a sequence of complex numbers? If $z_n = x_n + iy_n$ we could define $\lim_{n\to\infty} z_n = z$ where $z = x + iy$ to mean $\lim_{n\to\infty} x_n = x$ and $\lim_{n\to\infty} y_n = y$. In other words, we separate the real and imaginary parts of the complex numbers and require convergence of each. In fact, with most concepts involving complex numbers we will take this approach. But there is another ap-

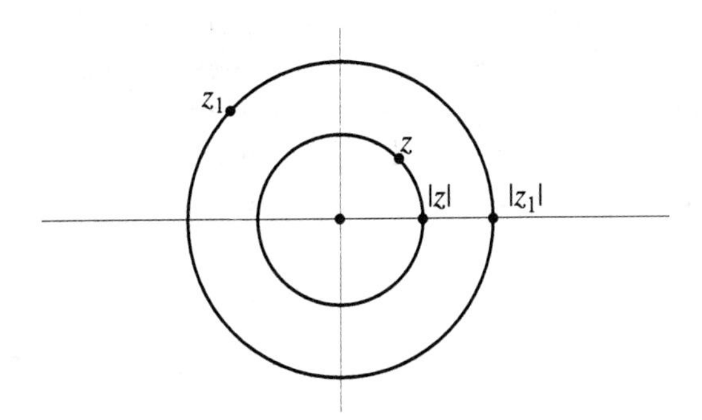

Figure 7.1.4:

proach, which is to replace the neighborhood $|x - x_0| < 1/n$ concept for reals by the neighborhood $|z - z_0| < 1/n$ concept for complex numbers. This would lead to the definition: $\lim_{k\to\infty} z_k = z$ means given any error $1/n$ there exists m such that $k \geq m$ implies $|z_k - z| < 1/n$.

We claim that the two definitions are equivalent. Indeed the second definition says the sequence eventually is entirely contained in any disc about z, while the first definition says the sequence eventually is entirely contained in any rectangle about z, as in Figure 7.1.5 ($\lim_{k\to\infty} x_k = x$ means eventually $|x_k - x| < 1/n$ and $\lim_{k\to\infty} y_k = y$ means eventually $|y_k - y| < 1/m$, and these two conditions mean z_k lies eventually in a rectangle).

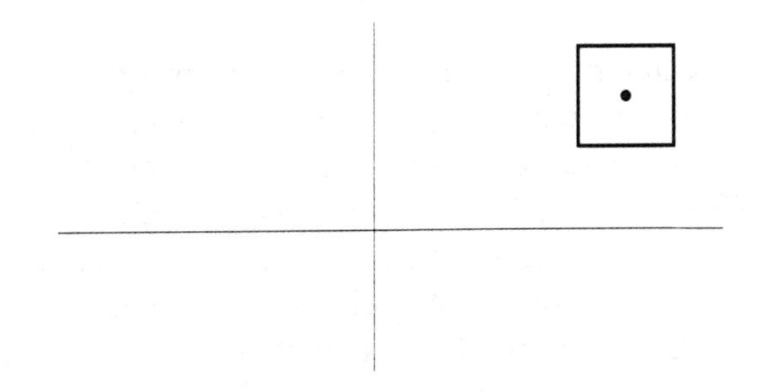

Figure 7.1.5:

The equivalence of the two definitions is thus a consequence of the fact that for every such rectangle there exists a disc about z entirely contained in the rectangle (so z_k in the disc implies z_k is in the rectangle), and conversely for every disc about z there exists such a rectangle entirely contained in the disc (so z_k in the rectangle implies z_k is in the disc). These facts are illustrated in Figure 7.1.6. We leave the details of the algebraic verification to the exercises.

Figure 7.1.6:

In the same way we can define a *Cauchy sequence* of complex numbers $z_1, z_2, \ldots$ by the condition: for all $1/n$ there exists m such that $|z_k - z_j| < 1/n$ if $j \geq m$ and $k \geq m$. We can show this is equivalent to the condition that $x_1, x_2, \ldots$ and $y_1, y_2, \ldots$ be Cauchy sequences of real numbers and hence obtain the *completeness* of the complex numbers: every Cauchy sequence of complex numbers converges to a complex number limit. (Note that the word *completeness* is used here in an entirely different sense than the algebraic completeness mentioned before.)

The complex number system has no direct intuitive interpretation. For this reason it met with some difficulty in being accepted into the mainstream of mathematics—such terminology as "imaginary" and "complex" reflects the resistance and even hostility of the imaginations of many mathematicians to accepting the complex numbers as a legitimate number system. After all, what do these numbers *mean*? Fortunately, such objections did not prevail, as the complex number system proved extraordinarily useful, in almost every branch of pure and applied mathematics. As far as the question of "meaning" is concerned, we can be content with no answer because there are so many different "interpretations" of the complex number system, each of which gives some significance—if not meaning—to the abstract mathematical system. Intuition is not completely excluded from the picture; one simply has to develop an intuition for the complex number system.

7.1.2 Complex-Valued Functions

We will be considering mostly functions whose domain lies in the real numbers and whose range lies in the complex numbers. Functions with domain and range in the complex numbers involve the theory of *complex analysis*, which requires a book of its own. We will only give a hint of this theory in the section on power series.

Functions whose range lies in the complex numbers, also called *complex-valued* functions, can be dealt with by splitting the real and imaginary parts. If $F(x)$ is a complex number for each x in the domain, then $F(x) = f(x) + ig(x)$ for uniquely determined real numbers $f(x)$ and $g(x)$; and this defines f and g as real-valued functions on the same domain. Conversely, if f and g are real-valued functions on a common domain, then $F(x) = f(x) + ig(x)$ is a complex-valued function on the same domain. We call f and g the *real* and *imaginary parts* of F, respectively. Properties such as continuity, differentiability, and integrability can be defined for complex-valued functions by modifying the definition for real-valued functions, and it can easily be shown that the complex-valued function has the property if and only if its real and imaginary parts both have the property. For example, if F is a complex-valued function on $[a, b]$, then F is said to be differentiable at x_0, a point in (a, b), if $\lim_{h\to 0}(F(x_0 + h) - F(x_0))/h$ exists. For integrability we do not have a notion of Riemann upper and lower sums, but we can still define F to be integrable if all the Cauchy sums $S(F, P)$ converge to a limit as the maximum length of subintervals of P goes to zero. We leave the details as exercises. Again we must stress that the complex values appear in the *range* and not the domain of the functions. For functions defined on the complex numbers there is *no* corresponding splitting into functions defined on the real numbers.

By and large, all the theorems we have established so far for real-valued functions are also valid for complex-valued functions. The proofs can either be modified simply or one may break up the complex-valued function $F = f + ig$ and apply the real-valued theorem to the real and imaginary parts. We discuss here only the few exceptions to this rule, where either the analogous theorem is false or a new idea is needed for the proof.

The *mean value theorem* does not hold for complex-valued functions. If we apply the theorem to the real and imaginary parts we obtain

$$\frac{f(b)-f(a)}{b-a}=f'(x_1)$$

and

$$\frac{g(b)-g(a)}{b-a}=g'(x_2),$$

but there does not have to be any connection between the points x_1 and x_2. It is easy to construct a counterexample, say $F(x)=x^2+ix^3$ on $[0,1]$ where $F'(x)=2x+i3x^2$ and $(F(1)-F(0))/(1-0)=1+i$, and ne'er the twain shall meet. Similarly, the Lagrange remainder formula for Taylor's theorem does not hold for complex-valued functions, but Taylor's theorem itself is true, as is the integral remainder formula. There is also no intermediate value property for continuous complex-valued functions, since there is no notion of "between" for complex numbers.

The next result is an example of a theorem that is true for both real- and complex-valued functions, but the complex version requires a more intricate proof.

Theorem 7.1.1 *Let F be a complex-valued integrable function. Then the real-valued function $|F|$ is integrable, and $|\int_a^b F(x)\,dx| \le \int_a^b |F(x)|\,dx$.*

Proof: We first show $\text{Osc}(|F|,P) \le \text{Osc}(f,P)+ \text{Osc}(g,P)$. This proves the integrability of $|F|$ since F integrable implies f and g are integrable, so $\text{Osc}(f,P)$ and $\text{Osc}(g,P)$ can be made as small as desired.

Now the oscillation over a partition is just the sum of the oscillations over each subinterval multiplied by the length of the subintervals, so it suffices to show that on each subinterval the oscillation of $|F|$ is less than or equal to the sum of the oscillations of f and g. Recall that the oscillation of a real-valued function f on an interval was defined to be the difference of the sup and inf, and this is clearly the sup of the values $|f(x)-f(y)|$ as x and y vary over the interval. Thus we are trying to prove

$$\sup_{x,y} \big|\,|F(x)|-|F(y)|\,\big| \le \sup_{x,y}|f(x)-f(y)|+\sup_{x,y}|g(x)-g(y)|,$$

and for this it suffices to prove

$$||F(x)| - |F(y)|| \le |f(x) - f(y)| + |g(x) - g(y)|$$

for then we may take the sup over all x and y in the interval and use the fact that the sup of a sum is less than or equal to the sum of the sups (the worst possible case is when the sups of $|f(x)-f(y)|$ and $|g(x)-g(y)|$ are assumed for the same value of x and y, when $\sup_{x,y}(|f(x)-f(y)|+|g(x)-g(y)|) = \sup_{x,y}|f(x)-f(y)| + \sup_{x,y}|g(x)-g(y)|$ whereas if $|f(x)-f(y)|$ and $|g(x)-g(y)|$ are largest for different values of x and y, then there will be inequality). But this follows from the inequality involving complex numbers $||z| - |z_1|| \le |x - x_1| + |y - y_1|$, where $z = x + iy$ and $z_1 = x_1 + iy_1$. The proof is left to the exercise set 7.1.3, number 10b.

Altogether, then, we have shown that $\mathrm{Osc}(|F|, P) \le \mathrm{Osc}(f, P) + \mathrm{Osc}(g, P)$. (We could also prove $\mathrm{Osc}(|F|, P) \le \mathrm{Osc}(F, P)$ if we defined $\mathrm{Osc}(F, P)$ appropriately, taking the definition of the oscillation of a complex-valued function on an interval to be the sup of $|F(x) - F(y)|$ over all x and y in the interval.) Thus F integrable implies $|F|$ integrable. The inequality $|\int_a^b F(x)\,dx| \le \int_a^b |F(x)|\,dx$ then follows from the corresponding inequality $|S(F,P)| \le S(|F|, P)$ for Cauchy sums (evaluated at the same points), which is just a consequence of the triangle inequality (iterated to sums of n numbers)

$$|S(F,P)| = \left|\sum_{j=1}^{n} F(y_j)(x_j - x_{j-1})\right| \le \sum_{j=1}^{n} |F(y_j)(x_j - x_{j-1})| = S(|F|, P)$$

since

$$|F(y_j)(x_j - x_{j-1})| = |F(y_j)|(x_j - x_{j-1}),$$

$x_j - x_{j-1}$ being positive. QED

7.1.3 Exercises

1. Prove that a complex-valued function is Riemann integrable (in the sense of convergence of Cauchy sums) if and only if its real and imaginary parts are Riemann integrable.

2. Prove the integral remainder formula for Taylor's theorem for a complex-valued function.

3. If $f(x)$ is a continuous complex-valued function and $f(x) \neq 0$ for any x in the domain, prove that $1/f(x)$ is continuous.

4. Find a countable dense subset of $\mathbb{C}$.

5. If z is a complex number define the *complex conjugate* $\bar{z}$ to be $x - iy$ if $z = x + iy$. a) Show that $\overline{z_1 + z_2} = \bar{z}_1 + \bar{z}_2, \overline{z_1 z_2} = \bar{z}_1 \bar{z}_2$, and $|z| = \sqrt{z\bar{z}}$. b) Show $1/z = \bar{z}/|z|^2$. c) Show that z is real if and only if $z = \bar{z}$.

6. If f is a complex-valued C^1 function and $f(x) \neq 0$ on the domain, prove that $|f|$ is C^1. Can you find a formula for $|f|'$? (**Hint:** use exercise 5.)

7. Show that $z^2 = i$ has solutions $z = 1/\sqrt{2} \pm i/\sqrt{2}$.

8. State and prove the fundamental theorem of the calculus (both forms) for complex-valued functions.

9. Prove that every disc $\{z : |z - z_0| < 1/n\}$ is contained in a square $\{z : |x - x_0| < 1/n \text{ and } |y - y_0| < 1/n\}$ and that every square $\{z : |x - x_0| < 1/n \text{ and } |y - y_0| < 1/n\}$ is contained in a disc $\{z : |z - z_0| < \sqrt{2}/n\}$.

10. a. Prove the inequality $||z| - |z_1|| \leq |z - z_1|$ for complex numbers. (See Figure 7.1.4.)

    b. Prove $||z| - |z_1|| \leq |x - x_1| + |y - y_1|$ for $z = x + iy$ and $z_1 = x_1 + iy_1$.

7.2 Numerical Series and Sequences

7.2.1 Convergence and Absolute Convergence

We have already discussed the meaning of the convergence of an infinite sequence of real or complex numbers. We will also want to consider infinite series, written $x_1 + x_2 + \cdots$ or $\sum_{k=1}^{\infty} x_k$. To each infinite series corresponds the infinite sequence of partial sums $s_n = \sum_{k=1}^{n} x_k$, and *we define $\sum_{k=1}^{\infty} x_k$ to be convergent with $\sum_{k=1}^{\infty} x_k = s$ if $s_1, s_2, \ldots$ is convergent with limit s.* This definition applies to real or complex-valued series. This convergence is sometimes called *ordinary convergence* to distinguish it from the more stringent absolute convergence we will define later. A series that is not convergent is called *divergent*, and we will sometimes speak of series that *diverge to* $+\infty$ or $-\infty$ (or write $\sum_{k=1}^{\infty} x_k = \pm\infty$) if $s_1, s_2, \ldots$ converges to $+\infty$ or $-\infty$ in the extended real numbers.

From the sequence of partial sums $s_1, s_2, \ldots$ we can recover the terms x_n of the series by the difference formula $x_1 = s_1$ and $x_n = s_n - s_{n-1}$ for $n > 1$. Note that if $s_1, s_2, \ldots$ is any infinite sequence of numbers we can obtain by the same formula an infinite series $\sum_{k=1}^{\infty} x_k$ that has $s_1, s_2, \ldots$ as partial sums. In this way we have a naturally defined one-to-one correspondence between infinite series and infinite sequences—and so in a sense the theory of the two is the same. Nevertheless, it is sometimes more convenient to think about certain problems in one or the other form, so we will maintain both perspectives. An example of a concept that is natural for series but not for sequences is that of *absolute convergence.* We say $\sum_{k=1}^{\infty} x_k$ is *absolutely convergent* if $\sum_{k=1}^{\infty} |x_k|$ is convergent. This definition applies to real- or complex-valued series. The idea of absolute convergence in the real case is that the convergence should not be caused by cancellation of positive and negative terms.

Later in this chapter we will discuss infinite series and sequences of functions. The material in this section lays the foundation for that more complicated theory. The reader is probably familiar with most of the ideas—at least on an informal level—from calculus.

Before beginning the discussion of general series, we start with an important and familiar example, the *geometric series* $\sum_{k=1}^{\infty} r^k$ for $0 < r < 1$. Since we have

$$(1-r)(r + r^2 + \cdots + r^n)$$
$$= r + r^2 + \cdots + r^n - r^2 - \cdots - r^n - r^{n+1} = r - r^{n+1},$$

we have

$$s_n = \sum_{k=1}^{n} r^k = \frac{r}{1-r} - \frac{r^{n+1}}{1-r},$$

so

$$s_n - \frac{r}{1-r} = -\frac{r^{n+1}}{1-r}.$$

We claim for $0 < r < 1$ that this has limit zero as $n \to \infty$, which will show $\sum_{k=1}^{\infty} r^k$ is convergent and equals $r/(1-r)$. Since $r < 1$, the factor $1/(1-r)$ is harmless, so we need to show $\lim_{n\to\infty} r^n = 0$. If $r < 1$, then $1/r > 1$, so by the axiom of Archimedes $1/r \geq 1 + 1/k$ for some integer k. Then

$$\frac{1}{r^n} \geq \left(1 + \frac{1}{k}\right)^n = 1 + \frac{n}{k} + \cdots \frac{n}{k^{n-1}} + \frac{1}{k^n}$$

by the binomial theorem. Since all the terms in the binomial expansion are positive, we have $1/r^n \geq 1 + n/k$, so $r^n \leq k/(n+k)$ and this can be made $\leq 1/m$ by taking $n \geq k(m-1)$.

Note that the rate of convergence of r^n to 0 (hence the rate of convergence of $\sum_{n=1}^{\infty} r^n$) depends on how close r is to 1. The closer r is to 1, the larger k is in the estimate $1/r \geq 1 + 1/k$ and the slower the convergence. The estimate we have given is rather crude; the geometric series actually converges much more rapidly for fixed r. Nevertheless it is true that there is no uniform rate of convergence for all $r < 1$. We leave this as an exercise (note that the argument we gave for the convergence does not suffice to *prove* that the rate of convergence is not uniform—it only shows that one attempt to estimate the rate of convergence leads to dependence on r).

We begin the discussion of general properties of series with the elementary observations of linearity and order preservation. If $\sum_{k=1}^{\infty} x_k$ and $\sum_{k=1}^{\infty} y_k$ are convergent, then $\sum_{k=1}^{\infty}(x_k + y_k)$ is convergent and $\sum_{k=1}^{\infty} ax_k$ is convergent with $\sum_{k=1}^{\infty}(x_k + y_k) = \sum_{k=1}^{\infty} x_k + \sum_{k=1}^{\infty} y_k$ and $\sum_{k=1}^{\infty} ax_k = a\sum_{k=1}^{\infty} x_k$. Also if $x_k \geq y_k$ for every k, then $\sum_{k=1}^{\infty} x_k \geq \sum_{k=1}^{\infty} y_k$. These follow from the corresponding results for the sequences of partial sums and the fact that like operations prevail (the partial sums of the sum series $\sum_{k=1}^{\infty}(x_k + y_k)$ are the sums of the partial sums of the summands, etc.). Incidentally, there is no such result for products; $\sum_{k=1}^{\infty} x_k y_k$ need not be convergent (see exercises). Indeed there is no relation between $\sum_{k=1}^{n} x_k y_k$ and $\sum_{k=1}^{n} x_k$ and $\sum_{k=1}^{n} y_k$.

Another relatively simple observation is that the convergence of an infinite series (but not the limit) does not depend on any finite number of terms. Indeed changing the terms x_k for $k \leq n$ means that all the partial sums s_m beyond s_n are changed to $s_m + c$ for a fixed number c and so converge or not as with s_m. A particular form of modifying a finite number of terms is to rearrange the first n terms. In this case the constant c is zero, so the limit is unchanged. Rearranging an infinite number of terms is another matter, which we will return to later.

The Cauchy criterion for convergence of sequences translates easily into a criterion for convergence of series that again does not involve the limit.

Theorem 7.2.1 *$\sum_{k=1}^{\infty} x_k$ converges if and only if for every error $1/n$ there exists m such that $|\sum_{k=p}^{q} x_k| < 1/n$ for all p, q satisfying $q \geq p \geq m$.*

Proof: If $s_1, s_2, \ldots$ denotes the partial sums, then $s_q - s_{p-1} = \sum_{k=p}^{q} x_k$, so this is exactly the Cauchy criterion for $s_1, s_2, \ldots$. QED

A special case of this is the observation that convergence of $\sum_{k=1}^{\infty} x_k$ implies $\lim_{k\to\infty} x_k = 0$, since $x_k = s_k - s_{k-1}$. Of course the converse is false (we have already discussed the fact that the smallness of the difference between neighboring terms s_{k-1} and s_k in a sequence is not the same thing as the Cauchy criterion). As we will show shortly, $\sum_{k=1}^{\infty} 1/k$ diverges.

A series $\sum_{k=1}^{\infty} x_k$ is said to be *absolutely convergent* if the series of absolute values $\sum_{k=1}^{\infty} |x_k|$ is convergent. To justify the terminology let us observe that absolute convergence implies convergence. This is an immediate consequence of the Cauchy criterion, since $|\sum_{k=p}^{q} x_k| \leq \sum_{k=p}^{q} |x_k|$ by the triangle inequality. Intuitively we can argue that the possibility of cancellation of positive and negative values (or the more complicated cancellation of positive and negative values in the real and imaginary parts in the complex-valued case) can only help with convergence. An example of a convergent series that is not absolutely convergent is $1 - 1/2 + 1/3 - 1/4 + 1/5 - 1/6 + \cdots = \sum_{k=1}^{\infty}(-1)^k/k$ (we will prove this later).

Absolute convergence is an extremely useful concept, and most tests for convergence actually establish absolute convergence. The most fundamental test is the comparison test.

Theorem 7.2.2 (*Comparison test*) *Let $\sum_{k=1}^{\infty} x_k$ and $\sum_{k=1}^{\infty} y_k$ be infinite series with the y_k's non-negative and $|x_k| \leq y_k$ (it suffices to have this for all but a finite number of terms). If $\sum_{k=1}^{\infty} y_k$ is convergent, then $\sum_{k=1}^{\infty} x_k$ is absolutely convergent.*

Proof: Note that for a series of non-negative terms, convergence and absolute convergence are the same. The theorem is a trivial consequence of the Cauchy criterion as $\sum_{k=p}^{q} |x_k| \leq \sum_{k=p}^{q} y_k$ from the hypothesis (if there is a finite number of exceptions to $|x_k| \leq y_k$ we must take p sufficiently large). From the convergence of $\sum_{k=1}^{\infty} y_k$ we know we can make $\sum_{k=p}^{q} y_k < 1/n$ if $q \geq p \geq m$ and so $\sum_{k=p}^{q} |x_k| < 1/n$ also, so $\sum_{k=1}^{\infty} |x_k|$ converges. QED

The comparison test can also be used in contrapositive form to prove divergence of series of non-negative terms (see exercises).

Many applications of the comparison test use the geometric series; in fact the familiar root and ratio tests are proved in this fashion. We state the results here but leave the proofs to the exercises.

Theorem 7.2.3

a. (*Ratio test*) *If* $|x_{n+1}/x_n| < r$ *for all sufficiently large* n *and some* $r < 1$, *then* Σx_n *converges absolutely; while if* $|x_{n+1}/x_n| \geq 1$ *for all sufficiently large* n, *then* Σx_n *diverges.*

b. (*Root test*) *If* $\sqrt[n]{|x_n|} < r$ *for all sufficiently large* n *and some* $r < 1$, *then* Σx_n *converges absolutely.*

Next we consider the important class of series $\sum_{n=1}^{\infty} 1/n^a$ where a is a positive real number. (Strictly speaking we have not yet defined n^a unless a is rational, so we should either restrict the discussion to rational numbers $a = p/q$ when $n^{p/q} = \sqrt[q]{n^p}$ or else observe that the argument we give depends only on the familiar properties of powers, so the results will ultimately be justified in the next chapter when we establish these properties in general.)

We want to show $\sum_{n=1}^{\infty} 1/n^a$ converges for $a > 1$ and diverges for $a \leq 1$. If this reminds you of the result concerning the improper integrals $\int_1^{\infty} 1/x^a \, dx$, it should. In fact one can use the results about the integrals to establish the results about the series (see exercises). We shall give a more direct argument here. Notice that we will not evaluate $\sum_{n=1}^{\infty} 1/n^a$ exactly, as a function of a; this is the notorious Riemann zeta function, which is related to questions about the distribution of prime numbers in ways that are too mysterious to explain here. You might also be intrigued with the identity $\sum_{n=1}^{\infty} 1/n^2 = \pi^2/6$, which we will establish in the chapter on Fourier series.

The key idea for deciding the convergence or divergence of the series $\sum_{n=1}^{\infty} 1/n^a$ is to break the sum into *dyadic* pieces, meaning the range $2^k \le n \le 2^{k+1} - 1$. For n in this range the values of $1/n^a$ vary between $1/2^{ka}$ at the largest to $1/2^{(k+1)a} = 1/2^a \cdot 1/2^{ka}$ at the smallest—so they are roughly the same order of magnitude. For the proof of convergence we use the upper bound, $1/n^a \le 1/2^{ka}$ if $2^k \le n < 2^{k+1}$, and for the proof of divergence we use the lower bound, $1/n^a \ge 1/2^a \cdot 1/2^{ka}$ if $2^k \le n < 2^{k+1}$. By the comparison test we thus need to determine the convergence of the series $\sum_{n=1}^{\infty} b_n$ where $b_n = 1/2^{ka}$ where k is related to n by $2^k \le n < 2^{k+1}$. Thus

$$\begin{aligned}\sum_{n=1}^{\infty} b_n &= \frac{1}{2^0} + \underbrace{\frac{1}{2^{1a}} + \frac{1}{2^{1a}}}_{2^1 \text{ terms}} \\ &\quad + \underbrace{\frac{1}{2^{2a}} + \frac{1}{2^{2a}} + \frac{1}{2^{2a}} + \frac{1}{2^{2a}}}_{2^2 \text{ terms}} \\ &\quad + \cdots + \underbrace{\frac{1}{2^{ka}} + \cdots + \frac{1}{2^{ka}}}_{2^k \text{ terms}} + \cdots .\end{aligned}$$

Note that each block of 2^k terms (all equal to $1/2^{ka}$) adds up to exactly $2^k \cdot 1/2^{ka} = 2^{(1-a)k}$, so $\sum_{n=1}^{2^m-1} b_n = \sum_{k=0}^{m-1} 2^{(1-a)k}$ and this is a geometric series with $r = 2^{1-a}$. If $a > 1$, then $r < 1$, so the geometric series converges; while if $a \le 1$, then $r \ge 1$ and the geometric series diverges. Strictly speaking we have only analysed the behavior of the partial sums $\sum_{n=1}^{k} b_n$ for $k = 2^m - 1$; but since the terms b_n are non-negative, this is enough to decide the convergence $(a > 1)$ and divergence $(a \le 1)$ of the series.

In particular, we have established the divergence of the harmonic series $\sum_{n=1}^{\infty} 1/n$. (As an interesting computer "experiment", try evaluating the sum of the harmonic series on your favorite computer or programmable calculator. You may come up with a finite answer because $1/k$ is rounded off to zero before the partial sums become very large. This illustrates how slowly the series diverges. It also raises a profound philosophical question: can there be any natural phenomenon whose existence depends on the divergence of the harmonic series?) Why, then, does the series $\sum_{n=1}^{\infty}(-1)^n/n$ converge? One way to see this is to combine consecutive terms, say $n = 2k - 1$ and $n = 2k$,

$$\frac{-1}{2k-1} + \frac{1}{2k} = \frac{-1}{(2k-1)2k}.$$

The series $\sum_{k=1}^{\infty} -1/(2k-1)2k$ converges absolutely by comparison with $\sum_{k=1}^{\infty} 1/k^2$, so the partial sums $\sum_{n=1}^{m}(-1)^n/n$ with m even converge and if m is odd we have $\sum_{k=1}^{m}(-1)^n/n = (-1)^m/m+\sum_{k=1}^{m-1}(-1)^n/n$, so the odd partial sums converge to the same limit.

We will discuss another way to establish the convergence of $\sum_{n=1}^{\infty}(-1)^n/n$ in section 7.2.3.

7.2.2 Rearrangements

If $\sum_{n=1}^{\infty} a_n$ is an infinite series, we say $\sum_{n=1}^{\infty} b_n$ is a *rearrangement* if there is a one-to-one correspondence between the terms; or, put another way, $b_n = a_{m(n)}$ where m is a function from the natural numbers to the natural numbers that is one-to-one and onto. Thus $\sum_{n=1}^{\infty} b_n$ is a series consisting of the same terms as $\sum_{n=1}^{\infty} a_n$ but with a different order.

At first we might think that if $\sum_{n=1}^{\infty} a_n$ is convergent, then $\sum_{n=1}^{\infty} b_n$ should also be convergent with the same limit, but this turns out to be the case only for absolutely convergent series. To understand why this should be so, we consider the intuitive idea that the tail of the series is small. That is, $\sum_{n=1}^{\infty} a_n = \sum_{n=1}^{N} a_n + \sum_{n=N+1}^{\infty} a_n$ and the finite sum $\sum_{n=1}^{N} a_n$ is close to the limit, so $\sum_{n=N+1}^{\infty} a_n$ is small. In the absolutely convergent case $\sum_{n=N+1}^{\infty} |a_n|$ can also be made small, but in the nonabsolutely convergent case $\sum_{n=N+1}^{\infty} |a_n|$ is divergent no matter how large N is. This means $\sum_{n=N+1}^{\infty} a_n$ is small only because of cancellation.

Now consider a partial sum of the rearrangement $\sum_{n=1}^{m} b_n$. If we take m large enough, then all the terms $a_1, a_2, \ldots, a_N$ will show up (here we fix N and m will depend on N) so that

$$\sum_{n=1}^{m} b_n = \sum_{n=1}^{N} a_n + \sum_{N+1}{}' a_n$$

where $\sum'_{N+1} a_n$ just represents a finite sum of *some* of the a_n with $n \geq N+1$—a selection of the tail $\sum_{n=N+1}^{\infty} a_n$. For the absolutely convergent case

$$\left| \sum_{N+1} {}' a_n \right| \leq \sum_{n=N+1}^{p} |a_n|$$

for p large enough so that each a_n in the selection of the tail $\sum'_{N+1} a_n$ has $n \leq p$. This can be made as small as desired by taking N large (hence m depending on N also large), so $\sum_{n=1}^{m} b_n$ differs from $\sum_{n=1}^{N} a_n$ by as little as desired, showing that $\sum_{n=1}^{\infty} b_n$ converges to the same limit. But in the nonabsolutely convergent case we do not know that the selection of the tail $\sum'_{N+1} a_n$ is small just because the tail is small. On the contrary, we expect that by an especially nasty selection of the tail we should be able to get something large, since $\sum_{n=N+1}^{\infty} |a_n|$ diverges. For example, in the case that all the terms are real, the divergence of $\sum_{n=N+1}^{\infty} |a_n|$ can only occur if either the sum of all the positive a_n's or the sum of all the negative a_n's diverges. Let us say $a_{n_k} > 0$ for some sequence $n_1 < n_2 < n_3 < \cdots$ and $\sum_{k=1}^{\infty} a_{n_k} = +\infty$. Then if the selection of the tail $\sum'_{N+1} a_n$ is made entirely from the values of n equal $n_1, n_2, \ldots$, then the selection of the tail can be made as large as desired! This is the idea that lies behind the next theorem.

Theorem 7.2.4

a. *Let $\sum_{n=1}^{\infty} a_n$ be absolutely convergent. Then any rearrangement is also absolutely convergent and has the same limit.*

b. *Suppose every rearrangement of $\sum_{n=1}^{\infty} a_n$ is convergent. Then $\sum_{n=1}^{\infty} a_n$ is absolutely convergent.*

Proof:

a. Given any error $1/m$, we have to show that $\sum_{n=1}^{k} b_n$ can be made to differ from $\sum_{n=1}^{\infty} a_n$ by at most $1/m$, by taking k large enough. Here $\sum_{n=1}^{\infty} b_n$ denotes a fixed rearrangement. First we choose N large enough so that $\sum_{n=N+1}^{\infty} |a_n| \leq 1/m$ and then choose k large enough so that $a_1, a_2, \ldots, a_N$ occur in $b_1, b_2, \ldots, b_k$. Then $\sum_{n=1}^{k} b_n - \sum_{n=1}^{N} a_n$ is a selection of the tail $\sum_{n=N+1}^{\infty} a_n$, hence

$$\left| \sum_{n=1}^{k} b_n - \sum_{n=1}^{N} a_n \right| \leq \frac{1}{m},$$

so $\sum_{n=1}^{\infty} b_n$ converges to the same limit as $\sum_{n=1}^{\infty} a_n$. The absolute convergence of $\sum_{n=1}^{\infty} b_n$ follows because $\sum_{n=1}^{\infty} |b_n|$ is a rearrangement of $\sum_{n=1}^{\infty} |a_n|$.

b. Suppose $\sum_{n=1}^{\infty} |a_n|$ diverges. We need to construct a rearrangement of $\sum_{n=1}^{\infty} a_n$ that diverges. Assume first that the terms a_n are real. We can also assume without loss of generality that there exists a sequence $n_1 < n_2 < n_3 < \cdots$ such that $a_{n_k} > 0$ and $\sum_{k=1}^{\infty} a_{n_k} = +\infty$ (if not, then we can find $a_{n_k} < 0$ with $\sum_{k=1}^{\infty} a_{n_k} = -\infty$, and the argument is essentially the same). To simplify the notation let $c_k = a_{n_k}$ and let $d_1, d_2, \ldots$ be the remaining a_n's (they may be finite in number). To describe a rearrangement of $\sum_{n=1}^{\infty} a_n$ we will tell how to pick all the c_k's and d_k's in some order.

Now the idea is that we would like to take $c_1, c_2, \ldots$ for this will clearly give a divergent series. But this is not a rearrangement, because we have omitted the d_k's. So we need to fix this by sprinkling the d_k's very thinly among the c_k's. Look at d_1. Wait until $c_1 + c_2 + \cdots + c_{N_1}$ exceeds $|d_1| + 1$, so then $c_1 + c_2 + \cdots + c_{N_1} + d_1 \geq 1$. Next look at d_2. Wait until $c_1 + \cdots + c_{N_1} + d_1 + c_{N_1+1} + \cdots + c_{N_2}$ exceeds $|d_2| + 2$, so then $c_1 + \cdots + c_{N_1} + d_1 + c_{N_1+1} + \cdots + c_{N_2} + d_2 \geq 2$. We can continue in this fashion to sprinkle in d_k after c_{N_k} so that the sum up to d_k exceeds k. In this way we obtain a rearrangement that diverges.

In the case where the a_n are complex, say $a_n = x_n + iy_n$, the divergence of $\sum_{n=1}^{\infty} |a_n|$ implies that either $\sum_{n=1}^{\infty} |x_n|$ or $\sum_{n=1}^{\infty} |y_n|$ diverges (else $\sum_{n=1}^{N} |a_n| \leq \sum_{n=1}^{N} |x_n| + \sum_{n=1}^{N} |y_n|$ by the triangle inequality, hence $\sum_{n=1}^{\infty} |a_n|$ converges by comparison). Then by the previous argument we can rearrange $\sum_{n=1}^{\infty} a_n$ so that either the real or imaginary parts of the series diverge, and this implies the complex series diverges (convergence of a complex series is equivalent to convergence of the real and imaginary parts). QED

We say that $\sum_{n=1}^{\infty} a_n$ converges *unconditionally* if every rearrangement converges. The theorem we have just proved shows that $\sum_{n=1}^{\infty} a_n$ *converges unconditionally if and only if it converges absolutely.*

A series is said to converge *conditionally* if it converges but some rearrangement diverges. We can generalize the previous argument to show that a conditionally convergent series of real numbers can be rearranged so that it converges to any prescribed real number! The idea is that now we can divide the terms a_n into those that are non-negative, $c_1, c_2, c_3, \ldots$ and those that are negative, $d_1, d_2, \ldots$, and we must have $\sum_{n=1}^{\infty} c_n = +\infty$ and $\sum_{n=1}^{\infty} d_n = -\infty$ (if both were finite the series would converge absolutely, while if only one were finite the series would diverge). Also, $\lim_{n\to\infty} c_n = 0$ and $\lim_{n\to\infty} d_n = 0$ since the series $\sum_{n=1}^{\infty} a_n$ converges. Now suppose we want to rearrange the series to converge to A. Say $A \geq 0$. Take $c_1, c_2, \ldots, c_{N_1}$ until $c_1 + c_2 + \cdots + c_{N_1} > A$ for the first time. Then take $d_1, \ldots, d_{N_2}$ until $c_1 + \cdots + c_{N_1} + d_1 + \cdots + d_{N_2} < A$ for the first time. Keep switching back and forth between c's and d's to make the partial sums oscillate above and below A. This is always possible since $\sum_{n=1}^{\infty} c_n = +\infty$ and $\sum_{n=1}^{\infty} d_n = -\infty$, and we eventually use all the c's and d's so that we have a genuine rearrangement. Finally the conditions $\lim_{n\to\infty} c_n = 0$ and $\lim_{n\to\infty} d_n = 0$ imply that the limit of the rearranged series is A, because we switch directions just when the partial sums cross the value A. We leave the details as an exercise.

Returning to the positive results, we note also that absolute convergence implies the possibility of rearrangement of multiply indexed series. For example, let a_{mn} denote a real or complex number for each natural number n and m. Then we can sum all the a_{mn} in either order,

$$\sum_{m=1}^{\infty}\left(\sum_{n=1}^{\infty} a_{mn}\right) \quad \text{or} \quad \sum_{n=1}^{\infty}\left(\sum_{m=1}^{\infty} a_{mn}\right).$$

Thinking of a_{mn} as an infinite matrix

$$\begin{matrix} a_{11} & a_{12} & a_{13} & \cdots \\ a_{21} & a_{22} & a_{23} & \cdots \\ a_{31} & a_{32} & a_{33} & \cdots \\ \vdots & & & \\ \vdots & & & \end{matrix}$$

these correspond to summing by rows or by columns. We could also consider summing along diagonals to get an ordinary infinite series. In general, the fact that one of these procedures yields a finite number does not imply that any of the others will or that even if they are all finite that they must be equal. However, if any of these procedures yields a finite number when applied to $|a_{mn}|$, then we say the double infinite series $\sum\sum a_{mn}$ is absolutely convergent, and we can prove that all three summing procedures yield the same finite number. The idea of the proof is the same as in Theorem 7.2.4. We leave the details as an exercise. An analogous result holds for multiple integrals; we will cover this in Chapter 15.

7.2.3 Summation by Parts*

In this section we establish a general method of proving convergence of non-absolutely convergent series, called summation by parts, which is an analog of the integration by parts formula. It can be used to give another proof of the convergence of $\sum_{n=1}^{\infty}(-1)^n/n$.

Suppose a series can be written in the form $\sum_{n=1}^{\infty} A_n b_n$ (of course this is always possible in many ways, the idea being that a clever choice of A_n and b_n will be required to get anything out of the method). We then want to do the analog of integrating b_n and differentiating A_n. So we let $B_n = \sum_{k=1}^{n} b_k$ and $a_n = A_{n+1} - A_n$. Then

$$\begin{aligned}\sum_{n=1}^{m} a_n B_n &= a_1B_1 + a_2B_2 + \cdots + a_mB_m \\ &= (A_2 - A_1)b_1 + (A_3 - A_2)(b_1 + b_2) \\ &\quad + (A_4 - A_3)(b_1 + b_2 + b_3) \\ &\quad + \cdots + (A_{m+1} - A_m)(b_1 + b_2 + b_3 + \cdots + b_m).\end{aligned}$$

If we collect all the terms that contain b_1, then all the terms that contain b_2, and so on, we obtain

$$\begin{aligned}\sum_{n=1}^{m} a_n B_n &= b_1((A_2 - A_1) + (A_3 - A_2) + \cdots + (A_{m+1} - A_m)) \\ &\quad + b_2((A_3 - A_2) + (A_4 - A_3) + \cdots + (A_{m+1} - A_m)) \\ &\quad + \cdots + b_m(A_{m+1} - A_m) \\ &= b_1(A_{m+1} - A_1) + b_2(A_{m+1} - A_2) \\ &\quad + \cdots + b_m(A_{m+1} - A_m) = -\sum_{n=1}^{m} b_nA_n + A_{m+1}B_m.\end{aligned}$$

Thus, if the term $A_{m+1}B_m$ goes to zero, the question of convergence and the limit of $\sum_{n=1}^{\infty} A_n b_n$ is the same as the question of convergence and the limit of $-\sum_{n=1}^{\infty} a_n B_n$ (note that the absolute convergence of these two series is a different question, because the above manipulations would not work with absolute values). By clever choice of A_n and b_n we may well find it easier to prove convergence of $\sum a_n B_n$. For the series $\sum_{n=1}^{\infty}(-1)^n/n$ we take $A_n = 1/n$ and $b_n = (-1)^n$. Then

$$a_n = \frac{1}{n+1} - \frac{1}{n} = \frac{-1}{(n+1)n}$$

and $B_n = 0$ or -1 depending on whether n is even or odd. Since $A_{n+1}B_n \to 0$ as $n \to \infty$, we conclude $\sum_{n=1}^{\infty}(-1)^n/n$ is convergent since $\sum_{n-1}^{\infty} a_n B_n$ is absolutely convergent by comparison with $\sum_{n=1}^{\infty} 1/n^2$. Of course this is essentially the same argument as before, but we can now generalize it.

Theorem 7.2.5

a. *Let $A_1, A_2, \ldots$ be a sequence of positive numbers converging monotonically to zero (so $A_1 \geq A_2 \geq A_3 \ldots$ and $\lim_{n\to\infty} A_n = 0$). Then $\sum_{n=1}^{\infty}(-1)^n A_n$ is convergent.*

b. *Suppose also $b_1, b_2, \ldots$ is any sequence of real numbers with $B_n = b_1 + \cdots + b_n$ bounded, say $|B_n| \leq M$ for all n. Then $\sum_{n=1}^{\infty} A_n b_n$ is convergent.*

Proof: Part a is a special case of part b with $b_n = (-1)^n$, so we prove part b. Note that if we form $a_n = A_{n+1} - A_n$, then $-a_n \geq 0$ since $A_{n+1} \leq A_n$ and

$$\begin{aligned} \sum_{n=1}^{m} -a_n &= -(A_2 - A_1) - (A_3 - A_2) - \cdots - (A_{m+1} - A_m) \\ &= A_1 - A_{m+1}, \end{aligned}$$

which converges as $m \to \infty$ to A_1 since $A_m \to 0$. Thus $\sum_{n=1}^{\infty} -a_n$ is an absolutely convergent series. Now if we multiply the terms of an absolutely convergent series by a bounded sequence the resulting series is still absolutely convergent. In this case $|B_n| \leq M$, so $\sum_{n=p}^{q} |a_n B_n| \leq M \sum_{n=p}^{q} |a_n| \to 0$ as $p, q \to \infty$, proving $\sum_{n=1}^{\infty} a_n B_n$ is absolutely convergent by the Cauchy criterion. Since we also have $\lim_{n\to\infty} A_{n+1}B_n = 0$ since $\lim_{n\to\infty} A_{n+1} = 0$ and $|B_n| \leq M$, we conclude that $\sum_{n=1}^{\infty} A_n b_n$ is convergent by summation by parts. QED

7.2.4 Exercises

1. Give an example of two convergent series $\sum_{k=1}^{\infty} x_k$ and $\sum_{k=1}^{\infty} y_k$ such that $\sum_{k=1}^{\infty} x_k y_k$ diverges. Can this happen if one of the series is absolutely convergent?

2. State a contrapositive form of the comparison test that can be used to show divergence of a series.

3. Show that it is not true that for every error $1/m$ there exists n such that $|\sum_{k=1}^{n} r^k - r/(1-r)| < 1/m$ for all r in $0 < r < 1$.

4. Prove the ratio test (Theorem 7.2.3a). What does this tell you if $\lim_{n\to\infty} |x_{n+1}/x_n|$ exists?

5. Show $\sum_{n=2}^{N} 1/n^a \leq \int_1^N 1/x^a\, dx$, and use this to prove the convergence of the series for $a > 1$.

6. Prove that every conditionally convergent series of real numbers that does not converge absolutely can be rearranged to have any prescribed real limit.

7. Prove that a series of complex numbers is absolutely convergent if and only if the series of real and imaginary parts are absolutely convergent.

8. Prove that if $(\sum_{n=1}^{\infty} |a_{mn}|)$ is finite for every m and $\sum_{m=1}^{\infty}(\sum_{n=1}^{\infty} |a_{mn}|)$ is finite, then $\sum_{m=1}^{\infty}(\sum_{n=1}^{\infty} a_{mn}) = \sum_{n=1}^{\infty}(\sum_{m=1}^{\infty} a_{mn})$.

9. Give an example of a doubly indexed series a_{mn} such that $\sum_{m=1}^{\infty}(\sum_{n=1}^{\infty} a_{mn}) \neq \sum_{n=1}^{\infty}(\sum_{m=1}^{\infty} a_{mn})$.

10. Prove the root test (Theorem 7.2.3b).

11. Suppose $|a_n| \leq b_n - b_{n+1}$ where b_n decreases monotonically to zero. Prove that $\sum_{n=1}^{\infty} a_n$ converges absolutely.

12. *Show that if $\sum_{n=1}^{\infty} a_n$ is absolutely convergent, there exists an absolutely convergent series $\sum_{n=1}^{\infty} b_n$ such that $\lim_{n\to\infty} a_n/b_n = 0$. Explain why this result shows that there is no "universal" comparison series for testing absolute convergence.

13. Give an example of a divergent series whose partial sums are bounded.

14. *Show that $\sum_{n=1}^{\infty} 1/a_n$ converges, where $a_1, a_2, \ldots$ is the Fibonacci sequence $1, 1, 2, 3, 5, 8, 13, \ldots$

7.3 Uniform Convergence

7.3.1 Uniform Limits and Continuity

The results discussed in this section are valid for both sequences and series of functions. We will usually state the result for sequences $\{f_n(x)\}$ and leave as an exercise for the reader the formulation of the analogous result for series $\sum_{k=1}^{\infty} g_k$, since this amounts to stating the sequence result for the partial sums $f_n = \sum_{k=1}^{n} g_k$ of the series.

Let $f_1(x), f_2(x), \ldots$ be a sequence of real- or complex-valued functions defined on a common domain D. We say the sequence *converges* to a function $f(x)$, written $\lim_{n\to\infty} f_n(x) = f(x)$, if for each x_0 in the domain the sequence of numbers $f_n(x_0)$ converges to the number $f(x_0)$. This notion of convergence is sometimes called *pointwise* or simple convergence in order to distinguish it from other notions of convergence we will have to consider. In fact, we will see that pointwise convergence is not always a very useful notion.

Consider, for example, the infinite series $\sum_{n=1}^{\infty} x^n$. From our discussion of the geometric series we recognize that this is a pointwise converging series of functions on the domain $-1 < x < 1$, and the limit is $x/(1-x)$. However, we observed that the rate of convergence gets slower as x approaches 1. Thus if we select an error $1/m$, we cannot say how many terms N we have to take to make $\sum_{n=1}^{N} x^n$ differ from $x/(1-x)$ by at most $1/m$ without first specifying x. In other words, the order of quantifiers in the definition of pointwise convergence is universal-existential (for every x there exists N). If we want the existential-universal form—which is a stronger condition—then we come up with a stronger notion of convergence, called *uniform convergence*: *a sequence $f_n(x)$ of functions on a common domain D is said to converge uniformly to a function $f(x)$ (equivalently, $f(x)$ is said to be the uniform limit of $f_n(x)$) if for every error $1/m$ there exists N (depending on $1/m$) such that for all x in the domain D, $|f_n(x) - f(x)| < 1/m$ if*

$n \geq N$. The definition of uniform convergence of a series of functions is analogous. In particular, $\sum_{n=1}^{\infty} x^n$ does not converge uniformly on $-1 < x < 1$. On the other hand it does converge uniformly on a smaller domain $-x_0 < x < x_0$ for fixed $x_0 < 1$.

If a convergent sequence fails to converge uniformly, there may be serious consequences in that the limit function may fail to share properties with the approximating functions. For example, the pointwise limit of continuous functions may not be continuous! Let $f_n(x)$ be the function whose graph is shown in Figure 7.3.1

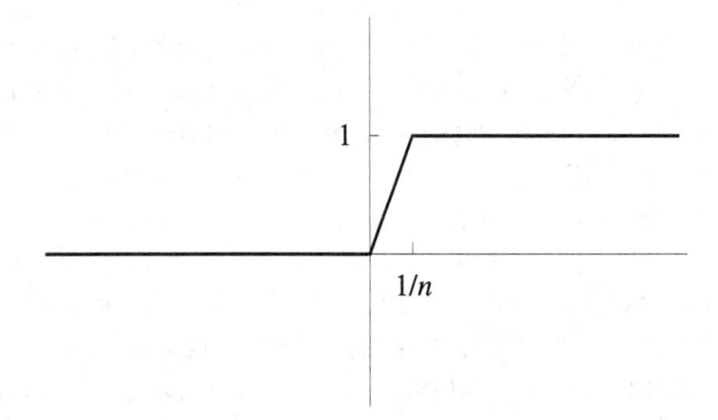

Figure 7.3.1:

so that $f_n(x) = 0$ if $x \leq 0$, $f_n(x) = 1$ if $x \geq 1/n$, and $f_n(x)$ is linear in between, $f_n(x) = nx$ if $0 < x < 1/n$. Then clearly $\lim_{n\to\infty} f_n(x)$ exists for every x. If $x \leq 0$, then $f_n(x) = 0$ for every n, so the limit is zero; while if $x > 0$, then $f_n(x) = 1$ once $n > 1/x$, so the limit is one. The limit function (shown in Figure 7.3.2) has a jump discontinuity at $x = 0$, but the functions $f_n(x)$ are all continuous. Of course the convergence is not uniform—the closer x is to zero the longer it takes for $f_n(x)$ to approach one. Since $f_n(1/2n) = 1/2$ and $f(1/2n) = 1$, we can never make $|f_n(x) - f(x)| < 1/2$ for all x. We will see shortly that this must always be the case: a uniform limit of continuous functions is continuous. This example also shows that compactness of the domain will *not* suffice to turn convergence into uniform convergence. (It is important not to confuse uniform convergence with uniform *continuity*, where compactness of the domain does suffice!)

The notion of uniform convergence can be rephrased in terms of

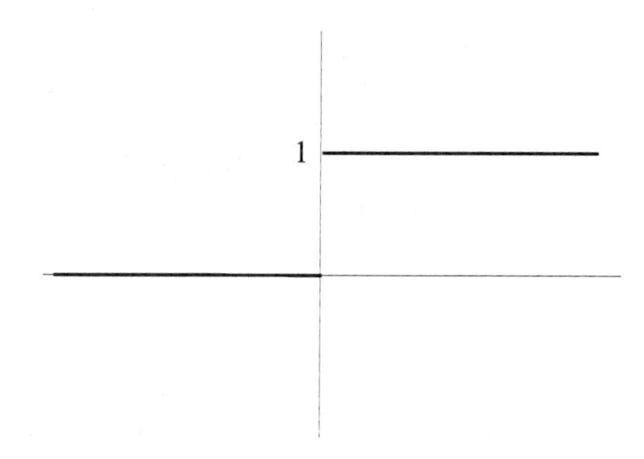

Figure 7.3.2:

the quantities $\sup_x\{|f_n(x) - f(x)|\}$, which can be thought of as a kind of "distance" between the functions f_n and f (we will return to this interpretation in a later chapter). We claim $f_n \to f$ uniformly if and only if $\lim_{n\to\infty} \sup_x\{|f_n(x) - f(x)|\} = 0$. Note that the limit here is the limit of a sequence of real numbers. Writing out the meaning of $\lim_{n\to\infty} \sup_x\{|f_n(x) - f(x)|\} = 0$ we have: given any error $1/m$ there exists N such that $n \geq N$ implies $\sup_x\{|f_n(x) - f(x)|\} \leq 1/m$. Of course $\sup_x\{|f_n(x) - f(x)|\} \leq 1/m$ is the same as $|f_n(x) - f(x)| \leq 1/m$ for every x, so we are back to the definition of uniform convergence.

There is a Cauchy criterion for uniform convergence:

Theorem 7.3.1 (*Cauchy criterion*) *A sequence of functions $f_n(x)$ converges uniformly to some limit function if and only if given any error $1/m$ there exists N such that $k, n \geq N$ imply $|f_n(x) - f_k(x)| \leq 1/m$ for all x.*

Proof: It is easy to show that a uniformly convergent sequence satisfies the Cauchy criterion using the estimate

$$|f_n(x) - f_k(x)| \leq |f_n(x) - f(x)| + |f(x) - f_k(x)|.$$

We leave the details to the reader.

Conversely, suppose the Cauchy criterion is satisfied. It follows that at each point x, the numerical sequence $\{f_n(x)\}$ satisfies the Cauchy criterion, hence it must converge to a limit. We then define $f(x)$ to be this limit. To complete the proof we need to show that the convergence of $f_n(x)$ to $f(x)$ is uniform. To see this, we take the Cauchy criterion as stated and let one of the indices, say k, go to infinity. That is, given the error $1/m$, we find N such that $k, n \geq N$ imply $|f_n(x) - f_k(x)| \leq 1/m$ for all x; then $n \geq N$ implies $|f_n(x) - f(x)| \leq 1/m$ for all x since non-strict inequalities are preserved in the limit. But this is exactly the statement of uniform convergence. QED

We leave it as an exercise to formulate the Cauchy criterion for pointwise convergence. The next theorem makes precise the idea that uniform limits preserve continuity.

Theorem 7.3.2 *Let f_n converge to f uniformly on the domain D. If all the f_n are continuous at a point x_0 in D, then f is also continuous at x_0. If all the f_n are continuous on D, then f is continuous on D. If all the f_n are uniformly continuous on D, then f is uniformly continuous on D.*

Proof: The idea of the proof is that since we can make $f_n(x)$ close to $f(x)$ for all points x, we can turn questions of continuity about f into questions of continuity about f_n. More precisely, let an error $1/m$ be given, and choose n large enough so that $|f_n(x) - f(x)| \leq 1/3m$ for all x. Then we can compare $f(x)$ with $f(y)$ by first comparing $f(x)$ with $f_n(x)$, then comparing $f_n(x)$ with $f_n(y)$, and finally comparing $f_n(y)$ with $f(y)$, using the triangle inequality, as shown in Figure 7.3.3:

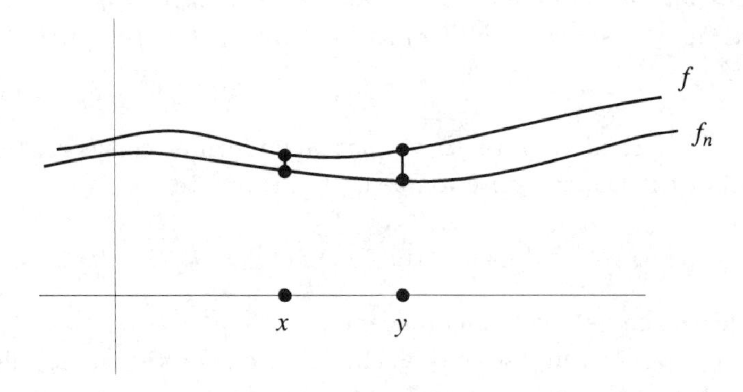

Figure 7.3.3:

$$|f(x) - f(y)| \leq |f(x) - f_n(x)| + |f_n(x) - f_n(y)| + |f_n(y) - f(y)|.$$

The first and third terms are at most $1/3m$ by the uniform closeness of f and f_n, so $|f(x) - f(y)| \leq 2/3m + |f_n(x) - f_n(y)|$ for all x and y. Note that we still have $1/3m$ to play around with when we want to make $|f(x) - f(y)| \leq 1/m$. The value of n and hence the function f_n is now fixed, but the choice of n did depend on the error $1/m$ that we were given.

Consider the continuity at the point x_0. After choosing n as above, we can then use the continuity of f_n to assert that there exists a neighborhood $|x - x_0| < 1/N$ of x_0 in which $|f_n(x) - f_n(x_0)| \leq 1/3m$. This will then make $|f(x) - f(x_0)| \leq 2/3m + |f_n(x) - f_n(x_0)| \leq 1/m$. This is the continuity of f at x_0.

Notice that the order of choice is crucial: given $1/m$ we first choose n, then based on the continuity of the particular f_n we choose the neighborhood $|x - x_0| < 1/N$. If we did not have uniform convergence we would not know that a single function f_n could make $|f_n(x) - f(x)|$ small for all x. We would only know that for each x some f_n would do this, and we would be unable to use the continuity of f_n to carry out the proof.

The continuity of f on D follows from the continuity of f_n on D because of the pointwise result.

Finally suppose the f_n are uniformly continuous. Then given the error $1/m$ we need to find $1/N$ independent of the point so that $|x-y| < 1/N$ implies $|f(x)-f(y)| < 1/m$. Choosing n as before, we already have $|f(x)-f(y)| \leq 2/3m+|f_n(x)-f_n(y)|$. By the uniform continuity of this f_n we can find $1/N$ so that $|x-y| < 1/N$ implies $|f_n(x)-f_n(y)| \leq 1/3m$, and this $1/N$ then does the trick. QED

This theorem is a two-edged sword: on the one hand it shows the usefulness of uniform convergence; but on the other hand it shows that we frequently can expect uniform convergence to fail. For example, when we consider Fourier series, we will want discontinuous functions to have Fourier series. But these cannot converge uniformly because the sines and cosines in the Fourier series are continuous functions, so the theorem says that uniformly convergent Fourier series can only represent continuous functions.

The theorem just proved is notorious because Cauchy got it wrong—he claimed to prove that the limit of continuous functions is continuous. At least part of the reason why he went wrong was that he couched his proof in the language of infinitesmals. For an interesting discussion of this see the Appendix 1 to *Proofs and refutations* by Imre Lakatos, Cambridge University Press, 1976.

Uniform convergence depends on the domain of the functions. Frequently we encounter the situation of a sequence of functions $f_n(x)$ that converges pointwise but not uniformly on an open domain D, but the restrictions of f_n to all compact subsets converge uniformly. In such a case we say that f_n *converges uniformly on compact sets.* For example, $\sum_{n=1}^{\infty} x^n$ converges uniformly on the domain $|x| \le 1 - \epsilon$ for any $\epsilon > 0$.

7.3.2 Integration and Differentiation of Limits

Uniform convergence can also be used to establish integrability and differentiability of limits of functions and the interchange of the operation and the limit. We begin with integration because it is simpler, and we will need it in the discussion of differentiation.

Theorem 7.3.3 *Let f_n converge uniformly to f on a finite interval $[a, b]$. If all the f_n are Riemann integrable on $[a, b]$, then so is f and*

$$\lim_{n \to \infty} \int_a^b f_n(x)\, dx = \int_a^b f(x)\, dx$$

(or more generally we can interchange the limit and integral over any subinterval).

Proof: The idea of the proof is that if $|f_n(x) - f(x)| \le 1/m$, then the Cauchy sums $S(f, P)$ and $S(f_n, P)$ evaluated at the same points can differ by at most $(b - a)/m$. Thus suppose the error $1/m$ is given. We want to show that the Cauchy sums $S(f, P)$ can be made to differ from each other by at most $1/m$ by making P sufficiently fine, since this will prove the integrability of f. Thus choose n large enough (say $n \ge N$) so that $|f_n(x) - f(x)| \le 1/3m(b - a)$. Then

$$\begin{aligned} |S(f_n, P) - S(f, P)| &= |\Sigma(f_n(y_j) - f(y_j))(x_{j+1} - x_j)| \\ &\le \Sigma|f_n(y_j) - f(y_j)|(x_{j+1} - x_j) \\ &\le \frac{1}{3m(b-a)}\Sigma(x_{j+1} - x_j) = \frac{1}{3m} \end{aligned}$$

for any partition P, where y_j denotes an arbitrary point in the subinterval $[x_j, x_{j+1}]$ that is the same for both Cauchy sums (we could also get the same estimate comparing Riemann upper and lower sums). Thus if P' is any other partition,

$$\begin{aligned}|S(f,P) - S(f,P')| &\leq |S(f,P) - S(f_n,P)| \\ &\quad +|S(f_n,P) - S(f_n,P')| \\ &\quad +|S(f_n,P') - S(f,P')| \\ &\leq \frac{2}{3m} + |S(f_n,P) - S(f_n,P')|.\end{aligned}$$

Now that n is fixed, we know from the integrability of f_n that $S(f_n, P)$ converges to $\int_a^b f_n(x)\,dx$ as the maximum interval length of the partition goes to zero. Therefore by taking the maximum interval length for P and P' sufficiently small we can make $|S(f_n,P) - S(f_n,P')| \leq 1/3m$, hence $|S(f,P) - S(f,P')| \leq 1/m$. This proves the integrability of f.

Finally we need to show that $\lim_{n\to\infty} \int_a^b f_n(x)\,dx$ exists and equals $\int_a^b f(x)\,dx$. From the integrability of f we know $|S(f,P) - \int_a^b f(x)\,dx| \leq 1/3m$ if the partition P is sufficiently fine. If we choose n as before, then $|S(f_n,P) - S(f,P)| \leq 1/3m$, so $|S(f_n,P) - \int_a^b f(x)\,dx| \leq 2/3m$. Note that for this to hold we only have to choose n large enough $(n \geq N)$ so that $|f_n(x) - f(x)| \leq 1/3m(b-a)$ and the partition P sufficiently fine. These conditions on n and P depend on $1/m$ but are independent of each other. Now for any fixed $n \geq N$, we can also require that the partition P be sufficiently fine so that $|S(f_n,P) - \int_a^b f_n(x)\,dx| \leq 1/3m$. Here the partition P does depend on n, but when we combine this with the previous estimate $|S(f_n,P) - \int_a^b f(x)\,dx| \leq 2/3m$ we obtain simply $|\int_a^b f_n(x)\,dx - \int_a^b f(x)\,dx| \leq 1/m$, so the partition P used for making the comparison drops out of the picture. Since this holds for *every* $n \geq N$, with N depending on $1/m$, we have the result $\lim_{n\to\infty} \int_a^b f_n(x)\,dx = \int_a^b f(x)\,dx$ as desired. QED

There are a few subtle points about the above proof that are worth observing. First note that in the proof that f is integrable we did not use the full strength of the uniform limit—we only needed $|f_n(x) - f(x)| \leq 1/m$ for one particular value of n for each $1/m$. For the proof of the limit and integral interchange we did need to use this estimate holding for all $n \geq N$. Here we had to introduce a partition P to make the comparisons $S(f,P)$ to $\int_a^b f(x)\,dx$, $S(f_n,P)$ to $S(f,P)$, and $S(f_n,P)$ to $\int_a^b f_n(x)\,dx$. The particular partition was restricted by the maximum length of subintervals in two ways, once depending on f and once depending on f_n. It is not true that one particular partition P can work for all $n \geq N$. Nevertheless, we still obtained the estimate

$$\left|\int_a^b f_n(x)\,dx - \int_a^b f(x)\,dx\right| \leq \frac{1}{m}$$

that we sought for all $n \geq N$.

This theorem may seem perfectly reasonable, but it turns out not to be as useful as one might like because there are many examples of non-uniform limits of functions (expecially in Fourier series) where one wants to interchange the limit and integral. Here is a particularly vexing example. Let $r_1, r_2, \ldots$ be an enumeration of the rational numbers in the interval $[0,1]$, and let f_n be defined on $[0,1]$ by $f_n(x) = 1$ if $x = r_1, r_2, \ldots, r_n$ and by $f_n(x) = 0$ otherwise. Clearly $f_n(x)$ is converging pointwise (but not uniformly) to Dirichlet's function $f(x) = 1$ if x is rational and $f(x) = 0$ if x is irrational. Now each f_n is zero except on a finite set of points, so f_n is integrable and $\int_0^1 f_n(x)\,dx = 0$. Thus $\lim_{n\to\infty} \int_0^1 f_n(x)\,dx$ exists and equals zero, so we would expect $\int_0^1 f(x)\,dx = 0$. But $f(x)$ is not integrable and $\int_0^1 f(x)\,dx$ is not defined. It is exactly this kind of failure of the Riemann integral to behave properly in limits that are not uniform that will motivate us to study a more general notion of integration—the Lebesgue theory—in a later chapter.

Next we consider the problem of interchanging limits and derivatives. First we note that the limit of differentiable functions need not be differentiable, even if the limit is uniform. The reason for this is that the uniform closeness of two functions, $|f_n(x) - f(x)| \leq 1/m$, does not imply *anything* about the relative smoothness of the graphs. The function $f(x)$ can be perfectly smooth, say $f(x) \equiv 0$, while $f_n(x)$ can have lots of bumps and wiggles, as in Figure 7.3.4. Perhaps the simplest example is the approximation of $f(x) = |x|$ by smooth functions obtained by rounding out the corner, as shown in Figure 7.3.5.

To get a positive result we need to assume the uniform convergence of the derivatives.

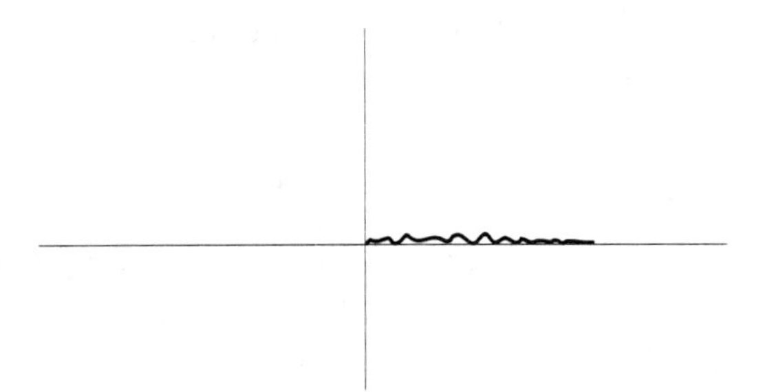

Figure 7.3.4:

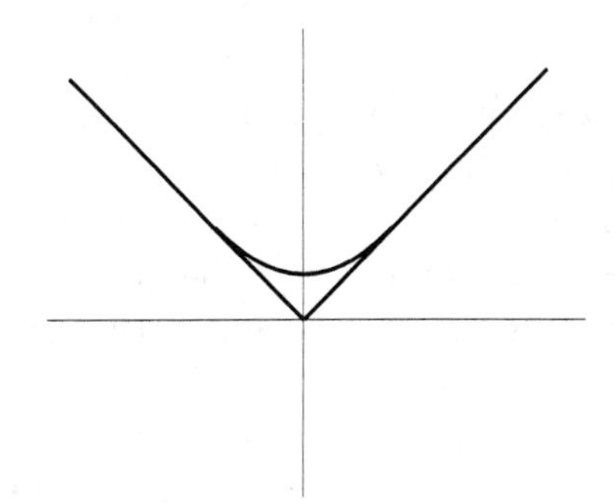

Figure 7.3.5:

Theorem 7.3.4 *Let f_n be defined on (a, b) and be C^1. If $f_n(x) \to f(x)$ pointwise and f_n' converges uniformly to $g(x)$, then f is C^1 and $f' = g$.*

Proof: A direct approach to the proof will not work, since there is no obvious way to estimate the difference of the difference quotients for f and f_n,

$$\frac{f(x+h)-f(x)}{h} - \frac{f_n(x+h)-f_n(x)}{h}$$

for all small h. We therefore seek an indirect method via the fundamental theorem of the calculus. This is frequently a good method for gaining information about derivatives.

We fix a point x_0 in the interval and write the integration of the derivative formula

$$f_n(x) = f_n(x_0) + \int_{x_0}^{x} f_n'(t)\,dt,$$

which holds because f_n is C^1. We now take the limit as $n \to \infty$, to obtain $f(x) = f(x_0) + \int_{x_0}^{x} g(t)\,dt$, where we have used the previous theorem to obtain the limit of the integral. Since we know g is continuous (it is the uniform limit of continuous functions), the differentiation of the integral theorem tells us that f is C^1 and $f'(x) = g(x)$ (the term $f(x_0)$ is just a constant whose derivative is zero). QED

There are a number of ways to improve this result. It suffices to have f_n' converge uniformly on every subinterval $(a + 1/m, b - 1/m)$ because then we can apply the argument on each subinterval and the conclusion $f' = g$ then holds on (a, b). It suffices to assume $f_n(x)$ converges at just one point $x = x_0$, for then $f_n(x) = f_n(x_0) + \int_{x_0}^{x} f_n'(t)\,dt$ implies $f_n(x)$ converges at every point since the right side converges. In fact, the convergence is uniform if the interval is bounded. We cannot, however, conclude that $f_n(x)$ converges merely from the uniform convergence of $f_n'(x)$, as the example $f_n(x) \equiv n$ shows. Also, the pointwise convergence of f_n' and f_n will not guarantee the differentiability of f. For a counterexample we need only integrate the functions in the example of a discontinuous limit of continuous functions.

It is also possible to show that the theorem is valid for differentiability in place of continuous differentiability. But the proof is trickier (using the mean value theorem in place of the fundamental theorem), and the result is not as useful.

7.3.3 Unrestricted Convergence*

We conclude this section with another way of looking at uniform convergence, at least for sequences of continuous functions on a compact interval. Since we are dealing with sequences of functions, it makes sense to vary the point as well as the function and to ask if $f_n(x_n)$ converges to $f(x)$ whenever x_n converges to x.

Theorem 7.3.5 *Let $f_n(x)$ be a sequence of continuous functions on a compact domain D. Then $f_n(x_n)$ converges to $f(x)$ for all sequences*

x_n in D convergent to x (for all x in D) if and only if f_n converges uniformly to f.

Proof: Suppose first $f_n \to f$ uniformly and $x_n \to x$ in D (this half of the proof does not require the compactness of D). We need to compare $f_n(x_n)$ with $f(x)$. Following the method of proof used in Theorem 7.3.2, there are two intermediate values, $f_n(x)$ and $f(x_n)$, that we might consider in making the comparison,

$$|f_n(x_n) - f(x)| \le |f_n(x_n) - f_n(x)| + |f_n(x) - f(x)|$$

or

$$|f_n(x_n) - f(x)| \le |f_n(x_n) - f(x_n)| + |f(x_n) - f(x)|.$$

The trouble with the first estimate is that we would need to invoke the continuity of f_n to estimate $|f_n(x_n) - f_n(x)|$, and this would require making $|x_n - x|$ small *depending* on f_n. Since we cannot vary the n separately for x_n and f_n, we would be in trouble. Thus we work with the second inequality.

Here we need to use the uniform convergence to estimate $|f_n(x_n)-f(x_n)|$ and the continuity of f to estimate $|f(x_n)-f(x)|$. Given any error $1/m$ we first use the uniform convergence to find N such that $n \ge N$ implies $|f_n(y) - f(y)| \le 1/2m$ for any y and, hence, in particular for $y = x_n$. Thus $|f_n(x_n) - f(x_n)| \le 1/2m$ for all $n \ge N$. Next by the continuity of f, a consequence of the fact f is the uniform limit of continuous functions, there exists $1/k$ such that $|y-x| \le 1/k$ implies $|f(y) - f(x)| \le 1/2m$ (here x is fixed and y is variable). In particular setting $y = x_n$ we have $|x_n - x| \le 1/k$ implies $|f(x_n) - f(x)| \le 1/2m$. Summing the two estimates we obtain $|f_n(x_n) - f(x)| \le 1/m$, the estimate we want, under the two conditions $n \ge N$ and $|x_n - x| \le 1/k$. But since $x_n \to x$ by assumption, we can make $|x_n-x| \le 1/k$ by taking n large enough (say $n \ge N'$), so the conclusion $|f_n(x_n) - f(x)| \le 1/m$ follows by taking n larger than both N and N'. This is the statement $\lim_{n\to\infty} f_n(x_n) = f(x)$.

Conversely, assume $f_n(x_n) \to f(x)$ if $x_n \to x$. To show f_n converges uniformly to f we consider the sets $A_{m,N} = \{x$ in D such that $|f_n(x) - f(x)| \le 1/m$ for all $n \ge N\}$. The condition for uniform convergence that we want to prove is that for all m there exists N such that $A_{m,N} = D$. Ordinary pointwise convergence would say that for all x and all m there exists N such that x is in $A_{m,N}$. But we are assuming more than pointwise convergence, and we claim the following stronger conclusion: *for each x in D and every m, there exists N such that $A_{m,N}$ contains a neighborhood of x.*

We prove this claim by contradiction. If it were not true, then there would exist a fixed x and m and a sequence of points $\{x_N\}$ converging to x such that each x_N is not in $A_{m,N}$. But x_N not in $A_{m,N}$ means there exists $k(N) \geq N$ such that $|f_{k(N)}(x_N) - f(x)| > 1/m$. By inserting some terms equal to x into the sequence $\{x_n\}$ we can obtain a new sequence $\{y_k\}$ converging to x such that $|f_k(y_k) - f(x)| > 1/m$ for infinitely many k. Specifically, we take $y_k = x_N$ if $k = k(N)$, choosing the smallest N if there is more than one choice and $y_k = x$ otherwise. But this contradicts the hypothesis $\lim_{k\to\infty} f_k(y_k) = f(x)$, proving the claim.

To complete the proof we apply the Heine-Borel theorem. Fix m. For each x in D there exists N and a neighborhood $(x - 1/n, x + 1/n)$ of x contained in $A_{m,N}$. This set of neighborhoods is an open cover of D, which is compact, so there exists a finite subcover. Thus D is a finite union of sets of the form $A_{m,N}$ for m fixed. But the sets $A_{m,N}$ increase with N, so the finite union is just $A_{m,N}$ for the largest value of N. Since $D = A_{m,N}$ for some N holds for every m, we have proved uniform convergence. QED

7.3.4 Exercises

1. State and prove a Cauchy criterion for pointwise convergence.

2. Suppose $f_n \to f$ and the functions f_n all satisfy the Lipschitz condition $|f_n(x) - f_n(y)| \leq M|x - y|$ for some constant M independent of n. Prove that f also satisfies the same Lipschitz condition.

3. Give an example of a sequence of continous functions on a compact domain converging pointwise but not uniformly to a continuous function.

4. Prove that a sequence of complex-valued functions converges uniformly if and only if the sequences of real and imaginary parts converge uniformly.

5. If $\lim_{n\to\infty} f_n = f$ and the functions f_n are all monotone increasing, must f be monotone increasing? What happens if f_n are all strictly increasing?

6. Give an example of a sequence of continuous functions converging pointwise to a function with a discontinuity of the second kind.

7. If $|f_n(x)| \le a_n$ for all x and $\sum_{n=1}^{\infty} a_n$ converges, prove that $\sum_{n=1}^{\infty} f_n(x)$ converges uniformly.

8. Give an example of a sequence of continuous functions on a non-compact domain D that does not converge uniformly, yet $\lim_{n\to\infty} f_n(x_n) = f(x)$ for every sequence $\{x_n\}$ converging to x in D.

9. If $f_n \to f$ uniformly and $\lim_{x\to x_0} f_n(x)$ exists for every n, then $\lim_{x\to x_0} f(x)$ exists and equals $\lim_{n\to\infty} \lim_{x\to x_0} f_n(x)$ (note: we are *not* assuming continuity of f_n or f).

10. If $f_n \to f$ uniformly and the functions f_n have only jump discontinuities, prove that f has only jump discontinuities.

11. *Give an example of a sequence of continuous functions f_n on $[0,1]$ that converge pointwise to zero but such that $\lim_{n\to\infty} \int_0^1 f_n(x)\,dx$ is not zero.

12. If $f_n \to f$ uniformly on $[a,b]$, prove that $F_n \to F$ uniformly on $[a,b]$ where $F_n(x) = \int_a^x f_n(t)\,dt$. Is the same true on the whole line?

13. Define a *step function* to be a function that is piecewise constant, $f(x) = \sum_{j=1}^{n} c_j \chi_{[a_j,b_j)}$ where $[a_j,b_j)$ are disjoint intervals. (χ denotes the characteristic function of the interval.)

    a. Prove that every continuous function on a compact interval is a uniform limit of step functions.

    b. Prove that a uniform limit of step functions (on a compact interval) is Riemann integrable.

14. Define a *linear spline* to be a continuous function on a compact interval that is piecewise linear (equal to an affine function on each

subinterval in a finite partition). Prove that every continuous function (on a compact interval) is the uniform limit of linear splines.

7.4 Power Series

7.4.1 The Radius of Convergence

An important class of infinite series of functions is the class of power series $\sum_{n=0}^{\infty} a_n(x-x_0)^n$ where x_0 is fixed (we refer to x_0 as the point *about* which the power series is *expanded*) and a_n are real (or complex) coefficients. By definition $(x-x_0)^0 = 1$, so the first term is merely the constant function a_0. Note that the partial sums of the power series are polynomials; $\sum_{n=0}^{N} a_n(x-x_0)^n$ is a polynomial of degree N. In many ways power series can be thought of as polynomials of infinite order. In fact, the theory of power series was developed during the eighteenth century in just such a spirit, with an unfortunate disregard for the question of convergence. We are in a position now to develop the basic properties of power series with complete rigor. Before beginning it is only fair to warn you that power series are very atypical of series of functions in general, so you should not expect the insights derived from the study of power series to extend to other classes of series. This will be especially important in a later chapter when we discuss Fourier series.

In discussing power series it is good to recall a nursery rhyme:

"There was a little girl
Who had a little curl
Right in the middle of her forehead
When she was good
She was very, very good
But when she was bad she was horrid."

The first question we need to answer is the question of convergence: given a specific power series $\sum_{n=0}^{\infty} a_n(x-x_0)^n$, that is, given x_0 and $a_1, a_2, \ldots$, for which values of x does it converge? Note that the answer depends only on $x - x_0$, so if we can answer the question for $x_0 = 0$ we can answer the question in general. In what follows we will frequently deal with the case $x_0 = 0$ only, in order to simplify notation.

The convergence of $\sum_{n=0}^{\infty} a_n x^n$ at a point x depends on the coefficients a_n. We can always arrange for $a_n x^n$ to be unbounded (if $x \neq 0$), and in fact by taking $a_n = n!$ we can make $a_n x^n$ unbounded for every $x \neq 0$, so the power series need not converge except at $x = 0$. However, if the power series does converge at some point $x \neq 0$, this has implications about convergence at other points. The idea is extremely simple. If $\sum_{n=0}^{\infty} a_n x^n$ converges at some point $x = x_1$, then the terms $a_n x_1^n$ must certainly be bounded. So there must exist a constant M such that $|a_n x_1^n| \leq M$ for all n. We are not saying that this condition is sufficient for the convergence; only that it is necessary. But it turns out to be sufficient for the convergence of the series for any x with $|x| < |x_1|$. In fact, if we rewrite the condition as $|a_n| \leq M|x_1|^{-n}$, then we have $|a_n x^n| \leq M|x/x_1|^n$ and so $\sum_{n=0}^{\infty} a_n x^n$ converges by comparison with $\sum_{n=0}^{\infty} M r^n$ for $r = |x/x_1| < 1$ if $|x| < |x_1|$. *In fact the convergence is absolute, and it is uniform in any interval* $|x| \leq R$ *for fixed* $R < |x_1|$, because we can compare with $\sum_{n=0}^{\infty} M\,(R/|x_1|)^n$ for all $x, |x| \leq R$. This implies that the power series represents a continuous function on $|x| < |x_1|$ since it is the uniform limit on compact subsets of the partial sums, which are continuous because they are polynomials. We will see shortly that the behavior of power series is much better than continuous.

This simple observation leads immediately to the definition of the *radius of convergence* of the power series $\sum_{n=0}^{\infty} a_n x^n$ as the unique number R such that the series converges for $|x| < R$ and diverges for $|x| > R$. We allow $R = +\infty$ if the series converges for all x and $R = 0$ if the series converges only for $x = 0$. If the series diverges for some value of x, then $R = \sup\{x : \text{series converges}\}$. At the values $x = \pm R$ the series may either converge or diverge. However, if $|x| > R$, then the series must diverge; in fact, the terms must be unbounded—otherwise by our observation we would have convergence for all values in $(-|x|, |x|)$, contradicting the definition of R. Thus the region $|x| < R$ is where the series is very, very good, and the region $|x| > R$ is where it is horrid. Again we must emphasize that this property is special to power series.

Here are some simple examples that illustrate the variety of behaviors of power series at the radius of convergence. The geometric series

$\sum_{n=0}^{\infty} x^n$ converges for $|x| < 1$ and diverges for $|x| > 1$, so the radius of convergence is 1. At $x = \pm 1$ the terms $(\pm 1)^n$ are bounded but the series diverges. The series $\sum_{n=0}^{\infty} nx^n$ also has radius of convergence 1 but the terms are unbounded at $x = \pm 1$. The series $\sum_{n=1}^{\infty}(1/n)x^n$ has radius of convergence 1, and at $x = 1$ it diverges while at $x = -1$ it converges, since $\sum_{n=1}^{\infty}(-1)^n/n$ converges, although not absolutely. Finally the series $\sum_{n=1}^{\infty}(1/n^2)x^n$ converges at both endpoints $x = \pm 1$. We will verify the computations of the radius of convergence for these series after the next theorem.

The root test for convergence gives a formula for the radius of convergence.

Theorem 7.4.1 *The radius of convergence R of a power series $\sum_{n=0}^{\infty} a_n x^n$ is given by $1/R = \limsup_{n\to\infty} \sqrt[n]{|a_n|}$.*

Proof: First we will establish the inequality $\limsup_{n\to\infty} |a_n|^{1/n} \le 1/R$. If $R = 0$ there is nothing to prove. If $R > 0$, choose any $r < R$. Then the terms $a_n x^n$ are bounded for $x = r$, $|a_n r^n| \le M$. This means $|a_n|^{1/n} \le M^{1/n}/|r|$ and so

$$\limsup_{n\to\infty} |a_n|^{1/n} \le (1/|r|) \limsup_{n\to\infty} M^{1/n} = 1/|r|$$

since $\lim_{n\to\infty} M^{1/n} = 1$ for any $M > 0$ (we will give a proof of this as a separate lemma below). This shows $\limsup_{n\to\infty} |a_n|^{1/n} \le 1/|r|$ for any value of r with $|r| < R$, so $\limsup_{n\to\infty} |a_n|^{1/n} \le 1/R$.

For the reverse inequality write $\limsup_{n\to\infty} |a_n|^{1/n} = 1/R_0$ (allowing $R_0 = 0$ or ∞). We need to show that $\sum_{n=0}^{\infty} a_n r^n$ converges if $|r| < R_0$, for that will imply $R \ge R_0$. Now for any *fixed* r with $|r| < R_0$ we can find R_1 satisfying $|r| < R_1 < R_0$ (if $R_0 = 0$ there is nothing to prove). Then $\limsup_{n\to\infty} |a_n|^{1/n} < 1/R_1$ so that for k large enough, $|a_n|^{1/n} \le 1/R_1$ for all $n \ge k$ (this is a consequence of the definition $\limsup_{n\to\infty} |a_n|^{1/n} = \lim_{k\to\infty} \sup_{n\ge k} |a_n|^{1/n}$). This means $|a_n| \le 1/R_1^n$ and, hence, $|a_n r^n| \le |r/R_1|^n$ for all $n \ge k$, which proves the convergence of $\sum a_n r^n$ by comparison with the geometric series since $|r/R_1| < 1$. One can also prove the convergence of $\sum a_n r^n$ by appealing directly to the root test—the above argument essentially incorporates the reasoning used to derive the root test. QED

Lemma 7.4.1 $\lim_{n\to\infty} M^{1/n} = 1$ *for any* $M > 0$.

Proof: It suffices to do this for $M > 1$ since $\lim_{n\to\infty}(1/M)^{1/n} = (\lim_{n\to\infty} M^{1/n})^{-1}$. One can give a quick proof using logarithms but since we have not discussed the properties of logarithms, we will give a longer proof using the binomial theorem. We use the identity

$$\left(1+\frac{1}{m}\right)^n = 1+\frac{n}{m}+\cdots+\left(\frac{1}{m}\right)^n.$$

All the terms are positive, so we have

$$\left(1+\frac{1}{m}\right)^n \geq 1+\frac{n}{m};$$

hence

$$1+\frac{1}{m} \geq \left(1+\frac{n}{m}\right)^{1/n}$$

since the function $f(x) = x^n$ preserves the order relation. Now if we fix $M > 1$, then given any error $1/m$ we can find k such that $M \leq 1 + k/m$, so $M \leq 1 + n/m$ for all $n \geq k$. Altogether $M^{1/n} \leq (1 + n/m)^{1/n} \leq 1 + 1/m$; and since the inequality $1 \leq M^{1/n}$ is an immediate consequence of $M > 1$, we have $|M^{1/n} - 1| \leq 1/m$ for all $n \geq k$, proving $\lim_{n\to\infty} M^{1/n} = 1$. QED

Next we look at some examples. In these, as is usually the case, the limit of $|a_n|^{1/n}$ will exist, so it is not necessary to invoke the limsup. First consider the case when the coefficients a_n are given by a rational function of n, $a_n = p(n)/q(n)$ where p and q are polynomials (perhaps modified at the finite number of zeroes of q). We claim the radius of convergence is 1. To see this it suffices to show $\lim_{n\to\infty} |p(n)|^{1/n} = 1$ for any non-zero polynomial. In the lemma we have already established this for $p(n) = M$, the constant polynomials. Since it is easy to show $M_0 n^k \leq |p(n)| \leq M n^k$ for all large n and some values of M_0, M and k (depending on the polynomial), it suffices to show $\lim_{n\to\infty} n^{1/n} = 1$. Then $\lim_{n\to\infty}(Mn^k)^{1/n} = (\lim_{n\to\infty} M^{1/n})(\lim_{n\to\infty} n^{1/n})^k = 1$ by the properties of limits ($f(x) = x^k$ is continuous, so we can interchange it with the limit). Finally we establish $\lim_{n\to\infty} n^{1/n} = 1$ without logarithms using the binomial theorem. Since $(1+1/m)^n = 1+n/m+\cdots+(1/m)^n$, we have $n/m \leq (1+1/m)^n$, so $n^{1/n} \leq m^{1/n}(1+1/m)$. Keeping m fixed and letting $n \to \infty$ we obtain $\limsup_{n\to\infty} n^{1/n} \leq (1 + 1/m)$

since we already have shown $\lim_{n\to\infty} m^{1/n} = 1$. Since this is true for any value of m, $\limsup_{n\to\infty} n^{1/n} \leq 1$ and the obvious inequality $n^{1/n} \geq 1$ implies $\lim_{n\to\infty} n^{1/n} = 1$.

To get an example of a power series with radius of convergence $+\infty$ or 0 we must do something more dramatic (to get a finite positive value for R we can merely take $\sum R^{-n}x^n$). The most important example is the exponential power series $\sum_{n=0}^{\infty} 1/n!x^n$. To see this has radius of convergence $+\infty$ we need to show $\lim_{n\to\infty} 1/(n!)^{1/n} = 0$ or, equivalently, $\lim_{n\to\infty}(n!)^{1/n} = +\infty$. This will also show that the series $\sum_{n=0}^{\infty} n!x^n$ has radius of convergence 0. Now $\lim_{n\to\infty}(n!)^{1/n} = +\infty$ is plausible because $n! = 1 \cdot 2 \cdots n$, a product of n factors, so that $(n!)^{1/n}$ is a kind of average (the geometric average) of the numbers from 1 to n, and this average should not remain bounded as $n \to \infty$. In fact, for any fixed m, once $n \geq m$ we can estimate $n! \geq m^{(n-m)}$ simply by ignoring the first m factors and noting that the last $n - m$ factors, $m+1, m+2, \ldots, n$, are all greater than m. Thus $(n!)^{1/n} \geq m^{1-m/n}$ and so $\liminf_{n\to\infty}(n!)^{1/n} \geq m \lim_{n\to\infty}(m^{-m})^{1/n} = m$ by our previous result $\lim_{n\to\infty} M^{1/n} = 1$. Since m is arbitrary, we have $\lim_{n\to\infty}(n!)^{1/n} = +\infty$.

Next we discuss differentiability of power series, which is based on the fact that in the interior of the interval of convergence the power series converges very rapidly. This will allow us to differentiate the series term-by-term. It is easy to verify that, on a formal level, the power series $\sum_{n=0}^{\infty} a_n x^n$ has derivative

$$\sum_{n=0}^{\infty} n a_n x^{n-1} = \sum_{n=0}^{\infty} (n+1) a_{n+1} x^n.$$

We can also verify that the derived power series $\sum(n+1)a_{n+1}x^n$ has the same radius of convergence as the original power series. Note that

$$\limsup_{n\to\infty} |(n+1)a_{n+1}|^{1/n} = \lim_{n\to\infty} (n+1)^{1/n} \limsup_{n\to\infty} |a_n|^{1/(n-1)}.$$

But $\lim_{n\to\infty}(n+1)^{1/n} = 1$ by the arguments already given, while $\limsup_{n\to\infty} |a_n|^{1/(n-1)} = \limsup_{n\to\infty} |a_n|^{1/n}$ (because $|a_n|^{1/(n-1)} = (|a_n|^{1/n})^{n/(n-1)}$ and $n/(n-1) \to 1$ as $n \to \infty$). One can also derive directly that the boundedness of the terms $a_n x^n$ for fixed $x = r$ implies the convergence of $\sum n a_n x^{n-1}$ for $|x| < r$ because the geometric decrease of $|x/r|^n$ swamps the growth of the factor n.

Theorem 7.4.2 *Let $\sum a_n x^n$ have radius of convergence R with $R \neq 0$. Then the function $f(x) = \sum a_n x^n$ is C^1 on $(-R, R)$ and $f'(x) = \sum n a_n x^{n-1}$ there. In fact $f(x)$ is C^∞ on $(-R, R)$, and the derivative $f^{(k)}$ of order k is given by the power series formally differentiated k times.*

Proof: We apply Theorem 7.3.4 on differentiating infinite series of functions. The individual terms $a_n x^n$ are all differentiable, so we need to show the convergence of $\sum a_n x^n$ and the uniform convergence of $\sum n a_n x^{n-1}$. But we have already shown that a power series with radius of convergence R converges uniformly on any smaller interval and that the power series $\sum n a_n x^{n-1}$ has radius of convergence R also. Thus we obtain $f'(x) = \sum n a_n x^{n-1}$ on any smaller interval and, hence, on $(-R, R)$. The results for $f^{(k)}$ then follows by induction. QED

This result is a two-edged sword. It shows that power series are really terrific—you can differentiate them term-by-term in the interior of the interval of convergence; but it also shows that only very special kinds of functions can be represented by power series—any such function must be C^∞. In the next chapter we will even give examples of C^∞ functions that do not have power-series expansions.

We can now relate the coefficients of a convergent power series with the Taylor expansions of the function equal to the sum, $f(x) = \sum_{n=0}^\infty a_n x^n$. By differentiating n times and setting $n = 0$ we obtain $f^{(n)}(0) = n! a_n$ (by convention $0! = 1$). In other words, the partial sums of the power series are the Taylor expansions of the function. From this we also obtain the uniqueness of the power series. If $\sum a_n x^n$ and $\sum b_n x^n$ converge in $|x| < r$ to the same function f, then $a_n = b_n = 1/n! f^{(n)}(0)$, so they are identical series. Notice that the same thing holds for power series about an arbitrary point x_0. If $\sum a_n (x - x_0)^n$ converges to $f(x)$ in $|x - x_0| < r$, then $a_n = 1/n! f^{(n)}(x_0)$; so if $\sum a_n (x - x_0)^n = \sum b_n (x - x_0)^n$ in $|x - x_0| < r$, then $a_n = b_n$ for all n. However, *we can have equality between different power series about different points.* As we will see in the next section, this is a rather important idea.

7.4.2 Analytic Continuation

Suppose $\sum a_n x^n$ converges to $f(x)$ in $|x| < R$ (we have expanded about $x = 0$ for simplicity). Now for any fixed x_0 in $|x_0| < R$, we might hope to find a power series $\sum b_n(x-x_0)^n$ converging to $f(x)$ in $|x-x_0|<r$ where $r = R - |x_0|$ (see Figure 7.4.1). In fact this is always possible, as we will now see. We know in fact that *if* there is such a series, it must be given by $b_n = 1/n! f^{(n)}(x_0)$, but surprisingly this remark does not help in establishing the convergence of $\sum b_n(x-x_0)^n$ to f. We therefore take a different approach, which is more direct.

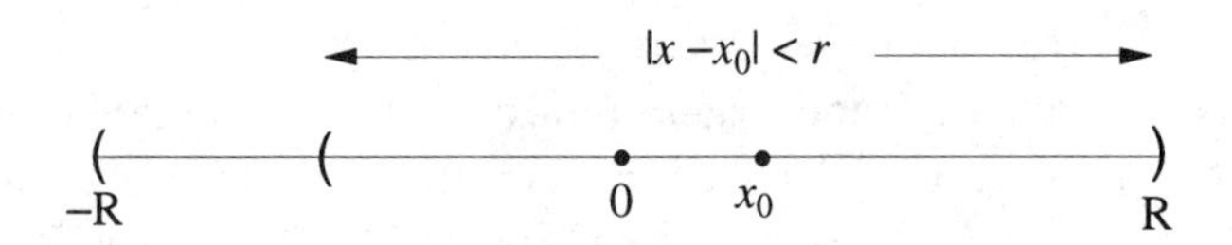

Figure 7.4.1:

We observe that if the series were finite (so f would be a polynomial), we could simply write $x = (x-x_0)+x_0$ and

$$x^n = ((x-x_0)+x_0)^n = \sum_{k=0}^{n} \binom{n}{k} (x-x_0)^k x_0^{n-k}$$

and substitute this into $f(x) = \sum_{n=0}^{N} a_n x^n$ to obtain

$$f(x) = \sum_{n=0}^{N} a_n \sum_{k=0}^{n} \binom{n}{k} (x-x_0)^k x_0^{n-k},$$

and by regrouping and rearranging terms, $f(x) = \sum_{k=0}^{N} b_k(x-x_0)^k$ where $b_k = \sum_{n=k}^{N} a_n \binom{n}{k} x_0^{n-k}$. This is just algebra. But it suggests that for the infinite series $f(x) = \sum_{n=0}^{\infty} a_n x^n$ we should have $f(x) = \sum_{k=0}^{\infty} b_k(x-x_0)^k$ with

$$b_k = \sum_{n=k}^{\infty} a_n \binom{n}{k} x_0^{n-k} = \sum_{n=0}^{\infty} a_{n+k} \binom{n+k}{k} x_0^n.$$

Notice that the expression we have guessed for the coefficient b_k is itself a power series in x_0 that converges also in $|x_0| < R$ (if $\sum a_n x^n$ converges in $|x| < R$) because the factor

$$\binom{n+k}{k} = (n+k)\cdot(n+k-1)\cdots(n+1)/k!$$

is a polynomial in n for k fixed, so $\lim_{n\to\infty}\binom{n+k}{k}^{1/n} = 1$ for each k. Thus the *rearrangement* (note that we are using this term in a somewhat different sense than in section 7.2.3) $\sum b_k(x - x_0)^k$ of $\sum a_n x^n$ where $b_k = \sum_{n=0}^{\infty} a_{n+k}\binom{n+k}{k}x_0^n$ is well defined for any x_0 in $(-R, R)$. Incidentally, the formula for b_k agrees with the previously derived $f^{(k)}(x_0)/k!$ as can be seen by differentiating $f(x) = \sum a_n x^n$ k times and setting $x = x_0$.

It remains to show that $\sum b_k(x - x_0)^k$ actually converges to $f(x)$, at least for $|x - x_0| < R - |x_0|$. To see this we want to invoke the unconditional nature of an absolutely convergent series, namely the expression

$$\sum_{n=0}^{\infty}\sum_{k=0}^{n} a_n \binom{n}{k} (x - x_0)^k x_0^{n-k}$$

(this is an infinite series, but it is not indexed by n; rather it is indexed by the pairs (n, k) with $k \le n$ in lexicographic order). If we put in absolute values for each term $a_n \binom{n}{k} (x - x_0)^k x_0^{n-k}$ we obtain the expression $\sum_{n=0}^{\infty}\sum_{k=0}^{n} |a_n| \binom{n}{k} |x - x_0|^k |x_0|^{n-k}$, since the binomial coefficients are positive. We can next evaluate the k-sum

$$\sum_{k=0}^{n} \binom{n}{k} |x - x_0|^k |x_0|^k = (|x - x_0| + |x_0|)^n = r^n$$

where $r = |x - x_0| + |x_0|$ satisfies $r < R$ by the conditions we have assumed for x and x_0. Thus

$$\sum_{n=0}^{\infty}\sum_{k=0}^{n} |a_n| \binom{n}{k} |x - x_0|^k |x_0|^{n-k} = \sum_{n=0}^{\infty} |a_n| r^n,$$

which converges because the power series is absolutely convergent in $(-R, R)$. Thus we are dealing with an absolutely convergent series. If we take the given order and sum on k first we obtain

$$\sum_{n=0}^{\infty}\sum_{k=0}^{n} a_n \binom{n}{k} (x - x_0)^k x_0^{n-k} = \sum_{n=0}^{\infty} a_n x^n = f(x).$$

On the other hand, if we rearrange the series, summing first on n with k fixed, we obtain

$$\sum_{n=0}^{\infty}\sum_{k=0}^{n} a_n \binom{n}{k} (x - x_0)^k x_0^{n-k} \quad = \quad \sum_{k=0}^{\infty}\left(\sum_{n=k}^{\infty} a_n \binom{n}{k} x_0^{n-k}\right)(x - x_0)^k$$

$$= \sum_{k=0}^{\infty} b_k (x - x_0)^k,$$

so $f(x) = \sum_{k=0}^{\infty} b_k (x - x_0)^k$ in $|x - x_0| < R - |x_0|$. We have thus proved the following theorem.

Theorem 7.4.3 *If $\sum a_n (x - x_1)^n$ is a convergent power series in $|x - x_1| < R$ converging to the function f, then f also has a power-series expansion about each point x_2 of the interval $|x - x_1| < R$, which converges at least in the largest symmetric interval $|x - x_2| < r$ lying entirely in the original interval $|x - x_1| < R$.*

Notice that the theorem does not preclude the convergence of the new power series in a larger interval. For example, $\sum_{n=0}^{\infty} x^n = 1/(1-x)$ converges in $|x| < 1$. If we fix a point x_0 in this interval, we can compute the power-series expansion $\sum_{n=0}^{\infty} b_n (x - x_0)^n$ by $b_n = f^{(n)}(x_0)/n!$ or more directly as

$$\begin{aligned}
\frac{1}{1-x} &= \frac{1}{1 - x_0 - (x - x_0)} = \frac{1}{1 - x_0} \frac{1}{1 - (x - x_0)/(1 - x_0)} \\
&= \frac{1}{1 - x_0} \sum_{n=0}^{\infty} \left(\frac{x - x_0}{1 - x_0} \right)^n = \sum_{n=0}^{\infty} (1 - x_0)^{-n-1} (x - x_0)^n,
\end{aligned}$$

provided $|(x - x_0)/(1 - x_0)| < 1$ using a power-series expansion about 0. Notice that we have a power-series expansion for $1/(1-x)$ convergent in the interval $|x - x_0| < |1 - x_0|$. If $0 < x_0 < 1$ this is the same interval predicted by the theorem, but for $-1 < x_0 < 0$ it is a larger interval.

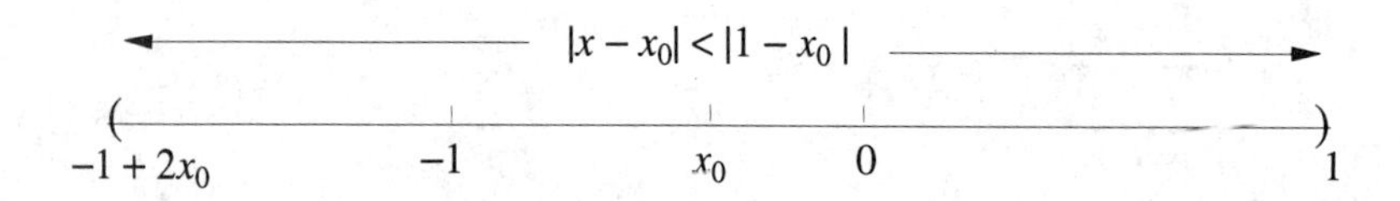

Figure 7.4.2:

In fact, this computation gives the power series of $1/(1 - x)$ about any point x_0 except 1.

Functions that have power-series expansions about all points in their domain are called *analytic functions.* The theorem shows that functions defined by power series about one point are analytic—they have convergent power series about the other points in their domain. Analytic functions have remarkable properties—the values of the function on any small interval determine the values of the function on any larger interval. This is just a consequence of the uniqueness of power series and the formula

$$a_n = \frac{1}{n!} f^{(n)}(x_0) \quad \text{if} \quad f(x) = \sum_{n=0}^{\infty} a_n (x - x_0)^n.$$

If we know $f(x)$ for $|x - x_0| < \epsilon$, then we can compute $f^{(n)}(x_0)$ for all n and so obtain the values of $f(x)$ on the interval of convergence from $\sum_{n=0}^{\infty} a_n (x - x_0)^n$. This in turn allows us to compute $f^{(n)}(x_1)$ for any point x_1 in the interval of convergence, hence we can obtain the power series for f about other points. This process of passing from power series to power series about new points is called *analytic continuation.* We leave as an exercise the proof that if f is defined in (a, b) and analytic, then we can obtain the value of $f(x_1)$ for any x_1 in (a, b) by analytic continuation from the power series $f(x) = \sum a_n (x - x_0)^n$ about any point x_0 in (a, b) in a finite number of steps. Note that analytic continuation of the power series $\sum_{n=0}^{\infty} x^n$ leads to the analytic function $1/(1 - x)$ on the domain $(-\infty, 1)$. However, analytic continuation will not enable us to extend $1/(1 - x)$ past the singularity at $x = 1$ to the region $(1, \infty)$. To do this, we need to move into the complex plane. This will be discussed briefly in the next section.

Most important functions in mathematics are analytic—at least if the domain is suitably restricted. Polynomials are the simplest examples, having power series with a finite number of non-zero terms. All the special functions—sine, cosine, exponential, and logarithm, and more exotic functions such as Bessel functions, hypergeometric functions, and so on, are analytic; indeed they are often defined by power-series. As we will see, the class of analytic functions is closed under arithmetic operations and compositions—so that any function for which you can write a formula is analytic (again with the domain suitably restricted). This means that power-series expansions should be extremely useful. However, it would be a mistake to think that only analytic functions are of interest. Many functions that are supposed to represent physical data from the real world are not analytic. Also, the procedure of analytic continuation is computationally unstable, involving high derivatives of the function that cannot be effectively controlled.

7.4.3 Analytic Functions on Complex Domains*

To understand the behavior of a power series, even if we are only interested in a real domain, requires that we consider its behavior in a complex domain. If $f(x)$ is a real- or complex-valued function of a real variable x, defined on some domain in $\mathbb{R}$, it is not a priori clear what we should mean by $f(z)$, where z varies in $\mathbb{C}$. Of course for some special functions we have a good candidate. If f is a polynomial, $f(x) = \sum_{n=0}^{N} a_n x^n$, then we simply take for granted that $f(z) = \sum_{n=0}^{N} a_n z^n$. This is not the unique possible choice—if $f(x) = 0$ we could consider $F(x+iy) = y$, which also extends the original function in that $F(z) = f(z)$ if $z = x$ is real. However, we are certainly justified in claiming that the first choice is somehow most natural. Now it turns out that we can do the same thing for power series. If $\sum_{n=0}^{\infty} a_n x^n = f(x)$ converges on the interval $|x| < R$, then $\sum_{n=0}^{\infty} a_n z^n = f(z)$ also converges in the circle $|x| < R$ (this explains why R is referred to as the *radius* of convergence). In fact the identical argument for the absolute convergence of the power series for real x proves also the absolute convergence for complex z. Thus analytic functions possess natural extensions to domains in $\mathbb{C}$. For example, $1/(1-x) = \sum_{n=0}^{\infty} x^n$ for $|x| < 1$, and also $\sum_{n=0}^{\infty} z^n = 1/(1-z)$ for $|z| < 1$. Clearly $f(z) = 1/(1-z)$ is the most natural candidate for the extension of $f(x) = 1/(1-x)$ to complex numbers—although you might be hard pressed to explain why without the use of power series. By using analytic continuation in $\mathbb{C}$ we can get around the obstacle at $z = 0$ that we could not get past in $\mathbb{R}$. However, analytic continuation in $\mathbb{C}$ leads to other types of complications that can't be discussed here.

The theory of analytic functions of a complex variable is beyond the scope of this book—it requires a book of its own. I will give one example that hints at some of the ways the theory of analytic functions of a complex variable can shed light on questions that only involve real numbers. If we are told that the function $1/(1-x)$ has a power-series expansion about $x = 0$, we can guess immediately that the power series could converge at most for $|x| < 1$ because the function $1/(1-x)$ has a singularity at $x = 1$. We might be led to guess, then, that a power series $\sum a_n x^n = f(x)$ should converge on the largest interval $|x| < R$ for which the function $f(x)$ has no singularities (never mind exactly what *singularity* means—just interpret it as any "irregular" behavior). Put another way, if $\sum a_n x^n = f(x)$ converges in $|x| < R$, then f is analytic in $|x| < R$; we can ask, conversely, if f being analytic in $|x| < R$ (remember this means it has a power-series expansion about each point in $|x| < R$) implies that the power series about 0 converges in $|x| < R$? The answer turns out to be no. The function $f(x) = 1/(1+x^2)$ is analytic on the whole real line. At any point x_0 we can compute its power series from

$$\begin{aligned} f(x) &= \frac{1}{1+x^2} = \frac{1}{1+((x-x_0)+x_0)^2} \\ &= \frac{1}{1+x_0^2+2x_0(x-x_0)+(x-x_0)^2} \\ &= \frac{1}{1+x_0^2}\frac{1}{1-(-2x_0(x-x_0)-(x-x_0)^2)/(1+x_0^2)} \\ &= \frac{1}{1+x_0^2}\sum_{n=0}^{\infty}\left(\frac{-2x_0(x-x_0)-(x-x_0)^2}{1+x_0^2}\right)^n \end{aligned}$$

with absolute convergence if $|2x_0(x-x_0)+(x-x_0)^2| < 1+x_0^2$. Now if $|x-x_0| < r$ for r sufficiently small we have $2x_0|x-x_0|$ and $(x-x_0)^2$ each $\le (1+x_0^2)/2$, so we not only have the absolute convergence of the above series (this is not quite a power series) but after substituting the binomial expansion

$$(-2x_0(x-x_0)-(x-x_0)^2)^n = (-1)^n\sum_{k=0}^{n}\binom{n}{k}(2x_0(x-x_0))^{n-k}(x-x_0)^{2k}$$

we still have the absolute convergence of

$$\frac{1}{1+x^2} = \frac{1}{1+x_0^2}\sum_{n=0}^{\infty}\sum_{k=0}^{n}\frac{(-1)^n\binom{n}{k}}{(1+x_0^2)^n}(2x_0(x-x_0))^{n-k}(x-x_0)^{2k}$$

because taking absolute values of $(-1)^n\binom{n}{k}(2x_0(x-x_0))^{n-k}(x-x_0)^{2k}$

leads to

$$\sum_{k=0}^{n} \left|(-1)^n \binom{n}{k} (2x_0(x-x_0))^{n-k}(x-x_0)^{2k}\right|$$
$$= (|2x_0(x-x_0)| + |x-x_0|^2)^n,$$

which is $< (1+x_0^2)^n$. Since we have an absolutely convergent series, we can rearrange it according to the powers of $(x-x_0)$ and so obtain the convergent power series of $f(x) = 1/(1+x^2)$ about $x = x_0$. The computation for general x_0 is quite messy, but for $x_0 = 0$ it is simply $1/(1+x^2) = \sum_{n=0}^{\infty}(-1)^n x^{2n}$. From this we see that the radius of convergence is exactly 1. This, despite the fact that the function $1/(1+x^2)$ does nothing unusual at $x = \pm 1$, destroys our conjecture.

Nevertheless, if we look at the function $1/(1+x^2)$ of a complex variable, we see immediately what is happening. We still have

$$\sum_{n=0}^{\infty}(-1)^n z^{2n} = 1/(1+z^2) \quad \text{for } |z| < 1,$$

but now there are singularities at $z = +i$ and $z = -i$ (because $1/(1+i^2) = 1/(1-1) = 1/0$, which is undefined). The presence of those singularities explains why the real power series cannot converge in any interval $|x| < R$ for $R > 1$. If it did, the complex power series would converge in $|z| < R$ and the function would be defined at $z = +i$ and $z = -i$. In fact one can prove—although we will not do so here—that the radius of convergence of a power series $\sum a_n(x-x_0)^n = f(x)$ is exactly equal to the distance from x_0 to the first complex singularity of $f(z) = \sum a_n(z-x_0)^n$.

7.4.4 Closure Properties of Analytic Functions*

In this section we show that analytic functions are preserved under operations of arithmetic and composition. This will involve understanding how power series behave under these operations. We work entirely in the real domain.

Theorem 7.4.4 *Let $\sum a_n(x-x_0)^n = f(x)$ and $\sum b_n(x-x_0)^n = g(x)$ converge in $|x-x_0| < R$. Then $f \pm g$ and $f \cdot g$ have power-series expansions about x_0 convergent in $|x-x_0| < R$. Furthermore, if $g(x_0) \neq$*

0, *then* f/g *has a power-series expansion about* x_0 *convergent in some (perhaps smaller) neighborhood of* x_0.

Proof: The theorem for $f \pm g$ is trivial since the sum or difference of the convergent power series gives a convergent power series for $f \pm g$, $(f \pm g)(x) = \sum_{n=0}^{\infty}(a_n \pm b_n)x^n$. We have to do a little work for the product. In the process we will find a formula for the power series of the product. If we formally multiply the two power series we obtain

$$\left(\sum_{n=0}^{\infty} a_n(x-x_0)^n\right)\left(\sum_{m=0}^{\infty} b_m(x-x_0)^m\right) = \sum_{n=0}^{\infty}\sum_{m=0}^{\infty} a_n b_m (x-x_0)^{n+m}.$$

(We must use a different label for the index of summation in each power series because the distributive law for multiplication requires that we take all products of terms—one from each series). Now the point is that the power series are absolutely convergent in $|x - x_0| < R$, and this implies that the double series is also absolutely convergent

$$\begin{aligned} &\sum_{n=0}^{N}\sum_{m=0}^{M} |a_n b_m (x-x_0)^{n+m}| \\ &= \left(\sum_{n=0}^{N} |a_n|\,|x-x_0|^n\right)\cdot\left(\sum_{m=0}^{M} |b_m|\,|x-x_0|^m\right) \\ &\le \left(\sum_{n=0}^{\infty} |a_n|\,|x-x_0|^n\right)\left(\sum_{m=0}^{\infty} |b_m|\,|x-x_0|^m\right) \end{aligned}$$

which is bounded (independent of N and M). This justifies the formal multiplication and allows us to rearrange terms to make a convergent power series:

$$f(x)g(x) = \sum_{n=0}^{\infty}\sum_{m=0}^{\infty} a_n b_m (x-x_0)^{n+m} = \sum_{k=0}^{\infty} c_k (x-x_0)^k$$

where $c_k = a_0 b_k + a_1 b_{k-1} + \cdots + a_k b_0$. Since everything we have done is valid in $|x - x_0| < R$, we may conclude that the power series for fg has radius of convergence at least R. (This could also be verified directly from the formula for c_k.)

Finally, we need to consider quotients. Since $f/g = f \cdot (1/g)$, it suffices to show that $1/g$ has a convergent power series about x_0 if

$g(x_0) \neq 0$. Note that $g(x_0) \neq 0$ is the same as $b_0 \neq 0$. We can then write

$$\begin{aligned} g(x) &= b_0 + \sum_{n=1}^{\infty} b_n(x-x_0)^n \\ &= b_0(1 + \sum_{n=1}^{\infty} d_n(x-x_0)^n) \end{aligned}$$

where $d_n = b_n/b_0$. We define $h(x) = \sum_{n=1}^{\infty} d_n(x-x_0)^n$. We note that this is a convergent power series in $|x - x_0| < R$, so h is continuous. Then since $h(x_0) = 0$, we can find some smaller neighborhood $|x-x_0| < r$ on which $|h(x) - h(x_0)| = |h(x)| < 1$, so that we can expand

$$\frac{1}{g(x)} = \frac{1}{b_0}\frac{1}{1+h(x)} = \frac{1}{b_0}\sum_{k=0}^{\infty}(-1)^k h(x)^k$$

with the series converging absolutely. Thus we have found

$$\frac{1}{g(x)} = \frac{1}{b_0}\sum_{k=0}^{\infty}(-1)^k \left(\sum_{n=1}^{\infty} d_n(x-x_0)^n\right)^k$$

where both the inner and outer series converge absolutely if $|x-x_0| < r$. This is not yet a power series, but it is very close to one. Remember that we have already shown how to multiply two power series, so by induction we know $h(x)^k = (\sum_{n=1}^{\infty} d_k(x-x_0)^n)^k$ has a power-series expansion obtained by formally multiplying out all terms and collecting the (finite) sums corresponding to each power of $x - x_0$. Let us say $h(x)^k = \sum_{n=k}^{\infty} h_{n,k}(x-x_0)^n$ (because there is no zero-order term in the $h(x)$ power series, the lowest order term in the $h(x)^k$ power series is $(x-x_0)^k$). Using this notation, we have

$$\frac{1}{g(x)} = \frac{1}{b_0}\sum_{k=0}^{\infty}(-1)^k h(x)^k = \frac{1}{b_0}\left(1 + \sum_{k=1}^{\infty}(-1)^k \sum_{n=k}^{\infty} h_{n,k}(x-x_0)^n\right).$$

It is merely a question of whether we can rearrange the sum to read

$$\frac{1}{g(x)} = \frac{1}{b_0}\left(1 + \sum_{n=1}^{\infty}\left(\sum_{k=1}^{n}(-1)^k h_{n,k}\right)(x-x_0)^n\right)$$

in order to obtain a convergent power series. We know that the condition for rearrangement is absolute convergence. Thus we need to show

$$\sum_{k=1}^{\infty} \sum_{n=k}^{\infty} |h_{n,k}|\,|x - x_0|^n$$

is finite.

At first glance this looks like an unpleasant task, since even to obtain a closed-form expression for the coefficients $h_{n,k}$ would require some formidable combinatorial notation. Fortunately we can argue the difficulties away. However, it is first necessary to shrink the neighborhood somewhat, say $|x - x_0| < r_0$, so that not only do we have $|h(x)| = |\sum_{n=1}^{\infty} d_n(x - x_0)^n| < 1$ but also $H(x) < 1$, where we define $H(x) = \sum |d_n|\,|x - x_0|^n$. This of course is possible since the convergence of $\sum_{n=1}^{\infty} d_n(x - x_0)^n$ implies the convergence of the power series defining H, and we still have $H(x_0) = 0$. Since we have $|H(x)| < 1$ for $|x - x_0| < r_0$, we may conclude $1/(1 - H(x)) = 1 + \sum_{k=1}^{\infty} H(x)^k$ converges absolutely, so

$$1 + \sum_{k=1}^{\infty} H(x)^k = 1 + \sum_{k=1}^{\infty} \left(\sum_{n=1}^{\infty} |d_n|\,|x - x_0|^n \right)^k$$

converges absolutely. In expanding the kth power we have

$$\left(\sum_{n=1}^{\infty} |d_n|\,|x - x_0|^n \right)^k = \sum_{n=k}^{\infty} H_{n,k}|x - x_0|^n.$$

Note that all the terms are non-negative (this uses the fact that the combinatorial factors are all natural numbers) and so the double series $\sum_{k=1}^{\infty} \sum_{n=k}^{\infty} H_{n,k}|x - x_0|^n$ converges in the order indicated. Finally, we have $|h_{n,k}| \le H_{n,k}$, since the $H_{n,k}$ are obtained in the same manner as $h_{n,k}$ except that d_n is replaced $|d_n|$. Thus we have proved the absolute convergence of the double series $\sum_{k=1}^{\infty}(-1)^k \sum_{n=k}^{\infty} h_{n,k}(x - x_0)^n$, hence its rearrangement as a power series in $|x - x_0| < r_0$ is justified. QED

The theorem implies immediately that if f and g are analytic functions on a domain D, then so are $f \pm g$, $f \cdot g$, and f/g (if $g \neq 0$ on D). The proof of the expansion for $1/g$ actually contains the germ of an important generalization concerning compositions ($1/g$ is the composition of g followed by $f(x) = 1/x$).

Theorem 7.4.5 *Let $f(x) = \sum_{n=0}^{\infty} a_n(x-x_0)^n$ converge in $|x-x_0| < r$, and let $g(x) = \sum_{n=0}^{\infty} b_n(x-x_1)^n$ converge in $|x-x_1| < r_1$ where $f(x_0) = x_1$ (so $x_1 = a_0$). Then $g \circ f(x)$ has a convergent power series in a sufficiently small neighborhood of x_0 that is obtained by rearrangement from $\sum_{n=0}^{\infty} b_n \left(\sum_{k=1}^{\infty} a_k(x-x_0)^k)^n\right)$.*

Proof: Since $f(x) - x_1 = \sum_{k=1}^{\infty} a_k(x-x_0)^k$ and this converges for $|x - x_0| < r_0$ (hence absolutely), we have

$$g \circ f(x) = \sum_{n=0}^{\infty} b_n \left(\sum_{k=1}^{\infty} a_k(x-x_0)^k \right)^n$$

converging if $|f(x) - x_1| < r_1$ also. Since $f(x_0) = x_1$ and f is continuous, this will be true if $|x - x_0|$ is small enough. To complete the proof we have to justify the rearrangement. We have

$$\left(\sum_{k=1}^{\infty} a_k(x-x_0)^k \right)^n = \sum_{k=n}^{\infty} a_{k,n}(x-x_0)^k$$

with convergence in $|x - x_0| < r_0$, so we need to show the absolute convergence of the double infinite series $\sum_{n=0}^{\infty} \sum_{k=n}^{\infty} b_n a_{k,n}(x-x_0)^k$. As in the previous proof we do this by comparison with

$$\sum_{n=0}^{\infty} \sum_{k=n}^{\infty} |b_n| A_{k,n} |x-x_0|^k$$

where $|a_{k,n}| \le A_{k,n}$ and $A_{k,n}$ is obtained in the same way as $a_{k,n}$ with $|a_n|$ replacing a_n,

$$\left(\sum_{k=1}^{\infty} |a_k| \, |x-x_0|^k \right)^n = \sum_{k=n}^{\infty} A_{k,n} |x-x_0|^k.$$

Finally the convergence of the dominating double series

$$\sum_{n=0}^{\infty} \sum_{k=n}^{\infty} |b_n| A_{k,n} |x-x_0|^k$$

follows if we restrict $x - x_0$ so that $\sum_{k=1}^{\infty} |a_k| \, |x-x_0|^k < r_1$, for then $\sum_{n=0}^{\infty} |b_n| \left(\sum_{k=1}^{\infty} |a_k| \, |x-x_0|^k\right)^n$ converges absolutely and expanding out $\left(\sum_{k=1}^{\infty} |a_k|\,|x-x_0|^k\right)^n$ does not change the sum since all terms are non-negative. QED

The theorems in this section justify the assertion that essentially all functions that can be written in closed form are analytic, provided the domain is suitably restricted to eliminate points where division by zero is called for at some stage of the definition. This of course assumes that all the special functions we introduce are analytic. (To show that general powers $f(x) = x^a$ are analytic for $x > 0$ we anticipate the results $x^a = e^{a \log x}$ and the analytic nature of e^x and $\log x$ from the next chapter.)

The proof of the rearrangement of the series for $g \circ f$ suggests an interesting question. If $f_1, f_2, \ldots$ are analytic functions with convergent power series $f_k = \sum_{n=0}^{\infty} a_{n,k}(x-x_0)^n$ in an interval $|x-x_0| < R$ and if $\sum_{k=1}^{\infty} f_k = f$ converges in $|x-x_0| < R$, does f have a convergent power series $\sum_{n=0}^{\infty} \left(\sum_{k=1}^{\infty} a_{n,k}\right)(x-x_0)^n$? The answer turns out to be no, even if we require the convergence of the series $\sum_{k=1}^{\infty} f_k$ to be uniform. We will see this in the next section, where we show that any continuous function can be obtained as a uniform limit of polynomials. A deeper theorem says that if $\sum f_k(z)$ converges uniformly for all *complex* z in a disc, then the limit is analytic. (For a proof, consult any book on complex variables.)

We now summarize what we have discovered about three closely related topics: power series, analytic functions, and Taylor series. Suppose we start with a C^∞ function $f(x)$ defined in a neighborhood of x_0. Then we can form the Taylor expansions $\sum_{n=0}^{N} f^{(n)}(x_0)(x-x_0)^n/n!$ to any order N at x_0. Taylor's theorem describes the accuracy of approximation to $f(x)$ as $x \to x_0$ for fixed N. It says nothing about limiting behavior as $N \to \infty$, and in fact we will see in the next chapter that in general nothing can be said. However, there is nothing to prevent us from considering the behavior as $N \to \infty$ and asking if the resulting power series $\sum_{n=0}^{\infty} a_n(x-x_0)^n$ converges (where we have set $a_n = f^{(n)}(x_0)/n!$). If it does converge to f (there are examples where it converges but to a different function) on an interval $|x-x_0| < R$, then we say that f is analytic on $|x-x_0| < R$. We have shown that f is also equal to a power-series expansion $\sum_{n=0}^{\infty} b_n(x-x_1)^n$ about an arbitrary point x_1 in the interval; the coefficients b_n are again given as in Taylor's theorem by $b_n = 1/n! f^{(n)}(x_1)$, and the interval of convergence is at least as large as the largest symmetric interval about x_1 contained in the original interval about x_0. Inside the interval of convergence, these power series are very well behaved. The convergence is absolute, uniform on any smaller interval, and the series may be differentiated term-by-term. Two power series about the same point converge to the same function only if they are identical series. Power series about a point x_0 may be combined by arithmetic operations, and power series may be composed if the expansion points match up appropriately. These properties make power series a powerful tool in the study of differential equations and other applications.

The class of analytic functions is wide enough to contain most important funtions. On the other hand, it is a rather special class of functions. An analytic function is determined by its values in an arbitrarily small neighbohood of a point since these suffice to determine all the derivatives $f^{(n)}(x_0)$ and, hence, the power series about x_0. It is even true that f is determined by its values on any sequence of points $x_1, x_2, \ldots$ converging to a point in the interior of its domain (see the exercises), but there is then no nice formula for $f(x)$ in terms of $f(x_1), f(x_2), \ldots$.

To learn more about analytic functions see *A primer of real analytic functions* by S. G. Krantz and H. R. Parks, Birkhauser-Verlag, 1992.

7.4.5 Exercises

1. Let f be defined in (a, b) and have a power-series expansion about every point x_0 in (a, b) that converges in a neighborhood of x_0. Show that the values of $f(x)$ on (a, b) are determined from the values of $f(x)$ on any neighborhood in (a, b).

2. If f is analytic in a neighborhood of x_0 and $f(x_0) = 0$, show that $f(x)/(x - x_0)$ is analytic in the same neighborhood.

3. If f is analytic on (a, b) and $f(x_k) = 0$ for a sequence of distinct points x_k in (a, b) with $\lim_{k\to\infty} x_k = x_0$ in (a, b) (note $x_0 = a$ or $x_0 = b$ is not allowed), prove that $f \equiv 0$. (**Hint**: show $f(x_0) = 0$ and divide the power series by $x - x_0$.)

4. Prove that the binomial series

$$(1+x)^a = 1 + ax + \frac{a(a-1)}{2!}x^2 + \cdots + \frac{a(a-1)\ldots(a-n)}{n!}x^n + \ldots$$

converges in $|x| < 1$, for any real a.

5. Expand $f(x) = \int_1^x (1/t)\,dt$ in a power series about $x = 1$ by expanding $1/t$ in a power series about $t = 1$ and integrating term-by-term. What is the radius of convergence of the series for $f(x)$?

6. Prove that if $f(x)$ is analytic on (a,b), then $F(x) = \int_c^x f(t)\,dt$ is also analytic on (a,b), where c is any point in (a,b).

7. Compute the power-series expansion of the following functions about $x = 0$:

   a. $f(x) = x^2/(1-x^2)$,
   b. $f(x) = 1/(1-x)^2$,
   c. $f(x) = \sqrt{1+x}$.

8. Compute the radius of convergence of the following power series:

   a. $\sum n^4/n!x^n$,
   b. $\sum \sqrt{n}\,x^n$,
   c. $\sum n^2 2^n x^n$.

9. *Compute the power-series expansion of $1/(1+x^2)$ about any point x_0 from the formula $a_n = f^{(n)}(x_0)/n!$.

10. If f is analytic on (a,b) prove that for every x_0 in (a,b) there exists a neighborhood of x_0 and constants M and r such that $|f^{(k)}(x)| \leq Mk!r^k$ for all k and x in the neighborhood.

11. *Prove the converse to 10, namely if f is C^∞ on (a,b) and if for every x_0 in (a,b) there exists a neighborhood of x_0 and constants M and r such that $|f^{(k)}(x)| \leq Mk!r^k$ for all k and x in the neighborhood, then f is analytic on (a,b). (**Hint**: show that the Taylor approximations about x_0 converge to f in a neighborhood of x_0.)

7.5 Approximation by Polynomials

7.5.1 Lagrange Interpolation

In this section we discuss the problem of approximating functions by sequences of polynomials. This is clearly a worthwhile goal, since a polynomial is a simpler object than a general continuous function, and many questions about polynomials can be answered easily by algebraic computations. One's first thoughts on the matter naturally turn to power series since the partial sums of a power series form a sequence of polynomials. However, using power series restricts us to analytic functions, excluding such simple functions as $|x|$, and even for analytic functions the interval of convergence may be too small. But we will show that an arbitrary continuous function on a compact interval can be approximated uniformly by polynomials—this is the famous *Weierstrass approximation theorem.* This does not in any way contradict the fact that some continuous functions do not have power-series expansions, because the partial sums of power series are very special ways of creating a sequence of polynomials. If the power series is expanded about the origin, for example, and $P_n(x) = \sum_{k=0}^{n} a_k x^k$, then the sequence $P_1(x), P_2(x), \ldots$ has the property that once the coefficient a_n of x_n is added in P_n, it never changes in the subsequent polynomials. When we construct sequences of polynomials $P_1, P_2, \ldots$ converging uniformly to f, we will not be requiring any relationship between the coefficients of P_n and the subsequent polynomials.

Before getting into the proof of the Weierstrass approximation theorem, we will discuss a related but simpler problem that was first solved by Lagrange—that of fitting a polynomial to any finite set of data. In other words, we want to find a polynomial $P(x)$ satisfying $P(x_k) = a_k$ for $k = 1, \ldots, n$ where x_k are arbitrary distinct real points and a_k are arbitrary real (or complex) values. Since an arbitrary polynomial of degree $n-1$, $c_0 + c_1 x + \cdots + c_{n-1}x^{n-1}$, has n arbitrary constants, we would hope to solve the problem with a polynomial of degree $n-1$. In fact, substituting this in the equations $P(x_n) = a_n$ leads to n linear equations in n unknowns, which may or may not have a unique solution. In the present case we can simply write down a solution. Note that the function $q_k(x) = \Pi_{j\neq k}(x - x_j)$ is a polynomial of degree $n-1$ that vanishes at every x_j except x_k. Note that $q_k(x_k) = \Pi_{j\neq k}(x_k - x_j)$ is a non-zero constant and the polynomials $Q_k(x) = [q_k(x_k)]^{-1}q_k(x)$ satisfy

$$Q_k(x_j) = \begin{cases} 0 & \text{if } j \neq k, \\ 1 & \text{if } j = k, \end{cases}$$

so $P(x) = \sum_{k=1}^{n} a_k Q_k(x)$ gives a solution to the problem that is called the *Lagrange interpolation polynomial.*

However, merely passing the graph of a polynomial through a finite set of points in the plane does not solve the problem of approximating a function because the formula for the Lagrange interpolation polynomial does not allow you to control $P(x)$ at points in between the x_k's. For example, for the function f in Figure 7.5.1, the polynomial $P(x) \equiv 0$ passes through all seven points where the function f crosses the x-axis, and yet P is a very poor approximation to f.

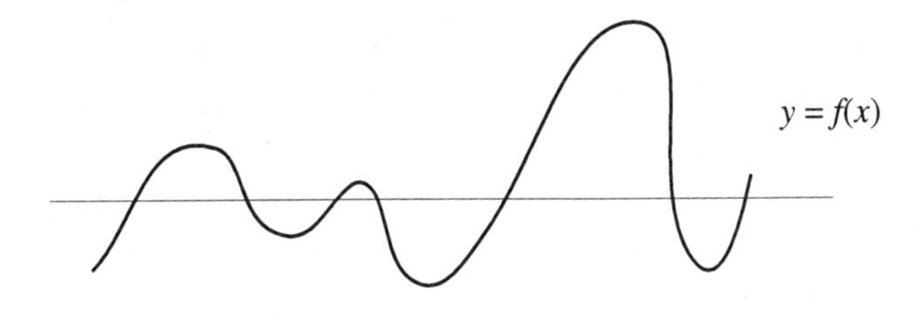

Figure 7.5.1:

Even if we were to choose additional points on the graph of f through which to pass a polynomial, we would not be sure of a better fit.

In section 7.5.3 we will give a constructive proof of the Weierstrass approximation theorem. The proof is instructive for two reasons. The first is that it is the prototype of a very general method for obtaining approximations to functions. The second is that it allows one to say more about the approximation if more is known about f (for example, if f is C^1, then the derivatives of the polynomials approximate f').

7.5.2 Convolutions and Approximate Identities

The method we are going to use is called *convolution with an approximate identity.* The term *convolution* refers to a kind of product between functions defined by

$$f * g(x) = \int_{-\infty}^{\infty} f(x-y)g(y)\,dy.$$

In the case we will consider both f and g will be continuous functions and one of them will be zero outside a bounded interval so that the integral will be a proper integral and $f*g(x)$ will be defined for each x. More generally one considers convolution products under much weaker assumptions on f and g. The key point to observe about the convolution is the commutativity, $f * g(x) = g * f(x)$. This follows from the simple change of variable in the integral—replace y by $x - y$ (hence $x - y$ gets replaced by y). The convolution product is also associative, $(f * g) * h = f * (g * h)$—a fact we leave as an exercise.

What is the significance of the convolution product? We can interpret $f(x - y)$ as a translate of f (the graph is translated to the right by y), and this is then "averaged" with the weight $g(y)$. So $f * g$ is a weighted average of translates of f. But writing $f * g(x) = g * f(x) = \int g(x-y)f(y)\,dy$ shows that the convolution product is also a weighted average of translates of g. Thus $f * g$ is a kind of hybrid, having the properties of both f and g (at least those properties that are preserved under translation and averaging). For example, if g is a polynomial and f is continuous and vanishes outside a bounded interval, then $f*g$ is a polynomial. Indeed if $g(x) = \sum_{k=0}^{n} a_k x^k$, then

$$\begin{aligned} f * g(x) &= \int g(x-y)f(y)\,dy \\ &= \int \left(\sum_{k=0}^{n} a_k (x-y)^k \right) f(y)\,dy \\ &= \int \sum_{k=0}^{n} \sum_{j=0}^{k} (-1)^{k-j} a_k \binom{k}{j} x^j y^{k-j} f(y)\,dy. \end{aligned}$$

Since this is a finite sum, we can interchange it with the integral to obtain $f * g(x) = \sum_{j=0}^{n} b_j x^j$ where

$$b_j = \sum_{k=j}^{n} (-1)^{k-j} a_k \binom{k}{j} \int y^{k-j} f(y)\,dy$$

are just constants (note the integral $\int y^{k-j} f(y)\,dy$ is a proper integral since $f = 0$ outside a bounded interval). Thus we can obtain polynomials to approximate f by taking $f * g$ where g is a polynomial.

We still need one further idea to make $f * g$ approximate f: that of an *approximate identity.* Looking at $f * g(x) = \int f(x-y)g(y)\,dy$, this will be close to $f(x)$ if only small values of y are emphasized (then $f(x-y)$ will be close to $f(x)$ by the continuity of f) and the average is "fair." Suppose for example that the graph of g were something like Figure 7.5.2.

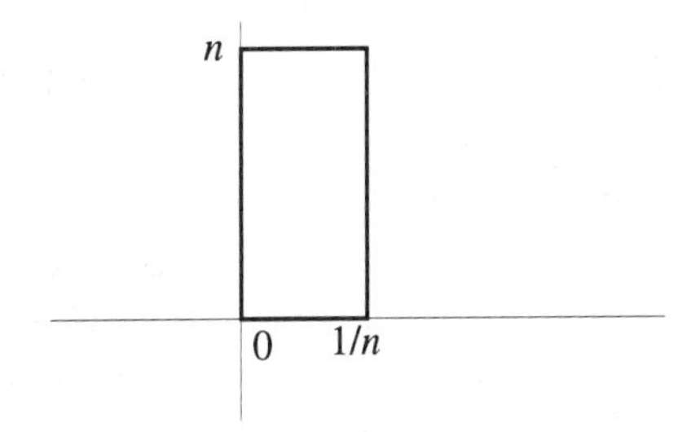

Figure 7.5.2:

Then $\int_0^{1/n} g(y)\,dy = 1$, so we are getting a fair average and $f*g(x) = n\int_0^{1/n} f(x-y)\,dy$ approximates f well since $f(x-y)$ is close to $f(x)$ for all y in $0 \le y \le 1/n$ by the uniform continuity of f. This can be seen most simply by writing

$$f * g(x) - f(x) = n\int_0^{1/n} (f(x-y) - f(x))\,dy,$$

so

$$\begin{aligned} |f * g(x) - f(x)| &\le n\int_0^{1/n} |f(x-y) - f(x)|\,dy \\ &\le n\int_0^{1/n} \frac{1}{m}\,dy = \frac{1}{m} \end{aligned}$$

if $|f(x-y) - f(x)| \le 1/m$ for all $|y| \le 1/n$.

Clearly the exact form of the function g is not crucial, but we do need the condition $\int g(y)\,dy = 1$ and some condition that makes g concentrate near zero. Any sequence of functions $\{g_n\}$ with such properties is called an *approximate identity.* This is not quite a definition because I have been vague as to what is meant by the condition that g_n gets more concentrated near zero as $n \to \infty$. In fact, depending upon the context, there are several ways of making this precise. Here is one: suppose $g_n \geq 0$ satisfies $\int g_n(y)\,dy = 1$ and $\int_{-\infty}^{-1/n} + \int_{1/n}^{\infty} g_n(y)\,dy \to 0$ as $n \to \infty$. Then

$$\begin{aligned} |f * g_n(x) - f(x)| &= \left| \int_{-\infty}^{\infty} (f(x-y) - f(x)) g_n(y)\,dy \right| \\ &\leq \int_{-\infty}^{-1/n} + \int_{1/n}^{\infty} |f(x-y) - f(x)| g_n(y)\,dy \\ &\quad + \int_{-1/n}^{1/n} |f(x-y) - f(x)| g_n(y)\,dy. \end{aligned}$$

We are assuming f is continuous and vanishes outside a bounded interval. This implies that f is bounded and uniformly continuous. Thus $|f(x)| \leq M$ for all x, and given any error $1/m$ there exists $1/n$ such that $|f(x-y) - f(x)| < 1/m$ for all $|y| \leq 1/n$. Substituting these estimates into the integrals (using $|f(x-y) - f(x)| \leq |f(x-y)| + |f(x)| \leq 2M$ in the $|y| > 1/n$ integral) we obtain

$$\begin{aligned} |f * g_n(x) - f(x)| &\leq 2M \int_{|y| \geq 1/n} g_n(y)\,dy + \frac{1}{m} \int_{-1/n}^{1/n} g_n(y)\,dy \\ &\leq 2M \int_{|y| \geq 1/n} g_n(y) dy + \frac{1}{m} \end{aligned}$$

since $\int_{-1/n}^{1/n} g_n(y)\,dy \leq \int_{-\infty}^{\infty} g_n(y)\,dy = 1$. But we are assuming $\int_{|y| \geq 1/n} g_n(y)\,dy \to 0$ as $n \to \infty$ (this is our concentrating hypothesis), so we can make $2M \int_{|y| \geq 1/n} g_n(y)\,dy \leq 1/m$ by taking n large enough and, hence, $|f * g(x) - f(x)| \leq 2/m$ if n is large enough. We can summarize this result as follows:

Definition 7.5.1 *A sequence of continuous functions on the line $\{g_n\}$ satisfying*

1. $g_n(x) \geq 0$,
2. $\int_{-\infty}^{\infty} g_n(x)\,dx = 1$,
3. $\lim_{n \to \infty} \int_{|x| \geq 1/n} g_n(x)\,dx = 0$

is called an approximate identity.

Lemma 7.5.1 (*Approximate Identity Lemma*) *Let $\{g_n\}$ be an approximate identity. Then if f is any continuous function on the line vanishing outside a bounded interval, $f * g_n$ converges uniformly to f.*

7.5.3 The Weierstrass Approximation Theorem

Theorem 7.5.1 (*Weierstrass Approximation Theorem*) *Let f be any continuous function on a compact interval $[a, b]$. Then there exists a sequence of polynomials converging uniformly to f on $[a, b]$.*

Proof: There are two obstacles to using the approximate identity lemma. The first is that f is not defined and continuous on the whole line, and the second is that it is impossible to find polynomials g_n to satisfy the approximate identity conditions.

The first obstacle is readily overcome in two different ways. Note that if it happens that $f(a) = f(b) = 0$, then we can extend the domain of f to the whole line by setting $f = 0$ outside $[a, b]$, and the extended function will be continuous. We can reduce the general case to this special case by enlarging the domain, say to $[a-1, b+1]$, adding "flaps" to the graph of f, as in Figure 7.5.3

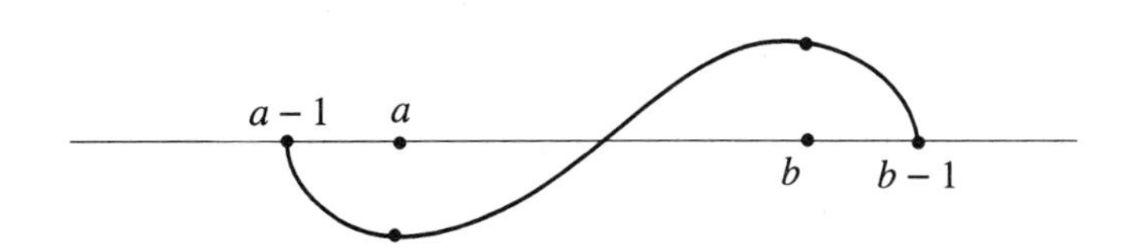

Figure 7.5.3:

and then extending f to be zero outside $[a-1, b+1]$. Any uniform approximation of f on $[a-1, b+1]$ will automatically yield uniform approximation on the smaller interval $[a, b]$. The other method is to subtract from f an appropriate affine function $Ax+B$ so that $f-Ax-B$ vanishes at the endpoints—since $Ax + B$ is a polynomial, we can add it back on to the polynomials approximating $f - Ax - B$ to obtain polynomials approximating f.

So now we assume that f is a continuous function on the whole line vanishing outside $[a, b]$, and we need to approximate f by polynomials on $[a, b]$. It is crucial that we observe that the approximation is only needed on $[a, b]$, since the growth of polynomials as $x \to \infty$ precludes approximation on the whole line. Also, by restricting attention to the interval $[a, b]$, we can overcome the problem that no polynomials satisfy the approximate identity properties. The idea is that if f vanishes outside $[a, b]$, then

$$f * g(x) = \int f(x - y)g(y)\, dy = \int_a^b f(y)g(x - y)\, dy$$

only involves the values of g on the interval $[x - b, x - a]$; and if x is also restricted to lie in $[a, b]$, then only the values of g on the compact interval $[a', b'] = [a - b, b - a]$ are involved. Therefore, if we take $g_n(x)$ to be equal to a polynomial on $[a', b']$ and zero elsewhere, then $f * g_n$ will be equal to a polynomial on $[a, b]$ (for x not in $[a, b]$ we will not have $f * g_n(x)$ equal to a polynomial, but we are only interested in what happens on $[a, b]$).

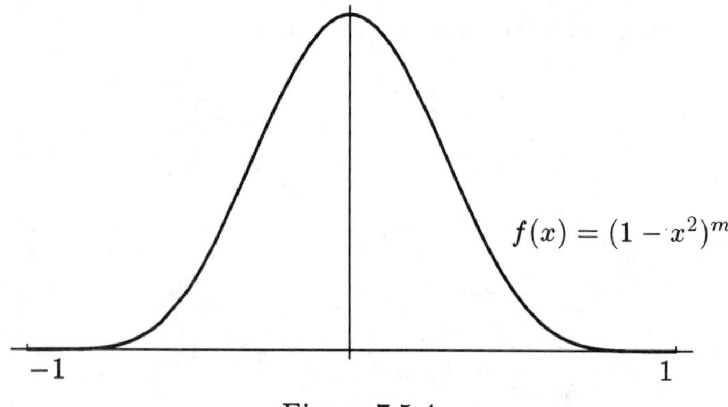

Figure 7.5.4:

Thus to complete the proof of the Weierstrass approximation theorem via the approximate identity lemma we need to show that there exists an approximate identity that consists of functions g_n that are equal to polynomials on $[a', b']$ and vanish outside. For simplicity we take the interval $[-1, 1]$, since the general case can be obtained by composing g_n with an appropriate affine function and multiplying by a constant.

We can actually write down a simple formula for the approximate identity we want. Consider the function $(1-x^2)^m$, shown in Figure 7.5.4. Note that it vanishes to high order at $x = \pm 1$ and its graph over the interval $[-1,1]$ appears concentrated near $x = 0$. The condition that the integral be equal to one is not satisfied, so we have to multiply by the appropriate constant, namely c_m^{-1}, where $c_m = \int_{-1}^{1}(1-x^2)^m\,dx$. Now while it is possible to compute c_m explicitly, the result is quite complicated (see exercise set 7.5.5, number 14) and it is more illuminating to get an estimate instead. Since we will want an estimate from above for c_m^{-1}, we need an estimate from below for c_m and, hence, an estimate from below for $(1-x^2)^m$. For x near zero it is natural to compare $(1-x^2)^m$ with $1-mx^2$, the first terms of the Taylor expansion about $x = 0$. (See Figure 7.5.5 where the two functions are graphed together.) Since they are equal at $x = 0$ and $d/dx(1-x^2)^m = -2mx(1-x^2)^{m-1}$ while $d/dx(1-mx^2) = -2mx$, we see $1-mx^2 \leq (1-x^2)^m$ for all x, the desired estimate.

Note that $1-mx^2$ vanishes at $x = \pm\sqrt{1/m}$, so the estimate $1-mx^2 \leq (1-x^2)^m$ is of interest only for $|x| \leq 1/\sqrt{m}$. In fact, if $|x| \leq 1/2\sqrt{m}$, then the estimate tells us

$$(1-x^2)^m \geq 1-mx^2 \geq 1-m/4m = 3/4$$

and so a lower bound for $\int_{-1}^{1}(1-x^2)^m dx$ is

$$\int_{-\frac{1}{2\sqrt{m}}}^{\frac{1}{2\sqrt{m}}}(1-x^2)^m\,dx \geq \frac{3}{4}m^{-1/2}$$

(the length of the interval $m^{-1/2}$ times the lower bound for the function).

So if we set $h_m(x) = c_m^{-1}(1-x^2)^m$ (on $[-1,1]$ and zero elsewhere) we will have $\int_{-1}^{1} h_m(x)\,dx = 1$ and, by the above estimate, $|h_m(x)| \leq (4/3)m^{1/2}(1-x^2)^m$. Clearly h_m satisfies the first two conditions for an approximate identity. To verify the third condition we need to show $\int_{|x|\geq 1/n} h_m(x)\,dx$ goes to zero as $m \to \infty$, for every fixed n. Since h_m vanishes outside $[-1,1]$, this is a proper integral, so it suffices to show $h_m(x) \to 0$ uniformly on $[1/n, 1]$ (since h_m is even, the behavior

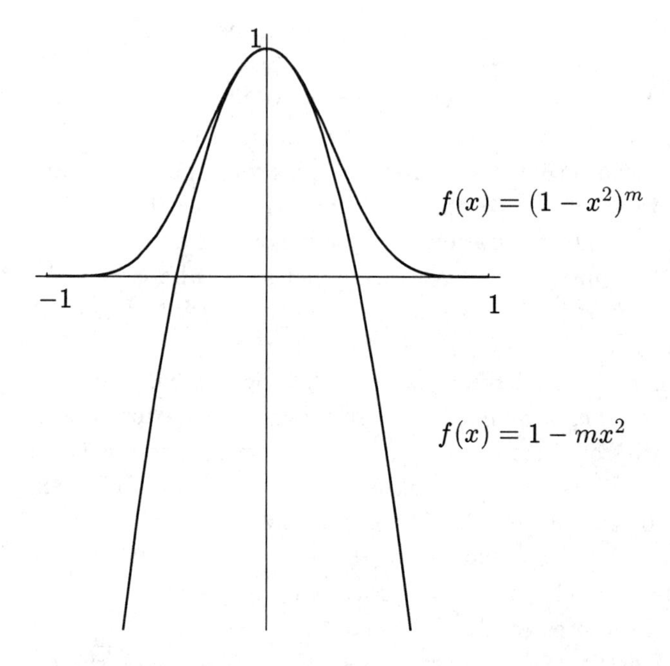

Figure 7.5.5:

is the same on $[-1, -1/n]$). But h_m is clearly decreasing on $[1/n, 1]$, so it assumes its maximum at $x = 1/n$. Thus we really need to show $\lim_{m\to\infty} h_m(1/n) = 0$ for each n. Because of our bound for $h_m(x)$, this will follow from $\lim_{m\to\infty} m^{1/2}\left(1 - 1/n^2\right)^m = 0$.

Notice what the issue is here. The factor $m^{1/2}$ goes to infinity, and the factor $(1 - 1/n^2)^m$ goes to zero because $1 - 1/n^2 < 1$. Which term dominates? In the next chapter we will discuss the general principle that exponential factors always dominate polynomial factors. Here we can finish the proof by recalling that our proof that $\lim_{m\to\infty} r^m = 0$ for $r < 1$ gave the estimate $(1 + c)^m \geq 1 + cm$; hence, $1/(1 + c)^m \leq 1/(1 + cm)$. In our case $1/(1 + c) = 1 - 1/n^2$; hence,

$$m^{1/2}\left(1 - \frac{1}{n^2}\right)^m \leq \frac{m^{1/2}}{1 + cm} \leq \frac{1}{cm^{1/2}} \to 0$$

as $m \to \infty$.

This completes the verification that $\{h_m\}$ is an approximate identity. The approximate identity lemma tells us that $h_m * f \to f$ uniformly on $[-1,1]$, and we have observed that $h_m * f$ is equal to a polynomial on $[-1,1]$. QED

A typical application of the Weierstrass approximation theorem is the following: if f is a continuous function on $[0,1]$ and all the *moments* $a_n = \int_0^1 f(x)x^n dx$ are known for $n = 0, 1, 2, \ldots$, then is f determined? By considering the difference of two functions with the same moments, the question reduces to: if the moments of f are all zero, is f zero? If the moments of f are all zero, then also $\int_0^1 f(x)P(x)\,dx = 0$ for any polynomial P by linearity of the integral. But by the Weierstrass approximation theorem we can find a sequence of polynomials P_n converging uniformly to f (if f is complex-valued take $\bar{f}$ instead). Then fP_n converges to $|f|^2$ uniformly (the difference $|fP_n - f^2| = |f|\,|P_n - f|$ is dominated by the maximum value of $|f|$, which is finite, times the value of $|P_n - f|$, which tends to zero uniformly). Thus we can interchange the limit and integral:

$$0 = \lim_{n\to\infty} 0 = \lim_{n\to\infty} \int_0^1 f(x)P_n dx = \int_0^1 |f(x)|^2\,dx,$$

so $|f|^2 \equiv 0$ and $f \equiv 0$. The problem of actually reconstructing a function from its moments is more difficult, and we will not discuss it here.

7.5.4 Approximating Derivatives

Next we show that if f is differentiable we can also approximate f' by the derivatives of the polynomials approximating f (this is by no means automatic, as we have seen). To be precise, let us assume that f is C^1 on $[a,b]$, meaning that one-sided derivatives exist at the endpoints and $f'(x)$ is continuous on $[a,b]$. As in the proof of the Weierstrass approximation theorem, we extend f to the whole line, but this time we need to do it in a more elaborate way so that the extension is C^1. This can be accomplished either by adding flaps that are adjusted to match up derivatives or by subtracting from f a higher order polynomial to make both f and f' vanish at the endpoints. After this more careful preparation, the same construction of $f * g_n$ will yield the desired approximation. Indeed it suffices to verify that $(f * g_n)' = f' * g_n$, for the proof of the Weierstrass approximation theorem shows $f' * g_n$ converges uniformly to f'. We state this as a general principle.

Theorem 7.5.2 *Let f be C^1 and vanish outside a bounded interval, and let g be continuous. Then $f * g$ is C^1 and $(f * g)' = f' * g$.*

Proof: We form the difference quotient

$$\begin{aligned}\frac{f * g(x) - f * g(x_0)}{x - x_0}\\ &= \frac{1}{x - x_0}\left(\int f(x-y)g(y)\,dy - \int f(x_0-y)g(y)\,dy\right)\\ &= \int \left(\frac{f(x-y) - f(x_0-y)}{x - x_0}\right) g(y)\,dy.\end{aligned}$$

Now we claim

$$\frac{f(x-y) - f(x_0-y)}{x - x_0}$$

converges uniformly to $f'(x_0 - y)$ as $x \to x_0$. Indeed by the mean value theorem it is $f'(x_1 - y)$ for some x_1 between x_0 and x (x_1 may depend on y), and by the uniform continuity of f' we have $f'(x_1 - y) \to f'(x_0 - y)$ uniformly. Since the integration only extends over a finite interval, we can interchange the integral with the uniform limit to obtain $\int f'(x_0 - y)g(y)\,dy$ as the limit of the difference quotient for $f * g$ at x_0, simultaneously proving that $f * g$ is differentiable and supplying the formula $f' * g$ for the derivative. QED

Because of the commutativity of the convolution product, we also have $(f * g)' = f * g'$. By induction we can extend the theorem to higher derivatives: if f is C^k, then so is $f * g$ and $(f * g)^{(k)} = f^{(k)} * g$. By choosing more sophisticated flaps we can extend C^k functions on a compact interval $[a, b]$ to C^k functions on the line that vanish outside a larger interval and so obtain in $\{f * g_n\}$ a sequence of polynomials that converges uniformly to f with all derivatives of orders $\leq k$ converging to the corresponding derivative of f. Finally it is even possible to have derivatives of all orders of $f * g_n$ simultaneously converge uniformly to the corresponding derivatives of f, provided f is C^∞. This requires adapting the flaps to match derivatives of all orders. We will see how to do this in the next chapter.

In obtaining a sequence of polynomials approximating f, we have not paid particular attention to the orders of the polynomials (of course the orders must increase to ∞ unless f is a polynomial). An interesting question that involes such considerations is the question of best approximation to f by polynomials of degree $\leq n$. For any fixed polynomial P, we define

$$E(f, P) = \sup\{|f(x) - P(x)| : x \text{ is in } [a, b]\},$$

the maximum error (the sup is achieved because $f - P$ is continuous and $[a, b]$ compact). We then define $E_n(f) = \inf\{E(f, P) : P \text{ is any polynomial of degree} \leq n\}$. It is not clear that this inf is attained nor that the polynomial attaining the inf—if it exists—is unique. Nevertheless, both statements are true. The Weierstrass approximation theorem implies $\lim_{n\to\infty} E_n(f) = 0$, but we can also ask at what rate $E_n(f)$ vanishes. It turns out that we can relate the rate of convergence to the smoothness of f. All these considerations are beyond the scope of this book.

7.5.5 Exercises

1. Show that there exists a polynomial of degree $2n - 1$ satisfying $f(x_k) = a_k$ and $f'(x_k) = b_k$ for $k = 1, \ldots, n$.

2. Let f be C^1 on $[a, b]$. Construct a C^1 extension of f to the line that vanishes outside $[a - 1, b + 1]$. (**Hint**: use exercise 1.)

3. If f and g are continuous on the line and f vanishes outside a bounded interval, prove $f * g$ is continuous.

4. Prove that $(f * g) * h = f * (g * h)$ if f, g, and h are continuous and two of them vanish outside a bounded interval. (Note: this requires interchanging the order of two integrations.)

5. If f is C^k and g is C^m and one of them vanishes outside a bounded interval, prove that $f * g$ is C^{k+m} and $(f * g)^{(k+m)} = f^{(k)} * g^{(m)}$.

6. Define the *support*indexsupport of f to be the closure of the set of points where $f \neq 0$. Prove that a continuous function f has compact support if and only if f vanishes outside a bounded interval. Prove that support $(f * g) \subseteq$ support $(f) +$ support (g) where the $+$ means the set of all sums of numbers from support (f) and support (g).

7. If f is C^1 on $[a, b]$ prove that there exists a cubic polynomial P such that $f - P$ and its first derivative vanish at the endpoints of the interval.

8. If $f \geq 0$, on $[a, b]$ show that the polynomials approximating f may be all taken ≥ 0 on $[a, b]$.

9. If $f(c) = 0$ for some point c in (a, b), prove that the polynomials approximating f on $[a, b]$ may be taken to vanish at c.

10. Let f be an even function ($f(x) = f(-x)$) on $[-1, 1]$. Prove that if $\int_{-1}^{1} f(x)x^{2k}dx = 0$ for $k = 0, 1, 2, \ldots$, then $f \equiv 0$.

11. Prove that none of the power-series expansions of $1/(1 + x^2)$ converge to it on $[-2, 2]$.

12. Let f be defined and C^1 on (a, b), and suppose one-sided limits of f' exist at a and b. Prove that one-sided limits of f exist at a and b and f can be extended to a C^1 function on $[a, b]$.

13. Let $P_n \to f$ uniformly on $[a, b]$ where $\{P_n\}$ is a sequence of polynomials of degree $\leq N$. Prove that f is a polynomial of degree $\leq N$. (**Hint**: for each $k \leq N$ find a continuous function $h_k(x)$ such that $\int_a^b h_k(x)x^j dx = 0$ for all $j \leq N$ such that $j \neq k$ but $\int_a^b h_k(x)x^k dx = 1$, and consider $\lim_{n\to\infty} \int_a^b h_k(x)P_n(x)\,dx$.)

14. a. For $c_m = \int_{-1}^{1}(1 - x^2)^m\,dx$, obtain the identity $c_m = c_{m-1} - (1/2m)c_m$ by integration by parts.

    b. Show that

$$c_m = \frac{1 \cdot 3 \cdot 5 \cdots (2m+1)}{2 \cdot 4 \cdot 6 \cdot \cdots (2m)} = \frac{(2m+1)!}{(2^m m!)^2}.$$

15. Compute $f * f$ for f equal to the characteristic function of $[0, 1]$ (equal to one on the interval, zero elsewhere). Explain why this is called a "hat function".

7.6 Equicontinuity

7.6.1 The Definition of Equicontinuity

Compactness is a powerful method for producing existence theorems—for example, the existence of a point where a continuous function achieves its maximum or minimum on a compact interval. In many problems we need to find functions, rather than points, that maximize or minimize certain quantities (often physical quantities such as energy or entropy are involved). For such problems we need a different notion of compactness. In order to describe one such notion we consider the compactness condition "every sequence of points in a set has a subsequence converging to a point in the set". If we replace the word "point" by "function", we are led to consider the problem of when a sequence of functions has a subsequence that converges. We will limit the discussion to continuous functions defined on a compact interval $[a, b]$, and we will demand uniform convergence. A typical way this problem arises is when we try to minimize some "functional" $E(f)$ over all continuous functions f (for example, $E(f) = \int_0^1 |f(x)|^2 dx$). If we can show that $E(f)$ is bounded from below, $E(f) \geq c$, then the inf of all the real values $E(f)$ exists and so we can find a sequence of functions f_n such that $E(f_n)$ converges to this inf. If we knew there were a subsequence that converged uniformly, $f_{n'} \to f$, then we could hope that $\lim_{n\to\infty} E(f_{n'}) = E(f)$ (in actual applications this is usually the most difficult step) and f would then be a continuous function (the uniform limit of continuous functions) minimizing E.

Not every sequence of continuous functions on $[a, b]$ has a uniformly convergent subsequence. The simplest example is $f_n(x) \equiv n$. Here the trouble is that the functions are unbounded. It is natural then to impose the condition that functions be *uniformly bounded,* meaning that $|f_n(x)| \leq M$ for some M for all n and all x. Note that the condition that the sequence of functions be bounded at each point (for every x there exists M_x such that $|f_n(x)| \leq M_x$ for all n) is enough to guarantee that for every point there is a subsequence converging at that point (for every x_0 there exists $\{n(k)\}$ depending on x_0 such that $\lim_{k\to\infty} f_{n(k)}(x_0)$ exists). The reason we require uniform boundedness (the bound M does not depend on the point x) is that it is a property of every uniformly convergent sequence of continuous functions. If $f_n \to f$

uniformly, then

$$|f_n(x) - f(x)| \le 1 \text{ for all } n \ge N,$$

so $|f_n(x)| \le 1 + \sup_x |f(x)|$ for $n \ge N$ and we can take for M the largest of $\sup_x |f_n(x)|$ for $n \le N$ and $1 + \sup_x |f(x)|$. Thus we are imposing on the original sequence a condition that must be met by the convergent subsequence.

But we are still far from our goal. It is easy to give examples of uniformly bounded sequences of continuous functions that possess no uniformly convergent subsequences—essentially because they oscillate too much. A simple example is the sequence $\{\sin nx\}$. Figure 7.6.1 shows a typical function from this sequence. We will not give a detailed proof that no convergent subsequence exists, but this is clear from the graph. In order to find the correct condition that will rule out this kind of oscillatory behavior, let us try to determine what kind of behavior a uniformly convergent sequence of continuous functions possesses that rules out such unrestrained agitation.

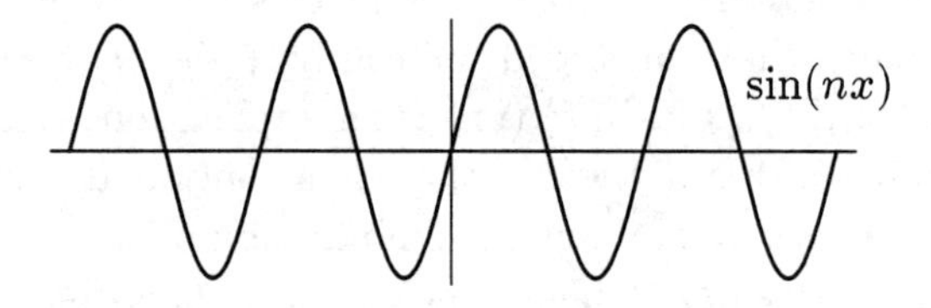

Figure 7.6.1:

An individual continuous function is prevented from jumping around by the continuity condition. Since the domain is assumed compact, we automatically have uniform continuity: for every error $1/m$ there exists $1/n$ such that $|x - y| < 1/n$ implies $|f(x) - f(y)| < 1/m$. If we switch to a different continuous function g, the same condition is satisfied, except that the value of $1/n$ may be different. If we look at functions like $\sin kx$ for large k we see that $1/n$ must be taken very small—the larger k the smaller $1/n$. However, if $f_k \to f$ uniformly, then the error estimate $1/n$ for the single function f can serve as well for all the functions f_k with k sufficiently large. More precisely, if we choose $1/n$ so that $|x-y| < 1/n$ implies $|f(x) - f(y)| < 1/3m$, then $|x-y| < 1/n$ also implies $|f_k(x) - f_k(y)| < 1/m$ once k is large enough that $|f_k(x) - f(x)| < 1/3m$ for all x. (The proof follows from the three-term estimate

$$|f_k(x) - f_k(y)| \le |f_k(x) - f(x)| + |f(x) - f(y)| + |f(y) - f_k(y)|$$

and the fact that each term is $\le 1/3m$). This still leaves a finite set of functions f_k for which we don't have $|f_k(x) - f(x)| < 1/3m$ for all x. But they do not cause much trouble because there are a finite number of them. For each f_k there is $1/n_k$ such that $|x-y| < 1/n_k$ implies $|f_k(x) - f_k(y)| < 1/m$ because each f_k is uniformly continuous. Thus by taking the minimum of $1/n_k$ for the finite number for f_k not covered above and the $1/n$ that works for all large k, we obtain a *single value* of $1/N$ such that $|x-y| < 1/N$ implies $|f_k(x) - f_k(y)| < 1/m$ for *all* k. This kind of continuity estimate that is valid for all f_k is referred to as *equicontinuity.*

Definition 7.6.1 *A sequence of functions $\{f_k\}$ defined on a domain D is said to be uniformly equicontinuous if for every $1/m$ there exists an error $1/n$ (depending only on $1/m$) such that $|x-y| < 1/n$ implies $|f_k(x) - f_k(y)| < 1/m$ for all k. Similarly we say any family (possibly uncountable) of functions is uniformly equicontinuous if for every $1/m$ there exists $1/n$ such that $|x-y| < 1/n$ implies $|f(x) - f(y)| < 1/m$ for every function of f in the family.*

We can also define equicontinuity at a point, but since we will not use this notion explicitly, we will not give the definition.

We can summarize the discussion so far by saying that any uniformly convergent sequence of continuous functions on a compact interval is

1. uniformly bounded and

2. uniformly equicontinuous.

We will now show that these two conditions imply the existence of a uniformly convergent subsequence. (Note: if you have difficulty pronouncing the words "equicontinuity" and "equicontinuous", try saying the words "continuity" and "continuous" and then prefix them with an unaccented "equi".)

7.6.2 The Arzela-Ascoli Theorem

Theorem 7.6.1 (*Arzela-Ascoli Theorem*) *Let* $\{f_k\}$ *be a sequence of functions on a compact interval that is uniformly bounded and uniformly equicontinuous. Then there exists a uniformly convergent subsequence.*

Proof: The idea of the proof is first to obtain a subsequence that converges at every point of a countable dense subset of the domain and then to show that this subsequence converges uniformly. The first step requires the uniform boundedness; the second step requires the uniform equicontinuity.

Let $x_1, x_2, \ldots$ be a countable dense subset of the domain (say all rational numbers in the interval). The sequence of numbers $f_1(x_k)$, $f_2(x_k), \ldots$ for each fixed k is bounded (by the uniform boundedness), so for each k there is a subsequence that converges. The problem is to obtain a subsequence that converges simultaneously at all the points x_k. We do this by a diagonalization procedure. First choose a subsequence of $\{f_n\}$ that converges at x_1. Call it $f_{11}, f_{12}, f_{13}, \ldots$. Then choose a subsequence of $f_{11}, f_{12}, f_{13}, \ldots$ that converges at x_2. Call it $f_{21}, f_{22}, f_{23}, \ldots$. Then this sequence also converges at x_1 because it is a subsequence of a sequence that converges at x_1. Next we choose a subsequence of $f_{21}, f_{22}, f_{23}, \ldots$ that converges at x_3. In this way we obtain a sequence of subsequences of the original sequence, each one a subsequence of the previous one, such that the subsequence $f_{k1}, f_{k2}, f_{k3}, \ldots$ converges at the points $x_1, x_2, \ldots, x_k$. We still do not have a subsequence converging at the infinite set of points $x_1, x_2, \ldots$. To get this we write all our subsequences in an infinite matrix

$$\begin{array}{cccc} f_{11} & f_{12} & f_{13} & \cdots \\ f_{21} & f_{22} & f_{23} & \cdots \\ f_{31} & f_{32} & f_{33} & \cdots \\ \vdots & & & \end{array}$$

and take the diagonal, $f_{11}, f_{22}, f_{33}, \ldots$. The diagonal is a subsequence of the original sequence; in fact, except for the first k terms it is a subsequence of the kth row $f_{k1}, f_{k2}, f_{k3}, \ldots$, because every row below the kth is a subsequence of the kth row. Therefore the diagonal sequence converges at x_k (a subsequence of a convergent sequence converges). This is true for all values of k, so $f_{11}, f_{22}, f_{33}, \ldots$ is the desired subsequence converging at a countable dense subset.

Now we come to the second step. We need to show that this subsequence actually converges uniformly. To do this we need to verify the uniform Cauchy criterion (given any error $1/m$ there exists N such that $j, k \geq N$ implies $|f_j(x) - f_k(x)| < 1/m$ for all x) since we know this implies uniform convergence. How can we accomplish this? The idea is that we need only nail down the Cauchy criterion at a finite set of points (depending on $1/m$) from our countable dense subset, and then the uniform equicontinuity holds down the other points (imagine a finite number of clothespins holding together two clotheslines that are not too wiggly). We will make the comparison

$$|f_j(x) - f_k(x)| \leq |f_j(x) - f_j(x_p)| + |f_j(x_p) - f_k(x_p)| + |f_k(x_p) - f_k(x)|$$

where x_p is chosen from the finite set of points and is near x. The middle term will be controlled by the Cauchy criterion at x_p, while the other two terms will be controlled by the uniform equicontinuity.

Let the error $1/m$ be given. By the uniform equicontinuity, there exists $1/n$ such that $|x - y| < 1/n$ implies $|f_j(x) - f_j(y)| < 1/3m$. Let $x_1, x_2, \ldots, x_r$ be chosen from the countable dense set such that every point x in the domain is within a distance of $1/n$ of at least one of $x_1, x_2, \ldots, x_r$. The number of such points will depend on $1/n$, but it will be finite because

1. the full sequence $x_1, x_2, \ldots$ is dense and

2. the interval is compact (if the countable dense set was all rational numbers in the interval, then we need only choose all points of the form k/n that lie in the interval).

Since f_k converges at each of the points $x_1, x_2, \ldots, x_r$, we can find N such that $j, k \geq N$ implies $|f_j(x_p) - f_k(x_p)| \leq 1/3m$ for $p = 1, 2, \ldots r$ (choose N to be the largest value of those required to make this statement true for each x_p).

We claim that $j, k \geq N$ implies $|f_j(x) - f_k(x)| \leq 1/m$ for any point in the interval. Indeed give any x we find x_p with $|x - x_p| < 1/n$, and then

$$\begin{aligned}|f_j(x) - f_k(x)| &\leq |f_j(x) - f_j(x_p)| + |f_j(x_p) - f_k(x_p)| \\ &\quad + |f_k(x_p) - f_k(x)| \\ &\leq 1/3m + 1/3m + 1/3m = 1/m,\end{aligned}$$

the first and third estimates following from $|x - x_p| < 1/n$ and the uniform equicontinuity, and the middle one from the Cauchy criterion at x_p. QED

In order to apply the Arzela-Ascoli theorem we need a criterion for equicontinuity. Fortunately there is one that is very simple to verify. Suppose all the functions in the sequence are C^1 and the derivatives are uniformly bounded, say $|f_k'(x)| \leq M$. Then $|f_k(x) - f_k(y)| = |f_k'(z)|\,|x-y| \leq M|x-y|$ by the mean value theorem and so the sequence is uniformly equicontinuous (choose $1/n = 1/Mm$). This is the most common condition used to verify uniform equicontinuity, so we state it separately:

Corollary 7.6.1 *Suppose $\{f_k\}$ is a sequence of C^1 functions on a compact interval such that $|f_k(x)| \leq M$ and $|f_k'(x)| \leq M$ for all k and x. Then there exists a uniformly convergent subsequence.*

Of course the condition of uniform boundedness of derivatives is not necessary for uniform equicontinuity, so there may be cases where the theorem applies but not the corollary. Unfortunately we cannot give here any honest applications of either, since they are all difficult and would take us far afield. (Some of the difficulties arise from the unbalanced nature of the corollary: although we make hypotheses about the derivatives of the functions, we do not obtain any conclusions concerning derivatives.) In a later chapter we will show how the Arzela-Ascoli theorem can be interpreted as a compactness theorem.

7.6.3 Exercises

1. If f_n is a uniformly equicontinuous sequence of functions on a compact interval and $f_n \to f$ pointwise, prove that $f_n \to f$ uniformly. (You should not assume that f is continuous, although this is a consequence of the result.)

2. If $|f_n(x) - f_n(y)| \leq M|x - y|^\alpha$ for some fixed M and $\alpha > 0$ and all x, y in a compact interval, show that $\{f_n\}$ is uniformly equicontinuous.

3. Let $\{f_n\}$ be a sequence of C^∞ functions on a compact interval such that for each k there exists M_k such that $|f_n^{(k)}(x)| \leq M_k$ for all n and x. Prove that there exists a subsequence converging uniformly, together with the derivatives of all orders, to a C^∞ function.

4. Prove that the sequence $f_n(x) = \sin nx$ is not uniformly equicontinuous on any non-trivial compact interval.

5. Give an example of a sequence that is uniformly equicontinuous but not uniformly bounded.

6. Prove that the family of all polynomials of degree $\leq N$ with coefficients in $[-1, 1]$ is uniformly bounded and uniformly equicontinuous on any compact interval.

7. Prove that the family of all polynomials $P(x)$ of degree $\leq N$ satisfying $|P(x)| \leq 1$ on $[0, 1]$ is uniformly equicontinuous on $[0, 1]$.

8. Let $f_1, f_2, \ldots, f_n$ be any finite set of continuous functions on a compact interval. Show that the family of all linear combinations $\sum_{j=1}^n a_j f_j$ with all $|a_j| \leq 1$ is uniformly bounded and uniformly equicontinuous.

9. Give an example of a uniformly bounded and uniformly equicontinuous sequence of functions on the whole line that does not have any uniformly convergent subsequences.

10. Give an example of a sequence of functions satisfying the hypotheses of the corollary that has no subsequence whose derivatives converge uniformly.

11. Let $\{f_k\}$ be a sequence of uniformly bounded uniformly equicontinuous functions on a bounded open interval (a, b). Show that the functions can be extended to the compact interval $[a, b]$ so that they are still uniformly bounded and uniformly equicontinuous.

12. Let $\{f_k\}$ be a sequence of functions defined an open interval (a, b) (not necessarily bounded) satisfying $|f_j(x)| \leq F(x)$ and $|f_j'(x)| \leq G(x)$ for all j, where F and G are continuous functions on (a, b). Prove that $\{f_k\}$ has a subsequence that converges uniformly on compact subsets of (a, b). (**Hint**: use the diagonalization argument after obtaining subsequences converging uniformly on $[a + 1/n, b - 1/n]$.)

13. Suppose $f_1, f_2, \ldots$ is a sequence of functions on a compact interval that is pointwise bounded ($|f_k(x)| \leq M(x)$ for all k) and pointwise equicontinuous (for each x in the interval and for all $1/m$, there exists $1/n$ such that $|f_k(x) - f_k(y)| \leq 1/m$ for all k provided $|x - y| \leq 1/n$). Prove that there is a pointwise convergent subsequence.

7.7 Summary

7.1 Complex Numbers

Definition *A complex number z is an expression $x + iy$ where x and y are real numbers. Under the operations $(x + iy) + (x' + iy') = x + x' + i(y + y')$ and $(x + iy) \cdot (x' + iy') = xx' - yy' + i(xy' + x'y)$ the complex numbers form a field, denoted $\mathbb{C}$. The modulus or absolute value of a complex number is $|z| = |x + iy| = \sqrt{x^2 + y^2}$.*

Theorem *$|z \cdot z_1| = |z|\,|z_1|$ and $|z + z_1| \leq |z| + |z_1|$ (triangle inequality).*

Definition *If $z_1, z_2, \ldots$ is a sequence of complex numbers ($z_k = x_k + iy_k$), then $z_k \to z$ if for every $1/n$ there exists m such that $k \geq m$ implies $|z_k - z| < 1/n$.*

Theorem *$z_k \to z$ if and only if $x_k \to x$ and $y_k \to y$.*

Theorem (*Completeness of $\mathbb{C}$*) *If $z_1, z_2, \ldots$ is a Cauchy sequence of complex numbers (for every $1/n$ there exists m such that $j, k \geq m$ implies $|z_j - z_k| \leq 1/n$), then there exists a complex number z such that $z_k \to z$.*

Theorem *If a complex-valued function F is integrable, then $|F|$ is integrable and $|\int_a^b F(x)\,dx| \le \int_a^b |F(x)|\,dx$.*

7.2 Numerical Series and Sequences

Definition *$\sum_{k=1}^{\infty} x_k$ is convergent if $s_1, s_2, \ldots$ is convergent, where s_n denotes the partial sums, $s_n = \sum_{k=1}^{n} x_k$, and $\sum_{k=1}^{\infty} x_k = \lim_{n\to\infty} s_n$. Otherwise, the series is said to be divergent. If $\lim_{n\to\infty} s_n = +\infty$, then $\sum_{k=1}^{\infty} x_k$ is said to diverge to $+\infty$. If $\sum_{k=1}^{\infty} |x_k|$ is convergent, $\sum_{k=1}^{\infty} x_k$ is said to be absolutely convergent.*

Example *The geometric series $\sum_{k=1}^{\infty} r^k = r/(1-r)$ converges for $0 < r < 1$ and diverges for $r \ge 1$.*

Theorem *If $\sum_{k=1}^{\infty} x_k$ and $\sum_{k=1}^{\infty} y_k$ are convergent, then so are $\sum_{k=1}^{\infty} ax_k$ and $\sum_{k=1}^{\infty}(x_k + y_k)$, and $\sum_{k=1}^{\infty} ax_k = a\sum_{k=1}^{\infty} x_k$ and $\sum_{k=1}^{\infty}(x_k + y_k) = \sum_{k=1}^{\infty} x_k + \sum_{k=1}^{\infty} y_k$. If $x_k \ge y_k$ for every k then $\sum_{k=1}^{\infty} x_k \ge \sum_{k=1}^{\infty} y_k$. If $\sum_{k=1}^{\infty} x_k$ is convergent and $y_k = x_k$ for all but a finite number of k, then $\sum y_n$ is convergent.*

Theorem 7.2.1 (*Cauchy criterion*) *$\sum_{k=1}^{\infty} x_k$ converges if and only if for every $1/n$ there exists m such that $q \ge p \ge m$ implies $\left|\sum_{k=p}^{q} x_k\right| < 1/n$.*

Theorem *Absolute convergence implies convergence.*

Theorem 7.2.2 (*Comparison test*) *If $|x_k| \le y_k$ and $\sum_{k=1}^{\infty} y_k$ converges, then $\sum_{k=1}^{\infty} x_k$ converges absolutely.*

Theorem 7.2.3

a. (*Ratio test*) *If $|x_{n+1}/x_n| < r$ for all sufficiently large n and some $r < 1$, then Σx_n converges absolutely, while if $|x_{n+1}/x_n| \ge 1$ for all sufficiently large n, then Σx_n diverges.*

b. (*Root test*) *If $\sqrt[n]{|x_n|} < r$ for all sufficiently large n and some $r < 1$, then Σx_n converges absolutely.*

Example $\sum_{n=1}^{\infty} 1/n^a$ *converges for* $a > 1$ *and diverges for* $0 \leq a \leq 1$.

Example $\sum_{n=1}^{\infty}(-1)^n/n$ *converges but not absolutely.*

Definition $\sum_{n=1}^{\infty} b_n$ *is a rearrangement of* $\sum_{n=1}^{\infty} a_n$ *if* $b_n = a_{m(n)}$ *for some one-to-one onto function* $m(n)$ *from the natural numbers to the natural numbers.*

Theorem 7.2.4 *A series is absolutely convergent if and only if every rearrangement is convergent, in which case the rearrangement is absolutely convergent and has the same limit.*

Definition *A series is said to converge unconditionally if every rearrangement converges, conditionally if it converges but some rearrangement diverges.*

Theorem *If* $\sum_{m=1}^{\infty} |a_{nm}|$ *is convergent for every* n *and* $\sum_{n=1}^{\infty} (\sum_{m=1}^{\infty} |a_{nm}|)$ *is convergent, then* $\sum_{n=1}^{\infty} (\sum_{m=1}^{\infty} a_{nm}) = \sum_{m=1}^{\infty} (\sum_{n=1}^{\infty} a_{nm})$.

Theorem (*Summation by Parts*) *Let* $A_1, A_2, \ldots$ *and* $b_1, b_2, \ldots$ *be two sequences and* $a_n = A_{n-1} - A_n$, $B_n = \sum_{k=1}^{n} b_k$; *and suppose* $A_{n+1}B_n \to 0$. *Then* $\sum_{n=1}^{\infty} A_n b_n$ *converges if and only if* $\sum_{n=1}^{\infty} a_n B_n$ *converges, and* $\sum_{n=1}^{\infty} A_n b_n = \sum_{n=1}^{\infty} a_n B_n$.

Theorem 7.2.5 *If* $A_1, A_2, \ldots$ *is a sequence of positive numbers,* $A_1 \geq A_2 \geq \ldots$, *and* $\lim_{n\to\infty} A_n = 0$, *then* $\sum_{n=1}^{\infty}(-1)^n A_n$ *converges. More generally* $\sum_{n-1}^{\infty} b_n A_n$ *converges if also* $|B_n| \leq M$ *for all* n, *where* $B_n = \sum_{k=1}^{n} b_k$.

7.3 Uniform Convergence

Definition *Let* $f_1(x), f_2(x), \ldots$ *be a sequence of functions on a domain* D. *We say* $f_n(x) \to f(x)$ *pointwise if for every* x *in* D, *the sequence of numbers* $\{f_n(x)\}$ *converges to the number* $f(x)$. *We say* $f_n(x) \to f(x)$ *uniformly if for every* $1/n$ *there exists* m *(not depending on* x*) such that for all* x *in* D *and all* $k \geq m$ *we have* $|f_k(x) - f(x)| \leq 1/n$.

Theorem 7.3.1 (*Cauchy criterion*) *$\{f_n(x)\}$ converges uniformly to some limit function if and only if for every $1/n$ there exists m such that $|f_j(x) - f_k(x)| \leq 1/n$ for all $j, k \geq m$ and all x in D.*

Theorem 7.3.2 *Uniform convergence preserves continuity at a point, continuity on the domain, or uniformly continuity.*

Theorem 7.3.3 $\lim_{n\to\infty} \int_a^b f_n(x)\,dx = \int_a^b f(x)\,dx$ *if $f_n(x) \to f(x)$ uniformly on $[a, b]$.*

Theorem 7.3.4 *If f_n are C^1 functions on (a, b), $f_n \to f$ pointwise and $f'_n \to g$ uniformly, then f is C^1 and $f' = g$.*

Theorem 7.3.5 *If f_n are continuous functions on a compact domain D, then $f_n(x_n) \to f(x)$ for all sequences $x_n \to x$ in D if and only if $f_n \to f$ uniformly.*

7.4 Power Series

Definition *A power series about x_0 is a series of the form $\sum_{n=0}^{\infty} a_n(x - x_0)^n$.*

Lemma *If a power series converges for $x = x_1$, then it converges absolutely for $|x - x_0| < |x_1 - x_0|$ and uniformly in $|x - x_0| \leq r$ for any $r < |x_1 - x_0|$.*

Definition *The radius of convergence of a power series is the unique number R in $[0, \infty]$ such that the series converges for $|x - x_0| < R$ and diverges for $|x - x_0| > R$.*

Example *The power series $\sum_{n=0}^{\infty} x^n$, $\sum_{n=1}^{\infty}(1/n)x^n$, and $\sum_{n=1}^{\infty}(1/n^2)x^n$ all have radius of convergence $R = 1$, but the first diverges at $x = \pm 1$, the second converges at $x = -1$ but diverges at $x = +1$, while the third converges at $x = \pm 1$.*

Theorem 7.4.1 *The radius of convergence R of $\sum_{n=0}^{\infty} a_n(x - x_0)^n$ is given by $1/R = \limsup_{n\to\infty} \sqrt[n]{|a_n|}$.*

Examples *If* $a_n = p(q)/q(n)$ *where* p *and* q *are polynomials, then* $R = 1$. $\sum_{n=0}^{\infty} x^n/n!$ *has* $R = +\infty$*, while* $\sum_{n=0}^{\infty} n!x^n$ *has* $R = 0$.

Theorem 7.4.2 *If* $f(x) = \sum_{n=0}^{\infty} a_n(x-x_0)^n$ *with radius of convergence* $R > 0$*, then* f *is* C^∞ *in* $|x-x_0| < R$ *and the series can be differentiated term-by-term.*

Theorem *If* $f(x) = \sum_{n=0}^{\infty} a_n(x - x_0)^n$ *with radius of convergence* $R > 0$*, then* $a_n = f^{(n)}(x_0)/n!$*. If* $\sum_{n=0}^{\infty} a_n(x-x_0)^n = \sum_{n=0}^{\infty} b_n(x-x_0)^n$ *for* $|x - x_0| < R$ *and* $R > 0$*, then* $a_n = b_n$ *for all* n.

Theorem 7.4.3 *If* $f(x) = \sum_{n=0}^{\infty} a_n(x - x_1)^n$ *in* $|x - x_1| < R$*, then for any* x_2 *satisfying* $|x_2 - x_1| < R$ *there exists a power series* $\sum_{n=0}^{\infty} b_n(x - x_2)^n$ *converging to* f *at least for* $|x - x_2| < R - |x_2 - x_1|$*, and*

$$b_k = \sum_{n=0}^{\infty} a_{n+k} \binom{n+k}{k} (x_2 - x_1)^n.$$

Definition *An analytic function* $f(x)$ *on an interval* (a, b) *is a function with a convergent power-series expansion (non-zero radius of convergence) about each point in the domain.*

Theorem 7.4.4 *If* f *and* g *are analytic, then so are* $f \pm g$*,* $f \cdot g$*, and* f/g *(provided* g *never vanishes), with the power series obtained by formal combination of the series for* f *and* g.

Theorem 7.4.5 *If* f *and* g *are analytic functions and the range of* f *lies in the domain of* g*, then* $g \circ f$ *is analytic with the power series obtained by formal substitution of the power series for* f *in the power series for* g.

7.5 Approximation by Polynomials

Theorem *Given distinct points* $x_1, \ldots, x_n$ *and values* $a_1, \ldots, a_n$*, there exists a unique polynomial* $P(x)$ *of degree* $\leq n-1$ *such that* $P(x_k) = a_k$*, namely* $P(x) = \sum_{k=1}^{n} a_k Q_k(x)$ *where* $Q_k(x) = (q_k(x_k))^{-1} q_k(x)$ *and* $q_k(x_k) = \Pi_{j \neq k}(x - x_j)$*.* $P(x)$ *is called the Lagrange interpolation polynomial.*

Definition *The convolution $f * g$ of two continuous functions on the line, one of them vanishing outside a bounded set, is given by*

$$f * g(x) = \int_{-\infty}^{\infty} f(x - y)g(y)\,dy.$$

Theorem *The convolution product is commutative.*

Definition 7.5.1 *An approximate identity is a sequence of continuous functions g_n on the line satisfying*

1. $g_n \geq 0$,
2. $\int_{-\infty}^{\infty} g_n(x)\,dx = 1$, *and*
3. $\lim_{n\to\infty} \int_{-\infty}^{-1/n} + \int_{1/n}^{\infty} g_n(x)\,dx = 0$.

Lemma 7.5.1 (*Approximate Identity*) *If g_n is an approximate identity and f is a continuous function vanishing outside a bounded interval, then $g_n * f \to f$ uniformly.*

Theorem 7.5.1 (*Weierstrass Approximation*) *Any continuous function on a compact interval is the uniform limit, on that interval, of polynomials.*

Corollary *If f is continuous on $[0,1]$ and $\int_0^1 f(x)x^n dx = 0$ for all $n = 0, 1, \ldots$, then $f \equiv 0$.*

Theorem 7.5.2 *If f is C^1 and vanishes outside a bounded interval and g is continuous, then $f * g$ is C^1 and $(f * g)' = f' * g$.*

Theorem *If f is C^k on $[a, b]$, then there exists a sequence of polynomials p_n with $p_n^{(j)} \to f^{(j)}$ uniformly on $[a, b]$ for all $j \leq k$.*

7.6 Equicontinuity

Definition 7.6.1 *A sequence of functions $\{f_k\}$ on a domain D is said to be uniformly bounded if there exists M such that $|f_k(x)| \leq M$ for all k and all x in D. It is said to be uniformly equicontinuous if for every*

$1/m$ there exists $1/n$ such that $|x - y| < 1/n$ implies $|f_k(x) - f_k(y)| < 1/m$ for all k.

Theorem 7.6.1 (*Arzela-Ascoli*) *A sequence of uniformly bounded and uniformly equicontinuous functions on a compact interval has a uniformly convergent subsequence.*

Corollary 7.6.1 *If $\{f_k\}$ is a sequence of C^1 functions on $[a,b]$ satisfying $|f_k(x)| \leq M$ and $|f_k'(x)| \leq M$ for all k and x, then it has a uniformly convergent subsequence.*

Chapter 8

Transcendental Functions

8.1 The Exponential and Logarithm

8.1.1 Five Equivalent Definitions

The exponential function, $\exp(x)$ or e^x, is one of the most important functions in mathematics. There are many ways to characterize it, such as:

1. the power series expansion $\exp(x) = \sum_{n=0}^{\infty} x^n/n!$,
2. the unique solution to the differential equation $f' = f$ with $f(0) = 1$,
3. the number e raised to the power x,
4. the inverse of the natural logarithm $\int_1^x 1/t\,dt$,
5. the limit of $(1 + x/n)^n$ as $n \to \infty$.

This list does not exhaust all the characterizations, but chances are any calculus book will use one of the above. One of the main goals of this section is to show that all five are equivalent, so it is not too important which one we take as the official definition. However, not all of these characterizations are of the same nature; for example, definitions 1 and 5 are directly algorithmic, giving a formula to compute $\exp(x)$ (although neither formula would be a particularly good choice for computing $\exp(100\pi)$). Many of these characterizations require that certain facts be proved first, for example, that the limit in definition 5 actually exists.

We will take definition 1 for our definition, use it to establish some of the basic properties of the exponential function, and then prove in turn that each of the other descriptions yields the same function.

Definition 8.1.1 *The exponential function* $\exp(x)$ *is defined for any real* x *by* $\exp(x) = \sum_{n=0}^{\infty} x^n/n!$.

We have observed in the last chapter that this power series has infinite radius of convergence, so $\exp(x)$ is well defined and is a C^{∞} function and the power series converges absolutely and uniformly on any bounded interval; in particular, the series can be rearranged and differentiated term-by-term.

Theorem 8.1.1

a. *The exponential function satisfies the differential equation* $f' = f$, *with* $f(0) = 1$.

b. $\exp(x + y) = \exp x \exp y$ *for any real* x *and* y.

c. $\exp(x) > 0$ *for any real* x.

Proof:

a. The differential equation follows by term-by-term differentiation of the power series, and $\exp(0) = 1$ since $0^n = 0$ for $n \geq 1$. Indeed, we could deduce the form of the power series from the differential equation: if $(\sum a_n x^n)' = \sum a_n x^n$, then $n\, a_n = a_{n-1}$, so $a_n = a_0/n!$ and $f(0) = 1$ yields $a_0 = 1$.

b. The identity may be established by multiplying the power series and rearranging terms:

$$\begin{aligned} \exp(x)\exp(y) &= \sum_{n=0}^{\infty} \frac{x^n}{n!} \sum_{m=0}^{\infty} \frac{y^m}{m!} \\ &= \sum_{k=0}^{\infty} \left(\sum_{n+m=k} \frac{x^n y^m}{n!m!} \right) = \sum_{k=0}^{\infty} \frac{(x+y)^k}{k!} \end{aligned}$$

by the binomial theorem. The rearrangement is justified by the absolute convergence of the doubly indexed series.

c. For $x > 0$ each term in the power series is positive, so $\exp(x) > 0$. For x negative just use part b to obtain $\exp(x)\exp(-x) = 1$, hence $\exp x = 1/\exp(-x)$ is positive. QED

Theorem 8.1.2 $\exp(x)$ *is the unique solution of the differential equation* $f' = f$ *with* $f(0) = 1$.

Proof: We have already established existence. The uniqueness is a special case of a more general theorem to be established in Chapter 11. We give here a simple but tricky proof of this special case.

Look at $g(x) = f(x)/\exp(x)$. Since $\exp(x)$ never vanishes, this is well defined and g is differentiable (it is implicit in writing $f' = f$ that f is differentiable; and since differentiability implies continuity, f is also continuous; hence f is C^1 since $f' = f$). We compute

$$g'(x) = \frac{f'(x)\exp(x) - f(x)\exp(x)}{(\exp x)^2}$$

using the fact that $d/dx \exp(x) = \exp(x)$. But $f' = f$ then gives $g' = 0$ for all x; hence g is constant, so $f(x) = g(x)\exp(x)$ is a constant multiple of $\exp(x)$. Clearly the constant is 1 if and only if $f(0) = 1$. QED

For the third characterization, we define the real number e to be $\exp(1) = \sum_{n=1}^{\infty} 1/n! \approx 2.7$. This number is not rational—a fact that is not too difficult to prove, but we will not give the proof here. It is not even algebraic (i.e., it does not satisfy any polynomial equation with rational coefficients)—although this is more difficult to establish. For this reason the number e is called *transcendental*. For a similar reason the function $\exp(x)$ is called transcendental (there is no polynomial $F(x, y)$ in two variables with rational coefficients such that $F(x, \exp x) = 0$, for then $F(1, e) = 0$ would imply that e is algebraic).

Theorem 8.1.3 *For every rational number* p/q *(*p *and* q *integers,* $q > 0$*) we have* $\exp(p/q) = (e^p)^{1/q}$. *For* x *real and* $x_k = p_k/q_k$ *a Cauchy sequence of rationals converging to* x, $\exp(x) = \lim_{k\to\infty} \exp(x_k) = \lim_{k\to\infty} (e^{p_k})^{1/q_k}$.

Proof: The identity $\exp(x + y) = \exp x \cdot \exp y$ implies $\exp(2) = e^2$, and by induction $\exp(k) = e^k$ for any non-negative integer k. From

$e = \exp(1/2 + 1/2) = \exp(1/2)^2$ we obtain $\exp(1/2) = e^{1/2}$ (remember exp is always positive). In a similar way we obtain $\exp(1/n) = e^{1/n}$ and $\exp(p/q) = (e^p)^{1/q}$ for every rational number p/q. Since exp is continuous, we have $\exp(x) = \lim_{k\to\infty} \exp(x_k)$ where x_k is any Cauchy sequence of rational numbers converging to x, so $\exp(x) = \lim_{k\to\infty}(e^{p_k})^{1/q_k}$. This shows the limit exists (it would be awkward, although not impossible, to prove this directly), and it justifies writing $\exp(x) = e^x$ even for irrational values of x. QED

Next we consider the fourth characterization in terms of the inverse function. For this we need to know that the image of the exponential function is the set of positive real numbers, which we denote $\mathbb{R}^+$.

Theorem 8.1.4 *The exponential function maps* $\mathbb{R}$ *one-to-one and onto* $\mathbb{R}^+$.

Proof: We have seen that $\exp(x)$ assumes only positive values. From the differential equation $d/dx \exp(x) = \exp(x)$ and the positivity of $\exp(x)$ we see that $\exp(x)$ is strictly increasing and, hence, one-to-one. To show that it assumes all positive values we observe that $\lim_{n\to\infty} e^n = +\infty$ and $\lim_{n\to\infty} e^{-n} = 0$, since $e > 1$ and $0 < e^{-1} < 1$. Since exp is continuous, the intermediate value theorem implies that it takes on all positive values. QED

The inverse function to exp is called the *natural logarithm*, denoted $\log x$ or $\ln x$ (or $\log_e x$). It is a function with domain $\mathbb{R}^+$ and range $\mathbb{R}$, increasing, one-to-one, and onto.

Theorem 8.1.5 $\log x = \int_1^x 1/t\, dt$.

Proof: By the inverse function theorem $\log x$ is differentiable since $d/dx \exp(x) \neq 0$, and the derivative of $\log x$ at $x = x_0$ is the reciprocal of the derivative of $\exp(x)$ at $x = y_0$ where $\exp(y_0) = x_0$. Thus

$$\frac{d}{dx}\log x\,\bigg|_{x=x_0} = \frac{1}{\frac{d}{dx}\exp x}\bigg|_{x=y_0} = \frac{1}{\exp y_0} = \frac{1}{x_0},$$

so $d/dx \log x = 1/x$. From the fundamental theorem of the calculus and $\log 1 = 0$ (from $e^0 = 1$) we deduce $\log x = \int_1^x 1/t\, dt$. QED

An alternative approach to defining the exponential and logarithm functions is to start with $\log x = \int_1^x 1/t\,dt$. From this definition it is a simple matter to deduce $\log(xy) = \log x + \log y$ by a change of variable argument. Then we can define exp as the inverse function to log and deduce the differential equation $\exp' = \exp$ from the inverse function theorem.

The fundamental identity $\exp(x+y) = \exp x \exp y$ and its equivalent $\log(xy) = \log x + \log y$ show that exp *and* log *establish an isomorphism between the additive group of* $\mathbb{R}$ *and the multiplicative group of* $\mathbb{R}^+$. Logarithms were invented by Napier to exploit this isomorphism for computational purposes—only recently have these applications become obsolete.

We come now to the fifth characterization of the exponential in terms of a limit.

Theorem 8.1.6 $\exp(x) = \lim_{n\to\infty}(1 + x/n)^n$ *for any real* x.

Proof: We start by writing

$$\lim_{n\to\infty}(1 + x/n)^n = \lim_{n\to\infty} \exp\log(1 + x/n)^n = \exp \lim_{n\to\infty} \log(1 + x/n)^n$$

since exp is continuous. Thus it suffices to show $\lim_{n\to\infty}\log(1+x/n)^n = x$. But

$$\log(1 + x/n)^n = n\log(1 + x/n) = x\left(\frac{\log(1 + x/n) - \log 1}{x/n}\right)$$

since $\log 1 = 0$. But with $x \neq 0$ fixed, x/n goes to zero as $n \to \infty$, so $(\log(1 + x/n) - \log 1)/(x/n)$ converges to the derivative of $\log x$ at $x = 1$, which is one. This establishes $\lim_{n\to\infty}\log(1 + x/n)^n = x$ as claimed. QED

This completes the proof of the equivalence of all five characterizations of exp.

An important property of the logarithm is that it also has power-series expansions, although the radius of convergence is finite. Starting from $\log x = \int_1^x 1/t\,dt$ we compute

$$\log x = \int_0^{x-1} \frac{1}{1+t}\,dt = \int_0^{x-1} \sum_{k=0}^{\infty} (-1)^k t^k\,dt = \sum_{k=0}^{\infty} (-1)^k \int_0^{x-1} t^k\,dt$$

$$= \sum_{k=0}^{\infty} \frac{(-1)^k}{k+1}(x-1)^{k+1}$$

where we assume $|x-1| < 1$ in order to have uniform convergence of the $1/(1+t)$ expansion, justifying the interchange of sum and integral. This is frequently written

$$\log(1+x) = x - \frac{x^2}{2} + \frac{x^3}{3} - \frac{x^4}{4} + \cdots.$$

For the power-series expansion of $\log x$ about an arbitrary point $x_0 > 0$ we use $\log x = \log x_0 + \log x/x_0 = \log x_0 + \log(1+x-x_0/x_0) = \log x_0 + \sum_{k=1}^{\infty}(-1)^{k+1}((x-x_0)/x_0)^k/k$, which converges for $|x-x_0| < x_0$. *Thus* $\log x$ *is an analytic function.*

We can use exp and log to define general powers, $a^b = \exp(b \log a)$ for $a > 0$ and any real b, and we can verify that $a^b = (a^p)^{1/q}$ if $b = p/q$ is rational. Also a^b is an analytic function of either variable. If we fix $a > 0$, then $a^x = \exp(x \log a)$ is analytic in x because exp is (here $\log a$ is just a constant). If we fix b, then $x^b = \exp(b \log x)$ is the composition of analytic functions and, hence, analytic in $x > 0$. Similarly $f(x)^{g(x)}$ is analytic if f and g are analytic and $f(x) > 0$. However, there may be points where $f(x) = 0$ and $f(x)^{g(x)}$ is still defined but not analytic (as in $(x^2)^{1/2} = |x|$). The familiar identities $(a^b)^c = a^{b \cdot c}$ and $\log(a^b) = b \log a$ can also be easily deduced—we leave these as exercises.

In addition to the identities for exp and log, we need to understand the asymptotic behavior of these functions. The significance of "exponential growth" is that $\exp(x)$ beats out any polynomial in x, that is, $\lim_{x\to\infty} x^{-n}\exp(x) = +\infty$ for any n. This is an immediate consequence of the power-series expansion, since $\exp(x) \geq x^n/n!$ for any n, if $x > 0$ (just throw away the other positive terms of the power series), and so (substituting $n+1$ for n) $x^{-n}\exp(x) \geq x^{-n}x^{n+1}/(n+1)! = x/(n+1)! \to \infty$ as $x \to \infty$. Similarly we have rapid exponential decay as $x \to -\infty$, $\lim_{x\to+\infty} x^n \exp(-x) = 0$ for any n, which is just the reciprocal of the exponential growth. For the logarithm we have slow growth, $\lim_{x\to\infty} \log x = +\infty$ but $\lim_{x\to\infty} x^{-a}\log x = 0$ for any $a > 0$. This follows from $\log x = \int_1^x 1/t\, dt \leq \int_1^x t^{a-1}\, dt = (x^a - 1)/a$ for any $a > 0$, hence (substituting $a/2$ for a) $\lim_{x\to\infty} x^{-a}\log x \leq \lim_{x\to\infty} x^{-a}\frac{x^{a/2}-1}{a/2} = 0$. Since $\log 1/x = -\log x$, we have similar estimates near zero: $\lim_{x\to 0} x^a \log x = 0$ for any $a > 0$ (substitute $1/x$ for x).

8.1.2 Exponential Glue and Blip Functions

With the aid of the exponential function, we can do gluing of functions with matching of all derivatives. The basic tool is the function $f(x) = \exp(-1/x^2)$, which is defined for all $x \neq 0$ and can be extended by defining $f(0) = 0$, since $\lim_{x\to 0} \exp(-1/x^2) = \lim_{x\to\infty} \exp(-x) = 0$. Thus $f(x)$ is a continuous function. We claim that f is in fact a C^∞ function. This is clear at every point $x \neq 0$ (in fact it is analytic on $x > 0$ and $x < 0$). At $x = 0$ the function $f(x)$ has a zero of infinite order, $f(x) = O(|x|^n)$ as $x \to 0$ for every n. Indeed the substitution $1/x^2 = t$ shows $\lim_{x\to 0} |x|^{-n} \exp(-1/x^2) = \lim_{t\to +\infty} t^{n/2} \exp(-t) = 0$. From this we can prove by induction that $f^{(n)}(0) = 0$. Since $f(0) = 0$, we have $(f(x) - f(0))/x = f(x)/x \to 0$ as $x \to 0$, so $f'(0) = 0$. Also $f'(x) = 2x^{-3} \exp(-1/x^2) = 2x^{-3} f(x)$ for $x \neq 0$, so $\lim_{x\to 0} f'(x) = 0$, proving that f is C^1. It is clear for $x \neq 0$ that $f^{(n)}(x) = Q_n(x) f(x)$ where Q_n is a polynomial in $1/x$, so $\lim_{x\to 0} f^{(n)}(x) = 0$; and if we assume by induction that $f^{(n)}(0) = 0$, then $(f^{(n)}(x) - f^{(n)}(0))/x = Q_n(x) f(x)/x \to 0$ as $x \to 0$ proving $f^{(n+1)}(0) = 0$. Thus f is C^∞ and all derivatives vanish at $x = 0$.

Notice that the Taylor expansions of f about $x = 0$ vanish identically for all orders. Thus f is not analytic at $x = 0$, for the Taylor expansions converge to zero and *not to* $f(x)$ (if $x \neq 0$). The graph of $f(x)$ is completely flat at $x = 0$, although the function is not constant in a neighborhood of $x = 0$. We can thus glue $f(x)$ for $x > 0$ to the zero function on $x \leq 0$ and have a C^∞ function. It is shown in Figure 8.1.1.

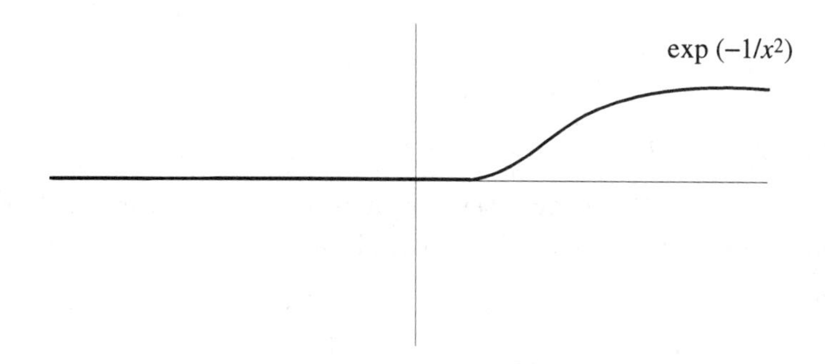

Figure 8.1.1:

Using this idea we can create C^∞ functions to suit every need. For example, the C^∞ function shown in Figure 8.1.2,

$$g(x) = \begin{cases} e^2 \exp\left(-1/(x-1)^2\right) \exp\left(-1/(x+1)^2\right), & -1 < x < 1, \\ 0, & x \geq 1 \text{ or } x \leq 1, \end{cases}$$

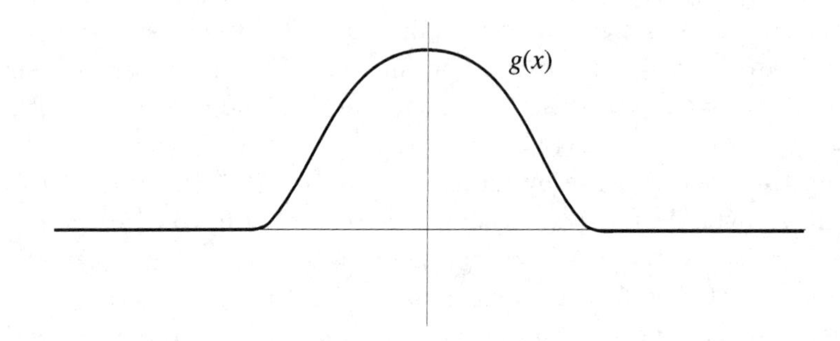

Figure 8.1.2:

vanishes outside $(-1,1)$ and satisfies $g(0) = 1$. We will refer to g as a *blip function* centered at $x = 0$. We can obtain other blip functions centered at other points $x = x_0$ by taking $g(\lambda^{-1}(x - x_0))$; this function vanishes outside $|x - x_0| < \lambda$ and satisfies $g(x_0) = 1$.

We can construct other C^∞ functions using blip functions. The idea is to use the convolution of a blip function with an arbitrary continuous function. Let $g(x)$ be a blip function satisfying

1. $g(x)$ is C^∞,
2. $g(x) = 0$ if $|x| \geq 1$,
3. $g(x) \geq 0$, and
4. $\int_{-\infty}^{\infty} g(x)\, dx = \int_{-1}^{1} g(x)\, dx = 1$.

This involves nothing more serious than multiplying the previous blip function by a constant to obtain condition 4. If we set

$$g_\lambda(x) = \lambda^{-1} g(\lambda^{-1} x),$$

then $g_\lambda(x)$ satisfies all of the above, except that in place of condition 2 we have $g_\lambda(x) = 0$ if $|x| \geq \lambda$. We call g_λ a C^∞ *approximate identity.*

If f is any continuous function, then we can form $f * g_\lambda$. As we saw in the discussion of the Weierstrass approximation theorem, $f * g_\lambda$ will be a C^∞ function ($(f * g_\lambda)^{(k)} = f * g_\lambda^{(k)}$) and $f * g_\lambda$ will converge to f as $\lambda \to 0$, uniformly on compact intervals. This procedure is called *regularization*. It differs from what we did in the proof of the Weierstrass approximation theorem only in that g_λ are C^∞ (the polynomials $(1 - x^2)^n$ were joined up to zero at $x = \pm 1$ creating discontinuities in the derivatives of order $\geq n$). Regularization is an important technique, although most of its applications are beyond the scope of this work.

Suppose we apply regularization to the function $f(x)$ whose graph is shown in Figure 8.1.3.

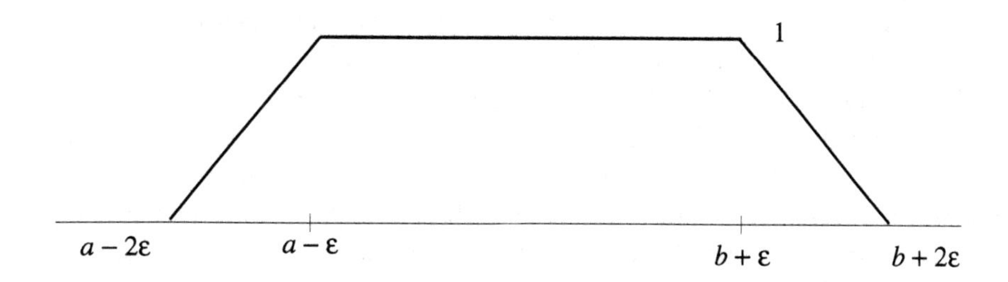

Figure 8.1.3:

Since $f \equiv 1$ on $[a - \varepsilon, b + \varepsilon]$, we have for $\lambda < \varepsilon$ and x in $[a, b]$

$$f * g_\lambda(x) = \int_{-\lambda}^{\lambda} f(x - y) g_\lambda(y)\, dy = \int_{-\lambda}^{\lambda} g_\lambda(y)\, dy = 1$$

since $f(x - y) = 1$ for all values of $x - y$ in the integral. We also have

$$f * g_\lambda(x) = \int_{-\lambda}^{\lambda} f(x - y) g_\lambda(y)\, dy = \int_{-\lambda}^{\lambda} 0 \cdot g_\lambda(y)\, dy = 0$$

if $x < a - 3\varepsilon$ or $x > b + 3\varepsilon$, again for $\lambda < \varepsilon$. Thus $F = f * g_\lambda$ is a C^∞ function satisfying $F(x) \equiv 1$ on $[a, b]$ and $F(x) \equiv 0$ outside $[a - 3\varepsilon, b - 3\varepsilon]$. We also observe that $0 \leq F(x) \leq 1$ for every x because f satisfies the same estimate. The graph of F is shown in Figure 8.1.4.

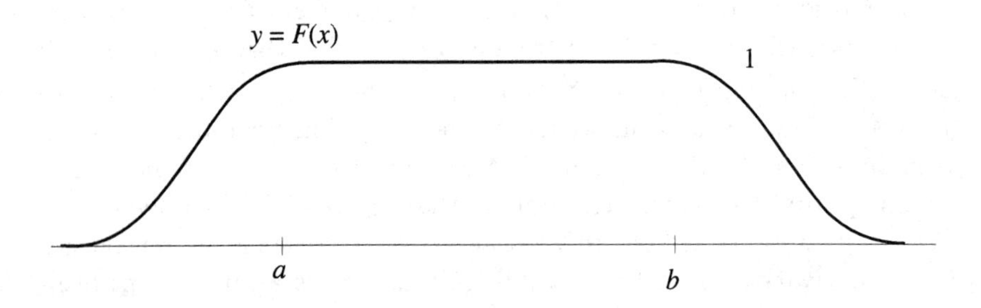

Figure 8.1.4:

8.1.3 Functions with Prescribed Taylor Expansions*

Using blip functions we can build C^∞ functions having prescribed Taylor expansions. The key observation is that the function $f_k(x) = g(\lambda^{-1}x)x^k/k!$ satisfies $(d/dx)^k f_k(x) = 1$ but $(d/dx)^j f_k(x)|_{x=0} = 0$ for $j < k$. The reason for this is that $(d/dx)^j x^k|_{x=0} = 0$ unless $j = k$, so by the product rule for derivatives each term of $(d/dx)^j g(\lambda^{-1}x)(x^k/k!)$ contains a factor vanishing at $x = 0$ if $j < k$ and when $j = k$ the only term not vanishing is $g(\lambda^{-1}x)(d/dx)^k(x^k/k!)$, which gives 1 at $x = 0$. Thus if we want a C^∞ function f vanishing outside $|x| < \lambda$ with $(d/dx)^k f(x)|_{x=0} = a_k$ for $k = 0, \ldots, n$ we need only take $f = \sum_{k=0}^n b_k x^k g(\lambda^{-1}x)$ and then successively solve the linear equations

$$\begin{aligned}
f(0) &= b_0 = a_0 \\
f'(0) &= b_1 + d/dx(b_0 g(\lambda^{-1}x))|_{x=0} = a_1 \\
f''(0) &= 2b_2 + d^2/dx^2(b_0 g(\lambda^{-1}x) + b_1 x g(\lambda^{-1}x)|_{x=0} = a_2 \\
&\vdots \\
f^{(n)}(0) &= n!b_n + (d/dx)^n \left(\sum_{k=0}^{n-1} b_k x^k g(\lambda^{-1}x) \right) |_{x=0} = a_n.
\end{aligned}$$

Since the kth equation expresses $k!b_k$ in terms of a_k and the previously determined $b_0, b_1, \ldots, b_{k-1}$, there is a unique solution. Notice that it would not hurt to vary λ with k (taking $\lambda_k \leq \lambda$). This is the important observation that enables us to control all derivatives simultaneously.

Theorem 8.1.7 (*Borel*) *Given any sequence $a_0, a_1, a_2, \ldots$ of real (or complex) numbers and given any point x_0 and neighborhood $|x - x_0| < \lambda$,* *there exists a C^∞ function f vanishing outside the neighborhood and satisfying $f^{(k)}(x_0) = a_k$ for all k.*

Proof: For simplicity we take $x_0 = 0$. (For the general case just replace x by $x - x_0$.) We want $f(x) = \sum_{k=0}^\infty b_k x^k g(\lambda_k^{-1}x)$ where b_k and $\lambda_k < \lambda$ are to be determined. The idea of the proof is that while we cannot control the values of b_k (they will be determined as above by solving linear equations to make $f^{(k)}(0) = a_k$), we can control the values of λ_k. Thus if we insist that $\lambda_k \to 0$ as $k \to \infty$, the sum defining $f(x)$ is actually finite for any fixed x. Indeed, if $x = 0$ only the first term is nonzero; while if $x \neq 0$, then $g(\lambda_k^{-1}x) = 0$ once $\lambda_k \leq |x|$ (because $g(\lambda_k^{-1}x)$ vanishes outside $|x| < \lambda_k$). Of course the number of non-zero terms increases without bound as $x \to 0$.

Assuming we can differentiate the series term-by-term at $x = 0$ (this is true for $x \neq 0$ because the series is finite, but it is not clear at $x = 0$, and indeed it won't be true unless we choose λ_k cleverly), we find the following equation for the condition $f^{(n)}(0) = a_n$:

$$n!b_n + (d/dx)^n \left(\sum_{k=0}^{n-1} b_k x^k g(\lambda_k^{-1}x) \right) |_{x=0} = a_n.$$

We will use this equation to define inductively $b_0, b_1, \ldots$. Notice that the value of λ_n does not enter into this equation or any earlier one, so we are free to choose λ_n after seeing what $b_0, \ldots, b_n$ are, and the choice of λ_n will not change $b_0, b_1, \ldots, b_n$. Therefore we choose λ_n to make $(d/dx)^k(b_n x^n g(\lambda_n^{-1}x))$ very small for all $k < n$ and all x, say $|(d/dx)^k(b_n x^n g(\lambda_n^{-1}x)| \leq 2^{-n}$ for all $k < n$ and all x. Why is this possible? Essentially because the function $x^n g(x)$ vanishes to order n at $x = 0$, so by taking λ_n very small we can nip it off before it has grown very much. This is easiest to see for $k = 0$, where there are no derivatives involved. Then we can simply estimate

$$|b_n x^n g(\lambda_n^{-1}x)| \leq |b_n| \lambda_n^n$$

since $g(\lambda_n^{-1}x) = 0$ if $|x| \geq \lambda_n$, $|g(\lambda_n^{-1}x)| \leq 1$ in general, and $|x^n| \leq \lambda_n^n$ if $|x| \leq \lambda_n$. Thus, no matter how large b_n may be, we can make $|b_n|\lambda_n^n \leq 2^{-n}$ by taking $\lambda_n \leq 1/2|b_n|^{1/n}$. But even when there are derivatives involved, there are less than n of them. If $k = 1$ we have

$$d/dx(b_n x^n g(\lambda_n^{-1}x)) = n\, b_n x^{n-1} g(\lambda_n^{-1}x) + b_n \lambda_n^{-1} x^n g'(\lambda_n^{-1}x)$$

and in general $(d/dx)^k(b_n x^n g(\lambda_n^{-1}x))$ is a sum of terms of the form

$$c_j b_n x^{n-j}(\lambda_n^{-1})^{k-j} g^{(k-j)}(\lambda_n^{-1}x)$$

where c_j are combinatorial coefficients and $j \leq k < n$. Each such term can be estimated by the same reasoning, using $g^{(k-j)}(\lambda_n^{-1}x) = 0$ for $|x| \geq \lambda_n$ and $|g^{(k-j)}(\lambda_n^{-1}x)| \leq M_{k-j}$ where M_{k-j} is the sup of $|g^{(k-j)}|$, some finite number independent of λ_n, and $|x^{n-j}| \leq \lambda_n^{n-j}$ on $|x| \leq \lambda_n$. We obtain

$$\begin{aligned} &|c_j b_n x^{n-j}(\lambda_n^{-1})^{k-j} g^{(k-j)}(\lambda_n^{-1})| \\ &\quad \leq \; c_j M_{k-j} b_n \lambda_n^{n-j}(\lambda_n^{-1})^{k-j} \\ &\quad = \; c_j M_{k-j} b_n \lambda_n^{n-k}. \end{aligned}$$

Since $n - k > 0$, we can make this as small as we like by taking λ_n small enough. There are only a finite number of such terms in $(d/dx)^k$ $(b_n x^n g(\lambda_n^{-1}x))$, so we can bound this uniformly by 2^{-n}. Finally, we can do this for all $k < n$ since there are only a finite number of k.

This completes the description of how to choose the b_n's and λ_n's. We still have to justify the term-by-term differentiation of $f(x) = \sum_{k=0}^{\infty} b_k x^k g(\lambda_k^{-1}x)$ in order to conclude that $f^{(n)}(0) = a_n$. To do this we use the estimates $|(d/dx)^k b_n x^n g(\lambda_n^{-1}x)| \leq 2^{-n}$ for all $k < n$ and all x. These estimates imply that the differentiated series converges uniformly, since for any fixed k the condition $k < n$ is satisfied for all but a finite number of n, so

$$\sum_{n=0}^{\infty} \left(\frac{d}{dx}\right)^k (b_n x^n g(\lambda_n^{-1}x))$$

converges uniformly. Thus f is C^∞ and term-by-term differentiation is justified, proving $f^{(n)}(0) = a_n$. Finally we note that the conditions on λ_n were all of the nature that λ_n must be sufficiently close to zero, so we can arrange to have $\lambda_n \leq \lambda$ for any prescribed $\lambda > 0$, proving that f vanishes outside $|x| < \lambda$. QED

Using this theorem we can glue flaps onto C^∞ functions on an interval. Suppose f is C^∞ on $[a, b]$ (by this we mean one-sided derivatives exist at the endpoints, and $f^{(n)}$ is continuous on $[a, b]$ for all n). Then by constructing C^∞ functions f_a and f_b to match up all derivatives with f at $x = a$ and $x = b$ and to vanish outside $[a-1, b+1]$, we can obtain in

$$F(x) = \begin{cases} f_a(x) & \text{if } x < a, \\ f(x) & \text{if } a \le x \le b, \\ f_b(x) & \text{if } x > b \end{cases}$$

a C^∞ function on the whole line extending f and vanishing outside $[a-1, b+1]$.

This theorem can be paraphrased as saying that there are no a priori restrictions on the infinite Taylor expansion of a C^∞ function, since the Taylor expansion is determined by the derivatives at the point. In particular, the infinite Taylor series expansion about a point x_0 may diverge at every point except x_0.

8.1.4 Exercises

1. Using $\log x = \int_1^x 1/t\, dt$ show $\log(xy) = \log x + \log y$.

2. Find $\lim_{x \to 0} x^x$.

3. Compute $d/dx(f(x)^{g(x)})$ if f and g are C^1, $f(x) > 0$.

4. Show directly that if $(1 + x/n)^n$ converges uniformly to a function f on a compact interval, then f is C^1 and $f' = f$.

5. Show that $f' = \lambda f$ for a real constant λ has only $ce^{\lambda x}$ as solutions.

6. For which values of a is the improper integral

$$\int_0^{1/2} \frac{1}{x|\log x|^a dx}\, dx$$

finite?

7. For which values of a is the improper integral

$$\int_2^\infty \frac{1}{x|\log x|^a\, dx}$$

finite?

8. For which values of a is the improper integral

$$\int_{1/2}^{2} \frac{1}{x|\log x|^a\, dx}$$

finite?

9. Show that for every closed set A there exists a C^∞ function on the line such that $f(x) = 0$ if and only if x is in A.

10. Show that there exists a C^∞ function on the line having prescribed derivatives of all orders on any sequence of distinct points $x_1, x_2, \ldots$ with no finite limit point.

11. Show that there exists a C^∞ function on (a, b) having prescribed derivatives of all order on any sequence of distinct points $x_1, x_2, \ldots$ with no limit points in (a, b).

12. Prove that any C^1 real-valued function satisfying $f(x + y) = f(x)f(y)$ must be $\exp(ax)$ for some real a. (**Hint**: differentiate the identity.)

13. Give an interpretation of definition 5 in terms of compound interest rates. What does it say about interest that is compounded instantaneouly?

14. For $f(t) = e^{-at}$ (a positive) define the *half life* as the time T such that $f(t+T) = f(t)/2$. Show that T is well defined, and find the relationship between a and T.

15. For $f(t) = e^{at}$ (a positive) define the *doubling time* as the time T such that $f(t+T) = 2f(t)$. Show that T is well defined, and find the relationship between a and T.

16. Show that any function of the form $f(x) = cx^a$ will have a straight line graph on log-log graph paper. What is the slope of the line?

8.2 Trigonometric Functions

8.2.1 Definition of Sine and Cosine

Our approach to defining the sine and cosine functions will be indirect: first we will obtain the inverse functions, which will enable us to define sine and cosine on an interval, and then we will extend the definition to the whole line. Along the way we will derive some important properties of these functions. We begin with the geometric idea that $(\cos\theta, \sin\theta)$ should be the x- and y-coordinates of the point obtained by measuring a length θ along the unit circle from the point $(1,0)$, with the counterclockwise direction taken as positive, as in Figure 8.2.1.

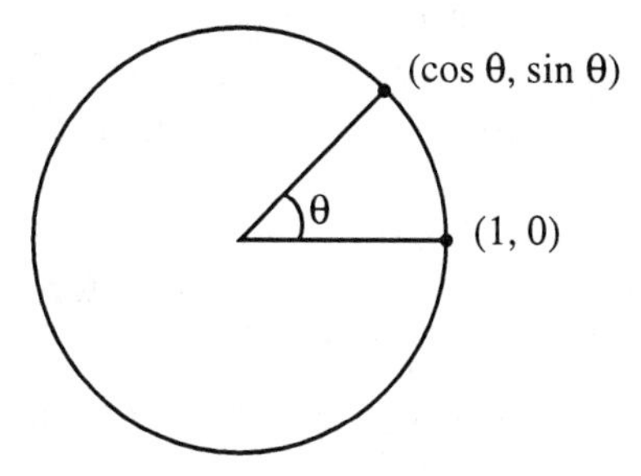

Figure 8.2.1:

For $0 \le \theta \le \pi/2$ this is consistent with the sides of a right triangle definition—taking the radian measurement of angles (equal to the length of the arc of the unit circle subtended), and for other values of θ it is consistent with the usual conventions. In order to turn this into an analytic definition we use the calculus formula for the arc length of a piece of the circle. We can compute the length of the piece of the circle between 0 and θ (for $\theta < \pi/2$) by considering it as the graph of the function $g(y) = \sqrt{1-y^2}$ as y varies between 0 and $\sin\theta$, as in Figure 8.2.2:

$$\theta = \int_0^{\sin\theta} \sqrt{1+g'(y)^2\,dy} = \int_0^{\sin\theta} \sqrt{1+\left(-\frac{y}{\sqrt{1-y^2}}\right)^2}\,dy$$

$$= \int_0^{\sin\theta} \frac{1}{\sqrt{1-y^2}}\,dy.$$

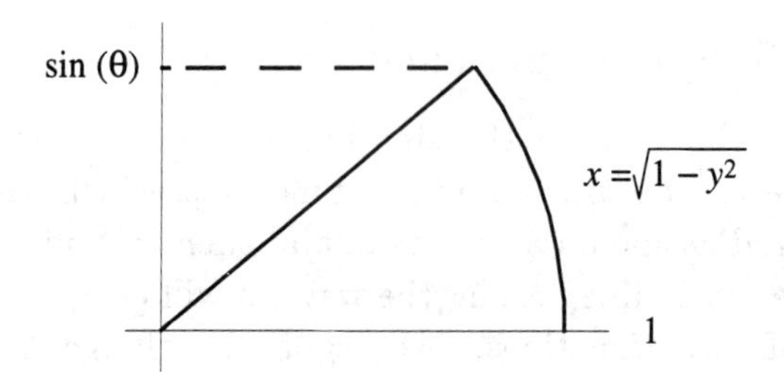

Figure 8.2.2:

Upon examining this formula we find that it does not in fact furnish a definition of $\sin\theta$ but rather of its inverse function arcsin. It tells us how to compute θ from $\sin\theta$. If we substitute $x = \sin\theta$ so that $\theta = \arcsin x$ we have

$$\arcsin x = \int_0^x \frac{1}{\sqrt{1-y^2}}\,dy, \quad \text{for} \quad -1 < x < 1$$

(because of our sign conventions this formula is also valid for negative values of x). For $x = 1$ we have an improper integral since the integrand tends to $+\infty$. However, note that

$$\frac{1}{\sqrt{1-y^2}} = \frac{1}{\sqrt{1+y}}\frac{1}{\sqrt{1-y}}$$

and the first term remains bounded as $y \to 1$. Thus,

$$\frac{1}{\sqrt{1-y^2}} \leq \frac{c}{\sqrt{1-y}}$$

in $[0, 1)$ and so the improper integral exists and is finite (recall that for a local singularity the cutoff point for integrability of $|x - x_0|^{-\alpha}$ is $\alpha = 1$).

We can now define the number π by

$$\frac{\pi}{2} = \arcsin 1 = \int_0^1 \frac{1}{\sqrt{1-y^2}}\,dy = \lim_{x\to 1}\int_0^x \frac{1}{\sqrt{1-y^2}}\,dy,$$

which is simply a translation of the geometric definition of π as half the length of the unit circle.

The function $\arcsin x$ is defined on $(-1,1)$ and is C^1 with derivative $1/\sqrt{1-x^2}$ by the fundamental theorem of the calculus (note the derivative does not exist at $x = \pm 1$). Since the derivative is strictly positive, we can invoke the inverse function theorem to define $\sin\theta$ on $(-\pi/2, \pi/2)$—the image of $\arcsin x$ on $(-1,1)$ (note that the continuity of $\arcsin x$ up to $x = \pm 1$ guarantees that $(-\pi/2, \pi/2)$ is the image of $(-1,1)$). The inverse function theorem tells us that there is a unique C^1 function $\sin\theta$ defined on $(-\pi/2, \pi/2)$ taking values in $(-1,1)$ satisfying

$$\theta = \int_0^{\sin\theta} 1/\sqrt{1-y^2}\, dy,$$

such that

$$\frac{d}{d\theta}\sin\theta = 1/\left(\frac{d}{dx}\arcsin x\right) = \sqrt{1-x^2} \quad \text{for } x = \sin\theta.$$

Thus on the right half circle we have recovered the geometric description of $\sin\theta$ as the y-coordinate of the point on the unit circle where the angle θ is measured by arc length along the circle from the point $(1,0)$, as shown in Figure 8.2.3.

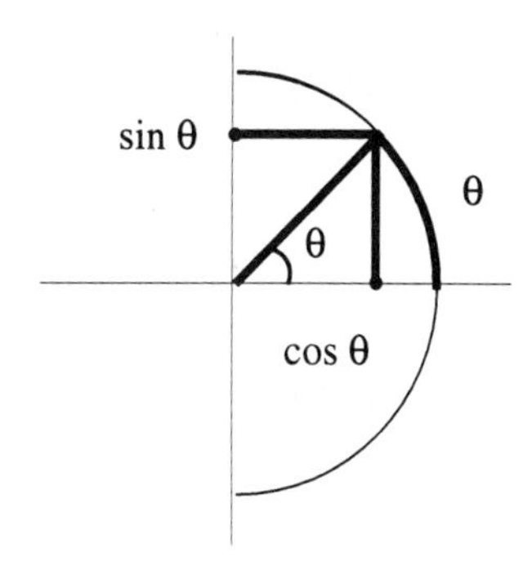

Figure 8.2.3:

Since the equation of the circle is $x^2 + y^2 = 1$ and the right half circle is given by $x > 0$, we can define $\cos\theta = +\sqrt{1-\sin^2\theta}$ for $-\pi/2 < \theta < \pi/2$. This gives the x-coordinate of the point on the circle with angle θ. The formula for the derivative of $\sin\theta$ now simplifies to

$$\frac{d}{d\theta}(\sin\theta) = \sqrt{1-\sin^2\theta} = \cos\theta,$$

and by differentiating the equation defining $\cos\theta$ we obtain

$$\begin{aligned}\frac{d}{d\theta}(\cos\theta) &= \frac{d}{d\theta}\sqrt{1-\sin^2\theta} \\ &= \frac{-\sin\theta(d/dx)\sin\theta}{\sqrt{1-\sin^2\theta}} = \frac{-\sin\theta\cos\theta}{\cos\theta} = -\sin\theta.\end{aligned}$$

The graphs of the functions $\sin\theta$ and $\cos\theta$ for θ in $(-\pi/2, \pi/2)$ are shown in Figure 8.2.4. We have the fundamental identities

$$\sin^2\theta + \cos^2\theta = 1, \quad \frac{d}{d\theta}\sin\theta = \cos\theta, \quad \frac{d}{d\theta}\cos\theta = -\sin\theta,$$

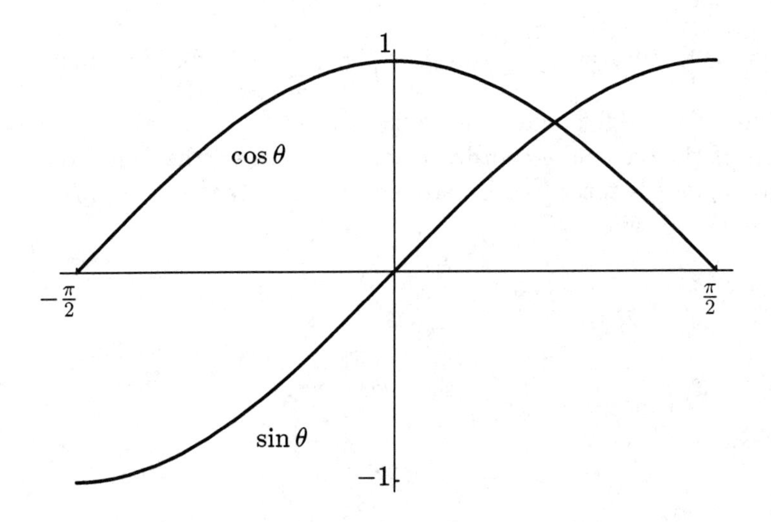

Figure 8.2.4:

holding on this interval.

Next we will extend the functions $\sin\theta$ and $\cos\theta$ to the whole line in accordance with our geometric definition so that these identities continue to hold. The easiest way to do this is first to establish the identities

$$\sin\theta = \cos\left(\theta - \frac{\pi}{2}\right), \quad \cos\theta = -\sin\left(\theta - \frac{\pi}{2}\right)$$

for $0 < \theta < \pi/2$. Indeed suppose $(x, y) = (\cos\theta, \sin\theta)$ is the point on the unit circle corresponding to angle θ in $0 < \theta < \pi/2$. Then

$(y, -x)$ is also a point on the unit circle (see Figure 8.2.5). It clearly lies in the fourth quadrant, so we must have $(\sin\theta, -\cos\theta) = (y, -x) = (\cos\phi, \sin\phi)$ for some ϕ in $-\pi/2 < \phi < 0$. We need to show $\phi = \theta - \pi/2$ in order to have the desired identities. In other words, we need to show that the arc length along the circle from $(y, -x)$ to (x, y) is $\pi/2$—one quarter of the circle.

Since (x, y) is in the first quadrant and $(y, -x)$ is in the fourth quadrant, $(1, 0)$ is an intermediate point.

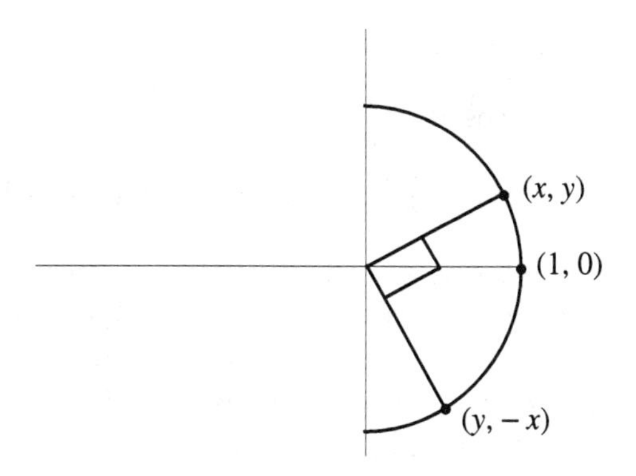

Figure 8.2.5:

The portion of the arc from $(1, 0)$ to (x, y) has length $\int_0^y 1/\sqrt{1-t^2}\,dt$ and the portion from $(y, -x)$ to $(1, 0)$ has length $\int_y^1 1/\sqrt{1-t^2}\,dt$ (here we are using the x-coordinate as parameter, or equivalently the identity $\int_y^1 1/\sqrt{1-t^2}\,dt = \int_0^{\sqrt{1-y^2}} 1/\sqrt{1-s^2}\,ds$, which follows from the substitution $s = \sqrt{1-t^2}$). Thus the total length is

$$\int_0^y \frac{1}{\sqrt{1-t^2}}\,dt + \int_y^1 \frac{1}{\sqrt{1-t^2}}\,dt = \int_0^1 \frac{1}{\sqrt{1-t^2}}\,dt = \frac{\pi}{2}$$

as desired. This result is clear on geometric grounds because the radii joining the origin to the points (x, y) and $(y, -x)$ are perpendicular and perpendicular radii cut the circle in quarters.

Now the identity $\sin\theta = \cos(\theta - \pi/2)$ allows us to define $\sin\theta$ in the interval $[\pi/2, \pi]$, since $\cos(\theta - \pi/2)$ will already be defined. Similarly we can define $\cos\theta = -\sin(\theta - \pi/2)$ for θ in $[\pi/2, \pi)$, and then we can extend sine and cosine for $[\pi, 3\pi/2)$, and so on. The extension to values of $\theta \leq -\pi/2$ is accomplished in a similar way. Notice what happens to $\sin\theta$ near $\theta = \pi/2$. We have $\sin\theta = \cos(\theta - \pi/2)$ (for $\theta < \pi/2$ because we proved it, and for $\theta \geq \pi/2$ because we defined it), so $\sin\theta$ is continuous and differentiable at $\theta = \pi/2$ and $d/d\theta \sin\theta|_{\theta=\pi/2} = d/d\theta \cos(\theta - \pi/2)|_{\theta=\pi/2} = -\sin(\theta - \pi/2)|_{\theta=\pi/2} = 0$.

It is easy to verify that the functions $\sin\theta$ and $\cos\theta$ so extended agree with the geometric definition in terms of $(\cos\theta, \sin\theta)$ giving the coordinates of a point on the unit circle a distance θ from $(1, 0)$ in the counterclockwise direction, and they continue to satisfy the identities

$$\begin{aligned}
&\sin^2\theta + \cos^2\theta = 1, \\
&\frac{d}{d\theta}\sin\theta = \cos\theta, \quad \frac{d}{d\theta}\cos\theta = -\sin\theta \\
&\sin\theta = \cos\left(\theta - \frac{\pi}{2}\right), \quad \cos\theta = -\sin\left(\theta - \frac{\pi}{2}\right).
\end{aligned}$$

Iterating the last two identities we find $\sin\theta = -\sin(\theta - \pi), \cos\theta = -\cos(\theta - \pi), \sin\theta = \sin(\theta - 2\pi), \cos\theta = \cos(\theta - 2\pi)$, so sine and cosine are periodic functions of period 2π. Their graphs are shown in Figure 8.2.6. Since sine and cosine are continuous, the derivative formulas imply they are C^∞, $(d/d\theta)^{2k}\sin\theta = (-1)^k \sin\theta$, etc.

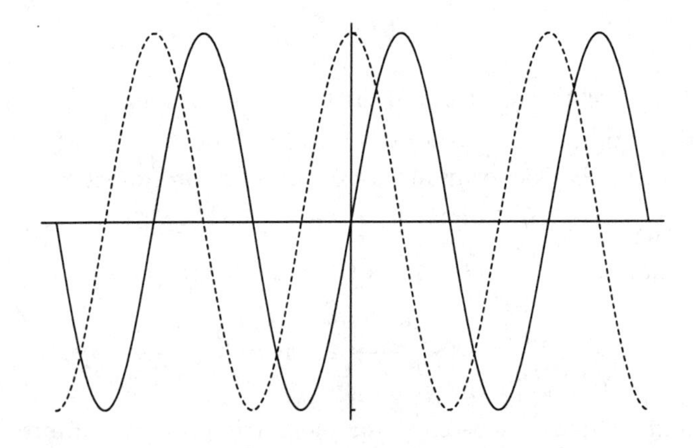

Figure 8.2.6:

Another way to obtain a global definition of sine and cosine is via power series. Since we can easily compute the derivatives of sine and cosine of all orders, we find the infinite Taylor expansions about the origin are

$$S(x) = x - \frac{x^3}{3!} + \frac{x^5}{5!} - \cdots = \sum_{k=0}^{\infty}(-1)^k \frac{x^{2k+1}}{(2k+1)!},$$

$$C(x) = 1 - \frac{x^2}{2!} + \frac{x^4}{4!} - \cdots = \sum_{k=0}^{\infty}(-1)^k \frac{x^{2k}}{(2k)!}.$$

We consider them as power series. From the size of the coefficients (they are comparable with the exponential power series) we see that they converge for all x. However, we cannot conclude from this that they converge to $\sin x$ and $\cos x$ (as we have seen the Taylor expansion of $\exp(-1/x^2)$ about $x = 0$ converges to zero). This will turn out to be true, but it will require a proof—and a rather indirect proof, since our definition of sine and cosine via the arcsine does not give us direct information about the power series. We do observe that $S(x)$ and $C(x)$ satisfy the same sort of differential equations as sine and cosine, $S'(x) = C(x)$ and $C'(x) = -S(x)$—these follow simply by differentiating the power series term-by-term. It would seem plausible that these differential equations essentially characterize sine and cosine, at least with the conditions $S(0) = 0$ and $C(0) = 1$, which are clearly satisfied. Thus we will have $S(x) = \sin x$ and $C(x) = \cos x$ for all x as a consequence of the following lemma.

Lemma 8.2.1 *If f and g are C^1 functions on the line satisfying $f' = g$ and $g' = -f$ and $f(0) = 0$, $g(0) = 1$, then $f(x) = \sin x$ and $g(x) = \cos x$.*

Proof: This result is a consequence of a general uniqueness theorem for ordinary differential equations that we will establish in a later chapter. However, there is a simpler proof of this special case, modeled on the proof of the uniqueness of solutions of the exponential differential equation. We observe that the complex-valued function $F(x) = g(x) + if(x)$ satisfies the differential equation $F'(x) = -f(x) + ig(x) = iF(x)$, as does the function $\operatorname{cis} x = \cos x + i \sin x$. Also, both F and cis take on the value 1 at $x = 0$. Therefore we want to form $F(x)/\operatorname{cis} x$ and show that the derivative is zero. We note that $|\operatorname{cis} x|^2 = \cos^2 x + \sin^2 x = 1$, so $\operatorname{cis} x \neq 0$; hence we can divide by it.

The computation

$$\begin{aligned}\frac{d}{dx}\left(\frac{F(x)}{\operatorname{cis} x}\right) &= \frac{\operatorname{cis} x \frac{d}{dx}F(x) - F(x)\frac{d}{dx}\operatorname{cis} x}{(\operatorname{cis} x)^2} \\ &= \frac{i\operatorname{cis} x F(x) - iF(x)\operatorname{cis} x}{(\operatorname{cis} x)^2} = 0\end{aligned}$$

is as easy as before. Thus $F(x) = a \operatorname{cis} x$ and $a = 1$ follows by setting $x = 0$. QED

We could also disguise the argument to remove all reference to the complex numbers by noting $1/(\cos x + i \sin x) = \cos x - i \sin x$, since $\cos^2 x + \sin^2 x = 1$, so that

$$\begin{aligned}\frac{F(x)}{\operatorname{cis} x} &= (\cos x - i \sin x)(g(x) + if(x)) \\ &= [\cos x g(x) + \sin x f(x)] + i[\cos x f(x) - \sin x g(x)].\end{aligned}$$

Thus we compute

$$\frac{d}{dx}[\cos x g(x) + \sin x f(x)] = 0 \quad \text{and} \quad \frac{d}{dx}[\cos x f(x) - \sin x g(x)] = 0$$

from the differential equations, hence

$$\cos x g(x) + \sin x f(x) = 1 \quad \text{and} \quad \cos x f(x) - \sin x g(x) = 0$$

(the values at $x = 0$ providing the constants 1 and 0). We get $f(x) = \sin x$ and $g(x) = \sin x$ by solving these linear equations (multiply the first by $\sin x$ and the second by $\cos x$ and add, using $\cos^2 x + \sin^2 x = 1$). Of course such a purely real-valued proof would seem quite magical (out of what hat did we draw $\cos x g(x) + \sin x f(x)$ and $\cos x f(x) - \sin x g(x)$?) if presented without motivation.

8.2.2 Relationship Between Sines, Cosines, and Complex Exponentials

The connection between sines and cosines and the exponential function is now clear from two perspectives. We have seen that $F(x) = \cos x + i \sin x$ satisfies $F'(x) = iF(x)$, the same differential equation that $\exp(ix)$ should satisfy. By $\exp(z)$ for z a complex number we mean the result of substituting z in the exponential power series, $\exp(z) = \sum_{n=0}^{\infty} z^n/n!$. For $z = ix$ we find

$$\exp(ix) = \sum_{n=0}^{\infty} \frac{(ix)^n}{n!} = \sum_{n=0}^{\infty} \frac{i^n x^n}{n!} = \sum_{k=0}^{\infty} (-1)^k \frac{x^{2k}}{(2k)!} + i \sum_{k=0}^{\infty} (-1)^k \frac{x^{2k+1}}{(2k+1)!}$$

by separating the real and imaginary parts (the rearrangement is justified by the absolute convergence of the series). Thus we have $\exp(ix) = \cos x + i \sin x$, the notorious *Euler identities.* In a like manner we find $\exp(-ix) = \cos x - i \sin x$, and so we can solve for sine and cosine:

$$\begin{aligned} \sin x &= \frac{\exp(ix) - \exp(-ix)}{2i}, \\ \cos x &= \frac{\exp(ix) + \exp(-ix)}{2}. \end{aligned}$$

These are truly remarkable identities. Although they are trivial consequences of the power-series expansions, they are totally unexpected from the geometric definitions of sine and cosine. We should also point out the rather remarkable nature of the power-series expansions for sine and cosine. For example, the periodicity of the cosine means that $\sum_{k=0}^{\infty} (-1)^k x^{2k}/k! = \sum_{k=0}^{\infty} (-1)^k (x + 2\pi)^{2k}/k!$. This identity is true in the sense that rearrangement of the second series—expanding $(x + 2\pi)^{2k}$ in the binomial theorem—gives the first series. But a direct proof of this fact is out of the question.

The Euler identities simplify the study of the trigonometric functions. For example, the addition formulas

$$\sin(x + y) = \sin x \cos y + \cos x \sin y,$$

$$\cos(x + y) = \cos x \cos y - \sin x \sin y$$

can be obtained by taking the real and imaginary parts of the identity

$$\exp(i(x + y)) = \exp(ix) \exp(iy)$$

after substituting the Euler identities:

$$\cos(x + y) + i \sin(x + y) = (\cos x + i \sin x)(\cos y + i\text{siny}).$$

The other trigonometric functions are definable in terms of sine and cosine, so we will not discuss their properties in detail. We will mention one interesting result, the arctangent integral: arctan: arctan $x = \int_0^x 1/(1+t^2)\,dt$. To derive this we compute

$$\frac{d}{d\theta}\tan\theta = \frac{d}{d\theta}\frac{\sin\theta}{\cos\theta} = \frac{\cos^2\theta + \sin^2\theta}{\cos^2\theta} = 1 + \tan^2\theta,$$

which shows $\tan\theta$ is increasing on $(-\pi/2, \pi/2)$; and since $\cos \pm\pi/2 = 0$, we see that $\tan\theta$ maps $(-\pi/2, \pi/2)$ onto the real line. The graph is shown in Figure 8.2.7.

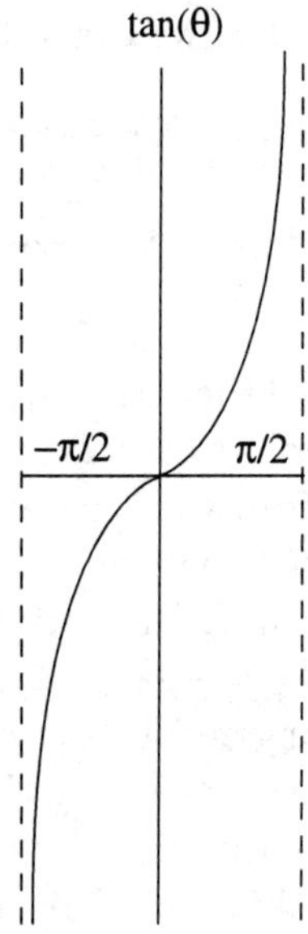

Figure 8.2.7:

By the inverse function theorem $\tan\theta$ has a C^1 inverse (denoted arctan x) on the whole line taking values in $(-\pi/2, \pi/2)$ with the derivative given by d/dx arctan $x = 1/(d/d\theta \tan\theta) = 1/(1+\tan^2\theta)$ if $x = \tan\theta$. Thus $d/dx(\arctan x) = 1/(1+x^2)$; and since arctan $0 = 0$, we obtain arctan $x = \int_0^x 1/(1+t^2)\,dt$ by the fundamental theorem of the calculus.

With the aid of the sine and cosine we can give another perspective on the complex number system. If z is a non-zero complex number, then $z/|z|$ is a complex number of absolute value 1, $|z/|z|| = |z|/|z| = 1$. Since $|z|^2 = x^2 + y^2$ for $z = x + iy$, a complex number of absolute value one corresponds to a point (x, y) in the plane lying on the unit circle. Thus, $(x, y) = (\cos\theta, \sin\theta)$ and $z/|z| = \cos\theta + i\sin\theta$. The value of the angle θ is only determined up to a multiple of 2π and is called the *argument* of z or $\arg z$. Thus, we have the polar coordinates representation of an arbitrary non-zero complex number $z = re^{i\theta} = r\cos\theta + i\,r\sin\theta$ where $r = |z|$ is the length of the line segment joining $(0, 0)$ to (x, y) and θ is the angle it makes with the positive x-axis. This gives us a better understanding of complex multiplication. If $z_1 = r_1e^{i\theta_1}$ and $z_2 = r_2e^{i\theta_2}$, then $z_1z_2 = r_1e^{i\theta_1}r_2e^{i\theta_2} = (r_1r_2)e^{i(\theta_1+\theta_2)}$, so the absolute values multiply and the arguments add. Similarly for powers: $z^n = (re^{i\theta})^n = r^ne^{in\theta}$. The absolute value is raised to the power n and the argument multiplied by n. Notice that in both these formulas the ambiguity in the argument does not matter. For example,

$$(re^{i(\theta+2\pi k)})^n = r^ne^{i(n\theta+2\pi kn)} = r^ne^{in\theta}$$

since kn is also an integer. However, for non-integer powers we find more than one value for z^a, since

$$(re^{i(\theta+2\pi k)})^a = r^ae^{i(a\theta+2\pi ak)}$$

and now ak need not be an integer. In fact, if a is irrational there are infinitely many values for z^a, while for a rational there are only finitely many. For example, there are two square roots, $(re^{i\theta})^{1/2} = r^{1/2}e^{i\theta/2}$ and $(re^{i(\theta+2\pi)})^{1/2} = r^{1/2}e^{i\theta/2}e^{\pi i} = -r^{1/2}e^{i\theta/2}$, exactly as we expect. Thus

$$(i)^{1/2} = (e^{\pi i/2})^{1/2} = \pm e^{\pi i/4} = \pm\left(\cos\frac{\pi}{4} + i\sin\frac{\pi}{4}\right) = \pm\left(\frac{1}{\sqrt{2}} + \frac{i}{\sqrt{2}}\right).$$

We can use these ideas to define an infinite-valued logarithm function for complex numbers. We want log to be the inverse function of exp, but exp is not one-to-one on $\mathbb{C}$, since $\exp(z + 2\pi ki) = \exp z$. To see how to define $\log z$ we look at the equation $\exp(\log z) = z$ (in the other direction we can only expect that one of the values of $\log(\exp z)$ will equal z). Writing $z = re^{i\theta}$ and $\log z = a + bi$ we are led to the equation $re^{i\theta} = e^ae^{ib}$ and, hence, $r = e^a$ and $e^{i\theta} = e^{ib}$, so $a = \log r$ and $b = \theta + 2\pi k$ for some integer k. Thus $\log z = \log r + i(\theta + 2\pi k) = \log|z| + i\arg z$. We can also express this in terms of the real and imaginary parts of z since $|z|^2 = x^2 + y^2$ and

$\arg z = \arctan y/x$, so $\log(x+iy) = 1/2\log(x^2+y^2) + i \arctan y/x$ (actually there is more ambiguity in $\arctan y/x$ than in $\arg z$, so this equation must be understood in the sense that only half the values of $\arctan y/x$ are allowed).

With the aid of the polar coordinate representation we can give a simple proof of the triangle inequality ($|z_1 - z_2| \le |z_1| + |z_2|$) for complex numbers. Writing $z_1 = r_1 e^{i\theta_1}$ and $z_2 = r_2 e^{i\theta_2}$ we compute

$$\begin{aligned} |z_1 - z_2|^2 &= (z_1 - z_2)(\bar{z}_1 - \bar{z}_2) = (r_1 e^{i\theta_1} - r_2 e^{i\theta_2})(r_1 e^{-i\theta_1} - r_2 e^{-i\theta_2}) \\ &= r_1^2 + r_2^2 - r_1 r_2 (e^{i(\theta_1-\theta_2)} + e^{-i(\theta_1-\theta_2)}) \\ &= r_1^2 + r_2^2 - 2r_1 r_2 \cos(\theta_1 - \theta_2), \end{aligned}$$

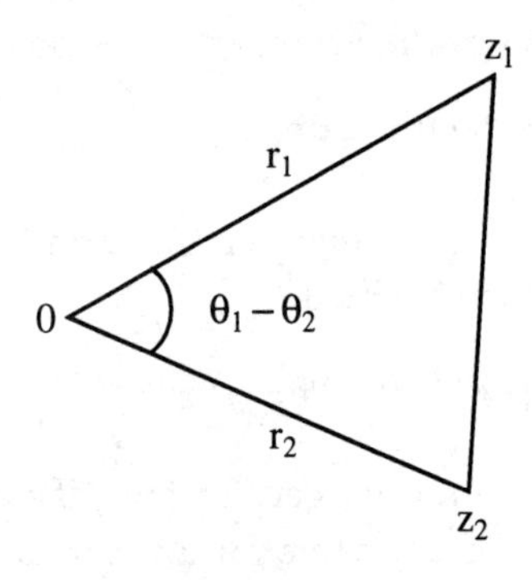

Figure 8.2.8:

which is the familiar *law of cosines* for the triangle (shown in Figure 8.2.8) with vertices at $0, z_1, z_2$. Since $\cos(\theta_1 - \theta_2)$ varies between $+1$ and -1, we have $|z_1 - z_2|^2$ lying between $r_1^2 + r_2^2 - 2r_1r_2 = (r_1 - r_2)^2$ and $r_1^2 + r_2^2 + 2r_1r_2 = (r_1 + r_2)^2$, with the maximum and minimum values assumed when the angle $\theta_1 - \theta_2$ is π and 0. Thus $|z_1 - z_2|^2 \le (r_1 + r_2)^2$, so $|z_1 - z_2| \le |z_1| + |z_2|$ with equality holding only when z_1 and z_2 are colinear.

In addition to the exponential, logarithm, and trigonometric functions, there are a number of other transcendental functions that have been carefully studied. Examples of these *special functions* are Bessel functions, hypergeometric functions, Legendre functions, and the gamma function. The main tools used in the study of these functions are those we have already discussed: representation as integrals, power-series expansions, and differential equations.

8.2.3 Exercises

1. Show by direct substitution ($t = x/\sqrt{1+x^2}$) that

$$\int_0^\infty 1/(1+x^2)\,dx = \int_0^1 1/\sqrt{1-t^2}\,dt.$$

2. Show by direct substitution that

$$\int_0^1 1/(1+x^2)\,dx = \int_1^\infty 1/(1+x^2)\,dx.$$

   Interpret this result in terms of the arctangent.

3. *Using $\pi = 4\int_0^1 1/(1+x^2)\,dx$ and the midpoint and trapezoidal rules with increments of $1/10$, compute the approximate value of π. Compare the predicted error with the actual error. (Use a calculator for this problem. You might also try programming the computation to allow different values of the increment and see if the actual error depends on the increment in the predicted manner.)

4. Verify the identity $\sin^2\theta + \cos^2\theta = 1$ by rearranging the power-series expansions.

5. Compute the power-series expansions of $\sin\theta$ and $\cos\theta$ about the point $\pi/2$.

6. Find all complex solutions of the equation $z^n = 1$. Can you give a geometric interpretation of the result?

7. Let $f(z) = \sum_{n=0}^\infty a_n z^n$ be convergent in $|z| < R$, and suppose the coefficients a_n are all real. Show $f(\bar{z}) = \overline{f(z)}$.

8. *Show that $e^z = z$ has no real solutions but that it has complex solutions. (**Hint**: write $z = x + iy$, and show that $e^z = z$ is equivalent to $x^2 + y^2 = e^{2x}$ and $y/x = \tan y$.)

9. By expanding $1/(1+x^2)$ in a power series about $x = 0$ prove $\pi/4 = \int_0^1 1/(1+x^2)\,dx = \lim_{r\to 1}\sum_{k=0}^\infty (-1)^k r^{2k+1}/(2k+1)$, and then show $\pi/4 = \sum_{k=0}^\infty (-1)^k/(2k+1)$. (**Hint**: combine neighboring terms in the series.)

10. Show $a \sin x + b \cos x = A \cos(x + B)$. How are a, b and A, B related?

11. Show that $\exp(z)$ assumes every complex value except zero and that $\exp(z_1) = \exp(z_2)$ if and only if $z_1 - z_2 = 2\pi k i$ for some integer k.

12. Express $(\sin x)^n$ as a linear combination of $\cos kx$ and $\sin kx$ for $0 \le k \le n$. (**Hint:** use $\sin x = (e^{ix} - e^{-ix})/2i$.)

13. Prove $\int_0^\pi \sin nx \sin mx \, dx = 0$ if $n \neq m$.

14. Prove $\cos x \cos y = 1/2 \cos(x + y) + 1/2 \cos(x - y)$. Explain how this formula could be used, in conjuction with a table of cosines, to simplify the process of multiplication of two numbers in $[0, 1]$. (This idea was proposed by Francois Viete in the 1590s under the name *prosthaphaeresis*. It became obsolete after the introduction of logarithms by Napier.)

8.3 Summary

8.1 The Exponential and Logarithm

Definition 8.1.1 *The exponential function* $\exp(x)$ *is defined for any real* x *by* $\exp(x) = \sum_{n=0}^{\infty} x^n/n!$.

Theorem 8.1.1

a. *The exponential function satisfies the differential equation* $f' = f$, *with* $f(0) = 1$.

b. $\exp(x + y) = \exp x \exp y$ *for any real* x *and* y.

c. $\exp(x) > 0$ *for any real* x.

Theorem 8.1.2 $\exp(x)$ *is the unique solution of the differential equation* $f' = f$ *with* $f(0) = 1$.

Theorem 8.1.3 *For every rational number* p/q *(*p *and* q *integers,* $q > 0$*) we have* $\exp(p/q) = (e^p)^{1/q}$. *For* x *real and* $x_k = p_k/q_k$ *a Cauchy sequence of rationals converging to* x, $\exp(x) = \lim_{k\to\infty} \exp(x_k) = \lim_{k\to\infty}(e^{p_k})^{1/q_k}$.

Theorem 8.1.4 *The exponential function maps* $\mathbb{R}$ *one-to-one and onto* $\mathbb{R}^+$.

Theorem 8.1.5 $\log x = \int_1^x 1/t\, dt$.

Theorem *The functions* exp *and* log *establish an isomorphism between the additive group of the reals and the multiplicative group of the positive reals.*

Theorem 8.1.6 $\exp(x) = \lim_{n\to\infty}(1 + x/n)^n$ *for any real* x.

Theorem

$$\log(1+x) = x - \frac{x^2}{2} + \frac{x^3}{3} - \frac{x^4}{4} + \cdots$$

for $|x| < 1$, *and more generally*

$$\log x = \log x_0 + \sum_{k=1}^{\infty} \frac{(-1)^{k+1}}{k} \left(\frac{x - x_0}{x_0}\right)^k$$

for $|x - x_0| < x_0$; *hence* $\log x$ *is analytic.*

Theorem $\lim_{x\to+\infty} x^{-n}\exp(x) = +\infty$ *and* $\lim_{x\to+\infty} x^n \exp(-x) = 0$, *for any* $n \geq 0$; $\lim_{x\to+\infty} x^{-a}\log x = 0$ *and* $\lim_{x\to 0} x^a \log x = 0$ *for any* $a > 0$.

Theorem *The function*

$$F(x) = \begin{cases} \exp(-1/x^2), & x \neq 0, \\ 0, & x = 0, \end{cases}$$

is C^∞ *and vanishes to infinite order at* $x = 0$; *hence, it is not analytic at* $x = 0$.

Theorem *The "blip function"*

$$g(x) = \begin{cases} e^2 \exp\left(-1/(x-1)^2\right) \exp\left(-1/(x+1)^2\right), & -1 < x < 1, \\ 0, & |x| \geq 1, \end{cases}$$

is C^∞, vanishes outside $|x| \geq 1$, and satisfies $g(0) = 1$.

Theorem 8.1.7 (*Borel*) *Given any sequence $a_0, a_1, \ldots$ of reals, any point x_0, and any neighborhood $|x - x_0| < \lambda$, there exists a C^∞ function f vanishing outside the neighborhood such that $f^{(k)}(x_0) = a_k$ for all k.*

Corollary *Given any C^∞ function on a compact interval $[a, b]$ (meaning one-sided derivatives exist at the endpoints) there exists a C^∞ extension to the line vanishing outside a larger interval $[a - \varepsilon, b + \varepsilon]$.*

8.2 Trigonometric Function

Definition $\arcsin x = \int_0^x 1/\sqrt{1-y^2}\, dy$ *for* $-1 \leq x \leq 1$, *and* $\pi = 2 \arcsin 1$.

Theorem $\arcsin x$ *is a C^1 function mapping $(-1, 1)$ onto $(-\pi/2, \pi/2)$ with derivative $1/\sqrt{1-x^2}$, and has a C^1 inverse function* $\sin\theta$ *mapping $(-\pi/2, \pi/2)$ onto $(-1, 1)$ with derivative $d/d\theta \sin\theta = \sqrt{1 - \sin^2\theta}$.*

Definition $\cos\theta = +\sqrt{1-\sin^2\theta}$ *for* $-\pi/2 < \theta < \pi/2$.

Theorem *There exist unique functions $\sin\theta$ and $\cos\theta$ defined for all real θ coinciding with the above definitions for $-\pi/2 < \theta < \pi/2$ and satisfying* $\sin\theta = \cos(\theta - \pi/2), \cos\theta = -\sin(\theta - \pi/2)$ *for all real θ.*

Theorem $\sin\theta$ *and* $\cos\theta$ *are C^1 functions periodic of period 2π, and satisfy the identities* $\sin^2\theta + \cos^2\theta = 1$, $d/d\theta \sin\theta = \cos\theta, d/d\theta \cos\theta = -\sin\theta$.

Theorem sin *and* cos *are analytic functions with power series*

$$\sin x = x - \frac{x^3}{3!} + \frac{x^5}{5!} - \cdots = \sum_{k=0}^{\infty} (-1)^k \frac{x^{2k+1}}{(2k+1)!}$$

$$\cos x = 1 - \frac{x^2}{2!} + \frac{x^2}{4!} - \ldots = \sum_{k=0}^{\infty} (-1)^k \frac{x^{2k}}{(2k)!}$$

converging for all real x.

Lemma 8.2.1 *The unique C^1 solutions to $f' = g$ and $g' = -f$ with $f(0) = 0$ and $g(0) = 1$ are $f(x) = \sin x$ and $g(x) = \cos x$.*

Theorem (*Euler identities*) $\exp(ix) = \cos x + i \sin x$,

$$\sin x = \frac{\exp(ix) - \exp(-ix)}{2i},$$
$$\cos x = \frac{\exp(ix) + \exp(-ix)}{2}.$$

Theorem $\arctan x = \int_0^x 1/(1+t^2)\, dt$.

Theorem *An arbitrary non-zero complex number z can be written $z = re^{i\theta}$, where θ (determined modulo 2π) is called the argument and r the modulus of z. In multiplying complex numbers the moduli are multiplied and the arguments added, while $z^n = r^n e^{in\theta}$ for integer n.*

Theorem (*Law of Cosines*) $|z_1 - z_2|^2 = r_1^2 + r_2^2 - 2r_1 r_2 \cos(\theta_1 - \theta_2)$ *if* $z_1 = r_1 e^{i\theta_1}, z_2 = r_2 e^{i\theta_2}$.

Chapter 9

Euclidean Space and Metric Spaces

9.1 Structures on Euclidean Space

9.1.1 Vector Space and Metric Space

We are now ready to begin the study of functions of several variables, $f(x_1, x_2, \ldots, x_n)$, where each x_k varies over $\mathbb{R}$. Since we live in a three-dimensional world, we can easily appreciate the importance of this subject, at least for $n = 3$. We allow the number n to be arbitrary because it is necessary to do so for many applications and also because the mathematics is not appreciably more difficult. We will frequently appeal to our three-dimensional geometric intuition to guide us to an understanding of the general case.

We begin by studying the Euclidean space $\mathbb{R}^n$ of n real variables. We can simply define $\mathbb{R}^n$ to be the set of all ordered n-tuples $(x_1, x_2, \ldots, x_n)$ of real numbers. Of course this is not the whole story. We also want to define on the set $\mathbb{R}^n$ various "structures": vector space, metric space, normed space (Banach space), and inner product space (Hilbert space). Each of these structures can be defined by rather simple formulas; however, we will also be interested in a more abstract description of these structures. For each type of structure, we will first describe the Euclidean version, then present some of the basic properties of the structure, and finally use these properties to define a general notion of the structure. In this way, the Euclidean version becomes just a special case of abstract structure; but it is the special case in which we are most interested, and it serves to motivate the abstract definition. All these structures play an important role in mathematics, although it is the metric space structure that will be most emphasized in this book. The reader who has not seen these ideas before should not expect to appreciate the signficance of these structures immediately; such an understanding can only come after the theory is developed.

In order to simplify the notation we will adopt the following *conventions.* Letters at the end of the alphabet x, y, z, etc., will be used to denote points in $\mathbb{R}^n$, so $x = (x_1, x_2, \dots, x_n)$ and x_k will always refer to the kth coordinate of x. We will reserve the letter n for the dimension of the space. Letters from the beginning of the alphabet, a, b, c, etc., will denote real numbers, also called *scalars.*

We begin with the vector space structure of $\mathbb{R}^n$. We define vector addition by $x + y = (x_1 + y_1, x_2 + y_2, \dots, x_n + y_n)$ and scalar multiplication by $ax = (ax_1, ax_2, \dots, ax_n)$. It is easy to verify that with these definitions $\mathbb{R}^n$ forms a *vector space* over the scalar field $\mathbb{R}$. We recall the vector space axioms: a set V with a vector addition and scalar multiplication is said to be a vector space over the scalar field $\mathbb{F}$ (in this book we will always take $\mathbb{R}$ or $\mathbb{C}$) provided

1. vector addition satisfies the commutative group axioms: commutativity ($x + y = y + x$), associativity ($(x + y) + z = x + (y + z)$), existence of zero ($x + 0 = x$ for all x), and existence of additive inverses ($x + (-x) = 0$); and
2. scalar multiplication is associative ($(ab)x = a(bx)$) and distributes over addition in both ways ($a(x + y) = ax + ay$ and $(a + b)x = ax + bx$).

The study of vector spaces is called *linear algebra.* We assume the reader has had some exposure to the elementary theory of linear algebra, at least in the concrete setting of $\mathbb{R}^n$. We recall that $\mathbb{R}^n$ has dimension n because the vectors $(1, 0, \dots, 0), (0, 1, 0, \dots, 0), \dots, (0, \dots, 0, 1)$ form a basis (a set of vectors that is linearly independent and spans) and that every basis must have n elements. We refer to this special basis as the *standard basis*, and we sometimes denote it by $e^{(1)}, e^{(2)}, \dots, e^{(n)}$. The vector x is written uniquely $x_1 e^{(1)} + x_2 e^{(2)} + \cdots + x_n e^{(n)}$ as a linear combination of the standard basis vectors, and the coordinate x_k is the coefficient of $e^{(k)}$ in this representation.

The vector space structure of $\mathbb{R}^n$ is not enough—on its own—to allow us to express geometric concepts. We are thus led to consider the *metric* structure in order to define length. We take the Pythagorean formula $d(x,y) = \sqrt{(x_1 - y_1)^2 + \cdots + (x_n - y_n)^2}$ as the definition of the Euclidean distance between x and y. Our geometric intuition validates this definition when $n = 1, 2, 3$. It is also the only reasonable choice in general if we want subspace consistency (if x and y happen to lie in an m-dimensional subspace defined by the vanishing of a specified set of $n - m$ coordinates, then the distance is the same measured in the subspace or in all of $\mathbb{R}^n$). This distance function satisfies three basic conditions:

1. $d(x,y) \geq 0$ with equality if and only if $x = y$ (positivity).

2. $d(x,y) = d(y,x)$ (symmetry).

3. $d(x,z) \leq d(x,y) + d(y,z)$ (triangle inequality).

The Euclidean distance clearly satisfies the first two properties; only the triangle inequality requires proof. We have already proved this in the case $n = 2$ under the guise of a triangle inequality for distances in the complex plane. Indeed we can regard $\mathbb{C}$ as $\mathbb{R}^2$ with some additional structure—complex multiplication—by identifying $x = (x_1, x_2)$ in $\mathbb{R}^2$ with $z = x_1 + ix_2$ in $\mathbb{C}$, and the distance functions are the same. We could generalize the proof given for $\mathbb{C}$ to the case of $\mathbb{R}^n$ but will obtain a simpler proof very shortly, so we postpone the discussion until the next section.

The abstract notion of *metric space* is that of a set M with a *distance function* (or *metric*) $d(x,y)$ taking real values for x and y in M and satisfying the above three conditions of positivity, symmetry, and triangle inequality. These are considered to be the minimum conditions needed to justify the crudest intuitions of distance (sometimes, however, a condition weaker than the triangle inequality, such as $d(x,z) \leq M(d(x,y) + d(y,z))$ for some constant M, can be substituted and one still obtains a useful notion of distance). Thus we have shown that $\mathbb{R}^n$ with the Pythagorean distance function forms a metric space. We will give many more examples in the next section.

9.1.2 Norm and Inner Product

The metric structure of $\mathbb{R}^n$ is related to the vector space structure, since $d(x,y)$ depends only on $x-y$. To make this explicit we introduce the Euclidean *norm*, defined by $|x| = \sqrt{x_1^2 + \cdots + x_n^2}$, so that $d(x,y) = |x-y|$. The basic properties of the norm are:

1. $|x| \geq 0$ with equality if and only if $x = 0$ (positivity).

2. $|ax| = |a|\,|x|$ for any scalar a (homogeneity).

3. $|x+y| \leq |x| + |y|$ (triangle inequality).

Note that in the statement of homogeneity the symbol $|\,|$ is used with two different meanings—$|a|$ referring to the absolute value of a. Of course the absolute value and the norm coincide for $\mathbb{R}^1$, so there is no danger of misinterpreting the formula. Nevertheless, it is sometimes preferable to use double bars $||x||$ to donate the norm. We have chosen to use the single bars for the norm on $\mathbb{R}^n$ so that we can use double bars to refer to other norms. The abstract definition of *norm* is of course any function *on a vector space* that satisfies the above conditions of positivity, homogeneity, and triangle inequality. Notice that a norm must be defined on a vector space in order for conditions 2 and 3 to make sense. For a metric there is no need to assume the space has a vector space structure.

The verification that the Euclidean norm on $\mathbb{R}^n$ actually satisfies the positivity and homogeneity conditions is trivial, and again we postpone the proof of the triangle inequality. What we want to show is that these conditions on the norm imply the defining conditions of a metric for the distance $d(x,y) = |x-y|$. In other words, *if we start with any vector space V with a norm $||x||$, then V becomes a metric space with the distance function $d(x,y) = ||x-y||$.* The proof of this is trivial: the positivity of the norm implies the positivity of the metric, $d(x,y) = ||x-y|| \geq 0$ with equality if and only if $x - y = 0$, or in other words $x = y$. The symmetry of the metric follows from the homogeneity of the norm with $a = -1$,

$$d(x,y) = ||x-y|| = ||(-1)(y-x)|| = |-1|\,||y-x|| = ||y-x|| = d(y,x).$$

Finally, the triangle inequality for the norm implies the triangle inequality for the metric—

$$\begin{aligned} d(x,z) &= ||x-z|| = ||(x-y)+(y-z)|| \\ &\leq ||x-y|| + ||y-z|| = d(x,y) + d(y,z). \end{aligned}$$

The metric $d(x,y) = ||x-y||$ is said to be the metric *associated with* (or *induced by*) the norm. Note that in our proof that $d(x,y)$ is a metric we used the homogeneity of the norm only in the special case $a = -1$. This means that the homogeneity condition is considerably stronger than is absolutely essential—so not every metric on a vector space that depends only on $x - y$ is associated with a norm (for an example see exercise set 9.1.4, number 15).

Here are two more examples of norms, with the underlying vector space being $\mathbb{R}^n$:

$$\begin{aligned} ||x||_1 &= \sum_{j=1}^{n} |x_j|, \\ ||x||_{\text{sup}} &= \max_j \{|x_j|\}. \end{aligned}$$

The verification of the norm axioms is straightforward for these—we leave it to the exercises. The associated distances are different from the Pythagorean distance—in the first case we can interpret the distance $||x-y||$ as the shortest distance between x and y along a broken line segment that moves parallel to the axes (taxicab distance in a city laid out on a square grid). The norms we are considering are special cases of the p-norm

$$||x||_p = \left(\sum_{j=1}^{n} |x_j|^p \right)^{1/p}$$

with p a constant satisfying $1 \leq p < \infty$ (the norm $||x||_{\text{sup}}$ can be thought of as $\lim_{p\to\infty} ||x||_p$ and is frequently denoted $||x||_\infty$). The proof of the triangle inequality for the p-norm in general is more difficult. If we draw the graph of the set of points in $\mathbb{R}^2$ satisfying $||x|| = 1$ we can get a picture of the differences between these norms, as in Figure 9.1.1.

For another important example we consider the space $C([a,b])$ of continuous real-valued functions on the compact interval $[a,b]$, with norm $||f||_{\text{sup}} = \sup_x |f(x)|$. The vector space structure of $C([a,b])$

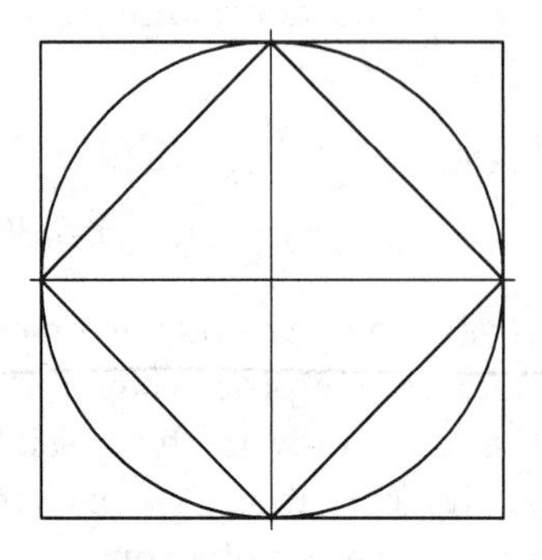

Figure 9.1.1:

is the obvious one, $f + g$ is the function $f(x) + g(x)$, and af is the function $af(x)$. The verification of the norm axioms for the sup-norm is again simple and left as an exercise. This vector space is not finite dimensional.

The last structure on $\mathbb{R}^n$ we want to consider is the *inner product.* We need this structure in order to express the geometric notion of angle. We define $x \cdot y = x_1 y_1 + \cdots + x_n y_n$—this is sometimes called the *scalar product* or *dot product.* The connection with angle is given by the formula $x \cdot y = |x|\,|y| \cos\theta$ defining the angle θ between the vectors x and y. Note that the angle is only defined if both x and y are nonzero, and then the sign of θ is not defined (in dimension $n \geq 3$ we cannot unambiguously choose a sign convention). For most applications we will only be interested in the condition $x \cdot y = 0$ characterizing perpendicular vectors. The geometric justification of the angle formula in $\mathbb{R}^2$ is familiar: if $x = (r\cos\theta, r\sin\theta)$ and $y = (R\cos\phi, R\sin\phi)$ in polar coordinates, then

$$x \cdot y = rR(\cos\theta\cos\phi + \sin\theta\sin\phi) = rR\cos(\theta - \phi),$$

and $r = |x|, R = |y|$, and $\theta - \phi$ is the angle between, as in Figure 9.1.2.

The basic properties of the inner product are:

1. $x \cdot y = y \cdot x$ (symmetry).

2. $(ax + by) \cdot z = ax \cdot z + by \cdot z$ and $x \cdot (ay + bz) = ax \cdot y + bx \cdot z$ (bilinearity).

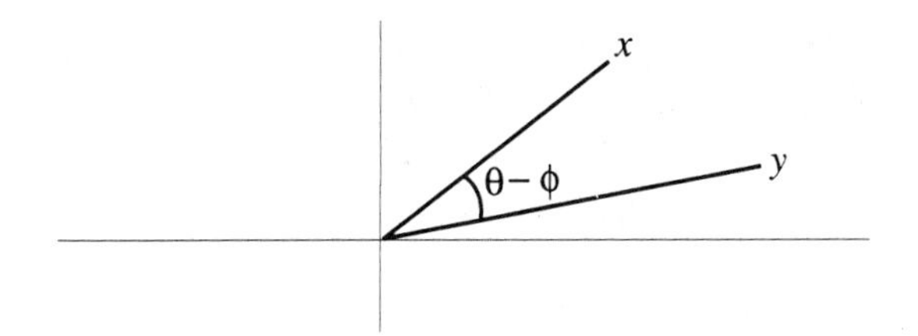

Figure 9.1.2:

3. $x \cdot x \geq 0$ with equality if and only if $x = 0$ (positive definiteness).

Notice that all three conditions are trivially verified. Any real-valued function $\langle x, y\rangle$ defined for x and y in a vector space and satisfying symmetry, bilinearity, and positive definiteness is said to be an *inner product.*

The connection between the Euclidean inner product and norm is evidently given by $|x| = \sqrt{x \cdot x}$. In general, if we are given an inner product $\langle x, y\rangle$ we define the *associated* (or *induced*) *norm* to be $||x|| = \sqrt{\langle x, x\rangle}$. This requires, of course, that we prove $||x|| = \sqrt{\langle x, x\rangle}$ actually is a norm. To do this we need an important estimate, known as the *Cauchy-Schwartz inequality*:

$$|\langle x, y\rangle| \leq \sqrt{\langle x, x\rangle}\sqrt{\langle y, y\rangle}.$$

This is one of the most important inequalities in all of analysis and appears under myriad guises. By stating it as we have in the abstract formulation we include all the special cases—corresponding to different choices of inner product. For the case of the Euclidean inner product the inequality reads

$$\left|\sum_{j=1}^{n} x_j y_j\right| \leq \left(\sum_{j=1}^{n} x_j^2\right)^{1/2} \left(\sum_{j=1}^{n} y_j^2\right)^{1/2}.$$

An integral version,

$$\left|\int_a^b f(x)g(x)\,dx\right| \leq \left(\int_a^b |f(x)|^2\,dx\right)^{1/2} \left(\int_a^b |g(x)|^2\,dx\right)^{1/2},$$

follows easily from the fact that $\langle f, g\rangle = \int_a^b f(x)g(x)\,dx$ is an inner product on $C([a,b])$. We will give a simple proof in the general setting. A good exercise would be to rewrite it in the special case of the Euclidean inner product.

Theorem 9.1.1 (*Cauchy-Schwartz Inequality*) *On any vector space with an inner product $\langle x, y\rangle$, we have $|\langle x, y\rangle| \leq \langle x, x\rangle^{1/2}\langle y, y\rangle^{1/2}$ (or equivalently, $|\langle x, y\rangle| \leq ||x||\,||y||$) for any x, y. Furthermore, equality occurs if and only if the vectors x and y are colinear ($x = ay$ or $y = bx$ for some scalar).*

Proof: The proof is based on the observation $\langle x+y, x+y\rangle \geq 0$ and $\langle x-y, x-y\rangle \geq 0$ by the positivity. Expanding these using the bilinearity and symmetry we obtain $\langle x, x\rangle + \langle y, y\rangle + 2\langle x, y\rangle \geq 0$ and $\langle x, x\rangle + \langle y, y\rangle - 2\langle x, y\rangle \geq 0$, so $|\langle x, y\rangle| \leq (\langle x, x\rangle + \langle y, y\rangle)/2$. This is almost the inequality we want. Notice that if $\langle x, x\rangle = 1$ and $\langle y, y\rangle = 1$, then this is just $|\langle x, y\rangle| \leq 1 = \langle x, x\rangle^{1/2}\langle y, y\rangle^{1/2}$ as desired. Thus we already have the Cauchy-Schwartz inequality for unit vectors (those satisfying $\langle x, x\rangle = \langle y, y\rangle = 1$). Finally we can reduce the general case to this special case. Leaving aside the trivial cases $x = 0$ or $y = 0$ (then both sides of the inequality are zero), we can always write $x = ax'$, $y = by'$ with $\langle x', x'\rangle = \langle y', y'\rangle = 1$ simply by choosing $a = \langle x, x\rangle^{1/2}$ and $b = \langle y, y\rangle^{1/2}$ and setting $x' = a^{-1}x$, $y' = b^{-1}y$ (a^{-1} and b^{-1} are defined because $x \neq 0, y \neq 0$, and the inner product is positive definite). Then

$$|\langle x, y\rangle| = |\langle ax', by'\rangle| = |ab\langle x', y'\rangle| = |ab|\,|\langle x', y'\rangle| \leq |ab|$$

by the special case $|\langle x', y'\rangle| \leq 1$, and $|ab| = \langle x, x\rangle^{1/2}\langle y, y\rangle^{1/2}$.

To see when equality can hold consider first the special case of unit vectors. Then $\langle x', y'\rangle = 1$ implies $\langle x' - y', x' - y'\rangle = 0$, while $\langle x', y'\rangle = -1$ implies $\langle x' + y', x' + y'\rangle = 0$. Thus equality holds if and only if $x' = \pm y'$. In the general case equality holds if and only if $|\langle x', y'\rangle| = 1$, so the condition of colinearity follows. QED

Notice that the first inequality obtained, $|\langle x, y\rangle| \leq (\langle x, x\rangle + \langle y, y\rangle)/2$, is in general weaker than the Cauchy-Schwartz inequality because the geometric mean $\langle x, x\rangle^{1/2}\langle y, y\rangle^{1/2}$ is less than the arithmetic mean $(\langle x, x\rangle + \langle y, y\rangle)/2$. Thus the "scaling" part of the proof is indispensible.

We can now prove that the formula $||x|| = \langle x, x\rangle^{1/2}$ really defines a norm—incidentally giving the promised proof of the triangle inequality for the Euclidean norm, since we have already verifed that the Euclidean inner product $x \cdot y$ is an inner product.

Theorem 9.1.2 *Let $\langle x, y\rangle$ be an inner product on a vector space. Then $||x|| = \langle x, x\rangle^{1/2}$ is a norm.*

Proof: The positive definiteness of the inner product implies the positivity of the norm, and the homogeneity

$$||ax|| = \langle ax, ax\rangle^{1/2} = (a^2\langle x, x\rangle)^{1/2} = |a|\langle x, x\rangle^{1/2} = |a|\,||x||$$

follows from the bilinearity. The only nontrivial property is the triangle inequality, which we prove in its squared version, $||x + y||^2 \leq (||x|| + ||y||)^2 = ||x||^2 + 2||x||\,||y|| + ||y||^2$, for we can then take the square root. Now

$$\begin{aligned} ||x + y||^2 &= \langle x + y, x + y\rangle = \langle x, x\rangle + 2\langle x, y\rangle + \langle y, y\rangle \\ &= ||x||^2 + 2\langle x, y\rangle + ||y||^2. \end{aligned}$$

Now by the Cauchy-Schwartz inequality $||x+y||^2 \leq ||x||^2 + 2||x||\,||y|| + ||y||^2$ as desired. QED

It is a curious, and sometimes useful, fact that we can also express the inner product in terms of the norm. This is called the *polarization identity*:

$$\langle x, y\rangle = \frac{1}{4}(||x + y||^2 - ||x - y||^2),$$

which follows immediately by expanding $||x \pm y||^2 = \langle x \pm y, x \pm y\rangle$ using the bilinearity. Of course not every norm is associated to an inner product (of the examples of norms on $\mathbb{R}^n$, only the Euclidean one is), so the polarization identity does not make sense for an arbitrary norm. Norms that are associated to inner products satisfy the *parallelogram law*

$$||x + y||^2 + ||x - y||^2 = 2(||x||^2 + ||y||^2).$$

Geometrically this can be interpreted to say the sum of the squares of the diagonals of a parallelogram equals the sum of the squares of the sides (see Figure 9.1.3). It is again a simple exercise to derive the parallelograph law for norms associated with an inner product. It is actually true that the parallelogram law characterizes such norms. If a norm satisfies the parallelogram law, then the polarization identity defines an inner product, and the norm associated with the inner product is equal to the original norm. We leave the details as an exercise.

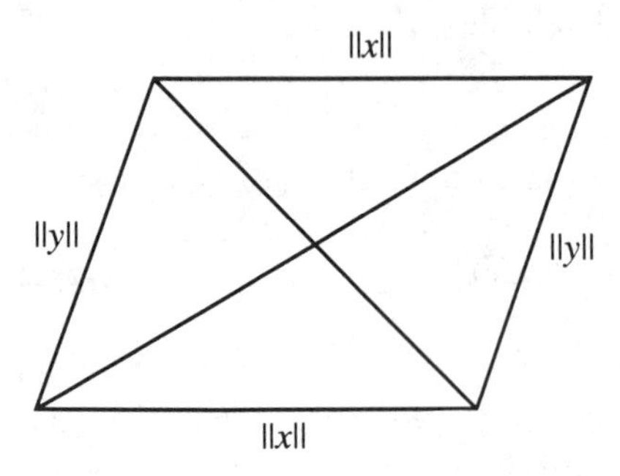

Figure 9.1.3:

9.1.3 The Complex Case

A discussion of the structure of vector space, norm, and inner product would be incomplete without mention of the complex analogs because they have many important applications. A *complex vector space* is a space satisfying the axioms of a vector space with $\mathbb{C}$ as the field of scalars. This means that ax is defined for complex numbers a. A complex vector space may be thought of as a real vector space with additional structure (multiplication by $i = \sqrt{-1}$). The simplest example is $\mathbb{C}^n$, the set of n-tuples $z = (z_1, z_2, \ldots, z_n)$ of complex numbers. This has complex dimension n since the basis vectors $e^{(1)}, \ldots, e^{(n)}$ of $\mathbb{R}^n$ also form a basis of $\mathbb{C}^n$, $z = z_1 e^{(1)} + \cdots + z_n e^{(n)}$. (Warning: regarded as a real vector space, $\mathbb{C}^n$ has dimension $2n$, with $e^{(1)}, \ldots, e^{(n)}, ie^{(1)}, \ldots, ie^{(n)}$ forming a basis. The paradox is explained because the definitions of linear independence and spanning involve the notion of linear combinations, which refers to the scalar field.)

The definition of *norm* on a complex vector space is the same as for a real vector space except that the homogeneity condition $||ax|| = |a|\,||x||$ must hold for all complex scalars, where $|a|$ is the absolute value of the complex number a. The proof that a complex norm defines a metric via $d(x,y) = ||x-y||$ is the same as in the real case.

The definition of *complex inner product* requires one important modification. In place of the bilinearity, we must introduce complex conjugates on one or the other side. (By convention mathematicians have chosen the right side and physicists the left.) This is referred to as Hermitian linearity, and it also spoils the symmetry. The conditions on $\langle x,y\rangle$ necessary for it to be an inner product in the complex case read as follows:

1. $\langle x,y\rangle = \overline{\langle y,x\rangle}$ (Hermitian symmetry).
2. $\langle ax+by,z\rangle = a\langle x,z\rangle + b\langle y,z\rangle$ and $\langle x,ay+bz\rangle = \bar{a}\langle x,y\rangle + \bar{b}\langle x,z\rangle$ (Hermitian linearity).
3. $\langle x,x\rangle$ is real and $\langle x,x\rangle \geq 0$ with equality if and only if $x=0$.

For $\mathbb{C}^n$ the usual inner product is $\langle z,w\rangle = z_1\bar{w}_1 + \cdots + z_n\bar{w}_n$. It satisfies these conditions, as is easily verified. The associated norm is $||z|| = \langle z,z\rangle^{1/2}$, which is still a positive real number. Note that the homogeneity of the norm, $||az|| = \langle az,az\rangle^{1/2} = (a\bar{a}\langle z,z\rangle)^{1/2} = |a|\,||z||$, depends on the Hermitian form of the linearity since $|a| = (a\bar{a})^{1/2}$ but $|a| \neq (a^2)^{1/2}$ for complex a.

Now the Cauchy-Schwartz inequality $|\langle z,w\rangle| \leq ||z||\,||w||$ is still valid for complex inner products, but the proof requires one additional twist. We begin, as in the real case, by expanding $\langle z+w,z+w\rangle \geq 0$ and $\langle z-w,z-w\rangle \geq 0$ to obtain $\langle z,z\rangle + \langle z,w\rangle + \langle w,z\rangle + \langle w,w\rangle \geq 0$ and $\langle z,z\rangle - \langle z,w\rangle - \langle w,z\rangle + \langle w,w\rangle \geq 0$. Thus, $\langle z,w\rangle + \langle w,z\rangle$ is real and satisfies

$$|\langle z,w\rangle + \langle w,z\rangle| \leq \langle z,z\rangle + \langle w,w\rangle.$$

However, we do not have $\langle w,z\rangle = \langle z,w\rangle$ but rather $\langle w,z\rangle = \overline{\langle z,w\rangle}$ because of the Hermitian symmetry, so

$$\langle z,w\rangle + \langle w,z\rangle = 2\mathrm{Re}\langle z,w\rangle$$

($a+\bar{a} = 2\,\mathrm{Re}\,a$ for any complex number a). Thus we have $|\mathrm{Re}\langle z,w\rangle| \leq (||z||^2 + ||w||^2)/2$ or, $|\mathrm{Re}\langle z,w\rangle| \leq 1$ for $||z|| = ||w|| = 1$. Now for the twist: if we knew that $\langle z,w\rangle$ were real, then we would have $\mathrm{Re}\langle z,w\rangle = \langle z,w\rangle$ and, hence, $|\langle z,w\rangle| \leq 1$. The point is that we can always make

$\langle z, w\rangle$ real by multiplying z by the appropriate complex number of absolute value one *without changing* $||z||$. In other words, $||az|| = ||z||$ if $|a| = 1$ and if $\langle z, w\rangle = re^{i\theta}$ in polar coordinates, then the choice $a = e^{-i\theta}$ makes $\langle az, w\rangle = a\langle z, w\rangle = e^{-i\theta}re^{i\theta} = r$ real, so

$$\begin{aligned} |\langle z, w\rangle| &= |\langle az, w\rangle| = |\mathrm{Re}\langle az, w\rangle| \leq \frac{1}{2}(||az|| + ||w||) \\ &= \frac{1}{2}(||z|| + ||w||). \end{aligned}$$

Thus $|\langle z, w\rangle| \leq 1$ for $||z|| = ||w|| = 1$ and we can complete the proof in general by scaling as in the real case. The triangle inequality then follows by the same token:

$$\begin{aligned} ||z + w||^2 &= \langle z + w, z + w\rangle \\ &= \langle z, z\rangle + 2\,\mathrm{Re}\langle z, w\rangle + \langle w, w\rangle \\ &\leq ||z||^2 + 2||z||\,||w|| + ||w||^2 \\ &= (||z|| + ||w||)^2 \end{aligned}$$

where we use the Cauchy-Schwartz inequality and the trivial inequality $\mathrm{Re}\langle z, w\rangle \leq |\langle z, w\rangle|$.

Incidentally, the polarization identity also must be modified in the complex case. It reads:

$$\langle z, w\rangle = \frac{1}{4}(||z + w||^2 - ||z - w||^2 + i||z + iw||^2 - i||z - iw||^2).$$

We leave the verification as an exercise.

9.1.4 Exercises

1. Let $u^{(1)}, u^{(2)}, \ldots, u^{(n)}$ be any orthonormal basis ($u^{(j)} \cdot u^{(k)} = 0$ if $j \neq k$, $u^{(j)} \cdot u^{(j)} = 1$) in $\mathbb{R}^n$. Prove that $x = \sum_{j=1}^n a_j u^{(j)}$ where $a_j = x \cdot u^{(j)}$ and $|x| = (\sum_{j=1}^n a_j^2)^{1/2}$.

2. Verify that the sup norm on $C([a, b])$ is a norm.

3. Verify that $||f||_1 = \int_a^b |f(x)|dx$ on $C([a, b])$ is a norm.

4. Prove that $||x||_{\sup} = \lim_{p\to\infty} ||x||_p$ on $\mathbb{R}^n$.

5. Verify that $\langle f, g\rangle = \int_a^b f(x)g(x)\,dx$ is an inner product on $C([a,b])$. What is the associated norm and metric?

6. Verify the polarization identity in both the real and complex cases.

7. Verify the parallelogram law.

8. *Prove that if a norm $||x||$ on a real vector space satisfies the parallelogram law, then the polarization identity defines an inner product and that the norm associated with this inner product is the original norm.

9. Prove that if $||x||$ is any norm on $\mathbb{R}^n$, then there exists a positive constant M such that $||x|| \leq M|x|$ for all x in $\mathbb{R}^n$ where $|x|$ is the Euclidean norm. (**Hint**: $M = (\sum_{j=1}^n ||e^{(j)}||^2)^{1/2}$ will do.)

10. Prove that the norm $||x||_1$ on $\mathbb{R}^n$ for $n > 1$ is not associated with an inner product. (**Hint**: violate the parallelogram law.) Do the same for $||x||_{\text{sup}}$.

11. Prove that a real $n \times n$ matrix A satisfies $Ax \cdot Ay = x \cdot y$ for all x and y in $\mathbb{R}^n$ if and only if $|Ax| = |x|$ for all x in $\mathbb{R}^n$. Such matrices are called *orthogonal.*

12. Prove that a real $n \times n$ matrix is orthogonal if and only if its columns form an orthonormal basis for $\mathbb{R}^n$.

13. Prove that $Ax \cdot y = x \cdot A^{\text{t}}y$ where A^{t} denotes the transpose matrix—obtained from A by interchanging rows and columns.

14. Let A denote any $n \times m$ matrix, and define $||A|| = \sup\{|Ax| : |x| \leq 1\}$. Show that this is indeed a norm on the space of $n \times m$ matrices (regarded as an $(n \cdot m)$-dimensional vector space).

15. Verify that $d(x,y) = |x-y|/(1+|x-y|)$ defines a metric on $\mathbb{R}^n$, but this metric is not induced by any norm. (**Hint**: homogeneity fails.)

9.2 Topology of Metric Spaces

9.2.1 Open Sets

In this section we want to discuss the generalizations of the concepts introduced in Chapter 3, including open set, closed set, compact set, limits, and completeness. It turns out that these concepts can be described in terms of the most rudimentary structure on $\mathbb{R}^n$, the metric space structure. In order to make this clear we will develop those concepts for an abstract metric space. This doesn't really involve any more work, and it has many rewards, for there are other contexts in which these concepts are very useful.

Recall that a *metric space* is simply a set M on which we have defined a distance function $d(x, y)$, real-valued, for x and y in M, satisfying

1. $d(x, y) \geq 0$ with equality if and only if $x = y$ (positivity),
2. $d(x, y) = d(y, x)$ (symmetry), and
3. $d(x, z) \leq d(x, y) + d(y, z)$ for any x, y, z in M (triangle inequality).

We begin by describing some examples so that we can refer the abstract concepts to concrete situations. Our basic example is Euclidean space $\mathbb{R}^n$ with the Pythagorean metric

$$d(x, y) = \sqrt{(x_1 - y_1)^2 + \cdots + (x_n - y_n)^2}.$$

We have also seen that $\mathbb{R}^n$ has other metrics, such as those associated with the norms $||x||_{\text{sup}}$ and $||x||_1$. Another important example is the space $C([a, b])$ of real-valued continuous functions on a compact interval $[a, b]$ with the sup-norm metric $d(f, g) = \sup_x |f(x) - g(x)|$.

Whenever we have one metric space we can immediately get many more by the simple device of restricting to a subset. If M is a metric space with distance function $d(x, y)$, then any subset $M' \subseteq M$ becomes a metric space, called a *subspace*, with the same distance function $d(x, y)$ (now x and y are restricted to lie in M'). Clearly there are many interesting metric subspaces of $\mathbb{R}^n$ (warning: the term "subspace" here is being used in a completely different way than in linear algebra, where a vector subspace must have the vector space properties). We will find in particular that it is very useful to consider metric space concepts for subspaces of Euclidean space. Sometimes the restricted distance function for a subspace may not be the most natural one to consider. For example, the unit circle in the plane as a subspace of $\mathbb{R}^2$ has a chord-length distance function, as illustrated in Figure 9.2.1. It is not difficult to verify that arc-length distance makes the circle a metric space—a different but perhaps more natural metric space. We will return to this idea in a later chapter. For a completely different example of a metric, see exercise set 9.2.5, number 12.

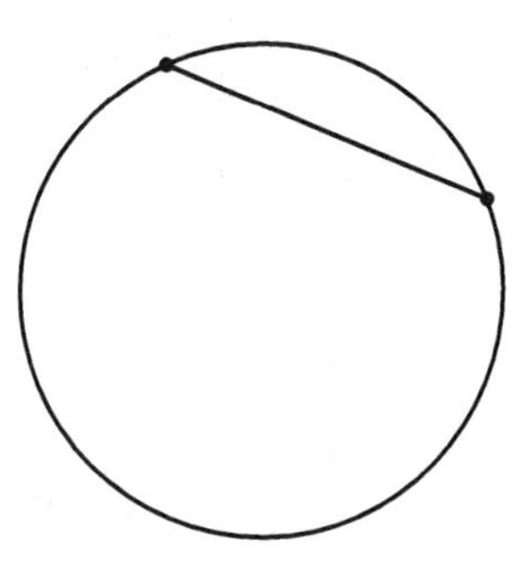

Figure 9.2.1:

Now that we have some examples, we begin the topological theory. The analog of open intervals in $\mathbb{R}$ will be played by the *open balls*

$$B_r(y) = \{x : d(x, y) < r\}$$

where the *radius* r is positive and the *center* y is an arbitrary point in the metric space. We follow the mathematical convention of using the word "ball" for the solid region and "sphere" for the boundary, $\{x : d(x, y) = r\}$. In the example of $\mathbb{R}^n$ for $n = 2$ or 3 these balls are what we normally think of as balls (or discs in $\mathbb{R}^2$) with radius r and center y. In a general metric space, however, we cannot expect the balls to have any "roundness". As a consequence of the triangle inequality, we can easily show that an open ball with center y also contains open balls centered at all its other points. To be precise, if x is a point in $B_r(y)$, then $d(x, y) = r_1 < r$ and so $B_{r-r_1}(x) \subseteq B_r(y)$ because if z is in $B_{r-r_1}(x)$, then $d(x, z) \leq r - r_1$, which implies

$$\begin{aligned} d(y, z) &\leq d(y, x) + d(x, z) \\ &\leq r_1 + r - r_1 = r \end{aligned}$$

by the triangle inequality; so z is in $B_r(y)$. (See Figure 9.2.2.) We will use this property to define a general notion of open set.

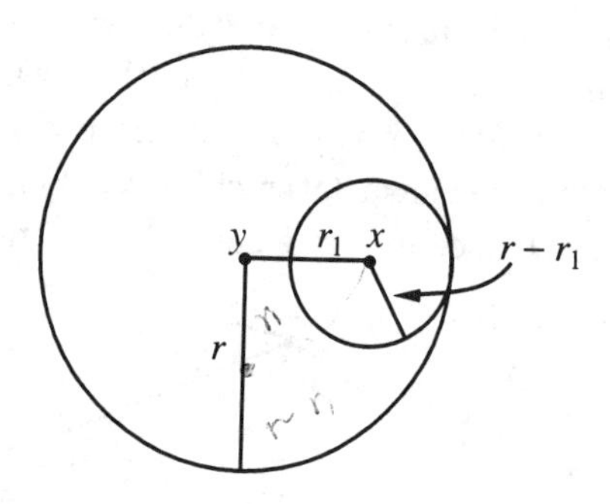

Figure 9.2.2:

Specifically, we define an *open set* in a metric space to be *a set A with the property that every point in A lies in an open ball contained in A*. This is equivalent to saying for every point y in A, $B_r(y) \subseteq A$ for some $r > 0$. Notice that it suffices to consider only balls with radius $1/m$ for some integer m, since $r > 1/m$ for some m by the axiom of Archimedes, so $B_{1/m}(y) \subseteq B_r(y)$. We define a *neighborhood* of a point to be an open set containing that point. For most applications we can restrict attention to the countable sequence $B_{1/m}(y)$ of neighborhoods of y. We define the *interior* of a set A to be the subset of all points in A contained in open balls contained in A. Thus A is open if and only if it equals its interior.

From the definition it is clear that a nonempty open set (the empty set is open by convention) is a union of open balls and conversely. However, the open sets in $\mathbb{R}^n$ are quite complicated—there is no nice structure theorem as there is in $\mathbb{R}^1$, where every open set is a disjoint union of open intervals. Even more distressing, if we consider subspaces of $\mathbb{R}^n$, the concept of open set changes. If M is a subspace of $\mathbb{R}^n$, then the open balls of M are the sets $\{x \text{ in } M : d(x,y) < r\}$ for center points y in M. The open subsets of M are thus unions of these kinds of balls and need not be open in $\mathbb{R}^n$. To make the distinction clear we sometimes refer to these sets as *open relative* to M or *relatively open* sets. The set M is always open relative to M but need not be open in $\mathbb{R}^n$. Of course if M is open in $\mathbb{R}^n$, then all open subsets of M are open in $\mathbb{R}^n$. But in general all we can say is that the open subsets of M are the intersections of M with open subsets of $\mathbb{R}^n$. None of this has to do specifically with $\mathbb{R}^n$, so we prove it in general.

Theorem 9.2.1 *Let M be a metric subspace of a metric space M_1. Then a subset A of M is open in M if and only if there exists an open subset A_1 of M_1 such that $A = A_1 \cap M$. If M is open in M_1, then for subsets A of M we have A is open in M if and only if A is open in M_1.*

Proof: If $A \subseteq M$ is open in M, this means for every y in A, $B_r(y) \subseteq A$ for small enough radius r. Here $B_r(y)$ refers to the metric space M and so is defined by

$$B_r(y) = \{x \text{ in } M : d(x, y) < r\}.$$

On the other hand we can also consider the ball of radius r centered at y in the larger metric space M_1,

$$B_r(y)_1 = \{x \text{ in } M_1 : d(x, y) < r\}.$$

Since the distance function is the same (this is the definition of metric subspace), we clearly have $B_r(y) = M \cap B_r(y)_1$. In other words, the open balls in M are exactly the intersections of M with open balls in M_1 whose centers happen to lie in M.

Now given an open set A in M, we construct the required set A_1. Simply take A_1 to be the union of all the balls $B_r(y)_1$ such that $B_r(y)$ lies in A. Since $B_r(y) \subseteq A \subseteq M$, we have $B_r(y)_1 \cap M = B_r(y)$; so $A_1 \cap M$ is the union of all the balls $B_r(y)$, which equals A from the defining property of A being open in M. Also A_1 is clearly open in M_1 because it is a union of open balls. Thus we have constructed an open subset A_1 of M_1, such that $A_1 \cap M = A$. Incidentally, we are not claiming that such a set A_1 is unique.

Conversely, if A_1 is an open subset of M_1, then $A = A_1 \cap M$ is an open subset of M. Indeed if y is in A, then $B_r(y)_1 \subseteq A_1$ for small enough radius r because A_1 is open and y is in A_1. Then

$$B_r(y) = M \cap B_r(y)_1 \subseteq M \cap A_1 = A,$$

so A is open in M.

Finally if M itself is open in M_1, then $B_r(y)_1 = B_r(y)$ for small enough r (depending on y) for each y in M. Thus the conditions that A be open in M and M_1 are the same: for all y in A, $B_r(y) = B_r(y)_1$ must lie in A for r small enough. QED

The basic closure properties of open sets on the line are true in general.

Theorem 9.2.2 *In any metric space, an arbitrary union of open sets is open and a finite intersection of open sets is open.*

Proof: These are immediate consequences of the definition. If $\mathcal{A}$ is any set of open subsets A of M, we want to show $\bigcup_{\mathcal{A}} A$ is open. Since the union consists of all points that lie in at least one of the set A, given any point x in the union there is a set A in $\mathcal{A}$ containing x. Since A is open, it contains $B_r(x)$ for small enough radius r; so $B_r(x)$ is contained in the union, proving the union is open.

Next consider a finite intersection $A = A_1 \cap A_2 \cap \cdots \cap A_m$, where each A_k is open. If x is in A, then x is in each A_k; hence, $B_{r_k}(x) \subset A_k$ for some positive r_k because A_k is open. Taking r to be the minimum of $r_1, \ldots, r_m$ (r is positive because there are only a finite number of r_k's) we have $B_r(x) \subseteq B_{r_k}(x) \subseteq A_k$ for every k; so $B_r(x) \subseteq A$, proving A is open. QED

In the more general theory of topological spaces, the closure properties of this theorem together with the trivial property that the empty set and the whole space are open are taken as axioms for the open sets. That is, a *topological space* is defined to be any set with a collection of open subsets satisfying those axioms. A metric space then becomes a special case of a topological space with the open sets given by the definition we have chosen. It is possible to introduce many of the concepts we are discussing in the still more abstract setting of general topological spaces, but we have chosen not to do this here because all the examples we need to deal with are in fact metric spaces. There are, however, topological spaces that are not metric spaces, and some of these are important in more advanced analysis.

9.2.2 Limits and Closed Sets

Next we discuss the concept of *limit* in a metric space. If $x_1, x_2, \ldots$ is a sequence of points in M, then we say the sequence has a *limit* x in M (or the sequence *converges* to x), written $x_n \to x$ or $\lim_{n\to\infty} x_n = x$, provided that for *any error* $1/m$ *there exists* N *such that* $n \geq N$ *implies* $d(x_n, x) \leq 1/m$. An equivalent way of saying this is that *every neighborhood of* x *contains all but a finite number of* x_n. This is the identical definition we gave for limits in $\mathbb{R}$; we have simply replaced the distance $|x_n - x|$ in $\mathbb{R}$ with the distance function $d(x_n, x)$ in M. Again we have the bull's eye picture of the neighborhoods $B_{1/m}(x)$, as shown in Figure 9.2.3, and the points in the sequence must eventually end up in each neighborhood.

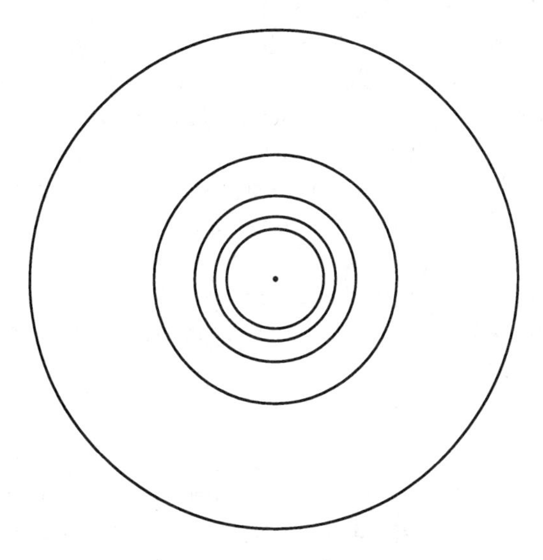

Figure 9.2.3:

In a similar way we define *limit point* of a sequence, this time requiring infinitely many x_n to be in each neighborhood of x. Just as in the real case, this is equivalent to a subsequence converging to x. A *limit point* of a set A is a point x such that every $B_{1/m}(x)$ contains a point of A different from x—hence, infinitely many points of A. A set A is said to be *closed* if it contains all its limit points. The *closure* of a set is the union of the set and all its limit points. The closure is always a closed set, and a set is closed if and only if it equals its closure. These are all straightforward to verify. We say a subset A of B is *dense* in B if the closure of A contains B.

Theorem 9.2.3 *A subset of a metric space is closed if and only if its complement is open.*

Proof: Let A be open and B be its complement. To show B is closed we need to show it contains its limit points. So let x be a limit point of B. To show x is in B we must show x is not in A since B is the complement of A. But if x were in A, then $B_r(x) \subseteq A$ for the radius r sufficiently small since A is open. But this contradicts the fact that x is a limit point of B, for every $B_r(x)$ must contain points of B. Therefore x is not in A, hence in B.

Now conversely assume B is closed and A is its complement. We want to show A is open. So given any point x in A, we want to show $B_r(x) \subseteq A$ for some r. If not, then $B_r(x)$ would contain points of B (not equal to x) for all r, hence x would be a limit point of B. Since B is assumed closed, it would have to contain x, contradicting the fact that x is in A. Thus $B_r(x) \subseteq A$ for some r, proving A is open. QED

From this we can conclude that finite unions and arbitrary intersections of closed sets are closed. We could also prove this directly from the definition.

9.2.3 Completeness

One very important fact that we learned about convergence in $\mathbb{R}$ is that a sequence converges if and only if it satisfies the Cauchy criterion. In an arbitrary metric space we can define a *Cauchy sequence* as follows: for every error $1/m$ there exists N such that $j, k \geq N$ implies $d(x_j, x_k) \leq 1/m$. It is easy to see that a convergent sequence is always a Cauchy sequence because once $d(x_k, x) \leq 1/2m$ for all $k \geq N$ it follows that $d(x_j, x_k) \leq 1/m$ for $j, k \geq N$ by the triangle inequality. However, the converse is *not* true for the general metric space. For example it is not true for the rational numbers, which is a metric subspace of $\mathbb{R}$. A metric space is called *complete* if every Cauchy sequence is convergent. We will show that $\mathbb{R}^n$ is a complete metric space. This will be easy once we show that convergence in $\mathbb{R}^n$ is equivalent to convergence in each coordinate. Here we follow the notational conventions of the previous section.

Theorem 9.2.4 *Let $x^{(1)}, x^{(2)}, \dots$ be a sequence of points in $\mathbb{R}^n$. Then $x^{(1)}, x^{(2)}, \dots$ converges to a limit x if and only if the sequence of real numbers $x_k^{(1)}, x_k^{(2)}, \dots$ converges to x_k for $k = 1, 2, \dots, n$.*

Proof: The idea of the proof is that we can fit a circle in a square and a square in a circle (or more precisely, the n-dimensional analog). We essentially proved this result for $n = 2$ when we showed convergence in $\mathbb{C}$ was equivalent to convergence of the real and imaginary parts.

Suppose first $\lim_{j\to\infty} x^{(j)} = x$ exists. Given any error $1/m$ there exists N such that $j \geq N$ implies $d(x^{(j)}, x) \leq 1/m$. Since $|a_k| \leq (\sum_{p=1}^n |a_p|^2)^{1/2}$, we have $|x_k^{(j)} - x_k| \leq d(x^{(j)}, x)$, so $|x_k^{(j)} - x_k| \leq 1/m$ for $j \geq N$ and so $\lim_{j\to\infty} x_k^{(j)} = x_k$. This is the circle in the square part.

Conversely, suppose $\lim_{j\to\infty} x_k^{(j)} = x_k$ for each k. Then given any error $1/m$ we can find N_k such that $j \geq N_k$ implies

$$\left| x_k^{(j)} - x_k \right| \leq \frac{1}{\sqrt{n}\, m}.$$

By taking N to be the largest of the N_k we have $j \geq N$ implies $|x_k^{(j)} - x_k| \leq 1/\sqrt{n}m$ for all k. Then—square in the circle—

$$d(x^{(j)}, x) \leq \left(\sum_{k=1}^n \left(\frac{1}{\sqrt{n}\, m} \right)^2 \right)^{1/2} = \frac{1}{m}$$

for $j \geq N$, so $\lim_{j\to\infty} x^{(j)} = x$. QED

Corollary 9.2.1 *$\mathbb{R}^n$ is a complete metric space.*

Proof: Let $x^{(1)}, x^{(2)}, \dots$ be a Cauchy sequence. Then by the theorem, $x_k^{(1)}, x_k^{(2)}, \dots$ is a Cauchy sequence of real numbers for each k. By the completeness of the reals each of these sequences has a limit, say $\lim_{j\to\infty} x_k^{(j)} = x_k$. Then $\lim_{j\to\infty} x^{(j)} = x = (x_1, \dots, x_n)$, also by the theorem. QED

Another interesting example of a complete metric space is $C([a, b])$.

Theorem 9.2.5 *$C([a,b])$ with the sup-norm metric is a complete metric space.*

Proof: This theorem is really just a reformulation of the Cauchy criterion for uniform convergence (Theorem 7.3.1). In fact, we claim that convergence in the sup-norm metric is the same as uniform convergence.

Indeed convergence in the sup-norm metric means given any error $1/m$ there exists N such that $k \geq N$ implies $\sup_x |f_k(x) - f(x)| \leq 1/m$, which is the same as saying there exists N independent of x such that $k \geq N$ implies $|f_k(x) - f(x)| \leq 1/m$ for every x, which is uniform convergence.

The same reasoning shows that the Cauchy criterion for a sequence $\{f_n\}$ in the sup-norm metric is identical to the uniform Cauchy criterion: given any error $1/m$ there exists N such that $|f_j(x) - f_k(x)| \leq 1/m$ for all x provided $j, k \geq N$. Thus any Cauchy sequence in the sup-norm metric converges uniformly to a function f that is continuous (being the uniform limit of continuous functions), which is the same as saying $f_n \to f$ in the sup-norm metric. QED

It is important to realize that this result concerns just the sup-norm metric on the space $C([a,b])$. There are other metrics, such as the L^1 metric $d(f,g) = \int_a^b |f(x) - g(x)|dx$ on the same set $C([a,b])$ for which completeness fails. In fact the sequence of continuous functions converging pointwise to a discontinuous function gives an example of a Cauchy sequence in this metric with no limit in $C([a,b])$. See the exercises for details. The L^1 metric is associated to a norm and will be discussed in detail in Chapter 14. In a finite-dimensional vector space every metric associated to a norm is complete, so it is the infinite-dimensional nature of the space $C([a,b])$ that is crucial in this example. The kind of incompleteness here is different in nature from the incompleteness of the rationals in that there are no "holes" in between points of the space.

Any metric space that is not complete can be *completed* by a process completely analogous to the procedure whereby we constructed the real numbers from the rationals. The *completion* $\overline{M}$ of M is defined to be the set of equivalence classes of Cauchy sequences of points in M—where equivalence is defined in exactly the same way as with numerical sequences. We regard M as a subset of $\overline{M}$ by identifying the point x in M with the equivalence class of the sequence $(x, x, \ldots)$—exactly the way we regarded the rationals as a subset of the reals. We can make $\overline{M}$ into a metric space by defining the distance between the equivalence class of $x_1, x_2, \ldots$ and the equivalence class of $y_1, y_2, \ldots$ to be $\lim_{n\to\infty} d(x_n, y_n)$. This definition requires that we verify: 1) the limit exists, 2) the limit is independent of the choice of sequences from the equivalence classes, and 3) the distance so defined satisfies the axioms for a metric. All these verifications are routine. A general metric space M has no further structure, so we can't say any more about $\overline{M}$. But if M happens to be a vector space and the metric is associated to a norm or an inner product on M, then we can extend the norm or inner product structure to $\overline{M}$. Finally we can prove that $\overline{M}$ is complete, justifying the terminology. Again the proof of the completeness of $\overline{M}$ is completely analogous to the proof of the completeness of $\mathbb{R}$. A complete normed vector space is called a *Banach space*, and a complete inner product space is called a *Hilbert space*. The study of these structures forms an important part of twentieth century analysis, but it is beyond the scope of this work.

It should be pointed out that the abstract construction of the completion $\overline{M}$ sketched above is not always very satisfactory. Frequently we want a more concrete description of $\overline{M}$. For example, in the L^1 metric $d(f,g) = \int_a^b |f(x) - g(x)| dx$ on $C([a,b])$, the completion is essentially the space of Lebesgue integrable functions. However, the construction of Lebesgue integration theory is much more difficult than the abstract completion—indeed the abstract completion does not enable you to identify the elements of the completion space as certain functions on $[a,b]$, while the Lebesgue theory does. For this reason we will not discuss completions in more detail.

9.2.4 Compactness

The idea of compactness is important in the general setting of metric spaces and $\mathbb{R}^n$ in particular. The definition of a *compact* subset A of a metric space is that every sequence $a_1, a_2, \ldots$ of points in A has a limit point in A (or, equivalently, has a subsequence that converges to a point in A). Notice that the definition refers only to the points in A and to the distance function between points of A, but it does not refer to points not in A. Thus it is the same thing to say A is a compact subset of M or A is a compact subset of N if N is any subspace of M containing A (remember that "subspace" implies the metric is the same). Thus compactness is an absolute concept, unlike "open" and "closed", which depend on the whole space. In particular we say that a metric space M is *compact* if M is a compact subset of itself or, in other words, if all sequences of points in M have limit points in M. Then A is a compact subset of M if and only if A as a subspace is a compact metric space.

Clearly there is a close connection between the concepts of compactness and completeness, since both refer to the existence of limits of sequences. In fact it is easy to see that *compactness implies completeness, but completeness does not imply compactness.* Indeed suppose M is compact. To show that M is complete, consider any Cauchy sequence $x_1, x_2, \ldots$. We need to show it has a limit in M. By the compactness it has a limit point in M, and it is easy to show that the Cauchy criterion then implies that the limit point is actually a limit (see exercises). On the other hand, $\mathbb{R}$ is already an example of a complete metric space that is not compact.

As in the case of subsets of $\mathbb{R}$, compactness is also equivalent to other conditions. The most important is the *Heine-Borel property: every open covering has a finite subcovering.* The definitions are the same as in the case of subsets of $\mathbb{R}$. If A is a subset of M, we say $\mathcal{B}$, a collection of subsets B of M, is a covering of A if $\bigcup_{\mathcal{B}} B \supseteq A$ and an *open covering* if all the sets B are open sets in M. A *subcovering* simply means a subcollection $\mathcal{B}'$ of $\mathcal{B}$ that still covers A.

Notice now that there are two ways in which we can interpret the Heine-Borel property—in the metric space M and in the metric subspace A. They are somewhat different but turn out to be equivalent. The reason is that "open" means different things for subsets of A and subsets of M. If we were to consider the Heine-Borel property with respect to A we would have to consider coverings of A by open *subsets of* A. If B is an open subset of A, it is not in general an open subset of M. Thus a covering of A by open subsets of A is not a special case of a covering of A by open subsets of M. Nevertheless, because of the relationship between open subsets of A and M, we know that if $\mathcal{B}$ is a covering by open subsets of A, then by extending each B in $\mathcal{B}$ to $\tilde{B}$ an open subset of M (with $\tilde{B} \cap A = B$), we obtain covering $\tilde{\mathcal{B}}$ by open subsets of M. Conversely we can intersect open subsets of M with the set A to obtain open subsets of A, and by doing this to each set in a covering of A by open subsets of M we can obtain a covering of A by open subsets of A. In this way we can go back and forth between the two Heine-Borel properties and show they are equivalent (see exercises).

Theorem 9.2.6 (*Abstract Heine-Borel Theorem*) *A metric space (or a subset of a metric space) is compact if and only if it has the Heine-Borel property.*

Proof: Many of the ideas of the proof have already been given in the proof of the analogous theorem for sets of real numbers. We start by proving that the Heine-Borel property implies compactness. Thus we need to show that an arbitrary sequence $x_1, x_2, \ldots$ of points in A has a limit point in A. We may assume without loss of generality that all the points $x_1, x_2, \ldots$ are distinct. We want to show that not having a limit point would contradict the Heine-Borel property. To do this we construct a cover of A by the sets $B_1, B_2, \ldots$ where B_1 is A with the sequence $x_1, x_2, \ldots$ removed, B_2 is A with $x_2, x_3, \ldots$ removed, and in general B_k is A with $x_k, x_{k+1}, \ldots$ removed. Clearly the sets $\{B_k\}$ cover A, and no finite subcovering exists. But are the sets B_k open in A? The answer is yes because we are assuming the sequence $x_1, x_2, \ldots$ (and hence $x_k, x_{k+1} \ldots$) has no limit points. Since the sequence $x_k, x_{k+1}, \ldots$ has no limit points, it is a closed subset of A (the definition of closed is vacuously satisfied), so its complement B_k is open. Thus the Heine-Borel property is not satisfied, proving half the theorem by contradiction.

Conversely, suppose A is compact. We want to show that the Heine-Borel property holds. Let us first show that every countable open cover has a finite subcover—and afterward we will reduce the general case to this one. Let the sets in the cover be $B_1, B_2, \ldots$. If there were no finite subcover, then each of the sets $B_1, B_1 \cup B_2, \ldots, B_1 \cup B_2 \cup \cdots \cup B_n, \ldots$ would fail to cover A. Thus there would be points $a_1, a_2, \ldots$ in A with a_n not in $B_1 \cup \cdots \cup B_n$. By the compactness of A there must be a limit point a of the sequence $a_1, a_2, \ldots$ in A. For any fixed n, the tail of the sequence $a_n, a_{n+1}, a_{n+2}, \ldots$ consists of points not in B_n; and since the complement of the open set B_n is closed, it must contain its limit point a (of course a separate—more trivial—argument must be given if infinitely many of the a_n equal a). Thus a is not in any B_n, contradicting the fact that $B_1, B_2, \ldots$ was supposed to be a cover of A.

To reduce the case of an arbitrary open cover to a countable one we need the fact that a compact metric space has a countable dense set (without the assumption of compactness this is not necessarily true of an arbitrary metric space—although it is true of an arbitrary subspace of the reals, or of $\mathbb{R}^n$). We will prove this as a separate lemma following this proof. Assuming it is true, let us denote by $x_1, x_2, \ldots$ this countable dense set and consider the countable collection of balls $B_{1/m}(x_n)$ of radius $1/m$ about x_n, where m and n vary over all positive integers. The idea is to reduce an arbitrary cover to a cover that is in one-to-one correspondence with a subset of this countable collection of balls—hence will be a countable (or finite) subcover. To do this we start with the original cover $\mathcal{B}$ and choose one set B containing $B_{1/m}(x_n)$, if any such sets exist. This requires the countable axiom of choice, one arbitrary selection from each of the sets $\{B \text{ in } \mathcal{B} \text{ containing } B_{1/m}(x_n)\}$. This gives us a countable (or finite) subcollection $\mathcal{B}'$ of $\mathcal{B}$. Why does it cover A? Let a be any point in A. Since $\mathcal{B}$ covers A, there must be a set B in $\mathcal{B}$ containing a. Since B is open, it must contain a neighborhood of a. Now since $x_1, x_2, \ldots$ is dense in A, we can find points in the sequence arbitrarily close to a, and it follows easily that one of the balls $B_{1/m}(x_n)$ contains a and is contained in B (if say $B_{2/m}(a) \subseteq B$, then by choosing $d(x_n, a) < 1/m$ we have a in $B_{1/m}(x_n)$ and $B_{1/m}(x_n) \subseteq B_{2/m}(a) \subseteq B$), as shown in Figure 9.2.4.

Since $B_{1/m}(x_n) \subseteq B$ for some B in $\mathcal{B}$, by the definition of $\mathcal{B}'$ there is a set in $\mathcal{B}'$ also containing $B_{1/m}(x_n)$ and, hence, a since a is in $B_{1/m}(x_n)$. Thus $\mathcal{B}'$ covers A. To complete the proof of the theorem we need only prove the following lemma.

Lemma 9.2.1 *Any compact metric space has a countable dense subset.*

Proof: Choose any point for x_1. For x_2 we choose any point not too close to x_1. We let R be the sup of $d(x_1, x)$ as x varies over the space and require that $d(x_1, x_2) \geq R/2$. The fact that R is finite follows from the compactness, since otherwise there would be a sequence $y_1, y_2, \ldots$ with $\lim_{n\to\infty} d(x_1, y_n) = +\infty$, and no subsequence of $y_1, y_n, \ldots$ could converge—for $y'_j \to y$ would imply

$$d(x_1, y'_j) \leq d(x_1, y) + d(y, y'_j)$$

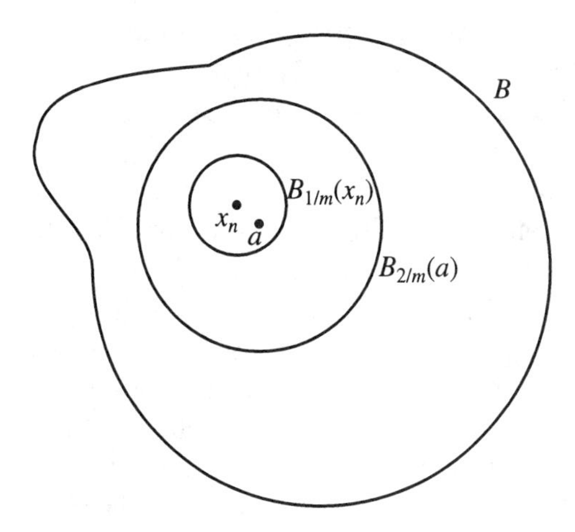

Figure 9.2.4:

remains bounded since $d(y, y'_j) \to 0$ and $d(x_1, y)$ is a fixed constant.

Having chosen x_1 and x_2, we choose x_3 so that both $d(x_1, x_3)$ and $d(x_2, x_3)$ exceed $R_2/2$, where R_2 is the sup of the minimum of $d(x_1, x)$ and $d(x_2, x)$ as x varies over the space. In this way x_3 is not too close to either x_1 or x_2. Proceeding inductively, having chosen $x_1, \ldots, x_n$ we let R_n be the sup of the minimum of $d(x_1, x), \ldots, d(x_n, x)$ as x varies over the space. As long as there are an infinite number of points in the space (if there were only a finite number there would be nothing to prove) there are always points x not equal to any of $x_1, \ldots, x_n$, so R_n is not zero. And R_n is finite since it is always $\leq R$. We can think of R_n as measuring the furthest one can possibly get away from all the points $x_1, \ldots, x_n$ in the space. We then choose for x_{n+1} any point such that $d(x_k, x_{n+1}) \geq R_n/2$ for all $k = 1, 2, \ldots, n$. This construction requires the countable axiom of choice.

So far we have merely obtained a sequence $x_1, x_2, \ldots$ that is "spread out" over the space. Our argument did not use the compactness of the space in any essential way (only boundedness). Now we are going to use the compactness to show that the set $x_1, x_2, \ldots$ is in fact dense. First we need to show that $\lim_{n\to\infty} R_n = 0$. Note that the choice of x_{n+1} required that $d(x_{n+1}, x_k) \geq R_n/2$ for $k \leq n$; so if we did not have $\lim_{n\to\infty} R_n = 0$, then there would exist $\varepsilon > 0$ such that $d(x_j, x_k) \geq \varepsilon$ for all j and k (just take $\varepsilon = 1/2 \lim_{n\to\infty} R_n$, the limit existing because R_n

is decreasing). But then no subsequence could converge, contradicting the compactness. Thus $\lim_{n\to\infty} R_n = 0$.

To show that $x_1, x_2, \ldots$ is dense, choose any point x in the space. Given any $1/m$ we need to find x_n with $d(x, x_n) \leq 1/m$. Choose n large enough so that $R_n < 1/m$. Then $d(x, x_k) \leq 1/m$ for some $k \leq n$ or else by the definition of R_n we would have $R_n \geq 1/m$. QED

In the course of the proof we have established several properties of compact metric spaces that are of interest on their own. The first is *boundedness*. We can express this in two different but equivalent ways: 1) there exists a point x in the space such that $d(x, y) < R$ for every y in the space, or $B_R(x)$ is the whole space. The inf of such R defines the *radius* of the space with respect to x. The triangle inequality then implies that the radius is finite with respect to every point in the space, $\leq 2R$ in fact. 2) Let $D = \sup_{x,y} d(x, y)$ be the *diameter* of the space. The diameter is finite if and only if the radius is finite, with $R \leq D \leq 2R$ no matter which point we use to compute the radius. These are immediate consequences of the triangle inequality.

In addition to boundedness, compact metric spaces have the property that given any $1/m$ there exists a finite set of points $x_1, \ldots, x_n$ such that every point x is within a distance $1/m$ of one of them. This was established in the course of the proof of the lemma. Notice that it says $B_{1/m}(x), \ldots, B_{1/m}(x_n)$ covers the space, so it is also an immediate consequence of the Heine-Borel property.

We have seen that a compact metric space must be bounded and complete. Here the completeness is playing the role of "closed" for subsets of the reals. In the abstract setting, saying that a set is closed may have no significance—if the whole space is being considered, then it is always closed. The connection between "complete" and "closed subset of the reals" is explained by the following simple fact, whose proof we leave as an exercise: *a subspace A of a complete metric space M is itself complete if and only if it is a closed set in M.* Since the reals are complete, the closed subsets (as metric subspaces) are the same as the complete subspaces.

It is natural to pose the question: if a metric space is complete and bounded, is it necessarily compact? For subsets of the reals we have seen this is true, but in general it is false. Before giving a counterexample, let us first show that the analogous statement in Euclidean space

$\mathbb{R}^n$ is true.

Theorem 9.2.7 *A subspace of $\mathbb{R}^n$ is compact if and only if it is closed and bounded.*

Proof: Since we have already seen that a compact set is closed and bounded, it remains to show that if A is a closed and bounded subset of $\mathbb{R}^n$, then it is compact. Thus let $x^{(1)}, x^{(2)}, \ldots$ be any sequence of points in A. Since A is bounded, the sequence of coordinates $x_k^{(1)}, x_k^{(2)}, \ldots$ must also be bounded, for each fixed k, $1 \leq k \leq n$. Thus each of these sequences of reals has a limit point and, hence, a convergent subsequence. By taking subsequences of subsequences we can arrange to find a single subsequence, call it $y^{(1)}, y^{(2)}, \ldots$ such that $y_k^{(1)}, y_k^{(2)}, \ldots$ converges, for each $k, 1 \leq k \leq n$. Since convergence in each coordinate implies convergence in $\mathbb{R}^n$, the subsequence $y^{(1)}, y^{(2)}, \ldots$ converges to some limit y in $\mathbb{R}^n$. Since A is closed, the limit must be in A. This shows A is compact. QED

The complete metric space $C([a,b])$ with sup-norm metric

$$d(f,y) = \sup_x |f(x) - g(x)|$$

provides us with examples of subspaces that are closed (hence complete) and bounded but not compact. To see this, let $f_1, f_2, \ldots$ be a sequence of continuous function converging pointwise to a discontinuous function; for example, on $[-1,1]$, we could take f_n as shown in Figure 9.2.5. Let A be the set $\{f_1, f_2, \ldots\}$ in $C([-1,1])$. A is bounded because $d(f_n, 0) \leq 1$. A is closed because it has no limit points—any sequence from A converges pointwise to a discontinuous function and so cannot converge uniformly (remember uniform convergence is the metric convergence) to a continuous function. But A is not compact, since as we have just seen $f_1, f_2, \ldots$ is a sequence from A with no convergent subsequence.

We can now give an interpretation of the Arzela-Ascoli theorem in terms of compactness in $C([a,b])$. Recall that the Arzela-Ascoli theorem says that a sequence of functions in $C([a,b])$ that is both uniformly bounded and uniformly equicontinuous has a uniformly convergent subsequence. Now suppose A is any subspace of $C([a,b])$ that is 1) closed,

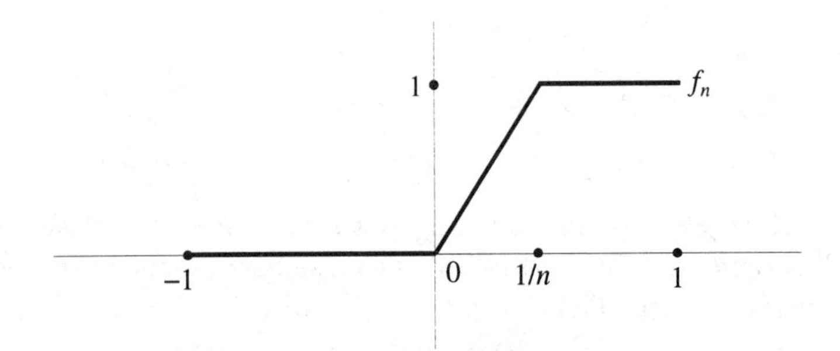

Figure 9.2.5:

2) bounded, and 3) uniformly equicontinuous (for every $1/m$ there exists $1/n$ such that for every f in A and x and y in $[a, b]$ with $|x - y| < 1/n$, we have $|f(x) - f(y)| < 1/m$). Given any sequence of functions in A, 2) and 3) imply by the Arzela-Ascoli theorem that there exists a uniformly convergent subsequence, and the limit is in A by 1). Thus A is compact. Conversely, if A is a compact subspace of $C([a, b])$, then it must satisfy 1), 2), and 3). Indeed, we have already seen that a compact set is closed and bounded; the uniform equicontinuity is a consequence of the Heine-Borel property. We leave the details as an exercise. Thus we have a complete characterization of the compact subspaces of $C([a, b])$.

9.2.5 Exercises

1. If $x_n \to x$ in a metric space and y is any other point in the space, show $\lim_{n\to\infty} d(x_n, y) = d(x, y)$.

2. If $x_n \to x$ and $y_n \to y$ in a metric space, show $\lim_{n\to\infty} d(x_n, y_n) = d(x, y)$.

3. Prove that the metric $d(f, g) = \int_a^b |f(x) - g(x)|dx$ on $C([a, b])$ is not complete. (**Hint**: consider the example of a sequence of continuous functions converging pointwise to a discontinuous function.)

4. Prove that the space of bounded sequences with metric $d(\{x_n\}, \{y_n\}) = \sup_n |x_n - y_n|$ is complete, and the same is true on the subspace of sequences converging to zero.

5. Prove that if a Cauchy sequence in a metric space (not assumed to be complete) has a limit point, then it has a limit.

6. Prove directly that if A is a subspace of M, then the Heine-Borel property for A as a subspace of M (open meaning open in M) is equivalent to the Heine-Borel property for A as a subspace of A.

7. Prove that a metric space is compact if and only if it is bounded, complete, and given any $1/m$ there exists a finite subset $x_1, \ldots, x_n$ such that every point x in the space is within $1/m$ of one of them ($d(x, x_k) \leq 1/m$ for some k, $1 \leq k \leq n$).

8. Prove that a subspace of a complete metric space is complete if and only if it is closed.

9. Construct a sequence of functions $f_1, f_2, \ldots$ in $C([0,1])$ with sup-norm metric such that $d(f_k, 0) = 1$ and $d(f_j, f_k) = 1$ for any j and k.

10. Show that any compact subspace of $C([a,b])$ is uniformly equicontinuous. (**Hint**: for each $1/m$ consider the covering by $B_n = \{f \text{ in } C([a,b]) : |x - y| \leq 1/n \text{ implies } |f(x) - f(y)| < 1/m\}$.)

11. Prove that if $A_1 \supseteq A_2 \supseteq A_3 \supseteq \cdots$ is a nested sequence of non-empty compact subsets of a metric space, then $\bigcap_{n=1}^{\infty} A_n$ is non-empty.

12. *a. Let $\mathbb{Z}$ denote the integers, and let p be any fixed prime. Every integer z can be written uniquely base p as $\pm a_N a_{N-1} \cdots a_1 a_0$ where $0 \leq a_j \leq p - 1$ and $z = \pm \sum_{j=0}^{N} a_j p^j$. Let $|z|_p = p^{-k}$, where k is the *smallest* integer for which $a_k \neq 0$. Prove that $d(x,y) = |x - y|_p$ is a metric. (It is called the *p-adic metric.*)

    b. Show that the p-adic metric satisfies

    $$d(x,z) \leq \max(d(x,y), d(y,z)).$$

    (Metrics with this property are called *ultra metrics.*)

    c. Show that the Euclidean metric on $\mathbb{R}^3$ is not an ultra metric.

    d. Show that the completion of the integers with the p-adic metric can be realized concretely by infinite base p integers

$\pm \cdots a_n a_{n-1} \cdots a_1$ where $0 \leq a_n < p-1$ and that the ordinary rules of addition and subtraction make the completion a group (called the p-adic integers).

13. Give explicitly a countable dense subset of $\mathbb{R}^n$.

14. Call two metrics d_1 and d_2 on the same set M *equivalent* if there exist positive constants c_1, c_2 such that $d_1(x, y) \leq c_2 d_2(x, y)$ and $d_2(x, y) \leq c_1 d_1(x, y)$ for all x and y in M. Prove that $x_n \to x$ in d_1-metric if and only if $x_n \to x$ in d_2-metric, if d_1 and d_2 are equivalent.

15. Prove that equivalent metrics have the same open sets. Give an example of two metrics on $\mathbb{R}$ that have the same open sets but are not equivalent.

16. Prove that the metrics on $\mathbb{R}^n$ associated to the norms $|x|_{\text{sup}}$ and $|x|_1$ are equivalent to the usual metric.

17. Which subsets of $\mathbb{R}^n$ are both open and closed?

18. Prove that the intersection of an open subset of $\mathbb{R}^2$ with the x-axis is an open subset of the line.

19. Let S denote the circle $x^2 + y^2 = 1$ in $\mathbb{R}^2$. For points on S define the distance to be the length of the shortest arc of the circle joining them. Prove this is a metric. Is it the same metric as that of S as a subspace of $\mathbb{R}^2$? Can you describe the distance function in terms of the angular parameter θ in the representation $(\cos\theta, \sin\theta)$ of points on S?

9.3 Continuous Functions on Metric Spaces

9.3.1 Three Equivalent Definitions

In this section we are going to discuss functions whose domains and ranges are metric spaces. Although most of the examples we deal with will concern only subspaces of Euclidean space, it is instructive to see the theory in the more abstract setting. The proofs are certainly no more difficult, and there are many applications in which other metric spaces are involved. Since we have had the experience of studying functions whose domains and ranges are subsets of the reals, we will find many of the concepts and proofs familiar. On the other hand there will also be a few new ideas.

We introduce the notation $f : M \to N$ to mean f is a function whose domain is M and whose range is N, where both M and N are metric spaces. The *image* $f(M)$ is the set of all values actually assumed by $f, f(M) = \{y$ in N: there exists x in M with $f(x) = y\}$. It is a subset of N, not assumed equal to all of N. If it is all of N we say f is *onto*. We are interested primarily in continuous functions, and as in the case of numerical functions there are several equivalent definitions. We list three important definitions, all familiar:

1. For every $1/m$ and every x_0 in M there exists $1/n$ such that $d(x, x_0) \leq 1/n$ implies $d(f(x), f(x_0)) \leq 1/m$.

2. For all sequences in M, if $x_1, x_2, \ldots$ converges in M, then $f(x_1)$, $f(x_2), \ldots$ converges in N.

3. The inverse image $f^{-1}(B)$ of any open set B in N is an open set in M.

Note that in definition 1 the distance function $d(x, x_0)$ refers to the metric on M, while $d(f(x), f(x_0))$ refers to the metric on N. We will not burden the notation with this distinction. Of course definition 1 is the continuity at the point x_0 for each point x_0 in the domain. We occasionally will need the notion of continuity at a point—the condition in definition 1 for an individual x_0. In definition 2 we do not need to add the condition $f(\lim_{n\to\infty} x_n) = \lim_{n\to\infty} f(x_n)$, but this will be an immediate consequence of shuffling $x = \lim_{n\to\infty} x_n$ into the sequence. We can paraphrase definition 2 as saying f preserves limits.

In the case of numerical functions we only discussed definition 3—inverse images of open sets are open—for functions whose domains were open subsets of $\mathbb{R}$. The reason for this is that we had not yet discussed the concept of open subset of M where M is a subspace of $\mathbb{R}$. Of course when M is an open subset of $\mathbb{R}$ the open subsets of M are the subsets of M that are open in $\mathbb{R}$, so there is no difficulty. When M is not open in $\mathbb{R}$, the meaning of "$f^{-1}(B)$ is open in M" is exactly what it says, *open in M, not necessarily open in* $\mathbb{R}$. Thus the general viewpoint enables us to improve our understanding even in the concrete setting of subsets of $\mathbb{R}$.

On the other hand, it is immaterial whether we take the range N as given, or reduce it to the image $f(M)$, or enlarge it to some space containing N, as long as we keep the same metric on the image. The reason for this is that $f^{-1}(B)$ is the same as $f^{-1}(B \cap f(M))$, because only points of B in the image of f contribute to the inverse image. Thus as B varies over the open subsets of N, $B \cap f(M)$ varies over the open subsets of the image $f(M)$—and the same inverse images occur.

We can take any one of the three conditions as the definition of f *is continuous on* M, since we will now show they are equivalent.

Theorem 9.3.1 *For a function $f : M \to N$, the three definitions above are equivalent.*

Proof: First we show the equivalence of definitions 1 and 2. Suppose definition 1 holds, and let $x_k \to x_0$ in M. We want to show $f(x_k) \to f(x_0)$ in N. Given any error $1/m$ we use definition 1 at x_0 to find $1/n$ such that $d(x, x_0) \leq 1/n$ implies $d(f(x), f(x_0)) \leq 1/m$. Then from $x_k \to x_0$ we know there exists j such that $k \geq j$ implies $d(x_k, x_0) \leq 1/n$. Thus $d(f(x_k), f(x_0)) \leq 1/m$ for $k \geq j$, which proves $f(x_k) \to f(x_0)$ and, hence, definition 2.

Conversely, assume definition 2 holds. By shuffling $x_0 = \lim_{k\to\infty} x_k$ into the original sequence $x_0, x_1, x_0, x_2, x_0, x_3, \ldots$ we still have a convergent sequence, and hence by definition 2 $f(x_0), f(x_1), f(x_0), f(x_2), \cdots$ is also convergent, which can only happen if $\lim_{k\to\infty} f(x_k) = f(x_0)$. Thus definition 2 implies the stronger statement "$x_k \to x_0$ implies $f(x_k) \to f(x_0)$". Now let's establish definition 1 at the point x_0. Suppose it were false. Then there would exist $1/m$ such that for every $1/n$ there exists x_n such that $|x_n - x_0| \leq 1/n$ but $|f(x_n) - f(x_0)| > 1/m$. The sequence $x_1, x_2, \ldots$ clearly violates definition 2, since $x_n \to x_0$ but $f(x_n)$ does not converge to $f(x_0)$.

Next we show the equivalence of definitions 1 and 3. Assume first definition 1 holds, and let B be any open set in N. We have to show $f^{-1}(B)$ is open. So let x_0 be in $f^{-1}(B)$; this simply means $f(x_0)$ is in B. We need to find a ball $B_r(x_0)$ contained in $f^{-1}(B)$. Since B is open, it contains a ball $B_{1/m}(f(x_0))$ about $f(x_0)$, and by definition 1 there exists $1/n$ (for that $1/m$ and x_0) such that $|x - x_0| < 1/n$ (x in

$B_{1/n}(x_0)$) implies $|f(x) - f(x_0)| < 1/m$ ($f(x)$ in $B_{1/m}(f(x_0))$). Thus the ball $B_{1/n}(x_0)$ lies in $f^{-1}(B)$.

Conversely, assume definition 3 holds. Given the point x_0 and $1/m$, we look at the inverse image of the ball $B_{1/m}(f(x_0))$. Since $B_{1/m}(f(x_0))$ is open in N, $f^{-1}(B_{1/m}(f(x_0)))$ is open in M by definition 3. Now x_0 belongs to this inverse image since $f(x_0)$ is in $B_{1/m}(f(x_0))$; so by the definition of open set in M there is a ball $B_{1/n}(x_0)$ contained in the inverse image $f^{-1}(B_{1/m}(f(x_0)))$, or x in $B_{1/n}(x_0)$ implies $f(x)$ is in $B_{1/m}(f(x_0))$. This is just another way of writing definition 1. QED

We note some simple properties of continuity that are easily deduced from the definition. If $f : M \to N$ and $g : N \to P$ are continuous, then $g \circ f : M \to P$ is continuous. If $f : M \to \mathbb{R}^n$ and $g : M \to \mathbb{R}^n$ are continuous, then $f + g : M \to \mathbb{R}^n$ is continuous. If $f : M \to \mathbb{R}^n$ and $g : M \to \mathbb{R}$ are continuous, then $g \cdot f : M \to \mathbb{R}^n$ is continuous. If $f : M \to N$ is continuous and $M_1 \subseteq M$ is any subspace, then the restriction of f to M_1 is continuous. If $f : M \to \mathbb{R}^n$ and we define the coordinate functions $f_k : M \to \mathbb{R}$ for $1 \leq k \leq n$ by $f(x) = (f_1(x), \ldots, f_n(x))$, then f is continuous if and only if all the f_k are continuous. We leave the verification of these facts as exercises.

In the special case where the domain and range are Euclidean s-paces, or subspaces of Euclidean spaces, it is important to have a rich collection of continuous functions. We have already observed that the range space can be split into coordinate components, so it is really enough to consider the case $f : \mathbb{R}^n \to \mathbb{R}$. The simplest nontrivial examples are the coordinate projection maps $f_k(x) = f_k(x_1, \ldots, x_n) = x_k$. These are easily seen to be continuous by the preservation of limits criterion. From these, using composition and arithmetic operations, we can establish the continuity of all "elementary" functions—functions for which we have a finite formula—provided we restrict the domain to all points for which the formula defining the function makes sense.

The class of *polynomials* on $\mathbb{R}^n$ is the class of functions built up by addition and multiplication from the coordinate projections and the constants. Writing these in concise notation is difficult, but the following *multi-index* convention seems to work extremely well. Let $\alpha = (\alpha_1, \ldots, \alpha_n)$ denote an n-tuple of non-negative integers (each α_k can equal $0, 1, 2, \ldots$), and let $x^\alpha = x_1^{\alpha_1} x_2^{\alpha_2} \cdots x_n^{\alpha_n}$. Then $p(x) = \sum c_\alpha x^\alpha$, where the sum is finite and c_α are constants, is the general polynomial on $\mathbb{R}^n$. We further let $|\alpha| = \alpha_1 + \alpha_2 + \cdots + \alpha_n$ (note this is a *different* notational convention from the usual Pythagorean metric on $\mathbb{R}^n$). We call x^α a *monomial* of *order* or *degree* $|\alpha|$, and we call the *order* of the polynomial the order of the highest monomial appearing in it with non-zero coefficient (frequently we are sloppy in saying "a polynomial of order m" when we really mean "a polynomial of order $\leq m$").

In dealing with functions defined in $\mathbb{R}^n$ or subsets of $\mathbb{R}^n$, it is tempting to think of the variable $x = (x_1, \ldots, x_n)$ as consisting of n distinct real variables $x_1, x_2, \ldots, x_n$. In particular, we can hold $n - 1$ of them fixed and vary just one, obtaining a function of one variable. We could hope to reduce questions about a single function of n variables to questions about the many functions of one variable that arise in this fashion. However, while there are some situations in which this technique proves useful, in general it is very misleading. To understand why, we have only to consider the case $n = 2$, where we can visualize the domain as the plane. By fixing one variable and varying the other, we sweep out all the horizontal and vertical lines in the plane. The function $f(x_1, x_2)$ considered as a function of x_1 with x_2 fixed is then the restriction of f to the horizontal line $x_2 = a$. Thus our seductive suggestion is to try to reduce questions about f to questions about its restriction to all horizontal and vertical lines. This may not be too helpful if we need to compare the values of f at two different points that are not on the same horizontal or vertical line.

Let's look at continuity from this point of view. It is easy to see that the continuity of $f : \mathbb{R}^n \to \mathbb{R}$ implies the continuity of the restriction of f to each of the lines obtained by holding $n-1$ variables constant—this is an immediate consequence of the preservation of limits criterion for continuity. We say that f is *separately* continuous if it has this property: for every k and every fixed value of all x_j with $j \neq k$, the function $g(x_k) = f(x_1, \ldots, x_n)$ is continuous. Thus continuity implies separate continuity. But the converse is not true. It is easy to give a counterexample in $\mathbb{R}^2$. Take the function $f(x, y) = \sin 2\theta$ where $\theta = \arctan(y/x)$ is the angular polar coordinate, and $f(0, 0) = 0$. The factor of 2 is chosen so that f is zero, hence continuous, on the coordinate axes. On every line not passing through the origin the function is continuous because $f(x, y) = 2xy/(x^2 + y^2)$ (this follows from $\sin 2\theta = 2 \sin\theta \cos\theta$ and $\sin\theta = y/\sqrt{x^2 + y^2}, \cos\theta = x/\sqrt{x^2 + y^2}$) and we don't encounter any zero divisions. Thus f is separately continuous. But f is discontinuous along any line through the origin not equal to one of the axes, for it assumes the value zero at the origin and is equal to the constant non-zero value $\sin 2\theta$ along the ray with θ constant. From this it is easy to see that f is not continuous.

9.3.2 Continuous Functions on Compact Domains

We continue our discussion of continuous functions $f : M \to N$ for general metric spaces. Just as in the case of numerical functions, we can define a notion of *uniform continuity* in which the error relations are uniform over the domain: *for every error $1/m$ there exists an error $1/n$ such that $d(x, y) \leq 1/n$ implies $d(f(x), f(y)) \leq 1/m$.* Once again we can show that a continuous function on a compact domain is automatically uniformly continuous. The proof is a nice application of the Heine-Borel property.

Theorem 9.3.2 *Let M be a compact metric space, and let $f : M \to N$ be continuous. Then f is uniformly continuous.*

Proof: Since f is continuous, for each $1/m$ and each point x_0 there exists $1/n$ such that $d(x, x_0) < 2/n$ implies $d(f(x), f(x_0)) < 1/2m$. By the triangle inequality $d(f(x), f(y)) < 1/m$ if both x and y are in $B_{2/n}(x_0)$. Here $1/n$ depends on x_0 and $1/m$. Keeping $1/m$ fixed and varying x_0 over M, consider the open covering by the smaller balls $B_{1/n}(x_0)$. By the Heine-Borel property, there exists a finite subcover. That means there exists a finite number of points $x_1, x_2, \ldots, x_N$ and radii $1/n_j$ such that every point lies in one of the balls $B_{1/n_j}(x_j)$. We now take $1/n$ to be the smallest value of $1/n_j$. Given any point x, we have x in $B_{1/n_j}(x_j)$ for some j. If $d(x, y) \leq 1/n$, then $d(y, x_j) \leq d(x, y) + d(x, x_j) \leq 1/n + 1/n_j \leq 2/n_j$ by the triangle inequality, so both x and y belong to $B_{2/n_j}(x_j)$. We have already observed that this implies $d(f(x), f(y)) \leq 1/m$. This proves the uniform continuity. QED

Continuous functions on compact domains have other special properties. For example, if the range is the reals, then we can assert that the sup and inf are attained.

Theorem 9.3.3 *Let $f : M \to \mathbb{R}$ be continuous and M compact. Then* $\sup f(x)$ *and* $\inf f(x)$ *are both finite, and there are points in M where these values are assumed.*

Proof: We give the proof just for $\sup_x f(x)$. There exists a sequence of values $\{f(x_n)\}$ converging to the sup (or to $+\infty$ if the sup is $+\infty$). By the compactness of M we can obtain a convergent subsequence $x'_n \to x_0$. Then $f(x'_n) \to f(x_0)$ by the continuity of f. Thus the sup is finite and equals $f(x_0)$. QED

More generally, if $f : M \to N$ is continuous and M is compact, then the image $f(M)$ is compact. This implies the previous result because a compact subset of $\mathbb{R}$ is bounded and contains its sup and inf.

Theorem 9.3.4 *The image of a compact set under a continuous function is compact.*

Proof: Let A be compact. To show $f(A)$ is compact we need to show every sequence of points in $f(A)$ has a subsequence converging to a point in $f(A)$. But a sequence of points in $f(A)$ must have the form $f(x_1), f(x_2), \ldots$ where $x_1, x_2, \ldots$ is a sequence of points in A. (The point x_k may not be uniquely determined by $f(x_k)$ if f is not one-to-one, but all that matters is that there is at least one such point). By the compactness of A there exists a convergent subsequence $x'_k \to x_0$ with x_0 in A. Then by the continuity of f, the subsequence $f(x'_k)$ converges to $f(x_0)$, which is in $f(A)$. Thus $f(A)$ is compact. QED

We mention now that for sequences or series of functions on metric spaces, results analogous to those of Chapter 7 for numerical functions can be obtained without difficulty. For example, if a sequence of continuous functions $f_n : M \to N$ converges uniformly to $f : M \to N$, then f is continuous. The definition of uniform convergence and the proof can be repeated almost word for word. We leave the details as an exercise. Similarly for the Arzela-Ascoli theorem: if $f_n : M \to \mathbb{R}$ is a sequence of real-valued functions on a compact metric space M that is uniformly bounded and uniformly equicontinuous, then there exists a uniformly convergent subsequence. The same result is even true for functions $f_n : M \to N$ as long as N is complete.

9.3.3 Connectedness

We consider next some analogs of the intermediate value property of numerical functions. If f is a continuous numerical function defined on an interval, then the intermediate value property says it assumes all values in between values it assumes—and this implies easily that the image of the interval must be an interval. The first question that faces us, then, is to find some intrinsic properties of intervals that have counterparts in a general metric space. The kind of properties for which we are looking will express the intuitive idea of being of one piece, or connectedness. There are actually two distinct concepts of connectedness. The first, which we will call *connectness,* involves the impossibility of splitting the space up into pieces. The second, which we call *arcwise connectedness,* involves being able to join any two points by a continuous curve; it turns out to be a stronger condition than connectness.

Let M be a metric space. We say M is *connected* if there does not exist a pair of disjoint nonempty open sets A and B with $M = A \cup B$. Equivalently, since the complement of an open set is closed, M *is connected if the only subsets of* M *that are both open and closed are the empty set and the whole space.* To justify—at least in part—this definition we should observe that an interval of $\mathbb{R}$ is connected, and the only connected subsets of $\mathbb{R}$ are intervals.

Let I be an interval of $\mathbb{R}$—it doesn't matter whether or not I contains its endpoints or whether or not the endpoints are finite or infinite. Suppose $I = A \cup B, A$ and B disjoint open (in I) subsets. If A and B are both nonempty, they must contain points a and b and we may assume $a < b$ (A and B are disjoint, so we can't have $a = b$). To produce a contradiction we look for a dividing point between A and B. Thus let $r = \sup\{x \text{ in } I : x \leq b \text{ and } x \text{ is in } A\}$. Since a is in A and $a < b$, we have $r \geq a$. By the definition of the sup we have a sequence of points $x_1, x_2, \ldots$ in A converging to r, so r is in A since A is closed in I. But A is also open in I. Since $r < b$, (we have $r \neq b$ since r is in A), r is in the interior of I, so A must contain an open neighborhood of r, which contradicts the definition of r as the sup of points of A. Thus I is connected.

We now show, conversely, that intervals are the only connected subsets of $\mathbb{R}$ (points are special cases of intervals, $[a, a]$). Suppose A is a connected set in $\mathbb{R}$ and $a < b < c$ with a and c in A. We want to show this implies b is in A. Suppose it were not; then by dividing A at b as $A = (A \cap (-\infty, b)) \cup (A \cap (b, \infty))$ we would obtain a forbidden decomposition into disjoint open sets. Thus we must have b in A. It is then a simple exercise to show that A must be an interval with endpoints $\inf A$ and $\sup A$.

For the second concept of connectedness, we need to define the notion of *curve*, also known as *path* or *arc*. A *curve* in M is defined to be a continuous function $f : I \to M$ where I is an interval in $\mathbb{R}$. The image $f(I)$ is a set of points in M, which we can think of as being traced out by $f(t)$ as t varies in I, interpreted as a time variable. Thus the curve is a "trajectory of a moving particle" in M. We do not assume f is one-to-one, so the image of the curve is allowed to intersect itself. It is tempting to identify the curve with the image $f(I)$ in M, with $f(t)$ supplying a parametric representation of it. Of course the same set of points in M can be the image of many different curves, so, strictly speaking, we are not justified in calling $f(I)$ a curve—nevertheless, we will do so when there is no danger of confusion. When M is a subspace of $\mathbb{R}^n$ the curve has the form $f(t) = (f_1(t), f_2(t), \ldots, f_n(t))$ where $f_k(t)$ are continuous numerical functions, giving the coordinates of the trajectory at each time t. For example, $f(t) = (\cos t, \sin t)$ is the curve whose image is the unit circle traced out in the usual manner (counterclockwise at constant angular velocity) infinitely often as t varies in $\mathbb{R}$. The graph of a continuous function $g : I \to \mathbb{R}$ is a curve in the plane given by $f(t) = (t, g(t))$ for t in I.

A metric space (or subspace of a metric space) is called *arcwise connected* (sometimes the term *pathwise* is used instead, but never *curvewise*) if there exists a curve connecting any two points. In more precise language, given any two points x and y in M there exists a curve $f : [a, b] \to M$ such that $f(a) = x$ and $f(b) = y$. The intuitive content of this definition is clear; on a technical level it allows us to reduce questions about M to questions about the intervals $[a, b]$. For example, if $g : M \to \mathbb{R}$ is any continuous real-valued function on an arcwise connected space M, we can show that g has the intermediate value property. Suppose $g(x) = a$ and $g(y) = b$, with $a < b$. Let $f : I \to M$ be a curve connecting x to y, as shown in Figure 9.3.1. Then $g \circ f$ is a continuous function from I to $\mathbb{R}$ that assumes the values a and b at the endpoints of I. By the intermediate value property for numerical functions, $g \circ f$ assumes all values in $[a, b]$; hence, g assumes all these values.

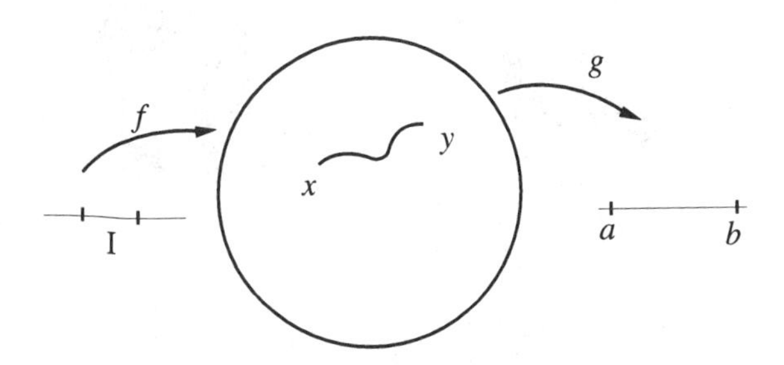

Figure 9.3.1:

Theorem 9.3.5 *An arcwise connected space is connected.*

Proof: Suppose M were arcwise connected but not connected. Then $M = A \cup B$ where A and B are disjoint non-empty open sets. Let x be in A and y be in B, and let $f : [a, b] \to M$ be a curve connecting x and y in M. We claim $f^{-1}(A)$ and $f^{-1}(B)$ give a forbidden decomposition of the connected interval $[a, b]$. Indeed they are open sets because f is continuous and the inverse image of open sets is open. Furthermore, for each t in $[a, b]$, either $f(t)$ is in A or B but not both. Thus $[a, b] = f^{-1}(A) \cup f^{-1}(B)$ and $f^{-1}(A)$ and $f^{-1}(B)$ are disjoint. This contradiction to the connectedness of $[a, b]$ proves that M is connected. QED

The converse of this theorem is not true, although it is true under extra assumptions on M. (See exercise set 9.3.7, number 7.)

From a practical point of view, the easiest way to show a space is connected is to show that it is arcwise connected. Indeed, most of the subspaces of $\mathbb{R}^n$ with which we will deal, such as balls and generalized rectangles ($\{x \text{ in } \mathbb{R}^n; a_j \leq x_j \leq b_j \text{ for } j = 1, \ldots, n\}$), are obviously arcwise connected. Any straight line segment in $\mathbb{R}^n$ is a curve, as is any broken line segment; so any subspace of $\mathbb{R}^n$ such that all points may be joined by broken line segments is connected, such as the region shown in Figure 9.3.2.

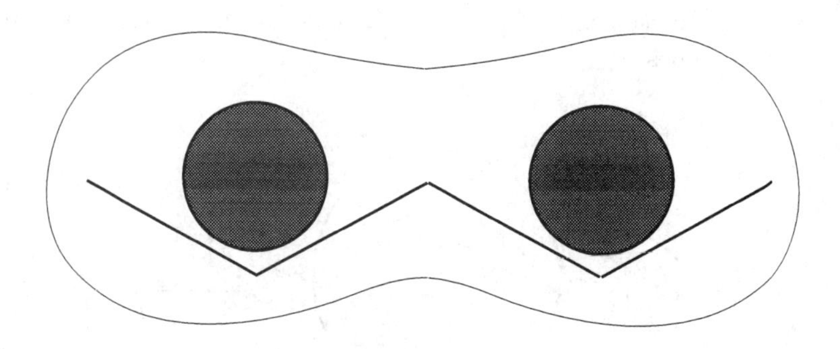

Figure 9.3.2:

The notion of connectedness allows us to attain the broadest generalization of the intermediate value property: the continuous image of a connected set is connected. Furthermore, since the only connected subsets of $\mathbb{R}$ are intervals, this implies immediately that a continuous function $f : M \to \mathbb{R}$ has the intermediate value property if M is connected.

Theorem 9.3.6 *Let $f : M \to N$ be continuous and onto, $f(M) = N$. If M is connected, then N is connected. If M is arcwise connected, then N is arcwise connected.*

Proof: Both results are simple consequences of the definitions. Suppose, first, that M is connected. If $N = A \cup B$ with A and B disjoint non-empty open subsets, then $M = f^{-1}(A) \cup f^{-1}(B)$ since $N = f(M)$ with $f^{-1}(A)$ and $f^{-1}(B)$ disjoint non-empty open subsets of M, a contradiction. So N is connected.

Next assume M is arcwise connected. Let x and y be points of $N = f(M)$. Then $x = f(u)$ and $y = f(v)$ for u and v in M. Since M is arcwise connected, there exists a curve $g : [a, b] \to M$ joining u to v. Then $f \circ g : [a, b] \to N$ is a curve joining x to y. Thus N is arcwise connected. QED

The implications in the theorem cannot be reversed. It is easy to give examples (see exercises) of $f : M \to N$ onto, where N is connected but M is not.

9.3.4 The Contractive Mapping Principle

The next theorem we discuss is the Contractive Mapping Principle. It has a very simple proof but many important applications. We will use it many times in the chapters to come. We consider a function whose domain and range are of the same metric space, which we assume is complete. The term *mapping* is sometimes used to denote a continuous function and suggests the intuitive idea of moving points around. For a mapping $f : M \to M$, we can ask if there are any *fixed points*, points for which $f(x) = x$. This is not merely idle speculation, because many problems can be cast in this form. The Contractive Mapping Principle gives a criterion for there to be a unique fixed point and a simple constructive means of finding the fixed point.

Definition 9.3.1 *We say $f : M \to M$ is a contractive mapping if there exists a contant $r < 1$ such that $d(f(x), f(y)) \leq r\, d(x, y)$ for all x and y in M.*

Note that this condition is just a Lipschitz condition with constant less than one. It says that under the mapping, all distances are reduced by at least a factor of r. If we apply f repeatedly we will shrink distances drastically, which makes the existence of a unique fixed point seem plausible.

We introduce the notation f^n for the iterated mapping $f \circ f \circ \cdots \circ f$ (n times).

Contractive Mapping Principle *Let M be a complete metric space and $f : M \to M$ a contractive mapping. Then there exists a unique fixed point x_0, and $x_0 = \lim_{n\to\infty} f^n(x)$ for any point x in M, with $d(x_0, f^n(x)) \leq c\, r^n$ for a constant c depending on x.*

Proof: Choose any x in M, and consider the sequence $f(x), f^2(x), f^3(x), \ldots$. We will show it is a Cauchy sequence; for then, by the completeness of M, it will have a limit. By the contractive mapping property,

$$d(f^{n+1}(x), f^n(x)) \leq r\, d(f^n(x), f^{n-1}(x)).$$

Hence, by induction,

$$d(f^{n+1}(x), f^n(x)) \leq r^n d(f(x), x).$$

Thus, if $m > n$, we have

$$\begin{aligned} d(f^m(x), f^n(x)) &\leq d(f^m(x), f^{m-1}(x)) + \cdots + d(f^{n+1}(x), f^n(x)) \\ &\leq (r^{m-1} + r^{m-2} + \cdots + r^n) d(f(x), x) \end{aligned}$$

by the triangle inequality. But $d(f(x), x)$ is just a constant, and we can make $(r^{m-1} + \cdots + r^n)$ as small as we please by making n large enough, because $\sum r^n$ converges (since $r < 1$). Thus, $\{f^n(x)\}$ is a Cauchy sequence and $x_0 = \lim_{n\to\infty} f^n(x)$ exists because M is complete. It is easy to see that x_0 is a fixed point since

$$f(x_0) = f(\lim_{n\to\infty} f^n(x)) = \lim_{n\to\infty} f^{n+1}(x) = x_0$$

by the continuity of f. The above estimates also give us the rate of convergence,

$$\begin{aligned} d(f^n(x), x_0) &= \lim_{m\to\infty} d(f^n(x), f^m(x)) \\ &\leq \sum_{k=n}^{\infty} r^k d(f(x), x) = \frac{r^n}{1-r} d(f(x), x). \end{aligned}$$

Finally the fixed point is unique, for if x_1 were another one, then

$$d(x_0, x_1) = d(f(x_0), f(x_1)) \leq r\, d(x_0, x_1)$$

with $r < 1$, which is impossible. QED

It would seem that contractive mappings are hard to come by, s-ince they have such strong properties. This is in fact the case, and in most applications we must first restrict the space. That is, we start with a mapping $f : M \to M$ that is not necessarily contractive and seek a closed subspace M_0 (this implies M_0 is complete) on which f is contractive. We then have to verify that $d(f(x), f(y)) \leq r\, d(x, y)$ for every x and y in M_0 and also that $f(x)$ is in M_0 for every x in M_0. This second condition is easy to overlook, but it is necessary if we are to apply the theorem to f restricted to M_0 and have a mapping $f : M_0 \to M_0$ with the same domain and range. We will discuss this in detail when we apply the Contractive Mapping Principle to obtain solutions of differential equations.

It is tempting to consider weakening the contractive property to allow $r = 1$, but this will not work. The simplest counterexample is the translation $f(x) = x+1$ on $\mathbb{R}$, for which $d(f(x), f(y)) = d(x, y)$ but there are no fixed points. It is not even possible to allow the condition $d(f(x), f(y)) < d(x, y)$ for every x and y, for there are examples of mappings without fixed points satisfying this condition.

There are many other fixed point theorems, such as the Brouwer fixed point theorem, which asserts that there is always a fixed point (not necessarily unique) if M is a closed ball in $\mathbb{R}^n$. But that theorem is non-constructive, and Brouwer himself had to renounce it when he became a constructivist! Incidentally, there does not have to be a fixed point if M is an open ball.

9.3.5 The Stone-Weierstrass Theorem*

We consider next a famous generalization of the Weierstrass approximation theorem. A straightforward generalization would simply be that any continuous real-valued function $f : M \to \mathbb{R}$ for M a compact subspace of $\mathbb{R}^n$ can be uniformly approximated by polynomials. This will follow from a more general theorem of Marshall Stone, which is known as the Stone-Weierstrass theorem. Stone asked the question: what are the properties of the polynomials that enable them to approximate arbitrary continuous functions on an interval? He observed that the collection of polynomials on an interval, call it $\mathcal{P}$, has the property that it forms a vector space: if f and g are in $\mathcal{P}$, then $af + bg$ is in $\mathcal{P}$ for constants a, b. Also it is closed under multiplication: f and g in $\mathcal{P}$ imply $f \cdot g$ is in $\mathcal{P}$. We summarize these properties by saying $\mathcal{P}$ forms an algebra. More generally, if $\mathcal{A}$ denotes any collection of real-valued functions on a set M, we say $\mathcal{A}$ forms an *algebra* if f and g in $\mathcal{A}$ imply $af + bg$ and $f \cdot g$ are in $\mathcal{A}$. If M is any subset of $\mathbb{R}^n$, then the polynomials on M form an algebra.

Stone discovered that any algebra of continuous functions on a compact metric space will suffice to approximate uniformly all continuous real-valued functions on M, provided that we impose some conditions—which are obviously necessary—to guarantee that $\mathcal{A}$ is large enough. We say that $\mathcal{A}$ *strongly separates points* on M if 1) given any x in M there exists f in $\mathcal{A}$ with $f(x) \neq 0$, and 2) given any distinct points x and y in M there exists f in $\mathcal{A}$ with $f(x) \neq f(y)$. If $\mathcal{A}$ failed to strongly separate points (say $f(x_0) = 0$ for all f in $\mathcal{A}$ or $f(x_1) = f(x_2)$ for all f in $\mathcal{A}$), then the same would be true of all uniform limits of functions in $\mathcal{A}$ and we could never approximate all continuous functions without these properties. Thus the condition that $\mathcal{A}$ strongly separates points is necessary. On the other hand the condition that $\mathcal{A}$ forms an algebra is not necessary; there are many interesting examples of vector spaces of functions that do not form an algebra and yet can uniformly approximate all continuous functions. Nevertheless, there are enough applications of the Stone-Weierstrass theorem to justify its fame. The proof of the Stone-Weierstrass theorem uses the one-dimensional Weierstrass theorem and so cannot be used to give an alternative proof of the Weierstrass theorem.

Theorem 9.3.7 (**Stone-Weierstrass Theorem**) *Let M be a compact metric space and $\mathcal{A}$ an algebra of continuous real-valued functions on M that strongly separates points. Then any continuous real-valued function on M can be uniformly approximated by functions in $\mathcal{A}$.*

Proof: We have to show that given any continuous function $f : M \to \mathbb{R}$ and any error $1/m$ there exists g in $\mathcal{A}$ with $|f(x) - g(x)| \leq 1/m$ for all x in M. We write this estimate as $f(x) - 1/m \leq g(x) \leq f(x) + 1/m$. We will first get g in $\mathcal{A}$ to satisfy $g(x) \geq f(x) - 1/m$, and then we will decrease g to get the other inequality. A key technical device is that if g_1 and g_2 are in $\mathcal{A}$, then $\max(g_1, g_2)$ and $\min(g_1, g_2)$ are "almost" in $\mathcal{A}$. More precisely, we have the following lemma:

Lemma 9.3.1 *Let g_1 and g_2 be in $\mathcal{A}$. Then for every error $1/m$ there exist functions g_3 and g_4 in $\mathcal{A}$ such that $|g_3 - \max(g_1, g_2)| \leq 1/m$ and $|g_4 - \min(g_1, g_2)| \leq 1/m$ for all points of M.*

Proof: Since $\max(g_1, g_2) = (g_1 + g_2 + |g_1 - g_2|)/2$ and $\min(g_1, g_2) = (g_1 + g_2 - |g_1 - g_2|)/2$, it suffices to prove the analogous statement for $|g|$: if g is in $\mathcal{A}$ there exists g_1 in $\mathcal{A}$ with $|g_1 - |g|\,| \leq 1/m$. Here we use the fact that $\mathcal{A}$ is an algebra. Let $A = \sup_x |g(x)|$, which is finite because M is compact and g is continuous. Then $|g(x)|$ is the composition of $g : M \to \mathbb{R}$ followed by the function $|x| : [-A, A] \to \mathbb{R}$. By the Weierstrass approximation theorem on the interval $[-A, A]$, there exists a polynomial p with $|p(t) - |t|\,| \le 1/m$ for every t in $[-A, A]$ (of course p depends on both $1/m$ and A). Then $|p(g(x)) - |g(x)|\,| \le 1/m$ (just take $t = g(x)$), and $p(g(x))$ is in $\mathcal{A}$ because $\mathcal{A}$ is an algebra. This proves the lemma.

Returning to the proof of the theorem, we begin by finding a solution of a simpler interpolation problem—to find a function g in $\mathcal{A}$ satisfying $g(x_k) = y_k$ for any finite distinct set of points $x_1, \ldots, x_n$ in M and any real values $y_1, \ldots, y_n$. This step in the proof uses the hypothesis that $\mathcal{A}$ strongly separates points. We will only need to use the result for pairs of points x_1, x_2 and so we give the proof in that case. The reader can easily supply the general case proof by induction. We know there is a function h_1 in $\mathcal{A}$ such that $h_1(x_1) \neq h_1(x_2)$, and we can assume without loss of generality that $h_1(x_1) \neq 0$ and even $h_1(x_1) = 1$ by multiplying by a suitable constant. We have to consider two cases, depending on whether or not $h_1(x_2) = 0$. If $h_1(x_2) \neq 0$, then $h_1^2(x_2) \neq h_1(x_2)$ (we know $h_1(x_2) \neq h_1(x_1) = 1$), and linear combinations of h_1 and h_1^2 will solve the interpolation problem. If $h_1(x_2) = 0$, then we use the existence of h_2 in $\mathcal{A}$ with $h_2(x_2) \neq 0$, say $h_2(x_2) = 1$, and then linear combinations of h_1 and h_2 will solve the interpolation problem.

Consider any continuous function $f : M \to \mathbb{R}$. For any two points x_1 and x_2 we can find h in $\mathcal{A}$ with $h(x) = f(x)$ for $x = x_1$ or x_2. Because both functions are continuous, there exists a neighborhood of each of these points on which $h(x) \ge f(x) - 1/m$. By holding one of the points fixed, say x_1, and varying x_2, we get an open covering of M by neighborhoods A for which there exists h in $\mathcal{A}$ with $h(x_1) = f(x_1)$ and $h(x) \ge f(x) - 1/m$ for all x in A. By the compactness of M there exists a finite subcovering. Thus $M \subseteq \bigcup_{j=1}^{k} B_j$ and for each B_j there exists h_j in $\mathcal{A}$ such that $h_j(x_1) = f(x_1)$ and $h_j(x) \ge f(x) - 1/m$ for x in B_j. Now the function $H = \max(h_1, \ldots, h_k)$ satisfies $H(x_1) = f(x_1)$ and $H(x) \ge g(x) - 1/m$ for every x in M. By the lemma, and induction, we can find g in $\mathcal{A}$ such that $|g(x) - H(x)| < 1/m$ for every x in M. Thus the function g in $\mathcal{A}$ satisfies $g(x) \ge f(x) - 2/m$ for every x in M but $g(x_1) \le f(x_1) + 1/m$, as well. (In Figure 9.3.3 the graph of g is constrained to lie above the lower line—the graph of $f(x) - 2/m$, but at $x = x_1$ it must pass through the vertical line—between $f(x_1) - 2/m$ and $f(x_1) + 1/m$.)

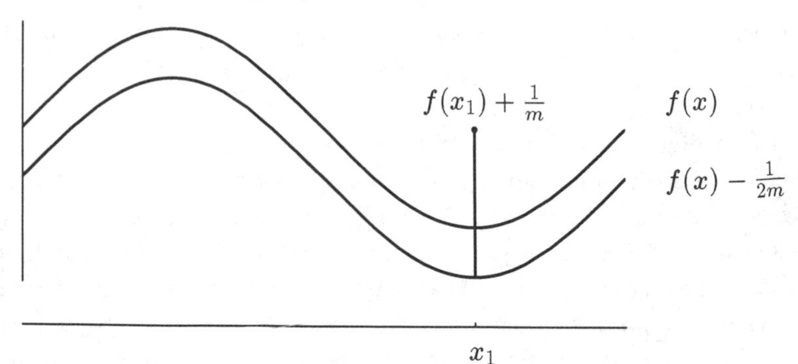

Figure 9.3.3:

Thus we have succeeded in both pinning down the value of g near that of f at one point x_1 and at the same time bounding the values of $g(x)$ below by $f(x) - 2/m$ at all points.

We now repeat the same procedure to decrease g, preserving the lower bound we have already obtained, and at the same time getting the desired upper bound. Since we have one such function for each point x_1 of M and by the continuity of g and f we have $g(x) \le f(x) + 2/m$ for x in a neighborhood of x_1, we can use the compactness of M to find a finite set of functions $g_1, \ldots, g_k$ in $\mathcal{A}$ and an open covering $B_1, \ldots, B_k$ such that $g_j(x) \le f(x) + 2/m$ on B_j. Of course we still have $g_j(x) \ge f(x) - 2/m$ on all of M. Thus $G(x) = \min\{g_1(x), \ldots, g_k(x)\}$ satisfies $f(x) - 2/m \le G(x) \le f(x) + 2/m$ for all x in M. By the lemma we can find g in $\mathcal{A}$ satisfying $|G(x) - g(x)| \le 1/m$ on M, so $|g(x) - f(x)| \le 3/m$. Thus any continuous function f can be uniformly approximated by functions in $\mathcal{A}$. QED

There is also a complex-valued version of this theorem, where we require the additional hypothesis that if $g(x)$ is in $\mathcal{A}$, then $\overline{g(x)}$ is in $\mathcal{A}$. The proof can be reduced to the real version by taking real and imaginary parts. We leave the details for the exercises. Without this additional hypothesis the theorem is not true, but a proper explanation of the counterexample requires the theory of complex variables.

If M is any compact subset of $\mathbb{R}^n$, then the Stone-Weierstrass theorem applies to the algebra of polynomials on M. Indeed the constant polynomial $f(x) \equiv 1$ gives a non-zero value at every point of M, and one of the coordinate monomials $f_k(x) = x_k$ will assume different values at two distinct points.

9.3.6 Nowhere Differentiable Functions, and Worse*

We turn now to certain examples of continuous functions that have variously been described as pathological, bizzare, monstrous, obscene, etc. The first such example is a continuous nowhere differentiable function, discovered by Bolzano around 1830. His discovery was not widely circulated and so when Weierstrass discovered a similar example some 40 years later it was regarded as new and shocking. The reason for this was that mathematicians had tacitly assumed that all continuous functions would be differentiable except for isolated exceptional points. It is tempting to react to these examples by saying, "very well, but let's just add some hypotheses to rule them out". It turns out, however, that nowhere differentiable functions have an important role to play in very down-to-earth problems. In the study of Brownian motion—one of the central topics in modern probability theory with widespread physical applications—one finds that nowhere differentiable functions are the rule, not the exception. (With probability one, every Brownian motion path is nowhere differentiable.)

Bolzano's example was very graphic. It is also a model for a kind of construction that occurs often in the theory of fractals. The function we seek will be the limit of a sequence of approximating functions. For simplicity, we take $[0, 1]$ for the domain and range. The first function we consider is $f_1(x) = x$. For the second function we add a zigzag to the graph, as shown in Figure 9.3.4. For the third function we take each of the three straight line segments of the graph of f_2 and add zigzags to them. Continuing in this fashion we pass from f_{n-1} to f_n by adding zigzags to all the straight line segments of the graph of f_{n-1}. This process is illustrated in Figure 9.3.5. By controlling the size of the zigzags appropriately we can make the sequence $f_1, f_2, \ldots$ converge uniformly to a limit function f that will be continuous. It is certainly plausible that f should fail to be differentiable at the countable set of points where we have corners in the graphs of the f_n. But why should $f(x)$ fail to have a derivative at every point?

To understand this we need a simple remark. Normally we compute the derivative at a point x_0 by taking the limit of a difference quotient

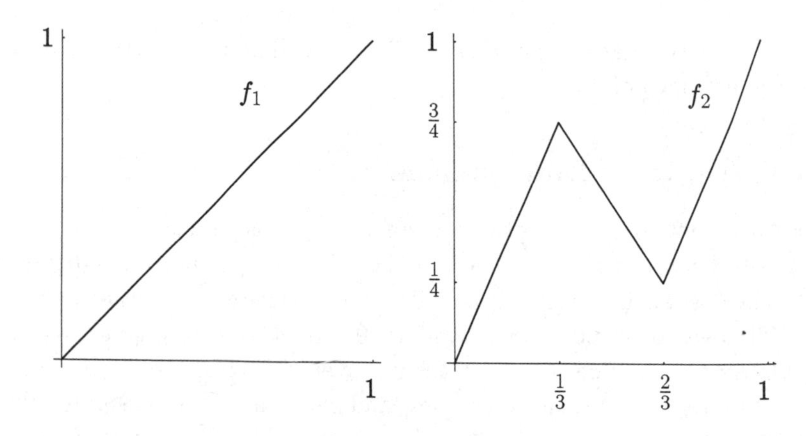

Figure 9.3.4:

$(f(x) - f(x_0))/(x - x_0)$ where x_0 is one of the points at which we evaluate f. However we could also hope to get $f'(x_0)$ as a limit of different quotients $(f(x) - f(y))/(x - y)$ where both points x and y approach x_0. It turns out that this is the case if x and y stay on opposite sides of x_0.

Lemma 9.3.2 *Suppose $f(x)$ is defined in a neighborhood of x_0 and differentiable at x_0. Then if $x_n \to x_0$ and $y_n \to x_0$ with $y_n \leq x_0 \leq x_n$ we have*

$$\lim_{n\to\infty} \frac{f(x_n) - f(y_n)}{x_n - y_n} = f'(x_0).$$

Proof: The idea is that the difference quotient $(f(x_n)-f(y_n))/(x_n- y_n)$ is an average of the difference quotients $(f(x_n) - f(x_0))/(x_n - x_0)$ and $(f(y_n) - f(x_0))/(y_n - x_0)$. In fact we compute

$$\frac{f(x_n) - f(y_n)}{x_n - y_n} = a_n \frac{f(x_n) - f(x_0)}{x_n - x_0} + b_n \frac{f(y_n) - f(x_0)}{y_n - x_0}$$

where $a_n = (x_n - x_0)/(x_n - y_n)$ and $b_n = (x_0 - y_n)/(x_n - y_n)$. Notice that $a_n + b_n = 1$ and a_n and b_n are ≥ 0 because of the assumption $y_n \leq x_0 \leq x_n$. Thus $0 \leq a_n, b_n \leq 1$ and since both

$$\frac{f(x_n) - f(x_0)}{x_n - x_0} - f'(x_0)$$

and

$$\frac{f(y_n) - f(x_0)}{y_n - x_0} - f'(x_0)$$

can be made as small as desired,

$$\begin{aligned} \frac{f(x_n) - f(y_n)}{x_n - y_n} - f'(x_0) &= a_n \left(\frac{f(x_n) - f(x_0)}{x_n - x_0} - f'(x_0) \right) \\ &\quad + b_n \left(\frac{f(y_n) - f(x_0)}{y_n - x_0} - f'(x_0) \right) \end{aligned}$$

can also be made as small as desired. QED

We use this result in contrapositive form. To show $f'(x_0)$ does not exist we need only find sequences $\{x_n\}$ and $\{y_n\}$ surrounding x_0 and converging to x_0 for which the difference quotients $(f(x_n) - f(y_n))/(x_n - y_n)$ do not converge. This makes life easy, for we need only know how to compute f at a countable dense set of points in order to compute difference quotients and show f' does not exist at every point.

We are now in a position to verify Bolzano's example. To be specific, we add the zigzags by dividing each interval of the domain in thirds. On the first and last thirds we cover 3/4 of the vertical distance in the same direction and in the middle third we cover 1/2 the vertical distance in the opposite direction, as shown in Figure 9.3.5. The exact values are not important. We note that $|f_1'(x)| \geq 3/2$ at all points except $x = 1/3$ and 2/3 where f_1' doesn't exist. By induction $|f_n'(x)| \geq (3/2)^n$ at all points except $x = k/3^n$ where f_n' doesn't exist, since adding the zigzag multiplies the slope of the line segment by 9/4 on the first and last thirds and $-3/2$ on the middle third. We also note that the value of $f_n(x)$ does not change for x of the form $k/3^m$ once $n \geq m$ so that

$$\frac{f\left(\frac{k+1}{3^m}\right) - f\left(\frac{k}{3^m}\right)}{\frac{k+1}{3^m} - \frac{k}{3^m}} = 3^m \left(f_m\left(\frac{k+1}{3^m}\right) - f_m\left(\frac{k}{3^m}\right)\right).$$

Note that f_m is linear on the interval $[k/3^m, (k+1)/3^m]$, so the estimate for f_m' shows $|3^m(f_m((k+1)/3^m) - f_m(k/3^m))| \geq (3/2)^m$. Thus the difference quotient for f computed at points $k/3^m$ and $(k+1)/3^m$ is large. But for any point x_0 in the unit interval and any m there is some

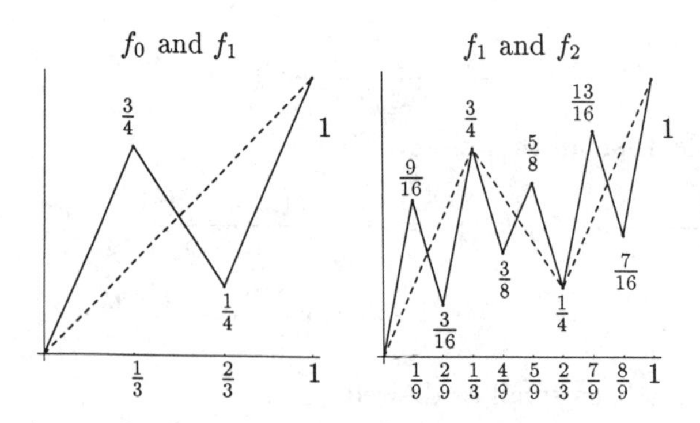

Figure 9.3.5:

value of k such that $k/3^m \leq x_0 \leq (k+1)/3^m$ and so, by the lemma, f is not differentiable at x_0. Thus f is nowhere differentiable.

It remains to verify that f is continuous—that the sequence $\{f_n\}$ converges uniformly. In fact we claim $|f_n(x) - f_{n-1}(x)| \leq (3/4)^n$, for all x, which will establish the uniform convergence by comparison with $\Sigma(3/4)^n$. Notice $|f_1(x) - f_0(x)| \leq 3/4$ by inspection. Furthermore, the maximum vertical height of any line segment in the graph of f_1 is at most $3/4$. Adding the zigzag to any line segment does not change the value of the function by more than $3/4$ times the vertical height of the segment, and each of the three segments of the zigzag are of height at most $3/4$ times the vertical height of the original segment. Thus by induction we have that all the vertical heights of the segments composing the graph of f_n are at most $(3/4)^n$, and $|f_n(x) - f_{n-1}(x)| \leq (3/4)^n$ as desired.

While Bolzano's example is geometrically appealing, it has the defect that there is no reasonable formula for the function. To obtain an example of a nowhere differentiable continuous function with a formula, we note that in Bolzano's example the differences $f_n - f_{n-1}$ have graphs that zigzag rapidly around the x-axis. Now if $g(x)$ denotes the function periodic of period 2 whose graph is as shown in Figure 9.3.6, then $g(2^n x)$ is also a function that zigzags rapidly about the x-axis and vanishes at $x = k/2^n$.

Thus we could try to get an example in the form $\sum_{n=0}^{\infty} a_n g(2^n x)$ for appropriate coefficients a_n. In fact it is not hard to show $f(x) = \sum_{n=0}^{\infty} (3/4)^n g(4^n x)$ will work. We leave the details for the exercises.

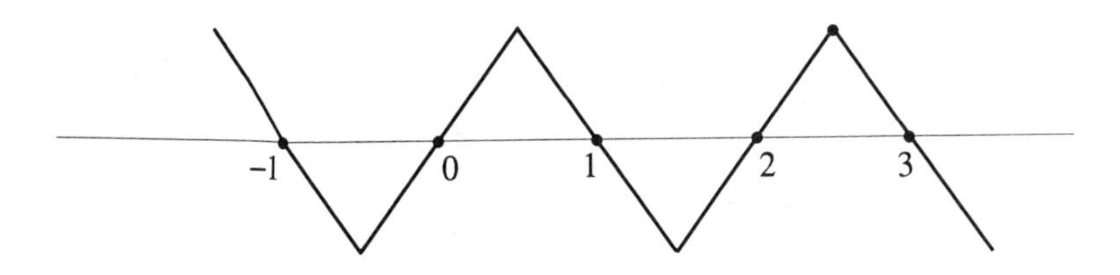

Figure 9.3.6:

The example of Weierstrass is similar, except $\sin \pi x$ is used in place of $g(x)$.

Our last example is that of a space-filling curve. The individual components of the curve are nowhere differentiable functions, so we can think of this curve as a higher species of monster. To be specific, we will construct a continuous function $f : [0,1] \to \mathbb{R}^2$ such that the image is the square $0 \le x, y \le 1$. This function is not one-to-one, and it is possible to prove that there is no one-to-one example. However, the proof requires a rather elaborate study of the topological meaning of dimension and is beyond the scope of this book. Think of a child scribbling on a piece of paper if you want an intuitive idea of how one might proceed to obtain an approximation to such a curve. We just have to make sure that the scribbling doesn't omit any region of the paper!

The original example is due to Peano, and such curves are usually called *Peano curves.* The example we give is due to I. J. Schonberg. We let t denote the parameter variable in $[0,1]$ and $x(t)$ and $y(t)$ the coordinates of the curve $(x(t), y(t))$. We need to show that $x(t)$ and $y(t)$ are continuous functions and that for every point (x_0, y_0) in the square there exists t_0 such that $(x(t_0), y(t_0)) = (x_0, y_0)$.

Let $f(t)$ and $g(t)$ be the functions whose graphs are sketched in Figure 9.3.7 and which are extended to the whole line to have period 1. The idea behind these functions is that the curve $(f(t), g(t))$ visits all four corners $(0,0), (0,1), (1,0)$, and $(1,1)$ of the square and remains in them during the intervals $[.1,.2], [.3,.4], [.5,.6], [.7,.8]$ respectively. In particular, if we write an arbitrary point (x, y) in the square by expressing x and y in binary expanion, $x = .x_1x_2\ldots, y = .y_1y_2\cdots$, then we can match the first digits $.x_1$ and $.y_1$ by $f(t)$ and $g(t)$ by controlling the first digit in the decimal expansion of t (here we are

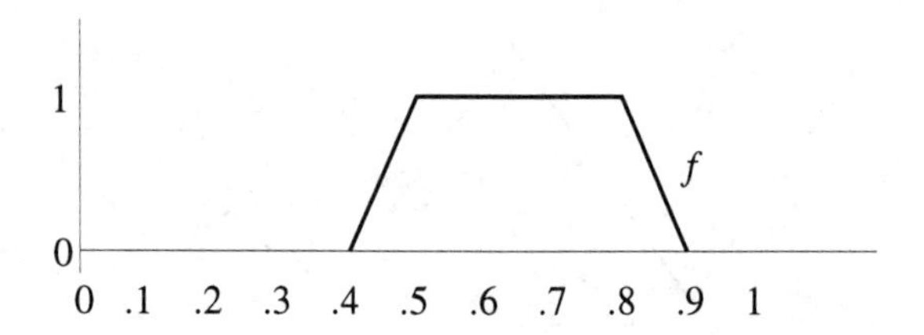

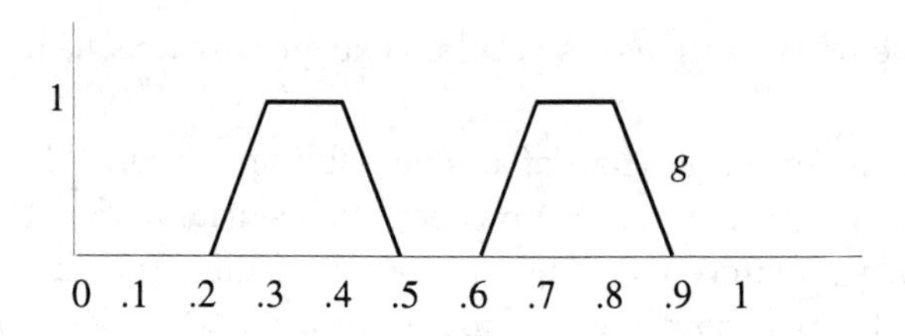

Figure 9.3.7:

thinking of $.111\ldots$ as the binary expansion of 1).

To control all the binary digits of x and y simultaneously by the decimal digits in t we take the infinite series

$$x(t) = \sum_{k=1}^{\infty} \frac{1}{2^k} f(10^{k-1}t),$$

$$y(t) = \sum_{k=1}^{\infty} \frac{1}{2^k} g(10^{k-1}t).$$

Notice that these series converge uniformly by comparison with $\Sigma 1/2^k$ and so the limits are continuous functions.

Now suppose t is a real number in $[0, 1]$ whose decimal expansion $t = .t_1t_2\ldots$ contains only the digits $1, 3, 5, 7$. Then $f(10^{k-1}t) = f(.t_kt_{k+1}\ldots)$ because f is periodic of period 1, and this is 0 if $t_k = 1$ or 3 and 1 if $t_k = 5$ or 7. Thus $x(t) = \sum_{k=1}^{\infty} \varepsilon_k/2^k = .\varepsilon_1\varepsilon_2\ldots$ in binary notation, where

$$\varepsilon_k = \begin{cases} 0 & \text{if } t_k = 1 \text{ or } 3, \\ 1 & \text{if } t_k = 5 \text{ or } 7. \end{cases}$$

Similarly we have $y(t) = \sum_{k=1}^{\infty} \delta_k/2^k = .\delta_1\delta_2\ldots$ in binary notation, where

$$\delta_k = \begin{cases} 0 & \text{if } t_k = 1 \text{ or } 5, \\ 1 & \text{if } t_k = 3 \text{ or } 7. \end{cases}$$

Thus by choosing t_k appropriately we can simultaneously obtain any binary expansions $.\varepsilon_1\varepsilon_2\ldots$ and $.\delta_1\delta_2\ldots$ and so $(x(t), y(t))$ takes on every value in the unit square for one such t.

9.3.7 Exercises

1. Prove that $f : M \to N$ is continuous if and only if $f^{-1}(A)$ is closed in M for every set A closed in N.

2. If M is a metric space and x_0 a point in M, prove that $d(x, x_0) : M \to \mathbb{R}$ is continuous.

3. If $f : M \to N$ and $g : N \to P$ are continuous, prove $g \circ f : M \to P$ is continuous. Give an example where $g \circ f$ is continuous but g and f are not.

4. If $f : M \to \mathbb{R}^n$ and $g : M \to \mathbb{R}$ are continuous, prove $g \cdot f : M \to \mathbb{R}^n$ is continuous.

5. If $f : M \to N$ is continuous and $M_1 \subseteq M$ is any subspace, prove that the restriction of f to M_1 is continuous.

6. If $f_k : M \to \mathbb{R}$ for $k = 1, \ldots, n$ and $f(x) = (f_1(x), \ldots, f_n(x)) : M \to \mathbb{R}^n$, prove f is continuous if and only if all the f_k are continuous.

7. Prove that if A is an open set in $\mathbb{R}^n$, then A is connected if and only if A is arcwise connected. (**Hint:** consider the set of points in A that can be joined to a given point x_0 by a curve.)

8. Give an example of a continuous mapping of a disconnected set onto a connected set.

9. Give an example of a continuous mapping of a noncompact set onto a compact set.

10. Formulate a definition of uniform convergence for a sequence of functions $f_n : M \to N$, and prove that the uniform limit of continuous functions is continuous.

11. State and prove a version of the Arzela-Ascoli theorem for sequences of functions $f_n : M \to N$ where N is complete and M is compact.

12. Prove that if $\mathcal{A}$ satisfies the hypotheses of the Stone-Weierstrass theorem then for any distinct points $x_1, \ldots, x_n$ in M and any real values $y_1, \ldots, y_n$ there exists a function f in $\mathcal{A}$ satisfying $f(x_k) = y_k, k = 1, \ldots, n$.

13. Prove that if $\mathcal{A}$ is an algebra of complex-valued functions on a compact metric space that strongly separates points and such that if f is in $\mathcal{A}$, then $\bar{f}$ is in $\mathcal{A}$, then any continuous complex-valued function on M can be uniformly approximated by functions in $\mathcal{A}$.

14. Show that the conclusion of the Stone-Weierstrass theorem is equivalent to saying $\mathcal{A}$ is dense in the metric space $C(M)$ of continuous real-valued functions on M with metric $d(f, g) = \sup_x |f(x) - g(x)|$.

15. Give an example of a continuous mapping of $(0, 1)$ onto $(0, 1)$ with no fixed points.

16. Let a be a fixed real number with $1 < a < 3$. Prove that the mapping $f(x) = (x/2) + (a/2x)$ satisfies the hypotheses of the contractive mapping principle on the domain $(1, \infty)$. What is the fixed point?

17. Let $T : C([0, 1]) \to C([0, 1])$ be defined by $Tf(x) = x + \int_0^x tf(t)\, dt$. Prove that T satisfies the hypotheses of the contractive mapping principle. Show that the fixed point is a solution to the differential equation $f'(x) = xf(x) + 1$.

18. Show that the set of trigonometric polynomials (functions of the form $\sum_{k=-n}^{n} a_k e^{ik\theta}$ for some n) satisfies the hypotheses of the Stone-Weierstrass theorem (complex-valued) on the circle.

19. Let $g(x)$ be the sawtooth function in Figure 9.3.6. Prove that $f(x) = \sum_{n=0}^{\infty} (3/4)^n g(4^n x)$ is a nowhere differentiable continuous function. (**Hint:** evaluate the difference quotient $4^n(f((k+1)/4^n) - f(k/4^n))$, and show that the contribution from $(3/4)^{n-1} g(4^{n-1}x)$ is the dominant term.)

20. Let $f : M \to N$ be continuous and onto, and let M be compact. Prove that A is closed in N if and only if $f^{-1}(A)$ is closed in M. Prove that if f is also one-to-one, then $f^{-1} : N \to M$ is continuous.

21. Show that $f : [0, 2\pi) \to S$ defined by $f(t) = (\cos t, \sin t)$, where S is the unit circle in $\mathbb{R}^2$, is one-to-one, onto, and continuous, but f^{-1} is not continuous.

22. Give an example of a metric space that is complete but not connected and one that is connected but not complete.

23. Give an example of a continuous function $f : M \to N$ that does not take Cauchy sequences in M to Cauchy sequences in N.

24. Give examples of continuous functions $f : M \to N$ that are onto such that M is complete and N is incomplete or M is incomplete and N is complete.

25. Give an example of a contractive mapping on an incomplete metric space with no fixed point (**Hint**: remove the fixed point from a complete metric space.)

26. Show that the graph of

$$f(x) = \begin{cases} \sin 1/x & x \neq 0 \\ 0 & x = 0 \end{cases}$$

is a connected subset of $\mathbb{R}^2$ even though f is discontinuous. Is it arcwise connected?

27. Prove that the graph of a continuous function $f : I \to \mathbb{R}$ for an interval I is a connected subset of $\mathbb{R}^2$.

28. If $A_1 \supseteq A_2 \supseteq \cdots$ is a nested sequence of compact, connected sets, show that $\bigcap_n A_n$ is also connected. Similarly, show that if the sets are arcwise connected, then so is the intersection. Would the same be true if we did not assume compactness?

29. *Let $\mathcal{K}$ denote the set of compact subsets of a complete metric space M. Define the *Hausdorff distance* on $\mathcal{K}$ as follows: $d_{\mathrm{H}}(A, B)$

is the smallest value of ε such that for every point a in A there exists a point b in B with $d(a,b) \leq \varepsilon$ and for every b in B there exists a in A with $d(a,b) \leq \varepsilon$.

a. Show that d_{H} is a metric on $\mathcal{K}$.

b. Show that $\mathcal{K}$ is complete in this metric.

30. *An *iterated function system* on a complete metric space M is defined to be a finite set $f_1, \ldots, f_m$ of contractive mappings. Prove that there exists a unique compact set K (called the *attractor*) such that $K = \bigcup_{j=1}^{m} f_j(K)$. (**Hint**: show that the mapping $A \to \bigcup_{j=1}^{m} f_j(A)$ on $\mathcal{K}$ satisfies the hypotheses of the contractive mapping principle, using the Hausdorff distance from the previous exercise.)

31. If f satisfies the hypotheses of the contractive mapping principle and x_1 is any point in M, show that $d(x_1, x_0) \leq d(x_1, f(x_1))/(1-r)$ where x_0 is the fixed point. Informally, this says that if $f(x_1)$ is close to x_1, then x_1 is close to the fixed point (but r must not be too close to 1 for this to be a good estimate).

32. Let T_1 and T_2 satisfy the hypotheses of the contractive mapping theorem on the same metric space M with the same contractive ratio r. Suppose T_1 and T_2 are close together, in the sense that $d(T_1 x, T_2 x) \leq \varepsilon$ for all x in M. Show that the fixed points x_1 and x_2 of T_1 and T_2 are also close together, namely $d(x_1, x_2) \leq \varepsilon/(1-r)$.

9.4 Summary

9.1 Structures on Euclidean Space

Definition *$\mathbb{R}^n$ is the set of ordered n-tuples $x = (x_1, x_2, \ldots, x_n)$ of real numbers.*

Definition *A metric space M is a set with a real-valued distance function $d(x,y)$ defined for x, y in M satisfying*

1. *$d(x,y) \geq 0$ with equality if and only if $x = y$ (positivity),*

2. $d(x, y) = d(y, x)$ *(symmetry)*,
3. $d(x, z) \leq d(x, y) + d(y, z)$ *(triangle inequality)*.

Definition *A norm on a real or complex vector space is a function* $||x||$ *defined for every* x *in the vector space satisfying*

1. $||x|| \geq 0$ *with equality if and only if* $x = 0$ *(positivity)*,
2. $||ax|| = |a|\,||x||$ *for any scalar* a *(homogeneity)*,
3. $||x + y|| \leq ||x|| + ||y||$ *(triangle inequality)*.

Theorem *If* $||x||$ *is a norm, then* $d(x, y) = ||x - y||$ *(called the induced metric) is a metric.*

Example *Let* $C([a, b])$ *denote the continuous functions on* $[a, b]$. *Then* $||f||_{\text{sup}} = \sup |f(x)|$ *is a norm on* $C([a, b])$, *called the sup norm.*

Definition *An inner product on a real vector space is a real-valued function* $\langle x, y\rangle$ *defined for all* x *and* y *in the vector space satisfying*

1. $\langle x, y\rangle = \langle y, x\rangle$ *(symmetry)*,
2. $\langle ax + by, z\rangle = a\langle x, z\rangle + b\langle y, z\rangle$ *and* $\langle x, ay + bz\rangle = a\langle x, y\rangle + b\langle x, z\rangle$ *for all real numbers* a, b *(bilinearity)*,
3. $\langle x, x\rangle \geq 0$ with equality if and only if $x = 0$ *(positive definiteness)*.

Theorem 9.1.1 *(Cauchy-Schwartz Inequality) On a real inner product space,* $|\langle x, y\rangle| \leq \sqrt{\langle x, x\rangle}\sqrt{\langle y, y\rangle}$, *with equality if and only if* x *and* y *are colinear.*

Theorem 9.1.2 *If* $\langle x, y\rangle$ *is an inner product, then* $||x|| = \sqrt{\langle x, x\rangle}$ *is a norm.*

Example *On* $\mathbb{R}^n$ $x \cdot y = \sum_{j=1}^n x_j y_j$ *is an inner product; hence,* $|x| = \sqrt{\sum_{j=1}^n x_j^2}$ *is a norm and* $|x - y|$ *is a metric.*

Theorem *On an inner product space the polarization identity* $\langle x, y\rangle = (||x + y||^2 - ||x - y||^2)/4$ *holds, and the associated norm satisfies the parallelogram law* $||x + y||^2 + ||x - y||^2 = 2||x||^2 + 2||y||^2$.

Definition *A complex inner product on a complex vector space is a complex-valued function $\langle x, y\rangle$ defined for all x and y in the space satisfying*

1. $\langle x, y\rangle = \overline{\langle y, x\rangle}$ *(Hermitian symmetry),*
2. $\langle ax + by, z\rangle = a\langle x, z\rangle + b\langle y, z\rangle$ *and* $\langle x, ay + bz\rangle = \bar{a}\langle x, y\rangle + \bar{b}\langle x, z\rangle$ *(Hermitian linearity),*
3. $\langle x, x\rangle$ *is real and* $\langle x, x\rangle \geq 0$ *with equality if and only if* $x = 0$ *(positive definiteness).*

9.2 Topology of Metric Spaces

Definition *A subspace M' of a metric space M is a subset of M with the same metric.*

Definition *The open ball $B_r(y)$ in a metric space with center y and radius r is $B_r(y) = \{x : d(x, y) < r\}$.*

Definition *A subset A of a metric space M is said to be open in M if every point of A lies in an open ball entirely contained in A. A neighborhood of a point is an open set containing the point. The interior of a set A is the subset of all points contained in open balls contained in A.*

Theorem 9.2.1 *Let M be a subspace of M_1. A set A is open in M if and only if there exists A_1 open in M_1 such that $A = A_1 \cap M$. If M is open in M_1, then a subset of M is open in M if and only if it is open in M_1.*

Theorem 9.2.2 *In a metric space, an arbitrary union of open sets, or a finite intersection of open sets, is open.*

Definition $\lim_{n\to\infty} x_n = x$ *means for all $1/m$ there exists N such that $n \geq N$ implies $d(x_n, x) \leq 1/m$. We say x is a limit point of a sequence $\{x_n\}$ if every neighborhood of x contains x_n for infinitely many n and x is a limit point of a set A if every neighborhood of x contains points* *of A not equal to x. A set is closed if it contains all its limit points. The closure of a set consists of the set together with all its limit points. If* $A \subseteq B$ *we say A is dense in B if the closure of A contains B.*

Theorem 9.2.3 *In a metric space, a set is closed if and only if its complement is open.*

Definition *We say* $\{x_n\}$ *is a Cauchy sequence if for every* $1/m$ *there exists N such that* $d(x_j, x_k) \leq 1/m$ *for all* $j, k \geq N$*. A metric space is complete if every Cauchy sequence has a limit.*

Theorem 9.2.4 *A sequence* $x^{(1)}, x^{(2)}, \ldots$ *in* $\mathbb{R}^n$ *converges to x if and only if the sequence of coordinates* $x_k^{(1)}, x_k^{(2)}, \ldots$ *converges to* x_k *for every* $k = 1, \ldots, n$.

Corollary 9.2.1 $\mathbb{R}^n$ *is complete.*

Theorem 9.2.5 $C([a, b])$ *with the sup-norm metric is complete.*

Definition *We say A is compact if every sequence of points in A has a limit point in A.*

Theorem *A compact metric space is complete.*

Theorem 9.2.6 (*Heine-Borel*) *A metric space is compact if and only if it has the Heine-Borel property: every open covering has a finite subcovering.*

Lemma 9.2.1 *A compact metric space has a countable dense subset, and given any* $1/m$ *there exists a finite set of points such that every point is within distance* $1/m$ *of one of them.*

Theorem *A subspace A of a complete metric space M is itself complete if and only if it is a closed set in M.*

Theorem 9.2.7 *A subspace of* $\mathbb{R}^n$ *is compact if and only if it is closed and bounded (this is not true of general metric spaces).*

9.3 Continuous Functions on Metric Spaces

Theorem 9.3.1 *Let M and N be metric spaces, $f : M \to N$ a function. The following three conditions are equivalent (and a function satisfying them is called continuous):*

1. *for every $1/m$ and x_0 in M there exists $1/n$ such that $d(x, x_0) = 1/n$ implies $d(f(x), f(x_0)) \leq 1/m$;*

2. *if $x_1, x_2, \ldots$ is any convergent sequence in M, then $f(x_1), f(x_2), \ldots$ is convergent in N;*

3. *if B is any open set in N, then $f^{-1}(B)$ is open in M.*

In part 2 it follows that $\lim_{n\to\infty} f(x_n) = f(\lim_{n\to\infty} x_n)$.

Theorem *Continuous functions are closed under restriction to a subspace, composition, addition (when the range is $\mathbb{R}^n$) and multiplication (when the range of one is $\mathbb{R}$ and the other $\mathbb{R}^n$). If $f(x) = (f_1(x), \ldots, f_n(x))$ is a function $f : M \to \mathbb{R}^n$, then f is continuous if and only if all $f_k : M \to \mathbb{R}$ are continuous.*

Example *The function $f : \mathbb{R}^2 \to \mathbb{R}$ given by*

$$f(x, y) = \begin{cases} \sin 2\theta & (x, y) \neq (0, 0), \\ 0 & (x, y) = (0, 0) \end{cases}$$

where $\theta = \arctan(y/x)$ is the polar coordinates angular variable, is not continuous at the origin, but is continuous in x for each fixed y and continuous in y for each fixed x.

Definition *$f : M \to N$ is said to be uniformly continuous if for every $1/m$ there exists $1/n$ such that $d(x, y) \leq 1/n$ implies $d(f(x), f(y)) \leq 1/m$.*

Theorem 9.3.2 *Let M be compact. Then $f : M \to N$ continuous implies it is uniformly continuous.*

Theorem 9.3.3 *If M is compact and $f : M \to \mathbb{R}$ is continuous, then $\sup_x f(x)$ and $\inf_x f(x)$ are finite and there are points in M where f attains these values.*

Theorem 9.3.4 *The image of a compact set under a continuous function is compact.*

Definition *M is said to be connected if there do not exist disjoint nonempty open sets A and B with $M = A \cup B$; or equivalently, the only sets both open and closed in M are the empty set and M.*

Theorem *A subspace of $\mathbb{R}$ is connected if and only if it is an interval.*

Definition *A curve (or arc) in M is a continuous function from an interval to M. A space M is said to be arcwise connected if for every pair of points x, y in M there exists a curve $f : [a, b] \to M$ with $f(a) = x, f(b) = y$.*

Theorem 9.3.5 *Arcwise connected implies connected.*

Theorem 9.3.6 *Let $f : M \to N$ be continuous and onto. If M is connected, then so is N. If M is arcwise connected, then so is N.*

Definition *A contractive mapping is a function $f : M \to M$ such that there exists $r < 1$ with $d(f(x), f(y)) \le r\, d(x, y)$ for all x, y in M.*

Theorem (*Contractive Mapping Principle*) *A contractive mapping on a complete metric space has a unique fixed point x_0, which is the limit of $f^n(x)$ (f^n denotes f iterated n times) for any x, and $d(x_0, f^n(x)) \le cr^n$.*

Theorem 9.3.7 (*Stone-Weierstrass*) *Let M be a compact metric space and $\mathcal{A}$ a collection of real-valued functions on M that forms an algebra (is closed under linear combinations and products) and strongly separates points (there are functions in $\mathcal{A}$ assuming distinct and non-zero values at distinct points). Then any continuous real-valued function on M can be uniformly approximated by functions in $\mathcal{A}$.*

Corollary *Any continuous function $f : M \to \mathbb{R}$ for $M \subseteq \mathbb{R}^n$ compact can be uniformly approximated by polynomials.*

Example (*Bolzano*) *There exists a continuous function $f : \mathbb{R} \to \mathbb{R}$*

such that f' fails to exist at every point.

Lemma 9.3.2 *If $f'(x_0)$ exists, then*

$$f'(x_0) = \lim_{n\to\infty} \left(f(x_n) - \frac{f(y_n)}{(x_n - y_n)} \right)$$

for any pair of sequences $x_n \to x_0$ and $y_n \to x_0$ such that $y_n \le x_0 \le x_n$.

Example (*Peano*) *There exists a continuous function from $[0, 1]$ onto the unit square in $\mathbb{R}^2$.*

Chapter 10

Differential Calculus in Euclidean Space

10.1 The Differential

10.1.1 Definition of Differentiability

Recall that the idea of the differential calculus for numerical functions was to approximate locally a general function by a special kind of function, an affine function $ax + b$, and to relate properties of the general function to properties of its affine approximation. The idea of the differential calculus for functions $f : \mathbb{R}^n \to \mathbb{R}^m$ is essentially the same. The affine functions will still have the form $ax + b$, but now a is an $m \times n$ matrix (m rows and n columns) and b is vector in $\mathbb{R}^m$. We adopt the convention that all vectors are considered as column vectors in any equation involving matrix multiplication. Thus

$$g(x) = ax + b = \begin{pmatrix} g_1(x) \\ \vdots \\ g_m(x) \end{pmatrix}$$

with $g_k(x) = \sum_{j=1}^n a_{kj}x_j + b_k$. Since these affine functions are more complicated than in the $n = m = 1$ case, the differential calculus in general will also be more complicated. However, we assume the reader has some familiarity with elementary linear algebra and so will feel comfortable in dealing with these affine functions.

Suppose f is a function $f : D \to \mathbb{R}^m$ with $D \subseteq \mathbb{R}^n$. For simplicity we will assume that the domain D is an open set. We want to define the best affine approximation to f at a point y. From the $n = m = 1$ case we expect this should be given by the condition $f(x) = g(x) + o(|x-y|)$ as $x \to y$, where g is an affine function. This means

$$\lim_{x \to y} (f(x) - g(x))/|x - y| = 0.$$

The presumption is that if such an affine function exists, then it is unique. To see why this is so suppose there were two such functions g_1 and g_2, and let $g = g_1 - g_2$ be their difference. Then

$$\begin{aligned} g(x) &= -(f(x) - g_1(x)) + (f(x) - g_2(x)) \\ &= o(|x - y|) + o(|x - y|) = o(|x - y|), \end{aligned}$$

and of course g is also affine. But we can show that zero is the only affine function such that $g(x) = o(|x - y|)$ as $x \to y$. We write $g(x) = a(x - y) + b$; since $g(y) = b$ we obtain $b = 0$ from $g(x) = o(|x - y|)$ as $x \to y$. So it suffices to show that $|a(x - y)| = o(|x - y|)$ implies $a = 0$. Setting $z = x - y$, the hypothesis is $|az|/|z| \to 0$ as $z \to 0$. But if we let z vary along a ray $z = tw$, for $t > 0$ and w in $\mathbb{R}^n$ with $w \neq 0$ fixed, then $|az| = t|aw|$ while $|z| = t|w|$, so $|az|/|z| = |aw|/|w|$ is a constant independent of t. Since this must tend to zero as $t \to 0$, it must already be zero, so $aw = 0$ for all $w \neq 0$, and this implies $a = 0$.

Thus the best affine approximation to f at a given point, if it exists, is uniquely defined. We can also assert that the best affine approximation, if it exists, must have the form $g(x) = a(x-y)+b$ where $b = f(y)$. This follows simply from the fact that $f(x)-g(x) = o(|x-y|)$ as $x \to y$ implies $f(y) - g(y) = 0$, and $b = g(y)$. In other words, the graphs of f and g pass through the same point at the value y of the independent variable. Thus it is only the matrix a that needs to be determined, and we shall call it the *differential* of f at y, written $df(y)$. (The terms *derivative* and *total derivative* are often used to mean the same thing.)

Definition 10.1.1 *We say f is differentiable at y if there exists an $m \times n$ matrix $df(y)$, called the differential of f at y, such that*

$$f(x) = f(y) + df(y)(x - y) + o(|x - y|)$$

as $x \to y$; or in other words, given any error $1/N$, there exists $1/k$ such that $|x - y| < 1/k$ implies $|f(x) - f(y) - df(y)(x - y)| \leq |x - y|/N$.

Notice that $|x-y|$ refers to the distance in $\mathbb{R}^n$ and $|f(x)-f(y)-df(y)(x-y)|$ refers to the distance in $\mathbb{R}^m$. Otherwise, the definition is the same as in the case of numerical functions. We do not specify in the definition what the entries of the matrix $df(y)$ should be. We will deal with this question in the next section, after we derive some basic properties that follow from the definition directly.

It is easy to show that the m variables in the range do not add anything significantly new. In fact, if we write

$$f(x) = \begin{pmatrix} f_1(x) \\ \vdots \\ f_m(x) \end{pmatrix}$$

in components, then f is differentiable at y if and only if all the functions $f_j(x)$ are differentiable at y and $df(y)$ is the $m \times n$ matrix whose jth row is $df_j(y)$. However, having n variables in the domain is a significant generalization.

If f is real-valued, then $df(y)$ is a $1 \times n$ matrix—a row vector. It is sometimes called the *gradient* of f and written $\nabla f(y)$. We will later give a geometric interpretation of this vector. We may also rearrange the row vector into a column and then write

$$f(x) = f(y) + \nabla f(y) \cdot (x-y) + o(|x-y|)$$

where the dot indicates the inner product of vectors. It will not be necessary for us to distinguish between row and column vectors. However, it is only fair to point out that in the study of differential geometry this distinction becomes meaningful and important.

The best affine approximation $f(y) + df(y)(x-y)$ has a graph that can be interpreted as a tangent plane (actually an n-dimensional plane) to the graph of f at y. This is most meaningful geometrically in the cases $f : \mathbb{R}^2 \to \mathbb{R}^1$ where the tangent plane is an honest plane in $\mathbb{R}^3$ and $f : \mathbb{R}^1 \to \mathbb{R}^2$ where the tangent "plane" is a line in $\mathbb{R}^3$.

Just as in the numerical case, it is easy to show that the differentiability of f at y implies the continuity of f at y, as follows. From $f(x) = f(y) + df(y)(x-y) + o(|x-y|)$ we take limits as $x \to y$. We obtain $\lim_{x\to y} f(x) = f(y)$ because $f(y)$ is constant and both $df(y)(x-y)$ and $o(|x-y|)$ tend to zero as $x \to y$. We can show a little more, namely that a pointwise Lipschitz condition holds: there exists a neighborhood of y on which $|f(x) - f(y)| \leq M|x - y|$ for some constant M. To see this, observe that

$$|f(x) - f(y)| = |df(y)(x - y) + o(|x - y|)| \leq |df(y)(x - y)| + |o(|x - y|)|.$$

Since $o(|x - y|)$ implies $O(|x - y|)$, we can make $|o(|x - y|)| \leq |x - y|$ on a neighborhood of y. Then $|df(y)(x - y)| \leq M|x - y|$ because $|Ax| \leq M|x|$ for any matrix A, with M depending on A (see exercises). This proves the pointwise Lipschitz condition $|f(x) - f(y)| \leq M|x - y|$, for x in a neighborhood of y. Note, however, that since the constant M may depend on y, this is not as strong as a uniform Lipschitz condition.

If f is differentiable at every point of the domain D, we say f is *differentiable on* D. We can then regard the differential $df(y)$ as a function of y, taking values in the space of $m \times n$ matrices $\mathbb{R}^{m \times n}$. If $df : D \to \mathbb{R}^{m \times n}$ is continuous we say f is *continuously differentiable* or f is C^1. This will turn out to be the more useful notion.

We note that differentiability and the differential are linear: if f and g (both mapping $D \to \mathbb{R}^m$) are differentiable at y, then so is $af + bg$ for scalars a, b and $d(af + bg) = adf + bdg$. These are immediate consequences of the linearity of matrix multiplication,

$$\begin{aligned} af(x) + bg(x) &= a(f(y) + df(y)(x - y) + o(|x - y|)) \\ &\quad + b(g(y)dg(y)(x - y) + o(|x - y|)) \\ &= af(y) + bg(y) + (adf(y) + bdg(y))(x - y) \\ &\quad + ao(|x - y|) + bo(|x - y|) \end{aligned}$$

and the error term $ao(|x - y|) + bo(|x - y|)$ is also $o(|x - y|)$. It also follows that if f and g are differentiable or C^1 on D, then so is $af + bg$.

There is also a product formula, in the case that $f : D \to \mathbb{R}^m$ and $g : D \to \mathbb{R}$ so that $g \cdot f : D \to \mathbb{R}^m$. Then the differentiability of f and g at y implies that $g \cdot f$ is differentiable at y and $d(g \cdot f)(y) = g(y)df(y) + f(y)dg(y)$. Here $df(y)$ is an $m \times n$ matrix and is multiplied by the scalar $g(y)$, while $f(y)$ is an $m \times 1$ matrix and $dg(y)$ is a $1 \times n$ matrix, so the matrix product $f(y)dg(y)$ is an $m \times n$ matrix. This follows from

$$\begin{aligned} g(x)f(x) &= (g(y) + dg(y)(x - y) + o(|x - y|)) \\ &\quad \cdot (f(y) + df(y)(x - y) + o(|x - y|)) \\ &= g(y)f(y) + [(dg(y)(x - y))f(y) \\ &\quad + g(y)df(y)(x - y)] + o(|x - y|) \end{aligned}$$

(the new remainder term includes $o(|x-y|)$ multiplied by bounded functions, and $(dg(y)(x-y))df(y)(x-y)$, which is $O(|x-y|^2)$ hence $o(|x-y|)$.) All that remains is to verify that $(dg(y)(x-y))f(y) = (f(y)dg(y))(x-y)$, and this is simply matrix algebra.

10.1.2 Partial Derivatives

We come now to the problem of determining the entries of the matrix df. If we choose $x = y + te_j$ where e_j is the unit vector in the jth direction, then

$$\begin{aligned} f(x) &= f(y+te_j) = f(y) + df(y)te_j + o(|te_j|) \\ &= f(y) + t\,df(y)e_j + o(|t|). \end{aligned}$$

The kth coordinate of this equation is

$$f_k(y+te_j) = f_k(y) + t\,df(y)_{kj} + o(|t|),$$

where $df(y)_{kj}$ denotes the kth coordinate of the vector $df(y)e_j$ or the kj entry of the matrix $df(y)$. But this equation simply expresses the fact that the function $f_k(y+te_j)$ as a function of t is differentiable at $t=0$ with derivative $df(y)_{kj}$. We recognize this as just the usual *partial derivative* $\partial f_k/\partial x_j$, which is obtained by keeping all the variables $x_1, \ldots, x_n$ except x_j fixed and differentiating f_k as a function of x_j.

Note that we can define the *partial derivatives* independent of the concept of the differential. We simply say $\partial f_k/\partial x_j$ exists at a point y if $f_k(y+te_j) = f_k(y) + at + o(|t|)$ for some constant a as $t \to 0$ and then set $a = \partial f_k/\partial x_j(y)$. This can also be expressed as the limit of the difference quotient

$$\frac{\partial f_k}{\partial x_j}(y) = \lim_{t\to 0} \frac{f_k(y+te_j) - f_k(y)}{t}.$$

Notice that in general the differential is not expressible as the limit of a difference quotient, and for this reason we have consistently downplayed difference quotients in our development of differential calculus.

More generally, if u is any non-zero vector in $\mathbb{R}^n$, we can define the *directional derivative in the u direction*:

$$d_u f(y) = \lim_{t\to 0}(f(y+tu) - f(y))/t.$$

Note that this means $f(y+tu) = f(y) + td_u f(y) + o(t)$ as $t \to 0$. If f takes values in $\mathbb{R}^m$, then so does $d_u f$; we can also consider $d_u f_k$, the kth component of $d_u f$. Then $\partial f_k/\partial x_j(y) = d_{e_j} f_k(y)$. We will also write $\partial f/\partial x_j$ for $d_{e_j} f$ without taking components and call this the partial derivative of f with respect to x_j. (In some calculus books the term "directional derivative" is reserved for the case that u is a unit vector, but there is no need for such a restriction. In fact, we will also allow $u = 0$ in $d_u f$, in which case we obviously have $d_u f = 0$.)

We have observed that the existence of the differential implies the existence of parital derivatives. This generalizes easily to the case of directional derivatives.

Theorem 10.1.1 *If $f : D \to \mathbb{R}^m$ with $D \subseteq \mathbb{R}^n$ is differentiable at y with differential $df(y)$, then $d_u f$ exists at y for any u in $\mathbb{R}^n$ and $d_u f(y) = df(y)u$.*

Proof: From $f(x) = f(y) + df(y)(x-y) + o(|x-y|)$ we obtain by substituting $x = y + tu$

$$\begin{aligned} f(y+tu) &= f(y) + df(y)tu + o(|tu|) \\ &= f(y) + tdf(y)u + o(t). \end{aligned}$$

QED

Notice that choosing $u = e_j$ shows $df(y)e_j = \partial f/\partial x_j(y)$ as we originally noted. Also we can write out $df(y)u = \sum_{j=1}^n \partial f/\partial x_j(y)u_j$ or $(df(y)u)_k = \sum_{j=1}^n \partial f_k/\partial x_j(y)u_j$ in components. Thus all directional derivatives are determined by the special ones in the coordinate directions, if f is differentiable. This is a somewhat surprising result, because the directional derivative $d_u f(y)$ is defined entirely in terms of the values of f along the line $y + tu$ passing through y in the u direction (these lines are shown in Figure 10.1.1.) Aside from the point y itself, these lines have no points in common, so there is no reason to believe there should be any connection between the derivatives of f along them. In fact it is easy to construct functions for which there is no connection. Take the plane $\mathbb{R}^2$ and $y = (0,0)$, and define $f(r\cos\theta, r\sin\theta) = rg(\theta)$ for any function $g(\theta)$ that is odd, $g(-\theta) = -g(\theta)$. Then $d_u f(0,0) = g(\theta)$ if $u = (\cos\theta, \sin\theta)$. The function f is linear along each line through the origin, but the slopes are unrelated.

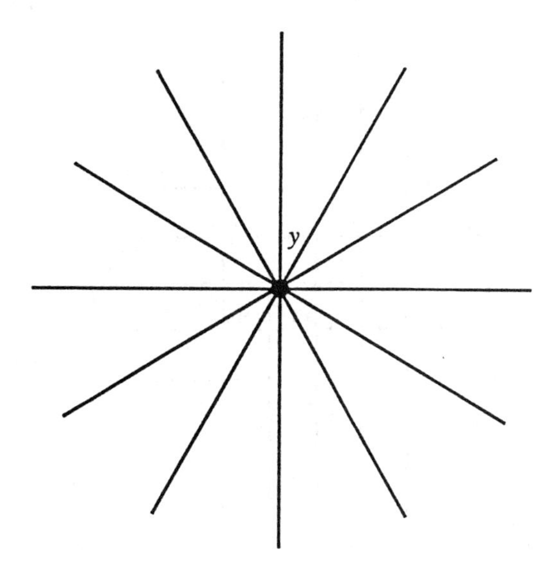

Figure 10.1.1:

The point is that these functions are not differentiable (unless $g(\theta)$ is very special). The graph of f near the origin has a folding fan appearance and is not smooth. We see that differentiability for a function of several variables is a rather strong condition. The existence of the partial derivatives, or even all directional derivatives, does not imply differentiability. In fact, in the above example if we take $g(\theta)$ smooth, say $g(\theta) = \sin 2\theta$, then $f(x, y) = xy/\sqrt{x^2 + y^2}$ has direction derivatives in all directions and at all points in the plane but is not differentiable at the origin. Thus the converse to the last theorem is false. It is a remarkable fact, however, that if the partial derivatives are continuous, then f has to be differentiable; this will provide a modified converse to the theorem. The hypothesis of continuity of $\partial f/\partial x_j(y)$ is strong enough to allow us to compare quantities computed from the values of f along different lines parallel to the axes and passing through points near y. Figure 10.1.2 illustrates the geometry in the plane. These lines fill up in a neighborhood of y.

Theorem 10.1.2 *Let $f : D \to \mathbb{R}^m$ for $D \subseteq \mathbb{R}^n$ have partial derivatives $\partial f/\partial x_j : D \to \mathbb{R}^m$ for $j = 1, \ldots, n$ that are continuous in a neighborhood of y. Then f is differentiable at y. Moreover, a necessary and*

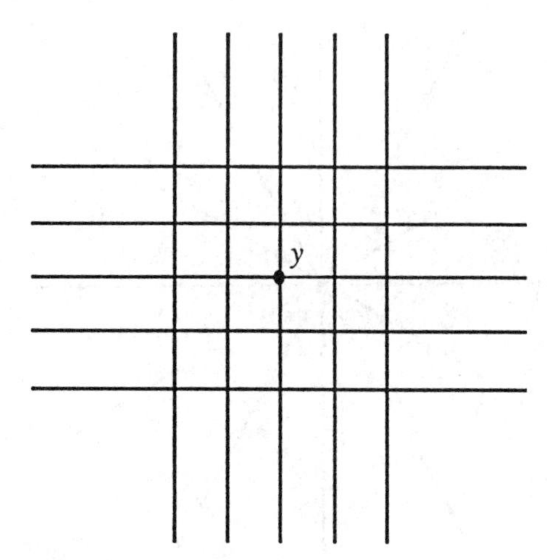

Figure 10.1.2:

sufficient condition that f be C^1 on D is that all the partial derivatives $\partial f / \partial x_j$ exist and are continuous.

Proof: The idea of the proof is to apply the mean value theorem n times, to express $f(x) - f(y)$ in terms of partial derivatives. The mean value theorem, however, involves evaluating derivatives at undetermined points, and we will use the continuity of the partial derivatives to make the change to the partial derivatives evaluated at y. For simplicity of notation we present the proof in the case $n = 2$ and $m = 1$.

Assume that $\partial f / \partial x_1$ and $\partial f / \partial x_2$ are continuous in a neighborhood of y. To prove the differentiability of f at y we write

$$f(x_1, x_2) - f(y_1, y_2) = [f(x_1, x_2) - f(x_1, y_2)] + [f(x_1, y_2) - f(y_1, y_2)]$$

and apply the mean value theorem to both terms. (These are the three corner points in Figure 10.1.3.) Notice that $f(x_1, x_2) - f(x_1, y_2)$ is the difference of the values of f along the horizontal line, so by the one-dimensional mean value theorem

$$f(x_1, x_2) - f(x_1, y_2) = \frac{\partial f}{\partial x_2}(x_1, z_2)(x_2 - y_2)$$

for some value z_2 between x_2 and y_2.

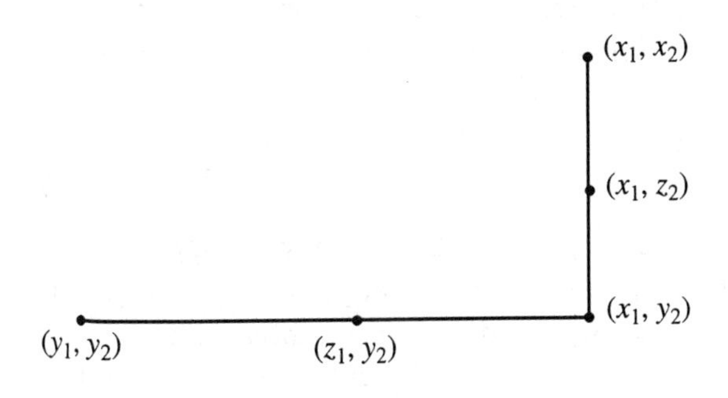

Figure 10.1.3:

Similarly, by the mean value theorem applied to f on the vertical line,

$$f(x_1, y_2) - f(y_1, y_2) = \frac{\partial f}{\partial x_1}(z_1, y_2)(x_1 - y_1)$$

for some z_1 between x_1 and y_2. This stage of the argument only requires the existence of the partial derivatives (not their continuity) to justify the use of the mean value theorem. Summing the two identities, we obtain

$$f(x_1, x_2) - f(y_1, y_2) = \partial f/\partial x_1(z_1, y_2)(x_1 - y_1) + \partial f/\partial x_2(x_1, z_2)(x_2 - y_2),$$

which is related to but not identical to the statement of differentiability

$$\begin{aligned} f(x_1, x_2) - f(y_1, y_2) &= \partial f/\partial x_1(y_1, y_2)(x_1 - y_1) \\ &\quad + \partial f/\partial x_2(y_1, y_2)(x_2 - y_2) + o(|x - y|), \end{aligned}$$

that we are trying to prove. Therefore, we rewrite our identity as

$$\begin{aligned} f(x_1, x_2) - f(y_1, y_2) &= \partial f/\partial x_1(y_1, y_2)(x_1 - y_1) \\ &\quad + \partial f/\partial x_2(y_1, y_2)(x_2 - y_2) + R, \end{aligned}$$

where

$$\begin{aligned} R &= [\partial f/\partial x_1(z_1, y_2) - \partial f/\partial x_1(y_1, y_2)](x_1 - y_1) \\ &\quad + [\partial f/\partial x_2(x_1, z_2) - \partial f/\partial x_2(y_1, y_2)](x_2 - y_2). \end{aligned}$$

To complete the proof we need to show $R = o(|x - y|)$ as $x \to y$ or $\lim_{x \to y} |R|/|x - y| = 0$. But $|x_1 - y_1| \leq |x - y|$ and $|x_2 - y_2| \leq |x - y|$, so

$$\frac{|R|}{|x-y|} \leq \left| \frac{\partial f}{\partial x_1}(z_1, y_2) - \frac{\partial f}{\partial x_1}(y_1, y_2) \right| + \left| \frac{\partial f}{\partial x_2}(x_1, z_2) - \frac{\partial f}{\partial x_2}(y_1, y_2) \right|.$$

Now we are ready to use the continuity of $\partial f/\partial x_1$ and $\partial f/\partial x_2$. We have only to note that as (x_1, x_2) approaches (y_1, y_2), so do (z_1, y_2) and (x_1, z_2) because z_1 and z_2 are intermediate values. Thus $\lim_{x \to y} R/|x - y| = 0$, proving the differentiability of f at y.

Next, assume that the partial derivatives are continuous on all of D. Then, by what we have shown, f is differentiable on all of D. Since the matrix $df(y)$ has entries $\partial f_k/\partial x_j(y)$ and these are continuous by hypothesis, it follows that $df : D \to \mathbb{R}^{m \times n}$ is continuous, hence f is C^1. Conversely, assume f is C^1; then $\partial f_k/\partial x_j(y)$ exists and is an entry of the matrix $df(y)$. But the continuity of the matrix function implies the continuity of the entries, so $\partial f_k/\partial x_j(y)$ is continuous. QED

If we examine the argument carefully, we observe an asymmetry among the variables. This was caused by the arbitrary choice of $f(x_1, y_2)$ over $f(y_1, x_2)$ for the intermediate comparison. As a result we ended up with a comparison of values of $\partial f/\partial x_1$ along a horizontal line, but the values of $\partial f/\partial x_2$ are compared between (x_1, z_2) and (y_1, y_2)—not along a vertical line. We will see in the next section that the proof of commutativity of partial derivatives is very similar.

This theorem is extraordinarily useful, since it provides a simple and expedient method of showing functions are differentiable. We even get continuity of the derivative in the bargain! We can immediately conclude from this theorem that every function $f : \mathbb{R}^n \to \mathbb{R}^m$ given by a finite formula involving arithmetic operations, roots, and special numerical functions such as exp, log, sin, cos, which are known to be differentiable, is continuously differentiable on its natural domain—all x for which all operations are defined (without dividing by zero or taking roots of zero, etc.).

10.1.3 The Chain Rule

While the previous theorem allows us to reduce many questions in the differential calculus of several variables to computations of partial derivatives, hence to the one-dimensional calculus, there are still situations in which we need to think in terms of the differential in its full multi-dimensional incarnation. One of these is the chain rule. The underlying idea is that the composition of two linear transformations corresponds to matrix multiplication. Since the differential calculus lets us approximate a differentiable function f by an affine function, which is a linear transformation plus a constant, the composition of differentiable functions should be differentiable and the differential should be obtained by appropriately multiplying the differentials of the functions being composed. As in the one-dimensional case, we need to specify the points at which the differentials are evaluated.

Let $f : D \to \mathbb{R}^m$ with $D \subseteq \mathbb{R}^n$ and $g : A \to \mathbb{R}^p$ with $A \subseteq \mathbb{R}^m$, and suppose the image $f(D)$ is contained in A; so $g \circ f : D \to \mathbb{R}^p$ is defined by $g \circ f(x) = g(f(x))$ for x in D. Now fix a point y in D, and let $z = f(y)$. Note that z is an image point of f, so z is in A.

Theorem 10.1.3 (*Chain Rule*) *If f is differentiable at y and g is differentiable at $z = f(y)$, then $g \circ f$ is differentiable at y and $d(g \circ f)(y) = dg(z)df(y)$ (matrix multiplication). Moreover, if f and g are differentiable (respectively C^1) on their domains, then so is $g \circ f$.*

Proof: We write

$$f(x) = f(y) + df(y)(x - y) + R_1(x, y)$$

and

$$g(w) = g(z) + dg(z)(w - z) + R_2(w, z)$$

where $R_1(x, y) = o(|x - y|)$ as $x \to y$ and $R_2(w, z) = o(|w - z|)$ as $w \to z$. Setting $w = f(x)$ and $z = f(y)$ in the second equation and then using the first equation we obtain

$$\begin{aligned} g(f(x)) &= g(f(y)) + dg(z)(f(x) - f(y)) + R_2(f(x), f(y)) \\ &= g(f(y)) + dg(z)df(y)(x - y) \\ &\quad + dg(z)R_1(x, y) + R_2(f(x), f(y)). \end{aligned}$$

This will prove both the differentiability of $g \circ f$ at y and the formula for the differential, if we can show that $dg(z)R_1(x, y) + R_2(f(x), f(y)) = o(|x - y|)$ as $x \to y$. But $dg(z)R_1(x, y) = o(|x - y|)$ because $R_1(x, y) =$

$o(|x-y|)$ and $dg(z)$ is a fixed matrix (so $|dg(z)R_1(x,y)| \le c|R_1(x,y)|$ where c depends only on $dg(z)$). To show $R_2(f(x), f(y)) = o(|x-y|)$ we need to appeal to the pointwise Lipschitz continuity of f at y, which we have shown to be a consequence of the differentiability of f at y.

Given any error $1/N$, we first use $R_2(z,w) = o(|z-w|)$ to find $1/k$ such that $|z-w| \le 1/k$ implies $|R_2(z,w)| \le |z-w|/NM$, where M is the constant in the pointwise Lipschitz condition $|f(x) - f(y)| \le M|x-y|$ holding for x in a neighborhood of y, say $|x-y| < 1/p$. Next we choose $1/q$ so that $1/q \le 1/p$ and $|x-y| < 1/q$ implies $|f(x)-f(y)| < 1/k$ since f is continuous (we may take $1/q =$ minimum $(1/p, 1/kM)$). Then if $|x-y| < 1/q$ we have

$$|R_2(f(x), f(y))| \le \frac{|f(x) - f(y)|}{MN} \le \frac{|x-y|}{N}.$$

This proves $R_2(f(x), f(y)) = o(|x-y|)$.

Thus we have simultaneously shown that $g \circ f$ is differentiable at y and computed its differential at y to be $dg(z)df(y)$, under the assumptions that f is differentiable at y and g is differentiable at $z = f(y)$. If f and g are differentiable on their domains this shows $g \circ f$ is differentiable on its domain, and if dg and df are continuous the formula for $d(g \circ f)$ shows that it, too, is continuous. QED

By taking components we can obtain the chain rule for partial derivatives. For simplicity of notation we assume the range of g is $\mathbb{R}^1$, so $g \circ f(x) = g(f_1(x), \ldots, f_m(x))$. Then

$$\partial(g \circ f)/\partial x_j(y) = \sum_{k=1}^{m} \partial g/\partial z_k(z) \partial f_k/\partial x_j(y).$$

This formula was derived under the assumption that f and g are differentiable—not just that the partial derivatives exist. We can easily interpret this formula if we think about how a small change in the x_j variable affects $g \circ f$. First it produces a change in each $f_k(x)$ roughly proportional to $\partial f_k/\partial x_j$, and each of these is transmitted to $g \circ f$ with a factor roughly equal to $\partial g/\partial z_k$. The fact that these terms are more or less independent, so we sum their contributions, can be justified by arguing that any interactions between terms will be of smaller order and hence vanish in the limit. This can be seen clearly if we take polynomial functions, say $f_1(x_1, x_2) = x_1x_2$, $f_2(x_1, x_2) = x_1^2 - x_2^2$, and $g(z_1, z_2) = z_1z_2$. Then giving x_1 an increment, $x_1 + h$ results in $f_1(x_1 + h, x_2) = (x_1+h)x_2 = x_1x_2+x_2h$ and $f_2(x_1+h, x_2) = (x_1+h)^2 - x_2^2 = x_1^2 - x_2^2 + 2x_1h + h^2$. Here we have $f_1(x_1+h, x_2) = f_1(x_1, x_2) + \partial f_1/\partial x_1(x_1, x_2)h$ exactly, while $f_2(x_1+h, x_2) = f_2(x_1, x_2) + \partial f_2/\partial x_1(x_1, x_2)h + h^2$. Substituting into g we find

$$\begin{aligned}
& g(f_1(x_1 + h, x_2), f_2(x_1 + h, x_2)) \\
&\quad = f_1(x_1 + h, x_2) \cdot f_2(x_1 + h, x_2) \\
&\quad = (x_1x_2 + x_2h)(x_1^2 - x_2^2 + 2x_1h + h^2) \\
&\quad = (x_1x_2)(x_1^2 - x_2^2) + ((x_1x_2)2x_1 + x_2(x_1^2 - x_2^2))h \\
&\qquad + (x_2 2x_1 + x_1x_2)h^2 + x_2h^3
\end{aligned}$$

and the h^2 and h^3 terms are discarded in the $o(h)$ remainder. Notice that the term $x_2 2x_1h^2$ comes from an interaction between $\partial f_1/\partial x_1(x_1, x_2)h$ and $\partial f_2/\partial x_1(x_1, x_2)h$ and it is, as expected, of smaller order than the significant terms.

The chain rule can also be interpreted as giving a formula for transforming partial derivatives under a change of variable. For this interpretation we think of $g(z_1, \ldots, z_m)$ as a given function of m variables whose partial derivatives $\partial g/\partial z_k$ are known. We then assume that $x_1, \ldots, x_n$ are new variables that are connected to the z variables by the equations $z_k = f_k(x_1, \ldots, x_n)$. (Usually we have $n = m$ in this interpretation, although strictly speaking this isn't necessary.) Then we may regard g as also a function of the x variables, $g(z_1, \ldots, z_m) = g(f_1(x_1, \ldots, x_n), f_2(x_1, \ldots, x_n), \ldots, f_m(x_1, \ldots, x_n))$. Of course this new function is just $g \circ f(x)$. Then the partial derivatives of the new g with respect to x_j are computed by the chain rule:

$$\frac{\partial g}{\partial x_j} = \sum_{k=1}^{n} \frac{\partial g}{\partial z_k}\left(\frac{\partial z_k}{\partial x_j}\right).$$

Of course this notation suppresses some of the evidence of what is going on, but it is undeniably convenient. There is, however, one grave danger of confusion. If some of the z variables are the same as the x variables, the corresponding partial derivatives may be different. In other words, $z_1 = x_1$ does not imply $\partial g/\partial x_1$ and $\partial g/\partial z_1$ are equal. Perhaps the simplest way to understand this is to realize that $\partial g/\partial x_1$

means the derivative of g obtained by varying x_1 and holding fixed $x_2, \ldots, x_n$, while $\partial g/\partial z_1$ involves varying $z_1 = x_1$ but holding fixed $z_2, \ldots, z_m$. Clearly these are different. The real problem is that when $z_1 = x_1$ one is tempted not to introduce a new name for the variable. Clearly this is a temptation to be resisted.

We conclude this section with a discussion of maximum and minimum problems in several variables. This is just a preliminary discussion, and we will return to the topic several times again. Here we simply want to observe that the vanishing of the gradient is a necessary condition for the existence of a max or min in the interior of the domain, which is almost an immediate consequence of the analogous one-dimensional result.

Theorem 10.1.4 *Let $f : D \to \mathbb{R}$ for $D \subseteq \mathbb{R}^n$, and let y be a point in the interior of D. If f assumes its maximum or minimum value at y and f is differentiable at y, then $df(y) = 0\,(i.e., \partial f/\partial x_1(y) = 0, \partial f/\partial x_2(y) = 0, \ldots, \partial f/\partial x_n(y) = 0)$.*

Proof: If y is in the interior of D, then $y + te_j$ must belong to D for t in a neighborhood of zero; and if $df(y)$ exists, then $g(t) = f(y + te_j)$ is differentiable at $t = 0$. Clearly g attains its max or min at $t = 0$, so $g'(0) = 0$ by the $n = 1$ case. Thus $\partial f/\partial x_j(y) = g'(0)$ for all j, so $df(y) = 0$. QED

10.1.4 Differentiation of Integrals

We return to the question of differentiating a general function defined by an integal. A function of a single variable x might be given by an expression

$$f(x) = \int_{a(x)}^{b(x)} g(x, y)\, dy$$

where $a(x)$ and $b(x)$ are C^1 functions of one variable and g is C^1 on $\mathbb{R}^2$. In Chapter 6, we stated without proof the theorem that f is C^1 and its derivative is given by

$$f'(x) = b'(x)g(x, b(x)) - a'(x)g(x, a(x)) + \int_{a(x)}^{b(x)} \frac{\partial g}{\partial x}(x, y)\, dy.$$

We are now in a position to give the proof.

To do this we introduce a function $F(x_1, x_2, x_3)$ of three variables that isolates the three appearances of x in the definition of f:

$$F(x_1, x_2, x_3) = \int_{a(x_2)}^{b(x_1)} g(x_3, y)\, dy.$$

Then $f(x) = F(x,x,x)$ and so

$$f'(x) = \frac{\partial F}{\partial x_1}(x,x,x) + \frac{\partial F}{\partial x_2}(x,x,x) + \frac{\partial F}{\partial x_3}(x,x,x)$$

by the chain rule (provided F is C^1). But the computation of $\partial F/\partial x_1$ and $\partial F/\partial x_2$ is easily accomplished by the one-dimensional chain rule and the fundamental theorem of the calculus (differentiation of the integral). For $F(x_1, x_2, x_3) = G(b(x_1), a(x_2), x_3)$ where $G(b, a, x_3) = \int_a^b g(x_3, y)\, dy$, so

$$\frac{\partial F}{\partial x_1}(x_1, x_2, x_3) = \frac{\partial G}{\partial b}\frac{\partial b}{\partial x_1} = b'(x_1)g(x_3, b)(x_1))$$

and

$$\frac{\partial F}{\partial x_2}(x_1, x_2, x_3) = \frac{\partial G}{\partial a}\frac{\partial a}{\partial} x_2 = -a'(x_2)g(x_3, a(x_2)).$$

This shows $\partial F/\partial x_1$ and $\partial F/\partial x_2$ are continuous, and $\partial F/\partial x_1(x,x,x)$ and $\partial F/\partial x_2(x,x,x)$ are the first two terms in the claimed formula for $f'(x)$.

It remains to compute $\partial F/\partial x_3$. Since $b(x_1)$ and $a(x_2)$ are held fixed, we can simplify the notation and show

$$\frac{\partial}{\partial x}\int_a^b g(x,y)\, dy = \int_a^b \frac{\partial g}{\partial x}(x,y)\, dy.$$

We also need to show that this a continous function. Thus to complete the proof of our differentiation formula we need to establish the following lemma.

Lemma 10.1.1

a. *If g is a continuous function on $\mathbb{R}^2$, then $\int_a^b g(x,y)\, dy$ is a continuous function of x.*

b. *If g is C^1 on $\mathbb{R}^2$, then $\int_a^b g(x,y)\,dy$ is C^1 on $\mathbb{R}$ and*

$$\frac{d}{dx}\int_a^b g(x,y)\,dy = \int_a^b \partial g/\partial x(x,y)\,dy.$$

Proof:

a. On any compact set, say the rectangle $a \le x \le b$, $c \le y \le d$, the function g is uniformly continuous. So given any $1/n$ there exists $1/m$ such that $|g(x,y) - g(x',y)| \le 1/n$ provided $|x - x'| \le 1/m$ for (x,y) and (x',y) in the rectangle (of course more is true, but this is all we need). If we consider two Cauchy sums approximating $\int_a^b g(x,y)\,dy$ and $\int_a^b g(x',y)\,dy$, evaluating at the same y values, they can differ by at most $(b-a)/n$, because

$$\begin{aligned}
&\left|\sum g(x,y_j)\Delta y_j - \sum g(x',y_j)\Delta y_j\right| \\
&\quad \le \sum |g(x,y_j) - g(x',y_j)|\Delta y_j \\
&\quad \le \frac{1}{n}\sum \Delta y_j = (b-a)/n.
\end{aligned}$$

Keeping x and x' fixed (with $|x-x'| \le 1/m$) and taking the limit as the maximum interval length of the partition goes to zero, the sums become integrals and we obtain $|\int_a^b g(x,y)\,dy - \int_a^b g(x',y)\,dy| \le (b-a)/n$. This is the desired continuity of the integral.

b. We form the difference quotient

$$\frac{1}{h}\left(\int_a^b g(x+h,y)\,dy - \int_a^b g(x,y)\,dy\right) = \int_a^b \frac{1}{h}(g(x+h,y) - g(x,y))\,dy.$$

We want to take the limit as $h \to 0$ and interchange the limit and the integral, which will be justified if we can show that $(g(x+h,y) - g(x,y))/h$ converges to $\partial g/\partial x(x,y)$ uniformly for y in $[c,d]$. But the mean value theorem shows that the difference quotient $(g(x+h,y) - g(x,y))/h$ is equal to $\partial g/\partial x(z,y)$ for some point z between x and $x+h$ (z depends on y also). The uniform convergence of $\partial g/\partial x(z,y)$ to $\partial g/\partial x(x,y)$ then follows from the uniform continuity of $\partial g/\partial x$ on the rectangle. QED

10.1.5 Exercises

1. If A is any $m \times n$ matrix prove that there exists a constant M such that $|Ax| \leq M|x|$ for every x in $\mathbb{R}^n$.

2. Prove that $f : D \to \mathbb{R}^m$ is differentiable at a point if and only if each of the coordinate functions $f_k : D \to \mathbb{R}$ is differentiable at that point.

3. If f is differentiable at y, show that $d_u f(y)$ is linear in u, meaning $d_{(au+bv)} f(y) = a d_u f(y) + b d_v f(y)$ for any scalars a and b.

4. Let $f : D \to \mathbb{R}^m$ and $g : D \to \mathbb{R}^m$ be differentiable at y. Let $f \cdot g : D \to \mathbb{R}$ be defined by the dot product in $\mathbb{R}^m$. Prove that $f \cdot g$ is differentiable at y, and find a formula for the differential $d(f \cdot g)(y)$.

5. Let $f : D \to \mathbb{R}^3$ and $g : D \to \mathbb{R}^3$ be differentiable at y, and let $f \times g$ be defined by the vector cross product in $\mathbb{R}^3$. Prove that $f \times g$ is differentiable at y and $d(f \times g)(y) = df(y) \times g(y) + f(y) \times dg(y)$.

6. Let $f : D \to \mathbb{R}$ be differentiable at y, and suppose $\nabla f(y) \neq 0$. Show that $d_u f(y)$ as u varies over all unit vectors ($|u| = 1$) attains its maximum value when $u = \lambda \nabla f(y)$ for some $\lambda > 0$ and $d_u f(y) = |\nabla f(y)|$ for that choice of u.

7. Let $f : D \to \mathbb{R}$ be differentiable at y, and suppose $\nabla f(y) \neq 0$. Show that $d_u f(y) = 0$ if u is orthogonal to $\nabla f(y)$.

8. Let $f : D \to \mathbb{R}$ for $D \subseteq \mathbb{R}^2$ be differentiable. Let (x, y) denote cartesian coordinates in $\mathbb{R}^2$ and (r, θ) denote polar coordinates in $\mathbb{R}^2$. Express $\partial f / \partial x$ and $\partial f / \partial y$ in terms of $\partial f / \partial r$ and $\partial f / \partial \theta$ and, conversely, at every point except the origin.

9. Let $f : \mathbb{R}^n \to \mathbb{R}$ be differentiable. Show that there exists $g : \mathbb{R}^{n-1} \to \mathbb{R}$ with $f(x_1, \ldots x_n) = g(x_2, \ldots, x_n)$ if and only if $\partial f / \partial x_1 \equiv 0$.

10. Let $g : [a, b] \to \mathbb{R}^n$ be differentiable. If $f : \mathbb{R}^n \to \mathbb{R}^1$ is differentiable, what is the derivative $(d/dt) f(g(t))$?

11. Let $x_1, \ldots, x_n$ and $y_1, \ldots, y_n$ be given real numbers with the x's distinct. Find the affine function $g(x) = ax + b$ such that $\sum_{j=1}^{n}(y_j - g(x_j))^2$ is minimized.

12. *Suppose $f : \mathbb{R}^n \to \mathbb{R}$ is C^1 and $g : \mathbb{R}^n \to \mathbb{R}$ is an affine function such that the graphs of f and g intersect at the point $(y, f(y))$ in $\mathbb{R}^{n+1}$ but do not intersect at any other point in a neighborhood of $(y, f(y))$, where $n \geq 2$. Prove that g is the best affine approximation to f at y.

13. Show that the following functions are differentiable and compute df:

    a. $f : \mathbb{R}^2 \to \mathbb{R}^1, \quad f(x_1, x_2) = x_1 e^{x_2}$.
    b. $f : \mathbb{R}^3 \to \mathbb{R}^2, \quad f(x_1, x_2, x_3) = (x_3, x_2)$.
    c. $f : \mathbb{R}^2 \to \mathbb{R}^3, \quad f(x_1, x_2) = (x_1, x_2, x_1 x_2)$.

14. *A contour map shows the curves $h(x, y) = c$ for values of c differing by fixed amounts (usually 50 feet or 100 feet), where h is the altitude function. The gradient ∇h is larger in regions where the contour curves are denser, and ∇h lies in a direction roughly perpendicular to the contour curves. Explain why this is so.

15. If $f : D \to \mathbb{R}$ is C^1 with $D \subseteq \mathbb{R}^n$ and D contains the line segment joining x and y, show that $f(y) = f(x) + \nabla f(z) \cdot (y - x)$ for some point z on the line segment. Explain why this is an n-dimensional analog of the mean value theorem.

16. Let $f : \mathbb{R}^n \to \mathbb{R}^m$ be C^1. Show that $df \equiv 0$ if and only if f is contant and that df is constant if and only if f is an affine function.

17. Let $f : \mathbb{R}^2 \to \mathbb{R}$ be C^1 and satisfy $f(0, y) = 0$ for all y. Prove that there exists $g : \mathbb{R}^2 \to \mathbb{R}^1$ that is C^1, such that $f(x, y) = xg(x, y)$

18. If $f : \mathbb{R} \to \mathbb{R}$ is C^1 and $g : \mathbb{R} \to \mathbb{R}$ is continuous and one of them has compact support, show that $f * g$ is C^1 and $(f * g)' = f' * g$.

19. If $g : \mathbb{R}^{n+1} \to \mathbb{R}$ is C^1 and $f(x) = \int_a^b g(x, y)\, dy$ (for $x \in \mathbb{R}^n$), show that f is C^1 and $\partial f / \partial x_j(x) = \int_a^b \partial g / \partial x_j(x, y)\, dy$.

10.2 Higher Derivatives

10.2.1 Equality of Mixed Partials

We begin by defining the second derivative. If $f : D \to \mathbb{R}^m$ (with $D \subseteq \mathbb{R}^n$) is differentiable, we can regard df as a function $df : D \to \mathbb{R}^{m\times n}$ taking values in the space of $m \times n$ matrices. We can then ask if this function is differentiable. If it is, its differential $d(df)(y)$ at a point will be an $(m \times n) \times n$ matrix, which we will call the *second derivative*, $d^2f(y)$. For simplicity we will usually deal with the case $m = 1$, since the general case reduces to this by considering coordinate functions. If $f : D \to \mathbb{R}$, then $d^2f(y)$, if it exists, is an $n \times n$ matrix called the *Hessian* of f at y. We can also define higher derivatives, but the notation becomes a bit awkward.

What are the entries of the Hessian matrix? To answer this we write

$$df = \begin{pmatrix} \dfrac{\partial f}{\partial x_1} \\ \vdots \\ \dfrac{\partial f}{\partial x_n} \end{pmatrix}$$

as a column vector. (We are thinking of df as taking values in $\mathbb{R}^n$ that we identify with $1 \times n$ matrices by transposing the row vector to a column vector.) If $df : D \to \mathbb{R}^n$ is differentiable, then we can again express $d(df)$ in terms of partial derivatives:

$$d(df) = \begin{pmatrix} \dfrac{\partial}{\partial x_1}\dfrac{\partial f}{\partial x_1} & \cdots & \dfrac{\partial}{\partial x_n}\dfrac{\partial f}{\partial x_1} \\ \vdots & & \vdots \\ \dfrac{\partial}{\partial x_1}\dfrac{\partial f}{\partial x_n} & \cdots & \dfrac{\partial}{\partial x_n}\dfrac{\partial f}{\partial x_n} \end{pmatrix}.$$

We write $\partial^2 f/\partial x_j \partial x_k$ for $\partial/\partial x_j\ (\partial f/\partial x_k)$ and so $(d^2f)_{jk} = \partial^2 f/\partial x_j \partial x_k$. We call $\partial^2 f/\partial x_j \partial x_k$ a *second-order partial derivative*, which is given by

$$\frac{\partial^2 f}{\partial x_j \partial x_k}(y) = \lim_{t\to 0} \frac{\partial f/\partial x_k(y + te_j) - \partial f/\partial x_k(y)}{t}.$$

Notice that this can be defined, independent of the existence of d^2f, as long as $\partial f/\partial x_k$ exists in a neighborhood of y.

The main result in which we are interested is the equality $\partial^2 f/\partial x_j \partial x_k = \partial^2 f/\partial x_k \partial x_j$. This is not true without additional hypotheses, but the counterexamples are not of great significance, so we leave them to the exercises. We can interpret the identity $\partial^2 f/\partial x_j \partial x_k = \partial^2 f/\partial x_k \partial x_j$ in two ways. First, it says that the Hessian is a symmetric matrix. This is a significant observation since there are a number of powerful theorems of linear algebra that apply to symmetric matrices. For example, a symmetric matrix always has a complete set of eigenvectors. This fact is especially valuable in studying maxima and minima.

The second interpretation of $\partial^2 f/\partial x_j \partial x_k = \partial^2 f/\partial x_k \partial x_j$ involves the commutativity of the "operators" $\partial/\partial x_j$ and $\partial/\partial x_k$. Here we are thinking of the partial derivatives $\partial/\partial x_j$ as functions ("operators") whose domain and range consist of spaces of functions $f : D \to \mathbb{R}$, with $\partial/\partial x_j$ mapping the "point" f to the "point" $\partial f/\partial x_j$. Without going into the details of the precise definition of $\partial/\partial x_j$ as an operator, it is clear that $\partial/\partial x_j \, (\partial f/\partial x_k) = \partial/\partial x_k \, (\partial f/\partial x_j)$ does in fact express a commutative law for partial derivatives.

If we simply substitute the definitions in terms of difference quotients we find

$$\begin{aligned}
&\frac{\partial^2 f}{\partial x_k \partial x_j}(x) \\
&= \frac{\partial}{\partial x_k}\left(\frac{\partial f}{\partial x_j}(x)\right) = \lim_{t\to 0} \frac{\dfrac{\partial f}{\partial x_j}(x+te_k) - \dfrac{\partial f}{\partial x_j}(x)}{t} \\
&= \lim_{t\to 0}\lim_{s\to 0} \frac{f(x+te_k+se_j) - f(x+te_k)}{st} - \lim_{s\to 0} \frac{f(x+se_j)-f(x)}{st} \\
&= \lim_{t\to 0}\lim_{s\to 0} \frac{f(x+te_k+se_j) - f(x+te_k) - f(x+se_j) + f(x)}{st}.
\end{aligned}$$

On the other hand, if we compute $\partial^2 f/\partial x_j \partial x_k$ we will obtain the double limit of the same expression (think of it as a mixed second difference quotient) with the order of the limits interchanged. Thus the identity of $\partial^2 f \partial x_k \partial x_j$ and $\partial^2 f \partial x_j \partial x_k$ is a statement about the interchange of two limits.

Theorem 10.2.1 *Let $f : D \to \mathbb{R}$ and all partial derivatives of order one and two be continous. Then $\partial^2 f/\partial x_j \partial x_k = \partial^2 f/\partial x_k \partial x_j$ for all j*

and k.

Proof: We will use the mean value theorem twice to replace the mixed second difference quotient by a second partial derivative evaluated at an undetermined point and then use the continuity of the second partial derivatives. Define the *difference operator,* Δ_u, by $\Delta_u f(x) = f(x+u) - f(x)$. In this notation the mixed second difference quotient is $\Delta_{te_k}\Delta_{se_j} f(x)/st$. A direct computation shows that the operators Δ_{te_k} and Δ_{se_j} commute, that is, $\Delta_{te_k}\Delta_{se_j} f(x) = \Delta_{se_j}\Delta_{te_k} f(x)$, since both are equal to $f(x + te_k + se_j) - f(x + te_k) - f(x + se_j) + f(x)$. Consider first $\Delta_{te_k}\Delta_{se_j} f(x)$, and think of it as $\Delta_{te_k} g(x)$ for the function $g(x) = \Delta_{se_j} f(x)$. Note that the mean value theorem for g (regarded as a function of x_k alone, with the other variables held fixed) can be written in the form $\Delta_{te_k} g(x)/t = \partial g/\partial x_k(x + t_1 e_k)$ for some t_1 between 0 and t. Thus

$$\begin{aligned}
\frac{1}{st}\Delta_{te_k}\Delta_{se_j} f(x) &= \frac{1}{s}\frac{\partial}{\partial x_k}\left(\Delta_{se_j} f\right)(x + t_1 e_k) \\
&= \frac{1}{s}\frac{\partial}{\partial x_k}(f(x + se_j + t_1 e_k) - f(x + t_1 e_k)) \\
&= \frac{1}{s}\left(\frac{\partial f}{\partial x_k}(x + t_1 e_k + se_j) - \frac{\partial f}{\partial x_k}(x + t_1 e_k)\right).
\end{aligned}$$

Then one more application of the mean value theorem yields

$$\frac{1}{st}\Delta_{te_k}\Delta_{se_j} f(x) = \frac{\partial^2 f}{\partial x_j \partial x_k}(x + t_1 e_k + s_1 e_j),$$

where s_1 lies between 0 and s. The same argument with the difference operators in the reverse order shows

$$\frac{1}{st}\Delta_{se_j}\Delta_{te_k} f(x) = \frac{\partial^2 f}{\partial x_k \partial x_j}(x + t_2 e_k + s_2 e_j)$$

for different values t_2 and s_2 in the same range. Since the difference operators commute, we have the equality of the mixed partial derivatives in reverse order at two different points, $\partial^2 f/\partial x_k \partial x_j(x') = \partial^2 f/\partial x_j \partial x_k(x'')$, where $x' = x + t_1 e_k + s_1 e_j$ and $x'' = x + t_2 e_k + s_2 e_j$ are both confined to the same small rectangle near x, as shown in Figure 10.2.1. Since this is true for all s and t, we have only to let $s \to 0$ and

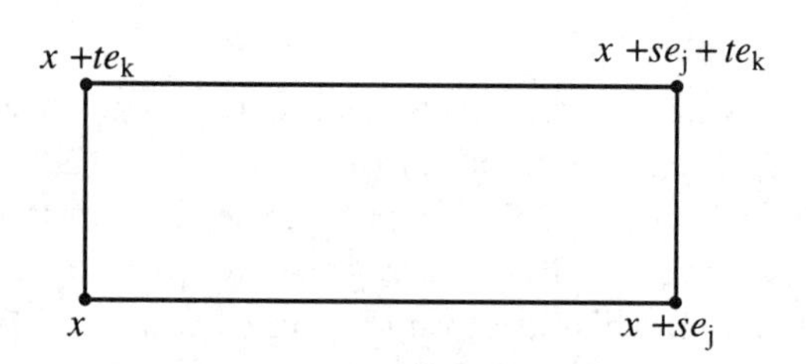

Figure 10.2.1:

$t \to 0$ in any order and appeal to the continuity of $\partial^2 f/\partial x_j \partial x_k$ and $\partial^2 f/\partial x_k \partial x_j$ to obtain $\partial^2 f/\partial x_j \partial x_k(x) = \partial^2 f/\partial x_k \partial x_j(x)$. QED

A careful reworking of the final limiting argument in the proof will allow you to deduce the equality from the assumption that just one of the mixed second partial derivatives is continuous—however, this is of minor interest.

Under the hypothesis that all partial derivatives of orders one and two are continuous we can apply Theorem 10.1.2 twice to conclude that df is differentiable. Such functions are said to be of class C^2. Similarly if all partial derivatives of orders up to k are continuous, the function is said to be C^k. This again implies that full derivatives up to order k exist and are continuous. If f is in C^k for all finite k, we say f is C^∞.

In dealing with partial derivatives of higher order, a good notation is very important. We will find the multi-index notation introduced for polynomials in the last chapter extends nicely to this context. We let

$$\left(\frac{\partial}{\partial x}\right)^\alpha f = \left(\frac{\partial}{\partial x_1}\right)^{\alpha_1} \cdot \left(\frac{\partial}{\partial x_2}\right)^{\alpha_2} \cdots \left(\frac{\partial}{\partial x_n}\right)^{\alpha_n} f$$

for any multi-index α. We will always assume the functions involved are C^k for $k = |\alpha|$ (recall $|\alpha| = \alpha_1 + \alpha_2 + \cdots + \alpha_n$) so that the order of the partial derivatives is irrelevant. Note that

$$\left(\frac{\partial}{\partial x}\right)^\alpha \left(\frac{\partial}{\partial x}\right)^\beta f = \left(\frac{\partial}{\partial x}\right)^{\alpha+\beta} f,$$

again assuming f is C^k for $k = |\alpha| + |\beta|$.

10.2.2 Local Extrema

We now consider further the question of local maxima and minima for C^2 functions. We have seen already that the vanishing of the differential is a necessary condition for a local extremum. A point where $df(x) = 0$ is called a *critical point.* In this section we will show how to use the second derivative d^2f to analyze the behavior of a function near a critical point. It will turn out that we can reduce the problem to the one-dimensional situation by considering the restriction of the function to all lines passing through the critical point.

Let $f : D \to \mathbb{R}$ be a C^2 function with a critical point at y, so $df(y) = 0$. Let $d^2f(y)$ denote the Hessian matrix $\{\partial^2 f(y)/\partial x_j \partial x_k\}$ at the point y. For any line passing through y, given as $y + tu$, the restriction of f to the line gives a function $g(t) = f(y + tu)$ of one variable, which is C^2. Note that $g'(t) = df(y + tu)u = \sum_{j=1}^n \partial f/\partial x_j(y + tu)u_j$ by the chain rule, so $g'(0) = 0$. Also $g''(t) = \sum_{j,k=1}^n \partial^2 f/\partial x_j \partial x_k(y+tu)u_j u_k = \langle d^2f(y+tu)u, u\rangle$, where $\langle\ ,\ \rangle$ denotes the inner product on $\mathbb{R}^n$. Thus it is not surprising that the expression $\langle d^2 f(y)u, u\rangle$ (regarded as a function of u in $\mathbb{R}^n$) is the key quantity to study, since it is $g''(0)$ and we know how $g''(0)$ relates to the behavior of $g(t)$ near the critical point $t = 0$. For example, if f has a local minimum at y, then so does g at 0 and so $g''(0) \geq 0$. On the other hand, if $g''(0) > 0$, then we know g has a strict local minimum at 0. If this is true for all lines through y, we expect to be able to prove that f has a strict local minimum at y.

This discussion points to the importance of conditions like $\langle d^2f(y)u, u\rangle \geq 0$ or $\langle d^2f(y)u, u\rangle > 0$ for all vectors $u \neq 0$ in $\mathbb{R}^n$. The first case we call *non-negative definite* and the second we call *positive definite.* These definitions come from the theory of *quadratic forms,* which we can define simply as functions on $\mathbb{R}^n$ of the form $\langle Au, u\rangle = \sum_{j,k=1}^n A_{jk}u_j u_k$ where A is any *symmetric* matrix. A quadratic form is said to be *non-negative definite* if $\langle Au, u\rangle \geq 0$ for all non-zero u in $\mathbb{R}^n$ and *positive definite* if $\langle Au, u\rangle > 0$ for all non-zero u in $\mathbb{R}^n$. We define *non-positive definite* and *negative definite* in the same way with the inequalities reversed.

We will need the following basic facts about positive definite quadratic forms:

Lemma 10.2.1

a. *A quadratic form $\langle Au, u\rangle$ is positive definite if and only if there exists $\varepsilon > 0$ such that $\langle Au, u\rangle \geq \varepsilon |u|^2$ for all u.*

b. *If $\langle Au, u\rangle$ is positive definite, then so is $\langle Bu, u\rangle$ for all symmetric matrices B sufficiently close to A.*

Proof:

a. It is clear that the inequality implies that the quadratic form is positive definite, so it suffices to prove the converse. Note that $\langle Atu, tu\rangle = t^2\langle Au, u\rangle$ for real t, so the quadratic form has the same sign along lines through the origin. Thus $\langle Au, u\rangle$ is positive definite if and only if it satisfies $\langle Au, u\rangle > 0$ for u in the unit sphere $|u| = 1$. (The setup is symbolized in Figure 10.2.2).

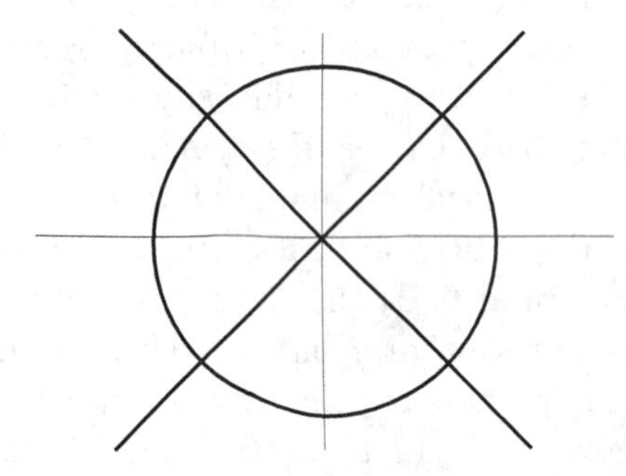

Figure 10.2.2:

But the sphere is compact, and $\langle Au, u\rangle$ is continuous on $\mathbb{R}^n$ and, hence, on the sphere (with the subspace metric). Thus the quadratic form attains its inf on the sphere, which must be positive since $\langle Au, u\rangle > 0$. Thus we must have $\langle Au, u\rangle \geq \varepsilon > 0$ on the sphere for some $\varepsilon > 0$. This then implies $\langle Au, u\rangle \geq \varepsilon|u|^2$ for general u (because $|tu| = 1$ if $t = |u|^{-1}$).

b. The symmetric matrices may be viewed as a Euclidean space of dimension $n(n+1)/2$ (as coordinate variables take all entries on or above the diagonal—the below diagonal entries are determined by symmetry). Given a positive definite quadratic form $\langle Au, u\rangle$, choose

$\varepsilon > 0$ as in part a. If we take any symmetric matrix B sufficiently close to A in the Euclidean metric, then all the entries of $B - A$ will be small and we can make $\langle (B - A)u, u\rangle \leq \varepsilon|u|^2/2$. Then

$$\langle Bu, u\rangle = \langle Au, u\rangle + \langle (B - A)u, u\rangle \geq \varepsilon|u|^2 - \varepsilon|u|^2/2 = \varepsilon|u|^2/2,$$

so B is positive definite. QED

Part b of the lemma says exactly that the positive definite matrices form an open set in the space of symmetric matrices. In particular, for a C^2 function f, if $\langle d^2f(x)u, u\rangle$ is positive definite for one value of x it must be positive definite in a neighborhood of x.

Theorem 10.2.2 *Let $f : D \to \mathbb{R}$ be C^2 with $D \subseteq \mathbb{R}^n$ an open set. Let y in D be a critical point.*

a. *If y is a local minimum (respectively maximum), then $d^2f(y)$ is non-negative definite (respectively nonpositive definite).*

b. *If $d^2f(y)$ is positive definite (respectively negative definite), then y is a strict local minimum (respectively maximum).*

Proof:

a. This is an immediate consequence of the one-dimensional case and the fact that $g''(0) = \langle d^2f(y)u, u\rangle$ for $g(t) = f(y + tu)$.

b. Here we have to proceed more cautiously, because we cannot deduce that f has a local minimum at y merely from the fact that $g(t) = f(y + tu)$ has a strict local minimum at 0 for every u. The problem is that $g(t)$ having a strict local minimum at 0 means there exists $\varepsilon > 0$ such that $g(t) > g(0)$ for $t \neq 0$ and $|t| < \varepsilon$. But this ε depends on g and, hence, on u. Thus we know only that $f(y+tu) > f(y)$ if $t \neq 0$ and $|t| < \varepsilon(u)$, and the set of such values $y + tu$ may not necessarily constitute a neighborhood of y, as in Figure 10.2.3. To show y is a strict local minimum we need to show $f(x) > f(y)$ for x in a neighborhood of y. We will show that this is true on a ball $|x - y| < \varepsilon$ on which $d^2f(x)$ is positive definite—we have already observed that such a ball exists by the lemma.

Fix u with $|u| = 1$, and look at $g(t) = f(y+tu)$ on $|t| < \varepsilon$ (this ε does not depend on u). We have $g'(0) = 0$ and $g''(t) = \langle d^2f(y+tu)u, u\rangle > 0$ as a consequence of the positive definiteness. We claim that this implies

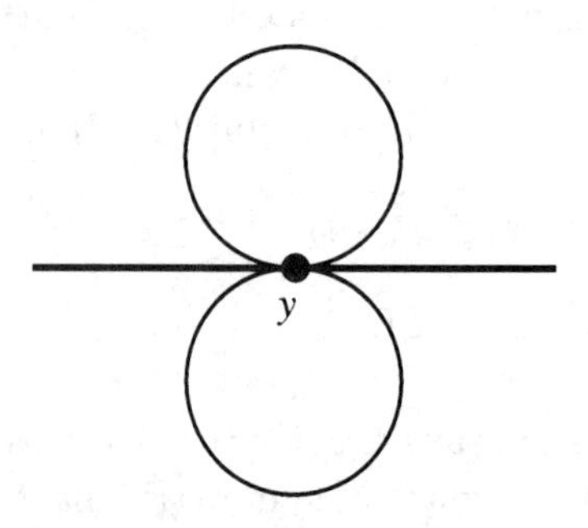

Figure 10.2.3:

0 is a strict minimum of g on the interval $|t| < \varepsilon$. (Indeed if $g(t_1) \leq g(0)$ for $0 < t_1 < \varepsilon$, then

$$\frac{g(t_1) - g(0)}{t_1} = g'(t_2) \leq 0$$

for $0 < t_2 < t_1$ by the mean value theorem and then

$$\frac{g'(t_2) - g'(0)}{t_2} = g''(t_3) \leq 0$$

for $0 < t_3 < t_2$ by another application of the mean value theorem (a similar argument works if $-\varepsilon < t_1 < 0$.) Thus $g(y) < g(y + tu)$ for $t \neq 0$, $|t| < \varepsilon$, and every u with $|u| = 1$; hence, $g(y) < g(x)$ for $x \neq y$ in the ball $|x - y| < \varepsilon$. QED

In view of this result it becomes important to know whether or not a quadratic form is positive definite. We can understand the problem better if we use a basic fact from linear algebra—the spectral theorem. Given any square matrix A, we say u is an *eigenvector* with *eigenvalue* λ if $u \neq 0$ and $Au = \lambda u$. Here λ is any scalar value, with $\lambda = 0$ allowed. We must insist $u \neq 0$, otherwise $A0 = \lambda 0$ for any λ. *The spectral theorem says that for a symmetric matrix A there exists a complete set of eigenvectors*; that is, there exists an orthonormal basis $u^{(1)}, \ldots, u^{(n)}$ of $\mathbb{R}^n$ with $Au^{(k)} = \lambda_k u^{(k)}$. This is sometimes expressed by saying A is diagonalizable by an orthogonal matrix, for the linear transformation

$x \to Ax$ is represented by the diagonal matrix

$$\begin{pmatrix} \lambda_1 & & 0 \\ & \vdots & \\ 0 & & \lambda_n \end{pmatrix}$$

with respect to the basis $u^{(1)}, \ldots, u^{(n)}$. We will not make use of this interpretation.

What does the spectral theorem say about the quadratic form $\langle Ax, x\rangle$? Since an arbitrary vector x can be expressed as a linear combination of the orthonormal basis elements $u^{(j)}$ as $x = \sum_{j=1}^{n} \langle x, u^{(j)}\rangle\, u^{(j)}$, we can substitute this into $\langle Ax, x\rangle$, noting that

$$Ax = \sum_{j=1}^{n} \langle x, u^{(j)}\rangle A u^{(j)} = \sum_{j=1}^{n} \langle x, u^{(j)}\rangle \lambda_j u^{(j)}$$

and so $\langle Ax, x\rangle = \sum_{j=1}^{n} \lambda_j \langle x, u^{(j)}\rangle^2$ because the $u^{(j)}$ are orthonormal ($\langle u^{(j)}, u^{(k)}\rangle = 0$ if $j \neq k$ and $\langle u^{(j)}, u^{(j)}\rangle = 1$). *Thus the spectral theorem shows us that the most general quadratic form is a weighted sum of squares.* The terms $\langle x, u^{(j)}\rangle$ are just the coordinates of x with respect to the orthonormal basis $\{u^{(j)}\}$, and the weighting factors λ_j are just the eigenvalues. It is clear that the sign of $\langle Ax, x\rangle$ will be determined by the sign of the eigenvalues λ_j. The quadratic form will be positive-definite if and only if all the eigenvalues are positive, non-negative definite if and only if the eigenvalues are all non-negative, and so on.

The problem of deciding the type of a quadratic form is thus reduced to the signs of the eigenvalues. Since the eigenvalues of a matrix A are the roots of the characteristic polynomial $p(\lambda) = \det(\lambda I - A)$, this reduces to an examination of the signs of the coefficients of $p(\lambda)$. In fact, $p(\lambda) = \prod_{j=1}^{n}(\lambda - \lambda_j) = \sum_{k=0}^{n} a_k \lambda^k$ with $a_n = 1$. If $\lambda_j > 0$ for all j, then the signs of a_k alternate; while if all $\lambda_j < 0$, then $a_k > 0$ for k. The converse statements are also true and so we have a criterion for positive or negative definiteness in terms of the characteristic polynomial. There is also a related criterion in terms of the signs of the determinants of submatrices of A.

In view of the importance of the spectral theorem, we will give a proof of it. This proof is not algebraic but rather uses the differential calculus in Euclidean space. It also has the virtue of being adaptable to certain infinite-dimensional situations. It is based on the observation that the function $\langle Ax, x\rangle / \langle x, x\rangle$, called the *Rayleigh quotient*, attains its maximum value when x is an eigenvector corresponding to the largest eigenvalue. If we assumed the spectral theorem this would be a simple exercise; but since we are trying to prove the spectral theorem, we will have to use this only for motivation. In fact, we observe that the Rayleigh quotient is homogeneous of degree zero—if we multiply x by a scalar it does not change the Rayleigh quotient. This means that the values of the Rayleigh quotient on $\mathbb{R}^n$ minus the origin (it is not defined for $x = 0$) are the same as its values on the unit sphere. But the unit sphere is compact and the Rayleigh quotient on the sphere is just $\langle Ax, x\rangle$, a continuous function, so it attains its maximum. Passing back to $\mathbb{R}^n$ minus the origin, we have shown the existence of a maximum for $\langle Ax, x\rangle / \langle x, x\rangle$, hence a critical point.

Lemma 10.2.2 *Let $x \neq 0$ in $\mathbb{R}^n$ be a critical point for $\langle Ax, x\rangle / \langle x, x\rangle$. Then x is an eigenvector for A.*

Proof: We compute

$$\frac{\partial}{\partial x_j}\left(\frac{\langle Ax, x\rangle}{\langle x, x\rangle}\right) = \frac{|x|^2 \dfrac{\partial}{\partial x_j}\langle Ax, x\rangle - 2x_j\langle Ax, x\rangle}{|x|^4}$$

and so the equations for a critical point are $|x|^2 \partial/\partial x_j \langle Ax, x\rangle = 2x_j\langle Ax, x\rangle$, for $j = 1, \ldots, n$. Now since A is symmetric,

$$\begin{aligned}\frac{\partial}{\partial x_j}\langle Ax, x\rangle &= \frac{\partial}{\partial x_j}\sum_{i=1}^{n}\sum_{k=1}^{n} A_{ik}x_i x_k = \sum_{k=1}^{n} A_{jk}x_k + \sum_{i=1}^{n} A_{ij}x_i \\ &= 2\sum_{k=1}^{n} A_{jk}x_k = 2(Ax)_j.\end{aligned}$$

Thus the critical point equations are $2|x|^2(Ax)_j = 2x_j\langle Ax, x\rangle$ or $Ax = \lambda x$ where $\lambda = \langle Ax, x\rangle / \langle x, x\rangle$. QED

Thus we have a way to get eigenvectors and so we know that there is at least one eigenvector, which we can normalize to have length one by multiplying by an appropriate scalar. Call it u. To complete the proof of the spectral theorem by induction we want to restrict attention to the orthogonal complement of u, the $(n-1)$-dimensional vector space of all x in $\mathbb{R}^n$ with $\langle x, u\rangle = 0$. Now the symmetry of the matrix A is equivalent to the condition $\langle Ax, y\rangle = \langle x, Ay\rangle$ for any x and y (choosing $x = e_j$ and $y = e_k$ gives $A_{jk} = A_{kj}$). Thus if x is orthogonal to u, then so is Ax, $\langle Ax, u\rangle = \langle x, Au\rangle = \langle x, \lambda_1 u\rangle = \lambda_1\langle x, u\rangle = 0$ since u is an eigenvector. Thus if we call this vector subspace $u^\perp$, we have A acting on $u^\perp$. Furthermore, A is still symmetric on $u^\perp$ because the condition $\langle Ax, y\rangle = \langle x, Ay\rangle$ is true when we restrict x and y to $u^\perp$. We can then repeat the argument to get an eigenvector of length one in $u^\perp$, and continuing in this fashion by induction we obtain the complete set of eigenvectors $u^{(1)}, \ldots, u^{(n)}$. The manner of choice shows that these form an orthonormal set, hence a basis for $\mathbb{R}^n$. This procedure also yields the formula

$$\lambda_k = \sup\left\{\frac{\langle Ax, x\rangle}{\langle x, x\rangle} : x \neq 0 \text{ and } x \text{ is orthogonal to } u^{(1)}, \ldots, u^{(k-1)}\right\}$$

for the kth largest eigenvalue. This formula is often used in obtaining estimates for the eigenvalues.

Example Let $f(x,y) = (1/32)x^4 + x^2y^2 - x - y^2$. The critical points are the solutions of $\partial f/\partial x = (1/8)x^3 + 2xy^2 - 1 = 0$ and $\partial f/\partial y = 2x^2y - 2y = 0$. A simple computation shows there are only the three critical points $(2,0), (1, \sqrt{7}/2)$ and $(1, -\sqrt{7}/2)$. The Hessian for f is

$$\begin{pmatrix} \frac{3}{8}x^2 + 2y^2 & 4xy \\ 4xy & 2x^2 - 2 \end{pmatrix}.$$

At the point $(2,0)$ this is

$$\begin{pmatrix} 3/2 & 0 \\ 0 & 6 \end{pmatrix},$$

which is clearly positive definite, so $(2,0)$ is a local minimum. At $(1, \pm\sqrt{7}/2)$ this is

$$\begin{pmatrix} \frac{31}{8} & \pm 2\sqrt{7} \\ \pm 2\sqrt{7} & 0 \end{pmatrix}.$$

Since this matrix has determinant -28 and the determinant is equal to the product of the eigenvalues, we conclude that the two eigenvalues have opposite signs and so these points are neither local maxima or minima (they are called *saddle points*).

We will return again to the theory of maxima and minima in Chapter 13, after we have discussed the implicit function theorem.

10.2.3 Taylor Expansions

We have been thinking of the differential df as giving the crucial piece of information needed to form the best approximation to f by an affine function in a neighborhood of a point. We can think of an affine function as a polynomial of degree one, and then the natural generalization is to consider the best local approximation to f by polynomials of higher degrees. For simplicity we take f to be real-valued, $f : D \to \mathbb{R}$ with $D \subseteq \mathbb{R}^n$ an open set. Motivated by the $n = 1$ case, we seek a polynomial $T_m(y, x)$ of degree m in x (here y is a fixed point in D) such that $f(x) = T_m(y, x) + o(|x - y|^m)$ as $x \to y$. Notice that this is truly an n-dimensional problem in that we need to show for every $1/N$ there exists $1/k$ such that $|x-y| < 1/k$ implies $|f(x) - T_m(y, x)| \leq |x - y|^m/N$. It is not enough to show that this is true along every line through y, for then the $1/k$ might depend on the choice of the line. Nevertheless, our method of proof will be to use Taylor's theorem in one dimension applied to the restriction of f to each line through y. This will tell us what $T_m(y, x)$ has to be, and we will be able to give a proof based on the proof of the one-dimensional Taylor's theorem that will enable us to control $1/k$ independent of the line.

Let f be C^m. This means all partial derivatives $(\partial/\partial x)^\alpha f$ for $|\alpha| \leq m$ (recall $|\alpha| = \alpha_1 + \cdots + \alpha_n$) exist and are continuous. It is easy to see that the restriction of f to every line in $\mathbb{R}^n$ (intersected with the domain D of f) is C^m. We will need a formula for the derivatives of the restriction. Let $y + tu$ be such a line, where y is in D, u is in $\mathbb{R}^n$ $(u \neq 0)$, and t is a real variable. Then $g(t) = f(y + tu)$ is the restriction.

Lemma 10.2.3 *If f is C^m, then $g(t) = f(y + tu)$ is C^m and*

$$\frac{1}{k!}\left(\frac{d}{dt}\right)^k g(t) = \sum_{|\alpha|=k} \frac{u^\alpha}{\alpha!}\left(\frac{\partial}{\partial x}\right)^\alpha f(y + tu)$$

where $\alpha! = \alpha_1!\alpha_2!\ldots\alpha_n!$, for any $k \leq m$ (note $0! = 1$ by convention).

Proof: This is essentially just an application of the chain rule k times. For $k = 1$ we have $d/dt f(y+tu) = \sum_{j=1}^{n} u_j \partial f/\partial x_j (y+tu)$, which is of the required form ($|\alpha| = 1$ implies all $\alpha_j = 0$ except for one $\alpha_j = 1$). For $k = 2$ we compute

$$\frac{d^2}{dt^2} f(y+ty) = \frac{d}{dt}\left(\sum_{j=1}^{n} u_j \frac{\partial f}{\partial x_j}(y+tu)\right) = \sum_{i=1}^{n}\sum_{j=1}^{n} u_i u_j \frac{\partial^2 f}{\partial x_i \partial x_j}(y+tu).$$

There are two kinds of multi-indices α with $|\alpha| = 2$: those with $n-2$ zeroes and two ones and those with $n-1$ zeroes and one two. The first kind correspond to $\partial^2 f/\partial x_i \partial x_j = \partial^2 f/\partial x_j \partial x_i$ (off-diagonal terms in $d^2 f$) with $j \neq i$ and occur twice in the sum. The second kind correspond to $\partial^2 f/\partial x_j \partial x_j$ (diagonal terms in $d^2 f$) and occur only once. But $\alpha!$ is equal to one for the first kind and two for the second kind, so

$$\sum_{i=1}^{n}\sum_{j=1}^{n} u_i u_j \frac{\partial^2 f}{\partial x_i \partial x_j}(y+tu) = 2 \sum_{|\alpha|=2} \frac{u^\alpha}{\alpha!}\left(\frac{\partial}{\partial x}\right)^\alpha f(y+tu),$$

which gives the result for $k = 2$.

The general case is proved by induction. So let us assume that

$$\frac{1}{(k-1)!}\left(\frac{d}{dt}\right)^{k-1} f(y+tu) = \sum_{|\alpha|=k-1} \frac{u^\alpha}{\alpha!}\left(\frac{\partial}{\partial x}\right)^\alpha f(y+tu)$$

and prove the result for k. We differentiate both sides of the induction hypothesis identity to obtain

$$\begin{aligned}\frac{1}{k!}\left(\frac{d}{dt}\right)^{k} f(y+tu) &= \sum_{|\alpha|=k-1} \frac{1}{k}\frac{u^\alpha}{\alpha!}\frac{d}{dt}\left(\frac{\partial}{\partial x}\right)^\alpha f(y+tu) \\ &= \sum_{|\alpha|=k-1}\sum_{j=1}^{n} \frac{1}{k}\frac{u^\alpha}{\alpha!} u_j \frac{\partial}{\partial x_j}\left(\frac{\partial}{\partial x}\right)^\alpha f(y+tu).\end{aligned}$$

But $(\partial/\partial x_j)(\partial/\partial x)^\alpha = (\partial/\partial x)^\beta$ where $\beta_i = \alpha_i$ for $i \neq j$ and $\beta_j = \alpha_j + 1$. Clearly every multi-index β with $|\beta| = k$ arises in this way n times, one for each j. Also $\beta! = \beta_j \alpha!$, so

$$\begin{aligned}&\sum_{|\alpha|=k-1}\sum_{j=1}^{n} \frac{1}{k}\frac{u^\alpha}{\alpha!} u_j \frac{\partial}{\partial x_j}\left(\frac{\partial}{\partial x}\right)^\alpha f(y+tu) \\ &\qquad = \sum_{|\beta|=k} \frac{1}{k}\sum_{j=1}^{n} \frac{\beta_j}{\beta!} u^\beta \left(\frac{\partial}{\partial x}\right)^\beta f(y+tu)\end{aligned}$$

and $(1/k)\sum_{j=1}^{n}\beta_j = 1$ (since $|\beta| = k$); hence, we have the desired identity. QED

We write out the one-dimensional Taylor expansion for $g(t) = f(y+tu)$, substituting in the above computation, to obtain

$$g(t) = \sum_{k=0}^{m}\sum_{|\alpha|=k}\frac{u^\alpha}{\alpha!}\left(\frac{\partial}{\partial x}\right)^\alpha f(y)t^k + o(t^m).$$

Choosing $u = (x-y)/|x-y|$ and $t = |x-y|$ this becomes

$$f(x) = \sum_{|\alpha|\le m}\frac{(x-y)^\alpha}{\alpha!}\left(\frac{\partial}{\partial x}\right)^\alpha f(y) + o(|x-y|^m).$$

This looks like the kind of result we want, with

$$T_m(y,x) = \sum_{|\alpha|\le m}\frac{(x-y)^\alpha}{\alpha!}\left(\frac{\partial}{\partial x}\right)^\alpha f(y).$$

Note that $T_m(y,x)$ is a polynomial of degree $\le m$ in x. We call it the *Taylor expansion of f at y of order m*. When $n = 1$ this agrees with our previous definition. Also note that $T_m(y,x) = f(x)$ exactly if f is a polynomial of degree $\le m$. The problem is that we do not yet have an honest proof that $f(x) = T_m(y,x) + o(|x-y|^m)$, because we have only proved it line by line. Since the $o(|x-y|^m)$ estimate on each line through y may vary, we may not be able to make $o(|x-y|^m)/|x-y|^m$ small uniformly on a neighborhood of y. Nevertheless, we will see that we can produce a valid argument by looking back at exactly what was proved in the one-dimensional case.

Taylor's Theorem *Let $f : D \to \mathbb{R}$ be C^m, where $D \subseteq \mathbb{R}^n$ is open, and let y be in D. Define the mth order Taylor expansion as*

$$T_m(y,x) = \sum_{|\alpha|\le m}\frac{(x-y)^\alpha}{\alpha!}\left(\frac{\partial}{\partial x}\right)^\alpha f(y).$$

Then $f(x) = T_m(y,x) + o(|x-y|^m)$ as $x \to y$.

Proof: We have to show that given any error $1/N$ there exists $1/k$ such that $|x-y| < 1/k$ implies $|f(x) - T_m(y,x) \le |x-y|^m/N$. Let

$h(x) = f(x) - T_m(y,x)$. Then we claim $(\partial/\partial x)^\beta h(y) = 0$ for any $|\beta| \le m$. This follows by direct computation since $(\partial/\partial x)^\beta (x-y)^\alpha = 0$ at $x = y$ unless $\beta = \alpha$, and then $(\partial/\partial x)^\alpha (x-y)^\alpha = \alpha!$. Thus $(\partial/\partial x)^\beta T_m(y,x)$ at $x = y$ consists of just the one term $(\partial/\partial x)^\beta f(y)$ and so $(\partial/\partial x)^\beta h(y) = 0$.

Now we use the assumption that f is C^m, hence h is C^m, to conclude that $(\partial/\partial x)^\beta h$ must be small in a neighborhood of y for all $|\beta| \le m$. Let $1/N$ be given. Then there exists $1/k$ such that $|x-y| < 1/k$ implies $|(\partial/\partial x)^\beta h(x)| \le 1/N$ for all $|\beta| \le m$. We want to prove that this in turn implies $|h(x)| \le c_n |x-y|^m/N$ where c_n is a constant depending only on the dimension n. If we can do this we will have shown $f(x) - T_m(y,x) = o(|x-y|^m)$ as desired.

But now we can apply the proof of the one-dimensional Taylor theorem to the function $g(t) = h(y + t(x-y))$. We have

$$\left(\frac{d}{dt}\right)^k g(t) = \sum_{|\alpha|=k} \frac{t^k (x-y)^k}{a!} \left(\frac{\partial}{\partial x}\right)^\alpha h(y + t(x-y))$$

by the lemma. So setting $t = 0$ we obtain $(d/dt)^k g(0) = 0$ for $k \le m$ since $(\partial/\partial x)^\alpha h(y) = 0$. We also have $|y + t(x-y) - y| = t|x-y| < 1/k$ for $0 \le t \le 1$ if $|x-y| < 1/k$. Thus $|(\partial/\partial x)^\alpha h(y + t(x-y))| \le 1/N$, which gives us the key estimate

$$(*) \qquad \left|\left(\frac{d}{dt}\right)^m g(t)\right| \le \frac{1}{N} \sum_{|\alpha|=m} \frac{t^m |(x-y)^\alpha|}{\alpha!} \le c_n |x-y|^m/N$$

for $0 \le t \le 1$, where $c_n = \sum_{|\alpha|=m} 1/\alpha!$. (We have also used the estimate $|(x-y)^\alpha| \le |x-y|^m$ for $|\alpha| = m$, which we leave as an exercise.) The proof of the one-dimensional Taylor's theorem (Theorem 5.4.5) shows that $|g(t)| \le t^m |(d/dt)^m g(t_1)|$ for some value of t_1 in $(0,t)$. Taking $t = 1$ and applying $(*)$ we obtain

$$|h(x)| = |g(1)| \le |\,(d/dt)^m\, g(t_1)| \le c_n |x-y|^m/N,$$

as desired. QED

It is worth pointing out the similarity of this proof with that of Theorem 10.2.2 ($d^2 f(y)$ being positive definite at a critical point implies the critical point is a local minimum). In both cases we reduced the n-dimensional result to the one-dimensional result along all lines passing through the point. However, in both cases it was not enough simply to quote the one-dimensional result; we needed further to use the continuity of the derivatives in n-dimensions to obtain estimates uniformly on the lines, and this in turn required that we re-examine the proof of the one-dimensional result.

There is also a Lagrange Remainder Formula for $f(x) - T_m(y, x)$, under the assumption that f is C^{m+1}. It has the form

$$f(x) - T_m(y, x) = \sum_{|\alpha|=m+1} \frac{(x-y)^\alpha}{\alpha!} \left(\frac{\partial}{\partial x}\right)^\alpha f(z)$$

where z is some point on the line joining x and y. From this we obtain $f(x) - T_m(y, x) = O(|x-y|^{m+1})$. The proof of the Lagrange Remainder Formula in n-dimensions is a direct consequence of the one-dimensional case and the lemma. We leave the details as an exercise.

10.2.4 Exercises

1. Let $f : \mathbb{R}^2 \to \mathbb{R}$ be defined by $f(x, y) = xy(x^2 - y^2)/(x^2 + y^2)$ for $(x, y) \neq (0, 0)$ and $f(0, 0) = 0$. Express f in polar coordinates. Show that $\partial f/\partial x, \partial f/\partial y, \partial^2 f/\partial x \partial y$, and $\partial^2 f/\partial y \partial x$ exist for all (x, y) in $\mathbb{R}^2$ but $\partial^2 f/\partial x \partial y(0, 0) \neq \partial^2 f/\partial y \partial x(0, 0)$.

2. Prove the equality of $\partial^2 f/\partial x_j \partial x_k$ and $\partial^2 f/\partial x_k \partial x_j$ under the hypothesis that one of them is continuous (and $\partial f/\partial x_j$ and $\partial f/\partial x_k$ are continuous).

3. Let A be a symmetric matrix. Prove that A is *nondegenerate* (for every $x \neq 0$ there exists $y \neq 0$ such that $\langle Ax, y \rangle \neq 0$), if and only if all the eigenvalues of A are non-zero.

4. Prove that a positive definite matrix has positive entries on the diagonal. Give an example of a symmetric matrix with positive entries on the diagonal that is not positive definite.

5. Prove $|x^\alpha| \leq |x|^{|\alpha|}$ for every α.

6. Prove the Lagrange Remainder Formula in n-dimensions.

7. *Find a formula for $(\partial/\partial x)^\alpha(f \cdot g)$ in terms of derivatives of f and g.

8. For the affine function $g(x) = ax + b$ that minimizes $\sum_{j=1}^{n}(y_j - g(x_j))^2$ (see problem 11 of section 10.1), show by direct computation that the Hessian is positive definite.

9. Show that if f is C^2 and $f(x,y) = g(r,\theta)$ where (r,θ) are polar coordinates in $\mathbb{R}^2$, then

$$\left(\frac{\partial^2}{\partial x^2} + \frac{\partial^2}{\partial y^2}\right) f(x,y) = \left(\frac{\partial^2}{\partial r^2} + \frac{1}{r}\frac{\partial}{\partial r} + \frac{1}{r^2}\frac{\partial^2}{\partial \theta^2}\right) g(r,\theta)$$

for all $(x,y) \neq (0,0)$.

10. Prove $d_u d_v f = d_v d_u f$ if f is C^2 for any vectors u and v.

11. Prove that if f is C^3 and $d^2 f(y) = 0$ at a critical point y but $(\partial/\partial x)^\alpha f(y) \neq 0$ for some α with $|\alpha| = 3$, then y is not a local maximum or minimum.

12. Let f and g be C^2 real-valued functions with $f(y) = g(y) = 0$, $df(y) = dg(y) = 0$, and $d^2 f(y) = \lambda d^2 g(y)$ where $d^2 g(y)$ is positive definite (or negative definite). Prove that $\lim_{x \to y} f(x)/g(x) = \lambda$. Give a counterexample to the naive generalization of l'Hôpital's rule to dimensions greater than one (i.e., find f and g that are C^1 with $f(y) = g(y) = 0$, $df(y) = \lambda dg(y) \neq 0$, but $\lim_{x \to y} f(x)/g(x)$ does not exist).

13. If f and g are C^m functions, prove that the Taylor expansion of order m about y of $f \cdot g$ is obtained by multiplying the corresponding Taylor expansions of f and g and retaining only the terms of order $\leq m$.

14. Prove that the Taylor expansion $T_m(y,x)$ is unique in that if g is any polynomial of degree $\leq m$ such that $f(x) = g(x) + o(|x-y|^m)$ as $x \to y$, then $g(x) = T_m(y,x)$.

15. Classify the critical points of the following functions:

    a. $f(x,y) = x^4 + x^2y^2 - y$.

b. $f(x, y) = \dfrac{x}{1 + x^2 + y^2}$.

c. $f(x, y) = x^4 + y^4 - x^3$.

16. Show that if $f : \mathbb{R}^n \to \mathbb{R}$ is C^2 and $d^2 f(y)$ is positive definite, then the graph of f locally lies above the graph of its tangent plane at y. Prove conversely that if the graph of f lies locally above its tangent plane at y that $d^2 f(y)$ is non-negative definite.

17. Let $f : \mathbb{R}^2 \to \mathbb{R}$ be C^2. Show that f satisfies $\partial^2 f / \partial x \partial y(x, y) \equiv 0$ if and only if there exist C^2 functions $g, h : \mathbb{R} \to \mathbb{R}$ such that $f(x, y) = g(x) + h(y)$. To what extent are g and h uniquely determined by f?

18. *Let $f : \mathbb{R}^2 \to \mathbb{R}$ be C^2. Show that f satisfies

$$\partial^2 / \partial t^2 f(x, t) = c^2 (\partial^2 / \partial x^2) f(x, t)$$

(vibrating string equation) if and only if there exist C^2 functions $g, h : \mathbb{R} \to \mathbb{R}$ such that $f(x, t) = g(x + ct) + h(x - ct)$. Here c is a constant (the speed of sound). (**Hint**: make a change of variable to reduce to problem 17.)

19. Define $Pf(x) = \sum_{|\alpha| \le m} c_\alpha (\partial / \partial x)^\alpha f(x)$ for $f : \mathbb{R}^n \to \mathbb{R}$ any C^m function, where c_α are constants. Show that if $Pf = 0$, then the same is true for any translate of f ($g(x) = f(x + y)$ for fixed y). P is called a *constant coefficient partial differential operator.*

20. If $f : \mathbb{R} \to \mathbb{R}$ is C^k and even ($f(-x) = f(x)$), show that $F : \mathbb{R}^n \to \mathbb{R}$ defined by $F(x) = f(|x|)$ is C^k.

10.3 Summary

10.1 The Differential

Definition *Let $f : D \to \mathbb{R}^m$ with $D \subseteq \mathbb{R}^n$ open. We say f is differentiable at y (a point in D) if there exists an $m \times n$ matrix $df(y)$, called the differential of f at y, such that*

$$f(x) = f(y) + df(y)(x - y) + o(|x - y|)$$

as $x \to y$ (or in other words, given any $1/N$ there exists $1/k$ such that $|x-y| < 1/k$ implies

$$|f(x) - f(y) - df(y)(x-y)| \le |x-y|/N).$$

The differential is uniquely determined by this condition. If $m = 1$ we also call $df(y)$ the gradient of f at y and write it $\nabla f(y)$. If f is differentiable at every point of D we say f is differentiable on D, and if $df : D \to \mathbb{R}^{m \times n}$ is also continuous we say f is continuously differentiable or C^1.

Theorem *If f is differentiable at y, then f is continuous at y; in fact, $|f(x) - f(y)| \le M|x-y|$ for x in a neighborhood of y, for some M.*

Theorem *If f and g are differentiable at y (or C^1, respectively), then so is $af + bg$ and the differential is linear: $d(af+bg)(y) = adf(y) + bdg(y)$. If $f : D \to \mathbb{R}^m$ and $g : D \to \mathbb{R}$ are differentiable, then so is $g \cdot f$ and $d(gf)(y) = g(y)df(y) + f(y)dg(y)$.*

Definition *The partial derivative $\partial f_k/\partial x_j$ is said to exist at a point y if $f_k(y + te_j) = f_k(y) + \partial f_k/\partial x_j(y)t + o(t)$ as $t \to 0$. More generally, if u is in $\mathbb{R}^n$, the directional derivative $d_u f$ is said to exist at y if $f(y+tu) = f(y) + td_u f(y) + o(t)$ as $t \to 0$.*

Theorem 10.1.1 *If f is differentiable at y, then all partial and directional derivatives exist at y and $df(y)$ is the matrix $\partial f_k/\partial x_j(y)$, while $d_u f(y) = df(y)u$.*

Example *$f(x,y) = xy/\sqrt{x^2+y^2}$ has directional derivatives in all directions at all points in the plane but is not differentiable at the origin.*

Theorem 10.1.2 *A function $f : D \to \mathbb{R}^m$ with $D \subseteq \mathbb{R}^n$ open is C^1 if and only if the partial derivatives exist and are continuous on D.*

Theorem 10.1.3 *(Chain Rule) Let $f : D \to \mathbb{R}^m$ with $D \subseteq \mathbb{R}^n$ open, and let $g : A \to \mathbb{R}^p$ with $A \subseteq \mathbb{R}^m$ open and $f(D) \subseteq A$. If f is differentiable at y and g is differentiable at $z = f(y)$, then $g \circ f$ is differentiable at y and $d(g \circ f)(y) = dg(z)df(y)$ (matrix multiplication).*

Theorem 10.1.4 *Let $f : D \to \mathbb{R}$ with $D \subseteq \mathbb{R}^n$ and y in the interior of D. If f assumes its maximum or minimum at y and f is differentiable at y, then $df(y) = 0$.*

Theorem *Let $g : \mathbb{R}^2 \to \mathbb{R}, a : \mathbb{R} \to \mathbb{R}$, and $b : \mathbb{R} \to \mathbb{R}$ be C^1. Then $f(x) = \int_{a(x)}^{b(x)} g(x,y)\,dy$ is C^1 and*

$$f'(x) = b'(x)g(x, b(x)) - a'(x)g(x, a(x)) + \int_{a(x)}^{b(x)} \frac{\partial g}{\partial x}(x,y)\,dy.$$

Lemma 10.1.1

a. *If $g : \mathbb{R}^2 \to \mathbb{R}$ is continuous, then $G(x) = \int_a^b g(x,y)\,dy$ is continuous.*

b. *If g is C^1, then G is C^1 with $G'(x) = \int_a^b \partial g/\partial x(x,y)\,dy$.*

10.2 Higher Derivatives

Definition *Let $f : D \to \mathbb{R}^m$ with $D \subseteq \mathbb{R}^n$ be differentiable. If $df : D \to \mathbb{R}^{n \times m}$ is differentiable at y, then its differential $d(df)(y)$ (if it exists) is called the second derivative, denoted $d^2 f(y)$. If $m = 1$, then $d^2 f(y)$ is an $n \times n$ matrix called the Hessian of f at y.*

Definition *If f is differentiable and $\partial f/\partial x_j$ has a partial derivative $\partial/\partial x_k(\partial f/\partial x_j)$ at a point y, we say the second-order partial derivative $\partial^2 f/\partial x_k \partial x_j$ exists at y and equals $\partial/\partial x_k(\partial f/\partial x_j)$.*

Theorem *If $f : D \to \mathbb{R}$ with $D \subseteq \mathbb{R}^n$ has a second derivative at a point y, then all second-order partial derivatives exist at y and the Hessian $d^2 f(y)$ matrix is $(d^2 f)_{jk} = \partial^2 f/\partial x_j \partial x_k$.*

Theorem 10.2.1 *Let $f : D \to \mathbb{R}$ be continuous together with all partial derivatives of order one and two. Then $\partial^2 f/\partial x_k \partial x_j = \partial^2 f/\partial x_j \partial x_k$; hence, the Hessian metrix is symmetric.*

Definition *A function is said to be of class C^k if all partial derivatives of orders up to k exist and are continuous.*

Notation $\alpha = (\alpha_1, \alpha_2, \ldots, \alpha_n)$ *is called a multi-index, each* α_j *being a non-negative integer;* $(\partial/\partial x)^\alpha f = (\partial/\partial x_1)^{\alpha_1}(\partial/\partial x_2)^{\alpha_2}\cdots(\partial/\partial x_n)^{\alpha_n} f$*; and* $|\alpha| = \alpha_1 + \cdots + \alpha_n$ *is the order of the partial derivative.*

Definition *A quadratic form on* $\mathbb{R}^n$ *is a function of the form* $\langle Au, u\rangle$ *where* A *is a symmetric* $n \times n$ *matrix. It is said to be non-negative definite if* $\langle Au, u\rangle \geq 0$ *for all* u *and positive definite if* $\langle Au, u\rangle > 0$ *for* $u \neq 0$. *(Similarly, we define non-positive definite and negative definite by reversing the inequalities).*

Lemma 10.2.1

a. *A quadratic form* $\langle Au, u\rangle$ *is positive definite if and only if there exists* $\varepsilon > 0$ *such that* $\langle Au, u\rangle \geq \varepsilon|u|^2$ *for all* u.

b. *If* $\langle Au, u\rangle$ *is positive definite, then so is* $\langle Bu, u\rangle$ *for all symmetric matrices* B *sufficiently close to* A.

Theorem 10.2.2 *Let* $f : D \to \mathbb{R}$ *be* C^2 *with* $D \subseteq \mathbb{R}^n$ *an open set. Let* y *in* D *be a critical point.*

a. *If* y *is a local minimum (respectively maximum), then* $d^2 f(y)$ *is non-negative definite (respectively non-positive definite).*

b. *If* $d^2 f(y)$ *is positive definite (respectively negative definite), then* y *is a strict local minimum (respectively maximum).*

Definition *An eigenvector* u *for a matrix* A *with eigenvalue* λ *is a non-zero solution of* $Au = \lambda u$.

Spectral Theorem *A symmetric matrix has a complete set of eigenvectors.*

Lemma 10.2.2 *Let* $x \neq 0$ *in* $\mathbb{R}^n$ *be a critical point for* $\langle Ax, x\rangle/\langle x, x\rangle$. *Then* x *is an eigenvector for* A.

Lemma 10.2.3 *If* f *is* C^m, *then* $g(t) = f(y + tu)$ *is* C^m *and*

$$\frac{1}{k!}\left(\frac{d}{dt}\right)^k g(t) = \sum_{|\alpha|=k} \frac{u^\alpha}{\alpha!}\left(\frac{\partial}{\partial x}\right)^\alpha f(y + tu)$$

where $\alpha! = \alpha_1!\alpha_2!\cdots\alpha_n!$, *for any* $k \leq m$ *(note* $0! = 1$ *by convention).*

Taylor's Theorem *Let* $f : D \to \mathbb{R}$ *be* C^m, *where* $D \subseteq \mathbb{R}^n$ *is open; and let* y *be in* D. *Define the* m*th order Taylor expansion as*

$$T_m(y,x) = \sum_{|\alpha| \leq m} \frac{(x-y)^\alpha}{\alpha!} \left(\frac{\partial}{\partial x}\right)^\alpha f(y).$$

Then $f(x) = T_m(y,x) + o(|x-y|^m)$ *as* $x \to y$.

Chapter 11

Ordinary Differential Equations

11.1 Existence and Uniqueness

11.1.1 Motivation

An *ordinary differential equation* is a relationship between a function and its derivatives, where the derivative is taken with respect to a single variable. Usually we think of this variable as time, although in some applications there may be other interpretations for it. We also want to consider *systems of ordinary differential equations* that involve more than one function. The term "ordinary" thus refers to the number of variables with respect to which we take derivatives—not to the number of variables in the problem. For example, the theory of celestial mechanics—describing the motions of any number of bodies through three-dimensional space under the influence of gravitational forces—is cast in the form of a system of ordinary differential equations since only time derivatives appear. On the other hand, Einstein's theory of General Relativity describing the same system involves partial derivatives.

We have already used ordinary differential equations in the study of transcendental functions. The exponential function satisfies $y' = y$, and up to a constant multiple it is the unique solution. Similarly $y(t) = \cos t$ and $z(t) = \sin t$ satisfy the system $y' = -z, z' = y$; and with the conditions $y(0) = 1, z(0) = 0$ they are again the unique solution to the system. These observations played a crucial role in our derivation of the properties of these functions. We will see that these results are special cases of the general theory that we are going to develop; however, the special tricks that made the proof of existence and uniqueness for these special equations so simple are not available in general, so we will have to work harder.

We let t denote the variable with respect to which we take derivatives, and we let $x(t) : [a, b] \to \mathbb{R}^n$ denote a function of t taking values in $\mathbb{R}^n$. The coordinate components $x_j(t)$ of $x(t)$ can be thought of as n separate real-valued functions that the system of equations describes. An *ordinary differential equation*, abbreviated *o.d.e.*, is an equation of the form $F(t, x(t), x'(t), \ldots, x^{(m)}(t)) = 0$ where F is a function defined on an open subset of $\mathbb{R}^{1+n(m+1)}$ and taking values in $\mathbb{R}^k$. Each of the k components of F may be thought of as a separate equation, so we are compressing k equations into one. (Usually $k = n$, but we will not insist on this here.) Choosing zero for the right side is merely a simplifying convention, for we could always absorb a constant into F (in place of $y' + y = 27$ write $y' + y - 27 = 0$). The *order* of the o.d.e. is defined to be m—the highest derivative that appears in it. Notice that all the functions in the o.d.e. are evaluated at the same time t. Sometimes it is necessary to consider relationships such as $x'(t)+x(t-1) = 1$, which involve simultaneously different time values; however, such equations are not included in the standard theory of o.d.e.'s; they are called *functional-differential equations* or *retarded differential equations.*

By a *solution* to the o.d.e. $F(t, x(t), x'(t), \ldots, x^{(m)}(t)) = 0$ on an interval I we mean a function $x(t) : I \to \mathbb{R}^n$ of class C^m such that the o.d.e. holds for every t in I (if the interval contains one or both endpoints we interpret the o.d.e. as referring to one-sided derivatives at those endpoints). This means in particular that for each t in I, the value $(t, x(t), x'(t), \ldots, x^{(m)}(t))$ must lie in the domain of F. The solutions we are considering may be thought of as "local" solutions, in that we do not require that the interval I be of maximal length in any sense. The function F may be defined for t outside the interval I, and it may even be possible to extend the solution beyond I. For example, consider the o.d.e. $x'(t) + x(t)^2 = 0$ for $x(t)$ taking values in $\mathbb{R}$. The function $F(t, x(t), x'(t))$ is defined on all of $\mathbb{R}^3$ (in terms of coordinates (y_1, y_2, y_3) for $\mathbb{R}^3$ the function is $F(y) = y_2^2 + y_3$). A solution for this o.d.e. is the function $x(t) = 1/t$ on any interval I not containing zero. Other solutions are $x(t) = 1/(t - t_0)$ where t_0

is any fixed constant and the interval I does not contain t_0. It is easy to verify that these are solutions since $x'(t) = -1/(t-t_0)^2$ and $x'(t) + x(t)^2 = -1/(t-t_0)^2 + 1/(t-t_0)^2 = 0$. As a consequence of the uniqueness theorem we will prove that there are no other solutions to this o.d.e. In particular there are no solutions defined on the entire line—despite the fact that the function F is defined everywhere and is C^∞. This example underlies the motivation for looking at merely "local" solutions—if we demanded "global" solutions defined for all t for which the o.d.e. makes sense, then we would put ourselves out of business for even this simple example. Later we will see that it is still possible to get global solutions for a certain important class of o.d.e.'s, the *linear* o.d.e.'s.

There is a fairly standard device for reducing an arbitrary o.d.e. to one of first order. In the process we will increase the number of variables (the dimension of the range of $x(t)$)—in particular, we will always end up with a system, even if we started out with a single equation. The trick is to introduce new names for the derivatives $x'(t), \ldots, x^{(m-1)}(t)$, up to but not including the highest order. Say $m = 3$; then let $y(t) = x'(t)$ and $z(t) = x''(t)$. If the original o.d.e. involves $x(t)$ taking values in $\mathbb{R}^n$ (n unknown functions) and F taking values in $\mathbb{R}^k$ (k equations), the new o.d.e. will involve $3n$ unknown functions, namely $x(t), y(t)$, and $z(t)$, and $k + 2n$ equations, namely

$$\begin{aligned} F(t, x(t), y(t), z(t), z'(t)) &= 0, \\ x'(t) - y(t) &= 0, \\ y'(t) - z(t) &= 0. \end{aligned}$$

The second and third sets of equations simply say what $y(t)$ and $z(t)$ are in terms of $x(t)$, and the first set of equations restates the original o.d.e. The claim is that $x(t) : I \to \mathbb{R}^n$ is a solution of the original o.d.e. if and only if $(x(t), x'(t), x''(t)) : I \to \mathbb{R}^{3n}$ is a solution of the new o.d.e. and furthermore any solution $(x(t), y(t), z(t)) : I \to \mathbb{R}^{3n}$ of the new o.d.e. must be of the above form. We leave the simple verification of this claim as an exercise. The new o.d.e. may appear more complicated than the original o.d.e., but it is of first order.

It is customary in discussions of o.d.e.'s to insist on this reduction to the first-order case since then one can proceed with a minimum of notational baggage. However, this has the disadvantage that whenever one wants to apply the theory to a higher order equation, it becomes necessary to go through a translation process. We will sometimes opt to deal directly with equations of higher order.

There is one other preliminary simplification that we will need to do; namely to solve the equations for the highest order derivatives. In other words, we want to rewrite the equation in the form $x^{(m)}(t) = G(t, x(t), \ldots, x^{(m-1)}(t))$. Usually this is easy to do by inspection. If the original o.d.e. is $x''(t)x'(t) + t^2 = 0$, then we can rewrite it as $x''(t) = -t^2/x'(t)$. However, if the original o.d.e. is $x'(t)^2 - 1 = 0$, then solving for $x'(t)$ yields two different o.d.e.'s, $x'(t) = 1$ and $x'(t) = -1$, which have different solutions, all of which are solutions of the original equation. Notice that by solving for the highest order derivative we may end up with a more singular equation. In going from $x'(t)^2 + 1 - x(t)^2 = 0$ to $x'(t) = \pm\sqrt{x(t)^2 - 1}$ we go from an everywhere defined C^∞ F to a partially defined, multivalued G that is not even differentiable when $x(t) = 1$. This is an important observation, since we will be making assumptions about the function G, and these are *not* consequences of the corresponding assumptions on F. The abstract question of when we can solve equations like $F(t, x(t), \ldots, x^{(m)}(t)) = 0$ for $x^{(m)}(t) = G(t, x(t), \ldots, x^{(m-1)}(t))$ is one we will take up in a later chapter with the aid of the Implicit Function Theorem. For now we will simply assume that this preliminary step has been accomplished, and all o.d.e.'s we consider will be in the *normal form* $x^{(m)}(t) = G(t, x(t), \ldots, x^{(m-1)}(t))$ where G is a function defined on an open set of $\mathbb{R}^{(mn+1)}$ taking values in $\mathbb{R}^n$. We will always assume G to be continuous, since we want $x(t)$ to be C^m and so $x^{(m)}(t)$ to be continuous. Notice that in the normal form the number of equations and unknown functions is the same (namely n). From now on we will always assume this.

Perhaps the simplest o.d.e. in normal form we might consider is $x'(t) = g(t)$ for a continuous function $g : \mathbb{R} \to \mathbb{R}$ (here $n = k = 1$). We know from the fundamental theorem of the calculus that this o.d.e. has solutions $x(t) = \int_0^t g(x)\, ds + c$ where c is an arbitrary constant. (The choice of 0 for the lower endpoint in the integral is also arbitrary; we could just as well have started the integration from another point.) We also know these are the only solutions. We can therefore expect in general that solutions to o.d.e.'s are not unique, but we can hope that they can all be described in terms of a finite number of constants.

Looked at another way, we can hope to adjoin a finite number of side conditions to the o.d.e. in order to make the solution unique. In the example $x'(t) = g(t)$, the condition $x(0) = a$ determines the unique solution $x(t) = \int_0^t g(s)\, ds + a$. Ideally we would like to adjoin just the correct number of conditions so that the solution is unique and so the solution always exists for all values of the parameters that are involved. Clearly, existence demands fewer conditions and uniqueness demands more conditions, so we hope to strike a happy balance in the middle. (In the example $x'(t) = g(t)$, the conditions $x(0) = a$ and $x(1) = b$ would also guarantee uniqueness, but there does not exist a solution for every choice of a and b.)

We would like to give an intuitive answer to the question: how much extra information do we need to determine a solution to a given o.d.e.? The first approach is to ask how many free constants—parameters—one can expect in the general solution. That is, we can expect all solutions to be given by a formula $x(t) = f(t, a, b, \ldots)$, each choice of the parameters $a, b, \ldots$ yielding a solution and all solutions corresponding to some choice of the parameters. Indeed, if you look at any elementary textbook on o.d.e.'s, you will find most of the book devoted to methods for explicitly obtaining such formulas in many special cases. Such an approach is often disparagingly described as "cookbook", but in fact many of the "recipes" turn out to be valuable in unexpected ways. Here we will use only the very imprecise observation that all the "recipes" for solving an o.d.e. of order m involve performing m integrations, and of course each integration picks up an arbitrary constant. Since there are n functions $x_1(t), \ldots, x_n(t)$ being integrated m times, we expect mn constants to be generated and so there should be mn parameters in the general solution. This also means that we will be looking for mn side conditions in order to determine a solution uniquely. This does not mean that we can expect *any* mn side conditions to work—we still have to examine the problem in detail for particular choices of side conditions. We are only claiming at this point to have a grasp of the number of conditions that need to be imposed.

We will now simplify the discussion to consider only conditions involving the unknown functions and their derivatives at a single value of t, say t_0. Such conditions are known as *initial value conditions*, with the interpretation that t_0 is some initial time at which we make some measurements on the system, and then we want to predict the behavior of the system at future times $t > t_0$. Actually our methods will also allow us to "predict the past" as well, $t < t_0$, so we should take "initial" with a grain of salt. It is certainly *not* the case that all interesting or natural problems involving o.d.e.'s lead to initial value conditions—many involve boundary conditions in which the values at two points, thought of as the endpoints of the interval on which the solution is expected to exist, are specified. However, initial value conditions are the easiest with which to deal, and form the basis for discussing more general problems.

Let us look at the simplest case: one first-order equation in one unknown function, $x'(t) = G(t, x(t))$; and consider an initial value condition at $t = t_0$. Since we expect only one such condition ($m = n = 1$), the simplest choice would be to specify the value $x(t_0) = a$. Is it reasonable to expect that this should determine a unique solution for each choice of the parameter a? One way to think about this is graphically. Suppose we consider the graph of the solution in the $t - x$ plane. What does the differential equation say about this graph? It says the slope of the curve $x'(t)$ at a point t is specified by $G(t, x(t))$—the value of G at the point in the plane on the graph of the solution, as shown in Figure 11.1.1.

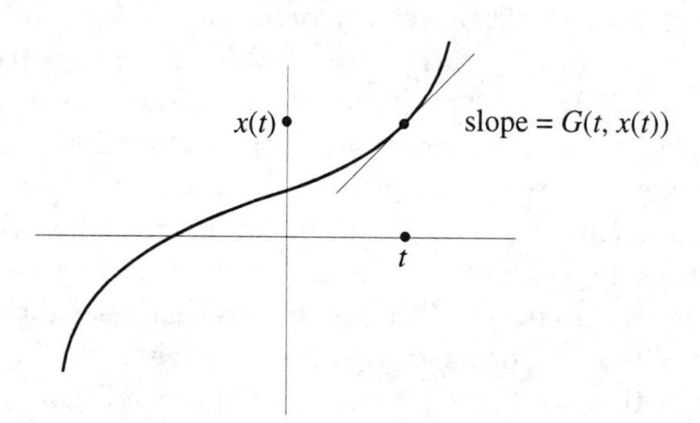

Figure 11.1.1:

We can thus think of $G(t, x)$ as a function assigning a slope to every point in the $t - x$ plane. The differential equation requires that we draw the graph of a function subject to the restriction that at each point on the graph the slope is the one prescribed by G. We can further imagine G as given by drawing a tiny (perhaps infinitesimal) line segment at each point with the given slope so that the plane is covered by a pattern of porcupine quills. The graph of a solution is then obtained by piecing together a curve out of the porcupine quills. The initial condition $x(t_0) = a$ asserts that the graph of the solution must pass through the point (t_0, a). Some people find that this graphical description makes the existence and uniqueness of a solution seem plausible.

Another way of looking at the question is in terms of the Taylor expansion of the solution at the point t_0. The initial condition $x(t_0) = a$ allows one to determine the value of the first derivative $x'(t)$ of the point t_0 via the differential equation $x'(t) = G(t, x(t))$, namely $x'(t_0) = G(t_0, x(t_0)) = G(t_0, a)$. (This computation reveals why it would be foolish to try to specify $x'(t_0)$ as well as $x(t_0)$.) Thus we have the first two terms of the Taylor expansion of x about $t_0, x(t) = a + G(t_0, a)(t - t_0) + o(|t - t_0|)$. But we can in fact go further, simply by differentiating the differential equation. Since $x'(t) = G(t, x(t))$ holds for t in an interval, we also have $(d/dt)x'(t) = (d/dt)G(t, x(t))$ or $x''(t) = \partial G/\partial t(t, x(t)) + \partial G/\partial x(t, x(t))x'(t)$ by the chain rule. This, of course, presupposes that $x(t)$ is C^2 and G is C^1. Under these additional hypotheses we can set $t = t_0$ and obtain

$$\begin{aligned} x''(t_0) &= \frac{\partial G}{\partial t}(t_0, x(t_0)) + \frac{\partial G}{\partial x}(t_0, x(t_0))x'(t_0) \\ &= \frac{\partial G}{\partial t}(t_0, a) + \frac{\partial G}{\partial x}(t_0, a)G(t_0, a) \end{aligned}$$

where we have substituted the previously determined $x'(t_0) = G(t_0, a)$. By induction we can continue to determine all the derivatives $x^{(n)}(t_0)$ at t_0 by differentiating the o.d.e. $n - 1$ times, $(d/dt)^{(n-1)}x'(t) = (d/dt)^{(n-1)}G(t, x(t))$ and setting $t = t_0$, for the right side of this equation only involves derivatives of x at t_0 of orders $< n$, which have previously been determined in the induction step. Again we need to assume that G is C^n and $x(t)$ is C^{n+1}. If the function $x(t)$ were actually analytic, then we would have a complete determination of the power series of $x(t)$ about t_0, which would uniquely specify $x(t)$. Thus we have another plausible reason why the initial condition $x(t_0) = a$ should specify a unique solution.

Despite these plausible reasons, the uniqueness of the solution is not assured without further assumptions. The way to obtain a counterexample is to work backward from the solution to the equation. We want a function whose graph looks like that in Figure 11.1.2

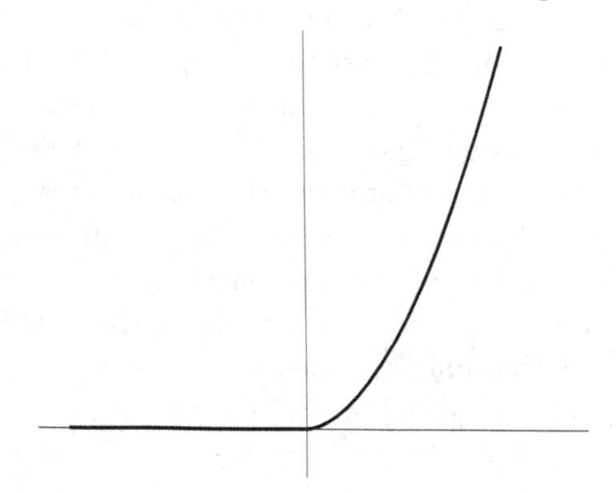

Figure 11.1.2:

so that it is identically zero for a while and then slowly lifts off, say

$$x(t) = \begin{cases} t^{3/2} & \text{if } t > 0, \\ 0 & \text{if } t \le 0. \end{cases}$$

We choose $t^{3/2}$ so that the second derivative will fail to exist, but the function is still C^1 and

$$x'(t) = \begin{cases} (3/2)t^{1/2} & \text{if } t > 0, \\ 0 & \text{if } t \le 0 \end{cases}$$

so that $x'(0) = 0$. Now it is easy to concoct the o.d.e., namely $x'(t) = (3/2)x(t)^{1/3}$, which this function satisfies. Note that the function $G(t, x) = (3/2)x^{1/3}$ is continuous and everywhere defined, although it does fail to be differentiable. This o.d.e. also has the solution $x(t) \equiv 0$, and both these solutions satisfy the initial condition $x(t_0) = 0$ for any choice of $t_0 \le 0$. Thus uniqueness fails for this example. It will turn out that the failure of G to be differentiable is the culprit here—we will be able to prove the uniqueness and existence under hypotheses on G that are slightly weaker than differentiability.

Before discussing the positive results, let's consider the case of general systems ($n \ge 1$) and general orders ($m \ge 1$). For first-order systems, the same reasoning as above leads us to choose the initial conditions $x(t_0) = a$, where this is now a vector equation, $x_k(t_0) = a_k$

for $k = 1, \ldots, n$, involving n parameters. (Incidentally, we can't reduce the general first-order system to n separate first-order equations, because they may be "coupled", as in the example

$$x_1'(t) = -x_2(t), \quad x_2'(t) = x_1(t),$$

which is familiar from the discussion of the sine and cosine functions.) When we reduce the general o.d.e. system of order m to one of first order, we introduce new variables for the derivatives of $x(t)$ up to order $m-1$, so the initial conditions for the first-order system translate to the conditions $x(t_0) = a^{(0)}, x'(t_0) = a^{(1)}, \ldots, x^{(m-1)}(t_0) = a^{(m-1)}$ where each $a^{(j)}$ is a vector in $\mathbb{R}^n$. Notice that there are exactly mn conditions. We will call these conditions the *Cauchy initial value conditions*, and we will refer to the specified values $a^{(j)}$ of the parameters as the *Cauchy data.* Then the *Cauchy problem* for the o.d.e. $x^{(m)}(t) = G(t, x(t), \ldots, x^{(m-1)}(t))$ consists of solving the o.d.e. with given Cauchy data. It is a simple matter to verify that the Cauchy data determines uniquely the values of all derivatives of all the functions $x_k(t)$ at the point $t = t_0$, under the assumption that the x_k and G are C^∞ functions.

11.1.2 Picard Iteration

We will now consider how to obtain a proof of the existence and uniqueness of solutions of an o.d.e. with Cauchy initial conditions. We will begin with first-order systems and then extend the results to higher order systems. The key idea of the proof is to integrate the o.d.e. to obtain an equivalent integral equation. That is, we take the equation $x'(t) = G(t, x(t))$ and integrate both sides from t_0 to t:

$$x(t) - x(t_0) = \int_{t_0}^{t} G(s, x(s))\, ds$$

or

$$x(t) = a + \int_{t_0}^{t} G(s, x(s))\, ds$$

using the initial condition $x(t_0) = a$. This is exactly the way we would obtain the solution of the simple o.d.e. $x'(t) = g(t)$. In the general case we do not obtain the solution in this manner, because after integration the unknown function $x(t)$ appears on both sides of the equation. But is is easy to show that the new equation is equivalent to the o.d.e with the Cauchy initial condition.

Lemma 11.1.1 *A C^1 function $x(t)$ defined on an interval I containing t_0 satisfying the Cauchy problem $x'(t) = G(t, x(t))$ and $x(t_0) = a$ also satisfies the integral equation*

$$x(t) = a + \int_{t_0}^{t} G(s, x(s))\, ds.$$

Conversely, any continuous function $x(t)$ satisfying the integral equation is C^1 and satisfies the Cauchy problem.

Proof: This is an immediate consequence of the two fundamental theorems of the calculus. We use integration of the derivative as above to go from the Cauchy problem to the integral equation. Conversely, given the integral equation, the differentiation of the integral theorem tells us that $f(t) = \int_{t_0}^{t} G(s, x(s))\, ds$ is a C^1 function with derivative $f'(t) = G(t, x(t))$—this uses the hypotheses that G and x are continuous functions, hence so is $G(t, x(t))$. But the integral equation says $x(t) = f(t)$, so we have that x is C^1 and $x'(t) = f'(t) = G(t, x(t))$. Thus the o.d.e. is satisfied. Finally we obtain the initial condition $x(t_0) = a$ by substituting $t = t_0$ in the integral equation. QED

Thus we have reduced the Cauchy problem to the solution of an integral equation. Have we gained anything by doing this? On a very superficial level we have managed to combine both the o.d.e. and the initial value condition into a single equation. But the real accomplishment is considerably deeper. We can see a hint of it in the asymmetric statement of the lemma. For the o.d.e. we had to assume $x(t)$ was C^1, but in the integral equation it was enough to assume $x(t)$ was continuous—it then followed that $x(t)$ was C^1. This gratuitous gain of a derivative is highly non-trivial—normally one doesn't expect a continuous function to be automatically differentiable. There is a reason for it in this case: the integral equation. Integration "smooths things out", while differentiation "roughens things up". Therefore it is easier to deal with an equation in which integrals appear than with an equation involving derivatives. There is also one technical point that we will exploit in the proof: aside from the constant a, the right side of the integral equation $\int_{t_0}^{t} G(s, x(s))\, ds$ is small if t is close to t_0.

It is also significant to consider that the process whereby we passed from the o.d.e. to the integal equation, integration, gave the exact solution to the simpler o.d.e. $x'(t) = g(t)$. In fact, the gist of our argument will be that the general o.d.e. is some kind of a "perturbation" of the simpler one—this will be especially clear in the case of linear equations where we can actually write down a perturbation series expansion of the solution. The idea of exploiting an explicit solution to a simpler problem to prove the existence of a less explicit solution to a more difficult problem is an important theme in the theory of differential equations and perhaps in all of mathematics.

The form of the integal equation suggests that we look for the solution as a fixed point of the mapping $x(t) \to a + \int_{t_0}^t G(s, x(s))\, ds$. That is, we think of $a + \int_{t_0}^t G(s, x(s))\, ds$ as a mapping from the function $x(t)$ to a new function of t—the integral equation then says that the solution we seek is a fixed point of this mapping. It is natural, then, to try to apply the Contractive Mapping Principle, since this will give us a unique fixed point. We consider the metric space $C(I)$ of continuous functions $x : I \to \mathbb{R}^n$, where I is some interval containing t_0, with the sup-norm metric

$$d(x, y) = \sup_I |x(t) - y(t)|$$

where $|\ |$ denotes the Euclidean norm in $\mathbb{R}^n$. We have seen that this is a complete metric space. We define a mapping $T : C(I) \to C(I)$ by $Tx(t) = a + \int_{t_0}^t G(s, x(s))\, ds$. For this to be defined we need only assume G to be continuous, but for this to be a contraction ($d(Tx, Ty) \leq \rho d(x, y)$ for some $\rho < 1$) we need to assume more. We have already seen that uniqueness can fail, and even if the solution exists it may not exist over the whole interval. Thus we must both put additional hypotheses on G and shrink the interval I. Since the contraction property is a kind of Lipschitz condition, it is not surprising that we will need to have G satisfy a Lipschitz condition. A quick glance at the form of

$$\begin{aligned} d(Tx, Ty) &= \sup_I \left| a + \int_{t_0}^t G(s, x(s))\, ds - \left(a + \int_{t_0}^t G(s, y(s)) ds \right) \right| \\ &= \sup_I \left| \int_{t_0}^t (G(s, x(s)) - G(s, y(s)))\, ds \right| \end{aligned}$$

indicates that we will have to estimate differences $G(s, x) - G(s, y)$ where the value of s is the same for both terms. Thus the kind of

Lipschitz condition we want is $|G(s,x) - G(s,y)| \le M|x-y|$ for all x and y in $\mathbb{R}^n$ and all s in I (here we are using the letters x and y to denote points in $\mathbb{R}^n$ rather than functions taking values in $\mathbb{R}^n$—although there is no harm in substituting these functions into the Lipschitz condition). One might think that we have to put some restriction on the size of the Lipschitz constant M because for the contraction property we need $\rho < 1$, but it turns out that this is unnecessary because we can obtain $\rho < 1$ by shrinking the interval. We will refer to this kind of Lipschitz condition as a *global Lipschitz condition*, because it must hold for all x and y in $\mathbb{R}^n$. This is usually too strong a condition—it rules out too many important examples—so we will later have to consider also local Lipschitz conditions.

Lemma 11.1.2 *Suppose $G(t,x)$ is a continuous function defined for t in I and x in $\mathbb{R}^n$ taking values in $\mathbb{R}^n$, which satisfies the global Lipschitz condition $|G(t,x) - G(t,y)| \le M|x-y|$ for all x and y in $\mathbb{R}^n$ and t in I, and some constant M. Let J be any subinterval of I contained in $|t - t_0| \le \rho/M$ for some $\rho < 1$. Then the mapping*

$$Tx(t) = a + \int_{t_0}^{t} G(s, x(s))\, ds$$

on $C(J)$ is contractive, $d(Tx, Ty) \le \rho d(x,y)$ for any $x(t)$ and $y(t)$ in $C(J)$.

Proof: As before, we have

$$d(Tx, Ty) = \sup_J \left| \int_{t_0}^{t} (G(s, x(s)) - G(s, y(s)))\, ds \right|,$$

except now we are restricting attention to the interval J. Here we need to use a basic inequality known as *Minkowski's inequality*, which says $|\int_a^b f(s)\, ds| \le \int_a^b |f(s)|\, ds$ for any continuous function $f(s)$ taking values in $\mathbb{R}^n$, where $|\ |$ denotes the Euclidean norm. We have already discussed this inequality in the case $n = 1$. For the general case it is really a generalization of the triangle inequality. The integral $\int_a^b f(s)\, ds$ is a limit of Cauchy sums $\sum f(t_j)(s_j - s_{j-1})$, for which we have $|\sum f(t_j)(s_j - s_{j-1})| \le \sum |f(t_j)|(s_j - s_{j-1})$ by the triangle inequality, and $\sum |f(t_j)|(s_j - s_{j-1})$ is a Cauchy sum for $\int_a^b |f(s)|\, ds$. Passing to the limit we obtain Minkowski's inequality.

We apply Minkowski's inequality to estimate

$$\left| \int_{t_0}^{t} (G(s,x(s)) - G(s,y(s)))\, ds \right| \leq \int_{t_0}^{t} |G(s,x(s)) - G(s,y(s))|\, ds.$$

We then substitute the global Lipschitz condition to obtain

$$\int_{t_0}^{t} |G(s,x(s)) - G(s,y(s))|\, ds \leq \int_{t_0}^{t} M|x(s) - y(s)|\, ds.$$

Thus we have altogether

$$d(Tx,Ty) \leq \sup_J M \int_{t_0}^{t} |x(s) - y(s)|\, ds.$$

Finally we make the crude estimate $|x(s) - y(s)| \leq d(x,y)$ for s in J, so $\int_{t_0}^{t} |x(s) - y(s)| ds \leq |t - t_0| d(x,y)$ for t in J. Since we chose J so that $|t - t_0| < \rho/M$ for t in J, we have

$$\begin{aligned} d(Tx,Ty) &\leq \sup_J M \int_{t_0}^{t} |x(s) - y(s)|\, ds \leq \sup_J M|t - t_0| d(x,y) \\ &\leq M \cdot (\rho/M) d(x,y) = \rho d(x,y), \end{aligned}$$

which is the desired contraction property. QED

Notice the crucial way we have used the observation that $\int_{t_0}^{t} G(s,x(s))\, ds$ is small for t near t_0 to overcome the lack of control over M.

We are now in a position to apply the contractive mapping principle to obtain the existence and uniqueness of the solution of the integral equation, hence the Cauchy problem, on the subinterval J. Notice that the contractive mapping principle gives us a constructive method for finding the solution by iteration: choose an arbitrary first guess $x_1(t)$ (it is simplest to take $x_1(t) \equiv a$) and then define inductively $x_{k+1}(t) = a + \int_{t_0}^{t} G(s,x_k(s))\, ds$. The solution $x(t)$ is just the limit of $x_k(t)$ as $k \to \infty$ in the metric, which means $x_k(t)$ converges to $x(t)$ uniformly on J. This construction of the solution is known as the *Picard iteration method.* It can be shown by using more careful estimates that the Picard iteration method actually converges to the solution on the whole interval I not just on J. However, the Picard iteration method is not really very practical for obtaining approximations to the solution— since doing successive integrations is very time consuming. (It is never used in numerical solutions.) Therefore we will not devote our energy here to obtain more information about the scope of this method, but we will simply take it as a method for proving existence and uniqueness over a small interval. We can then get existence and uniqueness over a larger interval by piecing together solutions over smaller intervals.

Theorem 11.1.1 (*Global Existence and Uniqueness*) *If $G(s,x)$ satisfies a global Lipschitz condition for s in I, then the Cauchy problem*

$$x'(t) = G(t, x(t)), \quad x(t_0) = a,$$

has a unique solution on I (the solution is even unique on any subinterval containing t_0).

Proof: By the lemmas and the Contractive Mapping Principle, there exists a unique solution on J. If J is not all of I we take a point t_1 of J and the value $x(t_1) = b$ and give the Cauchy initial conditions $x(t_1) = b$. Applying the argument to this Cauchy problem we obtain a solution to the o.d.e. on an interval J_1 containing t_1. Since the two solutions have the same Cauchy data at t_1, they must be equal on the overlap of the domains $J \cap J_1$ because of the uniqueness of the Cauchy problem on J_1. Thus we may combine the

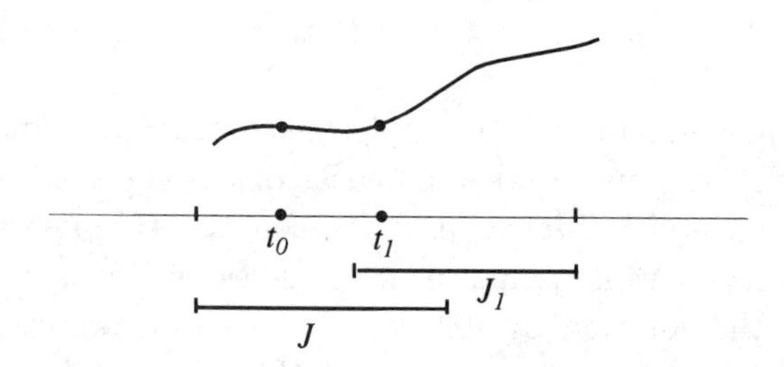

Figure 11.1.3:

two solutions to obtain a single solution of the o.d.e on the union $J \cup J_1$ of the two intervals, as shown in Figure 11.1.3. Repeating this process a finite number of times, we will eventually extend the solution to every point of I since the size of the interval of solution at each stage is restricted only by the condition $|t - t_k| \leq \rho/M$ for the same fixed ρ and M (t_k here denotes the point at which the Cauchy data is given). If the interval I is unbounded it will take an infinite number of steps to extend the solution to all of I but only a finite number of steps to get to any given point in I.

In this way we pass from local existence and uniqueness to global existence by piecing together solutions. To complete the proof it remains to show that the solution is unique. This is done by looking at the first point where two proposed solutions begin to differ and applying the local uniqueness result there. Suppose then that $x(t)$ and $y(t)$ are two solutions of the o.d.e. on a subinterval I_1, containing t_0, and suppose $x(t_0) = y(t_0)$. We want to show $x(t) = y(t)$ on I_1. Let $t_1 = \sup\{t \text{ in } I_1 : t \geq t_0 \text{ and } x(t) = y(t)\}$. Then, by the continuity of x and y, we have $x(t_1) = y(t_1)$ and so, by the local uniqueness of the Cauchy problem at t_1, we have $x(t) = y(t)$ in a neighborhood of t_1. This contradicts the definition of t_1, unless t_1 is the upper endpoint of I_1. In a similar way we show that $x(t) = y(t)$ down to the lower endpoint of I_1. QED

11.1.3 Linear Equations

The global existence and uniqueness theorem allows an immediate extension: if $I = \bigcup_{j=1}^{\infty} I_j$ and G satisfies a global Lipschitz condition $|G(s,x) - G(s,y)| \leq M_j|x - y|$ for s in I_j and all x, y in $\mathbb{R}^n$, where the Lipschitz constant M_j depends on the subinterval I_j, then existence and uniqueness holds on I. The reason for this is simply that existence and uniqueness on I is equivalent to existence and uniqueness on each I_j, which is a consequence of the theorem. This extension is valuable because such a function G may fail to satisfy a global Lipschitz condition on all of I.

An important example where this remark applies is the class of *linear* o.d.e.'s, $x'(t) = A(t)x(t) + b(t)$ where $A : I \to \mathbb{R}^{n \times n}$ is a continuous $(n \times n)$-matrix-valued function and $b : I \to \mathbb{R}^n$ is continuous. Strictly speaking we should call these "affine" equations, limiting the term "linear" to the case $b = 0$, but instead it is traditional to call $x'(t) = A(t)x(t)$ the *homogeneous linear* o.d.e. and $x'(t) = A(t)x(t) + b(t)$ the *inhomogeneous linear* o.d.e. Of course the homogeneous linear o.d.e. is "linear" in the sense that linear combinations of solutions are solutions, and the theory of linear algebra may be applied.

For a linear first-order o.d.e., the function $G(t,x)$ is $A(t)x + b(t)$. The assumptions that A and b are continuous imply that G is continuous. The global Lipschitz condition that needs to be verifed is

$$|G(t,x) - G(t,y)| = |A(t)x - A(t)y| = |A(t)(x-y)| \leq M|x-y|.$$

Now for any fixed matrix A we have $|A(x-y)| \leq M|x-y|$ where M depends on the entries of the matrix. The question of finding the smallest M is very delicate, but we can easily get a crude estimate with $M^2 = \sum_{j,k} |A_{jk}|^2$. Thus if all the entries $A_{jk}(t)$ of $A(t)$ are bounded on I, we will have a global Lipschitz condition for G holding on I. Since we are assuming that $A(t)$—hence the entries—are continuous, this is immediately the case if I is compact. But even if I is not compact, we can always write $I = \bigcup_{j=1}^{n} I_j$ with I_j compact, and G satisfies a global Lipschitz condition on each I_j. Thus we have

Corollary 11.1.1 *The Cauchy problem for the linear o.d.e.* $x'(t) = A(t)x(t) + b(t)$ *with* $A : I \to \mathbb{R}^{n \times n}$ *and* $b : I \to \mathbb{R}^n$ *continuous has a unique solution on* I.

It is instructive to examine the Picard iteration method in the special case of linear equations, for then we can represent it in terms of a perturbation series. We consider the equation $x'(t) = A(t)x(t) + b(t)$ as a perturbation of the simpler equation $x'(t) = b(t)$, which we know how to solve exactly. We can represent this in concise notation by writing Dx for $x'(t)$ and Ax for $A(t)x(t)$. Then $Dx = b$ is the simple equation and $Dx = Ax + b$ or $(D-A)x = b$ the perturbed equation. Under the initial condition $x(t_0) = a$, the simple equation $Dx = b$ has the explicit solution $x(t) = a + \int_{t_0}^{t} b(s)\,ds$. For simplicity we will set $a = 0$. We define $D^{-1}f(t) = \int_{t_0}^{t} f(s)\,ds$ so that $D^{-1}b$ is the solution of $Dx = b$ with initial condition $x(t_0) = 0$.

In the perturbed equation $(D-A)x = b$ with initial condition $x(t_0) = 0$, we will think of A as being small relative to D. This suggests writing the equation as $(D-A)x = D(I - D^{-1}A)x = b$, where I denotes the identity operator, $Ix = x$. Then, at least formally, we can solve $x = (I - D^{-1}A)^{-1}D^{-1}b$. We already know what D^{-1} means, but what does $(I - D^{-1}A)^{-1}$ mean? We are thinking of A as small compared to

D, so $D^{-1}A$ should be small compared to I. Now if r is a real number that is small compared to 1 (if $|r| < 1$), then $(1-r)^{-1} = 1+r+r^2+\cdots = \sum_{k=0}^{\infty} r^k$. We could thus hope that

$$(I - D^{-1}A)^{-1} = I + D^{-1}A + (D^{-1}A)^2 + \cdots = \sum_{k=0}^{\infty}(D^{-1}A)^k.$$

This leads to the "solution"

$$\begin{aligned} x &= \sum_{k=0}^{\infty}(D^{-1}A)^k D^{-1}b \\ &= D^{-1}b + D^{-1}AD^{-1}b + D^{-1}AD^{-1}AD^{-1}b + \cdots. \end{aligned}$$

Of course we have only derived this *perturbation series solution* in a formal manner.

Now, what does this have to do with the Picard iteration method? Since we are free to choose any initial approximation $x_0(t)$ to the solution, let's take $x_0(t) = D^{-1}b(t)$, the solution to the simpler equation. Then the Picard iteration method defines

$$x_1(t) = \int_{t_0}^{t} (A(s)x_0(s) + b(s))\, ds,$$

and inductively

$$x_k(t) = \int_{t_0}^{t} (A(s)x_{k-1}(x) + b(s))\, ds.$$

Since this is just $x_k = D^{-1}Ax_{k-1} + D^{-1}b$, we find

$$\begin{aligned} x_1 &= D^{-1}AD^{-1}b + D^{-1}b, \\ x_2 &= D^{-1}AD^{-1}AD^{-1}b + D^{-1}AD^{-1}b + D^{-1}b, \end{aligned}$$

and in general

$$x_k = \sum_{j=0}^{k}(D^{-1}A)^j D^{-1}b.$$

Thus the Picard interation method produces the partial sums of the perturbation series. If we take the general case $a \neq 0$ and choose $x_0(t) = a + D^{-1}b(t)$, then we find $x_k = \sum_{j=0}^{k}(D^{-1}A)^j(a + D^{-1}b)$. We leave the details as an exercise. In particular, since we know the Picard iterations approximate the solution on some interval about t_0, we can conclude that the perturbation series converges on that interval (in fact, that the convergence is uniform, since we are using the sup-norm metric in the Contractive Mapping Principle). There are many other contexts in which the method of perturbation series can be applied.

11.1.4 Local Existence and Uniqueness*

Aside from the linear equations, there are very few o.d.e.'s for which the function G satisfies a global Lipschitz condition. We have already observed that the o.d.e. $x'(t) = -x(t)^2$ has solutions $x(t) = 1/(t - t_1)$ that fail to exist for all t, so the global existence and uniqueness theorem can't apply. In this case it is easy to see that $G(t, x) = -x^2$ fails to satisfy a global Lipschitz condition. Nevertheless, it satisfies a local Lipschitz condition, $|G(t, x) - G(t, y)| \leq M|x - y|$, if we restrict x and y suitably, say $|x| \leq N$ and $|y| \leq N$ (then the Lipschitz constant M will depend on N). Such a local Lipschitz condition is rather easy to obtain—in fact it is true in general if we merely assume that G is C^1. We could hope, then, that the local Lipschitz condition would suffice to prove a local existence and uniqueness theorem. This is indeed the case and provides us with a very valuable theorem.

The idea of the proof is again to use the contractive mapping principle, not on the whole space $C(I)$, but rather on a part of it—a part on which $x(t)$ is bounded—so the local Lipschitz condition applies. The difficult part of the proof turns out not to be the contractive estimate—that argument is the same as before—but the verification that the part of $C(I)$ is mapped into itself. As before we will take a subinterval J of I, but now we will also impose the condition $|x(t) - a| \leq N$ for all t in J. We let $C_0(J)$ denote the metric subspace of $C(J)$ of all continuous functions $x : J \to \mathbb{R}^n$ satisfying $|x(t) - a| \leq N$ for all t in J, with the sup-norm metric $d(x, y) = \sup_J |x(t) - y(t)|$. Note that the condition $|x(t) - a| \leq N$ for all t in J is the same as $d(x, a) \leq N$, where a denotes the function that is identically equal to a. Thus the subspace $C_0(J)$ is actually the closed ball of radius N and center a in $C(J)$. Being a closed subspace of a complete space it is complete, so we can apply the contractive mapping principle to mappings of $C_0(J)$ to itself. We want to apply it in particular to the mapping $Tx(t) = a + \int_{t_0}^{t} G(s, x(s))ds$.

It will not be hard to show that T is a contractive mapping from $C_0(J)$ to $C(J)$, but first we show that the image lies in $C_0(J)$.

Lemma 11.1.3 *Let $G(t,x)$ be defined and continuous for t in I and x in $|x-a| \leq N$, and let M_0 be the* sup *of $|G(t,x)|$ for t in I and $|x-a| \leq N$. Then $Tx(t) = a + \int_{t_0}^t G(s,x(s))\,ds$ maps $C_0(J)$ into $C_0(J)$ provided J is contained in $|t-t_0| \leq N/M_0$.*

Proof: Note that $Tx(t)$ is a continous function (even C^1) because it is a constant plus an integral of the continuous function $G(s,x(s))$. Thus the image of T lies in $C(J)$. We need to show $|Tx(t)-a| \leq N$ for every t in J. But

$$|Tx(t)-a| = \left|\int_{t_0}^t G(s,x(s))\,ds\right| \leq \int_{t_0}^t |G(s,x(s))|\,ds$$

by Minkowski's inequality, and then we can substitute $|G(s,x(s))| \leq M_0$ to obtain $|Tx(t)-a| \leq M_0|t-t_0| \leq N$ if $|t-t_0| \leq N/M_0$. QED

Notice that the local Lipschitz condition did not play a role in the above argument—it will only be used to get the contractive estimate.

Theorem 11.1.2 (*Local Existence and Uniqueness*) *Let $G(t,x)$ be defined and continuous for t in I and $|x-a| \leq N$, and let it satisfy the local Lipschitz condition $|G(t,x) - G(t,y)| \leq M|x-y|$ for t in I and $|x-a| \leq N, |y-a| \leq N$. Then for t_0 in I there exists a subinterval J containing t_0 on which the Cauchy problem $x'(t) = G(t,x(t)), x(t_0) = a$, has a unique solution.*

Proof: By the lemma, if we take J small enough, T will map $C_0(J)$ into $C_0(J)$. But by the same argument as in the proof of the global theorem, we will have the contractive estimate $d(Tx,Ty) \leq \rho d(x,y)$ for some $\rho < 1$ and for x and y in $C_0(J)$, if J is small enough. This is because that argument only involved using the Lipschitz condition for $G(s,x(s)) - G(s,y(s))$, and the condition that x and y be in $C_0(J)$ says exactly that the Lipschitz condition applies. Thus once again the Contractive Mapping Principle yields the existence and uniqueness of solutions to the integral equation and, hence, to the Cauchy problem. QED

Corollary 11.1.2 *Let $G(t,x)$ be a C^1 function defined for t in I and for x in some open set D in $\mathbb{R}^n$, taking values in $\mathbb{R}^n$. Then the Cauchy problem $x'(t) = G(t, x(t)), x(t_0) = a$ for t_0 in I and a in D has a unique solution on some subinterval J containing t_0.*

Proof: By taking N small enough we can make the ball $B = \{|x-a| \le N\}$ lie in D since D is open. We then can apply the theorem to this ball and a subinterval J, provided we verify

$$|G(t,x) - G(t,y)| \le M|x-y|$$

for x and y in B. But this follows from the hypothesis that G is C^1 by considering the restriction of G to the line segment joining x and y. We can parameterize this segment by $x + s(y-x)$ for $0 \le s \le 1$. Then for fixed t, x, and y, $g(s) = G(t, x + s(y-x))$ is a C^1 function (for x and y in the ball B the line segment stays inside B because the ball is convex, as shown in Figure 11.1.4).

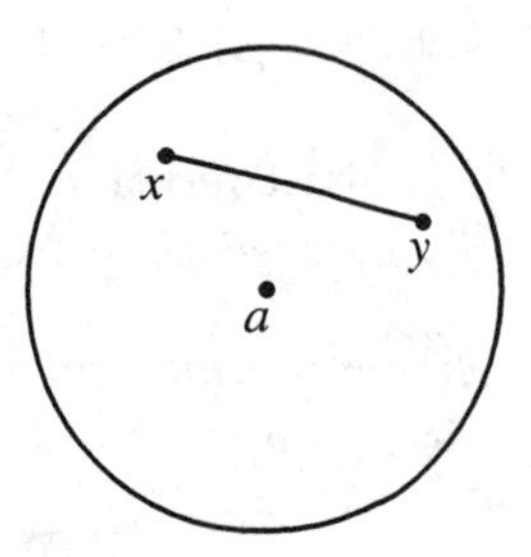

Figure 11.1.4:

Also $g'(s) = \sum_{k=1}^n (y_k - x_k)\partial G/\partial x_k(t, x + s(y-x))$ by the chain rule. We have

$$G(t,y) - G(t,x) = g(1) - g(0) = \int_0^1 g'(s)\, ds$$

by the fundamental theorem of the calculus, so

$$|G(t,y) - G(t,x)| \le \int_0^1 |g'(s)|\, ds$$

by Minkowski's inequality. But

$$|g'(s)| = \left|\sum_{k=1}^n (y_k - x_k)\partial G/\partial x_k(t, x + s(y-x))\right|,$$

and this can be dominated by $M|x-y|$ where M depends only on the maximum of $|\partial G/\partial x_k|$ on $J \times B$. (This is finite because B and J are compact and G is C^1.) The most efficient way to do this is to apply the Cauchy-Schwartz inequality in $\mathbb{R}^n$ to obtain

$$\begin{aligned}
&\left|\sum_{k=1}^{n}(y_k - x_k)\frac{\partial G}{\partial x_k}(t, x+s(y-x))\right| \\
&\quad \le \left(\sum_{k=1}^{n}(y_k - x_k)^2\right)^{1/2}\left(\sum_{k=1}^{n}\frac{\partial G}{\partial x_k}(t, x+s(y-x))^2\right)^{1/2} \\
&\quad = |x-y|\left(\sum_{k=1}^{n}\frac{\partial G}{\partial x_k}(t, x+s(y-x))^2\right)^{1/2}.
\end{aligned}$$

One could also use the triangle inequality and $|x_k - y_k| \le |x-y|$ to obtain

$$\begin{aligned}
&\left|\sum_{k=1}^{n}(y_k - x_k)\frac{\partial G}{\partial x_k}(t, x+s(y-x))\right| \\
&\quad \le \sum_{k=1}^{n}|y_k - x_k|\left|\frac{\partial G}{\partial x_k}(t, x+s(y-x))\right| \\
&\quad \le |x-y|\sum_{k=1}^{n}\left|\frac{\partial G}{\partial x_k}(t, x+s(y-x))\right|.
\end{aligned}$$

Either approach yields the desired Lipschitz condition with slightly different M. (One could also use the mean value theorem in place of the fundamental theorem of the calculus.) Thus the theorem applies. QED

It is natural to ask how big the interval J, on which we have existence and uniqueness, can be taken. In particular, why can't we use the idea of piecing together local solutions as we did in the proof of the global existence and uniqueness theorem? The answer is that of course we can use the idea, although it will not take us quite as far. The uniqueness of the solution—if it exists—is a global fact. If two solutions $x(t)$ and $y(t)$ exist on any subinterval I_0 of I and $x(t_0) = y(t_0)$ for t_0 in I_0 then they must be equal on I_0. The same argument as in the global theorem works—just apply the local uniqueness at the points where the two solutions stop being equal.

Regarding the existence of the solution on larger intervals, the best we can say is that *the solution exists until it becomes unbounded, or leaves the domain of the function* G. To treat the simplest case, let us assume that G is defined and C^1 for all t in $\mathbb{R}$ and all x in $\mathbb{R}^n$. Starting with the initial condition $x(t_0) = a$, we have the existence of the solution on an interval J. Then, by starting at points near the endpoints of J we can extend the solution. Let I denote the largest interval on which the solution exists. By the local existence and uniqueness this must be an open interval. If it is not the whole line, then what is the behavior of the solution as we approach an endpoint of I? We claim that the solution must become unbounded. The reason for this is a bit subtle. Suppose to the contrary that $x(t)$ remained bounded, say $|x(t)| \leq N$, as $t \to c$ from below, where c is the upper endpoint of I. Then the function $G(t, x)$ is bounded and satisfies a Lipschitz condition under the restrictions $t_0 \leq t \leq c+\varepsilon$ and $|x| \leq 2N$. Now an examination of the proof of the local existence and uniqueness theorem shows that starting with Cauchy data in the region $t_0 \leq t \leq c$ and $|x| \leq N$, the solution will exist over an interval of fixed length δ, where δ depends only on the bounds for G and the Lipschitz constant over the larger region $t_0 \leq t \leq c + \varepsilon, |x| \leq 2N$.

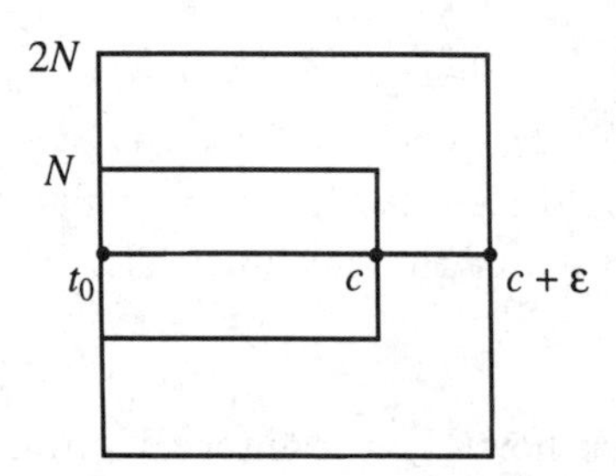

Figure 11.1.5:

Therefore, if we start at the point $t = c - \delta/2$ we will be able to extend the solution past c, contrary to the assumption that the solution could not be extended past c. Of course, if $x(t)$ becomes unbounded as $t \to c$, then the interval of existence gets smaller and smaller as $t \to c$ (because the bounds on G get larger and larger), and there is no contradiction. This argument shows that functions like $\sin 1/x$ with oscillating discontinuities cannot be solutions to the kind of o.d.e.'s we are considering.

11.1.5 Higher Order Equations*

Finally, we turn to the case of higher order equations,

$$x^{(m)}(t) = G(t, x(t), \ldots, x^{(m-1)}(t)),$$

where G is a continuous function defined on an open set in $\mathbb{R}^{1+mn}$ taking values in $\mathbb{R}^n$. We will assume that either G satisfies a global Lipschitz condition

$$|G(t,y) - G(t,z)| \leq M|y - z|$$

for all t in I and all y and z in $\mathbb{R}^{nm}$ or that G satisfies a local Lipschitz condition

$$|G(t,y) - G(t,z)| \leq M|y - z|$$

for all t in I and all y and z in some ball B in $\mathbb{R}^{nm}$. In the first case we have global existence and uniqueness and in the second case local existence and uniqueness for the Cauchy problem

$$\begin{aligned} x^{(m)}(t) &= G(t, x(t), \ldots, x^{(m-1)}(t)), \\ x(t_0) &= a^{(0)}, \ldots, x^{(m-1)}(t_0) = a^{(m-1)} \end{aligned}$$

(with the Cauchy data $a^{(0)}, \ldots, a^{(m-1)}$ in the ball B in the second case). This can be proved by reducing to the first-order case as indicated before, but it can also be proved directly. The idea is that we now want to integrate m times to obtain an equivalent integral equation. To see what this should look like let's consider the case $m = 2$. Then integrating $x''(t) = G(t, x(t), x'(t))$ once yields

$$x'(t) - x'(t_0) = \int_{t_0}^{t} x''(s)\, ds = \int_{t_0}^{t} G(s, x(s), x'(s))\, ds$$

or

$$x'(t) = a^{(1)} + \int_{t_0}^{t} G(s, x(s), x'(s))\, ds.$$

Integrating again we obtain

$$x(t) = a^{(0)} + (t - t_0)a^{(1)} + \int_{t_0}^{t} \int_{t_0}^{r} G(s, x(s), x'(s))\, ds\, dr.$$

We can simplify the iterated integral by interchanging the order (we will give a proof that this is valid in Chapter 15),

$$\begin{aligned}\int_{t_0}^{t}\int_{t_0}^{r} G(s,x(s),x'(s))\,ds\,dr &= \int_{t_0}^{t}\int_{s}^{t} G(s,x(s),x'(s))\,dr\,ds \\ &= \int_{t_0}^{t}(t-s)G(s,x(s),x'(s))ds.\end{aligned}$$

Thus the integral equation for $m=2$ is

$$x(t) = a^{(0)} + (t-t_0)a^{(1)} + \int_{t_0}^{t}(t-s)G(s,x(s),x'(s))\,ds,$$

and it is easy to verify directly that any solution of the integral equation is a solution to the Cauchy problem. By induction, it follows that the Cauchy problem in general is equivalent to the integral equation

$$\begin{aligned}x(t) &= \sum_{j=0}^{m-1}\frac{(t-t_0)^j}{j!}a^{(j)} \\ &+ \int_{t_0}^{t}\frac{(t-s)^{m-1}}{(m-1)!}G(s,x(s),\ldots,x^{(m-1)}(s))ds.\end{aligned}$$

We leave the details as an exercise. Notice that the integral equation also involves derivatives—up to order $m-1$—but there is still a gain of one derivative: assuming that $x(t)$ is C^{m-1} and satisfies the integral equation we obtain that $x(t)$ is C^m by the fundamental theorem of the calculus applied m times to the right side of the equation.

To solve the integral equation by the Picard iteration method we choose an initial approximation $x_0(t)$ and define inductively

$$\begin{aligned}x_k(t) &= \sum_{j=0}^{m-1}\frac{(t-t_0)^j}{j!}a^{(j)} \\ &+ \int_{t_0}^{t}\frac{(t-s)^{m-1}}{(m-1)!}G(s,x_{k-1}(s),\ldots,x_{k-1}^{(m-1)}(s))\,ds.\end{aligned}$$

Again we prove the sequence $\{x_k\}$ converges to a solution of the integral equation on a sufficiently small interval by appealing to the Contractive Mapping Principle. This time we have to take the metric space $C^{(m-1)}(J)$ with metric

$$d(x,y) = \sup_J \max_{j=0,\ldots,m-1} |x^{(j)}(t) - y^{(j)}(t)|$$

(or a ball in $C^{(m-1)}(J)$ in the local case). Of course, it is necessary to show that this is a complete metric space and that the Lipschitz condition on G implies that the mapping

$$\begin{aligned} Tx(t) &= \sum_{j=0}^{m-1} \frac{(t-t_0)^j}{j!} a^{(j)} \\ &+ \int_{t_0}^{t} \frac{(t-s)^{m-1}}{(m-1)!} G(s, x(s), \ldots, x^{(m-1)}(s))\, ds. \end{aligned}$$

is a contraction on that metric space. The details are not too much different than before, so we leave them for the exercises. It is interesting to observe that the Picard iterations described above are somewhat different from the iterations obtained if you first reduce the equation to first order. Of course either method produces approximations that converge to the same solution.

11.1.6 Exercises

1. Put the following o.d.e's into normal form by solving for the highest order derivatives:

   a. $x'(t)x(t) + \sin x(t) = 27.$

   b. $\begin{cases} x'(t)y(t) + t^2 y'(t) = 0, \\ x(t)y'(t) + x(t)^2 = t^4. \end{cases}$

   c. $\exp\,(x'') - x'(t)^2 = t^2.$

2. Reduce the following o.d.e.'s to first-order systems by introducing new variables:

   a. $x''(t) = t^2 x'(t) + \sin t.$

   b. $x'''(t) = x''(t)x'(t) + x(t).$

   c. $\begin{cases} x''(t) = \sin y'(t) + x(t)y(t), \\ y''(t) = y'(t)^2 + x'(t)^2. \end{cases}$

3. Verify that $x(t)$ is a solution of the mth order o.d.e. $x^{(m)}(t) = G(t, x(t), \ldots, x^{(m-1)}(t))$ if and only if $(x_0, \ldots, x_{m-1}) =$

$(x, x', \ldots, x^{(m-1)})$ is a solution of the first-order system

$$\begin{aligned} x'_{m-1}(t) &= G(t, x_0(t), \ldots, x_{m-1}(t)), \\ x'_k(t) &= x_{k+1}(t) \quad k = 0, 1, \ldots, m-2. \end{aligned}$$

Also verify that $x(t)$ satisfies the Cauchy initial conditions $x^{(k)}(t_0) = a^{(k)}$ for $k = 0, 1, \ldots, m-1$ if and only if $(x_0, \ldots, x_{m-1})$ satisfies the Cauchy initial conditions $(x_0(t_0), \ldots, x_{m-1}(t_0)) = (a^{(0)}, \ldots, a^{(m-1)})$.

4. Show that all solutions of $x''(t) = -x(t)$ are of the form $x(t) = A\cos t + B\sin t$. Using this, decide for which values of t_1 and t_2 the o.d.e. $x''(t) = -x(t)$ with boundary conditions $x(t_1) = a_1, x(t_2) = a_2$ has a unique solution on $[t_1, t_2]$, for any choice of a_1, a_2.

5. Prove that a homogeneous linear o.d.e. $x'(t) = A(t)x(t)$, where $x(t)$ takes values in $\mathbb{R}^m$ and $A(t)$ is a continuous $m \times m$ matrix valued function, has an m-dimensional vector space of solutions.

6. Write out Picard iterations explicitly as a perturbation series for the inhomogeneous linear o.d.e. $x'(t) = A(t)x(t) + b(t)$ with Cauchy data $x(t_0) = a$ choosing $x_0(t) = a + D^{-1}b(t)$. What would happen with a different choice of $x_0(t)$?

7. Prove that $x(t)$ is a C^m solution to the o.d.e.

$$x^{(m)}(t) = G(t, x(t), \ldots, x^{(m-1)}(t))$$

with Cauchy data

$$x^{(j)}(t_0) = a^{(j)}, \quad j = 0, 1, \ldots, m-1,$$

if and only if $x(t)$ is a C^{m-1} solution of the integral-differential equation

$$\begin{aligned} x(t) &= \sum_{j=0}^{m-1} \frac{(t-t_0)^j}{j!} a^{(j)} \\ &\quad + \int_{t_0}^{t} \frac{(t-s)^{m-1}}{(m-1)!} G(s, x(s), \ldots, x^{(m-1)}(s))\, ds. \end{aligned}$$

8. Compare the Picard iterations for the above integral-differential equation when $m = 2$ with the Picard iterations obtained by first reducing the system to first order.

9. The first-order o.d.e. with $n = 1$ is called *separable* if $G(s, x) = g_1(s)g_2(x)$ for some functions g_1 and g_2. Assume they are C^1 and g_2 is never zero. The usual technique for solving a separable o.d.e. is to write formally $dx/dt = g_1(t)g_2(x)$ hence $dx/g_2(x) = g_1(t)dt$, so $\int dx/g_2(x) = \int g_1(t)dt + c$. Justify this method by showing that for $G_1(t) = \int_{t_0}^{t} g_1(s)\,ds$ and $h(x) = \int_a^x dy/g_2(y)$, the solution to the o.d.e. with initial value $x(t_0) = a$ is given by $x(t) = h^{-1}(G_1(t))$.

11.2 Other Methods of Solution*

11.2.1 Difference Equation Approximation

In this section we discuss two alternate approaches to obtaining solutions of the Cauchy problem. The first, which is sometimes called *Euler's method*, involves obtaining approximations to the solution by replacing the differential equation by a difference equation. This is analogous to approximating a definite integral by a Cauchy sum. Euler's method actually yields a proof of existence under weaker conditions than Picard's method—we need only assume the function G in the o.d.e. is continuous. Of course we have seen that under such weak assumptions the uniqueness of the solution is not assured. Quite frankly, I don't know of any applications in which existence without uniqueness is of any value, but I will present the proof anyway because it uses the Arzela-Ascoli theorem in a nice way. The real importance of Euler's method is that it forms the basis of efficient numerical algorithms for approximating solutions of o.d.e.'s. There are many improvements on Euler's method, essentially based on the idea of using an improvement on the Cauchy sum approximation to an integral, such as the trapezoidal rule or Simpson's rule. We will not discuss these here.

The second method involves expansions into power series of both the solution and all the functions involved in the o.d.e. This method is evidently limited to o.d.e.'s for which power-series expansions are available but within this narrower class of equations gives a computational algorithm of some practical importance, and it yields some information that is not readily available through the other approaches.

We will consider Euler's method in the case of a single first-order o.d.e. $x'(t) = G(t, x(t))$. Recall the interpretation of this equation as saying that the graph of $x(t)$ in the t-x plane has its slope prescribed by $G(t, x)$ and so can be imagined as being pieced together out of infinitesmal line segments of slope $G(t, x)$ through the point (t, x). Euler's method takes small, rather than infinitesmal, line segments and pieces them together to obtain an approximate solution. Let $x(t_0) = a$ be the given initial data, and suppose we want the solution on the interval $[t_0, T]$. We partition the interval as $t_0 < t_1 < \cdots < t_N = T$. On the first interval $[t_0, t_1]$ we take the affine linear function $y(t)$ that satisfies the same initial condition $y(t_0) = y_0$ and the condition $y'(t) = G(t_0, y_0)$ where we set $y_0 = a$ for consistency of notation.

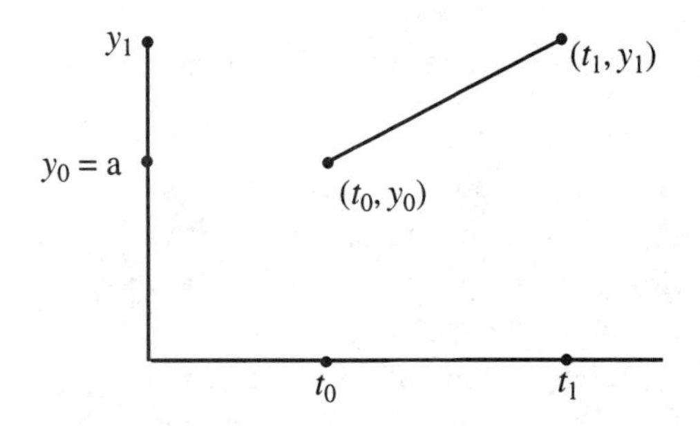

Figure 11.2.1:

Clearly this determines $y(t)$ on the interval $[t_0, t_1]$ to be $y(t) = y_0 + G(t_0, y_0)(t - t_0)$, as shown in Figure 11.2.1. We set $y_1 = y(t_1)$. The function $y(t)$ on the interval $[t_0, t_1]$ differs slightly from the solution $x(t)$ of the o.d.e. $x'(t) = G(t, x(t))$ because $G(t, x(t))$ will in general be different from $G(t_0, y_0)$. But if the interval is small the difference will not be great (assuming G is continuous), so we should have a good approximation. We then repeat the process on the interval $[t_1, t_2]$ starting with the initial data $y(t_1) = y_1$. We now require $y'(t) = G(t_1, y_1)$ and so $y(t) = y_1 + G(t_1, y_1)(t - t_1)$. We continue by induction to define $y(t)$ on each of the intervals $[t_k, t_{k+1}]$ by $y(t) = y_k + G(t_k, y_k)(t - t_k)$ where $y_k = y(t_k)$. In this way we obtain a continuous piecewise affine function $y(t)$ on the interval $[t_0, T]$ where each affine piece of the graph of y is a line segment of the prescribed slope at the left end, as illustrated in Figure 11.2.2.

Before we get into the question of whether (or how well) $y(t)$ approximates the solution $x(t)$, we should consider some alternate inter-

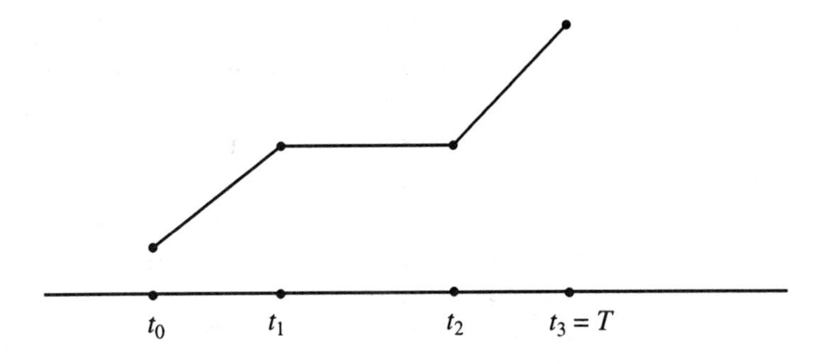

Figure 11.2.2:

pretations of $y(t)$. Since the graph of $y(t)$ is a broken line, it clearly suffices to compute the values $y_k = y(t_k)$—for then we can "connect the dots" to obtain the intermediate values. Notice that the equations $y(t) = y_k + G(t_k, y_k)(t - t_k)$ for t in $[t_k, t_{k+1}]$ evaluated at $t = t_{k+1}$ yield the relations $y_{k+1} - y_k = G(t_k, y_k)(t_{k+1} - t_k)$ or

$$\frac{y_{k+1} - y_k}{t_{k+1} - t_k} = G(t_k, y_k),$$

which is clearly a difference equation analog of the o.d.e. $x'(t) = G(t, x)$. Thus Euler's method can be construed as solving the differential equation approximately by replacing it with a difference equation.

Another interpretation of Euler's method derives from looking at the integral equation

$$x(t) = a + \int_{t_0}^{t} G(s, x(s))\, ds$$

and replacing the integral by a Cauchy sum. If we take $t = t_{k+1}$ and we partition the interval $[t_0, t_{k+1}]$ in the obvious way as $t_0 < t_1 < t_2 < \ldots < t_{k+1}$, this becomes

$$y_{k+1} = y_0 + \sum_{j=0}^{k} G(t_j, x(t_j))(t_{j+1} - t_j)$$

(evaluating $G(s, x(s))$ at the left endpoint $s = t_j$). Of course we don't know $x(t_j)$ exactly, but we could approximate it by y_j; this yields

$$y_{k+1} = y_0 + \sum_{j=0}^{k} G(t_j, y_j)(t_{j+1} - t_j),$$

which is just another way of writing Euler's method. This interpretation shows clearly that there are two sources of error, one from replacing the integral by a Cauchy sum, and one from evaluating G at (t_j, y_j) rather than $(t_j, x(t_j))$. Also there is the possibility that errors will compound, since the y_j are determined inductively and any error in one particular y_j gets passed on to all subsequent ones. Thus we can't simply conclude that y_k approximates $x(t_k)$ from the fact that the Cauchy sum approximates the integral.

Nevertheless, it is possible to show that if G is C^1, then on a sufficiently small interval $[t_0, T]$ Euler's method does approximate the solution. In fact if h denotes the maximum length of the subintervals $[t_k, t_{k+1}]$, then $y(t) - x(t) = O(h)$; in other words, $|y(t) - x(t)| \leq ch$ for $t_0 \leq t \leq T$ where c is a constant depending only on the o.d.e. From a practical point of view, the $O(h)$ error is not very good, and more sophisticated methods—based on better approximations to the integral—are preferred.

To avoid technicalities we assume the partition intervals are of equal length h, so $t_k = t_0 + kh$, and that G and its first derivatives are globally bounded. Under this hypothesis the solution exists for all t and the Euler approximation is valid on any interval.

Theorem 11.2.1 *Let G be C^1 function that is bounded together with its first derivatives. Then there exist constants M_1 and M_2 (depending on the bounds for G and dG) such that*

$$|y(t) - x(t)| \leq M_2(e^{M_1|t-t_0|} - 1)h,$$

where $x(t)$ is the exact solution and $y(t)$ is the Euler approximation using step size h.

Notice that if we extend the size of the interval on which we do the approximation, the constant $e^{M_1|t-t_0|} - 1$ that multiplies h in our estimate grows exponentially. This is exactly what we should expect because the errors in the method may compound. Of course, if we fix any interval, we can choose h small enough to make the error estimate as small as we like *on that interval.* But it is only fair to point out that the smaller we take h, the more points we have to take in the partition and, hence, the more computations we have to perform. Also, any numerical implementation of this algorithm will involve round-off error; and when a large number of calculations need to be performed, the round-off error has to be taken into account.

Proof: For simplicity we only prove the error estimate at the partition point t_k. Thus we need to show

$$(*) \qquad |y_k - x(t_k)| \leq M_2(e^{M_1 kh} - 1)h.$$

We will prove this by induction. Notice that for $k = 0$ we have $y_0 = x(t_0) = a$ as required. Thus we may assume that $(*)$ holds, and we need to show the analogous estimate holds for $k+1$. (In the course of the argument we will specify the constants M_1 and M_2.) Euler's method gives us $y_{k+1} = y_k + G(t_k, y_k)h$, and the integral equation gives $x(t_{k+1}) = x(t_k) + \int_{t_k}^{t_{k+1}} G(s, x(s))\, ds$ for the exact solution. Then $y_{k+1} - x(t_{k+1}) = y_k - x(t_k) + G(t_k, y_k)h - \int_{t_k}^{t_{k+1}} G(s, x(s))\, ds$ and the first difference $y_k - x(t_k)$ is controlled by $(*)$. In comparing the integral $\int_{t_k}^{t_{k+1}} G(s, x(s))\, ds$ with $g(t_k, y_k)h$ we are on slippery footing because the value y_k is not exactly $x(t_k)$. Therefore we will write $G(t_k, y_k)h - \int_{t_k}^{t_{k+1}} G(s, x(s))\, ds = [G(t_k, y_k)h - G(t_k, x(t_k))h] + [G(t_k, x(t_k))h - \int_{t_k}^{t_{k+1}} G(s, x(s))\, ds]$ and estimate each of the bracketed differences separately.

The first is easy. Using the mean value theorem we have $|G(t_k, y_k)h - G(t_k, x(t_k)h| \leq M_1 h |y_k - x(t_k)|$ where M_1 (this will be the constant M_1 in $(*)$) is an upper bound for the derivatives of G (actually just the x-derivatives, since the point t_k is the same in both $G(t_k, y_k)$ and $G(t_k, x(t_k))$). Notice that this is just what the doctor ordered, because we have $|y_k - x(t_k)|$ on the right, which is controlled by $(*)$.

For the second difference, notice that we have the integral over an interval of length h of the function $G(s, x(s))$ being compared with the value of this function at the left endpoint $s = t_k$ multiplied by the length of the interval. If we call this function $g(s) = G(s, x(s))$, then

$$\begin{aligned} g(t_k)h - \int_{t_k}^{t_{k+1}} g(s)\, ds &= \int_{t_k}^{t_{k+1}} (g(t_k) - g(s))\, ds \\ &= \int_{t_k}^{t_{k+1}} (t_k - s) g'(r(s))\, ds \end{aligned}$$

by the mean value theorem. Thus

$$\left| g(t_k)h - \int_{t_k}^{t_{k+1}} g(s)\,ds \right| \le M \int_{t_k}^{t_{k+1}} (s - t_k)\,ds = (1/2)Mh^2$$

where M is a bound for g'. But

$$g'(s) = \frac{\partial G}{\partial s}(s, x(s)) + \frac{\partial G}{\partial x}(s, x(s))x'(s)$$

by the chain rule and $x'(s) = G(s, x(s))$ by the o.d.e., so g' can be bounded in terms of the bounds for G and its derivatives.

If we combine all three estimates we have altogether (triangle inequality)

$$|y_{k+1} - x(t_{k+1})| \le |y_k - x(t_k)| + M_1 h |y_k - x(t_k)| + (1/2)Mh^2.$$

Then substituting $(*)$, we have

$$|y_{k+1} - x(t_{k+1})| \le (1 + M_1 h)M_2(e^{M_1 kh} - 1)h + (1/2)Mh^2.$$

We now can specify that $M_2 = M/2M_1$. It then follows that

$$\begin{aligned} |y_{k+1} - x(t_{k+1})| &\le M_2 h[(1 + M_1 h)(e^{M_1 kh} - 1) + M_1 h] \\ &= M_2 h[(1 + M_1 h)e^{M_1 kh} - 1] \le M_2 h(e^{M_1(k+1)h} - 1) \end{aligned}$$

because $1 + M_1 h \le e^{M_1 h}$ (remember the power series for exp). Thus we have the desired analog of $(*)$ for $k + 1$. QED

11.2.2 Peano Existence Theorem

We will now show how Euler's method can be used to establish the existence of solutions under the hypothesis that G is continuous. Since we don't yet know that solutions exist under such weak assumptions, we cannot refer to "the solution" $x(t)$—in fact we know by example ($x' = (3/2)x^{1/3}$) that there may be more than one solution to the Cauchy problem. Nevertheless, the formulas defining Euler's method,

$$(*) \qquad y(t) = y_k + G(t_k, y_k)(t - t_k)$$

for $t_k \le t \le t_{k+1}$, make sense without reference to any solution. We can use these to obtain a solution via the Arzela-Ascoli theorem.

Theorem 11.2.2 *(Peano Existence Theorem) Let $G(t,x)$ be a continuous real-valued function for t in some interval I containing t_0 and x in some neighborhood of a. Then there exists an interval J containing t_0 and a solution $x(t)$ to the Cauchy problem $x'(t) = G(t,x(t))$ and $x(t_0) = a$ on the interval J. The solution need not be unique.*

Proof: If the domain of G is not compact we can always make it so by shrinking it. Then let M be the sup of $|G|$ over this compact domain, which is finite because G is continuous. We want to shrink the domain still further to a rectangle (t in J and $|x-a| \le r$) in which the approximations $y(t)$ and the solution will be trapped, as shown in Figure 11.2.3.

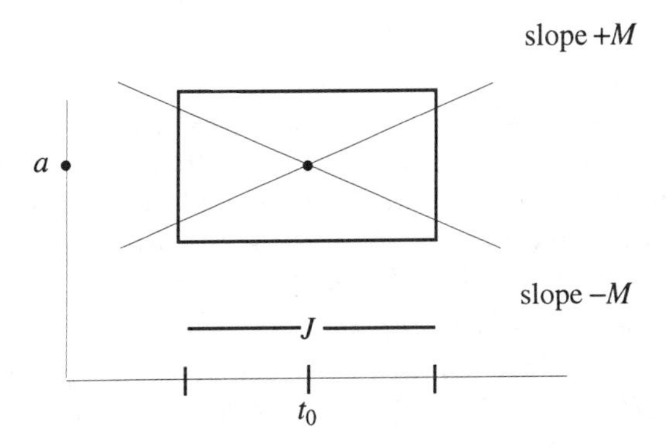

Figure 11.2.3:

The condition for trapping the solution is that $M|t-t_0| \le r$ for t in J, which can be achieved by making J small enough while holding r fixed (the constant M, which is the sup of $|G|$ over a larger domain, will still be an upper bound for $|G|$ over this smaller domain). The reason the condition $M|t-t_0| \le r$ for t in J traps the solution is that the derivative $x'(t)$ is trapped between $-M$ and $+M$, so any solution must lie between the lines through (t_0, a) with slope $-M$ and $+M$.

For simplicity we assume J is $[t_0, T]$. Consider any partition $t_0 < t_1 < \cdots < t_N = T$ of this interval, and let's attempt to define inductively the Euler method function

$$y(t) = y_k + G(t_k, y_k)(t - t_k)$$

for t in $[t_k, t_{k+1}]$ where $y_0 = a$ and $y_k = y(t_k)$. For this to make sense we need to verify that (t_k, y_k) is in the domain of G; in fact we will show $|y_k - a| \le r$ or more precisely $|y(t) - a| \le M|t - t_0|$ for t in J. We do this by induction. Since $y_0 = a$, we have trivially $|y_0 - a| \le r$ and

$$|y(t) - a| = |G(t_0, y_0)(t - t_0)| \le M|t - t_0|$$

for t in $[t_0, t_1]$. Since we are assuming $M|t - t_0| \le r$ in J, we have $|y_1 - a| \le r$. Continuing by induction, assuming $|y_k - a| \le r$ and $|y(t) - a| \le M|t - t_0|$ for $t_0 \le t \le t_k$, we have

$$\begin{aligned} |y(t) - a| &= |y_k - a + G(t_k, y_k)(t - t_k)| \\ &\le |y_k - a| + |G(t_k, y_k)(t - t_k)| \\ &\le M(t_k - t_0) + M(t - t_k) = M(t - t_0) \end{aligned}$$

for $t_k \le t \le t_{k+1}$. Thus also $|y_{k+1} - a| \le r$ since $M(t_{k+1} - t_0) \le r$ by assumption. This completes the induction argument.

Now we want to vary the partition, letting the maximum length of the subintervals go to zero. Let $y^{(1)}, y^{(2)}, \ldots$ denote the sequence of functions obtained by the Euler method from the partitions. We would hope to obtain a solution as the limit of this sequence, but the nonuniqueness means the limit may not exist. Instead we must settle for something weaker, a limit of a subsequence. For this we invoke the Arzela-Ascoli theorem. Thus we need to verify the hypotheses of that theorem, that the sequence is uniformly bounded and uniformly equicontinous (they are obviously continuous functions defined on a compact domain). The estimate $|y(t) - a| \le M|t - t_0| \le r$ established above shows that the functions are uniformly bounded. The uniform equicontinuity is essentially a consequence of the fact that the functions are piecewise affine with derivative $G(t_k, y_k)$ bounded by M. We claim that any function $y(t)$ constructed by Euler's method will satisfy $|y(t) - y(s)| \le M|t - s|$ for t and s in J—this just follows by summing the same estimate over all the subintervals connecting t and s, since

$$|y(t) - y(s)| = |G(t_k, y_k)(t - s)| \le M|t - s|$$

if t and s belong to the same subinterval $[t_k, t_{k+1}]$. Since the constant M does not depend on the partition, we obtain the uniform equicontinuity.

Thus the Arzela-Ascoli theorem applies and so there exists a uniformly convergent subsequence. For simplicity of notation let us also denote this subsequence $y^{(1)}, y^{(2)}, \ldots$. Let $x(t) = \lim_{j \to \infty} y^{(j)}(t)$, the limit converging uniformly on J. To complete the proof we will show that $x(t)$ satisfies the integral equation

$$x(t) = a + \int_{t_0}^{t} G(s, x(s))\, ds$$

for we have already seen that the integral equation is equivalent to the Cauchy problem. We will obtain the integral equation if we can show that

$$y^{(j)}(t) = a + \int_{t_0}^{t} G(s, y^{(j)}(s))\, ds + R_j(t)$$

where $R_j(t) \to 0$ as $j \to \infty$, for

$$\lim_{j\to\infty} \int_{t_0}^{t} G(s, y^{(j)}(s))\, ds = \int_{t_0}^{t} G(s, x(s))\, ds$$

because the convergence of $y^{(j)}$ to x is uniform.

The definition of $y^{(j)}(t)$ by Euler's method shows that $y^{(j)}(t) - a$ is a certain Cauchy sum for the integral $\int_{t_0}^{t} G(s, y^{(j)}(s))\, ds$ corresponding to the partition $t_0 < t_1^{(j)} < \cdots < t_k^{(j)} < t$ of $[t_0, t]$ and evaluation of $G(s, y^{(j)}(s))$ at the left endpoint of each subinterval. Here $t_0 < t_1^{(j)} < \cdots < t_N^{(j)} = T$ is the partition corresponding to $y^{(j)}$ and $t_k^{(j)}$ is the largest value for which $y_k^{(j)} < t$ (thus k depends on t and j, but we have not burdened the notation with this dependence). This is just the equation

$$\begin{aligned} y^{(j)}(t) &= y^{(j)}(t_k^{(j)}) + G(t_k^{(j)}, y^{(j)}(t_k^{(j)}))(t - t_k^{(j)}) \\ &= a + \sum_{p=0}^{k-1} G(t_p^{(j)}, y^{(j)}(t_p^{(j)}))(t_{p+1}^{(j)} - t_p^{(j)}) \\ &\quad + G(t_k^{(j)}, y^{(j)}(t_k^{(j)}))(t - t_k^{(j)}). \end{aligned}$$

Therefore the error R_j is just the difference between a Cauchy sum and an integral. Notice that both the partition and the function $G(s, y^{(j)}(s))$ being integrated vary with j. Since the maximum interval lengths of the partitions are going to zero, we know that the Cauchy sums for any fixed continuous function f are converging to the integral of f. An examination of that proof, however, shows that the rate of convergence depends only on the modulus of continuity, which is defined to be $\omega(\delta) = \sup\{|f(s) - f(t)|$ given that $|s-t| \le \delta\}$. In the present case, the uniform equicontinuity of the functions $y^{(j)}$ implies that a single function $\omega(\delta)$ dominates the modulus of continuity for all the integrands $G(s, y^{(j)}(s))$ so that the difference between $\int_{t_0}^{t} G(s, y^{(j)}(s))\, ds$ and a Cauchy sum is small for all j provided the partition is small. This implies $R_j \to 0$ as $j \to \infty$ and so the integral equation holds. QED

Although we have stated the theorem only for single equations, it is clear that the proof goes over almost without change to first-order systems and, hence, to systems of arbitrary order

$$x^{(m)}(t) = G(t, x(t), x'(t), \ldots, x^{(m-1)}(t))$$

for G continuous.

11.2.3 Power-Series Solutions

We turn now to power-series methods. We will consider only linear o.d.e.'s of second order, because this special case includes many important examples. Making the assumption of linearity vastly simplifies the problem, although it is still true that the Cauchy problem for the general equation

$$x^{(m)}(t) = G(t, x(t), x'(t), \ldots, x^{(m-1)}(t))$$

has a power-series solution that converges in a small neighborhood of t_0 provided the function G has a convergent multiple power series in a neighborhood of the Cauchy data. The proof of this, however, would take us too far afield. We should also point out that the result has a generalization to partial differential equations, the famous Cauchy-Kovalevsky Theorem.

The o.d.e.'s we consider are of the form

$$x''(t) = p(t)x'(t) + q(t)x(t) + r(t)$$

where p, q, r are the analytic functions

$$p(t) = \sum_{k=0}^{\infty} p_k (t - t_0)^k,$$

$$\begin{aligned} q(t) &= \sum_{k=0}^{\infty} q_k(t-t_0)^k, \\ r(t) &= \sum_{k=0}^{\infty} r_k(t-t_0)^k \end{aligned}$$

with power series converging at least in $|t-t_0| < \varepsilon$. We wish to solve the Cauchy problem $(x(t_0) = a, x'(t_0) = b)$ by an analytic function with power series $x(t) = \sum_{k=0}^{\infty} c_k(t-t_0)^k$, also convergent in $|t-t_0| < \varepsilon$. Notice then that the Cauchy initial data $x(t_0) = a, x'(t_0) = b$ just means $c_0 = a, c_1 = b$. Thus the problem can be restated: find an analytic solution of the o.d.e. $x(t) = \sum_{k=0}^{\infty} c_k(t-t_0)^k$ with c_0 and c_1 specified. For simplicity of notation we will set $t_0 = 0$.

We proceed backward, assuming the solution exists. What does the o.d.e. say about the coefficients c_k? Inside the radius of convergence we can perform all the operations called for in the o.d.e. by operating formally on the power series. We obtain

$$\begin{aligned} x''(t) &= \sum_{k=2}^{\infty} k(k-1)c_k t^{k-2} \\ &= \sum_{k=0}^{\infty} (k+1)(k+2)c_{k+2}t^k \end{aligned}$$

and also (by making a change of variable in the k summations)

$$\begin{aligned} & p(t)x'(t) + q(t)x(t) + r(t) \\ &= \left(\sum_{j=0}^{\infty} p_j t^j\right)\left(\sum_{k=0}^{\infty}(k+1)c_{k+1}t^k\right) \\ &\quad + \left(\sum_{j=0}^{\infty} q_j t^j\right)\left(\sum_{k=0}^{\infty} c_k t^k\right) + \sum_{k=0}^{\infty} r_k t^k \\ &= \sum_{k=0}^{\infty}\left(\sum_{j=0}^{k} p_j(k-j+1)c_{k-j+1}\right)t^k \\ &\quad + \sum_{k=0}^{\infty}\left(\sum_{j=0}^{k} q_j c_{k-j}\right)t^k + \sum_{k=0}^{\infty} r_k t^k \end{aligned}$$

$$= \sum_{k=0}^{\infty}\left(\left(\sum_{j=0}^{k} p_j(k-j+1)c_{k-j+1} + q_j c_{k-j}\right) + r_k\right)t^k.$$

All these power series will be convergent in $|t| < \varepsilon$. By the uniqueness of power-series expansions we can equate the coefficients of $x''(t)$ and $p(t)x'(t)+q(t)x(t)+r(t)$. This leads to a system of algebraic equations

$$(*) \qquad (k+1)(k+2)c_{k+2} = \left(\sum_{j=0}^{k} p_j(k-j+1)c_{k-j+1} + q_j c_{k-j}\right) + r_k$$

for $k = 0, 1, 2, \ldots$. Notice that the right side of the equation contains $c_0, \ldots, c_{k+1}$ but no other coefficients and that the left side is just c_{k+2} multiplied by a positive constant. Thus, given values for c_0 and c_1, these algebraic equations have a unique solution; in fact they are inhomogeneous linear equations in triangular form. It is the simplicity of the solution of these equations that makes the use of power series so appealing here.

So far we have seen that if the problem has a power-series solution, then we have the recursion formulas $(*)$ to find the coefficients. The next question is can we reverse the process? That is, we start by solving the equations $(*)$ to obtain the formal power series $\sum_{k=0}^{\infty} c_k t^k$. If we can show that this power series converges in $|t| < \varepsilon$, we will be done, for then $(*)$ will show that the power series satisfies the o.d.e. However, this seems a formidable task, since the coefficients c_k are only given indirectly as solutions of $(*)$ and because in order to show that the power series converges we will need some estimates for the size of $|c_k|$. While it is possible to derive such estimates directly from the equations $(*)$, there is a simpler indirect approach known as *Cauchy's method of majorants.* The idea of this method is to show there exists a *majorant* series $\sum_{k=0}^{\infty} d_k t^k$, with $|c_k| \le d_k$ (so the d_k are all positive) and yet $\sum_{k=0}^{\infty} d_k t^k$ converges in $|t| < \varepsilon$. This implies the convergence of $\sum_{k=0}^{\infty} c_k t^k$ by the comparison test. In order to produce the majorant series we will systematically increase all the coefficients of the power series of p, q, r. Because of the form of $(*)$, it is easy to show this produces a majorant series. Finally, we will obtain the convergence of the majorant series not by estimating its coefficients but because we will be able to write down the function $\sum_{k=0}^{\infty} d_k t^k$ explicitly—it will be the solution of an o.d.e. that can be solved by inspection. This is the trick that saves us the tedious work of estimating coefficients.

We begin by proving the Majorant lemma:

Lemma 11.2.1 (*Majorant lemma*) *Suppose*

$$P(t) = \sum_{k=0}^{\infty} P_k t^k, \quad Q(t) = \sum_{k=0}^{\infty} Q_k t^k, \quad \text{and } R(t) = \sum_{k=0}^{\infty} R_k t^k$$

are majorants of $p(t) = \sum_{k=0}^{\infty} p_k t^k, q(t) = \sum_{k=0}^{\infty} q_k t^k$, *and* $r(t) = \sum_{k=0}^{\infty} r_k t^k$ *respectively, meaning* $|p_k| \le P_k, |q_k| \le Q_k$, *and* $|r_k| \le R_k$. *Suppose* c_0 *and* c_1 *are given and* $\sum_{k=0}^{\infty} c_k t^k$ *is defined by the equation*

$$(*) \qquad (k+1)(k+2)c_{k+2} = \left(\sum_{j=0}^{k} p_j (k-j+1) c_{k-j+1} + q_j c_{k-j} \right) + r_k,$$

and similarly suppose d_0 *and* d_1 *are given and* $\sum_{k=0}^{\infty} d_k t^k$ *is defined by the analogous equation*

$$(**) \qquad (k+1)(k+2)d_{k+2} = \left(\sum_{j=0}^{k} P_j (k-j+1) d_{k-j+1} + Q_j d_{k-j} \right) + R_k.$$

Finally assume $|c_0| \le d_0$ *and* $|c_1| \le d_1$. *Then* $\sum d_k t^k$ *is a majorant for* $\sum c_k t^k$, *meaning* $|c_k| \le d_k$ *for every* k.

Proof: Notice that this is a purely algebraic fact—we assume nothing and conclude nothing about the convergence of any of the formal power series. The proof is by induction and is almost completely trivial. We are assuming $|c_0| \le d_0$ and $|c_1| \le d_1$, so let us assume we know $|c_j| \le d_j$ for $0 \le j \le k+1$ and then prove $|c_{k+2}| \le d_{k+2}$. But from $(*)$ and the triangle inequality

$$(k+1)(k+2)|c_{k+2}| \le \left(\sum_{j=0}^{k} |p_j| (k-j+1) |c_{k-j+1}| + |q_j| \, |c_{k-j}| \right) + |r_k|$$

since the coefficients $(k-j+1)$ are all positive. Substituting the majorants for p_j, q_j, r_k and the assumed estimates $|c_j| \le d_j$ for $0 \le j \le k+1$ we obtain

$$(k+1)(k+2)|c_{k+2}| \le \left(\sum_{j=0}^{k} P_j (k-j+1) d_{k-j+1} + Q_j d_{k-j} \right) + R_k$$

and the right side is just $(k+1)(k+2)d_{k+2}$. QED

Theorem 11.2.3 *Let the power series for p, q, r converge in $|t| < \varepsilon$. Then the power series $x(t) = \sum_{k=0}^{\infty} c_k t^k$, where c_k are given by $(*)$ and c_0 and c_1 are arbitrarily chosen, also converges in $|t| < \varepsilon$ to a solution of the Cauchy problem*

$$x''(t) = p(t)x'(t) + q(t)x(t) + r(t), \quad x(0) = c_0, x'(0) = c_1.$$

Proof: In order to put the lemma to work we have to make a clever choice of the majorants P, Q, R. Fix a value $\delta < \varepsilon$. From the convergence of $\sum_{k=0}^{\infty} p_k t^k$ in $|t| \le \delta$ we know that $|p_k| \le M\delta^{-k}$ for some M and all k (recall that this followed from the boundedness of $p_k t^k$ for $t = \delta$). Thus the choice $P_k = M\delta^{-k}$ would seem natural. Notice then that

$$\begin{aligned} P(t) &= \sum_{k=0}^{\infty} P_k t^k = \sum_{k=0}^{\infty} M\delta^{-k} t^k \\ &= M(1-\delta^{-1}t)^{-1} = \frac{M\delta}{\delta - t} = \frac{M_1}{\delta - t} \end{aligned}$$

is an elementary function and its power series converges in $|t| < \delta$.

We might similarly be tempted to try $Q(t) = M_2/(\delta - t)$ and $R(t) = M_3/(\delta - t)$ for suitable constants M_2 and M_3, but this choice leads to the o.d.e.

$$x''(t) = P(t)x'(t) + Q(t)x(t) + R(t),$$

which can't be solved by inspection. However, a slight modification of the choice of Q and R will lead to a linear o.d.e. of homogeneous type

$$x''(t) = \frac{M_1}{\delta - t}x'(t) + \frac{M_2}{(\delta - t)^2}x(t) + \frac{M_3}{(\delta - t)^3},$$

which can easily be solved. This means we want to choose

$$Q(t) = \frac{M_2}{(\delta - t)^2} \quad \text{and} \quad R(t) = \frac{M_3}{(\delta - t)^3}.$$

Since

$$Q(t) = \frac{M_2}{(\delta - t)^2} = \frac{d}{dt}\frac{M_2}{\delta - t} = \sum_{k=0}^{\infty} \delta^{-2} M_2 (k+1) \delta^{-k} t^k$$

and

$$R(t) = \frac{M_3}{(\delta - t)^3} = \frac{1}{2}\left(\frac{d}{dt}\right)^2 \frac{M_3}{\delta - t}$$
$$= \sum_{k=0}^{\infty} \delta^{-3} M_3 (k+2)(k+1)\delta^{-k} t^k,$$

we can always choose M_2 and M_3 large enough to make these majorants of $q(t)$ and $r(t)$—the point being that the factors $(k+1)$ and $(k+2)(k+1)$ do not spoil the estimates.

Choosing $d_0 = |c_0|$ and $d_1 = |c_1|$, we can apply the lemma to conclude $|c_k| \le d_k$ where $\sum d_k t^k$ is the power series for the solution to

$$x''(t) = \frac{M_1}{\delta - t} x'(t) + \frac{M_2}{(\delta - t)^2} x(t) + \frac{M_3}{(\delta - t)^3}$$

$x(0) = |c_0|, x'(0) = |c_1|$. But this o.d.e. has solutions of the form

$$a(\delta - t)^{\lambda_1} + b(\delta - t)^{\lambda_2} + c(\delta - t)^{-1}.$$

Indeed by substituting this in to the o.d.e. and doing the routine calculations (see exercises) we find that λ_1 and λ_2 must satisfy the quadratic equation $\lambda(\lambda - 1) = M_1\lambda + M_2$ and c must satisfy $2c = M_1 c + M_2 c + M_3$. Thus we have determined a unique value of c (if $M_1 + M_2 = 2$ we can increase M_1 to avoid this problem), and the quadratic equation for λ has two distinct real roots

$$-\left(\frac{1 + M_1}{2}\right) \pm \sqrt{\left(\frac{1 + M_1}{2}\right)^2 + M_2}.$$

Thus by adjusting the two constants a and b we can meet the initial conditions $x(0) = |c_0|, x'(0) = |c_1|$ (this requires only the distinctness of the two roots λ_1 and λ_2). The solution

$$a(\delta - t)^{\lambda_1} + b(\delta - t)^{\lambda_2} + c(\delta - t)^{-1}$$

is an analytic function with a power series about $t = 0$ convergent in $|t| < \delta$; the explicit power series is given by the binomial theorem (see exercise 4 of section 7.4). But this power series must be the same as $\sum d_k t^k$ because the coefficients must satisfy (**) of the majorant lemma. Thus $\sum d_k t^k$ converges in $|t| < \delta$ and so $\sum c_k t^k$ converges in $|t| < \delta$ by the majorant lemma. Since we could take any $\delta < \varepsilon$, it follows that $\sum c_k t^k$ converges in $|t| < \varepsilon$ as desired. QED

Of course one could also compute d_k explicitly (first computing $M_1, M_2, M_3, \lambda_1, \lambda_2, a, b, c$) from the binomal theorem and then prove by induction $|c_k| \leq d_k$ directly from $(*)$, thereby replacing the above indirect argument by a mass of incomprehensible computation.

For many applications it is necessary to extend this technique to the case when the functions $p(t), q(t), r(t)$ do not have power-series expansions about the point t_0 but rather have singularities of specified nature (for example, Bessel's o.d.e.

$$t^2 x''(t) + t x'(t) + (t^2 - \lambda^2) x(t) = 0,$$

where λ is a fixed constant, becomes

$$x''(t) = \frac{-1}{t} x'(t) + \left(\frac{\lambda_2}{t^2} - 1 \right) x(t),$$

so $p(t)$ and $q(t)$ have singularities at $t = 0$). This is the theory of *regular singular points.*

11.2.4 Exercises

1. Solve the Cauchy problem $x'(t) = x(t)^2, x(t_0) = a$ by power series. What is the radius of convergence of the solution?

2. Write out the explicit recursion formulas for Euler's method for appoximating solutions of the Cauchy problem

$$x''(t) = G(t, x(t), x'(t)), \quad x(t_0) = a, x'(t_0) = b,$$

for a second-order o.d.e. by reducing to a first-order system.

3. Prove that the first-order linear o.d.e. $x'(t) = p(t)x(t) + q(t)$ with initial condition $x(t_0) = a$ is solved in closed form by

$$x(t) = au(t) + \int_{t_0}^{t} \frac{u(t)q(s)}{u(s)}\, ds$$

where $u(t) = \exp(\int_{t_0}^{t} p(s)\, ds)$. (**Hint**: show first $u'(t) = p(t)u(t)$.)

4. Substitute $x(t) = a(\delta - t)^{\lambda_1} + b(\delta - t)^{\lambda_2} + c(\delta - t)^{-1}$ into the o.d.e.

$$x''(t) = \frac{M_1}{\delta - t}x'(t) + \frac{M_2}{(\delta - t)^2}x(t) + \frac{M_3}{(\delta - t)^3}$$

and derive the conditions on a, b, c that are necessary and sufficient for this to be a solution. Then show (with $M_1 + M_2 - 2 \neq 0$) that any initial conditions $x(0) = d_0$ and $x'(0) = d_1$ can be met by the appropriate choice of a and b.

5. Write out a proof that the first-order linear o.d.e. in problem 3 can also be solved by power-series methods if $p(t) = \sum_{k=0}^{\infty} p_k(t - t_0)^k$ and $q(t) = \sum_{k=0}^{\infty} q_k(t - t_0)$ converge in $|t - t_0| < \varepsilon$.

6. Show that if G is C^1, then there exists an interval $[t_0, T]$ and constants M_1, M_2 such that $|y(t) - x(t)| \leq M_2(e^{M_1|t-t_0|} - 1)h$ on $[t_0, T]$ where y is the Euler approximation to the exact solution x with stepsize h.

7. Show that

$$J_k(t) = \sum_{j=0}^{\infty} \frac{(-1)^j (t/2)^{k+2j}}{j!(k+j)!}$$

has infinite radius of convergence and satisfies Bessel's differential equation $x''(t) + (1/t)x'(t) + (1 - k^2/t^2)x(t) = 0$.

11.3 Vector Fields and Flows*

11.3.1 Integral Curves

We want to consider now some elegant applications of the existence and uniqueness theorem for o.d.e.'s. We want to discuss the notion of *vector field*, which intuitively is the assignment of an arrow to every point in space. More precisely, a *continuous vector field* on an open domain D in $\mathbb{R}^n$ is defined to be a continuous function $F : D \to \mathbb{R}^n$, but we want to think of the value $F(x)$ as giving a vector sitting at the point x (pictorially, the arrow joining x to $x + F(x)$, as in Figure 11.3.1).

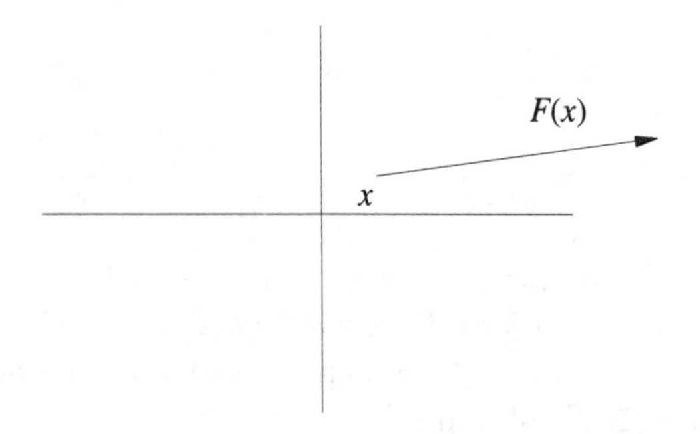

Figure 11.3.1:

Of course we cannot draw in too many of the arrows at once or they will criss-cross unintelligibly, but we can certainly imagine the vector field as a well-groomed head of hair on a flat-headed person. You can also think of a weather map showing wind velocities. A common way to get a vector field is to take the gradient of a scalar function, $f : D \to \mathbb{R}$ where f is assumed C^1, so $F = \nabla f$. This is called a *gradient vector field*. Not all vector fields are of this form, however (see exercises).

We want to think of the vector field as giving directional instructions: when at x, move with velocity $F(x)$. If we obey the instructions we will trace out an *integral curve* of the vector field, which is defined to be a C^1 curve $x : [a, b] \to \mathbb{R}^n$ such that $x'(t) = F(x(t))$ for all t in $[a, b]$. If we think of $x'(t)$ as the tangent vector to the curve, then we are saying that the curve has tangent vectors prescribed by the vector field. (*Warning*: This is slightly different from, although related to, our interpretation of the o.d.e. $x'(t) = G(t, x(t))$ as prescribing the slope of the graph of the solution.) It is clear that the equation for an integral curve is just a first-order o.d.e. that is already in normal form, so we are ready to apply the existence and uniqueness theorem. We will assume that the vector field F is C^1 so that the local theorem can be applied. The o.d.e. for the integral curve is sometimes called an *autonomous* o.d.e. because it does not depend on t (thought of as time).

Theorem 11.3.1 *Let $F : D \to \mathbb{R}^n$ with D an open subset of $\mathbb{R}^n$ be a C^1 vector field. Then through every point y of D there passes a unique integral curve $x(t)$ with $x(0) = y$. Furthermore $x(t)$ has a natural maximal domain of definition I, an interval of $\mathbb{R}$, such that as t approaches a finite endpoint of I (if I is not all of $\mathbb{R}$) the curve $x(t)$ either becomes unbounded or leaves the domain D. Two different integral curves either* *coincide as subsets of* $\mathbb{R}^n$ ($x^{(1)}(t) = x^{(2)}(t+s)$ *for some fixed* s) *or are disjoint.*

Proof: The existence and uniqueness of $x(t)$ follow from the local existence and uniqueness of $x'(t) = F(x(t))$ with $x(0) = y$. As we observed following the proof of that theorem, the solution can be continued beyond $t = a$ if $\lim_{t\to a} x(t)$ exists and is in D. Thus the integral curves cannot simply disappear but must move out of D or toward infinity if they don't exist for all t.

The fact that two integral curves must either coincide or be disjoint follows from the fact that the o.d.e. is autonomous. If $x(t)$ satisfies the o.d.e., then so does $x(t+s)$ for any fixed s, for $x'(t+s) = dx/dt(t+s) = F(x(t+s))$. Now suppose $x^{(1)}(t)$ and $x^{(2)}(t)$ are two integral curves and $x^{(1)}(t_1) = x^{(2)}(t_2)$ for some values t_1 and t_2. Then $x^{(3)}(t) = x^{(1)}(t+s)$ for $s = t_1 - t_2$ satisfies the o.d.e. and $x^{(3)}(t_2) = x^{(1)}(t_1) = x^{(2)}(t_2)$. By the uniqueness for the Cauchy problem at t_2 we must have $x^{(3)}(t) = x^{(2)}(t)$ and, hence, $x^{(1)}(t+s) = x^{(2)}(t)$ for all t. QED

Notice that the integral curves are actually C^2; one derivative exists via the o.d.e. and $x''(t) = dF(x(t))x'(t)$ by the chain rule, which is continuous because F is assumed C^1. You should imagine a picture of the domain D filled with disjoint smooth integral curves, such as the example shown in Figure 11.3.2. Incidentally, nothing prevents the curves from being closed. We can have $x(t_1) = x(0)$ for some t_1, in which case the curve repeats periodically with period t_1, $x(t+t_1) = x(t)$ for all t, again by the uniqueness. The curves may in fact consist of a single point, $x(t) \equiv y$ if $F(y) = 0$.

Next, we observe that we can follow the integral curves to obtain a *flow*. We think of t as a time parameter, and we move the point $x(0)$ to $x(t)$ in time t. Putting all the integral curves together we define the flow f_t by $f_t(y) = x(t)$ where x is the unique integral curve such that $x(0) = y$. The flow is possibly only locally defined, as $x(t)$ may not exist for all values of t; but for each fixed y it is defined for some interval of time containing zero. The flow has a local group property, $f_t(f_s(y)) = f_{t+s}(y)$, whenever both sides are defined. The reason is again the uniqueness. If we let $z = f_s(y)$ and start the flow at z, say $x^{(1)}(t)$ is the integral curve with $x^{(1)}(0) = z$, then $x(t+s) = x^{(1)}(t)$ because this is true for $t = 0$ and so $f_{s+t}(y) = f_t(z) = f_t(f_s(y))$.

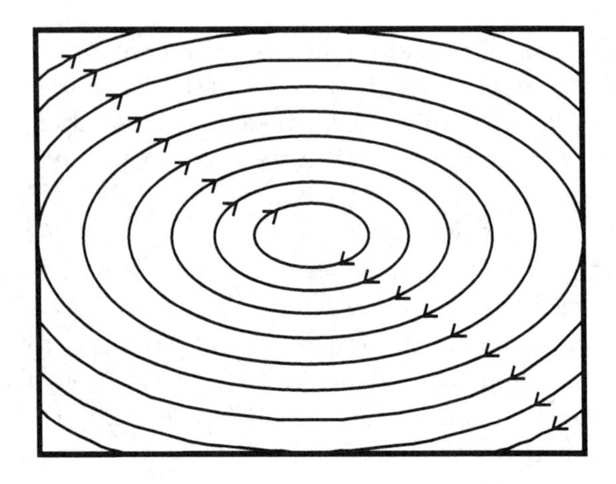

Figure 11.3.2:

Conversely, from the flow we can recover the vector field, $F(y) = (\partial/\partial t) f_t(y)|_{t=0}$. This is an immediate consequence of the definition, for if $x(t)$ is the integral curve with $x(0) = y$, then $(\partial/\partial t) f_t(y) = x'(t) = F(x(t))$, so $(\partial/\partial t) f_t(y)|_{t=0} = F(x(0)) = F(y)$. In fact we have actually shown $(\partial/\partial t) f_t(y) = F(f_t(y))$ for any value of t since $x(t) = f_t(y)$.

Even if we don't start with a vector field but are simply given a flow, a function $f_t(x)$ defined for x in D and t in some neighborhood of 0 (depending on x) with $f_t(x)$ also in D, satisfying the local group law $f_t(f_s(x)) = f_{t+s}(x)$ whenever both sides are defined, then we can define a vector field F by the equation $F(x) = (\partial/\partial t) f_t(x)|_{t=0}$ (we must assume that $f_t(x)$ is differentiable in t, of course). We can again show that $(\partial/\partial t) f_t(x) = F(f_t(x))$, this time by appealing to the local group law: $F(f_t(x)) = (\partial/\partial s) f_s(f_t(x))|_{s=0}$ by definition and $(\partial/\partial s) f_s(f_t(x))|_{s=0} = (\partial/\partial s) f_{s+t}(x)|_{s=0} = (\partial/\partial t) f_t(x)$. From this it follows that $f_t(x)$ for any fixed x is an integral curve for the vector field F. In this way we can establish a one-to-one correspondence between vector fields and flows. We will refrain from stating this as a formal theorem because the comparison of smoothness properties of the vector field and the flow requires some more careful work.

11.3.2 Hamiltonian Mechanics

We can describe classical mechanics in Hamiltonian form using vector fields and flows. Following the standard physicist's conventions, we let $q = (q_1, \ldots, q_n)$ denote a variable in a domain D in $\mathbb{R}^n$ called *position* or *configuration space* and let $p = (p_1, \ldots, p_n)$ denote a variable in a domain in $\mathbb{R}^n$ (usually all of $\mathbb{R}^n$) called *momentum space.* The domain $D \times \mathbb{R}^n$ in $\mathbb{R}^{2n}$ of all q, p variables is called *phase space.* Usually the position variables q are determined by the location of certain masses in space, and p_j is the coordinate of the momentum associated with the mass located at q_j. For k particles in three-dimensional space we take $n = 3k$.

Now we assume we are given a real-valued function on phase space $H(q,p)$ called the *Hamiltonian.* This function is interpreted as giving the energy of the system when the positions and momenta are given by q, p. The Hamiltonian is determined by the physics of the system under study. We are making the special assumption that the Hamiltonian is time-independent. That means that any external influences on the system cannot vary with time.

Each point in phase space represents a possible description of the system at a moment of time. The basic idea of Hamiltonian mechanics (which is, of course, equivalent to Newtonian mechanics) is that the system is completely determined by a point in phase space and that the future evolution of the system is also determined; in fact there is a Hamiltonian vector field determined by H, and the flow associated with this vector field, called the Hamiltonian flow, gives the evolution of the system. If the system is at the point x in phase space at time t_0, then it will be at the point $f_t(x)$ in phase space at time $t_0 + t$.

The Hamiltonian vector field is given by

$$\frac{\partial H(q,p)}{\partial p}, -\frac{\partial H(q,p)}{\partial q},$$

so the o.d.e. for the integal curves $x(t) = (q(t), p(t))$ is

$$\frac{dq_j(t)}{dt} = \frac{\partial H}{\partial p_j}(q, (t), p(t)), \qquad \frac{dp_j(t)}{dt} = -\frac{\partial H}{\partial q_j}(q(t), p(t)).$$

Note that this is really an o.d.e. because the partial derivatives that appear are taken of a known function H; the unknown functions are

$q(t)$ and $p(t)$. This o.d.e. system is known as the *Hamilton-Jacobi equations*—they play an important role in the modern theory of partial differential equations and in mechanics. Clearly our existence and uniqueness theorem applies if we assume H is C^2.

Conservation of energy is an immediate consequence of the Hamilton-Jacobi equations. Since $H(q(t), p(t))$ is the energy of the system at time t, we need to show $(d/dt)H(q(t), p(t)) = 0$ to conclude that the energy remains the same. But by the chain rule

$$\frac{d}{dt}H(q(t), p(t)) = \sum_{j=1}^{n} \frac{\partial H}{\partial p_j}(q(t), p(t))\frac{dp_j(t)}{dt} + \sum_{j=1}^{n} \frac{\partial H}{\partial q_j}(q(t), p(t))\frac{dq_j}{dt}(t)$$

and

$$\frac{\partial H}{\partial p_j}(q(t), p(t))\frac{dp_j(t)}{dt} = -\frac{\partial H}{\partial q_j}(q(t), p(t))\frac{dq_j}{dt}$$

by the Hamilton-Jacobi equations. It is usually possible to deduce the existence of solutions for all time from the conservation of energy and some special features of H that prevent the solution from becoming unbounded in a finite time with H constant.

A typical example of a Hamiltonian system is a single particle of mass m moving under the influence of a force vector field $F(q)$ that is assumed to be a gradient vector field, $F(q) = -\nabla V(q)$ where $V(q)$ is called the *potential energy*. The configuration space is all of $\mathbb{R}^3$, and the phase space is $\mathbb{R}^6$. The Hamiltonian for this system is $H(q, p) = (1/2m)|p|^2 + V(q)$, interpreted as a sum of kinetic and potential energy. The Hamilton-Jacobi equations are

$$\begin{aligned} \frac{dq_j(t)}{dt} &= \frac{p_j(t)}{m}, \\ \frac{dp_j(t)}{dt} &= -\frac{\partial V}{\partial q_j}(q(t)) = F_j(q(t)). \end{aligned}$$

The first set of equations defines the momentum $p_j(t) = m(dq_j(t)/dt)$ as mass times velocity, and the second set of equations can then be interpreted as Newton's $F = ma$ law.

11.3.3 First-Order Linear P.D.E.'s

As a final application of the existence and uniqueness theorem for o.d.e.'s, we show how to obtain all solutions to a first-order linear partial differential equation $\sum_{j=1}^{n} p_j(x)\partial f/\partial x_j(x) = q(x)f(x) + r(x)$ where

p_j, q, r are given real-valued continuous functions on an open domain D in $\mathbb{R}^n$ and f is the unknown function. (No such simple method works for higher order equations.) Let us consider first the case $q \equiv 0$ and $r \equiv 0$. Think of $p = (p_1, \ldots, p_n)$ as a vector field on D. Then $\sum_{j=0}^n p_j(x)\partial f/\partial x_j(x)$ says that f has directional derivative equal to zero along the vector field and so should be constant along the integral curves. Indeed if $x(t)$ is an integral curve for the vector field $p(t)$, then

$$\begin{aligned}\frac{d}{dt}f(x(t)) &= \sum_{j=1}^n \frac{dx_j(t)}{dt}\frac{\partial f}{\partial x_j}(x(t)) \\ &= \sum_{j=1}^n p_j(x(t))\frac{\partial f}{\partial x_j}(x(t)),\end{aligned}$$

so f being constant along the integral curves is necessary and sufficient for f to be a solution of $\sum_{j=1}^n p_j(x)\partial f/\partial x_j(x) = 0$. Here we must assume p is C^1 in order to have the domain D covered by integral curves.

If we want to consider the more general equation

$$(*) \qquad \sum_{j=1}^n p_j(x)\frac{\partial f}{\partial x_j}(x) = q(x)f(x)+r(x)$$

we can use the same computation to reduce the problem to a first-order linear o.d.e. along the integral curves as

$$(**) \qquad \frac{d}{dt}f(x(t)) = q(x(t))f(x(t))+r(x(t)).$$

A function f is a solution of the p.d.e. $(*)$ if and only if for each integral curve the restriction $f(x(t))$ is a solution of the o.d.e. $(**)$. Of course the o.d.e. $(**)$ is linear and so can be solved in closed form (see exercise 3 of section 11.2). However the o.d.e. for the integral curves $x(t)$ is in general non-linear and rarely can we give an explicit formula for $x(t)$.

11.3.4 Exercises

1. Prove that a C^1 vector field F that is a gradient vector field must satisfy the curl $F = 0$ equations $\partial F_j/\partial x_k - \partial F_k/\partial x_j = 0$ for all $j \neq k$.

2. Prove that if F is a C^1 vector field on $\mathbb{R}^n$ satisfying curl $F = 0$, then F is a gradient vector field (**Hint**: integrate components of F along lines parallel to the axes.)

3. Let $F = \operatorname{grad} \theta$ where θ is the angular polar coordinate in $\mathbb{R}^2$. Write out an explicit formula for F in terms of rectilinear coordinates. Show that F is single-valued (even though θ is not) and C^1 in $\mathbb{R}^2 \backslash \{0\}$. Show that curl $F = 0$ but F is not a gradient vector field.

4. Suppose $F = \nabla V$ is a gradient vector field and $x(t)$ an integral curve. Show that $V(x(t))$ is an increasing function of t.

5. Describe explicitly the integral curves and the flow for the vector field

$$F(x, y) = \begin{pmatrix} -y \\ x \end{pmatrix} \quad \text{in } \mathbb{R}^2.$$

6. The n-body problem is described by the Hamiltonian

$$H = \sum_{k=1}^{n} \frac{1}{2m_k} |p^{(k)}|^2 - \sum_{j \neq k} \left| \frac{G m_j m_k}{q^{(j)} - q^{(k)}} \right|$$

where $q^{(k)} = (q_1^{(k)}, q_2^{(k)}, q_3^{(k)})$ denotes the position, $p^{(k)} = (p_1^{(k)}, p_2^{(k)}, p_3^{(k)})$ denotes the momentum of a particle of mass m_k, and G is the universal gravitational constant. Show that the Hamilton-Jacobi equations reduce to Newton's $F = ma$ law and the universal law of gravitation. Show that the total momentum $\sum_{k=1}^{n} p^{(k)}$ and total angular momentum $\sum_{k=1}^{n} q^{(k)} \times p^{(k)}$ (here $\times$ is the vector cross product in $\mathbb{R}^3$) are conserved—that is, constant on integral curves.

7. Find all solutions to the p.d.e. $-y(\partial f/\partial x)(x, y) + x(\partial f/\partial y)(x, y) = f(x, y)$ in the quadrant $x > 0$, $y > 0$.

11.4 Summary

11.1 Existence and Uniqueness

Definition *An ordinary differential equation (or system of equations), abbreviated o.d.e., of order m is an equation*

$$F(t, x(t), x'(t), \ldots, x^{(m)}(t)) = 0$$

where F is a given function defined on an open subset of $\mathbb{R}^{1+n(m+1)}$ taking values in $\mathbb{R}^k$ and the unknown function $x(t)$ is defined on an interval I of $\mathbb{R}$ taking values in $\mathbb{R}^n$. We say $x(t)$ is a solution of the o.d.e. if the equation holds for every t in I.

Example *The o.d.e. $x'(t) + x(t)^2 = 0$ has solutions $x(t) = 1/(t - t_0)$ on any interval not containing t_0.*

Theorem *Every o.d.e. of order m may be reduced to an equivalent o.d.e. of order one by introducing new variables equal to the derivatives of x of orders $\leq m - 1$.*

Definition *An o.d.e. of order m is said to be in normal form if it is written $x^{(m)}(t) = G(t, x(t), \ldots, x^{(m-1)}(t))$ where G is a function defined on an open set in $\mathbb{R}^{1+nm}$ taking values in $\mathbb{R}^n$.*

Example *The o.d.e. $x'(t) = (3/2)x(t)^{1/3}$ has the solution*

$$x(t) = \begin{cases} t^{3/2} & \text{if } t \geq 0, \\ 0 & \text{if } t \leq 0 \end{cases}$$

as well as $x(t) = 0$, both with $x(0) = 0$, despite the fact that $G(x) = (3/2)x^{1/3}$ is continuous.

Definition *For an o.d.e. of order m in normal form, the conditions $x(t_0) = a^{(0)}, x'(t_0) = a^{(1)}, \ldots, x^{(m-1)}(t_0) = a^{(m-1)}$, where each $a^{(j)}$ is a vector in $\mathbb{R}^n$, are called the Cauchy initial value conditions, the vectors $a^{(j)}$ are called the Cauchy data, and the problem of solving the o.d.e. subject to the Cauchy initial value conditions on an interval containing t_0 is called the Cauchy problem.*

Lemma 11.1.1 *A C^1 function $x(t)$ solving the Cauchy problem $x'(t) = G(t, x(t)), x(t_0) = a$, also solves the integral equation $x(t) = a + \int_{t_0}^{t} G(s, x(s))\, ds$. Conversely, any continuous solution of the integral equation solves the Cauchy problem.*

Theorem (*Minkowski's Inequality*) *If $f : [a, b] \to \mathbb{R}^n$ is continuous, then $|\int_a^b f(t)dt| \leq \int_a^b |f(t)|\, dt$ where $|\ |$ denotes the Euclidean norm.*

Lemma 11.1.2 *Let $G(t, x)$ be continuous for t in I and x in $\mathbb{R}^n$; taking values in $\mathbb{R}^n$; and let G satisfy the global Lipschitz condition $|G(t, x) - G(t, y)| \leq M|x - y|$ for all x and y in $\mathbb{R}^n$ and t in I. Then the Contractive Mapping Principle applies to the mapping $Tx(t) = a + \int_{t_0}^{t} G(s, x(s))\, ds$ on the space $C(J)$ of continuous functions on J, where J is a sufficiently small interval containing t_0.*

Theorem 11.1.1 (*Global Existence and Uniqueness*) *Under the hypothesis of the lemma, the Cauchy problem has a unique solution on I. The same is true if we can write $I = \bigcup_{j=1}^{\infty} I_j$ and for every subinterval I_j there exists M_j such that $|G(s, x) - G(s, y)| \leq M_j|x - y|$ for all x and y in $\mathbb{R}^n$ and all s in I_j.*

Corollary 11.1.1 *The Cauchy problem for the linear o.d.e. $x'(t) = A(t)x(t) + b(t)$ where $A : I \to \mathbb{R}^{n\times n}$ and $b : I \to \mathbb{R}^n$ are continuous has a unique solution on I. In fact, the solution with $x(t_0) = 0$ can be written as a convergent perturbation series $x = \sum_{k=0}^{\infty}(D^{-1}A)^k D^{-1}b = D^{-1}b + D^{-1}AD^{-1}b + D^{-1}AD^{-1}AD^{-1}b + \cdots$ where D^{-1} denotes the solution to the problem with $A = 0$, namely $D^{-1}x(t) = \int_{t_0}^{t} x(s)\, ds$.*

Theorem 11.1.2 (*Local Existence and Uniqueness*) *Let $G(t, x)$ be defined and continuous for t in I and $|x - a| \leq N$, and satisfy $|G(t, x) - G(t, y)| \leq M|x - y|$ for t in I and $|x - a| \leq N, |y - a| \leq N$. Then for a sufficiently small interval J containing t_0 there exists a solution of the Cauchy problem $x' = G(t, x), x(t_0) = a$, which is unique on as large an interval as the solution exists.*

Corollary 11.1.2 *Let $G(t, x)$ be a C^1 function defined for t in I and x in D (an open set in $\mathbb{R}^n$), taking values in $\mathbb{R}^n$. Then the Cauchy* *problem* $x' = G(t,x), x(t_0) = a$ *for* t_0 *in* I *and* a *in* D *has a unique solution on a sufficiently small interval* J *containing* t_0.

Theorem *If* $G(t,x)$ *satisfies the hypotheses of the local existence and uniqueness theorem, then the solution can be extended uniquely until* $x(t)$ *either becomes unbounded or leaves the domain of* G.

Theorem *Let* $G(t,y)$ *be a continuous function defined for* t *in* I *and* y *in an open set in* $\mathbb{R}^{nm}$. *Assume* $|G(t,y) - G(t,z)| \leq M|y-z|$ *either* a) *for all* t *in* I *and all* y *and* z *in* $\mathbb{R}^{nm}$ *or* b) *for all* t *in* I *and* y *and* z *in some open ball* B *in* $\mathbb{R}^{nm}$. *Then the Cauchy problem* $x^{(n)} = G(t,x,x',\ldots,x^{(m-1)}), x(t_0) = a^{(0)},\ldots,x^{(m-1)}(t_0) = a^{(m-1)}$, *has a unique solution* a) *on* I *or* b) *on some sufficiently small interval* J *containing* t_0 *if the Cauchy data* $(a^{(0)},\ldots,a^{(m-1)})$ *belongs to* B.

11.2 Other Methods of Solution*

Definition *The Euler approximation to the solution of the Cauchy problem* $x'(t) = G(t,x), x(t_0) = a$, *associated to the partition* $t_0 < t_1 < \cdots < t_n = T$, *is the piecewise affine function* $y(t)$ *on* $[t_0,T]$ *defined inductively by* $y(t) = y_k + G(t_k,y_k)(t-t_k)$ *on* $[t_k,t_{k+1}]$ *where* $y_0 = a$ *and* $y_k = y(t_k)$ *for* $k \geq 1$.

Theorem 11.2.1 *If* G *and its first derivatives are bounded and* $t_k = t_0 + kh$ *is a partition with uniform stepsize, then there exist constants* M_1 *and* M_2 *(depending only on the bounds for* G *and* dG*) such that* $|y(t)-x(t)| \leq M_2(e^{M_1|t-t_0|}-1)h$ *where* $y(t)$ *is the Euler approximation.*

Theorem 11.2.2 (*Peano Existence*) *Let* $G(t,x)$ *be continuous for* t *in* I *and* x *in a neighborhood of* a. *Then there exists a solution (not necessarily unique) to the Cauchy problem* $x' = G(t,x), x(t_0) = a$, *on a sufficiently small interval* J *containing* t_0. *The solution is a limit of Euler approximations.*

Definition *If* $\sum_{k=0}^{\infty} a_k t^k$ *and* $\sum_{k=0}^{\infty} A_k t^k$ *are formal power series, we say* $\sum A_k t^k$ *is a majorant of* $\sum a_k t^k$ *if* $|a_k| \leq A_k$ *for every* k.

Lemma *If $\sum A_k t^k$ is a majorant of $\sum a_k t^k$ and $\sum A_k t^k$ converges in $|t| < \varepsilon$, then so does $\sum a_k t^k$.*

Lemma 11.2.1(*Majorant Lemma*) *Suppose $P(t) = \sum_{k=0}^{\infty} P_k t^k, Q(t) = \sum_{k=0}^{\infty} Q_k t^k, R(t) = \sum_{k=0}^{\infty} R_k t^k$ are formal power series that are majorants of $p(t) = \sum_{k=0}^{\infty} p_k t^k, q(t) = \sum_{k=0}^{\infty} q_k t^k, r(t) = \sum_{k=0}^{\infty} r_k t^k$, respectively. Suppose c_0 and c_1 are given and $\sum_{k=0}^{\infty} c_k t^k$ is defined by the equations*

$$(k+1)(k+2)c_{k+2} = \left(\sum_{j=0}^{\infty} p_j(k-j+1)c_{k-j+1} + q_j c_{k-j}\right) + r_k,$$

and suppose d_0 and d_1 are given and $\sum_{k=0}^{\infty} d_k t^k$ is defined by

$$(k+1)(k+2)d_{k+2} = \left(\sum_{j=0}^{k} P_j(k-j+1)d_{k-j+1} + Q_j d_{k-j}\right) + R_k.$$

If $|c_0| \leq d_0$ and $|c_1| \leq d_1$, then $\sum_{k=0}^{\infty} d_k t^k$ is a majorant for $\sum_{k=0}^{\infty} c_k t^k$.

Theorem 11.2.3 *Let p, q, r have convergent power series in $|t| < \varepsilon$. Then the Cauchy problem $x'' = px' + qx + r, x(0) = c_0, x'(0) = c_1$, has a solution given by a power series $x(t) = \sum_{k=0}^{\infty} c_k t^k$ convergent in $|t| < \varepsilon$.*

11.3 Vector Fields and Flows*

Definition *A continuous vector field on an open set D in $\mathbb{R}^n$ is a continuous function $F : D \to \mathbb{R}^n$. It is said to be a gradient vector field if there exists a C^1 function $f : D \to \mathbb{R}$ such that $F = \nabla f$. An integral curve of a vector field F is a C^1 curve $x : [a, b] \to \mathbb{R}^n$ such that $x'(t) = F(x(t))$ for all t in $[a, b]$.*

Theorem 11.3.1 *If a vector field F is C^1, then there exist unique integral curves satisfying $x(0) = y$ for any y in D. Two integral curves either coincide (as subsets of D) or are disjoint.*

Definition *A flow on an open set D in $\mathbb{R}^n$ is a continuous function $f_t(x)$ defined for x in D and t in a neighborhood of 0 (depending on x)*

taking values in D, satisfying $f_t(f_s(x)) = f_{t+s}(x)$ whenever both sides are defined. The flow associated to a C^1 vector field F is given by $f_t(y) = x(t)$ where $x(t)$ is the integral curve satisfying $x(0) = y$. The vector field associated to a C^1 flow f_t is $F(x) = \partial/\partial t f_t(x)|_{t=0}$.

Example (*Hamiltonian Mechanics*) *Let q and p denote variables in $\mathbb{R}^n$ (interpreted as position and momentum); and let $H(q,p)$ be a real-valued function, called the Hamiltonian (interpreted as energy). The vector field $(\partial H(q,p)/\partial p, -\partial H(q,p)/\partial q)$ is called the Hamiltonian vector field, and the o.d.e. for an integral curve comprise the Hamilton-Jacobi equations*

$$\frac{dq_j}{dt} = \frac{\partial H}{\partial p_j}(q,p), \quad \frac{dp_j}{dt} = -\frac{\partial H}{dt}(q,p).$$

The Hamiltonian is constant on integral curves (conservation of energy). If $H(q,p) = (1/2m)|p|^2 + V(q)$, the Hamilton-Jacobi equations are equivalent to Newton's laws of motion for a particle of mass m moving under the influence of a force $F = -\nabla V$ and V is interpreted as potential energy.

Theorem *Let p_j for $j = 1, \ldots, n$ be C^1 and q and r be continuous functions on an open set D in $\mathbb{R}^n$. Then $f(x)$ is a solution of the p.d.e.*

$$\sum_{j=1}^{n} p_j(x)\partial f/\partial x_j(x) = q(x)f(x) + r(x)$$

if and only if for each integral curve $x(t)$ of the vector field p, the restriction $f(x(t))$ is a solution of the o.d.e. $u'(t) = q(x(t))u(t) + r(x(t))$.

Chapter 12

Fourier Series

12.1 Origins of Fourier Series

12.1.1 Fourier Series Solutions of P.D.E.'s

Fourier series were first stumbled upon by Daniel Bernoulli in the 1750s while studying the equation of a vibrating string, which is the partial differential equation (p.d.e.)

$$\frac{\partial^2 y(x,t)}{\partial t^2} = c^2 \frac{\partial^2 y(x,t)}{\partial x^2}.$$

Here $y(x,t)$ denotes the vertical displacement of a string at the horizontal point x at time t for $0 \leq x \leq L$, as in Figure 12.1.1.

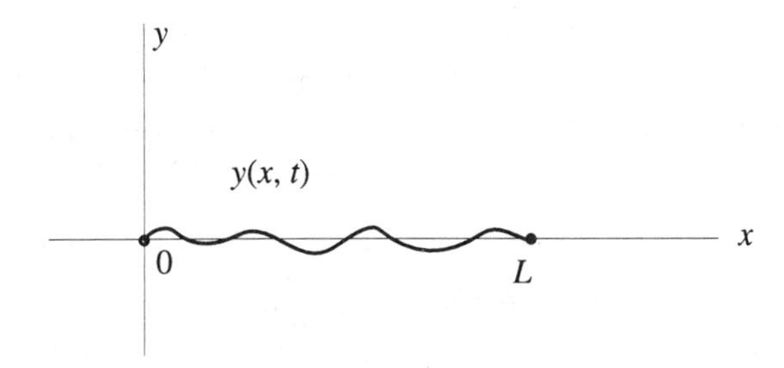

Figure 12.1.1:

Also L is the length of the string and the constant c is determined by the density and tension of the string ($c^2 = \dfrac{\text{tension}}{\text{density}}$ for appropriate units). We will not dwell on the derivation of this equation, which is a combination of Newtonian mechanics, simplfying assumptions, a limiting argument, and a little hocus pocus. We can give a plausible explanation of the equation, however, as follows: $\partial^2 y/\partial t^2$ is the acceleration of a point on the string located horizontally at x. According to Newtonian mechanics, the acceleration should be proportional to the force acting on that point on the string. This force is caused by the tension on the string, which acts to straighten it out. But $\partial^2 y/\partial x^2$ measures the concavity of the string shape and has the correct sign to impart an acceleration in the direction of straightening out the string, as shown in Figure 12.1.2.

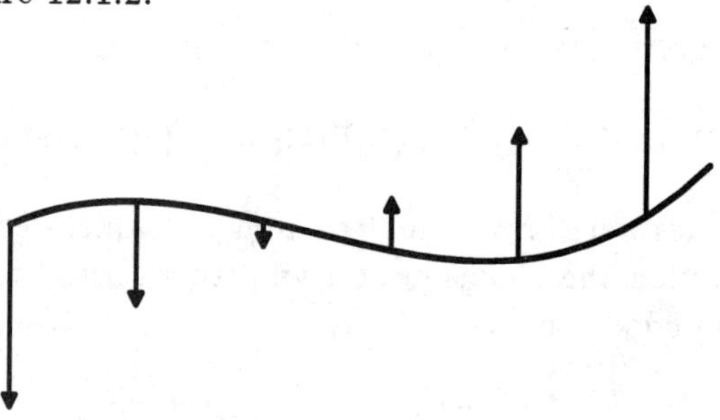

Figure 12.1.2:

In the problem considered by Bernoulli, the string is held down at the endpoints, yielding the boundary conditions $y(0,t) = y(L,t) = 0$ for all t. Also the string is initially ($t = 0$) held at rest in some position and then is released to follow the time evolution dictated by the equation (as in all Newtonian mechanics, the initial position and velocity are supposed to determine the future evolution of the system). This means the initial conditions

$$y(x,0) = f(x), \qquad \frac{\partial y}{\partial t}(x,0) = 0$$

are imposed, where $f(x)$ is the initial shape of the string (for consistency we need $f(0) = f(T) = 0$).

Bernoulli observed that there were some particularly simple solutions to the p.d.e., namely

$$y(x,t) = \sin \frac{k\pi}{L} x \cos \frac{ck\pi}{L} t$$

where k is a postive integer, that satisfy the boundary conditions and the initial conditions with $f(x) = \sin(k\pi/L)x$. These solutions have a musical interpretation in terms of the sound produced by the string—they give the overtone series for the fundamental tone ($k = 1$). Since the p.d.e. is linear, linear combinations of these simple solutions are also solutions. In Bernoulli's day it was presumed that infinite linear combinations would also satisfy the p.d.e., so

$$y(x,t) = \sum_{k=1}^{\infty} a_k \sin \frac{k\pi}{L} x \cos \frac{ck\pi}{L} t$$

would satisfy the p.d.e. and the initial conditions for

$$f(x) = \sum_{k=1}^{\infty} a_k \sin \frac{k\pi}{L} x.$$

We can in fact verify that this is the case if we make suitable assumptions on the coefficients to justify differentiating the infinite series term-by-term (see exercise set 12.1.4).

Bernoulli went further, however, and claimed that in this manner he obtained all solutions. This claim can be made plausible in terms of the musical interpretation, but Bernoulli rested his claim on the vague and dubious grounds that with an infinite number of variables (the a_k) at his disposal, he could do anything at all.

Euler immediately attacked Bernoulli's claim, pointing out that it would imply that an arbitrary function $f(x)$ on $0 \leq x \leq L$ with $f(0) = f(L) = 0$ could be expanded in an infinite sine series

$$f(x) = \sum_{k=1}^{\infty} a_k \sin \frac{k\pi}{L} x.$$

Euler thought this was patently absurd, since the sine series possessed special properties, such as being an odd function in x and periodic of period $2L$, which a function such as $f(x) = x(L - x)$ does not possess.

We will return to discuss this objection later. Of course, Euler had an axe to grind: he (and D'Alembert, independently) had found a completely different solution to the problem, namely

$$y(x,t) = \frac{1}{2}(\tilde{f}(x+ct) + \tilde{f}(x-ct))$$

where $\tilde{f}$ denotes the extension of f to the whole line, which is odd, $\tilde{f}(-x) = -f(x)$, and periodic of period $2L$, $\tilde{f}(x+2L) = \tilde{f}(x)$. It is a routine exercise to verify that Euler's solution actually works. Euler pointed out that Bernoulli's solution was a special case of his, which is clear from the trigonometric identity

$$\sin\frac{k\pi}{L}x\cos\frac{ck\pi}{L}t = \frac{1}{2}\left(\sin\frac{k\pi}{L}(x+ct) + \sin\frac{k\pi}{L}(x-ct)\right).$$

Bernoulli was not able to answer Euler well, except to repeat his lame argument that the equation

$$f(x) = \sum_{k=1}^{\infty} a_k \sin\frac{k\pi}{L}x$$

was like an algebraic system of an infinite number of linear equations in an infinite number of unknowns. One of the most significant weaknesses in his argument was that he could not produce a formula for the coefficients a_k in terms of f. This formula was actually first discovered by Euler several years later in the course of an unrelated investigation (Euler had cosines instead of sines), but as Euler was predisposed to reject Bernoulli's claim he never pointed out the possible relevance. Thus, did the two of them botch the opportunity of developing Fourier series a full half century before Fourier.

Before leaving this interesting historic episode behind, it is worth commenting on one line of thought that Bernoulli might have pursued to great advantage. If we modify the vibrating string equation to include a damping term for friction,

$$\frac{\partial^2 y}{\partial t^2} = c^2\frac{\partial^2 y}{\partial x^2} - a\frac{\partial y}{\partial t},$$

which is plausible on physical grounds (no real string vibrates forever like Bernoulli's simple solutions), then Bernoulli's method of solution is easily modified (see exercise set 12.1.4), but Euler's fails completely. If there had been as many mathematicians in those days as there are today, no doubt some graduate student looking for a thesis topic would have observed this, which might have been enough to push Euler and Bernoulli onto the right track.

Fourier came to consider Fourier series by way of a related p.d.e., the heat equation

$$\frac{\partial u}{\partial t}(x,t) = c^2 \frac{\partial^2 u}{\partial x^2}(x,t)$$

where $u(x,t)$ represents the temperature of a one-dimensional object (think of an insulated thin bar) at the point x at time t where c is a constant that depends on the physical properties of the object. Imposing boundary conditions $u(0,t) = u(L,t) = 0$ (the physical interpretation of this kind of boundary condition is that the ends of the bar are attached to reservoirs held at a constant temperature, which is taken to be zero) and initial conditions $u(x,0) = f(x)$ for $0 \leq x \leq L$, where $f(x)$ is the initial temperature distribution, Fourier found the simple solution (for $t > 0$)

$$u(x,t) = \exp\left(\frac{-c^2k^2\pi^2 t}{L^2}\right) \sin\frac{k\pi}{L}x$$

if $f(x) = \sin(k\pi/L)x$ and hence

$$u(x,t) = \sum_{k=1}^{\infty} a_k \exp\left(\frac{-c^2k^2\pi^2 t}{L^2}\right) \sin\frac{k\pi}{L}x$$

if $f(x) = \sum_{k=1}^{\infty} a_k \sin(k\pi/L)x$ by the linearity of the equation. Fourier went farther than Bernoulli and produced the formula

$$a_k = \frac{2}{L}\int_0^L f(x)\sin\frac{k\pi}{L}x\,dx$$

for the coefficients in terms of f. This is easily verified by multiplying by $\sin(j\pi/L)x$ and integrating term-by-term (assuming $\sum_{k=1}^{\infty} |a_k|$ is finite, so the series converges uniformly). We then need the elementary calculation

$$\int_0^L \sin\frac{k\pi}{L}x \sin\frac{j\pi}{L}x\,dx$$

$$= \frac{1}{2}\int_0^L \left[\cos\frac{(k-j)\pi}{L}x - \cos\frac{(k+j)\pi}{L}x\right] dx$$

$$= \begin{cases} L/2 & \text{if } k = j, \\ 0 & \text{if } k \neq j. \end{cases}$$

(Curiously, neither Euler nor Fourier presented this simple derivation immediately but rather first gave long-winded and largely incomprehensible derivations using power-series expansions and clever but invalid manipulations—presumably the method they used to discover the result—and then added the simple computation above as an after thought.)

Armed with a formula for the coefficients, Fourier was able to make a more convincing claim that every function $f(x)$ on $0 \leq x \leq L$ with $f(0) = f(L) = 0$ has a sine-series expansion. Actually this was quite a bold claim for his time, since neither the concept of function nor the meaning of the integral in the formula for a_k with a general function f (in Fourier's time the integral was usually defined as an anti-derivative) were at all clear. In fact it was the interest in understanding Fourier's claim that helped spur a good deal of the work we have discussed in earlier chapters. Fourier devoted a good deal of effort to computing the sine-series expansions of explicit functions and examining the convergence of the partial sums. This "empirical" evidence helped bolster his claim.

12.1.2 Spectral Theory

Before we begin the detailed study of Fourier series, it is worth remarking on some of the special features of the trigonometric functions that enter into the expansions. Bluntly stated, we want to know why we should care about Fourier series and into what general framework they fit. It turns out that there are two distinct perspectives available, one leading to *spectral theory* and the other to *harmonic analysis.*

If we ask why the sines appeared in the simple solutions to the vibrating string or heat equation, the answer can be given in terms of the "operator" $\partial^2/\partial x^2$ on the space of C^2 functions on $0 \leq x \leq L$ with boundary conditions $f(0) = f(L) = 0$. The sines are the solutions (eigenvectors) of the eigenvalue problem

$$\frac{d^2}{dx^2}\psi(x) = \lambda\psi(x) \quad \text{on} \quad 0 \le x \le L, \;\; \psi(0) = \psi(L) = 0.$$

This is completely analogous to the eigenvalue equation $A\psi = \lambda\psi$ where A is an $n \times n$ matrix and ψ a vector in $\mathbb{R}^n$. In the present case, in place of A we take d^2/dx^2 and in place of $\mathbb{R}^n$ we take ψ in the vector space of C^2 functions on $0 \le x \le L$ satisfying $\psi(0) = \psi(L) = 0$. The space of C^2 functions is an infinite-dimensional vector space, but d^2/dx^2 is linear:

$$\frac{d^2}{dx^2}(af(x) + bg(x)) = a\frac{d^2}{dx^2}f(x) + b\frac{d^2}{dx^2}g(x).$$

Since the o.d.e. has explicit solutions for every possible λ (real or complex), it is simple to verify, because of the boundary conditions, that the only solutions to the eigenvalue problem are multiples of $\sin(k\pi/L)x$, which correspond to $\lambda = -k^2\pi^2/L^2$ (see exercise set 12.1.4). The eigenvectors ψ then yield simple solutions of the above problems (and many more) by separation of variables: try $u(x,t) = g(t)\psi(x)$ in the heat equation and obtain $g'(t) = c^2\lambda g(t)$, hence g is a multiple of exp $(c^2\lambda t)$.

So we can restate Fourier's claim as follows: every function has an expansion in eigenvectors for the operator d^2/dx^2 on the space V of C^2 functions on $0 \le x \le L$ satisfying $f(0) = f(L) = 0$, called the *Dirichlet boundary conditions.* (The fact that the range of d^2/dx^2 on V is not equal to V creates some technical problems, but the eigenvectors will be in V as a consequence of the eigenvalue equation.) We note one special property of this operator. If $\langle f, g\rangle$ denotes the inner product $\int_0^L f(x)g(x)\,dx$ (or $\int_0^L f(x)\overline{g(x)}\,dx$ if we want to allow complex-valued functions), then

$$\left\langle \frac{d^2 f}{dx^2}, g\right\rangle = \left\langle f, \frac{\partial^2 g}{\partial x^2}\right\rangle$$

for f and g in V. This is easily seen by integrating by parts twice, using the boundary conditions $f(0) = f(L) = 0$ and $g(0) = g(L) = 0$ (from f and g being in V) to obtain the vanishing of the boundary terms in the integration by parts formula (see exercise set 12.1.4). The condition $\langle Af, g\rangle = \langle f, Ag\rangle$ for a linear operator on an inner product space is called *symmetry*, and it is a direct generalization of the symmetry of an $n \times n$ matrix considered as a linear operator on the inner product space $\mathbb{R}^n$. Recall that for a symmetric matrix we proved a spectral theorem asserting that $\mathbb{R}^n$ has an orthonormal basis of eigenvectors. The Fourier sine expansion can be expressed as a generalization of this spectral theorem, and both are special cases of a very general spectral theorem of von Neumann.

In the spectral theorem for symmetric matrices, the eigenvectors can be taken to be orthogonal. This is true in general for symmetric operators. If $A\psi_1 = \lambda_1\psi_1$ and $A\psi_2 = \lambda_2\psi_2$ for $\lambda_1 \neq \lambda_2$, then

$$\langle A\psi_1, \psi_2\rangle = \langle \lambda_1\psi_1, \psi_2\rangle = \lambda_1 \langle \psi_1, \psi_2\rangle$$

and

$$\langle \psi_1, A\psi_2\rangle = \langle \psi_1, \lambda_2\psi_2\rangle = \lambda_2 \langle \psi_1, \psi_2\rangle ;$$

so the symmetry $\langle A\psi_1, \psi_2\rangle = \langle \psi_1, A\psi_2\rangle$ implies $\lambda_1\langle \psi_1, \psi_2\rangle = \lambda_2\langle \psi_1, \psi_2\rangle$, hence $\langle \psi_1, \psi_2\rangle = 0$ if $\lambda_1 \neq \lambda_2$. This is just

$$\int_0^L \sin \frac{k\pi}{L} x \sin \frac{j\pi}{L} x \, dx = 0$$

if $j \neq k$ in the special case under consideration, which is the key formula for finding the Fourier coefficients. In the general case, if $f = \sum c_k\psi_k$ where the ψ_k are eigenvectors with distinct eigenvalues, then

$$c_k = \frac{\langle f, \psi_k\rangle}{\langle \psi_k, \psi_k\rangle}$$

(if there exist linearly independent eigenvectors with the same eigenvalue, then we must use the Gram-Schmidt orthogonalization process to make them orthogonal to each other).

Another closely related example is the operator d^2/dx^2 on the vector space V_1 of C^2 functions satisfying the boundary conditions $f'(0) = f'(L) = 0$ (called the *Neumann boundary* conditions). Notice that we have not changed the formula for the operator but that we have changed the vector space of functions on which it operates. We take the same formula for the inner product as before, $\langle f, g\rangle = \int_0^L f(x)g(x)\, dx$. Once again we find that d^2/dx^2 is symmetric, $\langle d^2/dx^2 f, g\rangle = \langle f, d^2/dx^2 g\rangle$ by integration by parts, this time using the Neumann boundary conditions to show that the boundary terms vanish. In case you are beginning to suspect that d^2/dx^2 will be symmetric for any boundary conditions, this turns out to be false (see exercise set 12.1.4).

The eigenvectors for d^2/dx^2 on V_2 are $\cos(k\pi/L)x$ with eigenvalue $-k^2\pi^2/L^2$ for $k = 0, 1, \ldots$. From the symmetry we can conclude immediately that $\int_0^L \cos(k\pi/L)x \cos(j\pi/L)x\, dx = 0$ if $j \neq k$. Using the standard trigonometric identities as in the sine case we could obtain the same result, and also

$$\int_0^L \left(\cos \frac{k\pi}{L} x\right)^2 dx = \begin{cases} L & \text{if } k = 0, \\ L/2 & \text{if } k \neq 0. \end{cases}$$

Thus if

$$f(x) = \frac{a_0}{2} + \sum_{k=1}^{\infty} a_k \cos \frac{k\pi}{L} x,$$

then

$$a_k = \frac{2}{L} \int_0^L f(x) \cos \frac{k\pi}{L} x\, dx$$

(the choice of the factor 1/2 in front of a_0 is purely conventional, so the formula for a_k is the same for $k = 0$ as for $k \neq 0$). This is called the *Fourier cosine expansion*, and Fourier also conjectured that every function $f(x)$ has such an expansion. Fourier used it to solve the heat equation with Neumann boundary conditions

$$\frac{\partial u}{\partial x}(0, t) = \frac{\partial u}{\partial x}(L, t) = 0,$$

which can be interpreted physically as saying the ends of the bar are insulated, hence no heat flows across them.

We can also combine the Fourier sine and cosine expansions by doubling the interval. If $f(x)$ is defined on $[-L, L]$, then we can write $f(x) = f_e(x) + f_o(x)$ where $f_e(x) = 1/2(f(x) + f(-x))$ is even and $f_o(x) = 1/2(f(x) - f(-x))$ is odd about the point $x = 0$. Since the cosines are even and the sines odd, if $f_e(x)$ restricted to $[0, L]$ has a Fourier cosine series

$$\frac{a_0}{2} + \sum_{k=1}^{\infty} a_k \cos \frac{k\pi}{L} x,$$

then the same series will give $f_e(x)$ on $[-L, L]$ and, similarly, if $f_o(x)$ restricted to $[0, L]$ has a Fourier sine expansion $\sum_{k=1}^{\infty} b_k \sin(k\pi/L)x$, then the same series will give $f_o(x)$ on $[-L, L]$. Adding, we have the full Fourier series

$$f(x) = \frac{a_0}{2} + \sum_{k=1}^{\infty} \left(a_k \cos \frac{k\pi}{L} x + b_k \sin \frac{k\pi}{L} x \right)$$

where

$$a_k = \frac{1}{L} \int_{-L}^{L} f(x) \cos \frac{k\pi}{L} x \, dx$$

and

$$b_k = \frac{1}{L} \int_{-L}^{L} f(x) \sin \frac{k\pi}{L} x \, dx,$$

the factor 2 having disppeared because we extended the integration to the full interval (notice also that

$$\int_{-L}^{L} f_o(x) \cos \frac{k\pi}{L} x \, dx = 0 \quad \text{and} \quad \int_{-L}^{L} f_e(x) \sin \frac{k\pi}{L} x \, dx = 0$$

because the integrands are odd functions).

Now recall that for the Fourier sine series we were assuming that $f(0) = f(L) = 0$. What does this mean for the odd function $f_o(x)$ above? The condition $f_o(0) = 0$ is automatic from the definition $f_o(x) = 1/2(f(x) - f(-x))$, but the condition $f_o(L) = 0$ is the same as $f(L) = f(-L)$ (for the even part the condition $f_e(L) = f_e(-L)$ is automatic). We can interpret this as a periodicity condition if we extend f to be periodic of period $2L$. This is a natural thing to do since the sines and cosines in the series already are periodic of period $2L$. Thus we define $f(x)$ on the whole line by saying $f(x + 2L) = f(x)$ for all x. The condition $f(-L) = f(L)$ is then necessary and sufficient for this condition to be fulfilled. We will usually discuss Fourier series in terms of the periodic extension—thus the condition that f be continuous means that f is continuous on $[-L, L]$ and $f(L) = f(-L)$, the condition that f be C^1 means that f is C^1 on $[-L, L]$ and $f(L) = f(-L)$ and $f'(L) = f'(-L)$, and so on.

Thus Fourier's conjecture is that every continuous function $f(x)$ on the line that is periodic of period $2L$ has a full Fourier series expansion. This follows from the two conjectures concerning Fourier sine and cosine series, and it implies them by considering even and odd extensions across $x = 0$ (see exercise set 12.1.4). The conjecture is actually not quite correct as we will see in the next section, but it can be modified in a number of ways to give a correct theorem. Notice already that we have successfully countered Euler's objection that the function $f(x) = x(L-x)$ cannot have a Fourier sine-series expansion because it is not odd and periodic. Fourier's claim is only that

$$x(L-x) = \sum_{k=1}^{\infty} b_k \sin \frac{k\pi}{L} x$$

for $0 \le x \le L$. For other values of x the sine series converges to the function $f(x)$, which is obtained by extending $x(L-x)$ to be odd and periodic of period $2L$, but $f(x) \neq x(L-x)$ outside the interval $[0, L]$. Euler could not conceive of this possibility because he was experienced in dealing with power series and knew that one cannot have a power series converge to different analytic expressions on different intervals. Euler naively assumed that trigonometric series would behave like power series—in fact nothing could be further from the truth!

12.1.3 Harmonic Analysis

In dealing with the full Fourier series we can simplify matters considerably by introducing complex numbers. Recall the Euler relations $e^{ix} = \cos x + i \sin x$, $\cos x = (e^{ix} + e^{-ix})/2$, and $\sin x = (e^{ix} - e^{-ix})/2i$. Using these it is easy to show that

$$\begin{aligned} \frac{a_0}{2} &+ \sum_{k=1}^{\infty} \left(a_k \cos \frac{k\pi}{L} x + b_k \sin \frac{k\pi}{L} x \right) \\ &= \sum_{n=-\infty}^{\infty} c_n e^{(in\pi/L)x} \end{aligned}$$

where $c_n = 1/2L \int_{-L}^{L} f(x) e^{(in\pi/L)x} dx$ (see exercise set 12.1.4). The condition that $f(x)$ be real-valued is then equivalent to the condition $c_{-n} = \bar{c}_n$ on the Fourier coefficients. Note then that the functions $e^{(in\pi/L)x}$ are exactly the eigenfunctions of the operator d/dx on the space of C^1 functions with the periodic boundary conditions $f(-L) = f(L)$ (note that the operator d/dx is now skew-symmetric instead of symmetric but it can be made symmetric by multiplying by $\pm i$).

Now we are in a position to discuss the second interpretation of Fourier series. Let τ_y denote the operation of translation by y, $\tau_y f(x) =$

$f(x+y)$. We can apply this to any function on the line, but for this discussion we restrict attention to periodic functions of period $2L$. We can relate the operators τ_y and the operator d/dx by the formula $d/dx = (\partial/\partial y)\tau_y|_{y=0}$. In a sense that we need not make precise, d/dx is a sort of infinitesmal translation, and conversely the translations may be obtained by integrating d/dx. The point of this vague discussion is the following: if the functions $e^{(ik\pi/L)x}$ are eigenvectors for d/dx, they should also be eigenvectors for τ_y. Having said this, it is immediately obvious that they are:

$$\tau_y(e^{(ik\pi/L)x}) = e^{(ik\pi/L)y}e^{(ik\pi/L)x}.$$

It is not hard to show (see exercise set 12.1.4) that (up to constant multiples) these are the only periodic functions that are eigenvectors for all the operators τ_y.

We can go a little further with the eigenvalue equation for τ_y and write it as

$$\psi(x+y) = \psi(x)\psi(y).$$

This equation, called the *character identity*, has a beautiful symmetry. It is not hard to show that any complex-valued C^1 solution must be of the form exp (λx) where λ is a complex number (see exercise set 12.1.4), and with more work it is possible to show the same is true under the hypothesis that ψ is continuous. The condition that ψ be periodic of period $2L$ then implies $\lambda = ik\pi/L$ for some integer k. Notice that the character identity determines $\exp(ik\pi/L)x$ exactly, whereas the eigenvalue equation allows an arbitrary multiple of $\exp(ik\pi/L)x$.

The significance of the character identity is that it has a group-theoretic interpretation. We think of the real numbers mod $2L$ as forming a commutative group under addition (the elements of this group are the sets $\{x+2Lk\}$ as k varies over the integers, and $\{x+2Lk\} + \{y+2Lk\} = \{x+y+2Lk\}$ is well defined). A periodic function of period $2L$ may be thought of as a function on this group. The character identity then says that ψ is a homomorphism of this group into the multiplicative group of the non-zero complex numbers. We note that in this case all the solutions of the character identity $e^{(ik\pi/L)x}$ take on only complex values of absolute value one, $|\psi(x)| = 1$. Recall that the complex numbers of absolute value one, $|z| = 1$, can be written $z = e^{i\theta}$

for θ real and θ is determined mod 2π. Since

$$e^{i\theta_1}e^{i\theta_2} = e^{i(\theta_1+\theta_2)},$$

we see that the complex numbers of absolute value one under multiplication give an isomorphic model of the group we are considering for $L = \pi$. Let us denote this group T.

We can summarize the above discussion (for $L = \pi$) by saying that the functions e^{ikx} are continuous homomorphisms of the group T into itself. It turns out that these are the only ones (see exercise set 12.1.4, number 18 for a slightly weaker result). They are called the *characters* of T. Fourier's conjecture then says that an arbitrary complex-valued function on T is an infinite linear combination of characters. This is a special case of what is called *harmonic analysis* (or sometimes *Fourier analysis*). There are far-reaching generalizations to other groups. If a group G is commutative we again consider characters, which are defined to be homomorphisms of G into T (usually there is a metric or more generally a topology on G, and the characters are assumed continuous). If a group is non-commutative, then the character identity must be further generalized to the theory of *group representations.* In the case where the group G is the additive group of the line, the harmonic analysis leads to the theory of Fourier transforms.

If we had not expanded our perspective to the complex numbers, the group-theoretic significance of the sines and cosines would not be so apparent. Nevertheless, there is still something we can say. Look at the two-dimensional vector space $\{a\cos(k\pi/L)x + b\sin(k\pi/L)x\}$ of linear combinations of the two functions $\cos(k\pi/L)x$ and $\sin(k\pi/L)x$. Then applying τ_y to any function in the space produces another function in this space:

$$\begin{aligned}
&\tau_y\left(a\cos\frac{k\pi}{L}x + b\sin\frac{k\pi}{L}x\right)\\
&= a\cos\frac{k\pi}{L}(x+y) + b\sin\frac{k\pi}{L}(x+y)\\
&= a\left(\cos\frac{k\pi}{L}x\cos\frac{k\pi}{L}y - \sin\frac{k\pi}{L}x\sin\frac{k\pi}{L}y\right)\\
&\quad + b\left(\sin\frac{k\pi}{L}x\cos\frac{k\pi}{L}y + \cos\frac{k\pi}{L}x\sin\frac{k\pi}{L}y\right)\\
&= A\cos\frac{k\pi}{L}x + B\sin\frac{k\pi}{L}x
\end{aligned}$$

where

$$\begin{pmatrix} A \\ B \end{pmatrix} = \begin{pmatrix} \cos(k\pi/L)y & \sin(k\pi/L)y \\ -\sin(k\pi/L)y & \cos(k\pi/L)y \end{pmatrix} \begin{pmatrix} a \\ b \end{pmatrix}.$$

The constant functions of course form a one-dimensional vector space of functions that is preserved by all translations. It turns out that any finite-dimensional vector space of functions (periodic of period $2L$) that is preserved by all translations must be a kind of sum of these basic building blocks. Thus the sines and cosines that appear in the Fourier series are in this sense the functions that behave most simply under translation.

We have now given three reasons why Fourier series are so important: 1) they are useful in solving p.d.e.'s, 2) they are a special case of the spectral theorem, 3) they are a special case of harmonic analysis.

12.1.4 Exercises

1. Assuming $\sum_{k=1}^{\infty} k^2|a_k| < \infty$, verify that

$$\sum_{k=1}^{\infty} a_k \sin \frac{k\pi}{L} x \cos \frac{ck\pi}{L} t$$

converges uniformly and absolutely to a C^2 solution of the vibrating string equation with boundary conditions $y(0,t) = y(L,t) = 0$ and initial conditions $y(x,0) = f(x)$, $(\partial/\partial t)y(x,0) = 0$ for

$$f(x) = \sum_{k=1}^{\infty} a_k \sin \frac{k\pi}{L} x.$$

2. Show that it is impossible to have $\sin x = \sum_{k=2}^{\infty} a_k \sin kx$ on $0 \leq x \leq \pi$ with the series converging uniformly, for any choice of the a_k, even though there are an infinite number of parameters in the problem. (**Hint**: multiply by $\sin x$ and integrate.)

3. Verify that the Euler-D'Alembert solution

$$y(x,t) = \frac{1}{2}(\tilde{f}(x+ct) + \tilde{f}(x-ct)),$$

where $\tilde{f}$ denotes the extension of f to the whole line satisfying $\tilde{f}(-x) = -f(x)$ and $\tilde{f}(x+2L) = \tilde{f}(x)$, actually solves the initial value problem for the vibrating string equation for any function $f(x)$ such that $\tilde{f}$ is C^2. What conditions on f are necessary and sufficient for $\tilde{f}$ to be C^2?

4. Solve the damped vibrating string equation

$$\frac{\partial^2 y}{\partial t^2} = c^2 \frac{\partial^2 y}{\partial x^2} - a\frac{\partial y}{\partial t}$$

with boundary conditions $y(0,t) = y(L,t) = 0$ and initial conditions $y(x,0) = f(x)$, $\partial y/\partial t(x,0) = 0$ assuming

$$f(x) = \sum_{k=1}^{\infty} a_k \sin \frac{k\pi}{L} x$$

with $\sum k^2|a_k| < \infty$. (**Hint:** look for special solutions $g(t)\sin(k\pi/L)x$ and obtain an o.d.e. for g. You will have to distinguish three cases depending on whether a is greater than, equal to, or less than $2ck\pi/L$.)

5. Verify that

$$u(x,t) = \sum_{k=1}^{\infty} a_k \exp\left(\frac{-c^2k^2\pi^2 t}{L^2}\right) \sin \frac{k\pi}{L} x$$

satisfies the heat equation for $t > 0$ with boundary conditions $u(0,t) = u(L,t) = 0$ and initial condition $u(x,0) = f(x)$ if $f(x) = \sum_{k=1}^{\infty} a_k \sin(k\pi/L)x$ with $\sum |a_k| < \infty$.

6. Show that all solutions of $\psi''(x) = \lambda\psi(x)$ on $0 \le x \le L$ with $\psi(0) = \psi(L) = 0$ are of the form $c\sin(k\pi/L)x$. (**Hint:** write down all solutions of the o.d.e. and impose the boundary conditions.)

7. Show that all solutions of $\psi''(x) = \lambda\psi(x)$ on $0 \le x \le L$ with $\psi'(0) = \psi'(L) = 0$ are of the form $c\cos(k\pi/L)x$.

8. Prove the symmetry of d^2/dx^2 on the vector space of C^2 functions with a) Dirichlet ($f(0) = f(L) = 0$) or b) Neumann ($f'(0) = f'(L) = 0$) boundary conditions.

9. For which values of a and b is the operator d^2/dx^2 symmetric on the vector space of C^2 functions on $[0, L]$ satisfying $af(0) + bf'(0) = af'(L) + bf'(L) = 0$? Prove the symmetry in the cases it is true, and give counterexamples to the symmetry in the cases it is false.

10. Find all eigenvectors for the operator d^2/dx^2 on the space of C^2 functions on $[0, L]$ satisfying the boundary conditions $f(0) = f'(L) = 0$.

11. Prove the symmetry of the *Sturm-Liouville operator* $Af = pf'' + p'f' + qf$, where p and q are real-valued functions, on the space of C^2 functions with a) Dirichlet boundary conditions, or b) Neumann boundary conditions.

12. Show how any result about convergence of the full Fourier series on $[-L, L]$ implies a result about Fourier sine and cosine series on $[0, L]$.

13. Express the a's and b's in terms of the c's and vice versa in the two forms

$$\frac{a_0}{2} + \sum_{k=1}^{\infty} \left(a_k \cos \frac{k\pi}{L} + b_k \sin \frac{k\pi}{L} x \right)$$

and

$$\sum_{n=-\infty}^{\infty} c_n e^{(ik\pi/L)x}$$

of the full Fourier series.

14. Show that $\sum_{n=-\infty}^{\infty} c_n e^{(ik\pi/L)x}$ is real-valued if and only if $c_{-n} = \overline{c_n}$.

15. Show that all eigenfunctions of d/dx on the space of C^1 functions on $[-L, L]$ satisfying $f(-L) = f(L)$ are of the form $ce^{(ik\pi/L)x}$.

16. Show that $\langle Af, g \rangle = -\langle f, Ag \rangle$ on a complex inner product space if and only if iA is symmetric.

17. Show that if $\tau_y f(x) = g(y)f(x)$ for every x and y and some C^1 function $g(y)$, where f is a C^1 periodic function of period $2L$, then $f(x) = ce^{(ik\pi/L)x}$. (**Hint**: differentiate and use exercise 15.)

18. Show that any complex-valued C^1 function satisfying the character identity must be of the form $\exp(\lambda x)$ for λ complex.

19. Show that any bounded C^1 function on the line satisfying the character identity must be of the form $f(x) = e^{itx}$ for real t.

20. a. "Sketch" the graph of $\sin 10,000x$. (Note: Your ear is capable of hearing a tone with this frequency, a little more than an octave above the highest note on a piano.)

    b. "Sketch" the graph of $\sin 10,000x + \sin 10,001x$. (**Hint**: use the trigometric identity $\sin(a+b) + \sin(a-b) = 2\sin a \cos b$.)

21. Solve the differential equation

$$\left(\frac{\partial^2}{\partial x^2} + \frac{\partial^2}{\partial y^2}\right) u(x,y) = 0$$

in the rectangle $0 \le x \le a, 0 \le y \le b$ with boundary conditions that $u(x,0) = f(x)$ on the bottom side and $u = 0$ on the other three sides (i.e., $u(x,b) = 0, u(0,y) = 0$ and $u(a,y) = 0$) by expanding $u(x,y)$ in a Fourier sine series in x, for each fixed y.

12.2 Convergence of Fourier Series

12.2.1 Uniform Convergence for C^1 Functions

Dirichlet was the first mathematician to give a good proof of the convergence of Fourier series. His work on the question was important also because it stimulated Riemann to investigate integration theory. Dirichlet proved that a continuous function is equal to a pointwise convergent Fourier series if it has only a finite number of maxima and minima. Dirichlet also proved a similar result for a function that has a finite number of discontinuities, and he remarked that he only used that property to ensure the existence of the integral. He had the integrity to admit that he saw no way to remove the hypothesis that the number of extrema be finite, even though he could not see any natural reason why the hypothesis should be needed. We will follow Dirichlet's argument part way, and use it to establish the uniform convergence for a C^1 function.

Our starting point will be a function $f(x)$ that is continuous and periodic of period 2π (we now follow the convention of taking the period equal to 2π in order to simplify the notation—clearly the arguments will go through in the general case). We form the Fourier coefficients

$$c_n = \frac{1}{2\pi}\int_{-\pi}^{\pi} f(x)e^{-inx}\,dx$$

and the Fourier series

$$\sum_{n=-\infty}^{\infty} c_n e^{inx}.$$

To investigate the convergence we form the partial sums $S_N f(x) = \sum_{n=-N}^{N} c_n e^{inx}$. Notice that the order we are taking corresponds to the natural order in terms of sines and cosines,

$$S_N f(x) = \frac{a_0}{2} + \sum_{k=1}^{N}(a_k \cos kx + b_k \sin kx).$$

We want to show $S_N f(x) \to f(x)$ as $N \to \infty$. Dirichlet's key idea is simply to substitute the definition of the Fourier coefficients in order to write $S_N f(x)$ as a convolution. To see how this works, we need the elementary observation that if f is 2π-periodic, then

$$\int_{-\pi}^{\pi} f(x)\,dx = \int_{a}^{a+2\pi} f(x)\,dx$$

for any real a—for any interval of length 2π can be cut in two pieces and translated back by multiples of 2π to the interval $[-\pi,\pi]$. Note that this remark also applies to $f(x)e^{-inx}$, since this function is again 2π-periodic.

Using these ideas, we find

$$\begin{aligned}
S_N f(x) &= \sum_{n=-N}^{N} c_n e^{inx} = \sum_{n=-N}^{N} \frac{1}{2\pi}\left(\int_{-\pi}^{\pi} f(y)e^{-iny}\,dy\right) e^{inx} \\
&= \frac{1}{2\pi}\sum_{n=-N}^{N}\int_{-\pi}^{\pi} f(y)e^{in(x-y)}\,dy \\
&= \frac{1}{2\pi}\int_{-\pi}^{\pi} f(y)\left(\sum_{n=-N}^{N} e^{in(x-y)}\right) dy \\
&= \frac{1}{2\pi}\int_{-\pi}^{\pi} f(y)D_N(x-y)\,dy,
\end{aligned}$$

where we have defined the *Dirichlet kernel* D_N by

$$D_N(t) = \sum_{n=-N}^{N} e^{int}.$$

Since this is a geometric progession, we can evaluate it exactly. We have

$$e^{it} D_N(t) = \sum_{n=1-N}^{N+1} e^{int} = D_N(t) + e^{i(N+1)t} - e^{-iNt},$$

so

$$D_N(t) = \frac{e^{i(N+1)t} - e^{-iNt}}{e^{it} - 1} = \frac{e^{i(N+1/2)t} - e^{-i(N+1/2)t}}{e^{it/2} - e^{-it/2}} = \frac{\sin(N+1/2)t}{\sin(1/2)t}.$$

We call the expression $\int_{-\pi}^{\pi} f(y)g(x-y)\,dy$ the *periodic convolution* of f and g, written $f * g(x)$, for any 2π-periodic functions f and g. We observe that it is a commutative product, for by the change of variable $y \to x - y$ we obtain

$$\int_{-\pi}^{\pi} f(y)g(x-y)\,dy = \int_{x-\pi}^{x+\pi} f(x-y)g(y)\,dy = \int_{-\pi}^{\pi} f(x-y)g(y)\,dy.$$

Thus we have

$$S_N f(x) = \frac{1}{2\pi} f * D_N(x) = \frac{1}{2\pi} \int_{-\pi}^{\pi} f(x-y) D_N(y)\,dy.$$

If $(1/2\pi)D_N$ behaves like an approximate identity we can hope to show $S_N f \to f$.

Let us examine the graph of $D_N(t)$ (see Figure 12.2.1 for the case $N = 3$). Notice that $D_N(0) = 2N + 1$, so D_N has a peak at the origin. However, the function $\sin(N + 1/2)x$ in the numerator oscillates frequently between -1 and $+1$, and the function $\sin x/2$ in the denominator has only the single zero at $x = 0$. Thus the quotient does not actually become small for x away from zero. However, there is a great deal of oscillation away from $x = 0$ so that "on the average" D_N is small away from the origin. We do have $(1/2\pi)\int_{-\pi}^{\pi} D_N(t)\,dt = 1$, which can be seen quite easily from the expression $D_N(t) = \sum_{n=-N}^{N} e^{int}$, since only the term $n = 0$ survives the integration.

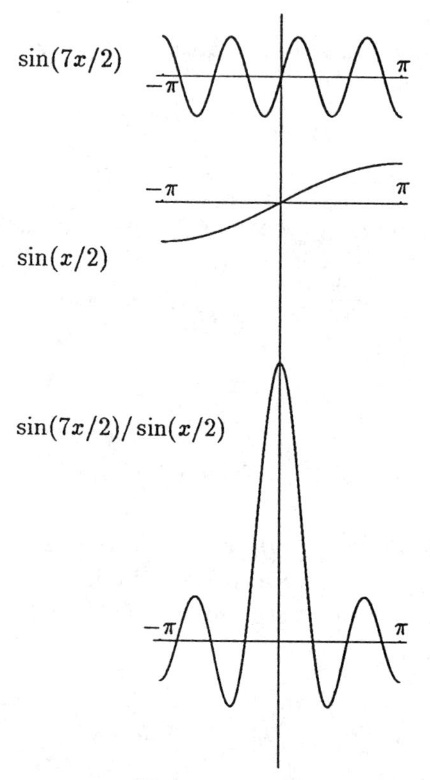

Figure 12.2.1:

Since we are interested in the convergence of $S_N f$ to f, we write

$$S_N f(x) - f(x) = \frac{1}{2\pi} \int_{-\pi}^{\pi} [f(x-y) - f(x)] D_N(y)\, dy.$$

But now we cannot take absolute values inside the integral without courting disaster, since we have no control over $|D_N(y)|$. In fact it can be shown that $\int_{-\pi}^{\pi} |D_N(y)| dy \geq c \log N$ (see exercise set 12.2.6). It follows that a direct appeal to the approximate identity method will not work. Nevertheless, the cancellation implied by the oscillation of D_N away from the origin should make the contribution to the integral from $\delta \leq |y| \leq \pi$ very small. The way to see this is by the appropriate integration by parts. We begin by proving the convergence under stronger hypotheses, namely that f is C^2. The proof is simpler in this case. As a bonus we get a faster rate of convergence.

Theorem 12.2.1 *Let f be a C^2 periodic function of period 2π. Then the Fourier series of f converges uniformly to f, and in fact $S_N f(x) - f(x) = O(1/N)$ uniformly as $N \to \infty$.*

Proof: We have seen that we can write

$$S_N f(x) - f(x) = \frac{1}{2\pi}\int_{-\pi}^{\pi} (f(x-y) - f(x)) D_N(y)\, dy.$$

The trick is to substitute the explicit formula for $D_N(y)$ and to group the terms as follows:

$$S_N f(x) - f(x) = \frac{1}{2\pi}\int_{-\pi}^{\pi} \left(\frac{f(x-y) - f(x)}{\sin y/2}\right) \sin\left(N + \frac{1}{2}\right) y\, dy.$$

We write $g(y) = (f(x-y) - f(x))/\sin y/2$, and observe that this is a C^1 function on $[-\pi, \pi]$. Indeed, there is no problem away from $y = 0$, and near $y = 0$ we can use the Taylor expansion of $f(x-y)$ about x, as in our discussion of L'Hôpital's rule. It is here that we use the hypothesis that f is C^2. Furthermore, we can find a bound $|g'(y)| \leq M$ for all x and y.

Now we are ready to do integration by parts, differentiating g and integrating $\sin(N + 1/2)y = (-d/dy)\cos(N + 1/2)y/(N + 1/2)$. There are no boundary terms because the cosine of $(N + 1/2)y$ vanishes when $y = \pm\pi$. Thus

$$S_N f(x) - f(x) = \frac{1}{2\pi}\int_{-\pi}^{\pi} g'(y) \frac{\cos(N + \frac{1}{2})y}{N + \frac{1}{2}}\, dy;$$

hence, $|S_N f(x) - f(x)| \leq M/(N + 1/2)$ as claimed. QED

In fact, by integrating by parts $k - 1$ times, you can show that if f is C^k, then the rate of convergence is $O(1/N^{k-1})$. We leave the details to the exercises.

Theorem 12.2.2 *Let f be a C^1 periodic function of period 2π. Then the Fourier series of f converges uniformly to f; in fact, $S_N f(x) - f(x) = O(N^{-1/2})$ uniformly as $N \to \infty$.*

Proof: Once again we write

$$S_N f(x) - f(x) = \frac{1}{2\pi}\int_{-\pi}^{\pi} g(y) \sin\left(N + \frac{1}{2}\right) y\, dy,$$

for $g(y) = (f(x-y) - f(x))/\sin y/2$. This time we divide the interval $[-\pi, \pi]$ of integration up into two pieces, the interval $[-\delta, \delta]$ and its complement $\delta \le |y| \le \pi$, where δ will be chosen later. In $[-\delta, \delta]$ we do nothing, and in its complement we integrate by parts. The hypothesis that f is C^1 implies $g(y)$ is continuous away from $y = 0$, but even at $y = 0$ we have the limit

$$\lim_{y\to 0} \frac{f(x-y) - f(x)}{\sin \frac{1}{2}y} = \lim_{y\to 0} \frac{y}{\sin \frac{1}{2}y} \frac{f(x-y) - f(x)}{y} = -2f'(x).$$

By the mean value theorem $f(x-y) - f(x) = -yf'(z)$ for some z, which gives the estimate

$$|g(y)| \le \left| \frac{yf'(z)}{\sin \frac{1}{2}y} \right| \le \pi \sup_z |f'(z)|,$$

using the fact that $|y| \le \pi |\sin y/2|$ on $|y| \le \pi$ (see exercise set 12.2.6). Thus

$$\left| \int_{-\delta}^{\delta} g(y) \sin\left(N + \frac{1}{2}\right) y \, dy \right| \le \int_{-\delta}^{\delta} |g(y) \, dy| \le 2\pi\delta \sup_z |f'(z)|.$$

This is a good estimate, because we can make the integral small by taking δ small.

Next we consider the integral over the interval $[\delta, \pi]$ (we treat $[-\pi, -\delta]$ analogously). We integrate by parts, differentiating $g(y)$ and integrating $\sin(N + 1/2)$, to obtain

$$\int_{\delta}^{\pi} g(y) \sin(N+1/2) y \, dy = \int_{\delta}^{\pi} g'(y) \frac{\cos(N + \frac{1}{2})y}{N + \frac{1}{2}} \, dy + g(\delta) \frac{\cos(N + \frac{1}{2})\delta}{N + \frac{1}{2}}$$

and now can take absolute values and estimate

$$\begin{aligned} &\left| \int_{\delta}^{\pi} g(y) \sin(N + \frac{1}{2}) y \, dy \right| \\ &\le \frac{1}{N + \frac{1}{2}} \left(\int_{\delta}^{\pi} |g'(y)| \left| \cos\left(N + \frac{1}{2}\right) y \right| dy + |g(\delta)| \cos\left(N + \frac{1}{2}\right) \delta \right). \end{aligned}$$

Now $|g(\delta)| \le \pi \sup_z |f'(z)|$ as above, and

$$g'(y) = \frac{-f'(x-y)}{\sin \frac{1}{2}y} - \frac{\cos \frac{1}{2}y}{2 \sin \frac{1}{2}y} g(y),$$

so $|g'(y)| \le (\pi/\delta)\sup_z |f'(z)| + (\pi/2\delta)\sup_z |f'(z)|$ using $\sin y/2 \ge \delta/\pi$ on $[\delta, \pi]$ (this follows from $|y| \le \pi|\sin y/2|$). Thus

$$\left|\int_\delta^\pi g(y)\sin\left(N+\frac{1}{2}\right)y\,dy\right| \le \frac{1}{N+\frac{1}{2}}\left(\frac{3}{2}\frac{\pi}{\delta}+\pi\right)\sup_z |f'(z)|.$$

The factor of $N+1/2$ in the denominator is helpful, but the δ in the denominator is a potential problem. Adding the estimates for $[-\delta,\delta]$ and its complement we obtain

$$\begin{aligned} |S_N f(x) - f(x)| &= \frac{1}{2\pi}\left|\int_{-\pi}^\pi g(y)\sin\left(N+\frac{1}{2}\right)y\,dy\right| \\ &\le \left[\delta + \frac{1}{N+\frac{1}{2}}\left(\frac{3}{2\delta}+1\right)\right]\sup_z |f'(z)|. \end{aligned}$$

Is this good enough? Yes, if we choose $\delta = (N+1/2)^{-1/2}$ we obtain the estimate

$$|S_N f(x) - f(x)| \le c\left(N+\frac{1}{2}\right)^{-1/2}\sup_z |f'(z)|$$

where the constant c does not depend on f or x or N (we use here $(N+1/2)^{-1} \le (N+1/2)^{-1/2}$ for $N \ge 1$). Thus we have the uniform convergence of the partial sums to f. We have actually shown that the rate of convergence is $O(N^{-1/2})$, with the constant depending only on $\sup_z |f'(z)|$. QED

It would be nice to know that the convergence is also absolute so that the order of terms is immaterial. This is indeed the case as we will show later. However, there are conditions on f that yield the convergence of $S_N f$ but not the absolute convergence. There are also many variants of this theorem that give convergence under weaker conditions and local versions in which only the convergence at a single point is proved. Some of these are given in exercise set 12.2.6.

12.2.2 Summability of Fourier Series

We have now seen that Fourier's conjecture is true if we add the additional hypothesis that the function be C^1. However, there is something very unsatisfactory about this result, because the hypothesis is stronger than the conclusion. That is, the natural hypothesis for convergence of the Fourier series is that f be continuous; if f is C^1 we would like to know that the differentiated Fourier series $\sum_{n=-\infty}^{\infty} inc_n e^{inx}$ also converges to f'. But neither statement is true. Still, there is a way to impove matters by changing the question. Instead of asking for the convergence of the Fourier series, we ask for a weaker condition called *summability.* This notion exploits the great amount of cancellation that is to be expected due to the oscillations of the functions e^{inx}.

The point is that by simply adding the terms of the Fourier series in pairs $(c_N e^{iNx} + c_{-N} e^{-iNx})$ in passing from $S_{N-1}f$ to $S_N f$, we are doing things too abruptly, giving too much emphasis to the new terms. It is like dropping a pebble in a lake—even if the pebble is small it can make a big splash. However, there is a way to ease the pebble into the water gently. What we do with the Fourier series is to multiply each term $c_n e^{inx}$ by a constant that depends on both n and N. That is, we look at

$$\sigma_N f(x) = \sum_{n=-N}^{N} A_{N,n} c_n e^{inx}$$

instead of $S_N f(x)$. We choose the coefficients $A_{N,n}$ to be small when n is near $\pm N$ so that the new terms are gently eased in, but we let $A_{N,n} \to 1$ for fixed n as $N \to \infty$ so that each term is eventually counted toward the sum. We can then ask if $\sigma_N f$ converges to f. Each choice of coefficients $A_{N,n}$ (with $0 \leq A_{N,n} \leq 1$ and $\lim_{N\to\infty} A_{N,n} = 1$ for each fixed n) is called a *summability method*, and if $\sigma_N f \to f$ we say the Fourier series is *summable* to f by the particular summability method. In fact the idea of summability methods can be applied to any sort of series, not just Fourier series. A key fact is that if the series converges, then it is summable to the same limit. If this were not true, the use of summability methods would be very suspicious. Fortunately, it is simple to prove, and we leave it to the exercises.

The simplest summability method is *Cesaro summability*, where we take the coefficients $A_{N,n}$ to be linear in n,

$$A_{N,n} = (N + 1 - |n|)/(N + 1) = 1 - |n|/(N + 1),$$

as shown in Figure 12.2.2. For this choice of coefficients $\sigma_N f$ is just the arithmetic mean of the first $N + 1$ partial sums, $\sigma_N f = (S_0 f + S_1 f + \cdots + S_N f)/(N + 1)$.

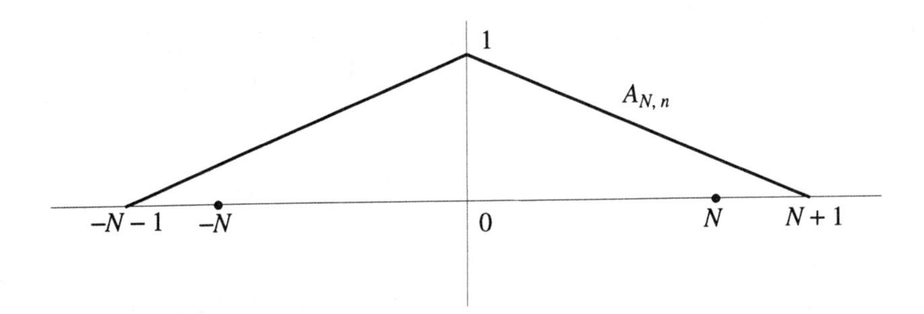

Figure 12.2.2:

The idea of using Cesaro summability on Fourier series is due to Fejér, and it works splendidly.

Following Fejér, we write

$$\begin{aligned}
\sigma_N f(x) &= \frac{1}{N+1}(S_0 f(x) + \cdots + S_N f(x)) \\
&= \frac{1}{N+1}\sum_{n=0}^{N}\frac{1}{2\pi}\int_{-\pi}^{\pi} f(x-y)D_N(y)\,dy \\
&= \frac{1}{2\pi}\int_{-\pi}^{\pi} f(x-y)K_N(y)\,dy
\end{aligned}$$

where the *Fejér kernel* $K_N(y)$ is given by

$$K_N(y) = \frac{1}{N+1}\sum_{n=0}^{N} D_n(y) = \frac{1}{N+1}\sum_{n=0}^{N}\frac{\sin(n+1/2)y}{\sin(1/2)y}.$$

Now the idea is that the Fejér kernel should be better behaved than the Dirichlet kernel, because it involves an averaging process that exploits the cancellations in the Dirichlet kernel. To see this more clearly we have to simplify the expression for K_N, using familiar trigonometric identities in an especially clever way. We observe that

$$\begin{aligned}
&-\cos(n+1)y + \cos ny \\
&= -\cos\left[\left(n+\frac{1}{2}\right)y + \frac{1}{2}y\right] + \cos\left[\left(n+\frac{1}{2}\right)y - \frac{1}{2}y\right] \\
&= -\left(\cos\left(n+\frac{1}{2}\right)y\cos\frac{1}{2}y - \sin\left(n+\frac{1}{2}\right)y\sin\frac{1}{2}y\right)
\end{aligned}$$

$$+\left(\cos\left(n+\frac{1}{2}\right)y\cos\frac{1}{2}y+\sin\left(n+\frac{1}{2}\right)y\sin\frac{1}{2}y\right)$$
$$= 2\sin\left(n+\frac{1}{2}\right)y\sin\frac{1}{2}y.$$

Thus

$$\begin{aligned}\sum_{n=0}^{N}\sin\left(n+\frac{1}{2}\right)y &= \frac{1}{2\sin\frac{1}{2}y}\sum_{n=0}^{N}(-\cos(n+1)y+\cos ny)\\ &= \frac{1-\cos(N+1)y}{2\sin\frac{1}{2}y}=\frac{(\sin(\frac{N+1}{2})y)^2}{\sin\frac{1}{2}y}\end{aligned}$$

and so finally

$$K_N(y)=\frac{1}{N+1}\left(\frac{\sin(\frac{N+1}{2})y}{\sin\frac{1}{2}y}\right)^2.$$

Examining this expression carefully, we notice two important improvements over the Dirichlet kernel

$$D_N(y)=\frac{\sin(N+\frac{1}{2})y}{\sin\frac{1}{2}y}.$$

The first is the appearance of the factor $1/(N+1)$, which means $K_N(y)$ gets small as $N\to\infty$, except near $y=0$ where $\sin y/2$ vanishes. This is the concentration of $K_N(y)$ near $y=0$, which gives it the properties of an approximate identity. This fact is not surprising in view of our strategy of averaging oscillations of $D_n(y)$. The second fact is that K_N is non-negative, $K_N(y)\geq 0$ for all y. This is rather startling, and I know of no explanation for it other than the computation itself. However, it is not essential for the theorems we will prove. The sketches of the graphs of $D_3(x)$ and $K_3(x)$ in Figure 12.2.3 show a marked contrast.

Theorem 12.2.3 (*Fejér*) *Let f be any continuous function periodic of period 2π. Then the Fourier series of f is uniformly Cesaro summable to $f, \sigma_N f\to f$ uniformly as $N\to\infty$ where*

$$\sigma_N f(x)=\sum_{n=-N}^{N}\frac{N+1-|n|}{N+1}c_n e^{inx}.$$

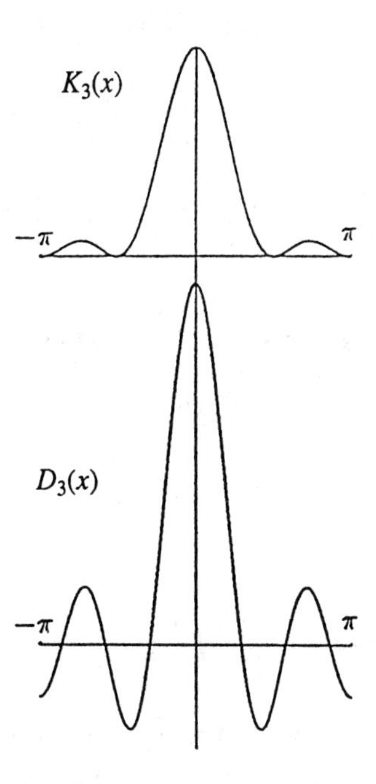

Figure 12.2.3:

Proof: This is essentially a consequence of the approximate identity lemma, since we have seen

$$\sigma_N f(x) = \frac{1}{2\pi} \int_{-\pi}^{\pi} f(x-y) K_N(y)\, dy.$$

Thus we need to verify that

1. $K_N(y) \geq 0$,
2. $(1/2\pi) \int_{-\pi}^{\pi} K_N(y)\, dy = 1$, and
3. $\lim_{N\to\infty} K_N(y) = 0$ uniformly on $|y| \geq \delta$ for any $\delta > 0$.

From the expression

$$K_N(y) = \frac{1}{N+1} \left(\frac{\sin(\frac{N+1}{2})y}{\sin \frac{1}{2}y} \right)^2$$

we obtain property 1, and then property 3 follows since

$$|K_N(y)| \leq \frac{1}{N+1}\frac{1}{(\sin\frac{1}{2}y)^2} \leq \frac{c}{\delta^2(N+1)}$$

for $\delta \leq |y| \leq \pi$. To verify property 2 we integrate the expression $K_N(y) = \sum_{n=0}^N D_n(y)/(N+1)$, using the fact that $\int_{-\pi}^{\pi} D_n(y)\,dy = 1$ for all n.

Repeating the argument of the approximate identity lemma, we write (using property 2)

$$\sigma_N f(x) - f(x) = \frac{1}{2\pi}\int_{-\pi}^{\pi} [f(x-y) - f(x)]K_N(y)\,dy$$

and so

$$|\sigma_N f(x) - f(x)| \leq \frac{1}{2\pi}\int_{-\pi}^{\pi} |f(x-y) - f(x)|K_N(y)\,dy$$

by Minkowski's inequality. Given $\varepsilon > 0$ we first find $\delta > 0$ such that $|y| \leq \delta$ implies $|f(x-y) - f(x)| \leq \varepsilon/2$ for all x. This follows from the uniform continuity of f, which is a consequence of periodicity. Indeed it suffices to prove the result for $|x| \leq \pi$ by periodicity, and then we have the uniform continuity of f on the compact interval $|x| \leq \pi + 1$ (keeping $\delta \leq 1$ to stay inside it). This enables us to estimate

$$\int_{-\delta}^{\delta} |f(x-y)-f(x)|K_N(y)\,dy \leq \frac{\varepsilon}{2}\int_{-\delta}^{\delta} K_N(y)\,dy \leq \frac{\varepsilon}{2}\int_{-\pi}^{\pi} K_N(y)\,dy = \frac{\varepsilon}{2}$$

for all N. Given δ we choose N large enough so that

$$\int_{-\pi}^{-\delta} + \int_{\delta}^{\pi} K_N(y)\,dy \leq \frac{\varepsilon}{4M}$$

where $M = \sup_x |f(x)|$. This is possible using property 3 if we take $2\pi c/\delta^2(N+1) \leq \varepsilon/4M$. Then

$$\int_{-\pi}^{-\delta} + \int_{\delta}^{\pi} |f(x-y)-f(x)|K_N(y)\,dy \leq 2M\left(\int_{-\pi}^{-\delta} + \int_{\delta}^{\pi} K_N(y)\,dy\right) \leq \frac{\varepsilon}{2}$$

and so $|\sigma_N f(x) - f(x)| \leq \varepsilon$ for all x. QED

There are many variants of this result. For example, if f is just Riemann integrable but continuous at a point x_0, a simple modification of the same proof shows $\sigma_N f(x_0) \to f(x_0)$. Even if f has a jump discontinuity at $x_0, \sigma_N f(x_0)$ will converge to the average value

$$\frac{1}{2}\left(\lim_{x \to x_0^+} f(x) + \lim_{x \to x_0^-} f(x)\right) = \lim_{s \to 0} \frac{1}{2}(f(x_0 + s) + f(x_0 - s)).$$

Notice that we are then approximating a discontinuous function by continuous functions, so the convergence cannot be uniform. For the same reason, the family $\sigma_N f$ cannot be uniformly equicontinuous.

Fejér's theorem gives an explicit proof that every periodic continuous function can be uniformly approximated by trigonometric polynomials. A *trigonometic polynomial* is defined to be any finite Fourier series, $\sum_{n=-N}^{N} c_n e^{in\theta}$. Since the trigonometric polynomials form an algebra, this fact is also obtainable from the Stone-Weierstrass theorem, although in less explicit form. In the opposite direction, we can obtain the Weierstrass approximation theorem as a consequence of Fejér's theorem by replacing the exponentials e^{inx} in $\sigma_N f(x)$ by partial sums of their power-series expansion, which we know converge uniformly on bounded intervals.

An immediate corollary of Fejér's theorem is the *uniqueness of Fourier series*: if f and g are two continuous functions with the same Fourier coefficients, then $f = g$. Indeed if the Fourier coefficients are equal, then $\sigma_N f = \sigma_N g$ for every N and, letting $N \to \infty$, we obtain $f = g$.

12.2.3 Convergence in the Mean

Although Fejér's theorem suggests that the Cesaro sums $\sigma_N f$ may do a better job than the partial sums $S_N f$ in approximating f, there is a criterion by which $S_N f$ is the best approximation among all trigonometric polynomials $\sum_{-N}^{N} a_n e^{inx}$ of order N. This involves measuring the error in the mean-square sense as

$$d_2(f, g) = \left(\frac{1}{2\pi}\int_{-\pi}^{\pi} |f(x) - g(x)|^2 \, dx\right)^{1/2}.$$

This is the metric associated to the L^2-norm

$$||f||_2 = \left(\frac{1}{2\pi}\int_{-\pi}^{\pi} |f(x)|^2\, dx\right)^{1/2}$$

associated to the inner product

$$\langle f, g\rangle = \frac{1}{2\pi}\int_{-\pi}^{\pi} f(x)\overline{g(x)}\, dx.$$

Recall that the exponential functions e^{inx} are orthogonal vectors with respect to this inner product, $\langle e^{inx}, e^{ikx}\rangle = 0$ if $k \neq n$; and that the choice of the factor $1/2\pi$ in the inner product makes them normalized, $\langle e^{inx}, e^{inx}\rangle = 1$. The normalization is purely conventional, and everything would work as well without it.

We will need to use the following basic fact about orthonormal vectors in an inner product space (real or complex), which generalizes the formula for projecting a vector onto a subspace of $\mathbb{R}^n$. We state and prove the result in the abstract context because it clarifies the ideas.

Theorem 12.2.4 (*Projection Theorem*) *Let $v_1, \ldots, v_N$ be orthonormal vectors in an inner product space V and W denote the subspace they span, $W = \{\sum_{n=1}^{N} a_n v_n\}$. Given any vector u in V, the problem of finding the vector w in W that minimizes the distance $d(u, w)$ is solved by $w = Pu = \sum_{n=1}^{N}\langle u, v_n\rangle v_n$. Pu is called the orthogonal projection of u onto W and is characterized by the condition that Pu is in W and $u - Pu$ is orthogonal to W. The distance $d(u, Pu)$ is given by*

$$d(u, Pu)^2 = ||u||^2 - \sum_{n=1}^{N} |\langle u, v_n\rangle|^2,$$

hence, we have Bessel's inequality

$$\sum_{n=1}^{N} |\langle u, v_n\rangle|^2 \leq ||u||^2.$$

Proof: We define $Pu = \sum_{n=1}^{N}\langle u, v_n\rangle v_n$ and observe $\langle Pu, v_k\rangle = \langle u, v_k\rangle$ for any k by the orthonormality condition. Therefore $\langle u - Pu, v_k\rangle = 0$ and so $\langle u - Pu, \sum_{k=1}^{N} a_k v_k\rangle = 0$, showing $u - Pu$ is orthogonal to W. Of course Pu is in W from the definition. Now if

w is any vector in W, then $u - w = (u - Pu) + (Pu - w)$ and $Pu - w$ is also in W. Thus

$$\begin{aligned} d(u,w)^2 &= \langle u - w, u - w \rangle \\ &= \langle (u - Pu) + (Pu - w), (u - Pu) + (Pu - w) \rangle \\ &= \langle u - Pu, u - Pu \rangle + \langle Pu - w, Pu - w \rangle \\ &= d(u, Pu)^2 + d(Pu, w)^2 \end{aligned}$$

because $u - Pu$ is orthogonal to W, as in Figure 12.2.4. This is a generalization of the Pythagorean theorem.

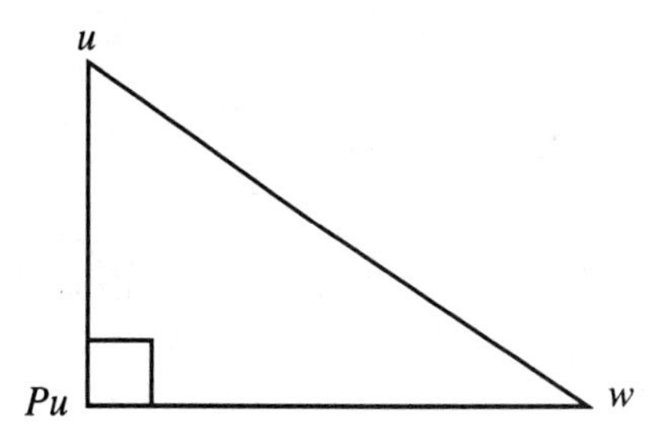

Figure 12.2.4:

Notice that the first term $d(u, Pu)^2$ is independent of w, and the second term $d(Pu, w)^2$ is non-negative. It is obvious that the sum is minimized if the second term $d(Pu, w)^2$ vanishes, which happens exactly when $w = Pu$. Thus $w = Pu$ is the unique minimizer. To obtain the formula for $d(u, Pu)^2$ we compute

$$\begin{aligned} d(u, Pu)^2 &= \langle u - Pu, u - Pu \rangle \\ &= \langle u, u \rangle + \langle Pu, Pu \rangle - \langle u, Pu \rangle - \langle Pu, u \rangle \end{aligned}$$

and (using the definitions of Pu)

$$\langle Pu, Pu \rangle = \langle u, Pu \rangle = \langle Pu, u \rangle = \sum_{n=1}^{N} |\langle u, v_n \rangle|^2.$$

QED

Returning to the concrete context of the L^2 inner product on 2π-periodic continuous functions, we consider the set of orthonormal vectors e^{inx} for $-N \leq n \leq N$. Note that Pf is exactly $S_N f$. Thus we have the following corollary.

Corollary 12.2.1 *Among all trigonometric polynomials $\sum_{n=-N}^{N} a_n e^{inx}$ of degree $N, S_N f$ minimizes the L^2 distance to f, for any 2π-periodic continuous function f. Furthermore, we have Bessel's inequality*

$$\sum_{n=-N}^{N} |c_n|^2 \leq \frac{1}{2\pi}\int_{-\pi}^{\pi} |f(x)|^2\, dx$$

and

$$\frac{1}{2\pi}\int_{-\pi}^{\pi} |f(x) - S_N f(x)|^2\, dx + \sum_{n=-N}^{N} |c_n|^2 = \frac{1}{2\pi}\int_{-\pi}^{\pi} |f(x)|^2\, dx$$

where $c_n = (1/2\pi)\int_{-\pi}^{\pi} f(x)e^{-inx}\, dx$.

In particular,

$$\frac{1}{2\pi}\int_{-\pi}^{\pi} |f(x) - S_N f(x)|^2\, dx$$

is less than

$$\frac{1}{2\pi}\int_{-\pi}^{\pi} |f(x) - \sigma_N f(x)|^2\, dx.$$

On the other hand, $f - \sigma_N f$ goes to zero uniformly, so

$$\lim_{N\to\infty} \frac{1}{2\pi}\int_{-\pi}^{\pi} |f(x) - \sigma_N f(x)|^2\, dx = 0.$$

Thus combining Fejér's theorem and the projection theorem we obtain

$$\lim_{N\to\infty} \frac{1}{2\pi}\int_{-\pi}^{\pi} |f(x) - S_N f(x)|^2\, dx = 0,$$

the mean-square convergence of Fourier series. We can then take the limit as $N \to \infty$ in the identity

$$\frac{1}{2\pi}\int_{-\pi}^{\pi} |f(x) - S_N f(x)|^2\, dx + \sum_{n=-N}^{N} |c_n|^2 = \frac{1}{2\pi}\int_{-\pi}^{\pi} |f(x)|^2\, dx$$

to obtain *Parseval's identity*:

$$\sum_{n=-\infty}^{\infty} |c_n|^2 = \frac{1}{2\pi} \int_{-\pi}^{\pi} |f(x)|^2 \, dx.$$

This identity gives us very useful information concerning the rate of growth of the Fourier coefficients. In particular, it implies $\lim_{n \to \pm\infty} c_n = 0$, a fact that is referred to as the *Riemann-Lebesgue Lemma*. Incidentally, both the Riemann-Lebesgue lemma and Parseval's identity are valid under much more general hypotheses than continuity on f, but it requires the Lebesgue theory of integration to explain these hypotheses. We will return to this in Chapter 14. There is also a bilinear form of Parseval's identity:

$$\sum_{n=-\infty}^{\infty} c_n(f)\overline{c_n(g)} = \frac{1}{2\pi} \int_{-\pi}^{\pi} f(x)\overline{g(x)} \, dx$$

where $c_n(f)$ and $c_n(g)$ denote the Fourier series coefficients of f and g. This is obtainable by applying the polarization indentity to both sides of Parseval's identity.

The mean-square convergence of Fourier series implies a property of the orthonormal system e^{inx} called *completeness* (*warning*: this use of the term is somewhat different from the previous usages we have encountered): there is no non-zero continuous function $f(x)$ that is orthogonal to all the functions e^{inx}, $-\infty < n < \infty$. For if there were such a function, then all $S_N f$ would be identically zero and so

$$\frac{1}{2\pi} \int_{-\pi}^{\pi} |f(x)|^2 \, dx = \frac{1}{2\pi} \int_{-\pi}^{\pi} |f(x) - S_N f(x)|^2 \, dx \to 0$$

as $N \to \infty$; hence

$$\frac{1}{2\pi} \int_{-\pi}^{\pi} |f(x)|^2 \, dx = 0,$$

so $f \equiv 0$ (the same conclusion also follows from Fejér's theorem since $\sigma_N f$ would also be identically zero). The completeness can be interpreted as saying we have not omitted any terms in forming the Fourier series expansion—in retrospect, this is the idea Bernoulli should have advanced.

One striking property of the mean-square convergence is that it does not depend on the order of the terms. If A is any finite subset of integers, let $P_A f(x) = \sum_A c_n e^{inx}$. Then

$$\frac{1}{2\pi}\int_{-\pi}^{\pi} |f(x) - P_A f(x)|^2\, dx = \sum_B |c_n|^2$$

where B denotes the complement of A, by Parseval's identity (the Fourier coefficients of $f - P_A f$ are c_n if n is in B and 0 if n is in A). If A_N is any sequence of finite subsets increasing to all of the integers and B_N is the complement of A_N, then

$$\lim_{N\to\infty} \frac{1}{2\pi}\int_{-\pi}^{\pi} |f(x) - P_N f(x)|^2\, dx = \lim_{N\to\infty} \sum_{B_N} |c_n|^2 = 0.$$

We are now in a position to show the absolute convergence of Fourier series of C^1 functions. Because $|e^{inx}| \equiv 1$, the absolute convergence is only a question of the size of the coefficients, $\sum_{n=-\infty}^{\infty} c_n e^{inx}$ converges absolutely if and only if $\sum_{n=-\infty}^{\infty} |c_n| < \infty$. Now if f is C^1, then f' is continuous, and the Fourier coefficients of f' are

$$\frac{1}{2\pi}\int_{-\pi}^{\pi} f'(x)e^{-inx}\, dx = in\frac{1}{2\pi}\int_{-\pi}^{\pi} f(x)e^{-inx}\, dx = inc_n$$

by integration by parts (there are no boundary terms because f is assumed periodic, so $f(-\pi) = f(\pi)$). Note this is the same result that we would obtain by formally differentiating the Fourier series. The Parseval identity for f' is

$$\sum_{n=-\infty}^{\infty} n^2|c_n|^2 = \frac{1}{2\pi}\int_{-\pi}^{\pi} |f'(x)|^2\, dx,$$

which implies

$$\sum_{n=-\infty}^{\infty} |c_n| = |c_0| + \sum_{n\neq 0} n^{-1}(n|c_n|) \le |c_0| + \left(\sum_{n\neq 0} \frac{1}{n^2}\right)^{1/2} \left(\sum_{n\neq 0} n^2|c_n|^2\right)^{1/2}$$

is finite by the Cauchy-Schwartz inequality. Thus the order of terms in the Fourier series of a C^1 function is immaterial, $P_{A_N} f \to f$ uniformly. We should point out, however, that this argument only shows that $S_N f$

converges; it does not show that the limit is f. For that we have to refer to the original proof.

Another application of Parseval's identity (actually the Riemann-Lebesgue lemma) is the principle of *localization*: the convergence of a Fourier series at a point x_0 depends only on the behavior of $f(x)$ for x in a neighborhood of x_0.

Theorem 12.2.5 (*Localization*) *Let f and g be two continuous periodic functions such that $f(x) = g(x)$ for x in $[x_0 - \delta, x_0 + \delta]$, for some fixed $\delta > 0$. Then the Fourier series for f and g either both converge or both diverge at x_0.*

Proof: As in the proof of convergence of the Fourier series for C^1 functions, we write

$$\begin{aligned} S_N f(x_0) - f(x_0) &= \frac{1}{2\pi}\int_{-\pi}^{\pi}[f(x_0 - y) - f(x_0)]D_N(y)\,dy \\ &= \frac{1}{2\pi}\int_{-\delta}^{\delta}[f(x_0 - y) - f(x_0)]D_N(y)\,dy \\ &\quad + \frac{1}{2\pi}\int_{-\pi}^{-\delta} + \int_{\delta}^{\pi}[f(x_0 - y) - f(x_0)]D_N(y)\,dy. \end{aligned}$$

The first term will be the same for g in place of f, so it suffices to show that the second term goes to zero as $N \to \infty$. Using a trigonometric identity for $\sin(N + 1/2)y$ we compute

$$\begin{aligned} &(f(x_0 - y) - f(x_0))D_N(y) \\ &= \left(\frac{f(x_0 - y) - f(x_0)}{\sin \frac{1}{2}y}\right)\sin\left(N + \frac{1}{2}\right)y \\ &= (f(x_0 - y) - f(x_0))\sin Ny + \frac{(f(x_0 - y) - f(x_0))}{\tan\frac{1}{2}y}\cos Ny. \end{aligned}$$

When we integrate over $[\pi, -\delta]$ and $[\delta, \pi]$ we avoid the zero of $\tan y$, so the limit is 0 as $N \to \infty$ by the Riemann-Lebesgue lemma. (Actually we have to extend the Riemann-Lebesgue lemma to discontinuous functions because the integral cuts off at $y = \pm\delta$. Another way to get around this is to use a continuous cutoff of the integral, multiplying by a continuous function equal to 1 on $|y| \geq \delta$ and zero on $|y| \leq \delta/2$. We leave the details to the exercises.) QED

12.2.4 Divergence and Gibbs' Phenomenon*

We come now to our first negative result, an example of a continuous function whose Fourier series diverges at a point. The first such example was given by du Bois Reymond in 1876. The example we give is due to Fejér. In a sense these examples show that Fourier's conjecture was false. Nevertheless, the three positive results that we have established—mean-square convergence and Cesaro summability for continuous functions and uniform convergence for C^1 functions—more than compensate. In this regard we should also mention three more recent results that are beyond the scope of this work:

1. A "typical" continuous function has a Fourier series that diverges on a countable dense set of points.

2. In 1926 Kolmogorov gave an example of a function (integrable in the sense of Lebesgue) whose Fourier series diverges at every point. This function is not continuous, however.

3. In 1966 Carleson showed that for a large class of functions (L^2 integrable in the sense of Lebesgue) including all continuous functions, the Fourier series converges at "almost every" point (the set of points at which the series diverges has Lebesgue measure zero).

The first result is relatively easy to demonstrate using some functional analysis, the second result is very difficult to demonstrate, and the third result is horrendously difficult. These results should give you an inkling of the complexity and subtlety of the issues raised by Fourier's conjecture. However, the essential validity of Fourier's point of view is firmly established.

The idea of Fejér's example of a divergent Fourier series is to construct the Fourier coefficients c_n in such a way that a certain subsequence of partial sums converges uniformly (this will give us the continuous function f) but other subsequences of partial sums will be unbounded at $x = 0$. The basic building block is the sum

$$Q_{n,m}(x) = \frac{\cos mx}{n} + \frac{\cos(m+1)x}{n-1} + \frac{\cos(m+2)x}{n-2} + \cdots + \frac{\cos(m+n-1)x}{1}$$

$$-\frac{\cos(m+n+1)x}{1}-\frac{\cos(m+n+2)}{2}-\cdots-\frac{\cos(m+2n)x}{n}.$$

Notice that if we sum just the first n positive terms at $x=0$ we get $1/n+1/(n-1)+\cdots+1>\log n$ by comparison with $\log n=\int_1^n 1/t\,dt$. This sum is $S_{n+m}Q_{n,m}(0)$. On the other hand, $Q_{n,m}(0)=0$, and in fact we will show that $Q_{n,m}(x)$ is bounded by a constant independent of n and m. Thus $Q_{n,m}(x)$ is a function with the property that a certain partial sum of its Fourier series is very much larger at $x=0$ than any value of $Q_{n,m}(x)$. Once we have established this it will be a simple matter to create $f(x)$ by taking a suitable infinite linear combination of such functions.

Lemma 12.2.1 *There exists a constant c such that $|Q_{n,m}(x)|\le c$ for all n,m, and x.*

Proof: We group together the two terms

$$\frac{\cos(m+n-k)x}{k}-\frac{\cos(m+n+k)x}{k}$$

and use $\cos(m+n\pm k)x=\cos(m+n)x\cos kx\mp\sin(m+n)x\sin kx$ to obtain

$$Q_{n,m}(x)=2\sin(m+n)x\sum_{k=1}^{n}\frac{\sin kx}{k}.$$

To complete the proof we will show that $\sum_{k=1}^{n}(\sin kx)/k$ is uniformly bounded.

We use an interpretation for $\sum_{k=1}^{n}(\sin kx)/k$ in terms of integrals of the Dirichlet kernel. Recall that

$$D_n(x)=\sum_{k=-n}^{n}e^{ikx}=1+2\sum_{k=1}^{n}\cos kx$$

and so

$$\sum_{k=1}^{n}\frac{\sin kx}{k}=\int_0^x\frac{1}{2}(D_n(t)-1)\,dt=\frac{1}{2}\int_0^x D_n(t)\,dt-\frac{1}{2}x.$$

Thus we need to verify the uniform boundedness of

$$\int_0^x D_n(t)\,dt=\int_0^x\frac{\sin(n+\frac{1}{2})t}{\sin\frac{1}{2}t}\,dt.$$

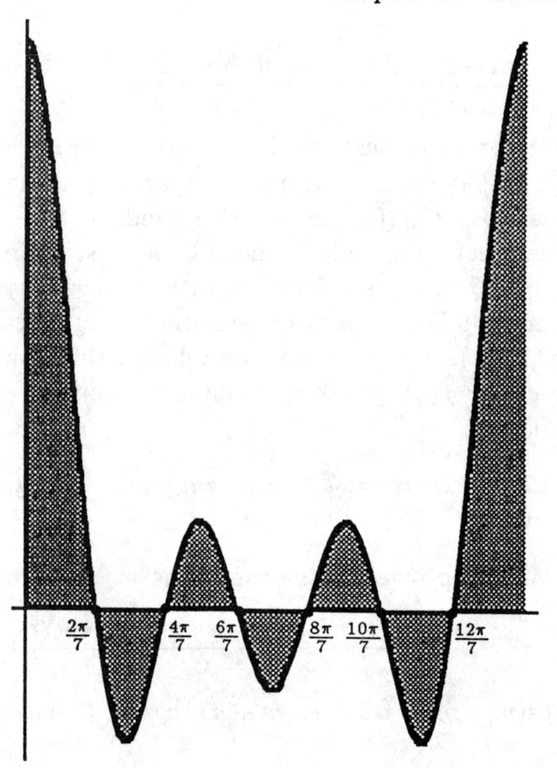

Figure 12.2.5:

This can easily be seen if we recall that the graph of $D_n(t)$ oscillates between positive and negative values, changing sign at the zeros of $\sin(n+1/2)t$,

$$t = \frac{\pi}{n+1/2}, \frac{2\pi}{n+1/2}, \dots, \frac{n\pi}{n+1/2}.$$

(See Figure 12.2.5). The function $\int_0^x D_n(t)\,dt$ thus increases on $[0, \pi/(n+1/2)]$, then decreases on $[\pi/(n+1/2), 2\pi/(n+1/2)]$, and continues to alternate between intervals of increase and decrease. Furthermore, the absolute value of

$$\int_{\frac{k\pi}{n+1/2}}^{\frac{(k+1)\pi}{n+1/2}} D_n(t)\,dt$$

decreases with k because the numerator $\sin(n+1/2)t$ is the same in absolute value and the denominator $\sin(1/2)t$ increases with k.

Thus the function $\int_0^x D_n(t)\,dt$ attains its absolute maximum at $x = \pi/(n+1/2)$. But we can easily estimate

$$\int_0^{\pi/(n+1)/2} D_n(t)\,dt \leq \frac{\pi}{n+\frac{1}{2}} \cdot 2\left(n+\frac{1}{2}\right) = 2\pi$$

because $D_n(t)$ attains its maximum value $2(n+1/2)$ at $t = 0$. QED

Concerning the functions $Q_{n,m}(x)$, we have shown $|Q_{n,m}(x)| \leq c$ for all n and m while $S_{n+m}Q_{n,m}(0) \geq \log n$. Now we select sequences $\{n_k\}$ and $\{m_k\}$ so that there is no overlap between the exponentials occurring in the different Q_{n_k,m_k}. We can accomplish this by requiring $m_k > m_{k-1} + 2n_{k-1}$ since $\cos jx$ occurs in $Q_{n,m}$ only for $m \leq j \leq m+2n$. We then choose a sequence $\{a_k\}$ of positive coefficients such that $\sum_{k=1}^{\infty} a_k$ is finite and set $f(x) = \sum_{k=1}^{\infty} a_k Q_{n_k,m_k}(x)$. The series converges uniformly by Lemma 12.2.1, so f is continuous. Also, the series that defines f is the Fourier series of f, because

$$c_j = \frac{1}{2\pi}\int_{-\pi}^{\pi} f(x)e^{-ijx}\,dx = \sum_{k=1}^{\infty} a_k \frac{1}{2\pi}\int_{-\pi}^{\pi} Q_{n_k,m_k}(x)e^{-ijx}\,dx,$$

the sum commuting with the integral because the limit is uniform. At most one of the terms $\int_{-\pi}^{\pi} Q_{n_k,m_k}(x)e^{-ijx}\,dx$ is non-zero because of the non-overlapping condition. In particular $S_{n_k+m_k}f(0) = a_k S_{n_k+m_k}f(0)$ because $S_{n_k+m_k}Q_{n_j,m_j} = Q_{n_j,m_j}$ for $j < k$ and $Q_{n_j,m_j}(0) = 0$ while $S_{n_k+m_k}Q_{n_j,m_j} \equiv 0$ for $j > k$. Thus $S_{n_k+m_k}f(0) \geq a_k \log n_k$, and if we choose a_k and n_k so that $a_k \log n_k \to \infty$ we have the divergence of the Fourier series for f at 0. A particular choice of the m_k, n_k, and a_k that meet all the above conditions is $a_k = 1/k^2$ (so $\sum a_k < \infty$), $n_k = 2^{(k^3)}$ (so $a_k \log n_k \to \infty$), and $m_k = 2^{(k^3)}$ (so $m_k > m_{k-1} + 2n_{k-1}$).

An interesting fact about the sums $\sum_{k=1}^{n}(\sin kx)/k$ is that they are the partial sums of the Fourier series of the function

$$g(x) = \begin{cases} (\pi - x)/2 & \text{if } 0 \leq x \leq \pi, \\ -(\pi - x)/2 & \text{if } -\pi \leq x < 0, \end{cases}$$

whose graph is shown in Figure 12.2.6. To see this, we compute

$$\frac{2}{\pi}\int_0^{\pi} \frac{(\pi - x)}{2} \sin kx\,dx = \frac{(\pi - x)}{\pi}\left(\frac{-\cos kx}{k}\right)\Bigg|_0^{\pi} - \frac{1}{\pi}\int_0^{\pi} \frac{\cos kx}{k}\,dx = \frac{1}{k}$$

and observe that there are no cosine terms since g is odd. Note that $g(x)$ has a jump discontinuity at $x = 0$. What is the behavior of the partial sums of the Fourier series of g in a neighborhood of this discontinuity? Of course they must approximate the jump, but in fact they do something more. At the point $x = \pi/(n + 1/2)$ we computed

$$\sum_{k=1}^{n} \frac{\sin kx}{k} = \frac{1}{2}\int_0^{\pi/(n+1/2)} D_n(t)\, dt - \frac{\pi}{2n+1}.$$

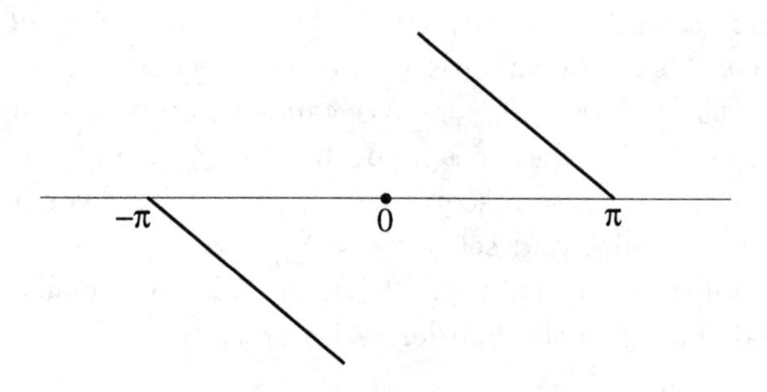

Figure 12.2.6:

Now the claim is that this value actually exceeds $\pi/2$, which is the maximum value of g, by an amount that does not go to zero as $n \to \infty$. In fact

$$\lim_{n\to\infty} \sum_{k=1}^{n} \frac{\sin kx}{k}\bigg|_{x=\frac{\pi}{n+1/2}} = \lim_{n\to\infty} \frac{1}{2}\int_0^{\frac{\pi}{n+1/2}} D_n(t) - \frac{\pi}{2n+1}$$

$$= \lim_{n\to\infty} \frac{1}{2}\int_0^{\frac{\pi}{n+1/2}} \frac{\sin(n+1/2)t}{\sin t/2}\, dt$$

$$= \lim_{n\to\infty} \frac{1}{2}\int_0^{\frac{\pi}{n+1/2}} \left(\frac{t}{\sin t/2}\right) \frac{\sin(n+1/2)t}{t}\, dt$$

$$= \lim_{n\to\infty} \int_0^{\frac{\pi}{n+1/2}} \frac{\sin(n+1/2)t}{t}\, dt = \lim_{n\to\infty} \int_0^{\pi} \frac{\sin t}{t}\, dt = \int_0^{\pi} \frac{\sin t}{t}\, dt$$

because $(1/2)t/\sin(1/2)t \to 1$ uniformly on $0 \le t \le \pi/(n+1/2)$. Now we can compute $\int_0^{\pi}(\sin t/t)\, dt \ge 1.08\,\pi/2$ by numerical integration (we can also argue that $\int_0^{\pi}(\sin t/t)\, dt > \int_0^{\infty}(\sin t/t)\, dt$ by an argument similar to the proof that $\int_0^{x} \sin(n+1/2)t/\sin(t/2)\, dt$ attains its maximum at $t = \pi/(n+1/2)$; and we compute $\int_0^{\infty}(\sin t)/t\, dt = \pi/2$ exactly, giving

$\int_0^\pi (\sin t)/t\, dt > \pi/2$). This means the partial sums of the Fourier series of $g(x)$ make a little leap before they jump across the discontinuity, as can be seen in Figure 12.2.7,

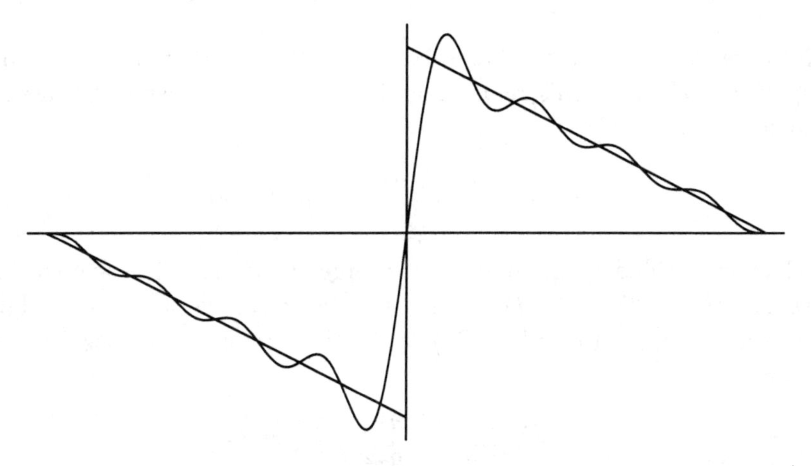

Figure 12.2.7:

and the height of the leap is independent of the number of terms in the partial sum. This is referred to as *Gibbs' phenomenon*, and it is true of any function with a jump discontinuity that is C^1 up to the jump, for such a function can be written as the sum of a C^1 function plus a variant of g. It may be regarded as a negative result because it shows that the behavior of Fourier series near jump discontinuities is worse than it has to be.

12.2.5 Solution of the Heat Equation*

To conclude this chapter on a positive note, let's return to the heat e-quation that Fourier considered and see what we can say. For simplicity of notation we choose $L = \pi$ and $c = 1$, so the p.d.e. is $\partial u/\partial t = \partial^2 u/\partial x^2$ on $0 \le x \le \pi$ with boundary conditions $u(0,t) = u(\pi,t) = 0$. We will reflect u oddly about $x = 0$, $u(-x,t) = -u(x,t)$, and then extend it to be periodic of period 2π. Thus instead of the original boundary conditions we will assume periodicity: $u(x + 2\pi, t) = u(x)$. We will assume that $u(x,t)$ is a C^2 function on $t > 0$ and that u is continuous on $t \ge 0$ with $u(x,0) = f(x)$.

For each fixed $t > 0$ we consider the Fourier series of the continuous function $u(x,t)$. The Fourier coefficients are

$$c_n(t) = (1/2\pi)\int_{-\pi}^{\pi} u(x,t)e^{-inx}\,dx.$$

Now we claim that the p.d.e. for u implies that $c_n(t)$ satisfies the o.d.e. $c_n'(t) = -n^2 c_n(t)$. This requires first differentiating inside the integral to show $c_n(t)$ is C^1 and

$$c_n'(t) = \frac{1}{2\pi}\int_{-\pi}^{\pi} \frac{\partial u}{\partial t}(x,t)e^{-inx}\,dx,$$

which is justified by the uniform convergence of the difference quotient $(u(x,t+s)e^{-inx} - u(x,t)e^{-inx})/s$ to the derivative as $s \to 0$. Then we use the p.d.e. to replace $\partial u/\partial t$ by $\partial^2 u/\partial x^2$ and integrate by parts twice:

$$\begin{aligned} c_n'(t) &= \frac{1}{2\pi}\int_{-\pi}^{\pi} \frac{\partial^2 u}{\partial x^2}(x,t)e^{-inx}dx \\ &= (-in)^2\frac{1}{2\pi}\int_{-\pi}^{\pi} u(x,t)e^{-inx}\,dx \\ &= -n^2 c_n(t), \end{aligned}$$

the boundary terms always cancelling because all the functions are periodic.

Now we know the o.d.e. has only the solutions $c_n(t) = c_n e^{-n^2 t}$ where c_n is a constant. The fact that $u(x,t)$ is continuous on the compact set $|x| \le \pi, 0 \le t \le 1$ implies that it is uniformly bounded there, so $c_n(t)$ must satisfy $|c_n(t)| \le M$ for some fixed constant M for $0 \le t \le 1$ and so $|c_n| \le M$. This means that the Fourier series $\sum_{n=-\infty}^{\infty} c_n e^{-n^2 t}e^{inx}$ converges absolutely if $t > 0$ (by comparison with $\sum e^{-n^2 t}$). We could also observe that since $u(x,t)$ is assumed differentiable, the absolute and uniform convergence $u(x,t) = \sum_{n=-\infty}^{\infty} c_n(t)e^{inx}$ has already been established. Finally we can assert that c_n are the Fourier coefficients of f because the uniform continuity of $u(x,t)$ on $0 \le t \le 1$ (by compactness) implies the uniform convergence of $u(x,t)$ to $f(x)$ as $t \to 0^+$, so

$$\begin{aligned} \frac{1}{2\pi}\int_{-\pi}^{\pi} f(x)e^{-inx}\,dx &= \lim_{t\to 0+}\frac{1}{2\pi}\int_{-\pi}^{\pi} u(x,t)e^{-inx}\,dx = \lim_{t\to 0+} c_n e^{-n^2 t} \\ &= c_n. \end{aligned}$$

Thus we have shown that if the heat equation with periodic boundary conditions and initial condition $u(x,0) = f(x)$ has a solution, then this solution is

$$u(x,t) = \sum_{n=-\infty}^{\infty} c_n e^{-n^2 t} e^{inx} \quad \text{where} \quad c_n = (1/2\pi)\int_{-\pi}^{\pi} f(x) e^{-inx}\,dx.$$

This establishes the uniqueness of the solution and justifies Fourier's formula.

But now we have to confront the question: does this formula always produce a solution? We cannot use the above argument to answer this question because we assumed the existence of the solution in deriving the formula. What we have to do is start from the formula and see if it solves the problem. In other words, we are given the continuous function $f(x)$, compute its Fourier coefficients $c_n = (1/2\pi)\int_{-\pi}^{\pi} f(x)e^{-inx}\,dx$ and then write the series

$$u(x,t) = \sum_{n=-\infty}^{\infty} c_n e^{-n^2 t} e^{inx}.$$

What can we say about it? First we observe that the sequence c_n is bounded, $|c_n| \le (1/2\pi)\int_{-\pi}^{\pi} |f(x)|\,dx$, so that the series defining $u(x,t)$ converges if $t > 0$; in fact the convergence is absolute and uniform in $t \ge \varepsilon$ for any $\varepsilon > 0$ by comparison with $\sum e^{-\varepsilon n^2}$. Thus the function $u(x,t)$ is well defined and continuous in $t > 0$, and it is clearly periodic in x. Furthermore, we can differentiate term-by-term any number of times with respect to either variable, because $\sum_{n=-\infty}^{\infty} |n|^k e^{-n^2 t}$ converges for any fixed k. This enables us to conclude that $u(x,t)$ is C^2 in $t > 0$ (it is also C^∞) and satisfies the p.d.e. by direct computation. There remains only the question of what happens as $t \to 0$. We would like to assert that $u(x,t) \to f(x)$ uniformly as $t \to 0^+$, for this will show that by setting $u(x,0) = f(x)$ we obtain a continuous function on $t \ge 0$ with the correct initial conditions.

This turns out to be more difficult to do. We do not know that $\sum_{n=-\infty}^{\infty} c_n e^{inx}$ converges in general, so the formula defining $u(x,t)$ does not necessarily make sense for $t = 0$. If we are willing to assume that f is C^1, then we know the Fourier series for f converges absolutely, $\sum |c_n| < \infty$, and from this it is straightforward to prove

$$\lim_{t\to 0+} \sum_{n=-\infty}^{\infty} c_n e^{-n^2 t} e^{inx} = \sum_{n=-\infty}^{\infty} c_n e^{inx}$$

uniformly since

$$\left|\sum_{n=-\infty}^{\infty}(c_n e^{-n^2 t} e^{inx} - c_n e^{inx})\right| \leq \sum_{n=-\infty}^{\infty} |c_n|\,|e^{-n^2 t} - 1|$$
$$\leq |e^{-t} - 1| \sum_{n=-\infty}^{\infty} |c_n|.$$

However, this hypothesis is not natural, and in fact the conclusion is true without it. The idea of the proof is very similar to the proof of Fejér's theorem. We substitute the definition of c_n in the formula for u and interchange the integral and sum to obtain

$$\begin{aligned} u(x,t) &= \sum_{n=-\infty}^{\infty} \frac{1}{2\pi}\int_{-\pi}^{\pi} f(y)e^{-iny}\,dy\, e^{-n^2 t} e^{inx} \\ &= \frac{1}{2\pi}\int_{-\pi}^{\pi} f(y)\left(\sum_{n=-\infty}^{\infty} e^{-n^2 t} e^{in(x-y)}\right) dy, \end{aligned}$$

this being justified by the uniform convergence of the sum for $t > 0$. We thus have $u(x,t)$ written as a periodic convolution operator applied to f:

$$u(x,t) = \frac{1}{2\pi}\int_{-\pi}^{\pi} f(y)H_t(x-y)\,dy$$

where the *heat kernel* H_t is given by

$$H_t(x) = \sum_{n=-\infty}^{\infty} e^{-n^2 t} e^{inx}.$$

It turns out that the heat kernel is positive and behaves like an approximate identity (with continuous parameter $t \to 0$ in place of the discrete parameter $n \to \infty$ before). This implies $u(x,t) \to f(x)$ uniformly as $t \to 0^+$. We will not give the proof of the properties of the heat kernel, since this would take us too far afield.

This example shows the utility of the Fourier series technique for the problems that gave rise to it, but it also indicates that it is by no means a trivial exercise to carry out the details of the applications.

During the eighteenth century, the concept of "function" was often confused with that of "analytic expression" or "formula". The concept of "function" that we now accept was introduced in the nineteenth century—Dirichlet is reputed to be the first to state it explicitly, although Fourier also seemed to have it in mind. On the other hand, if we accept a Fourier series as an "analytic expression", then we can close the circle of ideas: at least for periodic continuous functions, there always is an analytic expression!

12.2.6 Exercises

1. Prove there exists a constant $c > 0$ such that

$$\int_{-\pi}^{\pi} |D_N(t)|\, dt \geq c\left(1 + \frac{1}{2} + \cdots + \frac{1}{N}\right) \geq c \log N.$$

(**Hint**:

$$\int_{\frac{k\pi}{N+1/2}}^{\frac{(k+1)\pi}{N+1/2}} |D_N(t)|\, dt \geq \int_{\frac{k\pi}{N+1/2}}^{\frac{(k+1)\pi}{N+1/2}} \left|\sin\left(N + \frac{1}{2}\right) t\right| dt \Bigg)\, dt / \sin \frac{(k+1)\pi}{2N+1}.$$

2. Let f be continuous and let $f'(x_0)$ exist. Prove $S_N f(x_0) \to f(x_0)$ as $N \to \infty$.

3. Let f be Riemann integrable and continuous at the point x_0. Prove $\sigma_N f(x_0) \to f(x_0)$ as $N \to \infty$.

4. Let f be continuous except for the point x_0 where f has a jump discontinuity. Prove $\sigma_N f(x_0) \to (\lim_{x \to x_0^+} f(x) + \lim_{x \to x_0^-} f(x))/2$.

5. Let f be Riemann integrable. Prove $\int_{-\pi}^{\pi} |\sigma_N f(x) - f(x)|\, dx \to 0$ as $N \to \infty$.

6. Apply Parseval's identity to the function

$$g(x) = \begin{cases} (\pi - x)/2, & 0 \leq x \leq \pi, \\ -(\pi - x)/2, & -\pi \leq x < 0, \end{cases}$$

to evaluate $\sum_{n=1}^{\infty} 1/n^2$.

7. Compute the Fourier sine series for $\cos x$ and the Fourier cosine series for $\sin x$ on $[0, \pi]$.

8. Give a complete proof of the localization theorem.

9. Prove there exists a continuous function whose Fourier series diverges at a countable set of points. (**Hint**: if the example of a divergent Fourier series given in the text only diverges at a finite set of points, take an infinite linear combination of translates of it.)

10. If $\{f_k\}$ is a sequence of continuous functions converging uniformly to f, prove that the Fourier coefficients of f_k converge to the Fourier coefficients of f.

11. Let $\tau_y f(x) = f(x+y)$. What is the relationship between the Fourier coefficients of f and $\tau_y f$?

12. Let $f * g(x) = \int_{-\pi}^{\pi} f(x-y)g(y)\,dy$ for continuous periodic functions. What is the relationship between the Fourier coefficients of f, g, and $f * g$?

13. Prove that $|y| \le \pi |\sin y/2|$ on $|y| \le \pi$.

14. *An alternative summability method, invented by John Hubbard, uses the factors

$$A_{N,n} = \binom{2N}{N+n} \Big/ \binom{2N}{N}.$$

a. Show that

$$\sum_{n=-N}^{N} A_{N,n} e^{inx} = 2^{-2N} \binom{2N}{N} (\cos x/2)^{2N}.$$

(**Hint**: use the Euler identity for $\cos x/2$ and expand $(\cos x/2)^{2N}$ using the binomial theorem.)

b. Show that

$$2^{-2N} \binom{2N}{N} (\cos x/2)^{2N}$$

satisfies the same approximate identity properties as the Fejér kernel K_N (properties 1-3 in the proof of Fejér's theorem).

c. Conclude that $\sum_{n=-N}^{N} A_{N,n} c_n e^{inx}$ converges to f uniformly as $N \to \infty$ for any continuous, periodic function f.

15. *Let f be a C^k periodic function of period 2π for $k \geq 2$.

   a. Show that $g(y) = (f(x-y) - f(x))/\sin y/2$ is C^{k-1} and there exists M such that $|g^{(k-1)}(y)| \leq M$ for all x and y.

   b. Show that $g(y+2\pi) = -g(y)$.

   c. Show that

$$S_N f(x) - f(x) = \pm\frac{1}{2\pi}\int_{-\pi}^{\pi} g^{(k-1)}(y)\frac{\cos(N+1/2)y}{(N+1/2)^{k-1}}\,dy$$

   if k is even and a similar formula holds with cosine replaced with sine if k is odd. (**Hint**: use b to show that the boundary terms at $\pm\pi$ cancel.)

   d. Conclude that $S_N f(x) - f(x) = O(1/N^{k-1})$ uniformly as $N \to \infty$.

16. *Let $\Delta = \partial^2/\partial x^2 + \partial^2/\partial y^2$ in $\mathbb{R}^2$. For this problem you will have to use the polar coordinates representation

$$\Delta = \frac{\partial^2}{\partial r^2} + \frac{1}{r}\frac{\partial}{\partial r} + \frac{1}{r^2}\frac{\partial^2}{\partial \theta^2}$$

(section 10.2, exercise 9).

   a. Show that $r^{|k|}e^{ik\theta}$ is the unique solution of $\Delta u = 0$ in the disc $r < 1$ of the form $g(r)e^{ik\theta}$ with $g(1) = 1$. (**Hint**: the o.d.e. for g has a two-dimensional space of solutions, but some of these solutions are singular at $r = 0$.)

   b. Assuming $u(r,\theta) = \sum_{k=-\infty}^{\infty} g_k(r)e^{ik\theta}$ has a Fourier series expansion for each fixed $r < 1$ and that the series can be differentiated twice and also the limit as $r \to 1$ can be interchanged with the series, show that the Dirichlet problem $\Delta u = 0$ in $r < 1$ and $u(1,\theta) = f(0)$ has the unique solution $u(r,\theta) = \sum_{k=-\infty}^{\infty} c_k r^{|k|} e^{ik\theta}$ where c_k are the Fourier coefficients of f.

   c. Show that the solution can also be written in convolution form, $u(r,\theta) = (1/2\pi)\int_{-\pi}^{\pi} P_r(\theta - \varphi)f(\varphi)\,d\varphi$ where $P_r(\theta) = \sum_{k=-\infty} r^{|k|}e^{ik\theta}$.

d. Evaluate the infinite series for $P_r(\theta)$ to obtain

$$P_r(\theta) = \frac{1 - r^2}{1 - 2r\cos\theta + r^2}.$$

Note that $P_r(\theta - \varphi) = (1 - |A|^2)/|A - B|^2$ where $A = re^{i\theta}$ and $B = e^{i\varphi}$.

e. Show by direct differentiation that $\Delta u = 0$ if

$$u = (1/2\pi) \int_{-\pi}^{\pi} P_r(\theta - \varphi) f(\varphi)\, d\varphi$$

for any continuous function f on the circle.

f. Show by an approximate identity argument that $u(r, \theta) \to f(\theta)$ uniformly as $r \to 1$ if f is continuous.

12.3 Summary

12.1 Origins of Fourier Series

Theorem *The solutions to the eigenvalue problem $d^2\psi/dx^2(x) = \lambda\psi(x)$ on $0 \le x \le L$ with $\psi(0) = \psi(L) = 0$ (Dirichlet boundary conditions) are multiples of the functions $\sin(k\pi/L)x, k = 1, 2, \ldots$, with $\lambda = -k^2\pi^2/L^2$.*

Theorem *If $f(x) = \sum_{k=1}^{\infty} b_k \sin(k\pi/L)x$ with uniform convergence on $[0, L]$, then $b_k = (2/L)\int_0^L f(x)\sin(k\pi/L)x\, dx$. Similarly, if $f(x) = a_0/2 + \sum_{k=1}^{\infty} a_k \cos(k\pi/L)x$ with uniform convergence on $[0, L]$, then $a_k = (2/L)\int_0^L f(x)\cos(k\pi/L)x\, dx$. If*

$$f(x) = \frac{a_0}{2} + \sum_{k=1}^{\infty}\left(a_k \cos\frac{k\pi}{L}x + b_k \sin\frac{k\pi}{L}x\right)$$

with uniform convergence on $[-L, L]$, then $a_k = (1/L)\int_{-L}^{L} f(x)\cos(k\pi/L)x$ and $b_k = (1/L)\int_{-L}^{L} f(x)\sin(k\pi/L)x$, and in this case we also have $f(x) = \sum_{n=-\infty}^{\infty} c_n e^{i(n\pi/L)x}$ with $c_n = (1/2L)\int_{-L}^{L} f(x)e^{-i(n\pi/L)x}\, dx$.

Theorem *The solutions to the character identity $\psi(x+y) = \psi(x)\psi(y)$ that are C^1 and periodic of period $2L$ are the functions $e^{i(n\pi/L)x}$, for n an integer.*

Theorem *The two-dimensional real vector space of linear combinations of $\cos(k\pi/L)x$ and $\sin(k\pi/Lk)x$ is preserved under translation.*

12.2 Convergence of Fourier Series

Definition *If $f(x)$ is periodic of period 2π, the Fourier coefficients c_n are defined by $c_n = (1/2\pi)\int_{-\pi}^{\pi} f(x)e^{-inx}\,dx$ for n an integer and the formal series $\sum_{-\infty}^{\infty} c_n e^{inx}$ is called the Fourier series of f. The partial sums $S_N f(x)$ of the Fourier series are defined by $S_N f(x) = \sum_{n=-N}^{N} c_n e^{inx}$. The Fourier series is said to converge (pointwise, uniformly, or absolutely), if $S_N f(x) \to f(x)$ in the desired sense.*

Lemma *If f is periodic of period 2π, then $\int_{-\pi}^{\pi} f(x)\,dx = \int_a^{a+2\pi} f(x)\,dx$ for any a.*

Theorem $S_N f(x) = (1/2\pi)\int_{-\pi}^{\pi} f(y)D_N(x-y)\,dy$ *where D_N is the Dirichlet kernel*

$$D_N(t) = \sum_{n=-N}^{N} e^{int} = \frac{\sin(N+1/2)t}{\sin t/2}.$$

Theorem 12.2.1 and 12.2.2 *If f is periodic of period 2π and C^1, then the Fourier series converges uniformly to f and the rate of convergence is $O(N^{-1/2})$. If f is C^2, then the rate of convergence is $O(N^{-1})$.*

Definition *The Fourier series of f is said to be Cesaro summable to f if $\sigma_N f$ converges to f, where*

$$\sigma_N f(x) = \sum_{n=-N}^{N} \frac{N+1-|n|}{N+1} c_n e^{inx} = \frac{1}{N+1}(S_0 f(x) + \cdots + S_N f(x)).$$

Theorem $\sigma_N f(x) = (1/2\pi)\int_{-\pi}^{\pi} f(x-y)K_N(y)\,dy$ *where the Fejér ker-*

nel $K_N(y)$ is given by

$$K_N(y) = \frac{1}{N+1}\sum_{n=0}^{\infty} D_n(y) = \frac{1}{N+1}\left(\frac{\sin(\frac{N+1}{2})\, y}{\sin y/2}\right)^2.$$

Theorem 12.2.3 (*Fejér*) *If f is continuous and periodic of period 2π, then the Fourier series of f is uniformly Cesaro summable to f.*

Theorem (*Uniqueness of Fourier Series*) *If f and g are continuous periodic functions with the same Fourier series, then $f = g$.*

Theorem 12.2.4 (*Projection*) *Let $v_1, \ldots, v_N$ be orthonormal vectors in an inner product space V with span W. Then $Pu = \sum_{k=1}^{N}\langle u, v_n\rangle v_n$ for any vector u in V is the orthogonal projection of u onto W in the sense that Pu is the unique vector in W such that $u - Pu \perp W$. Furthermore Pu is the unique minimizer in W of the distance to u and $d(u, Pu)^2 = ||u||^2 - \sum_{n=1}^{N} |\langle u, v_n\rangle|^2$; hence, we have Bessel's inequality $\sum_{n=1}^{N} |\langle u, v_n\rangle|^2 \leq ||u||^2$.*

Definition *A trigonometric polynomial of degree N is a function of the form $\sum_{n=-N}^{N} a_n e^{inx}$.*

Corollary 12.2.1 (*Mean-Square Convergence of Fourier Series*) *Let f be continuous and periodic. Then $S_N f$ minimizes the L^2 distance to f among all trigonometric polynomials of degree N and*

$$(1/2\pi)\int_{-\pi}^{\pi} |f(x) - S_N f(x)|^2\, dx \to 0 \text{ as } N \to \infty.$$

Furthermore, Parseval's identity $\sum_{n=-\infty}^{\infty} |c_n|^2 = (1/2\pi)\int_{-\pi}^{\pi} |f(x)|^2\, dx$ holds or, more generally,

$$\sum_{n=-\infty}^{\infty} c_n(f)\overline{c_n(g)} = \frac{1}{2\pi}\int_{-\pi}^{\pi} |f(x)\overline{g(x)}\, dx$$

for two such functions.

Lemma (*Riemann-Lebesgue*) *If f is continuous and periodic, $\lim_{n\to\pm\infty} c_n = 0$.*

Theorem (*Completeness of Trigonometric System*) *The functions* e^{inx} *as* n *varies over the integers form an orthonormal set with respect to the inner product* $\langle f, g\rangle = (1/2\pi)\int_{-\pi}^{\pi} f(x)\overline{g(x)}\,dx$, *which is complete in the sense that no other non-zero continuous function is orthogonal to all* e^{inx}.

Theorem *The mean-square convergence of Fourier series does not depend on the order of the terms; i.e.,*

$$\lim_{n\to\infty} (1/2\pi)\int_{-\pi}^{\pi} |f(x) - P_N f(x)|^2\,dx = 0$$

if $P_N f(x) = \sum_{A_N} c_n e^{inx}$ *where* A_N *is any increasing sequence of subsets of integers such that* $\bigcup_{N=1}^{\infty} A_N$ *is all integers.*

Theorem *If* f *is* C^1 *and periodic with Fourier series* $\sum c_n e^{inx}$, *then* f' *has Fourier series* $\sum inc_n e^{inx}$.

Theorem *If* f *is* C^1 *and periodic its Fourier series converges absolutely.*

Theorem 12.2.5 (*Localization*) *The convergence or divergence of the Fourier series of a continuous function at a point* x_0 *depends only on the function on any neighborhood of* x_0.

Lemma *The trigonometric polynomials*

$$\begin{aligned} Q_{n,m}(x) &\\ = \quad & \frac{\cos mx}{n} + \frac{\cos(m+1)x}{n-1} + \cdots \\ & + \frac{\cos(m+n-1)x}{1} - \frac{\cos(m+n+1)x}{1} \\ & - \frac{\cos(m+n+2)x}{2} - \cdots - \frac{\cos(m+2n)x}{n} \end{aligned}$$

are uniformly bounded for all x, n, m, *but* $S_{n+m}Q_{n,m}(0) > \log n$.

Example *For the appropriate choice of sequence* a_k, n_k, m_k (*e.g.,* $a_k = k^{-2}, n_k = m_k = 2^{k^3}$) *the function* $\sum_{k=1}^{\infty} a_k Q_{n_k,m_k}$ *is continuous but has divergent* (*in fact unbounded*) *Fourier series at* $x = 0$.

Example (*Gibbs' Phenomena*) *Let*

$$g(x) = \begin{cases} (\pi - x)/2 & \text{if } 0 \leq x \leq \pi, \\ -(\pi + x)/2 & \text{if } -\pi \leq x < 0. \end{cases}$$

Then g has a jump discontinuity at $x = 0$ with jump π, but the partial sums of the Fourier series $S_N g$ jump by more than 1.08π between $\pm\frac{\pi}{N+1/2}$ for all N.

Theorem (*Fourier*) *If $u(x,t)$ is a C^2 function for $t > 0$, periodic of period 2π in x, satisfying the heat equation $\partial u/\partial t = \partial^2 u/\partial x^2$ together with the initial condition $\lim_{t\to 0} u(x,t) = f(x)$ uniformly, then*

$$(*) \qquad u(x,t) = \sum_{n=-\infty}^{\infty} c_n e^{-n^2 t} e^{inx}$$

where c_n are the Fourier coefficients of f. Conversely, for every C^1 function f (actually continuous will suffice), $(*)$ solves the heat equation and initial conditions.

Chapter 13

Implicit Functions, Curves, and Surfaces

13.1 The Implicit Function Theorem

13.1.1 Statement of the Theorem

It frequently happens that a function is not given explicitly as $y = f(x)$ but rather implicitly as the solution to an equation $F(x, y) = 0$. In this section we will describe the Implicit Function Theorem, which gives conditions under which such equations actually do define functions. This is an extremely useful and important theorem, but it is also quite subtle. We have already seen an important case, involving inverse functions. If $y = f(x)$ is given explicitly, the inverse function $f^{-1}(x)$ is defined implicitly as the solution of the equation $x = f(y)$, which we can write $x - f(y) = 0$. Note that this is of the form $F(x, y) = 0$ for $F(x, y) = x - f(y)$. We have proven an inverse function theorem in one dimension, and we exploited it in the definition of the sine and cosine. We will prove an n-dimensional version as a corollary of the Implicit Function Theorem.

Another situation in which implicit functions arise naturally is in some of the so-called exact (or cookbook) solutions of o.d.e.'s. For example, the o.d.e.

$$M(x, y)\frac{dy}{dx} + N(x, y) = 0$$

(often written $M(x,y)dy + N(x,y)dx$) is called *exact* if there exists $F(x,y)$ such that $M = \partial F/\partial y$ and $N = \partial F/\partial x$ (this will happen if $\partial M/\partial x = \partial N/\partial y$ and the region of definition is reasonable—see the exercises in section 11.3). Even if the o.d.e. is not exact, it can often be made exact by multiplying by an appropriate factor. Once the o.d.e. is exact, it is equivalent to $(\partial/\partial x)F(x, y(x)) = 0$ and, hence, $F(x, y(x)) = c$ for some constant c. Thus all solutions of the o.d.e. are solutions of the implicit function equation $F(x,y) - c = 0$. In order for this procedure to produce a solution to the o.d.e. we have to know how to solve the implicit equation.

A third natural situation in which implicit equations arise is in the study of curves and surfaces. We frequently describe such geometric objects as the solution sets of certain equations. For instance, the unit sphere in $\mathbb{R}^3$ is the solution set for $F(x,y,z) = 0$ where $F(x,y,z) = x^2 + y^2 + z^2 - 1$. We can think of this sphere as a two-dimensional surface because we can solve for one of the variables in terms of the others—at least in portions of the sphere, say $z = -\sqrt{1 - x^2 - y^2}$ in the lower hemisphere $z < 0$. Thus x and y provide a pair of coordinates for this hemisphere. But if the function F were more complicated, say

$$x^{27}y^5 - z^{16}x^9 + x^4y^4z^4 - 8xyz - 1,$$

so that we could not solve $F(x,y,z) = 0$ explicitly, could we still maintain that the solution set is a two-dimensional surface? We will see in a later section that the implicit function theorem is the perfect tool for answering this question.

The basic idea of the differential calculus is that if you have a question about a general function in a neighborhood of a point, ask the same question about the best affine approximation at that point and the answers should be qualitatively the same. This perspective will lead us to the correct statement of the implicit function theorem and will give us a hint at how to prove it. Notice that this approach limits us to a local theorem. This turns out to be just right because simple examples show there are no nice global theorems. In this respect the one-dimensional inverse function theorem is very misleading because there is a nice global result.

We want to consider a system of m equations for m unknown functions $y_1, \ldots, y_m$ (each being a function of n variables $x_1, \ldots, x_n$). We will write $y = (y_1, \ldots, y_m)$ and $x = (x_1, \ldots, x_n)$, and we will write the equations $F(x, y) = 0$ where F is a function defined in an open set in $\mathbb{R}^{n+m}$ taking values in $\mathbb{R}^m$. Notice that we are prejudging the outcome that the number of equations and unknowns should be equal if there is to be any hope of having unique solutions. We are looking for a local theorem, so we consider fixed points $\tilde{x}$ and $\tilde{y}$ that give one particular solution, $F(\tilde{x}, \tilde{y}) = 0$, and we ask if there is a way to define functions $y_1, \ldots, y_m$ of x (or $y(x)$ taking values in $\mathbb{R}^m$) in a neighborhood of $\tilde{x}$ so that $y(\tilde{x}) = \tilde{y}$ and $F(x, y(x)) = 0$ for all x in the domain of y. Because we want to use differential calculus we make the hypothesis that F be C^1 and we also demand that the solution functions $y(x)$ be C^1.

If we were to replace F by its best affine approximation at $(\tilde{x}, \tilde{y})$, then we would be looking at the system

$$\sum_{k=1}^{n} \frac{\partial F_j(\tilde{x}, \tilde{y})}{\partial x_k}(x_k - \tilde{x}_k) + \sum_{k=1}^{m} \frac{\partial F_j(\tilde{x}, \tilde{y})}{\partial y_k}(y_k - \tilde{y}_k) = 0$$

for $j = 1, \ldots, m$, since $F(\tilde{x}, \tilde{y}) = 0$. We abbreviate this system as

$$F_x(\tilde{x}, \tilde{y})(x - \tilde{x}) + F_y(\tilde{x}, \tilde{y})(y - \tilde{y}) = 0$$

where F_x stands for the $m \times n$ matrix

$$\left\{\frac{\partial F_j}{\partial x_k}\right\}, \quad 1 \le j < m, \ 1 \le k \le n,$$

and F_y stands for the $m \times m$ matrix

$$\left\{\frac{\partial F_j}{\partial y_k}\right\}, \quad 1 \le j \le m, \ 1 \le k \le m.$$

This is a system of affine linear (more commonly called inhomogeneous linear) equations for y, $Ay = b(x)$, where A is the $m \times m$ matrix $F_y(\tilde{x}, \tilde{y})$ and $b(x)$ is the vector $F_y(\tilde{x}, \tilde{y})\tilde{y} - F_x(\tilde{x}, \tilde{y})(x - \tilde{x})$. The implicit function theorem says essentially that if the affine linear sytem has a unique solution, then so does the original problem.

At this point we need to review some basic facts from linear algebra concerning matrix equations $Ay = b$. An $m \times m$ matrix is called *invertible* (or *nonsingular*) if there exists an $m \times m$ matrix B with $AB = I$, I denoting the identity matrix. If such a matrix exists it is unique, denoted A^{-1}, and it is a two-sided inverse, $A^{-1}A = AA^{-1} = I$.

If A^{-1} exists, then $y = A^{-1}b$ in the unique solution of $Ay = b$; and conversely if $Ay = b$ has a unique solution for each b, then A^{-1} exists. In fact if $Ay = 0$ has the unique solution $y = 0$ (otherwise said, zero is not an eigenvalue of A), then A is invertible. Finally there is the determinant criterion: A is invertible if and only if $\det A \neq 0$, in which case A^{-1} may be explicitly given by Cramer's rule in terms of ratios of determinants of $(m-1) \times (m-1)$ submatrices of A. This is not a particularly efficient method for computing A^{-1}, however.

One crucial fact that we will need is that the set of invertible matrices is an open set in $\mathbb{R}^{m \times m}$; if A is invertible, then any matrix sufficiently close to A is also invertible. One way to see this is via the determinant criterion: the set of matrices where $\det A \neq 0$ is the inverse image under the continuous function $\det : \mathbb{R}^{n \times n} \to \mathbb{R}^1$ of the open set $\mathbb{R}^1 \backslash \{0\}$. However, we prefer a different proof that is more in the spirit of what we will be doing.

Let us write a matrix close to A as $A + B$, where B is a small matrix. The most convenient measure of the "size" of a matrix M is the *norm* $||M||$, which is defined to be the least constant c such that $|Mx| \leq c|x|$ for all x in $\mathbb{R}^m$ where $|x|$ denotes the Euclidean norm (*warning*: $||M||$ is not the same as the Euclidean norm on $\mathbb{R}^{n \times n}$). We will frequently use the estimate $|Mx| \leq ||M||\,|x|$ in this section. Here we want to point out that every entry M_{jk} is dominated by the norm, $|M_{jk}| \leq ||M||$ (take $x = (0, \ldots, 0, 1, 0, \ldots, 0)$ with the 1 in the kth place, so $|x| = 1$ and $(Mx)_j = M_{jk}$; hence, $|M_{jk}| = |(Mx)_j| \leq |Mx| \leq ||M||\,|x| = ||M||$). On the ther hand $||M|| \leq (\sum_{j,k} M_{jk}^2)^{1/2}$ since $|Mx|^2 = \sum_j (\sum_k M_{jk} x_k)^2 \leq \sum_j (\sum_k M_{jk}^2 \sum_k x_k^2) = (\sum_{j,k} M_{jk}^2)|x|^2$ by the Cauchy-Schwartz inequality. Thus *the norm of a matrix is small if and only if all its entries are small.*

Now we will write a perturbation series for $(A+B)^{-1}$. We have

$$A + B = (I + BA^{-1})A,$$

so

$$\begin{aligned} (A+B)^{-1} &= A^{-1}(I + BA^{-1})^{-1} \\ &= A^{-1} \sum_{k=0}^{\infty} (-1)^k (BA^{-1})^k \\ &= A^{-1} - A^{-1}BA^{-1} + A^{-1}BA^{-1}BA^{-1} - \cdots . \end{aligned}$$

We need to verify that the infinite series converges, for then it is a simple matter to multiply by $A+B$ and get the identity matrix. But this is easy to show if $||BA^{-1}|| < 1$. In that case the series converges geometrically, so it gives an efficient method of computing $(A+B)^{-1}$ if A^{-1} is known. For the details, see the exercises. Note that we have actually established a quantitative result: *if A is invertible and $||BA^{-1}|| < 1$, then $A+B$ is invertible.* This method works in great generality, even in infinite-dimensional problems where there is no determinant.

We will also need to know that A^{-1} is a continuous function of the entries of A. This follows from Cramer's rule, and we can also deduce it from the perturbation series (see exercise set 13.1.3, number 2).

Let us now return to the general implicit equation $F(x,y) = 0$ and the linearized version

$$F_x(\tilde{x},\tilde{y})(x-\tilde{x}) + F_y(\tilde{x},\tilde{y})(y-\tilde{y}) = 0$$

near the point $(\tilde{x},\tilde{y})$. The condition we want in order that the linearized version have a unique solution is that $F_y(\tilde{x},\tilde{y})$ be invertible. This condition then implies that the equation $F(x,y) = 0$ has a local solution.

Theorem 13.1.1 (*Implicit Function Theorem*) *Let $F(x,y)$ be a C^1 function defined in a neighborhood of $\tilde{x}$ in $\mathbb{R}^n$ and $\tilde{y}$ in $\mathbb{R}^m$ taking values in $\mathbb{R}^m$, with $F(\tilde{x},\tilde{y}) = c$. Then if $F_y(\tilde{x},\tilde{y})$ is invertible there exists a neighborhood U of $\tilde{x}$ and a C^1 function $y : U \to \mathbb{R}^m$ such that $y(\tilde{x}) = \tilde{y}$ and $F(x,y(x)) = c$ for every x in U. Furthermore, y is unique in that there exists a neighborhood V of $\tilde{y}$ (V is the image $y(U)$) such that there is only one solution z in V of $F(x,z) = c$, namely $z = y(x)$. Finally, the differential of y can be computed by implicit differentiation as*

$$dy(x) = -F_y(x,y(x))^{-1}F_x(x,y(x)).$$

A special case is the following:

Theorem 13.1.2 (*Inverse Function Theorem*) *Let f be a C^1 function defined in a neighborhood of $\tilde{y}$ in $\mathbb{R}^n$ taking values in $\mathbb{R}^n$. If $df(\tilde{y})$ is invertible, then there exists a neighborhood U of $\tilde{x} = f(\tilde{y})$ and a C^1 function $g : U \to \mathbb{R}^n$ such that $f(g(x)) = x$ for every x in U. Furthermore g maps U one-to-one onto a neighborhood V of $\tilde{y}$ and*

$g(f(y)) = y$ for every y in V. The function g is unique in that for any x in U there is only one z in V with $f(z) = x$, namely $z = g(x)$. Finally $dg(x) = df(y)^{-1}$ if $f(y) = x$.

The inverse function theorem is a special case of the implicit function theorem with $F(x, y) = f(y) - x$ and $c = 0$. Notice that $F_y(x, y) = df(y)$, so the hypotheses are the same. There is also a trick for reducing the implicit function theorem to the inverse function theorem: given $F(x, y) = c$ we construct $f : U \to \mathbb{R}^{n+m}$, where U is a neighborhood of $(\tilde{x}, \tilde{y})$ in $\mathbb{R}^{n+m}$, by $f(x, y) = (x, F(x, y))$. Then

$$df(x, y) = \begin{pmatrix} I & 0 \\ F_x & F_y \end{pmatrix},$$

and this is invertible with inverse

$$\begin{pmatrix} I & 0 \\ -F_y^{-1}F_x & F_y^{-1} \end{pmatrix}$$

if and only if F_y is invertible. Thus the hypothesis of the implicit function theorem for F implies the hypothesis of the inverse function theorem for f. If g is the local inverse to f, then $g(x, c) = (x, y(x))$ where $y(x)$ is the solution to $F(x, y(x)) = c$. This is the way the implicit function theorem is proved in most texts. However, we will give a direct proof with an explicit algorithm for approximating the solution. Notice that in the above reduction the dimension was increased, so in order to have the one-dimensional implicit function theorem ($n = m = 1$) we need the two-dimensional inverse function theorem.

Before beginning the proof let's look at some examples.

Example 13.1.1 Consider the equation $x^2 + y^2 = 1$ of the unit circle in $\mathbb{R}^2$, as shown in Figure 13.1.1. Here $F(x, y) = x^2 + y^2$ and $F_y(x, y) = 2y$. This 1×1 matrix is invertible if and only if $y \neq 0$. Thus according to the implicit function theorem, at any point $(\tilde{x}, \tilde{y})$ on the circle except $(1, 0)$ and $(-1, 0)$, we can locally solve for y as a function of x. Of course there are two possible solutions, $y = \sqrt{1 - x^2}$ and $y = -\sqrt{1 - x^2}$, and the point $(\tilde{x}, \tilde{y})$ will determine which of these it is. At the points $(\pm 1, 0)$ this breaks down, as y can't be defined for $x < -1$ or $x > -1$ and the function $y = \pm\sqrt{1 - x^2}$ does not even have a one-sided derivative at $x = \pm 1$.

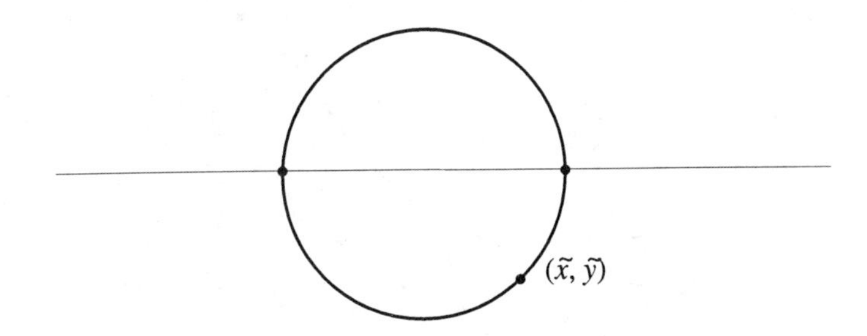

Figure 13.1.1:

Example 13.1.2 We consider what is essentially the function z^2 on $\mathbb{C}$. In terms of real variables alone this is the function $f : \mathbb{R}^2 \to \mathbb{R}^2$ given by $f(x, y) = (x^2 - y^2, 2xy)$. We would like to apply the inverse function theorem to this function. We compute

$$df(x,y) = \begin{pmatrix} 2x & -2y \\ 2y & 2x \end{pmatrix}$$

and det $df(x, y) = 4x^2 + 4y^2$, so $df(x, y)$ is invertible if $(x, y) \neq (0, 0)$. Thus the inverse function theorem says that there exist local complex square roots. But even if we cut away a neighborhood of the origin, we cannot define a global complex square root because the mapping f is not globally one-to-one. Indeed $f(x, y) = f(-x, -y)$. We can visualize the mapping f as follows: cut the plane along the positive real axis, then wrap it twice around (simultaneously stretching $r \to r^2$), and then glue it together again. This example shows that the local nature of the conclusion of the theorems is inherent in the situation.

13.1.2 The Proof*

We come now to the proof. The idea we use can be traced directly back to Isaac Newton, who used it to locate zeros of functions. *Newton's method* for solving $f(x) = 0$ is to obtain x as a limit of a sequence $x_0, x_1, \ldots$. The first value x_0 is chosen close to a root of $f(x) = 0$. We then replace f by its best affine approximation $f(x_0) + f'(x_0)(x - x_0)$, set this equal to zero, and solve $x = x_0 - f'(x_0)^{-1} f(x_0)$. The point we obtain is the intersection of the tangent line to f at x_0 with the x-axis, as shown in Figure 13.1.2.

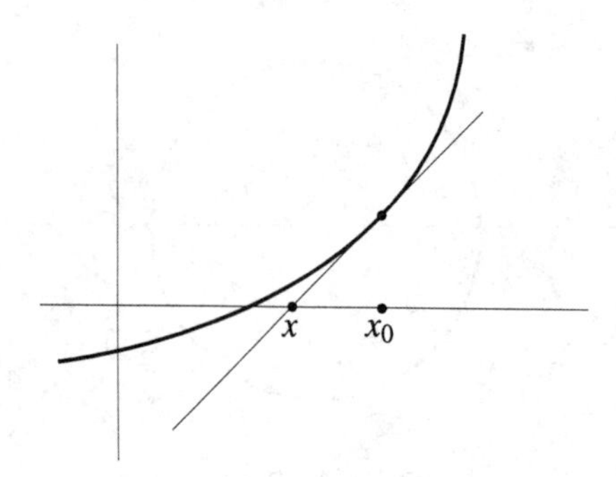

Figure 13.1.2:

This is not the solution to $f(x) = 0$ in general, but we take it to be x_1. Iterating this process, we define $x_n = x_{n-1} - f'(x_{n-1})^{-1} f(x_{n-1})$. For this to make sense we need f' to be non-zero at all the points considered. Notice also that the iterative equation for x_n in terms of x_{n-1} can be written concisely as $x_n = Tx_{n-1}$ where T is the transformation $Tx = x - f'(x)^{-1} f(x)$. We see that $f(x) = 0$ if and only if x is a fixed point of T. This suggests that we should apply the Contractive Mapping Principle. For technical reasons it will be easier to analyse the transformation $\tilde{T}x = x - f'(x_0)^{-1} f(x)$. Notice that it is also true of $\tilde{T}$ that we have $\tilde{T}x = x$ if and only if $f(x) = 0$. Furthermore, if f' doesn't vary too much the difference between T and $\tilde{T}$ will be slight. We will not carry this discussion of Newton's method further, since we intend it only as motivation for the proof.

Here is an outline of the proof:

1. For each point x in U we construct a map $T : V \to V$ with the property that $F(x, y) = c$ if and only if $Ty = y$. This mapping T is a slight variant of the one suggested by Newton's method.

2. We verify the hypotheses of the Contractive Mapping Principle. The fixed point, $y(x)$, is the implicitly defined function.

3. We show that $y(x)$ is differentiable and the derivative at $\tilde{x}$ is obtained by differentiating the identity $F(x, y(x)) = c$. This part of the proof only uses the identity $F(x, y(x)) = c$ itself and does not refer to the way we obtained $y(x)$.

4. By varying $\tilde{x}$, we establish $dy(x) = -F_y(x,y)^{-1}F_x(x,y(x))$ for x in a neighborhood of $\tilde{x}$. This formula shows that y is C^1.

There are unfortunately several small technical difficulties that make the proof long and pedantic. We mention one at the onset: the sets U and V in the implicit function theorem are required to be open sets, while the Contractive Mapping Principle requires a complete metric space and, hence, closed sets. (Actually this is only a problem for V, since U is in the parameter space.) To get around this mismatch, we have to consider both V and its closure $\bar{V}$. If we can show $T : \bar{V} \to V$ and is contractive on $\bar{V}$, then we can apply the Contractive Mapping Theorem to $\bar{V}$. This gives a unique fixed point in $\bar{V}$, but the fixed point must actually lie in V since T maps into V. Thus we may conclude the existence of a unique fixed point in V.

Proof of the Implicit Function Theorem: We want the solution y to $F(x,y) = c$ to be the limit of successive approximations obtained by iterating a contractive mapping. (The mapping will depend on x, which we treat as a constant.) Avoiding subscripts, let us suppose we have found y for which $F(x,y) \approx c$ and we want a better approximation z. Replacing F by its best affine approximation at (x,y), which is $F(x,y) + F_y(x,y)(z-y)$ (since we don't vary x, there is no F_x term), and setting this equal to c, we can solve $F(x,y)+F_y(x,y)(z-y) = c$ to obtain $z = y+F_y(x,y)^{-1}(c-F(x,y))$. Since our invertibility hypothesis concerned $F_y(\tilde{x},\tilde{y})$ and (x,y) is a nearby point, we will simplify this to $z = y + F_y(\tilde{x},\tilde{y})^{-1}(c - F(x,y))$. Thus we are led to consider the mapping $Ty = y + F_y(\tilde{x},\tilde{y})^{-1}(c - F(x,y))$, which is well defined since $F_y(\tilde{x},\tilde{y})^{-1}$ is assumed to exist, and has the property that $Ty = y$ if and only if $F(x,y) = c$. We would like to find a neighborhood V of $\tilde{y}$, say $V = \{y : |y - \tilde{y}| < \delta\}$, on which T is a contractive mapping; for then we will know that T has a unique fixed point and, hence, $F(x,y) = c$ has a unique solution for y in V. Presumably we will have to restrict x to lie in a neighborhood of $\tilde{x}$ in order for this to work, say $U = \{x : |x - \tilde{x}| < \varepsilon\}$.

In order to verify the amended hypotheses of the Contractive Mapping Principle we need to show that T maps $\bar{V}$ to V (y in $\bar{V}$ implies Ty is in V) and that $|Ty - Tz| \leq \rho|y - z|$ for some $\rho < 1$ for y and z in $\bar{V}$. We are free to take δ and ε as small as needed to obtain these results. First we show that T satisfies the contraction property. Since $F_y(\tilde{x}, \tilde{y})$ is invertible, we can write

$$\begin{aligned} Ty - Tz &= y - z + F_y(\tilde{x}, \tilde{y})^{-1}(F(x, z) - F(x, y)) \\ &= F_y(\tilde{x}, \tilde{y})^{-1}[F(x, z) - F(x, y) - F_y(\tilde{x}, \tilde{y})(z - y)]. \end{aligned}$$

Now the factor $F_y(\tilde{x}, \tilde{y})^{-1}$ is a fixed matrix, so if we can control the term in brackets it will only increase the value of ρ. The term in brackets looks very much like a comparison of $F(x, z)$ with the best affine approximation to F at (x, y), the only difference being that we are evaluating F_y at the wrong point. However, F_y is continuous so that should not be too serious.

Now if we were in one dimension we could use the mean value theorem on $F(x, z) - F(x, y)$, but in the general case we have to resort to a more technical argument involving the fundamental theorem of the calculus applied to the line segment joining y and z. We parametrize this segment $y + t(z - y)$ with $0 \le t \le 1$ and consider the restriction $h(t) = F(x, y + t(z - y))$ of F to this line (x remains constant throughout). Now $F(x, z) - F(x, y) = h(1) - h(0)$ and $h(1) - h(0) = \int_0^1 h'(t)\, dt$ by the fundamental theorem, while $h'(t) = F_y(x, y + t(z - y))(z - y)$ by the chain rule. Altogether $F(x, z) - F(x, y) = \int_0^1 F_y(x, y + t(z - y))(z - y)\, dt$, which enables us to compare it with $F_y(\tilde{x}, \tilde{y})(z - y)$ by writing

$$(*) \qquad \begin{aligned} &F(x, z) - F(x, y) - F_y(\tilde{x}, \tilde{y})(z - y) \\ &= \int_0^1 (F_y(x, y + t(z - y)) - F_y(\tilde{x}, \tilde{y}))(z - y)\, dt \end{aligned}$$

where we have used the linearity of matrix multiplication and the fact that $\int_0^1 dt = 1$ to bring $F_y(\tilde{x}, \tilde{y})$ inside.

Now we use the hypothesis that F_y is continuous. By taking ε small enough we can make x close to $\tilde{x}$, and by taking δ small enough we can make y and z close to $\tilde{y}$; hence, all points $y + t(z - y)$ on the line segment joining them will be close to $\tilde{y}$. By doing this we can make $F_y(x, y + t(z - y))$ close to $F_y(\tilde{x}, \tilde{y})$, say

$$||F_y(x, y + t(z - y)) - F_y(\tilde{x}, \tilde{y})|| \le \lambda$$

for any preassigned λ. Thus

$$|(F_y(x, y + t(z - y)) - F_y(\tilde{x}, \tilde{y}))(z - y)| \le \lambda |z - y|;$$

hence, by (*) and Minkowski's inequality (see section 11.1.2)

$$|F(x,z) - F(x,y) - F_y(\tilde{x},\tilde{y})(z-y)| \leq \int_0^1 \lambda|z-y|\,dt = \lambda|z-y|.$$

So

$$|Ty - Tz| = |F_y(\tilde{x},\tilde{y})^{-1}[F(x,z) - F(x,y) - F_y(\tilde{x},\tilde{y})(z-y)]| \leq M\lambda|z-y|$$

where $M = ||F_y(\tilde{x},\tilde{y})^{-1}||$ is a fixed constant. Thus by choosing λ so that $\rho = M\lambda < 1$ and ε and δ accordingly we obtain the contractive estimate for T and $\bar{V}$.

The argument that T maps $\bar{V}$ into V is quite similar. We need to show that for δ small enough, $|y - \tilde{y}| \leq \delta$ implies $|Ty - \tilde{y}| < \delta$. Now

$$\begin{aligned} Ty - \tilde{y} &= y - \tilde{y} + F_y(\tilde{x},\tilde{y})^{-1}(c - F(x,y)) \\ &= F_y(\tilde{x},\tilde{y})^{-1}[F(\tilde{x},\tilde{y}) - F(x,y) + F_y(\tilde{x},\tilde{y})(y-\tilde{y})] \end{aligned}$$

since $c = F(\tilde{x},\tilde{y})$. Again the terms in brackets resemble the comparison of $F(x,y)$ with the best affine approximation to F at the point $(\tilde{x},\tilde{y})$. We write it as

$$\begin{aligned} &[F(\tilde{x},\tilde{y}) + F_x(\tilde{x},\tilde{y})(x-\tilde{x}) + F_y(\tilde{x},\tilde{y})(y-\tilde{y}) - F(x,y)] \\ &\quad - F_x(\tilde{x},\tilde{y})(x-\tilde{x}). \end{aligned}$$

By the differentiability of F at $(\tilde{x},\tilde{y})$ we can arrange to make

$$|F(\tilde{x},\tilde{y}) + F_x(\tilde{x},\tilde{y})(x-\tilde{x}) + F_y(\tilde{x},\tilde{y})(y-\tilde{y}) - F(x,y)| \leq \lambda(\varepsilon + \delta)$$

for any given λ by making ε and δ small enough, and the second term we estimate by $|F_x(\tilde{x},\tilde{y})(x-\tilde{x})| \leq c\varepsilon$ for $c = ||F_x(\tilde{x},\tilde{y})||$. Combining these we obtain

$$\begin{aligned} |Ty - \tilde{y}| &\leq M|F(\tilde{x},\tilde{y}) - F(x,y) + F_y(\tilde{x},\tilde{y})(y-\tilde{y})| \\ &\leq M(\lambda(\varepsilon+\delta) + c\varepsilon) \end{aligned}$$

where $M = ||F_y(\tilde{x},\tilde{y})^{-1}||$ as before; and we can make this $< \delta$ by first taking λ so that $M\lambda < 1/2$, fixing the required δ, and then choosing ε so that $M(\lambda + c)\varepsilon < \delta/2$.

We have now completed the most difficult part of the proof. We have shown that there exist neighborhoods U of $\tilde{x}$ and V of $\tilde{y}$ such that $T : \bar{V} \to V$ is a contractive mapping for each x in U (T depends on x). Therefore $F(x, y) = c$ for fixed x in U has a unique solution $y(x)$ in V that is obtained algorithmically as the limit of iterating T on some initial first guess, say $\tilde{y}$, so $y(x) = \lim_{k\to\infty} T^k\tilde{y}$. All that remains is to show that $y(x)$ is a C^1 function and its differential is obtained by implicit differentiation of the identity $F(x, y(x)) = c$. We will obtain this information directly as a consequence of that identity and the differentiability of F.

Now both $F(\tilde{x}, \tilde{y}) = c$ and $F(x, y(x)) = c$, so $F(x, y(x)) - F(\tilde{x}, \tilde{y}) = 0$. On the other hand F is differentiable at $(\tilde{x}, \tilde{y})$, so

$$\begin{aligned} 0 &= F(x, y(x)) - F(\tilde{x}, \tilde{y}) \\ &= F_x(\tilde{x}, \tilde{y})(x - \tilde{x}) \\ &\quad + F_y(\tilde{x}, \tilde{y})(y(x) - \tilde{y}) + R(x) \end{aligned}$$

where $R(x) = o(|x - \tilde{x}| + |y(x) - \tilde{y}|)$. We can then solve

$$(*) \qquad \begin{aligned} y(x) - \tilde{y} &= -F_y(\tilde{x}, \tilde{y})^{-1} F_x(\tilde{x}, \tilde{y})(x - \tilde{x}) \\ &\quad - F_y(\tilde{x}, \tilde{y})^{-1} R(x). \end{aligned}$$

Notice that this will say that $y(x)$ is differentiable at $\tilde{x}$ with differential $-F_y(\tilde{x}, \tilde{y})^{-1} F_x(\tilde{x}, \tilde{y})$ once we show that

$$F_y(\tilde{x}, \tilde{y})^{-1} R(x) = o(|x - \tilde{x}|).$$

This is not immediately apparent because our estimate for $R(x)$ involves $|y(x) - \tilde{y}|$, but we can establish it in two steps.

The first step is to prove the Lipschitz condition $|y(x) - \tilde{y}| \le c|x - \tilde{x}|$ for some constant c and x near $\tilde{x}$. For this we use $(*)$ and estimate

$$\begin{aligned} |y(x) - \tilde{y}| &\le |F_y(\tilde{x}, \tilde{y})^{-1} F_x(\tilde{x}, \tilde{y})(x - \tilde{x})| \\ &\quad + |F_y(\tilde{x}, \tilde{y})^{-1} R(x)| \le M_1|x - \tilde{x}| + M_2|R(x)| \end{aligned}$$

where $M_1 = ||F_y(\tilde{x}, \tilde{y})^{-1} F_x(\tilde{x}, \tilde{y})||$ and $M_2 = ||F_y(\tilde{x}, \tilde{y})^{-1}||$ are fixed constants. Then by taking x close enough to $\tilde{x}$ we can make

$$|R(x)| \le (|x - \tilde{x}| + |y(x) - \tilde{y}|)/2M_2,$$

so

$$|y(x) - \tilde{y}| \le M_1|x - \tilde{x}| + \frac{1}{2}(|x - \tilde{x}| + |y(x) - \tilde{y}|).$$

We rearrange terms to obtain

$$\frac{1}{2}|y(x) - y| \leq \left(M_1 + \frac{1}{2}\right) |x - \tilde{x}|,$$

establishing the Lipschitz condition with constant $c = 2M_1+1$. The second step is then to argue that given any $\varepsilon > 0$ we can make $|F_y(\tilde{x}, \tilde{y})^{-1} \cdot R(x)| \leq \varepsilon |x - \tilde{x}|$, since if x is close enough to $\tilde{x}$ we can make

$$|R(x)| \leq \delta(|x - \tilde{x}| + |y(x) - \tilde{y}|),$$

and then

$$\begin{aligned} |F_y(\tilde{x}, \tilde{y})^{-1} R(x)| &\leq M_2|R(x)| \leq M_2\delta(|x - \tilde{x}| + |y(x) - \tilde{y}|) \\ &\leq M_2\delta(|x - \tilde{x}| + c|x - \tilde{x}|) \\ &\leq M_2\delta(1 + c)|x - \tilde{x}|. \end{aligned}$$

Then we need only choose δ so that $M_2\delta(1+c) = \varepsilon$. This completes the proof of the $o(|x - \tilde{x}|)$ estimate for the remainder in $(*)$, thus proving the differentiability of $y(x)$ at $\tilde{x}$ with the correct differential.

Finally, we need to vary the point $\tilde{x}$. We have seen that the invertibility of $F_y(\tilde{x}, \tilde{y})$ implies that all nearby matrices are also invertible, so $F_y(x, y(x))^{-1}$ exists for x near $\tilde{x}$. Thus the above argument can be repeated at x to show

$$dy(x) = -F_y(x, y)^{-1} F_x(x, y(x)),$$

so $y(x)$ is differentiable and, hence, continuous. But this expression for $dy(x)$ is clearly a continuous function. Thus $y(x)$ is C^1. QED

Notice that the formula for the differential shows that y is actually C^k if F is C^k. It can even be shown that y is analytic if F is analytic, although we will not do this here.

The reader may be struck with the resemblance of the proof of the implicit function theorem and the proof of the existence and uniqueness theorem for o.d.e.'s via Picard iteration. In a sense, the integral equation

$$x(t) = x(t_0) + \int_{t_0}^{t} G(s, x(s))\, ds$$

can be thought of as an infinite-dimensional implicit function equation.

We can now complete the discussion of the exact o.d.e.

$$M(x,y)\frac{dy}{dx} + N(x,y) = 0$$

where $M = F_y$ and $N = F_x$ for some C^1 function $F(x,y)$. We saw that the o.d.e. was equivalent to the implicit equation $F(x,y) = c$. We can now apply the implicit function theorem to assert the local existence of solutions of the implicit equation provided $F_y = M$ is non-zero. In other words, if $M(x_0, y_0) \neq 0$, then there exists a unique solution $y(x)$ of the o.d.e. with $y(x_0) = y_0$ defined in a neighborhood of x_0. The condition $M(x_0, y_0) \neq 0$ is quite natural, for it means we can put the o.d.e. in normal form,

$$\frac{dy}{dx} = -\frac{N(x,y)}{M(x,y)},$$

in a neighborhood of (x_0, y_0).

13.1.3 Exercises

1. Verify that if $\|BA^{-1}\| \leq r < 1$, then the series $A^{-1}\sum_{k=0}^{\infty}(-1)^k \cdot (BA^{-1})^k$ converges at a rate that depends only on r.

2. Show that $(A+B)^{-1} - A^{-1} = A^{-1}\sum_{k=1}^{\infty}(BA^{-1})^k \to 0$ as $B \to 0$.

3. Write out the proof of the inverse function theorem by modifying the given proof of the implicit function theorem.

4. In the following examples, decide at which points $(\tilde{x}, \tilde{y})$ the hypotheses of the implicit function theorem are satisfied:

   a. $x^4 + xy^6 - 3y^4 = c$.

   b. $\begin{cases} \sin(x + y_1) + y_2^2 = c_1, \\ y_1^2 + xy_2^2 = c_2. \end{cases}$

   c. $\begin{cases} y_1^5 + 3y_1y_2 = x_1, \\ y_2^6 + 4y_1^2y_2^2 = x_2. \end{cases}$

5. Prove that the o.d.e.

$$\left(\frac{3}{2}\sqrt{|y|} + 1 + x^2\right)\frac{dy}{dx} + 2xy = 0$$

   has unique local solutions with $y(x_0) = y_0$ for any x_0 and y_0. Does the existence and uniqueness theorem for o.d.e.'s apply?

6. a. Prove that for every $n \times n$ matrix M sufficiently close to the identity matrix there exists a square-root matrix (solution of $A^2 = M$) and the solution is unique if A is required to be sufficiently close to the identity matrix.

   b. Compute the derivative of the square-root function $M \to A$.

   c. Show that the binomial expansion can be used to compute A (write $M = I + B$ and substitute B into the power series for $\sqrt{1+x}$) for M sufficiently close to the identity.

7. a. Show that the autonomous o.d.e. $y' = F(y)$ has a unique solution with initial condition $y(t_0) = y_0$ in a neighborhood of t_0 provided F is continuous and $F(y_0) \neq 0$. (**Hint**: solve the o.d.e. that the inverse function satisfies.)

   b. *Let $F : \mathbb{R} \to \mathbb{R}$ be continuous, and let $A = \{y : F(y) = 0\}$. Since A is closed, its complement can be written uniquely as a (finite or countable) disjoint union of open intervals $\bigcup(a_j, b_j)$. Show that there exists a unique solution to $y' = F(y)$ with any initial condition $y(t_0) = y_0$ if and only if all the improper integrals $\int_{a_j}^{m_j} |F(y)|^{-1}\,dy$ and $\int_{m_j}^{b_j} |F(y)|^{-1}\,dy$ diverge, with m_j the midpoint of the interval (if one of the intervals is unbounded, say (a_j, ∞), the condition should read $\int_{a_j}^{a_j+1} |F(y)|^{-1}\,dy = +\infty$).

13.2 Curves and Surfaces

13.2.1 Motivation and Examples

We want to consider three ways to describe a subset A of $\mathbb{R}^n$:

1. *Parametrically*, as the image of a function $g : U \to \mathbb{R}^n$ where U is an open subset of $\mathbb{R}^m$, $A = g(U)$. The usual cartesian coordinates $(t_1, \ldots, t_m) = t$ in U give curvilinear coordinates for A. Usually the function g is assumed to be one-to-one. Frequently we can only describe part of the set A in this way, but we can describe all of A as a union $A = \bigcup_j g_j(U_j)$ where each $g_j : U_j \to \mathbb{R}^n$ is as above. It is natural to think of such sets A as being m-dimensional.

2. *Implicitly*, as the solution set of an equation or set of equations. That is, if $F : \mathbb{R}^n \to \mathbb{R}^k$, $A = \{x : F(x) = 0\}$. Here we expect k to be the codimension of A (the difference between the dimensions of the ambient space $\mathbb{R}^n$ and the subset A), so $m = n - k$ should be the dimension. The domain of F may also be an open subset of $\mathbb{R}^n$.

3. As the *graph* of a function. We split the variables $(x_1, \ldots, x_n)$ into two groups $t = (t_1, \ldots, t_m)$ and $s = (s_1, \ldots, s_k)$, where $m + k = n$, consider a function $f : \mathbb{R}^m \to \mathbb{R}^k$, and set

$$A = \{(t, s) \text{ in } \mathbb{R}^{m+k} : s = f(t)\}.$$

Again we may also allow the domain of f to be an open subset of $\mathbb{R}^m$. This is really a special case of part 1, since the t variables can serve as parameters, and $g(t) = (t, f(t))$. In this case the function g is obviously one-to-one.

We will loosely refer to the subsets A that can be described in one of the above manners as m-dimensional surfaces. When $m = 1$ we will use the term *curve*. We will give more precise definitions later. Our goal is to show that all three approaches yield the same class of surfaces. Also, we are interested only in smooth surfaces, so we will assume that all functions involved in the descriptions of A are C^1. The graph-of-function description is the simplest from this point of view, since we do not have to impose any further conditions on the function f. On the other hand, a graph-of-function description is not so easy to find. The main theorem in section 13.2.3 says that a parametric description leads to a graph-of-function representation, while the main theorem in section 13.2.4 says that an implicit description also leads to a graph-of-function representation. However, it is important to realize that both these theorems impose additional hypotheses on the defining functions. In fact, section 13.2.2 is devoted to explaining the condition we need to impose for parametric representations. Also, the theorems are local in nature; generally speaking, it is only the implicit representation that allows us to describe the whole surface at once.

The intuitive idea is that a smooth m-dimensional surface is a set that has an m-dimensional tangent space at each point. Unfortunately, there is an ambiguity in how one should define the tangent space at a point $\tilde{x}$: it can be 1) the vector subspace passing through the origin of all directions that are tangent to A at $\tilde{x}$; or 2) the affine subspace passing through $\tilde{x}$ lying tangent to A. The tangent space according to definition 2) is just the tangent space according to definition 1) translated by $\tilde{x}$ so as to sit tangent to A. We will use definition 1), but from time to time we will indicate how to obtain the affine tangent space of definition 2). One thing to keep in mind is that with whatever description of A we start—parametric, implicit, or graph-of-function—we will obtain the same kind of description of the tangent space.

Now we look at some examples that reveal some of the difficulties with which we have to deal.

1. The unit circle in $\mathbb{R}^2$ can be described implicitly as the solution set of $x^2 + y^2 - 1 = 0$. We can describe it parametrically by the angular variable θ and the function $g(\theta) = (\cos\theta, \sin\theta)$. Notice that we cannot find an open interval on which g is both one-to-one and onto (we could take a half-open interval such as $[0, 2\pi)$, but this just hides the difficulties at the endpoint 0). We can, however, cover the circle with two patches such that the parameter function g is one-to-one on each, say $g : (0, 2\pi) \to \mathbb{R}^2$ and $g : (-\pi/2, \pi/2) \to \mathbb{R}^2$, as indicated in Figure 13.2.1.

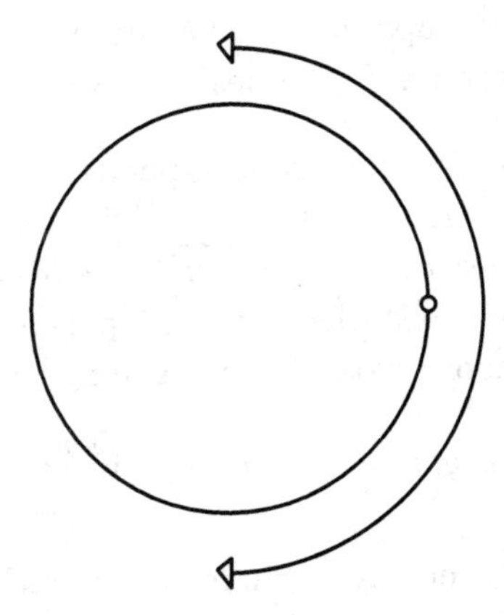

Figure 13.2.1:

Portions of the circle can be given as graphs of functions, $x = \pm\sqrt{1-y^2}$ or $y = \pm\sqrt{1-x^2}$.

2. Next consider the solution set in $\mathbb{R}^2$ of the equation $y^3 - x^2 = 0$, as shown in Figure 13.2.2. This can be represented parametrically by $g(t) = (t^3, t^2)$ or as the graph of $y = x^{2/3}$.

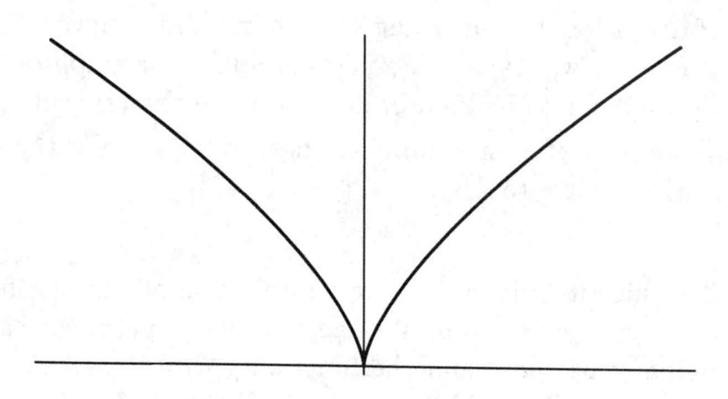

Figure 13.2.2:

Notice that there appears to be a sharp cusp at the point $(0,0)$ on the curve, even though the functions in the implicit and parametric representations are smooth. We will see that it is the vanishing of certain derivatives that is to be blamed for this. Notice that this problem is apparent in the representation as the graph of $f(x) = x^{2/3}$ because $f'(0)$ does not exist.

3. The unit sphere in $\mathbb{R}^3$ is given implicitly by the equation $x^2+y^2+z^2-1=0$. We can describe each of the six hemispheres as graphs of the functions $x = \pm\sqrt{1-y^2-z^2}, y = \pm\sqrt{1-x^2-z^2}, z = \pm\sqrt{1-x^2-y^2}$. Another popular parametric representation is the spherical coordinates (latitude–longitude)

$$(x,y,z) = (\cos\theta\sin\phi, \sin\theta\sin\phi, \cos\phi).$$

4. If we intersect the unit sphere in $\mathbb{R}^3$ with the $x-y$-plane, we obtain the unit circle in $\mathbb{R}^2$ again, but this time as a subset of $\mathbb{R}^3$. Implicitly this means considering the pair of equations $x^2 + y^2 + z^2 - 1 = 0$ and $z = 0$, or in other words, $F(x,y,z) = 0$ where $F(x,y,z) = (x^2+y^2+z^2-1, z)$ is a function from $\mathbb{R}^3$ to $\mathbb{R}^2$. Parametrically we can represent this circle by adding a zero

third coordinate to the $\mathbb{R}^2$ parametrization, $g(\theta) = (\cos\theta, \sin\theta, 0)$. We can also use the graph of $(y, z) = (\pm\sqrt{1 - x^2}, 0)$ or $(x, z) = (\pm\sqrt{1 - y^2}, 0)$ to represent parts of the circle, but we cannot use a graph of a function of z.

5. As a final example consider the implicit equation $xy = 0$ in $\mathbb{R}^2$, which defines the set consisting of the x-and y-axes. In a neighborhood of the origin where the axes cross, it is impossible to describe this set parametrically in a one-to-one fashion, although we can write the set as a union of the axes, each represented as a graph of a function.

These examples reveal some of the problems with which we are going to have to come to grips in creating a nice theory of surfaces. We can see from the start that we are going to have to describe surfaces locally and that there are really two distinct notions of "local". In the more stringent notion, we take a neighborhood U of a point on the surface in $\mathbb{R}^n$ and demand some property of the intersection of U with the surface—in other words we take a piece of the surface as it sits in $\mathbb{R}^n$. In this sense the crossing of the axes in the last example cannot be cut apart. In the other, less stringent, sense of "local", we ask only that the surface be a union of pieces with certain properties. In this sense the two axes are the union of axes, and each axis is well behaved. The first, stricter notion is the one we shall adopt, since we can obtain a nice result about implicitly defined surfaces with this concept. The technical name for the two concepts are *embedded surfaces* and *immersed surfaces.* Although we will not emphasize immersed surfaces, we should point out that they are important in topology and in some ways easier to study. For example, there is a famous theorem of Smale that says the unit sphere in $\mathbb{R}^3$ can be turned inside out—*everted*—by continuously deforming it through a sequence of immersed surfaces. It is intuitively clear that no such result would be possible using embedded surfaces.

13.2.2 Immersions and Embeddings

The next problem with which we have to deal is the possibility of cusps or other non-smooth behavior of the surface. We want to deal with smooth surfaces, so we can use differential calculus, but the second example shows that rough spots can develop even if the functions defining the surface are C^1. Of course we did notice that this problem doesn't seem to arise in the graph of function description. One way to evade the problem would be to make a definition in terms of the graph-of-function description: a C^1 surface in $\mathbb{R}^n$ is a subset that is locally the graph of a C^1 function. This approach seems overly restrictive, however. How do we know there aren't subsets of $\mathbb{R}^n$ that we would intuitively agree are smooth surfaces and yet cannot locally be expressed as a graph of a C^1 function? Actually it turns out that there aren't any; but this should be a theorem, not part of the definition. For the definition we should only require some sort of local parametric representation. But not every C^1 parametric representation will do, as the cusp example $g(t) = (t^3, t^2)$ shows. So our problem becomes how to restrict parametric representations to ensure "smoothness" of the surface.

To get a feel for the answer, let's look first at the simplest case: curves. The parameter is just a single variable t, which we can assume takes values in an open interval (a, b). Let $g : (a, b) \to \mathbb{R}^n$ be C^1. If we think of $g(t)$ as tracing out the curve in $\mathbb{R}^n$ with t as time, then $g'(t)$ represents the velocity of the motion. It will be a vector that is tangent to the curve, as in Figure 13.2.3. If the curve has a

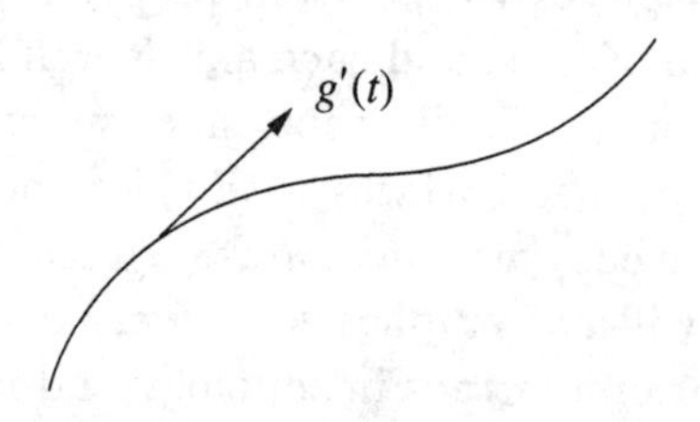

Figure 13.2.3:

tangent, then it is smooth, so what can go wrong? In the example $g(t) = (t^3, t^2)$, the thing that goes wrong is $g'(0) = (0, 0)$. If the velocity vector vanishes, then we don't necessarily have a tangent to the curve and the curve may not be smooth. Our interpretation of $g'(t)$ as velocity means that $g'(t_0) = 0$ is interpreted as saying the tracing comes to a stop at time t_0. In other words, *you can draw a rough curve in smooth motions if the motion is allowed to stop.* Therefore, the condition we want to impose is that g' is never zero. This is not to say that $g'(t_0) = 0$ necessarily implies that the curve is not smooth at t_0. There are many different parametric representations of a curve, and we can always stop along a smooth curve to admire the view (for example, $g(t) = (\cos t^3, \sin t^3)$ parameterizes the circle). Nevertheless, it is plausible that if a curve is smooth we can move along it at a non-zero velocity in a smooth manner.

Since we have seen that for a curve the condition we want to impose on g is the nonvanishing of the derivative, we might be tempted to guess that this is also the right condition in general. However, if we analyse the following simple "fold singularity" we will see that this is not strong enough. We simply take the cusp curve and add a dimension. The parameter space will have two variables t, s; and the two-dimensional surface in $\mathbb{R}^3$ will be parametrized by $g(t, s) = (t^3, t^2, s)$. Note that

$$dg(t, s) = \begin{pmatrix} 3t^2 & 0 \\ 2t & 0 \\ 0 & 1 \end{pmatrix}$$

is never zero. Nevertheless, the surface folds at $t = 0$ and so it is not smooth. The differential when $t = 0$ is

$$\begin{pmatrix} 0 & 0 \\ 0 & 0 \\ 0 & 1 \end{pmatrix};$$

and although it is not zero, it does appear less full of life than it might be. This gives us a good clue toward finding the condition we want.

It is time to appeal once again to the basic principle of differential calculus: *do unto the derivative what you would do unto the function.* In this case we want the parametrizing function g to be one-to-one, so we should demand that the best affine approximation be one-to-one. Clearly this doesn't depend on the constant term, but only on the linear part, the differential. The question is then whether matrix multiplication by dg is one-to-one. It is easy to see when $m = 1$ that this is the same as the non-vanishing of dg, but for $m > 1$ the question is more subtle. We need to recall some basic facts from the linear algebra.

Let A denote an $n \times m$ matrix, so Ax is a linear transformation from $\mathbb{R}^m$ to $\mathbb{R}^n$. The *rank* of A is defined to be the dimension of the image of A, which is the linear span of the columns of A. This is also equal to the dimension of the span of the rows of A and the size of the largest invertible square submatrix (a submatrix means we select certain rows and columns, not necessarily in any special order). The *kernel* of A, also called the *null-space*, is the set of solutions of $Ax = 0$; and the linear function Ax is one-to-one if and only if the kernel is $\{0\}$. The basic formula *dimension of domain* $=$ *dimension of image* $+$ *dimension of kernel* means that Ax is one-to-one if and only if the rank of A is equal to the dimension of the domain, which we are calling m. This is the largest possible value for the rank, since there are only m columns. Note that if A is one-to-one we necessarily have $m \leq n$, for there are only n rows. The matrix

$$A = \begin{pmatrix} 0 & 0 \\ 0 & 0 \\ 0 & 1 \end{pmatrix}$$

has rank 1 and Ax is clearly not one-to-one, as

$$A \begin{pmatrix} 1 \\ 0 \end{pmatrix} = 0.$$

The same is true of

$$A = \begin{pmatrix} 1 & 2 \\ 2 & 4 \\ 3 & 6 \end{pmatrix},$$

and we have

$$A \begin{pmatrix} -2 \\ 1 \end{pmatrix} = 0.$$

We now want to consider those C^1 functions $g : U \to \mathbb{R}^n$ where U is an open subset of $\mathbb{R}^m$ such that $dg(x)$ has rank m at every point x of U. Such functions are called *immersions*. We will insist that local parametrizations of C^1 surfaces be given by immersions. As expected, immersions will be locally one-to-one; we will not prove this now since it will be a consequence of a later result. An immersion does not have to be globally one-to-one, as the familiar circle parametrization shows. We define an *embedding* to be an immersion that has two additional properties: $g : U \to \mathbb{R}^n$ is one-to-one, and the inverse map

$g^{-1} : g(U) \to \mathbb{R}^m$ is continuous (here we are considering $g(U)$ as a metric subspace of $\mathbb{R}^n$). The following pictures of immersions that are not embeddings ($m = 1, n = 2$)

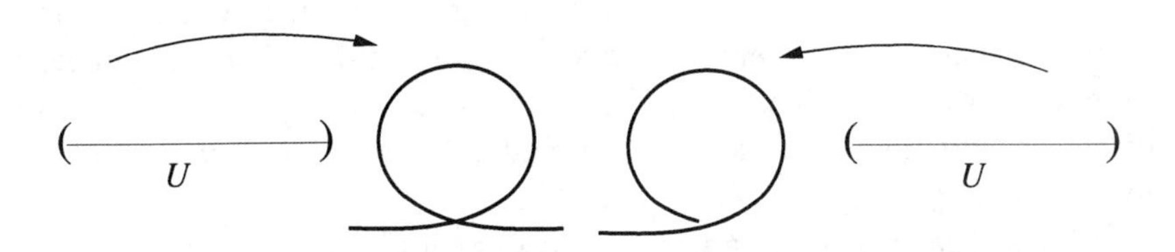

Figure 13.2.4:

should illustrate the bad behavior we are trying to avoid. In Figure 13.2.4 the map on the left fails to be one-to-one at the cross point. The map on the right is one-to-one, but the inverse fails to be continuous at the touching point (the map is defined on an open interval, so the touching point is only covered once). However, the map in Figure 13.2.5 does give an embedding

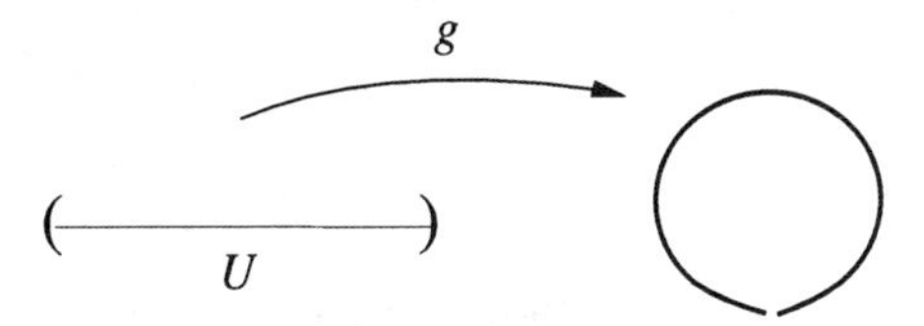

Figure 13.2.5:

because the touching point is no longer a point of $g(U)$.

If $g : U \to \mathbb{R}^n$ is an embedding, we take the usual Euclidean coordinates in $U \subseteq \mathbb{R}^m$ and carry them by g onto a set of coordinates for the image $g(U)$. Fix a point $\tilde{x}$ in U, and let $\tilde{y} = g(\tilde{x})$ be the corresponding point in the image $g(U)$. If we fix all the coordinates of $\tilde{x}$ except for one, say $\tilde{x}_2, \tilde{x}_3, \ldots, \tilde{x}_m$, and vary x_1, then we obtain a straight line in U that gets mapped under g into a curve in $g(U)$,

$$f(t) = g(t, \tilde{x}_2, \ldots, \tilde{x}_m),$$

passing through $\tilde{y}$. The tangent vector

$$\frac{df}{dt}(\tilde{x}_1) = \frac{\partial g}{\partial x_1}(\tilde{x}_1, \tilde{x}_2, \ldots, \tilde{x}_m)$$

is non-zero because it is a column of dg. In fact, if we vary each of the coordinates x_k in turn, we get curves in $g(U)$ with tangent vectors at $\tilde{y}$ that are the different columns of $dg(\tilde{x})$ and, hence, are linearly independent. This is indicated in Figure 13.2.6. They span an m-dimensional subspace of $\mathbb{R}^n$ that we will call the *tangent space* to $g(U)$ at the point $\tilde{y}$.

Notice that this gives a parametric description of the tangent space, namely all vectors $\sum_{k=1}^m c_k \partial g/\partial x_k(\tilde{x})$ for any choice of the parameters $c_1, \ldots, c_m$.

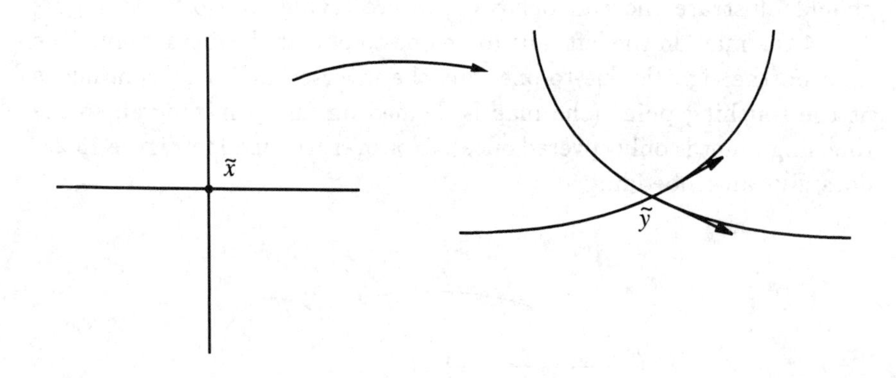

Figure 13.2.6:

If $h : (a, b) \to U$ is any C^1 curve passing through $\tilde{x}$, say $h(c) = \tilde{x}$, then $g \circ h : (a, b) \to g(U)$ is a C^1 curve in $g(U)$ passing through $\tilde{y}$, $g \circ h(c) = \tilde{y}$ and the tangent vector

$$\frac{d}{dt} g \circ h(c) = \sum_{k=1}^{m} \frac{dh_k}{dt}(c) \frac{\partial g}{\partial x_k}(\tilde{x})$$

is a linear combination of these vectors $\partial g/\partial x_k(\tilde{x})$ and, hence, belongs to the tangent space at $\tilde{y}$. Thus the tangent space is composed of tangent vectors to curves lying in $g(U)$ and passing through $\tilde{y}$. Also, because $g^{-1} : g(U) \to \mathbb{R}^m$ is continuous, any curve on $g(U)$ passing through $\tilde{y}$ is the image of a curve in U passing through $\tilde{x}$ as above.

13.2.3 Parametric Description of Surfaces

Definition 13.2.1 *A C^1 m-dimensional surface in $\mathbb{R}^n$ (with $m \leq n$) is a subset M_m of $\mathbb{R}^n$ such that for every point y in M_m there exists a neighborhood V of y in $\mathbb{R}^n$ and an embedding $g: U \to \mathbb{R}^n$ for U open in $\mathbb{R}^m$ such that $g(U) = V \cap M_m$. We say M_m is of class C^k if each of the embeddings is of class C^k.*

Our definition thus says that M_n is locally parametrized by C^1 embeddings. Next we want to show the fact hinted at before, that this implies that locally M_n is the graph of a C^1 function. By this we mean that for each V there is a partition of the variables $y_1, \ldots, y_n$ into two groups, $t_1, \ldots, t_m$ and $s_1, \ldots, s_{n-m}$, such that $V \cap M_m = \{(t, s) : s = f(t) \text{ for } t \text{ in } U\}$ where $f: U \to \mathbb{R}^{n-m}$ is a C^1 function. (Actually it may be necessary to decrease the size of the sets V for this to be true.) Note that this is a special case of an embedding, namely $g(t) = (t, f(t))$. Since the first m rows of dg are the $m \times m$ identity matrix, we have rank $dg = m$. Also g is clearly one-to-one, and the inverse is continuous because it is just the projection onto the t-variables.

Theorem 13.2.1 *Let $g: U \to \mathbb{R}^n$ with U an open subset of $\mathbb{R}^m$ be a C^1 immersion. Then for any point $\tilde{x}$ in U there exists a neighborhood $\tilde{U}$ of $\tilde{x}$ such that g is one-to-one on $\tilde{U}$ and $g(\tilde{U})$ is the graph of a C^1 function.*

The theorem holds for all immersions and is local in the domain of g. However, if g is actually an embedding, then $g(\tilde{U})$ is a neighborhood of $g(\tilde{x})$ in $g(U)$ and so the result is local on the surface. Thus the theorem shows that every surface is locally the graph of a function.

Proof: The idea of the proof is to solve for $x_1, \ldots, x_m$ (variables in the domain of g) as a function of some of the y variables (in the range of g) $t_1, \ldots, t_m$, using the inverse function theorem. Look at the matrix $dg(\tilde{x})$. By assumption it has rank m, so among the n rows $dg_j(\tilde{x})$, $j = 1, \ldots, n$, there are m linearly independent ones. The corresponding variables y_j may be taken as the t variables. The remaining $n - m$ variables will then be the s variables. In other words, among all possible partitions of the y variables we are allowed to choose any one for which the rows $dg_j(\tilde{x})$, corresponding to those variables y_j we have called

t variables, are linearly independent (or equivalently, the submatrix formed by taking the rows corresponding to t variables is invertible). To simplify the notation we now assume that the first m variables $y_1, \ldots, y_m$ are the t variables—we can always arrange this by relabeling the variables.

Now we use the inverse function theorem to solve for $x_1, \ldots, x_m$ as a function of $t_1, \ldots, t_m$. Indeed we are given the equations

$$\begin{aligned} t_1 &= g_1(x_1, \ldots, x_m) \\ &\vdots \\ t_m &= g_m(x_1, \ldots, x_m), \end{aligned}$$

which we can abbreviate $t = h(x)$ where

$$h = \begin{pmatrix} g_1 \\ \vdots \\ g_m \end{pmatrix},$$

so $dh(\tilde{x})$ consists of the first m rows of $dg(\tilde{x})$ and, hence, is invertible. By the inverse function theorem there exists a neighborhood $\tilde{U}$ of $\tilde{x}$ on which h has a C^1 inverse $h^{-1} : \tilde{V} \to \tilde{U}$ so that $x = h^{-1}(t)$ if and only if $t = h(x)$ for x in $\tilde{U}$ and t in $\tilde{V}$, where $\tilde{V}$ is an open set in the t space. This shows that g is one-to-one on $\tilde{U}$. Now the remaining equations

$$\begin{aligned} s_1 &= g_{m+1}(x_1, \ldots, x_m) \\ &\vdots \\ s_{n-m} &= g_n(x_1, \ldots, x_m), \end{aligned}$$

which we can abbreviate as $s = \phi(x)$ where

$$\phi(x) = \begin{pmatrix} g_{m+1}(x) \\ \vdots \\ g_n(x) \end{pmatrix},$$

allow us to write $s = \phi(h^{-1}(t)) = f(t)$ for $f = \phi \circ h^{-1}$; and f is a composition of C^1 functions and, hence, C^1. Thus $y = g(x)$ for x in $\tilde{U}$ if and only if $y = (t, s)$ with t in $\tilde{V}$ and $s = f(t)$, so $g(\tilde{U})$ is a graph of f. QED

Example 13.2.1 Consider

$$g(\theta) = \begin{pmatrix} \cos\theta \\ \sin\theta \end{pmatrix} = \begin{pmatrix} x \\ y \end{pmatrix},$$

an immersion of $\mathbb{R}^1$ into $\mathbb{R}^2$, the image being the unit circle. Since

$$dg(\theta) = \begin{pmatrix} -\sin\theta \\ \cos\theta \end{pmatrix},$$

we can solve for x as a function of y in a neighborhood of any point for which $\partial g_2/\partial\theta = \cos\theta \neq 0$ and for y as a function of x in a neighborhood of any point for which

$$\frac{\partial g_1}{\partial\theta} = -\sin\theta \neq 0.$$

Choosing a point

$$\begin{pmatrix} x_0 \\ y_0 \end{pmatrix} = \begin{pmatrix} \cos\theta_0 \\ \sin\theta_0 \end{pmatrix}$$

on the circle with $\sin\theta_0 \neq 0$ (hence $x_0 \neq \pm 1$), we can first solve for θ as a function of $x, \theta = \arccos x$ (choosing a branch of arccosine so that $\theta_0 = \arccos x_0$) and then $y = \sin(\arccos x)$ gives part of the circle as the graph of a function (depending on the branch of arccosine, we have $\sin(\arccos x) = \pm\sqrt{1-x^2}$).

Example 13.2.2 Consider the spherical coordinates map

$$\begin{pmatrix} x \\ y \\ z \end{pmatrix} = g\begin{pmatrix} \theta \\ \phi \end{pmatrix} = \begin{pmatrix} \cos\theta\sin\phi \\ \sin\theta\sin\phi \\ \cos\phi \end{pmatrix},$$

$g : \mathbb{R}^2 \to \mathbb{R}^3$, with the image of g equal to the unit sphere. We compute

$$dg = \begin{pmatrix} -\sin\theta\sin\phi & \cos\theta\cos\phi \\ \cos\theta\sin\phi & \sin\theta\cos\phi \\ 0 & -\sin\phi \end{pmatrix}.$$

When $\sin\phi = 0$ the first column is zero and dg has rank 1, which is less than $m = 2$; so g is not an immersion. However, at all other points dg

has rank 2, so g will be an immersion if we restrict it to any open set in (θ, ϕ) space for which $\sin\phi$ never vanishes, say the strip $0 < \phi < \pi$. Then g wraps this strip around the sphere with the north and south poles removed (θ is the longitude and $\phi - \pi/2$ the latitude). Suppose we want to solve for z as a function of x and y on the sphere. We need to have the first two rows of dg linearly independent or

$$\begin{pmatrix} -\sin\theta\sin\phi & \cos\theta\sin\phi \\ \cos\theta\sin\phi & \sin\theta\cos\phi \end{pmatrix}$$

invertible. Since the determinant of this 2×2 matrix is $-\sin\phi\cos\phi$, we need $\cos\phi \neq 0$—in other words we have to stay away from the equator. Then we can solve $x = \cos\theta\sin\phi, y = \sin\theta\sin\phi$ for θ and ϕ; indeed $x^2 + y^2 = \sin^2\phi$, so $\phi = \arcsin\sqrt{x^2 + y^2}$, and $y/x = \tan\theta$, so $\theta = \arctan y/x$, and then $z = \cos\phi = \cos(\arcsin\sqrt{x^2 + y^2})$, which simplifies to $z = \pm\sqrt{1 - x^2 - y^2}$ depending on the hemisphere. In this case the graph-of-function extends to the north and south poles, which had to be excluded in the spherical coordinate description.

At points on the equator ($\cos\phi = 0$) we have

$$dg = \begin{pmatrix} -\sin\theta & 0 \\ \cos\theta & 0 \\ 0 & -1 \end{pmatrix}$$

so that we can solve for x as a function of y and z if $\cos\theta \neq 0$ or we can solve for y as a function of x and z if $\sin\theta \neq 0$. In the first case we solve $y = \sin\theta\sin\phi, z = \cos\phi$ for θ and ϕ,

$$\phi = \arccos z,$$

and

$$\theta = \arcsin\left(\frac{y}{\sin\phi}\right) = \arcsin\left(\frac{y}{\sin\arccos z}\right).$$

Then $x = \cos\theta\sin\phi = \cos(\arcsin(y/\sin\arccos z))\sin(\arccos z)$, which as we know must simplify to $x = \pm\sqrt{1 - y^2 - z^2}$.

The theorem shows that we could just as well have defined a C^1 surface as a set that is locally the graph of a C^1 function. However, it is preferable to have the greater flexibility of general embeddings $g : U \to V \cap M_m$. We call the subsets $V \cap M_m$ of the surface *coordinate patches* and the function $g^{-1} : V \cap M_m \to U$ a *local coordinate map* since it assigns coordinates (the standard cartesian coordinates in U) to each point in the coordinate patch. The theorem asserts that if we take a small enough coordinate patch we can arrange for the coordinate map to be the projection onto m of the cartesian coordinates of $\mathbb{R}^n$, or in other words we can obtain local coordinates on the surface by selecting m of the cartesian coordinates of $\mathbb{R}^n$.

It is interesting to compare two local coordinate maps on coordinate patches that overlap. Suppose $V_1 \cap M_m$ and $V_2 \cap M_m$ are two patches for which $V_1 \cap V_2 \cap M_m$ is non-empty, as indicated in Figure 13.2.7. On the overlap we have two coordinate maps, g_1^{-1} and g_2^{-1}, and let $\tilde{U}_1$ and $\tilde{U}_2$ be the images of the overlap in U_1 and U_2. Then the maps $g_2^{-1} \circ g_1 : \tilde{U}_1 \to \tilde{U}_2$ and $g_1^{-1} \circ g_2 : \tilde{U}_2 \to \tilde{U}_1$

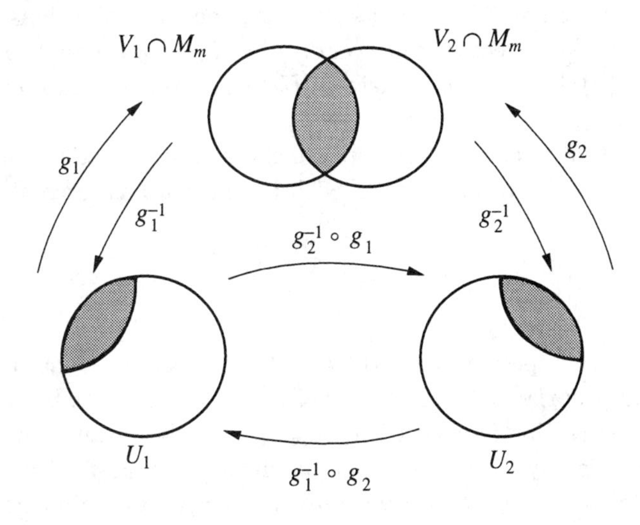

Figure 13.2.7:

tell us how to change from one coordinate system to another. We claim these functions are C^1. This is not obvious because g_1^{-1} and g_2^{-1} are not defined on an open subset of $\mathbb{R}^n$. However, if g_2 happens to be one of the special graph-of-function embeddings, then g_2^{-1} is just the projection onto some of the coordinates, so $g_2^{-1} \circ g_1$ is C^1 and then $g_1^{-1} \circ g_2 = (g_2^{-1} \circ g_1)^{-1}$ is C^1 by the inverse function theorem. Since by the theorem we can always interpose a third coordinate patch of this special form, it follows that these functions are C^1 in general. We can interpret this as saying that, despite the non-uniqueness of the local parametric representations, *any two local parametric representations of the same portion of the surface differ only by a C^1 change of variable in the parameters.*

Recall that an embedding $g : U \to \mathbb{R}^n$ allows us to define a tangent space to the image $g(U)$ at any point. We want similarly to define the *tangent space to a C^1 surface at a point* to be the tangent space given by one of the local embeddings $g : U \to V \cap M_m$. In order for this to make sense we must verify that we get the same tangent space in the overlap of coordinate patches. We do this by giving a characterization of the tangent space $TM_m(y)$ for y a point in M_m as the set of all tangent vectors to C^1 curves lying in M_m at the point y. This characterization is *intrinsic* in the sense that it does not refer to any coordinate patch. We will show that the two definitions coincide: the set of tangent vectors to C^1 curves in M_m at y is equal to the span of $\partial g/\partial x_k(x), k = 1, \ldots m$, where $g : U \to V \cap M_m$ is a local embedding and $g(x) = y$.

Recall that every C^1 curve passing through y that is obtained by composing a C^1 curve $f(t)$ in U with the map g has tangent vector

$$\frac{d}{dt} g \circ f(t) = \sum_{k=1}^{m} \frac{df_k}{dt} \frac{\partial g}{\partial x_k}(x)$$

that lies in this span, and clearly we get every vector in the span in this way for appropriate straight lines $f(t)$. What remains to be seen is that every C^1 curve in the patch $V \cap M_m$ is actually obtained in this way. But if $h(t)$ is any C^1 curve in the patch, we look at $f(t) = g^{-1} \circ h(t)$, which is clearly a curve in U with $g \circ f(t) = h(t)$, as indicated in Figure 13.2.8. We need to show that f is in fact C^1. But this is obvious if the coordinate map g^{-1} is a projection onto Euclidean coordinates, and it follows in general by the above observation that all changes of coordinates are C^1. This completes the proof of the equivalence of the two definitions of tangent space.

Using the local coordinate maps we can transfer much of the differential calculus in $\mathbb{R}^m$ to the surface M_m. A function f defined on M_m (say taking real values) is said to be C^1 if the coordinate version $f \circ g$ is C^1 for every coordinate patch. Questions about f can be pulled back

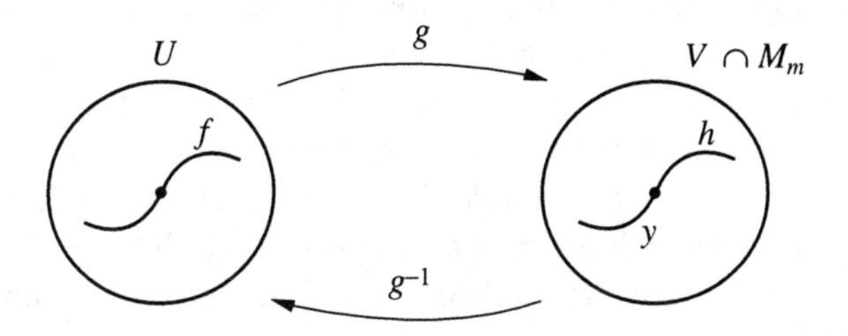

Figure 13.2.8:

to $f \circ g$, which is an ordinary C^1 function. It is obvious that if f is the restriction of a C^1 function on $\mathbb{R}^n$ to M_m, then f is C^1, say $f = F|M_m$, and then $f \circ g = F \circ g$, the composition of two C^1 functions. However, there are many instances when functions are defined on surfaces M_m without any obvious extension to $\mathbb{R}^n$ (it can be shown that such extensions must exist locally). We can similarly define what is meant by C^1 functions from one surface M_m to another surface, $M'_{m'}$; and it is possible to show that the differential $df(y)$ can be sensibly defined as a linear transformation from the tangent space $TM_m(y)$ to the tangent space $TM'_{m'}(f(y))$, but we will not go into the details of this.

13.2.4 Implicit Description of Surfaces

We have seen now some of the significance and versatility of the concept of C^1 surface from the point of view of parametric and graph-of-function descriptions. We now want to understand how C^1 surfaces may be given implicit descriptions as the solution set of $F(x) = 0$. Here

$$F(x) = \begin{pmatrix} F(x) \\ \vdots \\ F_{n-m}(x) \end{pmatrix}$$

is a C^1 function from $\mathbb{R}^n$ to $\mathbb{R}^{n-m}$, and we can think of the sets $\{x : F(x) = c\}$ as the *level sets* of F. Suppose for a moment that we knew that one of these level sets, say $M_m = \{x : F(x) = 0\}$, were a C^1 m-dimensional surface. Then if $x(t)$ is any curve lying in M_m we have $F_j(x(t)) \equiv 0$, so

$$0 = \frac{d}{dt} F_j(x(t)) = \nabla F_j(x(t)) \cdot \frac{dx}{dt}(t),$$

in other words the gradient of F_j is perpendicular to the tangent vector dx/dt for $j = 1, \ldots, m-n$. Since this is true for any curve in M_m, it follows that $\nabla F_j(x)$ is perpendicular to the tangent space $TM_m(x)$. We call the subspace of all vectors in $\mathbb{R}^n$ that are perpendicular to $TM_m(x)$ the *normal space* at x, denoted $NM_m(x)$. It is clear from linear algebra that the normal space has dimension $n-m$ and $TM_m(x)$ is the space of all vectors perpendicular to $NM_m(x)$. Thus the normal and tangent spaces determine one another.

We have observed that if a C^1 m-dimensional surface is given implicitly by $F(x) = 0$, then the gradients $\nabla F_j(x)$ give us $n-m$ vectors in the normal space $NM_m(x)$. Since the normal space has dimension $n-m$, it would be especially convenient if these gradients $\nabla F_j(x)$ were linearly independent because then they would span the normal space and hence determine the tangent space. Since these gradients are the rows of $dF(x)$, the condition that they be linearly independent is the same as saying rank $dF(x) = n-m$ (notice that $n-m$ is the maximal rank possible for an $(n-m) \times n$ matrix). If F was actually a linear transformation, then M_m would equal the tangent space, so this condition would be necessary and sufficient for M_m to be m-dimensional. Thus we expect that this condition should imply in general that the level set should be a C^1 m-dimensional surface, and the proof should involve the implicit function theorem.

Theorem 13.2.2 *Let $F : \mathbb{R}^n \to \mathbb{R}^{n-m}$ be a C^1 function, and suppose $dF(x)$ has rank $n-m$ at every point on a level set $\{x : F(x) = c\}$. Then this level set is a C^1 m-dimensional surface.*

Proof: Let x be a point on the level set. Since we are assuming $dF(x)$ has rank $n-m$, we can find $n-m$ variables x_j such that the columns $\partial F/\partial x_j(x)$ are linearly independent. Call these the s variables and the remaining variables the t variables. We want to show that in a neighborhood V of x the level set is the graph of a C^1 function,

$$V \cap \{x : F(x) = c\} = \{(t,s) : s = f(t) \text{ for } t \text{ in } U\}$$

for a C^1 function $f : U \to \mathbb{R}^{n-m}$, U an open set in $\mathbb{R}^m$. For simplicity we assume the first m variables $x_1, \ldots, x_m$ are the t variables. We write the equation of the level set $F(t,s) = c$. The condition on the linear independence of the vectors $\partial F/\partial s_j$ says that the $(n-m) \times (n-m)$

matrix F_s is invertible, so we can apply the implicit function theorem to solve for $s = f(t)$ as desired. QED

For example, the unit sphere in $\mathbb{R}^3$ is given implicitly by $x^2 + y^2 + z^2 = 1$. Here $F(x, y, z) = x^2 + y^2 + z^2$ and $dF(x, y, z) = (2x, 2y, 2z)$ has rank 1 provided $(x, y, z) \neq (0, 0, 0)$. Thus by the theorem all the spheres $x^2 + y^2 + z^2 = c$ for $c > 0$ are C^1 two-dimensional surfaces. The theorem does not apply to the level set $F(x, y, z) = 0$, which is in fact not a two-dimensional surface. If we intersect the sphere with the x-y-plane, we are looking at the pair of equations $x^2 + y^2 + z^2 = 1$ and $z = 0$, which can be written

$$F(x,y,z) = \begin{pmatrix} 1 \\ 0 \end{pmatrix} \quad \text{for } F(x,y,z) = \begin{pmatrix} x^2 + y^2 + z^2 \\ z \end{pmatrix}.$$

Then

$$dF = \begin{pmatrix} 2x & 2y & 2z \\ 0 & 0 & 1 \end{pmatrix},$$

which has rank 2 provided either $x \neq 0$ or $y \neq 0$. We can check that no point of the form $(0, 0, z)$ occurs on the given level set, so the theorem asserts it is a one-dimensional surface. In fact we know it is a circle.

The implicit representation is very convenient because with it we can describe the whole surface at once, whereas with the other representations we usually can only describe it in pieces. Also the implicit description allows us to compute easily the normal space and, hence, the tangent space at each point. For example, for the sphere $x^2 + y^2 + z^2 = 1$, the vector $(2x_0, 2y_0, 2z_0)$ is normal at the point (x_0, y_0, z_0); hence, the tangent plane at that point is given by the equation $2x_0x + 2y_0y + 2z_0z = 0$. Notice that by giving the tangent space as the orthogonal complement to the normal space we are giving an implicit description: the tangent space at x consists of all vectors v that satisfy the equations $v \cdot \nabla F_j(x) = 0,\ j = 1, \ldots, n - m$.

There is a more abstract concept, called *manifold*, which is essentially a surface that is not described as a subset of a Euclidean space. Roughly speaking, a C^1 m-dimensional manifold M_m is a metric space that is covered by coordinate patches V_j on which coordinate maps $h_j : V_j \to U_j$ are defined, where U_j are open subsets of Euclidean space. The condition for differentiability takes the form of an assumption that on overlapping patches the change of coordinates functions must be C^1. We will not go into the details here. There is a theorem to the effect that every C^1 manifold can be realized as a C^1 surface in a Euclidean space of sufficiently high dimension, so we don't get any new objects by considering manifolds rather than surfaces. Neverthless, we do gain a more intrinsic and flexible point of view by doing so.

13.2.5 Exercises

1. Decide which of the following maps are immersions, and for those that are decide which variables can be taken to be the independent variable(s) in the description of the image locally as a graph of a function:

   a. $g(t) = (\cos t, \sin t, t) \quad t$ in $\mathbb{R}^1$,
   b. $g(t, s) = (\cos t, \sin t, s) \quad (t, s)$ in $\mathbb{R}^2$,
   c. $g(t, s) = (s \cos t, s \sin t, s) \quad (t, s)$ in $\mathbb{R}^2$,
   d. $g(t, s) = (s \cos t, s \sin t, s) \quad s > 1, t$ in $\mathbb{R}^1$.

2. For which values of the constants do the following implicit equations define a C^1 surface? For those that do, decide which variables can be taken to be the independent variables in the description of the surface locally as a graph of a function:

   a. $x^2 + y^2 - z^2 = c$;
   b. $x^2 + y^2 + z^2 = c_1, \quad x^2 + y^2 - z^2 = c_2$;
   c. $xyz = c$.

3. For each of the surfaces in exercise 2, compute the normal space and tangent space at each point.

4. Let $f : M_m \to \mathbb{R}$ be a C^1 function, where $M_m \subseteq \mathbb{R}^n$ is a C^1 surface. Prove that every point in M_m lies in a neighborhood U such that there exists a C^1 function $F : U \to \mathbb{R}$ that extends f, i.e., $F(y) = f(y)$ for y in $M_m \cap U$. (**Hint**: use graph of function representation.)

5. For each point $(\tilde{x}, \tilde{y}, \tilde{z})$ on the unit sphere $x^2 + y^2 + z^2 = 1$ and each vector $v = (v_1, v_2, v_3)$ in the tangent space to the sphere at $(\tilde{x}, \tilde{y}, \tilde{z})$, construct a C^1 curve lying in the sphere whose tangent vector at the point $(\tilde{x}, \tilde{y}, \tilde{z})$ is (v_1, v_2, v_3).

6. Show that the set of $m \times n$ matrices of rank m (with $n \geq m$) is an open set in the space of matrices $\mathbb{R}^{m \times n}$.

7. Let M_2 be any C^1 two-dimensional surface in $\mathbb{R}^3$ that is compact. Show that for every two-dimensional vector space V in $\mathbb{R}^3$, there exists a point x on M_2 whose tangent space equals V. (**Hint**: if u is a vector perpendicular to V, what happens at points on M_2 where $x \cdot u$ achieves a maximum or minimum?)

8. Let M_2 be the surface of revolution in $\mathbb{R}^3$ obtained by rotating a circle in the x-z-plane, that does not intersect the z-axis, about the z-axis. Show that M_2 is a C^1 two-dimensional surface, and compute its tangent space at any point (Note: this surface is called a *torus*.)

9. Show that the Cartesian product $M_{m_1} \times M_{m_2}$ of two C^1 surfaces of dimensions m_1 and m_2 in $\mathbb{R}^{n_1}$ and $\mathbb{R}^{n_2}$ is a C^1 surface of dimension $m_1 + m_2$ in $\mathbb{R}^{n_1+n_2}$. Express the tangent space of $M_{m_1} \times M_{m_2}$ at a point (x, y) in terms of the tangent space of M_{m_1} at x and the tangent space of M_{m_2} at y.

10. Show that the equation $x_1^2 + x_2^2 + \cdots + x_n^2 = 1$ defines a C^1 $(n-1)$-dimensional surface in $\mathbb{R}^n$ (called the *unit sphere*). Compute its tangent space at every point.

11. *Show that the orthogonal $n \times n$ matrices form a C^1 surface of dimension $n(n-1)/2$ in $\mathbb{R}^{n^2}$.

12. Show that the $n \times n$ matrices with determinant equal to one form a C^1 surface of dimension $n^2 - 1$ in $\mathbb{R}^{n^2}$.

13. Let M_m be a C^1 m-dimensional surface in $\mathbb{R}^n$. Let TM_m be the set of points in $\mathbb{R}^{2n}$ of the form (x, y) with x in M_m and y in the tangent space to M_m at x. (This is called the *tangent bundle* of M_m.) Prove that TM_m is a C^1 $2m$-dimensional surface in $\mathbb{R}^{2n}$.

13.3 Maxima and Minima on Surfaces

13.3.1 Lagrange Multipliers

As an application of the material in section 13.2.4, we consider the problem of determining the local maxima and minima of a C^1 function f defined on a C^1 surface $M_m \subseteq \mathbb{R}^n$. Given a local parametric representation $g : U \to \mathbb{R}^n$ with U an open set in $\mathbb{R}^m$, we can find all local maxima and minima on $g(U)$ by finding the local maxima and minima of $f \circ g$ on U. This reduces the problem to one we have already solved. However, the procedure is often awkward and long, especially if we have to consider several local parametric representations to cover the surface or if the expressions for the functions g are complicated.

There is an alternate method, called *Lagrange multipliers*, that works in the situation where the surface is given implicitly and the function f is the restriction to the surface of a function defined on $\mathbb{R}^n$ (or at least a neighborhood of M_m in $\mathbb{R}^n$). Whenever these conditions are fulfilled, the method of Lagrange multipliers is almost always simpler and more direct. Frequently problems solved by Lagrange multipliers are stated in terms of *constraints*: maximize $f(x)$ defined on $\mathbb{R}^n$ subject to the constraints $G(x) = 0$. We recognize that the constraint equation restricts x to the surface $G(x) = 0$, and in fact the hypotheses on the constraint equation will be exactly what we needed to conclude that the solution set of $G(x) = 0$ is a C^1 surface.

The method can be described very elegantly. We assume f and G are C^1 functions, and we write

$$G(x) = \begin{pmatrix} G_1(x) \\ \vdots \\ G_k(x) \end{pmatrix}$$

where $G_j(x)$ are real-valued C^1 functions. We form the function $H(x, \lambda)$ of $n + k$ variables $H(x, \lambda) = f(x) + \lambda_1 G_1(x) + \cdots + \lambda_k G_k(x)$ and find all critical points of H (points where $\nabla H(x, \lambda) = 0$). Then the local maxima and minima of $f(x)$ on $G(x) = 0$ must occur at the x values of these critical points. The λ values are discarded. The method also finds some values of x that are neither maxima nor minima.

Why does the method work? We can break the $n + k$ equations $\nabla H = 0$ into two groups, namely the x-derivatives and the λ-derivatives.

The λ-derivatives yield $G_1(x) = 0, G_2(x) = 0, \ldots, G_k(x) = 0$, which are the constraint equations, so every solution of $\nabla H(x, \lambda) = 0$ automatically satisfies the constraint equations. The x-derivatives yield

$$\frac{\partial f}{\partial x_j}(x) = -\sum_{r=1}^{k} \lambda_r \frac{\partial G_r}{\partial x_j}(x), \quad j = 1, \ldots, n,$$

which we can abbreviate as

$$\nabla f(x) = -\sum_{r=1}^{k} \lambda_r \nabla G_r(x).$$

This equation says that ∇f lies in the span of the gradients of the functions $G_r(x)$, or in other words in the normal space to the surface $G(x) = 0$, provided we assume these gradients are linearly independent. This means the components of the gradient of f are zero in the tangential directions, which is what we expect if f has a maximum or minimum on the surface.

Theorem 13.3.1 (*Lagrange Multipliers*) *Let $f : \mathbb{R}^n \to \mathbb{R}$ and $G : \mathbb{R}^n \to \mathbb{R}^k$ be C^1 functions, and let $\tilde{x}$ be a point where $G(\tilde{x}) = 0$ and such that $dG(\tilde{x})$ has rank k. If $f(\tilde{x})$ is a local maximum or minimum for f on $\{x : G(x) = 0\}$, meaning there exists a neighborhood U of $\tilde{x}$ in $\mathbb{R}^n$ such that $f(\tilde{x}) \geq f(x)$ (or $f(\tilde{x}) \leq f(x)$) for all x in $U \cap \{x : G(x) = 0\}$, then there exists $\tilde{\lambda}$ in $\mathbb{R}^k$ such that $H(x, \lambda) = f(x) + \lambda \cdot G(x)$ has a critical point at $(\tilde{x}, \tilde{\lambda})$.*

Proof: The condition rank $dG(\tilde{x}) = k$ implies that the level set $G(x) = 0$ is a C^1 surface (dimension $m = n - k$) in a neighborhood of $\tilde{x}$, and since the result is purely local, we can disregard what happens away from $\tilde{x}$. We have seen that $\nabla_\lambda H(\tilde{x}, \tilde{\lambda}) = 0$ just says $G(\tilde{x}) = 0$, so it remains to show $\nabla_x H(\tilde{x}, \tilde{\lambda}) = 0$ for some $\tilde{\lambda}$. Now consider any C^1 curve $h(t)$ lying in the surface $G(x) = 0$ passing through $\tilde{x}$, say $h(t_0) = \tilde{x}$. Then $f \circ h(t)$ is a C^1 function (the composition of C^1 functions) with a local maximum or minimum at t_0. Hence $(d/dt) f \circ h(t_0) = 0$ by the one-dimensional theorem. But $(d/dt) f \circ h(t_0) = \nabla f(\tilde{x}) \cdot h'(t_0)$, so $\nabla f(\tilde{x})$ is perpendicular to the tangent vector $h'(t_0)$ to the curve. By varying the curves we see that $\nabla f(\tilde{x})$ is perpendicular to the tangent space of the surface at $\tilde{x}$; hence, $\nabla f(\tilde{x})$ lies in the normal space. But we know that the normal space is spanned by $\nabla G_1(\tilde{x}), \ldots, \nabla G_k(\tilde{x})$; hence, $\nabla f(\tilde{x}) = -\tilde{\lambda} \cdot \nabla G(\tilde{x})$ for some $\tilde{\lambda}$ in $\mathbb{R}^k$ and this says $\nabla_x H(\tilde{x}, \tilde{\lambda}) = 0$. QED

If the rank condition on G is not satisfied, the method may not work. For example, consider $G(x, y) = y^3 - x^2$ in $\mathbb{R}^2$, which gives a curve with a cusp at the origin, as shown in Figure 13.3.1.

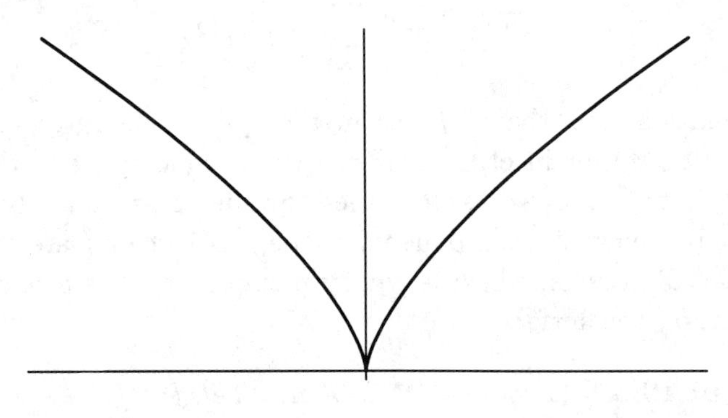

Figure 13.3.1:

Clearly the function $f(x, y) = y$ assumes a minimum at the origin, but

$$\nabla f(0,0) = \begin{pmatrix} 0 \\ 1 \end{pmatrix}$$

is not a linear multiple of

$$dG(0,0) = \begin{pmatrix} 0 \\ 0 \end{pmatrix}.$$

Thus in using the method of Lagrange multipliers, you must also search for maxima and minima among the points where rank $dG < k$. Fortunately these are usually either isolated points or themselves surfaces of lower dimension to which the method may be applied anew.

The method is also useful for finding maxima and minima of functions over regions given by inequalities—for interior points use the usual $\nabla f = 0$ test, and for boundary points use Lagrange multipliers.

For example, suppose we want to find the maximum and minimum values of $f(x,y) = x^2 - y^2$ on the unit disk $x^2 + y^2 \leq 1$. We first observe that

$$\nabla f(x,y) = \begin{pmatrix} 2x \\ -2y \end{pmatrix}$$

vanishes only at the origin $(0,0)$, so this is the only possible point in the interior. On the boundary $x^2 + y^2 - 1 = 0$ we solve $\nabla H(x,y,\lambda) = 0$ for $H(x,y,\lambda) = x^2 - y^2 + \lambda(x^2 + y^2 - 1)$, obtaining the equations

$$2x + 2x\lambda = 0, \quad -2y + 2y\lambda = 0, \quad x^2 + y^2 - 1 = 0.$$

From the first equation we obtain $x = 0$ or $\lambda = -1$ and from the second $y = 0$ or $\lambda = 1$, so all solutions (x,y,λ) are $(0,\pm 1,1), (\pm 1,0,-1)$. Discarding the λ values we have the four points $(\pm 1,0)$ and $(0,\pm 1)$. Finally we evaluate f at the five possible points and find that $(\pm 1,0)$ are maxima and $(0,\pm 1)$ are minima, but $(0,0)$ is neither.

As another example we consider the problem of maximizing and minimizing $f(x,y) = x + y$ on the lemniscate $(x^2 - y^2)^2 = x^2 + y^2$. Forming $H(x,y,\lambda) = x + y + \lambda((x^2 - y^2)^2 - x^2 - y^2)$ and setting $\nabla H = 0$ we obtain the equations

$$\begin{aligned} &1 + \lambda(4x(x^2 - y^2) - 2x) = 0, \\ &1 + \lambda(-4y(x^2 - y^2) - 2y) = 0, \\ &(x^2 - y^2)^2 = x^2 + y^2. \end{aligned}$$

Subtracting the first two equations we obtain

$$\lambda(4(x+y)(x^2 - y^2) - 2(x - y)) = 0.$$

Since $\lambda \neq 0$ by the first equation, we obtain either $x = y = 0$ or $4(x+y)^2 - 2 = 0$ and, hence, $x + y = \pm 1/\sqrt{2}$. The points on the lemniscate where $x + y = \pm 1/\sqrt{2}$ give the maximum and minimum values of $x + y$, since $f(0,0) = 0$ and $(0,0)$, is the only point at which the gradient of $(x^2 - y^2)^2 - x^2 - y^2$ vanishes.

13.3.2 A Second Derivative Test*

It is possible to give a second derivative test to distinguish between maxima, minima, and saddle points. Suppose $\tilde{x}$ is a point where

$\nabla H(\tilde{x}, \tilde{\lambda}) = 0$ and $dG(\tilde{x})$ has rank k. As in the proof of the theorem let $h(t)$ be a C^2 curve lying in the level set $G(x) = 0$ passing through $\tilde{x}$, $h(t_0) = \tilde{x}$. We assume that f and G are also C^2. We want to compute the second derivative $d^2/dt^2(f \circ h)$ at t_0. This is

$$(*) \qquad \sum_{j=1}^{n}\sum_{l=1}^{n} \frac{\partial^2 f}{\partial x_j \partial x_l}(\tilde{x})h'_j(t_0)h'_l(t_0) + \sum_{j=1}^{n} \frac{\partial f}{\partial x_j}(\tilde{x})h''_j(t_0).$$

Now we want to eliminate h'' from this expression. We can do this by observing that $G_r(h(t)) \equiv 0$, so $(d^2/dt^2)G_r(h(t_0)) = 0$ for $r = 1, \ldots, k$. Thus

$$(**) \qquad \sum_{j=1}^{n}\sum_{l=1}^{n} \frac{\partial^2 G_r}{\partial x_j \partial x_l}(\tilde{x})h'_j(t_0)h'_l(t_0) + \sum_{j=1}^{n} \frac{\partial G_r}{\partial x_j}(\tilde{x})h''_j(t_0) = 0.$$

Now recall that we have $\partial f/\partial x_j(\tilde{x}) = -\sum_{r=1}^{k} \tilde{\lambda}_r \partial G_r/\partial x_j(\tilde{x})$, so if we multiply each of the equations $(**)$ by $\tilde{\lambda}_r$ and sum we obtain

$$\begin{aligned}
\sum_{j=1}^{n} \frac{\partial f}{\partial x_j}(\tilde{x})h''(t_0) &= -\sum_{j=1}^{n}\sum_{r=1}^{k} \tilde{\lambda}_r \frac{\partial G_r}{\partial x_j}(\tilde{x})h''(t_0) \\
&= \sum_{j=1}^{n}\sum_{l=1}^{n}\sum_{r=1}^{k} \tilde{\lambda}_r \frac{\partial G_r}{\partial x_j \partial x_l}(\tilde{x})h'_j(t_0)h'_l(t_0).
\end{aligned}$$

Finally we can substitute this back into $(*)$ to obtain

$$\begin{aligned}
\frac{d^2}{dt^2} f \circ h(t_0) &= \sum_{j=1}^{n}\sum_{l=1}^{n} \left(\frac{\partial^2 f}{\partial x_j \partial x_l}(\tilde{x}) + \sum_{r=1}^{k} \tilde{\lambda}_r \frac{\partial^2 G_r}{\partial x_j \partial x_l}(\tilde{x}) \right) h'_j(t_0)h'_l(t_0) \\
&= \langle (d^2 f(\tilde{x}) + \tilde{\lambda} \cdot d^2 G(\tilde{x}))h'(t_0), h'(t_0) \rangle
\end{aligned}$$

with the obvious abbreviations. This may seem like a messy expression, but it is easy to interpret. Remember that $h(t)$ is just a curve lying in the surface passing through $\tilde{x}$ at $t = t_0$, so $h'(t_0)$ is just an arbitrary vector in the tangent space to the surface at $\tilde{x}$. Thus the signs of the second derivatives of f along curves in the surface through $\tilde{x}$ are given by the quadratic form associated to the matrix $d^2 f(\tilde{x}) + \tilde{\lambda} \cdot d^2 G(\tilde{x})$ restricted to the tangent space at $\tilde{x}$. Notice that the situation has changed in two ways from the unconstrained problem: first, we restrict attention to the tangent space to the surface at $\tilde{x}$; second, we "correct" the Hessian $d^2f(\tilde{x})$ by adding the term $\tilde{\lambda}\cdot d^2G(\tilde{x})$. Here $\tilde{\lambda}$ is the vector given by the equation $df(\tilde{x})+\tilde{\lambda}\cdot dG(\tilde{x})=0$. Also the tangent space at $\tilde{x}$ is the set of vectors u in $\mathbb{R}^n$ such that $dG(\tilde{x})u=0$. Thus everything we need is computable from G.

Now if f has a local maximum (or minimum) at $\tilde{x}$, then $(d^2/dt^2)f\circ h(t_0)$ is always ≤ 0 (or ≥ 0), so

$$\langle(d^2f(\tilde{x})+\tilde{\lambda}\cdot d^2G(\tilde{x}))u,u\rangle\le 0\ \ (\text{or}\ \ \ge 0)$$

for all u such that $dG(\tilde{x})u=0$ is a necessary condition for a maximum (or minimum). If we have strict inequality, then we have a local maximum (or minimum)—this requires an additional argument similar to the one in the unconstrained case to show that there is a neighborhood of $\tilde{x}$ on the surface (as opposed to a union of neighborhoods on the curves passing through $\tilde{x}$) such that $f(\tilde{x})$ is the maximum (or minimum).

Let's look at some examples. The parabolas $y=\pm x^2$ shown in Figure 13.3.2 have the same tangent space at the origin, but the function $f(x,y)=y$ has a maximum on one and a minimum on the other.

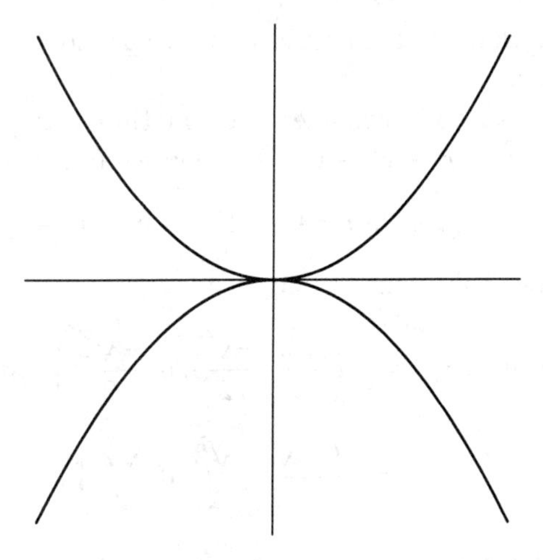

Figure 13.3.2:

Taking $G(x,y)=y\pm x^2$ and applying the method of Lagrange multipliers we set $H(x,y,\lambda)=y+\lambda(y\pm x^2)$. From $\nabla H(x,y,\lambda)=0$ we get $\pm 2x\lambda = 0, 1 + \lambda = 0, y \pm x^2 = 0$, which has the unique solution $(x, y) = (0, 0)$ and $\lambda = -1$. Now we compute the matrix

$$d^2 f + \lambda d^2 G = \begin{pmatrix} 0 & 0 \\ 0 & 0 \end{pmatrix} + \lambda \begin{pmatrix} \pm 2 & 0 \\ 0 & 0 \end{pmatrix}$$

and so for $\lambda = -1$ this is

$$\begin{pmatrix} \mp 2 & 0 \\ 0 & 0 \end{pmatrix}.$$

Now the tangent space at $(0, 0)$ is the x-axis (given by $(u, v) \cdot \nabla G(x, y) = \pm 2xu + v = 0$, which for $(x, y) = 0$ gives $v = 0$). Restricting the matrix

$$\begin{pmatrix} \mp 2 & 0 \\ 0 & 0 \end{pmatrix}$$

to vectors of the form

$$\begin{pmatrix} u \\ 0 \end{pmatrix}$$

gives

$$\left\langle \begin{pmatrix} \mp 2 & 0 \\ 0 & 0 \end{pmatrix} \begin{pmatrix} u \\ 0 \end{pmatrix}, \begin{pmatrix} u \\ 0 \end{pmatrix} \right\rangle = \mp 2u^2,$$

so we have a maximum in the first case (lower parabola) and a minimum in the second case.

For another example, consider the function $f(x, y, z) = x - y$ on the unit sphere $x^2 + y^2 + z^2 - 1 = 0$. Then we find from $\nabla H = 0$ that

$$1 + 2\lambda x = 0, \quad -1 + 2\lambda y = 0, \quad 2\lambda z = 0, \quad x^2 + y^2 + z^2 = 1,$$

which has the two solutions

$$\begin{aligned} (x, y, z, \lambda) &= \left(\frac{\sqrt{2}}{2}, \frac{-\sqrt{2}}{2}, 0, \frac{-\sqrt{2}}{2} \right) \quad \text{and} \\ &= \left(-\frac{\sqrt{2}}{2}, \frac{\sqrt{2}}{2}, 0, \frac{\sqrt{2}}{2} \right). \end{aligned}$$

The matrix $d^2 f + \lambda d^2 G$ is

$$\lambda \begin{pmatrix} 2 & 0 & 0 \\ 0 & 2 & 0 \\ 0 & 0 & 2 \end{pmatrix},$$

which is negative definite on all of $\mathbb{R}^3$ in the first case ($\lambda = -\sqrt{2}/2$) and positive definite in the second ($\lambda = \sqrt{2}/2$). It follows that it is negative and positive definite when restricted to the tangent space of the sphere at the appropriate points, so we have a maximum and a minimum at the points $(\sqrt{2}/2, -\sqrt{2}/2, 0)$ and $(-\sqrt{2}/2, \sqrt{2}/2, 0)$.

13.3.3 Exercises

1. Let M_m be a C^1 surface in $\mathbb{R}^n$, and let y be a point in $\mathbb{R}^n$ not on M_m. If x is a point on M_m that minimizes or maximizes the distance to y, prove that the line joining x and y is perpendicular to the surface at x (i.e., perpendicular to the tangent space at x). (**Hint**: it is easier to consider the square of the distance.)

2. If M_m and $M'_{m'}$ are two disjoint C^1 surfaces in $\mathbb{R}^n$ and if x in M_m and y in $M'_{m'}$ minimize or maximize the distance apart, prove that the line joining them is perpendicular to both surfaces at the intersections.

3. Let M_m be a C^1 surface in $\mathbb{R}^n$ that is compact (as a metric subspace of $\mathbb{R}^n$). Prove that there exist a pair of points x and y in M_m that maximize the distance among all such pairs and that the diameter joining them intersects the surface perpendicularly at both points.

4. Use the method of Lagrange multipliers to locate possible local maxima and minima of the function f subject to the conditions $G = 0$ in the following:

   a. $f(x, y, z) = x^2 + 4y^2 - z^2, G(x, y, z) = x^2 + y^2 + z^2 - 1.$

   b. $f(x, y, z) = zx + 2y, G_1(x, y, z) = x^2 + y^2 + 2z^2 - 1,$ $G_2(x, y, z) = x^2 + y + z.$

   c. $f(x, y, z) = x^2 + y^2 + z^2, G(x, y, z) = x^2 + 4y^2 - 2z^2 - 1.$

5. Apply the second derivative test to each of the critical points found in exercise 4.

6. Find the maximum and minimum values of $f(x, y) = x^3 - 3yx + y$ on the unit disk $x^2 + y^2 \leq 1$.

7. Show that the entropy $-\sum_{j=1}^{n} x_j \log x_j$ is maximized subject to the constraints $\sum_{j=1}^{n} x_j = 1$ and all $x_j > 0$, at the point $(1/n, 1/n, \ldots, 1/n)$.

13.4 Arc Length

13.4.1 Rectifiable Curves

Let $g : [a, b] \to \mathbb{R}^n$ be a continuous curve in $\mathbb{R}^n$. We want to talk about its length. The idea is very simple. If the curve consisted of broken s-traight line segments, we would simply add the lengths of the segments, measuring the length of a straight line segment by the Pythagorean distance from one endpoint to the other. For a general curve we can approximate the length by choosing a sequence of points lying along the curve and connecting them in order by straight line segments. Since we believe that the shortest distance between two points is a straight line, the length of the approximating broken line curve should be an underapproximation to the length of the curve, as in Figure 13.4.1.

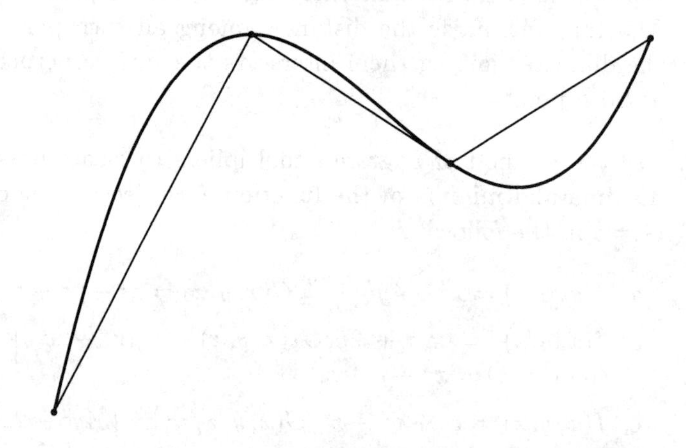

Figure 13.4.1:

We can think of this approximation in practical terms as the result of hammering nails along a curve and pulling a string taut between the nails and measuring the length of the string. As we increase the number of points along the curve, the length of the approximating broken line curves will increase (by the triangle inequality), and we expect the limit to be the length of the original curve. Because the lengths are increasing we can replace the limit with a sup.

Definition 13.4.1 *The length of a continuous curve* $g : [a, b] \to \mathbb{R}^n$ *is the* sup *of* $\sum_{j=0}^{N} |g(t_{j+1}) - g(t_j)|$ *taken over all partitions* $a = t_0 < t_1 < \ldots < t_N = b$ *of the interval* $[a, b]$, *where* $|g(t_{j+1}) - g(t_j)|$ *is the distance between the points* $g(t_{j+1})$ *and* $g(t_j)$ *in* $\mathbb{R}^n$. *If the length is finite, we say the curve is rectifiable.*

If a curve is rectifiable, then the length can be obtained as a limit of $\sum_{j=0}^{N} |g(t_{j+1}) - g(t_j)|$ along a sequence of partitions; and since adding points to the partition only increases the value of the sum, we can assume that the maximum of $t_{j+1} - t_j$ goes to zero. Conversely, *if for one sequence of partitions with the maximum of* $t_{j+1} - t_j$ *going to zero the limit of*

$$\sum_{j=0}^{N} |g(t_{j+1}) - g(t_j)|$$

exists and is finite, then the curve is rectifiable and the limit is the length. This is not immediately obvious from the definition but requires a proof. The idea is to use the uniform continuity of g to add points to the partition without appreciably increasing the sum.

Suppose $P_1, P_2, \ldots$ is the sequence of partitions of $[a, b]$, and denote by $L(P_k)$ the sum $\sum_{j=0}^{N} |g(t_{j+1}) - g(t_j)|$ for the partition P_k. (The number of points increases with the partition, so strictly speaking we should write N_k rather than N.) Since each $L(P_k) \leq L$ where L denotes the length of the curve, we have $\lim_{k\to\infty} L(P_k) \leq L$ and so it suffices to prove the reverse inequality. Since L is defined to be the sup of $L(Q)$ over all partitions Q, we need to show that for every Q and $\varepsilon > 0$ the inequality $L(Q) \leq L(P_k) + \varepsilon$ holds for all sufficiently large k; for then $L(Q) \leq \lim_{k\to\infty} L(P_k) + \varepsilon$, hence $L \leq \lim_{k\to\infty} L(P_k) + \varepsilon$ and, hence, $L \leq \lim_{k\to\infty} L(P_k)$. To do this we will throw the points of Q into the partition P_k—call the resulting partition $Q \cup P_k$ (we have to rearrange the points in their correct order and discard repeats, of course). Then $L(Q) \leq L(Q \cup P_k)$, so it suffices to show $L(Q \cup P_k) \leq L(P_k) + \varepsilon$ for all sufficiently large k.

We have $L(P_k) \leq L(Q \cup P_k)$, but to obtain the reverse inequality $L(Q \cup P_k) \leq L(P_k) + \varepsilon$ we will have to reason carefully. Let N denote the number of points in Q. This number remains fixed, while by varying k we can make the points in P_k increase indefinitely, with the distance between them very small. Let s denote a point in Q. If s is not in P_k, then it will fall between two points t_j and t_{j+1} of P_k. Thus the sum $L(Q \cup P_k)$ will contain

$$|g(s) - g(t_j)| + |g(t_{j+1}) - g(s)|$$

in place of $|g(t_{j+1}) - g(t_j)|$ in $L(P_k)$. (To avoid technicalities we assume that k is taken large enough so that at most one point of Q lies in each interval $[t_j, t_{j+1}]$. This is possible because the lengths of these intervals are going to zero.) For intervals $[t_j, t_{j+1}]$ of P_k containing no points of Q the same $|g(t_{j+1}) - g(t_j)|$ will occur in both $L(P_k)$ and $L(Q \cup P_k)$ sums. Thus we have an estimate $L(Q \cup P_k) \leq L(P_k) + N\delta$ where δ is any upper bound for

$$|g(s) - g(t_j)| + |g(t_{j+1}) - g(s)|$$

with s in $[t_j, t_{j+1}]$. By the uniform continuity of g (it is a continuous function on a compact interval) we can make δ as small as we like, say $\delta < \varepsilon/N$, by making all the intervals $[t_j, t_{j+1}]$ sufficiently small, and we do this by taking k large enough. Then $L(Q \cup P_k) \leq L(P_k) + \varepsilon$ and we have $\lim_{k\to\infty} L(P_k) = L$ as claimed. Incidentally, this argument also shows that $\lim_{k\to\infty} L(P_k) = L$ along any sequence of partitions $P_1, P_2, \ldots,$ such that the maximum length of subintervals goes to zero. The situation is analogous to that of the Riemann integral.

It is important to observe that the length of a curve is a property of the subset $g([a, b])$ of $\mathbb{R}^n$ and its ordering and not on the particular parametrization given by the function g. This is simplest to state if g is assumed to be one-to-one. If $h : [c, d] \to [a, b]$ is a continuous function that is one-to-one and onto, then $g \circ h : [c, d] \to \mathbb{R}^n$ is another one-to-one parametrization of the same subset of $\mathbb{R}^n$. Since h being one-to-one implies that h is either increasing or decreasing, the order of points along $g \circ h$ is either the same or the reverse of the order along g, so the lengths $L(P)$ for partitions of g and $g \circ h$ are in one-to-one correspondence and so the lengths of the curves g and $g \circ h$ are the same. It is possible to show further that any other one-to-one continuous parametrization $g_1 : [c,d] \to \mathbb{R}^n$ of the same set, $g_1([c,d]) = g([a,b])$, must be of the form $g_1 = g \circ h$ (see exercises). Thus for *simple curves*, those that have a one-to-one parametrization, the length is independent of the parametric representation. For curves that intersect themselves in complicated ways, the situation is more complicated and we will not attempt to describe it. There is, however, one case that is especially easy and important. Suppose $g : [a,b] \to \mathbb{R}^n$ with $g(a) = g(b)$ but otherwise $g(s) \neq g(t)$ for $s \neq t$ in $[a,b]$, as in Figure 13.4.2. The image $g([a,b)]$ is called a *simple closed curve*. For these it is again

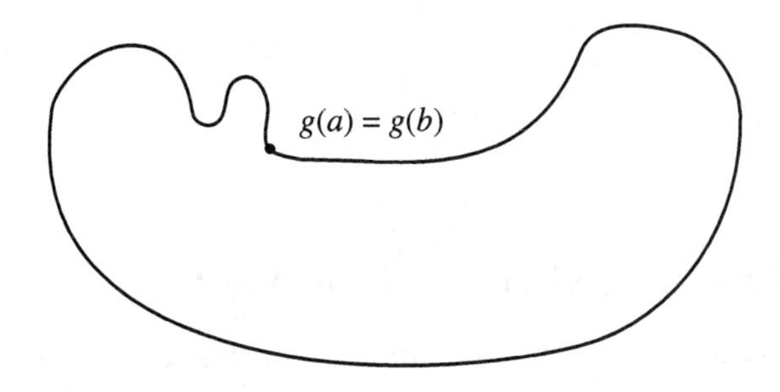

Figure 13.4.2:

true that the length is independent of the parametric representation, even of the point $g(a) = g(b)$ on the curve.

It is easy enough to give examples of reasonable curves that are not rectifiable. For example, Figure 13.4.3 shows the graph of $y = x \sin 1/x$ on $[0,1]$, given by the parametric representation $g(t) = (t, t\sin 1/t)$ for t in $[0,1]$. Note that $\lim_{t\to 0} t \sin 1/t = 0$, so g is continuous and, in fact, g is differentiable at every point of the interior although not at zero. However, if we take the partition points $t_k = 2/k\pi, k = 1, \ldots, N$, then $\sin 1/t_k = 0$ for k even and ± 1 for k odd; hence,

$$\begin{aligned} \sum |g(t_{k+1}) - g(t_k)| &\geq \sum \left| t_{k+1} \sin \frac{1}{t_{k+1}} - t_k \sin \frac{1}{t_k} \right| \\ &= 2 \sum_{k \text{ odd}} |t_k| = 2 \sum_{k \text{ odd}} \frac{2}{k\pi} = +\infty \end{aligned}$$

and so the length is infinite.

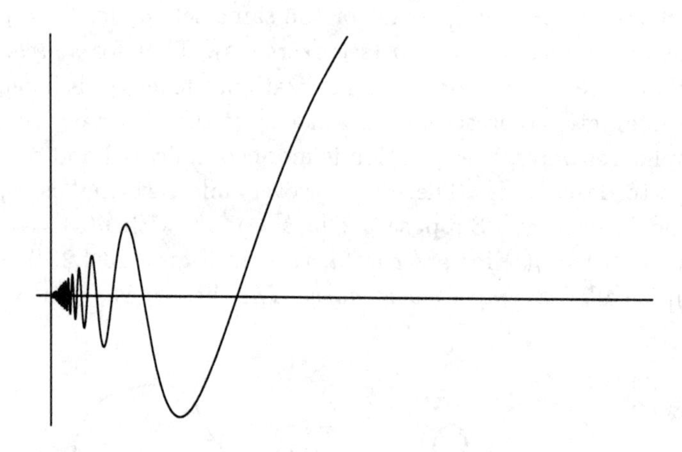

Figure 13.4.3:

13.4.2 The Integral Formula for Arc Length

Next, we would like to derive the familar calculus formula for the length of a curve under the assumption that $g : [a, b] \to \mathbb{R}^n$ is C^1. Actually, it suffices to have g piecewise C^1, which is quite handy for applications.

Theorem 13.4.1 *Let $g : [a, b] \to \mathbb{R}^n$ be a continuous curve, and assume that there exists a finite partition $a = t_0 < t_1 < \cdots < t_N = b$ of $[a, b]$ such that $g'(t)$ exists and is continuous on $[t_j, t_{j+1}]$ (at the endpoints this means one-sided derivatives) for each $j = 0, \ldots, N-1$. Then the curve is rectifiable and the length is given by $L = \int_a^b |g'(t)|\, dt$.*

If we recall the interpretation of $g'(t)$ as the velocity of the motion along the curve and $|g'(t)|$ as the speed, then this formula says that the distance traveled is obtained by integrating the speed.

Proof: It suffices to prove the formula if g is C^1 over the whole interval, since both L and the integral are additive over the finite partition. Then $\int_a^b |g'(t)|\, dt$ is the integral of a continuous function, so it is approximated by the Cauchy sums

$$\sum_{j=0}^{N} |g'(t_j)|(t_{j+1} - t_j)$$

for partitions of the interval (not to be confused with the single partition in the hypothesis of the theorem). We want to say that $|g'(t_j)|(t_{j+1} - t_j)$ is approximately equal to $|g(t_{j+1}) - g(t_j)|$. To do this we will show that the vectors $g'(t_j)(t_{j+1} - t_j)$ and $g(t_{j+1}) - g(t_j)$ are approximately equal and, hence, their lengths are approximately equal. But

$$g(t_{j+1}) - g(t_j) = \int_{t_j}^{t_{j+1}} g'(t)\, dt$$

by the fundamental theorem of the calculus (since $g(t)$ takes values in $\mathbb{R}^n$, we apply the theorem to each coordinate and put them all together). If the interval is small, then $g'(t)$ won't vary much over it, so $\int_{t_j}^{t_{j+1}} g'(t)\, dt$ will be approximately $g'(t_j)(t_{j+1} - t_j)$. More precisely,

$$\begin{aligned} \left| \int_{t_j}^{t_{j+1}} g'(t)\, dt - g'(t_j)(t_{j+1} - t_j) \right| &= \left| \int_{t_j}^{t_{j+1}} (g'(t) - g'(t_j))\, dt \right| \\ &\le \int_{t_j}^{t_{j+1}} |g'(t) - g'(t_j)|\, dt \end{aligned}$$

by Minkowski's inequality. Thus

$$|(g(t_{j+1}) - g(t_j)) - g'(t_j)(t_{j+1} - t_j)| \le \delta(t_{j+1} - t_j)$$

where δ is the sup of $|g'(t) - g'(t_j)|$ for t in $[t_j, t_{j+1}]$. By the uniform continuity of $g'(t)$, we can make δ as small as desired by taking the partition fine enough. Using the triangle inequality in the form $||u| - |v|| \le |u - v|$ for $u = g(t_{j+1}) - g(t_j)$ and $v = g'(t_j)(t_{j+1} - t_j)$, we also have $||g'(t_j)|(t_{j+1} - t_j) - |g(t_{j+1}) - g(t_j)|| \le \delta(t_{j+1} - t_j)$. Adding up, we obtain

$$\begin{aligned} &\left| \sum_{j=0}^{N} |g'(t_j)|(t_{j+1} - t_j) - \sum_{j=0}^{N} |g(t_{j+1}) - g(t_j)| \right| \\ &\le \sum_{j=0}^{N} ||g'(t_j)|(t_{j+1} - t_j) - |g(t_{j+1} - g(t_j)|| \\ &\le \delta \sum_{j=0}^{N} (t_{j+1} - t_j) = \delta(b - a). \end{aligned}$$

Thus the integral and the length are equal. QED

In particular, if the curve is given as the graph of a C^1 function of one of the variables, say $(x_1, \ldots, x_{n-1}) = h(x_n)$ for $a \leq x_n \leq b$, then the length is given by

$$\int_a^b \left(1 + \sum_{j=1}^{n-1} h_j'(t)^2\right)^{1/2} dt,$$

since $g(t) = (h_1(t), \ldots, h_{n-1}(t), t)$ is the associated parametrization.

13.4.3 Arc Length Parameterization*

The notion of arc length allows us to choose a particularly apt parametric representation for a simple rectifiable curve, in which arc length is the parameter. Let us say that the curve is initially given by $g : [a, b] \to \mathbb{R}^n$. We then define a function $s(t)$ on $[a, b]$ by setting $s(t_0)$ equal to the length of the curve for t restricted to $[a, t_0]$. Then $s(t)$ takes values in $[0, L]$ and is monotone increasing. Since we are assuming the curve is simple, g is one-to-one and so $s(t)$ is strictly increasing. It is not hard to show that $s(t)$ is continuous because the curve is rectifiable (see exercise set 13.4.4). Thus s is a one-to-one function taking $[a, b]$ onto $[0, L]$, so it has a continuous inverse because the intervals are compact (see exercises). We denote the inverse simply by $t(s)$. Then $g(t(s)) : [0, L] \to \mathbb{R}^n$ is a parametric representation of the curve and the parameter s equals the length of the curve from the initial point corresponding to $s = 0$ to the point with parameter s. We can interpret this parametric representation as tracing out the curve with speed equal to one, even if the derivative doesn't exist so that there is no tangent vector. For a simple curve it is easy to see that there are only two ways to parametrize by arc length—starting at either of the ends.

The arc length parametrization has several useful properties. If the curve is a C^1 one-dimensional surface so that it is given by a C^1 function $g : [a, b] \to \mathbb{R}^n$ with $g'(t) \neq 0$, then the arc length parametrization will be C^1 (in fact it will be C^k if g is C^k). The tangent vector will then have length equal to one. If the arc length parametrization $x(s)$ is C^2, then the acceleration vector $x''(s)$ will be normal to the curve. This follows by differentiating the identity $x'(s) \cdot x'(s) = 1$. This is called the *principal normal*. The plane determined by the tangent vector and the principal normal is called the *osculating plane*, and the reciprocal of the length of the principal normal is called the *radius of curvature*. The circle of that radius in the osculating plane tangent to the curve at the point (on the appropriate side) is thought of as the circle that best approximates the curve at the point.

13.4.4 Exercises

1. Prove that the shortest curve joining two points is a straight line segment. Can you also give a proof directly from the formula $L = \int_a^b |g'(t)|\, dt$ for those curves to which it applies?

2. Prove that arc length is additive: if the curve is divided into two parts the sum of the lengths of the parts equals the length of the whole.

3. Let $g : [a, b] \to \mathbb{R}^n$ be a rectifiable curve, and let $s(t) : [a, b] \to [0, L]$ be defined by $s(t_0) =$ length of the curve g restricted to $[a, b_0]$. Prove that $s(t)$ is continuous.

4. Let $g_1 : [a, b] \to \mathbb{R}^n$ and $g_2 : [c, d] \to \mathbb{R}^n$ be two one-to-one continuous function with the same image. Prove that there exists a continuous function $h : [a, b] \to [c, d]$ such that $g_1 = g_2 \circ h$.

5. Prove that a one-to-one onto map of compact intervals (or more generally compact metric spaces) has a continuous inverse.

6. Show that there are exactly two ways to parametrize a rectifiable curve so that the arc length is equal to the parameter.

7. Prove that if a curve is given by a C^1 parametrization g with $g'(t) \neq 0$, then the arc length parametrization is also C^1 and has tangent vector of length one. Show also that if g is C^k, then so is the arc length parametrization. Give an example to show that the condition $g'(t) \neq 0$ is essential.

8. Prove that if a C^2 curve lies in a plane in $\mathbb{R}^n$, then the osculating plane to the curve at any point is that plane.

9. Prove that the radius of curvature of any circle is equal to radius of the circle.

10. Prove that a continuous curve $g : [a,b] \to \mathbb{R}^n$ is rectifiable if and only if each coordinate function g_k is of *bounded variation* (a function $f : [a,b] \to \mathbb{R}$ is said to be of *bounded variation* if

$$\sup \sum_{j=0}^{N} |f(t_{j+1}) - f(t_j)|$$

is finite, the sup taken over all partitions of $[a,b]$).

11. Prove that a Lipschitz continuous curve $g : [a,b] \to \mathbb{R}^n$ is rectifiable.

12. Let $(r(t), \theta(t))$ for t in $[a,b]$ describe the polar coordinates of a curve in the plane. Assuming $r(t)$ and $\theta(t)$ are C^1 functions, find an integral fomula for the arc length of the curve.

13.5 Summary

13.1 Implicit Function Theorem

Definition *An o.d.e. in the form $M(x,y)dy/dx + N(x,y) = 0$ is called exact if there exists $F(x,y)$ such that $M = \partial F/\partial y$ and $N = \partial F/\partial x$.*

Theorem *$y(x)$ is a solution of the exact o.d.e. $M\partial y/\partial x + N = 0$ if and only if $F(x,y(x)) = c$ for some constant c, for all x.*

Definition *An $m \times m$ matrix A is called invertible (or nonsingular) if there exists an $m \times m$ matrix B with $AB = 1$, where I denotes the $m \times m$ identity matrix. We write $B = A^{-1}$. The norm $||A||$ of an $m \times m$ matrix is the least constant c such that $|Ax| \leq c|x|$ for all x in $\mathbb{R}^m$, where $|x|$ denotes the Euclidean norm.*

Theorem $|A_{jk}| \leq ||A|| \leq \left(\sum_{j,k} A_{jk}^2\right)^{1/2}$.

Theorem *If A is invertible and $||BA^{-1}|| < 1$, then $A + B$ is invertible and $(A+B)^{-1} = A^{-1}\sum_{k=0}^{\infty}(-1)^k(BA^{-1})^k$, the series converging.*

Theorem 13.1.1 (*Implicit Function Theorem*) *Let $F(x,y)$ be a C^1 function defined in a neighborhood of $\tilde{x}$ in $\mathbb{R}^n$ and $\tilde{y}$ in $\mathbb{R}^m$ taking values in $\mathbb{R}^m$, with $F(\tilde{x},\tilde{y}) = c$. Then if $F_y(\tilde{x},\tilde{y})$ is invertible there exists a neighborhood U of $\tilde{x}$ and a C^1 function $y : U \to \mathbb{R}^m$ such that $y(\tilde{x}) = \tilde{y}$ and $F(x,y(x)) = c$ for every x in U. Furthermore y is unique in that there exists a neighborhood V of $\tilde{y}$ (the image $y(U)$) such that there is only one solution z in V of $F(x,z) = c$, namely $z = y(x)$. Finally the differential of y can be computed by implicit differentiation,*

$$dy(x) = -F_y(x,y(x))^{-1}F_x(x,y(x)).$$

Theorem 13.1.2 (*Inverse Function Theorem*) *Let f be a C^1 function defined in a neighborhood of $\tilde{y}$ in $\mathbb{R}^n$ taking values in $\mathbb{R}^n$. If $df(\tilde{y})$ is invertible, then there exists a neighborhood U of $\tilde{x} = f(\tilde{y})$ and a C^1 function $g : U \to \mathbb{R}^n$ such that $f(g(x)) = x$ for every x in U. Furthermore, g maps U one-to-one onto an open neighborhood V of $\tilde{y}$ and $g(f(y)) = y$ for every y in V. The function g is unique in that for any x in U there is only one z in V with $f(z) = x$, namely $z = g(x)$. Finally $dg(x) = df(y)^{-1}$ if $f(y) = x$.*

Remark *The inverse function theorem is a special case of the implicit function theorem. Conversely, the implicit function theorem for $F(x,y)$ is a consequence of the inverse function theorem in $\mathbb{R}^{n+m}$ for $f(x,y) = (x,F(x,y))$.*

Theorem *If $M(x_0,y_0) \neq 0$, then the exact o.d.e. $M(dy/dx) + N = 0$ has a unique solution satisfying $y(x_0) = y_0$ in a neighborhood of x_0.*

13.2 Curves and Surfaces

Definition *A subset A of $\mathbb{R}^n$ is given parametrically if $A = \bigcup_j g_j(U_j)$ where $U_j \subseteq \mathbb{R}^m$ is open and $g : U_j \to \mathbb{R}^n$ is continuous; implicitly if $A = \{x : F(x) = 0\}$ where $F : \mathbb{R}^n \to \mathbb{R}^k$ is continuous; as the graph of a function if $A = \{(t,s)$ in $\mathbb{R}^{m+k} : s = f(t)\}$ where $f : \mathbb{R}^m \to \mathbb{R}^k$ is continuous, $m + k = n$, and the variables $t_1, \ldots, t_m, s_1, \ldots, s_k$ are a permutation of the variables $x_1, \ldots, x_n$.*

Example *The cusp curve in $\mathbb{R}^2$ given parametrically by $g(t) = (t^3, t^2)$*

for t in $\mathbb{R}^1$, or implicitly by $y^3 - x^2 = 0$, or as the graph of $y = x^{2/3}$, has a nonsmooth point at $(0,0)$.

Example *The unit sphere in $\mathbb{R}^3$ can be represented parametrically by spherical coordinates $(x,y,z) = (\cos\theta\sin\phi, \sin\theta\sin\phi, \cos\phi)$, implicitly by $x^2+y^2+z^2-1=0$, or locally as the graph of a function by $x = \pm\sqrt{1-y^2-z^2}$ or $y = \pm\sqrt{1-x^2-z^2}$ or $z = \pm\sqrt{1-x^2-y^2}$.*

Definition *If $g:(a,b) \to \mathbb{R}^n$ is a C^1 curve, then $g'(t)$ is the tangent vector to the curve at $g(t)$.*

Definition *The rank of an $m \times n$ matrix A is defined to be the dimension of the image of A (regarded as a linear transformation from $\mathbb{R}^m$ to $\mathbb{R}^n$). It is equal to the dimension of the span of the columns of A in $\mathbb{R}^n$ and is equal to the dimension of the span of the rows of A in $\mathbb{R}^m$ and equal to the size of the largest invertible square submatrix. A C^1 function $g: U \to \mathbb{R}^n$ where $U \subseteq \mathbb{R}^m$ is open is called an immersion if $dg(x)$ has rank m for every x in U. If g is also one-to-one and the inverse function $g^{-1}: g(U) \to \mathbb{R}^m$ is continuous it is called an embedding.*

Definition *If $g: U \to \mathbb{R}^n$ is an embedding, the tangent space to $g(U)$ at a point $\tilde{y} = \tilde{g}(x)$ in $g(U)$ is the span of the vectors $\partial g/\partial x_k(\tilde{x}), k = 1,\ldots,m$.*

Definition 13.2.1 *A C^1 m-dimensional surface in $\mathbb{R}^n$ (with $m \leq n$) is a subset M_m of $\mathbb{R}^n$ locally parametrized by embeddings, in the sense that every point in M_m lies in a neighborhood V in $\mathbb{R}^n$ such that there exists an embedding $g: U \to \mathbb{R}^n$ for $U \subseteq \mathbb{R}^m$ open such that $g(U) = V \cap M_m$.*

Theorem 13.2.1 *Let $g: U \to \mathbb{R}^n$ with $U \subseteq \mathbb{R}^n$ open be an immersion. For every point $\tilde{x}$ in U there exists a neighborhood $\tilde{U}$ of $\tilde{x}$ such that g is one-to-one on $\tilde{U}$ and $g(\tilde{U})$ is the graph of a C^1 function.*

Definition *In a C^1 surface, the subsets $V \cap M_m$ are called coordinate patches and the functions $g^{-1}: V \cap M_m \to U$ are called local coordinate maps.*

Theorem *If $V_1 \cap M_m$ and $V_2 \cap M_m$ are coordinate patches that intersect and $\tilde{U}_1$ and $\tilde{U}_2$ are the images of the intersection $V_1 \cap V_2 \cap M_m$ under g_1^{-1} and g_2^{-1} in U_1 and U_2, then $g_2^{-1} \circ g_1 : \tilde{U}_1 \to \tilde{U}_2$ and $g_1^{-1} \circ g_2 : \tilde{U}_2 \to \tilde{U}_1$ are C^1 functions.*

Theorem *The tangent space to $g_j(U_j)$ at a point $\tilde{y}$ on a surface M_m does not depend on the coordinate patch and can be taken as the definition of the tangent space $T_m M(\tilde{y})$ to M_m at $\tilde{y}$. It is equal to the set of tangent vectors $g'(t)$ at $g(t_0) = \tilde{y}$ as g ranges over all C^1 curves $g : (a, b) \to M_m$ lying in the surface and passing through $\tilde{y}$.*

Definition *The normal space $NM_m(x)$ to a surface M_m at a point x in M_m is the $(n-m)$-dimensional subspace of $\mathbb{R}^n$ of all vectors perpendicular to $TM_m(x)$.*

Theorem 13.2.2 *Let $F : \mathbb{R}^n \to \mathbb{R}^{n-m}$ be a C^1 function with rank $dF(x)$ equal to $n - m$ at every point on a level set $\{x : F(x) = c\}$. Then this level set is an m-dimensional C^1 surface M_m and $\nabla F_j(x)$ for $j = 1, \ldots, n-m$ span $NM_m(x)$.*

13.3 Maxima and Minima on Surfaces

Theorem 13.3.1 (*Lagrange Multipliers*) *Let $f : \mathbb{R}^n \to \mathbb{R}$ and $G : \mathbb{R}^n \to \mathbb{R}^k$ be C^1 functions, and let $\tilde{x}$ in $\mathbb{R}^n$ be a point where $G(\tilde{x}) = 0$ and $dG(\tilde{x})$ has rank k. If $f(\tilde{x})$ attains a local maximum or minimum for f on $\{x : G(x) = 0\}$, then there exists $\tilde{\lambda}$ in $\mathbb{R}^k$ such that $H : \mathbb{R}^{n+k} \to \mathbb{R}$ defined by $H(x, \lambda) = f(x) + \sum_{j=1}^{k} \lambda_j G_j(x)$ has a critical point at $(\tilde{x}, \tilde{\lambda})$.*

Example *If $G(x, y) = y^3 - x^2$ and $f(x, y) = y$, then f attains its minimum value on $G = 0$ at $(0, 0)$ but H does not have a critical point of the form $(0, 0, \lambda)$. In this case the rank condition of the theorem does not hold.*

Theorem (*Second Derivative Test*) *If f attains a local maximum (resp. minimum) on $G(x) = 0$ at $\tilde{x}$ and the conditions of the Lagrange Multiplier Theorem hold and if f and G are assumed C^2, then $d^2 f(\tilde{x}) + \tilde{\lambda} \cdot d^2 G(\tilde{x})$ is non-positive (resp. non-negative) definite on the tangent* *space at $\tilde{x}$ to $G(x) = 0$. Conversely, if f and G are C^2, if dG has rank k at $\tilde{x}$, if $(\tilde{x}, \tilde{\lambda})$ is a critical point of H, and if $d^2 f(\tilde{x}) + \tilde{\lambda} \cdot d^2 G(\tilde{x})$ is negative (resp. positive) definite on the tangent space to $G(x) = 0$ at $\tilde{x}$, then $\tilde{x}$ is a strict local maximum (resp. minimum) of f on $G(x) = 0$.*

13.4 Arc Length

Definition 13.4.1 *The length of a continuous curve $g : [a, b] \to \mathbb{R}^n$ is the* sup *of $\sum_{j=0}^{N} |g(t_{j+1}) - g(t_j)|$ over all partitions $a = t_0 < t_1 < \cdots < t_N = b$. If the length is finite the curve is called rectifiable.*

Theorem *If $\sum_{j=0}^{N} |g(t_{j+1}) - g(t_j)|$ has a finite limit for a sequence of partitions with the maximum of $t_{j+1} - t_j$ going to zero, then the curve is rectifiable and the limit equals the length.*

Definition *A curve $g : [a, b] \to \mathbb{R}^n$ is called simple if g is one-to-one. It is called a simple closed curve if g is one-to-one except $g(a) = g(b)$.*

Theorem *The length of a simple curve, or a simple closed curve, does not depend on the parametrization.*

Example *The graph of $y = x \sin 1/x$ on $[0, 1]$ is a continuous curve that is not rectifiable.*

Theorem 13.4.1 *A piecewise C^1 continuous curve $g : [a, b] \to \mathbb{R}^n$ (meaning there exists a finite partition $a = t_0 < t_1 < \cdots < t_N$ such that g is C^1 on each subinterval) is rectifiable, and the length equals $\int_a^b |g'(t)| dt$.*

Theorem *A simple rectifiable curve can be parametrized by arc length: if L denotes the length there exists $h : [0, L] \to \mathbb{R}^n$ onto the curve such that the length of h restricted to $[0, s]$ equals s, for $0 \leq s \leq L$, and there are exactly two possibilities for h (the other being $h(L - s)$).*

Chapter 14

The Lebesgue Integral

14.1 The Concept of Measure

14.1.1 Motivation

The idea of the Lebesgue integral is to enlarge the class of integrable functions so that $\int_a^b f(x)\,dx$ will be given a meaning for functions f that are not Riemann integrable. For functions that are Riemann integrable the Lebesgue theory will assign the same numerical value to $\int_a^b f(x)\,dx$ as the Riemann theory. Thus the Lebesgue integration theory can be thought of as a kind of completion of the Riemann integration theory. This can be given a precise sense in terms of the metric $d(f,g) = \int_a^b |f(x) - g(x)|dx$ on the continuous functions $C([a,b])$ so that the Lebesgue integrable functions are obtainable from the continuous functions by the same process as the real numbers are obtained from the rational numbers. However, it is best if we observe this fact after we have developed the Lebesgue theory in a more concrete way.

Before beginning on the rather difficult path of developing the Lebesgue theory we will recall some of the weak points of the Riemann theory that can serve as motivation for seeking a better theory. The Riemann theory works well for continuous functions and for uniformly convergent sequences, but one weakness is that it breaks down if we go too far beyond this tame territory. We have seen that a bounded function with a countable set of discontinuities is Riemann integrable, but Dirichlet's function

$$f(x) = \begin{cases} 0 & \text{if } x \text{ is irrational,} \\ 1 & \text{if } x \text{ is rational} \end{cases}$$

is not, even though it is a pointwise limit of functions

$$f_n(x) = \begin{cases} 0 & \text{if } x = r_1, r_2, \ldots, r_n, \\ 0 & \text{otherwise} \end{cases}$$

where $r_1, r_2, \ldots$ is an enumeration of the rationals, with $\int_0^1 f_n(x)\,dx = 0$. Now one might object that Dirichlet's example is somewhat artificial, but it is possible to conceive of more natural examples. For example, consider a Fourier series $\sum_{-\infty}^{\infty} a_n e^{inx}$ that is given by specifying the coefficients a_n in some way. Unless we know that $\sum_{-\infty}^{\infty} |a_n|$ is finite we cannot be sure that the series converges to a continuous function. Nevertheless it is tempting to integrate the series term-by-term as

$$\begin{aligned} \int_a^b \left(\sum_{\infty}^{\infty} a_n e^{inx} \right) dx &= \sum_{-\infty}^{\infty} \int_a^b a_n e^{inx}\,dx \\ &= \sum_{n \neq 0} a_n \left(\frac{e^{inb} - e^{ina}}{in} \right) + (b-a)a_0 \end{aligned}$$

because the factor n in the denominator helps make the series converge (if we assume $\sum |a_n|^2 < \infty$, which is a weaker condition than $\sum |a_n| < \infty$, we have

$$\sum_{n \neq 0} \left| a_n \left(\frac{e^{inb} - e^{ina}}{in} \right) \right| \leq \sum_{n \neq 0} \frac{|a_n|}{n} \leq 2 \left(\sum_{n \neq 0} |a_n|^2 \right)^{1/2} \left(\sum_{n \neq 0} \frac{1}{n^2} \right)^{1/2} < \infty$$

by the Cauchy-Schwartz inequality). The Riemann theory of integration, which was invented to handle similar problems in Fourier series, fails to give a meaning to this integration. Certainly the trend in mathematical analysis has been to consider more and more wildly behaved functions. Perhaps this trend has been encouraged by the Lebesgue theory, which allows us to deal with such functions, but it is still a worthwhile goal to pursue the integration of all functions that come along. (It would be nice to have a theory that provides an integral for all functions, but this turns out to be impossible.)

A second weakness of the Riemann theory of integration is the lack of a good convergence theorem. We have seen that the Riemann integral can be interchanged with a uniform limit, but in many applications this is not adequate. For example, with Fourier series we frequently do not have uniform convergence, even if the function is continuous. Of course even in the Lebesgue theory we will not be able to interchange all limits with integration. For example, if

$$f_n(x) = \begin{cases} n & \text{if } 0 < x < 1/n, \\ 0 & \text{otherwise,} \end{cases}$$

then $\int_0^1 f_n(x)\,dx = 1$ but $\lim_{n\to\infty} f_n(x) = 0$ at every point, so

$$\int_0^1 \lim_{n\to\infty} f_n(x)\,dx \neq \lim \int_0^1 f_n(x)\,dx.$$

Nevertheless we will find two rather useful criterion for interchanging limits and integrals—the monotone convergence theorem and the dominated convergence theorem.

A third weakness of the Riemann theory is that improper integrals have to be treated separately. In the Lebesgue theory we will be able to treat absolutely convergent improper integrals on the same footing as proper integrals.

A fourth weakness is that we have no reasonable criterion for deciding whether or not a function is Riemann integrable (Riemann did in fact give such a criterion, but I have not bothered to present it because it seems no easier to apply than to verify the definition of the Riemann integral). With the aid of the Lebesgue theory it is possible to give a criterion for the Riemann integral to exist (although it must be admitted that we don't have a very good criterion for the Lebesgue integral to exist).

Finally a fifth weakness involves the theory of multiple integrals. We have postponed discussing multiple integrals until after the Lebesgue theory because the Riemann theory yields only very awkward and incomplete results.

In addition to overcoming these weaknesses, the Lebesgue theory yields a remarkable bonus—it allows a very far-reaching and fruitful generalization of the concept of integration. We have already observed at least superficially an analogy between infinite series and integrals.

The Lebesgue theory allows us to say that the sum of an absolutely convergent series *is* a form of integration, and this conceptual framework allows us also to give a foundation to probability theory. Lebesgue's theory stands at the doorway to twentieth century mathematics, and all who enter must pass through this gate.

Much of the motivation we have given for the Lebesgue theory comes after the fact. Historically, Lebesgue was viewed by his contemporaries as almost a crack-pot—or at least as someone who was going very far out on a limb to study esoteric problems of little importance to the mainstream of mathematics. He was once asked why he bothered to study the problem of defining the area of very irregular surfaces since the usual calculus formula was valid for all surfaces that would ever arise in applications. Lebesgue replied by producing a crumpled handkerchief! However, once Lebesgue's results were published they were rapidly recognized as being of great importance.

Lebesgue explains the basic idea of his method by a parable. Suppose a merchant wishes to add the day's receipts. The most straightforward approach would be to add the amounts in the order in which they came, say $5+10+1+1+25+5+10+50+25+10$ (in Lebesgue's day you could buy things for those amounts). But a better approach would be to tally the number of coins of each denomination, 2 pennies, 2 nickles, 3 dimes, 2 quarters, 1 half-dollar, and then compute $2\times 1+2\times 5+3\times 10+2\times 25+1\times 50 = 2+10+30+50+50 = 142$. The Riemann integral is like the first approach; it adds the values of the function $f(x)$ in the order they occur—partitioning the domain. Lebesgue's integral is like the second approach; it first sorts the values of $f(x)$—partitioning the *range*, and then sums the values multiplied by the size of the set on which they occur. Of course if the function is continuous, or nearly so, the values of $f(x)$ will not vary much over a small interval and the Riemann integral will work well. But if the function $f(x)$ is wildly discontinuous—as the parable suggests it very well might be—then partitioning the domain really makes no sense. Why should $f(z_j)(x_j - x_{j-1})$ for z_j in $[x_{j-1}, x_j]$ represent a good approximation of $\int_{x_{j-1}}^{x_j} f(x)\,dx$ even if the interval is small if the values of $f(x)$ on the interval vary considerably? On the other hand, if A_j denotes the set of x such that $y_{j-1} < f(x) \le y_j$, then y_j times the size of the set A_j will be a good approximation to the contribution to the integral of f over the set A_j provided $y_j - y_{j-1}$ is small. If $y_0 < y_1 < y_2 < \cdots < y_n$ is a partition of the range of f (we assume for simplicity that f is bounded so we can do this), then the Lebesgue approximating sum $\sum_{j=1}^{n} y_j |A_j|$, where $|A_j|$ denotes the size of the set $A_j = \{x : y_{j-1} < f(x) \leq y_j\}$, will approximate the Lebesgue integral $\int_a^b f(x)\, dx$.

To complete the program we have to make precise the notion of the size of the set A_j. Here we come against a formidable difficulty. Even if the function f is quite smooth, the set A_j can be quite hairy; if $f(x) = x^2 \sin 1/x$ (as shown in Figure 14.1.1) the set where $0 < f(x) \leq 1$ is a countable union of intervals.

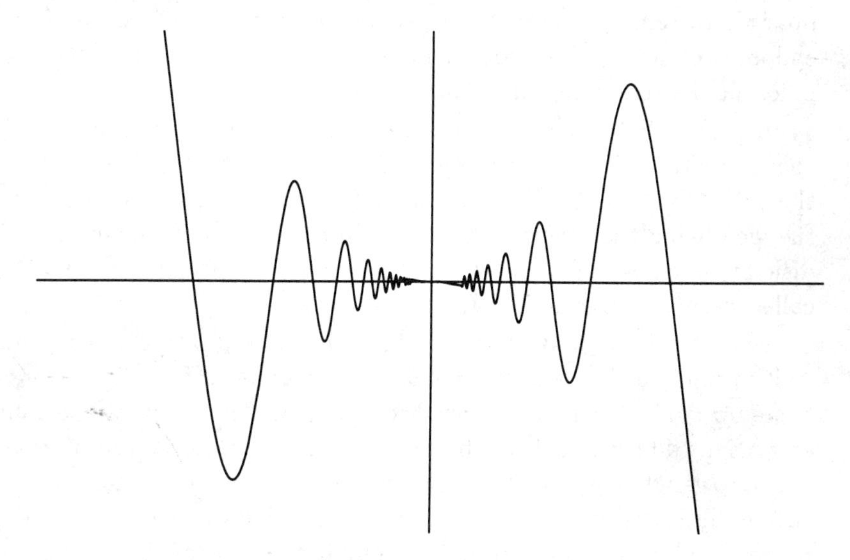

Figure 14.1.1:

If the function is wildly discontinuous, then we can expect even worse trouble. Thus we need to be able to give a numerical value to the size of a set for rather complicated sets, which generalizes the length of an interval. This leads to the concept of Lebesgue measure.

14.1.2 Properties of Length

Before we can discuss the concept of Lebesgue measure we need to discover the basic properties of the concept of length of an interval, for these will be the properties we can hope to generalize. If I denotes an interval (a, b) or $[a, b]$ or $(a, b]$ or $[a, b)$, then $b - a$ is its length, which we will denote $|I|$. Notice that if $a = -\infty$ or $b = +\infty$, then $|I| = +\infty$, so that $|I|$ is a non-negative extended real number. Perhaps the most obvious property of the length is its *additivity*: if I is the union of two disjoint intervals $I = I_1 \cup I_2$ (say $I = (a,b)$ and $I_1 = (a,c), I_2 = [c,b)$ with $a < c < b$), then $|I| = |I_1| + |I_2|$. Notice that this makes sense even if some of the lengths are zero or infinity. By induction we can obtain the finite additivity of length: if $I = I_1 \cup \cdots \cup I_n$ is a disjoint union of intervals, then $|I| = |I_1| + \cdots + |I_n|$. Actually the induction argument is not completely trivial since we need to observe that it is possible to remove one of the intervals I_k (say the one containing an endpoint or a neighborhood of an endpoint if I is open) so that the union of the remaining intervals is still an interval.

An immediate consequence of the finite additivity is the following *subadditivity*: if I is covered by $I_1, I_2, \ldots, I_n$, not necessarily disjoint, then $|I| \leq \sum_{j=1}^{n} |I_j|$. We leave the details of the proof as an exercise, the idea being that we can shrink the intervals I_j to I'_j so that I is the disjoint union $I = I'_1 \cup \cdots \cup I'_n$. An important special case, $n = 1$, is called *monotonicity*: if $I \subseteq J$, then $|I| \leq |J|$.

From the finite additivity it is tempting to jump to a general additivity principle that would include infinite unions. Here, however, we come up against a famous paradox that troubled many of the mathematicians who worked on the foundations of calculus: an interval of non-zero length, say $(0,1)$, is the union of the points it contains, yet each point c is an interval $[c,c]$ of zero length. How can a non-zero value be obtained by summing an infinite number of zeroes? There is no way to make sense out of this—although some have tried by arguing that the zeros are actually infinitesimals. The only way out is to conclude that there is no general principle for additivity of length. However, we observe that in the example we broke up the interval $(0,1)$ into an uncountable disjoint union of intervals. We know that uncountable sets are apt to cause more trouble than merely countable infinite sets, which are in effect limits of finite sets. Suppose we had a countable disjoint union $I = I_1 \cup I_2 \cup \cdots$ of intervals, with I also an interval. We could hope that $|I| = |I_1| + |I_2| + \cdots$. Such a principle does in fact hold, and we call it *countable additivity* or *σ-additivity* for short. Although it is plausible because a countable union is a limit of finite unions, we cannot prove it quite so easily because there are many complicated ways to write an interval as a countable disjoint union of intervals. Instead we must resort to trickery and the Heine-Borel theorem.

Lemma 14.1.1 *Let I be an interval, $I = I_1 \cup I_2 \cup \cdots$ where $I_1, I_2, \ldots$ are disjoint intervals. Then $|I| = \sum_{j=1}^{\infty} |I_j|$ (we interpret this to mean that if one side is $+\infty$, then so is the other, where $\sum_{j=1}^{\infty} |I_j|$ can be $+\infty$ either because one of the summands is $+\infty$ or because the series diverges).*

Proof: Consider first the case when $|I|$ is finite. If we break off the union after a finite number of terms, then $I_1 \cup I_2 \cup \cdots \cup I_n \subseteq I$ and the order of the terms can be rearranged if need be so that I_{k-1} lies to the left of I_k — for $2 \leq k \leq n$. We can then fill in other intervals $J_1, \cdots, J_m$ as needed in between (as indicated in Figure 14.1.2) so that $I = I_1 \cup \cdots \cup I_n \cup J_1 \cup \cdots \cup J_m$ (disjoint) so that $|I| = \sum_{j=1}^{n} |I_j| + \sum_{k=1}^{m} |J_k|$ by the finite additivity, hence $\sum_{j=1}^{n} |I_j| \leq |I|$.

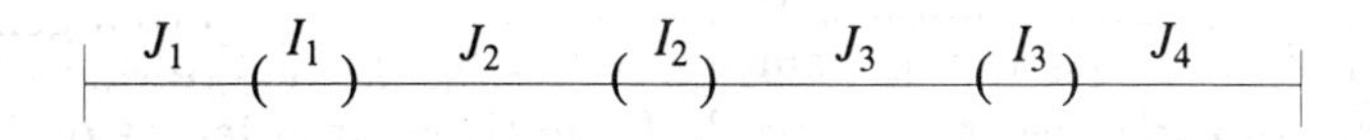

Figure 14.1.2:

Letting $n \to \infty$ we obtain $\sum_{j=1}^{\infty} |I_j| \leq |I|$.

The tricky step is obtaining the reverse inequality. We could try to argue directly that the J-intervals in the above picture must get smaller as we add more I-intervals, but it is hard to do this convincingly. Instead we resort to subterfuge. We shrink the interval I if need be to make it compact, and we expand the I_j intervals if need be to make them open. We claim we can do this while changing the values of $|I|$ and $\sum_{j=1}^{\infty} |I_j|$ by at most ε. Indeed if a and b are the endpoints of I, then set (if ε is small enough) $I' = [a + \varepsilon/2, b - \varepsilon/2]$; and if a_j and b_j are the endpoints of I_j, set $I_j' = (a - 2^{-j}\varepsilon/2, b + 2^{-j}\varepsilon/2)$. We then have I' compact, I_j' open, and $|I'| = |I| - \varepsilon$, $\sum_{j=1}^{\infty} |I_j'| = \varepsilon + \sum_{j=1}^{\infty} |I_j|$. Because the I_j cover I, it follows that the I_j' cover I' since we expanded the coverers and contracted the covered. The I_j' are no longer disjoint, but this doesn't matter. By the Heine-Borel theorem there is a finite subcover, $I' \subseteq \bigcup_{j=1}^{N} I_j'$ for some finite N.

Now the finite subadditivity implies $|I'| \leq \sum_{j=1}^{N} |I'_j|$. Finally, from $|I'| \leq \sum_{j=1}^{N} |I'_j|$ and $|I'| = |I| - \varepsilon$, $|I'_j| = |I_j| + 2^{-j}\varepsilon$ we obtain $|I| - \varepsilon \leq \varepsilon + \sum_{j=1}^{N} |I_j| \leq \varepsilon + \sum_{j=1}^{\infty} |I_j|$ and, since this holds for every $\varepsilon > 0$, $|I| \leq \sum_{j=1}^{\infty} |I_j|$. This completes the proof when $|I| < \infty$.

If $|I| = +\infty$ we simply intersect all the intervals with $[-N, N]$ and let $N \to \infty$. By the previous case we have

$$\begin{aligned} |I \cap [-N, N]| &= \sum_{j=1}^{\infty} |I_j \cap [-N, N]| \\ &\leq \sum_{j=1}^{\infty} |I_j| \end{aligned}$$

since $|I_j \cap [-N, N]| \leq |I_j|$ by the monotonicity. But

$$\lim_{N \to \infty} |I \cap [-N, N]| = +\infty \quad \text{if } |I| = +\infty$$

(for then $a = -\infty$ or $b = +\infty$), so $\sum_{j=1}^{\infty} |I_j| \geq +\infty$ hence $= +\infty$. QED

It is perhaps worth pointing out a technical aspect of the proof: when we expanded the intervals I_j we threw in a factor of 2^{-j} so that the sum of all the errors would remain small. This is a common theme in many arguments in Lebesgue integration theory. We can paraphrase it by saying *a countable number of small errors can be made small.*

When we give a formula for Lebesgue measure we will need the following corollary to the lemma.

Corollary 14.1.1 *If I is any interval, then*

$$|I| = \inf \left\{ \sum_{j=1}^{\infty} |I_j| : I \subseteq \bigcup_{j=1}^{\infty} I_j \right.$$
$$\left. \text{where } \{I_j\} \text{is any countable covering of } I \text{ by intervals.} \right\}$$

Proof: We can always cover I by itself ($I_1 = I$ and $I_j = \emptyset$ for $j > 1$), showing that the inf is at most $|I|$. Thus it suffices to show $I \subseteq \bigcup_{j=1}^{\infty} I_j$ implies $|I| \leq \sum_{j=1}^{\infty} |I_j|$. If the intervals I_j are not disjoint we can replace them by smaller intervals that are disjoint, in the process reducing $\sum_{j=1}^{\infty} |I_j|$. If I is then not equal to $\bigcup_{j=1}^{\infty} I$, we can make it so by replacing each interval with its intersection with I, again reducing $\sum_{j=1}^{\infty} |I_j|$. After these two reductions we apply the lemma to obtain equality $|I| = \sum_{j=1}^{\infty} |I_j|$, hence we must have had the desired inequality all along. QED

We have now completed a description of all the basic properties of the length of an interval that will be needed for the generalization of the concept of measure. We can summarize them succinctly in three simple statements:

1. $|I|$ is a value in $[0, \infty]$.
2. $|I| = 0$ if I is the empty interval.
3. $|I| = \sum_{j=1}^{\infty} |I_j|$ if $I = \bigcup_{j=1}^{\infty} I_j$ and the I_j are disjoint.

We are *not* claiming that these three statements determine $|I|$ uniquely—far from it, there are many different ways to define a measure of size $|I|$ for every interval in such a way that these three statements are valid. Rather we are claiming that these represent a sort of minimal distillation of the concept of length, or even of the general concept of measurement of "extent" that would include such things as area, volume, mass, and probability. We will eventually take them to be axioms for the abstract concept of *measure*, just as we took a few simple properties of distance and used them for axioms for the abstract concept of metric. But before doing this we need to extend our vision beyond the simple world of intervals and come to grips with the question of which kinds of sets we need to measure.

14.1.3 Measurable Sets

Our first choice would be to measure all sets. However this turns out to be impossible if we want to retain the properties listed above. (Actually the situation is more complicated: we need the uncountable axiom of choice to prove the impossibility, and Solovay has shown that without some such strongly nonconstructive axiom it is impossible to prove the impossibility.) This does not turn out to be too devastating a blow to the theory, since we can do analysis with a more restricted class of sets. It does mean, however, that we have to be more careful than we might prefer to be. Recall that the kind of sets we want to measure are of the form $A = \{x : a < f(x) \leq b\}$ for some function f we want to integrate. Because we cannot measure all sets, we cannot integrate all functions; but we would like to integrate as many functions as possible, so we want the collection of measurable sets, those A for which $|A|$ is defined, to be as versatile as possible so that we can manipulate the functions freely. This means we want these sets to be preserved under the usual operations of set theory: union, intersection, complement, difference. Any collection of sets with this property is called a *field* of sets. Technically it suffices to assume only the following axioms for a field $\mathcal{F}$ of sets:

1. The empty set is in $\mathcal{F}$.

2. If A is in $\mathcal{F}$, then the complement ${}^cA$ is in $\mathcal{F}$.

3. If A and B are in $\mathcal{F}$, then $A \cup B$ is in $\mathcal{F}$.

Here all the sets in $\mathcal{F}$ are subsets of a fixed universe X and the complement is defined with respect to this universe, ${}^cA = \{x \text{ in } X : x \text{ is not in } A\}$. For most of our applications X will be a subset of some Euclidean space. It is a simple exercise to show that these axioms imply that if A and B are in $\mathcal{F}$, then $A \cap B$ and $A \backslash B$ are also in $\mathcal{F}$. The term *algebra* of sets is sometimes used instead of *field*. The algebraic terminology derives from the fact that a field forms a *Boolean algebra* under the operations $A + B = A \Delta B$, $A \cdot B = A \cap B$ where $A \Delta B$ denotes the *symmetric difference* defined by $A \Delta B = (A \backslash B) \cup (B \backslash A) = (A \cup B) \backslash (A \cap B)$, as indicated by the shaded region in Figure 14.1.3.

An important example of a field of sets is the following: take for X any fixed interval of $\mathbb{R}$ and take for $\mathcal{F}$ all sets that are finite unions of intervals contained in X. We leave the simple verification of the axioms as an exercise. It is necessary to consider unions of intervals rather than merely intervals in order to have a field of sets.

So that we will be able to perform limit processes on functions, we have to assume more about the measurable sets—namely that they are preserved under countable set-theoretic operations. For technical reasons it suffices to assume the following additional axiom:

4. If $A_1, A_2, \ldots$ is a sequence of sets in $\mathcal{F}$, then $A = \bigcup_{j=1}^{\infty} A_j$ is in $\mathcal{F}$.

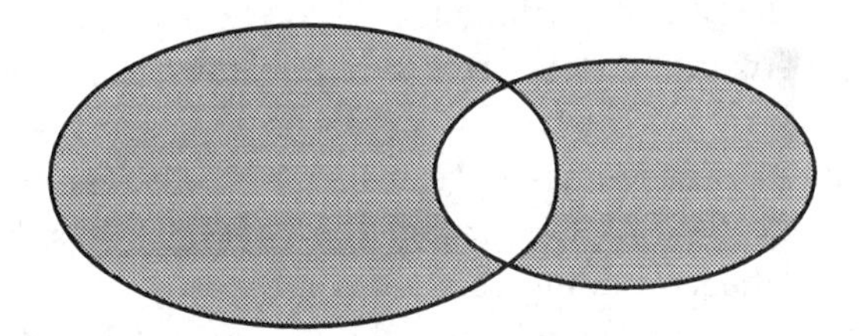

Figure 14.1.3:

Any field of sets that satisfies this condition is called a *σ-field.* The identity $\bigcap_{j=1}^{\infty} A_j = {}^{c}(\bigcup_{j=1}^{\infty} {}^{c}A_j)$ shows that a σ-field is closed under countable intersections as well. Notice that we only consider countable operations; we do *not* require that arbitrary uncountably infinite unions of sets in the σ-field must belong to the σ-field. Nevertheless a σ-field is in general a very large collection of sets.

For example, the field $\mathcal{F}$ of finite unions of intervals in X is *not* a σ-field. We could try to make it into a σ-field by considering $\mathcal{F}_1$, the collection of all countable unions of intervals of X. This will satisfy the last axiom, but it is not closed under complements (for example, the Cantor set is not in $\mathcal{F}_1$, but its complement is), so it isn't even a field. We could try to fix this by taking $\mathcal{F}_2$ to be all countable intersections of sets in $\mathcal{F}_1$ and then $\mathcal{F}_3$ to be all countable unions of sets in $\mathcal{F}_2$ and so on alternately taking countable intersections and unions and finally setting $\mathcal{F}_\infty = \bigcup_{j=1}^{\infty} \mathcal{F}_j$. It is not hard to show that $\mathcal{F}_\infty$ is again a field, but alas it too is not a σ-field. We need to repeat this process through a sophisticated transfinite induction up to the first uncountable ordinal in order to obtain a σ-field! Naturally such a complicated "construction" is of little use and so we rely on a non-constructive description. If $\mathcal{F}$ is any field of sets we define the σ-field *generated* by $\mathcal{F}$, denoted by $\mathcal{F}_\sigma$, to be the intersection of all σ-fields containing $\mathcal{F}$. We leave it as an exercise to verify that this is actually a σ-field and is the smallest σ-field containing $\mathcal{F}$, in the sense that any σ-field containing $\mathcal{F}$ must contain $\mathcal{F}_\sigma$. When $\mathcal{F}$ is the field of finite unions of intervals contained in a fixed interval X of $\mathbb{R}$ we call $\mathcal{F}_\sigma$ the σ-field of *Borel subsets* of X and we call sets in $\mathcal{F}_\sigma$ *Borel sets.* The Borel sets can equally well be described as the smallest σ-field containing all the open sets (or all the closed sets), and this definition makes sense in $\mathbb{R}^n$ or any metric space. The point of the above discussion is that there is no really satisfactory description of what a Borel set is like. However, it is a sufficiently large category of set so that it contains any set that is describable in conventional mathematical terms. (Actually, many authors use a slightly larger σ-field of sets, called the *Lebesgue sets*, but there is not much to be gained by doing this.)

The point of view we will take in the remainder of this chapter is that only the Borel sets are important. In this sense we can say that Lebesgue made Cantor obsolete: Cantor wanted to make set theory the foundation of mathematics, while Lebesgue showed that just about all mathematics can be done within the smaller confines of the Borel sets.

14.1.4 Basic Properties of Measures

We can now state our first main goal as follows: to extend the length measure $|I|$ from the intervals to the σ-field of Borel sets so as to preserve the basic properties. We will call the extended length measure *Lebesgue measure*. More generally, we will define a *measure* to be a function $|A|$ defined on a σ-field $\mathcal{F}$ of sets (called the *measurable sets*) satisfying:

1. (non-negativity) $|A|$ is in $[0, \infty]$;
2. $|\emptyset| = 0$ where $\emptyset$ is the empty set;
3. (σ-additivity) if $A = \bigcup_{j=1}^{\infty} A_j$ with A_j disjoint, $|A| = \sum_{j=1}^{\infty} |A_j|$. Note that this implies finite additivity simply be taking all but a finite number A_j equal to the empty set.

It is by no means clear that there exists such an extension as called for in the definition of Lebesgue measure. We will give a proof that such an extension exists in section 14.2. For the remainder of this section we will derive properties of measures that follow from the above three axioms. These properties will hold for Lebesgue measure, but the proofs are just as easy for general measures. It will also turn out that these properties lead to a formula for Lebesgue measure, which is given in section 14.1.5.

Next we observe some simple properties of measures that are consequences of the defining properties:

1. A measure is *monotone*: If A and B are measurable sets with $A \subseteq B$, then $|A| \leq |B|$. To see this observe that B is a disjoint union of A and $B \backslash A$, as shown in Figure 14.1.4, so $|B| = |A| + |B \backslash A|$ by additivity and, hence, $|B| \geq |A|$ since $|B \backslash A| \geq 0$.

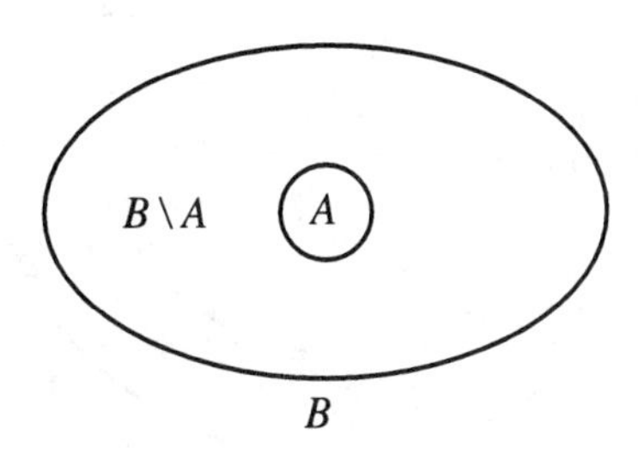

Figure 14.1.4:

2. *Continuity from below*: If $A_1 \subseteq A_2 \subseteq A_3 \subseteq \cdots$ is an increasing sequence of measurable sets and $A = \bigcup_{j=1}^{\infty} A_j$, then $|A| = \lim_{j\to\infty} |A_j|$. To see this we define the difference sets $B_k = A_k \backslash A_{k-1}$ for $k = 2, 3, \ldots$ and observe that A is the disjoint union $A = A_1 \cup B_2 \cup B_3 \cup \cdots$, so $|A| = |A_1| + |B_2| + |B_3| + \cdots$ by the σ-additivity. On the other hand if we terminate the union we have $A_1 \cup B_2 \cup \cdots \cup B_n = A_n$ a disjoint union, so $|A_1| + |B_2| + \cdots + |B_n| = |A_n|$ and so $|A| = \lim_{n\to\infty} |A_n|$.

3. *Conditional continuity from above*: If $B_1 \supseteq B_2 \supseteq B_3 \supseteq \cdots$ is a decreasing sequence of measurable sets and $B = \bigcap_{j=1}^{\infty} B_j$ and if the measures $|B_j|$ are finite, then $|B| = \lim_{j\to\infty} |B_j|$. To see this we define the difference sets $A_k = B_k \backslash B_{k+1}$ for $k = 1, 2, \ldots$ and observe that B_1 can be written as a disjoint union $B_1 = B \cup A_1 \cup A_2 \cup \cdots$ (see Figure 14.1.5); hence, $|B_1| = |B| + |A_1| + |A_2| + \cdots$, which we can write as $|B_1| - |B| = \sum_{j=1}^{\infty} |A_j|$ since $|B_1|$ and, hence, $|B|$ are finite. On the other hand $B_1 = B_n \cup A_1 \cup A_2 \cup \cdots \cup A_{n-1}$ is a disjoint union, so $|B_1| = |B_n| + |A_1| + \cdots + |A_{n-1}|$. Thus $|B_1| - |B_n| = \sum_{j=1}^{n-1} |A_j| \to |B_1| - |B|$ as $n \to \infty$, so $\lim_{n\to\infty} |B_n| = |B|$. The requirement that $|B_j|$ be finite was used in the argument to avoid meaningless manipulations with $\infty - \infty$ expressions. To see that it is actually necessary for the result consider the example of Lebesgue measure on $\mathbb{R}$ with $B_n = (n, \infty)$. Then $\bigcap B_n = \emptyset$, but $\lim_{n\to\infty} |B_n| = +\infty$.

4. *Subadditivity*: If $A_1, \ldots, A_n$ are measurable sets, not necessarily disjoint, then $|A_1 \cup \cdots \cup A_n| \leq \sum_{j=1}^{n} |A_n|$. To see this we replace the

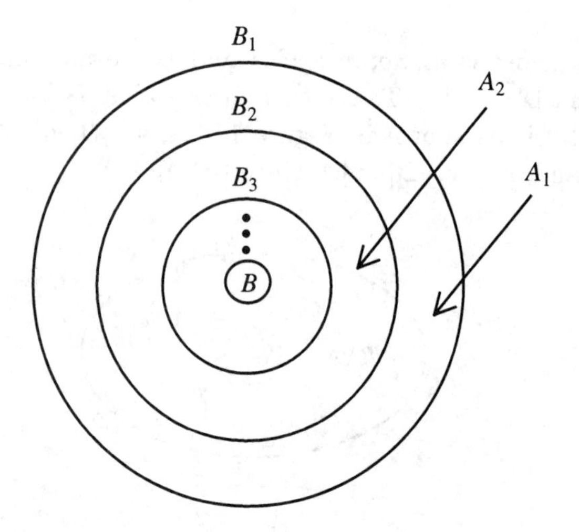

Figure 14.1.5:

union with an equivalent disjoint union $A_1 \cup \cdots \cup A_n = B_1 \cup B_2 \cup \cdots \cup B_n$ by defining $B_1 = A_1, B_2 = A_2 \backslash A_1, B_2 = A_3 \backslash (A_1 \cup A_2), \ldots, B_n = A_n \backslash (A_1 \cup \cdots \cup A_{n-1})$. Then the B_j are disjoint and $B_j \subseteq A_j$, so $|B_j| \leq |A_j|$ by monotonicity. Thus $|A_1 \cup \cdots \cup A_n| = \sum_{j=1}^n |B_j| \leq \sum_{j=1}^n |A_j|$.

5. *σ-Subadditivity*: If $A_1, A_2, \ldots$ is a sequence of measurable sets, then $|\bigcup_{j=1}^\infty A_j| \leq \sum_{j=1}^\infty |A_j|$. By essentially the same argument as in the finite case we write $\bigcup_{j=1}^\infty A_j = \bigcup_{j=1}^\infty B_j$ where the B_j are disjoint and $B_j \subseteq A_j$. Then $|\bigcup_{j=1}^\infty A_j| = \sum_{j=1}^\infty |B_j|$ by σ-additivity, and the result follows.

. Frequently we need to combine subadditivity and monotonicity to obtain the following statement: if $B \subseteq A_1 \cup \cdots \cup A_n$, then $|B| \leq \sum_{j=1}^n |A_j|$. A similar statement holds for countable unions.

14.1.5 A Formula for Lebesgue Measure

Let us return to the special case of Lebesgue measure. Although the proof of the existence of this measure is very difficult, it is not too hard to derive a formula for the measure. The remaining task will then be to show that the three axioms for a measure are indeed satisfied. Suppose that a set B is covered by a countable union of *intervals*, $B \subseteq \bigcup_{j=1}^\infty I_j$. Then if B is to be measurable, the σ-subadditivity would imply $|B| \leq \sum_{j=1}^\infty |I_j|$. Notice that here we know what $|I_j|$ is:

just the length of the interval I_j. Thus any countable covering of a set by intervals gives us some information about the measure of that set. Of course the same would be true about finite coverings of B by intervals. However it is a key observation of the Lebesgue theory that the information obtained from countable coverings is much more precise than that obtained from finite coverings. To understand this we should look at an example. Let B be the set of rational numbers in $[0,1]$. Then if we want to cover B by a finite union of intervals it is not hard to see that we must cover the entire interval $[0,1]$, so the lengths must add up to at least one. Thus we obtain only the estimate $|B| \le 1$. However, using a countable cover gives us greater flexibility. If $r_1, r_2, \ldots$ is an enumeration of the rational numbers in $[0,1]$ set $I_j = (r_j - 2^{-j}\varepsilon, r_j + 2^{-j}\varepsilon)$. Then $\bigcup_{j=1}^{\infty} I_j \supseteq B$ and $\bigcup_{j=1}^{\infty} |I_j| = \varepsilon$, so $|B| \le \varepsilon$. Since this is true for every $\varepsilon > 0$, we have $|B| \le 0$ and, hence, $|B| = 0$ (if we are willing to use closed intervals in the cover we can take $I_j = [r_j, r_j]$ and get $|B| \le 0$ immediately). In this case we have determined the value of $|B|$ exactly by considering countable coverings by intervals. It turns out that this is true if B is any Borel set.

Suppose B is an arbitrary set. We have seen that if B is to be measurable we must have $|B| \le \sum_{j=1}^{\infty} |I_j|$ where $\bigcup_{j=1}^{\infty} I_j$ is any countable covering of B by intervals. Similarly we must have $|B| \le \inf \bigcup_{j=1}^{\infty} |I_j|$ where the infininum is taken over all countable coverings of B by intervals. *The heart of the proof of the existence of Lebesgue measure is to take this to be an equality, to define*

$$|B| = \inf \left\{ \sum_{j=1}^{\infty} |I_j| : B \subseteq \bigcup_{j=1}^{\infty} I_j \right\}.$$

We can see immediately some consequences of this definition. First note that the arbitrary countable union of intervals can be replaced by a disjoint union to cover the same set B while reducing the value of $\sum_{j=1}^{\infty} |I_j|$. To do this we consider the differences $I_1, I_2 \backslash I_1, I_3 \backslash (I_1 \cup I_2), \ldots$ as in the proof of the subadditivity. These differences are not necessarily intervals, but they are clearly finite unions of intervals, so altogether we end up with a countable disjoint covering by intervals. Since the definition $|B|$ involves an infinum, we clearly obtain the same value if we restrict to disjoint converings. We may also assume that the intervals I_j are open, since we can always expand I_j to make it open, increasing its length by $\varepsilon 2^{-j}$; and the increase in $\sum_{j=1}^{\infty} |I_j|$ is at most ε.

In particular, if B happens to be an interval, then by Corollary 14.1.1 we know that the Lebesgue measure of B coincides with the length of B. Thus Lebesgue measure does extend the length measure of intervals.

We now want to look at two other special cases when the definition of Lebesgue measure can be given a significant interpretation. First suppose $\inf\{\bigcup_{j=1}^{\infty} |I_j| : B \subseteq \cup I_j\} = 0$. Then we have no choice by the σ-subadditivity but to set $|B| = 0$. Such sets are said to have *measure zero*. They will play an important role—that of "negligible" sets that can be ignored—in the integration theory. It is not difficult to show from the definition that sets of measure zero have the properties they should have: every Borel subset of a set of measure zero has measure zero, and any finite or countable union of sets of measure zero has measure zero. In particular we have a proof of a trivial kind of σ-additivity for sets of measure zero.

Second, we consider the case of an open set. Recall that we proved a structure theorem for open sets, $B = \bigcup_j I_j$ where I_j are disjoint open intervals, the union being finite or countable. Furthermore, this decomposition is unique. If we believe the countable additivity of Lebesgue measure, we must have $|B| = \sum_j |I_j|$. Now the definition of Lebesgue measure is somewhat different in that it requires that we take the inf of all such sums over countable covers of B by intervals; the union given by the structure theorem is just one of these covers. It certainly seems plausible that this is the best cover so that the infinum is actually achieved. This can be proved directly without using the σ-additivity (see exercise set 14.1.7).

Consider a general Borel set B. As we observed before, in the definition of $|B|$ we can restrict attention to disjoint open coverings, so $B \subseteq \bigcup I_j$ says exactly $B \subseteq A$ where A is open ($A = \bigcup I_j$) and so $\sum_j |I_j| = |A|$. Thus the definition becomes simply

$$|B| = \inf\{|A| : B \subseteq A, A \text{ open}\}$$

if we adopt the above definition for the measure of open sets. This property is called *outer regularity* (see exercises for *inner regularity*). In other words, once we know what the Lebesgue measure is for open sets, the outer regularity property tells us what it is for all Borel sets.

(Outer regularity is also an interesting property in its own right, and many other measures also share this property.)

Next we observe that we can replace the infinum by a limit: there must exist a sequence of open sets $A_1, A_2, \ldots$ such that $|B| = \lim_{j\to\infty} |A_j|$. (This is just a consequence of the properties of the in-f over any set of real numbers.) By the monotonicity we can obtain the same limit using $A_1, A_1 \cap A_2, A_1 \cap A_2 \cap A_3, \ldots$. Indeed we write $A'_n = A_1 \cap A_2 \cap \cdots \cap A_n$. It is easy to see that A'_n is also a countable disjoint union of open intervals and $B \subseteq A'_n$. Since $A'_n \subseteq A_n$, we have $|A'_n| \leq |A_n|$, so $|B| \leq \lim_{n\to\infty} |A'_n| \leq \lim_{n\to\infty} |A_n| = |B|$. The point of this is that the sequence $A'_1, A'_2, \ldots$ is decreasing. Finally set $A = \bigcap_{j=1}^{\infty} A'_j$. Since $B \subseteq A'_j$ for all j, we have $B \subseteq A$. In summary, for every Borel set B, there exists a decreasing sequence A'_1, A'_2 of coverings by countable disjoint unions of open intervals, such that $\lim_{j\to\infty} |A'_j|$ is the value we have defined for $|B|$. If we had $B = A$ we would certainly want to take $|B| = \lim_{j\to\infty} |A'_j|$; in fact we would have to take this value by the conditional continuity from above if the measures $|A'_j|$ are finite (this is always the case if the universe X is a bounded interval). In the general case we are merely asserting that $A \backslash B$ has measure zero.

A set that is the countable intersection of open sets is called a G_δ set (the G stands for "open" and the δ stands for "intersection"). As the above discussion shows, the definition of the measure of a G_δ set is perfectly natural. Now the class of G_δ sets is quite large, but it is not a σ-field. But if we modify the G_δ sets by sets of measure zero we do obtain the Borel σ-field: *every Borel set B can be covered by a G_δ set A such that $A \backslash B$ has measure zero, and $|B| = |A|$.*

14.1.6 Other Examples of Measures

Although Lebesgue measure on an interval of $\mathbb{R}$ is the principal example of a measure in which we will be interested, there are a few other examples worth keeping in mind:

1. We can define a Lebesgue measure on $\mathbb{R}^n$. In place of the intervals we consider rectangles $I = I_1 \times I_2 \times \cdots \times I_n = \{x \text{ in } \mathbb{R}^n : x_1 \text{ is in } I_1, x_2 \text{ is in } I_2, \ldots, x_n \text{ is in } I_n\}$ where $I_1, \ldots, I_n$ are intervals in $\mathbb{R}$, with measure (volume) equal to $|I_1| \cdot |I_2| \cdot \cdots \cdot |I_n|$. The Lebesgue measure of a set A that is a countable disjoint union of rectangles is the sum of the measures of the rectangles, and $|B| = \inf\{|A| : B \subseteq A\}$ with A as

above for a general Borel set B. We will use this measure in the theory of multiple integrals.

2. Let the universe X consist of the positive integers $\{1, 2, 3, \ldots\}$, and define $|A|$ to be the number of points in A ($+\infty$ if A is an infinite set). This is called *counting measure*, and it is defined on all subsets of X. It is trivial to verify that the axioms for a measure are satisfied. We will see that the integration theory associated with this measure is the theory of absolutely convergent series. More generally, counting measure can be defined on any universe X.

3. Let the universe X be a finite set $(x_1, \ldots, x_n)$. Let $p_1, p_2, \ldots, p_n$ be any non-negative values ($+\infty$ is allowed); and define $|A|$ to be the sum of p_j for all points x_j in A, where A is any subset of X. Again it is trivial to verify the axioms for a measure. In this case we can also easily show that any measure defined on all subsets of the universe X must have this form. If the values p_j also satisfy $\sum_{j=1}^{n} p_j = 1$ (or equivalently, $|X| = 1$), then we can interpret them as the probabilities. Thus p_j is the probability that x_j occurs; $|A|$ is the probability that one of the x's in A occurs.

4. More generally, for any universe X and any measure such that $|X| = 1$ we can interpret the measure as giving probabilities for "random" events with outcomes in X, $|A|$ being the probability that the outcome lies in A. Conversely, in the point of view pioneered by Kolmogorov and now almost universally accepted by mathematicians, every description involving probabilities may be cast into this form. To give one important illustration, let us give such a decription for the random tossing of a fair coin in an infinite sequence of independent trials. The universe X consists of all possible outcomes, i.e., sequences of heads (H) and tails (T), as $HTHT\ldots$. If A is a subset of X that is measurable (i.e., belongs to a certain σ-field of subsets that we will not describe explicitly), then $|A|$ will be interpreted as the probability that a random sequence of tossings will lie in A. For example, if A consists of all sequences beginning with H, then $|A| = 1/2$; and more generally if A consists of all sequences whose first n outcomes are specified, then $|A| = 2^{-n}$. However, for more general sets it is not immediately apparent how to define $|A|$. There is a device for doing this, however. We interpret H as 0 and T as 1, and each infinite sequence $HT\ldots$ as a binary expansion $.01\ldots$. In this way we obtain a mapping of the universe X onto the unit interval $[0, 1]$; this mapping is not

quite one-to-one because of identifications such as $.0111\ldots = .100\ldots$, but there are only a countable set of such exceptions and this set will have measure zero. Furthermore, if we consider the set A in X of all sequences starting with H, this is mapped into the interval $[0, 1/2]$, so $|A|$ is equal to the Lebesgue measure of $[0, 1/2]$. More generally, if A is the set of all sequences whose first n outcomes are specified, this is mapped into an interval of the form $[k/2^n, (k+1)/2^n]$ and $|A|$ still agrees with the Legesgue measure of the image of A. We can thus define $|A|$ for a general subset of X as the Lebesgue measure of the image of A under the mapping (A will be measurable if its image is a Borel set). In this way we obtain a new interpretation of Lebesgue measure on $[0, 1]$ as the probability measure associated with an infinite sequence of independent tosses of a fair coin.

14.1.7 Exercises

1. Show that a field of sets is closed under finite intersections, and differences.

2. Show that the collection of all finite unions of intervals forms a field. Show the same is true for finite unions of intervals in $\mathbb{R}$ that are left open and right closed, $(a, b]$.

3. Prove that the intersection of all σ-fields containing a field $\mathcal{F}$ is a σ-field and that it is the smallest σ-field containing $\mathcal{F}$.

4. Prove that if $A = \bigcup_{j=1}^n I_j$, a disjoint union of intervals, then $\sum_{j=1}^n |I_j|$ is independent of the particular decomposition. Show that if we define $|A| = \sum_{j=1}^n |I_j|$, then all the axioms for a measure are satisfied on the field of finite unions of intervals, where σ-additivity means $|\bigcup_{j=1}^\infty A_j| = \sum_{j=1}^\infty |A_j|$ if $A_1, A_2, \ldots$ and $\bigcup_{j=1}^\infty A_j$ are all in the field. (**Hint**: use the σ-additivity on intervals proved in the text.) Why doesn't this argument establish the existence of Lebesgue measure?

5. Prove that the Cantor set (delete middle thirds) has Lebesgue measure zero.

6. Prove that a countable union of sets of Lebesgue measure zero has

measure zero, directly from the definition $|B| = 0$ if

$$\inf\left\{\sum_{j=1}^{\infty}|I_j| : B \subseteq \bigcup_{j=1}^{\infty} I_j\right\} = 0.$$

7. What is the Lebesgue measure of the set of irrational numbers in $[0,1]$?

8. What is the Lebesgue measure of a countable set? Is the same true of a general measure?

9. Prove that a countable intersection of countable unions of intervals is a G_δ set.

10. Prove that the class of finite unions of rectangles in $\mathbb{R}^n$ is equal to the class of finite disjoint unions of rectangles in $\mathbb{R}^n$ and forms a field.

11. Prove that every open subset of $\mathbb{R}^n$ is a countable union of rectangles. Can the rectangles in the union be taken to be disjoint?

12. Prove that any measure on the field of all subsets of a finite set X has the form $|A| = \sum_{x_j \mathrm{in} A} p_j$ for some values p_j in $[0,\infty]$. What if X is countable?

13. Prove directly (without assuming σ-additivity) that $|\bigcup_{j=1}^{\infty} I_j| = \sum_{j=1}^{\infty}|I_j|$ if $\{I_j\}$ are disjoint intervals and $|I|$ denotes Lebesgue measure.

14. Prove the following *inner regularity* for Lebesgue measure: $|B| = \sup\{|F| : F \subseteq B \text{ is closed}\}$, for all Borel sets B. (**Hint**: if B is contained in $(-N,N)$ use the outer regularity for $(-N,N)\backslash B$.)

15. Prove that for compact sets A, the Lebesgue measure can be computed using only finite coverings,

$$|A| = \inf\left\{\sum_{j=1}^{N}|I_j| : A \subseteq \bigcup_{j=1}^{N} I_j\right\}.$$

16. Let μ be a measure on a σ-field $\mathcal{F}$, and let F be a set in $\mathcal{F}$. Define the restriction of μ to F, denoted $\mu|_F$, by $\mu|_F(A) = \mu(F \cap A)$. Prove that $\mu|_F$ is a measure on $\mathcal{F}$.

14.2 Proof of Existence of Measures*

14.2.1 Outer Measures

In this section we will give a proof of the existence of Lebesgue measure, following a method of Carathéodory. As a bonus we will also obtain the existence of other measures, including the Hausdorff measures that are used in the theory of fractals. The strategy of the proof is as follows. First, we weaken the axioms for a measure to conditions that are easy to verify. The resulting object will be called an *outer measure.* (This is truly dreadful terminology, because an outer measure is *not* a special case of a measure but something more general.) It will be easy to obtain examples of outer measures; in particular, the definition of Lebesgue measure yields an outer measure.

The second step in the proof is to show that an outer measure does yield a measure if we restrict the σ-field of sets appropriately. That is, we start with a σ-field $\mathcal{F}$ and consider a possibly smaller collection $\mathcal{F}_0$ of sets that satisfy a certain "splitting" condition. We prove a general theorem to the effect that $\mathcal{F}_0$ is a σ-field and the restriction of the outer measure to $\mathcal{F}_0$ is indeed a measure. Such a general theorem could yield a vacuous result in special cases because $\mathcal{F}_0$ could be very small, perhaps consisting of just the empty set and the whole space. To give significance to the general theorem we need the third step in the program, which gives a criterion for concluding $\mathcal{F}_0 = \mathcal{F}$. Fortunately, this criterion is easy to verify for Lebesgue measure and for other measures as well.

To begin the first step of the proof we give the axioms for an *outer measure* μ defined on σ-field of sets $\mathcal{F}$:

1. (non-negativity) $\mu(A)$ is in $[0, \infty]$ for every A in $\mathcal{F}$;

2. $\mu(\emptyset) = 0$;

3. (σ-subadditivity) if $A = \bigcup_{j=1}^{\infty} A_j$ with all A_j in $\mathcal{F}$, then $\mu(A) \leq \sum_{j=1}^{\infty} \mu(A_j)$;

4. (monotonicity) if $A \subseteq B$ are in $\mathcal{F}$, then $\mu(A) \leq \mu(B)$.

Notice that axioms 1 and 2 are the same as the corresponding axioms for measures and that axiom 3 is a weakening of σ-additivity.

Since we have an inequality, it is not necessary to assume the sets A_j are disjoint. Ordinary additivity is not being assumed, but of course we have subadditivity for finite unions as a consequence of axioms 2 and 3 by taking all but a finite number of sets equal to the empty set. We have already noted that axioms 3 and 4 are properties that hold for measures, *so every measure is an outer measure.* (Note: Most books require that $\mathcal{F}$ be the σ-field of all subsets of X in the definition of outer measure. This is totally unnecessary and contrary to the philosophy that non-measurable sets should play no role in analysis.)

Now, as promised, I will show that Lebesgue measure, defined by $\mu(A) = \inf\{\sum_{j=1}^{\infty} |I_j| : A \subseteq \bigcup_{j=1}^{\infty} I_j\}$, is an outer measure on the Borel sets (or any σ-field, for that matter). Indeed, axioms 1, 2, and 4 are immediate consequences of the definition; and axiom 3 follows easily because if we have countable coverings of each of the sets A_j by intervals, we can take the union of these coverings to obtain a countable covering of A by intervals. Thus if $A_j \subseteq \bigcup_k I_{jk}$, then $A \subseteq \bigcup_j \bigcup_k I_{jk}$, so $\mu(A) \leq \sum_j \sum_k |I_{jk}|$; and if we take the coverings of A_j so that $\sum_k |I_{jk}| \leq \mu(A_j) + \varepsilon 2^{-j}$ (note that if $\mu(A_j) = +\infty$ for any j, then there is nothing to prove) we have $\mu(A) \leq \sum_j (\mu(A_j) + \varepsilon 2^{-j}) = \varepsilon + \sum_j \mu(A_j)$. Since this is true for any $\varepsilon > 0$, it is true for $\varepsilon = 0$ and we have the required σ-subadditivity.

Since we have not worked very hard to establish the outer measure axioms for Lebesgue measure, we do not deserve to obtain very much as a consequence. Still, we get a little more than we deserve out of the following theorem. First we need to define the *splitting condition*, which will enable us to carry out the second step in the proof.

Definition 14.2.1 *Let μ be an outer measure on a σ-field $\mathcal{F}$. We say that a set A in $\mathcal{F}$ satisfies the splitting condition if we have*

$$\mu(B) = \mu(B \cap A) + \mu(B \backslash A)$$

for every set B in $\mathcal{F}$.

Since the splitting condition is an exact additivity statement for the disjoint sets $B \cap A$ and $B \backslash A$, we cannnot expect it to hold very often, unless μ is a measure, in which case it would always hold. It is obviously true for the empty set and the whole space but may not hold for any other sets. What the next theorem says is that things are very good for the sets that do satisfy the splitting condition.

Theorem 14.2.1 *Let μ be an outer measure on the σ-field $\mathcal{F}$, and let $\mathcal{F}_0$ denote the sets in $\mathcal{F}$ that satisfy the splitting condition. Then $\mathcal{F}_0$ is a σ-field, and μ restricted to $\mathcal{F}_0$ is a measure.*

Proof: It is clear from the definition that A is in $\mathcal{F}_0$ if and only if the complement of A is in $\mathcal{F}_0$ because the splitting condition is the same for A and ${}^cA$ $(B \backslash A = B \cap {}^cA)$. We first show that $\mathcal{F}_0$ is a field and that μ is finitely additive on $\mathcal{F}_0$. So let A_1 and A_2 be in $\mathcal{F}_0$, and take any set B in $\mathcal{F}$. Write $B = B_1 \cup B_2 \cup B_3 \cup B_4$ (disjoint), according to the Venn diagram shown in Figure 14.2.1.

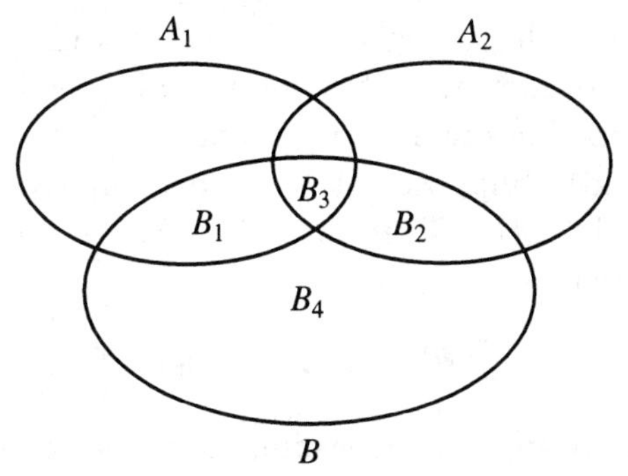

Figure 14.2.1:

All four sets B_j are in $\mathcal{F}$. The splitting condition for $A_1 \cup A_2$ and B that we need to prove is

$$\mu(B) = \mu(B_1 \cup B_2 \cup B_3) + \mu(B_4).$$

Now A_1 satisfies the splitting condition, so splitting B yields $\mu(B) = \mu(B_1 \cup B_3) + \mu(B_2 \cup B_4)$. Also A_2 satisfies the splitting condition, so we can use it to split the sets $B_1 \cup B_3$ and $B_2 \cup B_4$ to obtain $\mu(B_1 \cup B_3) = \mu(B_1) + \mu(B_3)$ and $\mu(B_2 \cup B_4) = \mu(B_2) + \mu(B_4)$. Combining these results yields

$$\mu(B) = \mu(B_1) + \mu(B_2) + \mu(B_3) + \mu(B_4).$$

If instead of starting with B we start with $B_1 \cup B_2 \cup B_3$ and repeat the above arguments we obtain

$$\mu(B_1 \cup B_2 \cup B_3) = \mu(B_1) + \mu(B_2) + \mu(B_3).$$

This proves the desired splitting of B,

$$\begin{aligned}\mu(B) &= (\mu(B_1)+\mu(B_2)+\mu(B_3))+\mu(B_4)\\ &= \mu(B_1\cup B_2\cup B_3)+\mu(B_4).\end{aligned}$$

So $\mathcal{F}_0$ is a field. Furthermore, if A_1 and A_2 in $\mathcal{F}_0$ are disjoint, then using A_1 to split $A_1 \cup A_2$ yields

$$\mu(A_1\cup A_2)=\mu(A_1)+\mu(A_2).$$

Thus μ is finitely additive on $\mathcal{F}_0$.

Next we show that $\mathcal{F}_0$ is a σ-field. Since we have shown that it is a field, it suffices to show that it is preserved by countable disjoint unions (since any countable union can be replaced by a disjoint one without leaving the field). Suppose $\{A_j\}$ is a disjoint sequence of sets in $\mathcal{F}_0$, and let $A = \bigcup_{j=1}^{\infty} A_j$. Given any B in $\mathcal{F}$, we need to show that the splitting formula

$$\mu(B)=\mu(B\cap A)+\mu(B\backslash A)$$

holds. Now μ is an outer measure, so by the σ-subadditivity, which implies finite sub-additivity, we already know

$$\mu(B)\le\mu(B\cap A)+\mu(B\backslash A);$$

so it suffices to establish the reverse inequality.

Consider the sets $F_n = \bigcup_{j=1}^{n} A_j$. Since $F_n \subseteq A$, we have $B\backslash A \subseteq B\backslash F_n$ and so $\mu(B\backslash A) \le \mu(B\backslash F_n)$ by the monotonicity of μ. On the other hand, F_n is in $\mathcal{F}_0$ because $\mathcal{F}_0$ is a field; so F_n splits B, yielding

$$\mu(B)=\mu(B\cap F_n)+(B\backslash F_n)$$

and furthermore

$$\mu(B\cap F_n)=\sum_{j=1}^{n}\mu(B\cap A_j)$$

by induction on the splitting property since the sets A_j are in $\mathcal{F}_0$ and disjoint. By the σ-subadditivity (applied to $B \cap A = \bigcup_{j=1}^{\infty}(B\cap A_j)$) we have $\mu(B\cap A) \le \sum_{j=1}^{\infty} \mu(B\cap A_j)$. Altogether we have shown

$$\mu(B\cap A)+\mu(B\backslash A)$$

$$\begin{aligned}
&\leq \sum_{j=1}^{\infty} \mu(B \cap A_j) + \mu(B \backslash A) \\
&= \lim_{n\to\infty} \sum_{j=1}^{n} \mu(B \cap A_j) + \mu(B \backslash A) \\
&\leq \lim_{n\to\infty} \left(\sum_{j=1}^{n} \mu(B \cap A_j) + \mu(B \backslash F_n) \right) \\
&= \lim_{n\to\infty} (\mu(B \cap F_n) + \mu(B \backslash F_n)) \\
&= \lim_{n\to\infty} \mu(B) = \mu(B).
\end{aligned}$$

This completes the verification that $\mathcal{F}_0$ is a σ-field. But if we look at the last string of inequalities for $B = A$ we find

$$\mu(A) = \sum_{j=1}^{\infty} \mu(A_j),$$

which is exactly the σ-additivity of μ on $\mathcal{F}_0$. This is the only axiom for a measure not contained in the axioms for an outer measure, so μ restricted to $\mathcal{F}_0$ is a measure. QED

14.2.2 Metric Outer Measure

We now come to the third stage in our process. We are going to give a simple criterion for $\mathcal{F}_0$ to equal $\mathcal{F}$ in the last theorem. This will allow us to conclude that μ is a measure on all of $\mathcal{F}$. We assume that X is a metric space and $\mathcal{F}$ is the σ-field of Borel sets. Recall that this is defined to be the smallest σ-field containing the open sets (or closed sets, since these are the complements of open sets).

Definition 14.2.2 *We say two sets A and B in X are separated if the distance from A to B (the* inf *of $d(x, y)$ for all x in A and y in B) is positive. We define a metric outer measure to be an outer measure that is additive on separated sets,*

$$\mu(A \cup B) = \mu(A) + \mu(B),$$

if A and B are separated.

Of course, separated sets are disjoint, but disjoint sets need not be separated (think of adjacent open intervals). It is usually not difficult to verify the metric condition in specific cases. For example, take Lebesgue measure. Notice that in the definition of $\mu(A)$ we can always assume that the covering intervals have length less than ε, for any fixed ε. (If not, break any interval of length greater then ε into a disjoint union of small intervals; this does not change the sum $\sum |I_j|$.) Given two separated sets A and B, say of distance ε apart, cover $A \cup B$ by $\bigcup I_j$ with $|I_j| < \varepsilon$. Then each I_j can meet only one of the sets A or B (or else the distance apart would be less than ε). Thus we can pick apart the covering of $A \cup B$ into a covering of A and B, say $A \subseteq \bigcup I'_j$ and $B \subseteq \bigcup I''_j$ with $\{I'_j\} \cup \{I''_j\} = \{I_j\}$ so that $\sum |I_j| = \sum |I'_j| + \sum |I''_j|$. If we chose the covering $A \cup B \subseteq \bigcup I_j$ so that $\mu(A \cup B) \geq \sum |I_j| - \delta$, then $\mu(A \cup B) \geq \mu(A) + \mu(B) - \delta$. Since this is true for any $\delta > 0$, it is true for $\delta = 0$. This completes the verification that Lebesgue measure is a metric outer measure, since the reverse inequality is automatic by subadditivity. Essentially the same argument shows that Lebesgue measure on $\mathbb{R}^n$ is also a metric outer measure.

Thus the next theorem will complete the proof of the existence of Lebesgue measure.

Theorem 14.2.2 (*Carathéodory*) *A metric outer measure is in fact a measure on the Borel sets.*

Proof: By the previous theorem it suffices to show that $\mathcal{F}_0 = \mathcal{F}$. Since $\mathcal{F}$ is generated by the closed sets and $\mathcal{F}_0$ is a σ-field, it suffices to show that every closed set A is in $\mathcal{F}_0$, which means

$$\mu(B) = \mu(B \cap A) + \mu(B \backslash A)$$

for every B in $\mathcal{F}$. Notice that we already know this is true if $B \cap A$ and $B \backslash A$ are separated. But in general, $B \cap A$ and $B \backslash A$ are disjoint but not separated.

Now we want to use the fact that A is closed and we are in a metric space. We break up $B \backslash A$ into "rings", as indicated in Figure 14.2.2, that is

$$B \backslash A = \sum_{j=1}^{\infty} R_j \text{ (disjoint)}$$

where the ring R_j is defined to be the set of points in $B \backslash A$ whose distance to the set $A(d(x, A) = \inf\{d(x, y) : y \text{ in } A\})$ satisfies

$$1/j \leq d(x, A) < 1/(j-1).$$

The fact that A is closed is equivalent to the statement that $d(x, A) > 0$ if x is not in A because $d(x, A) = 0$ says that x is a limit-point of A.

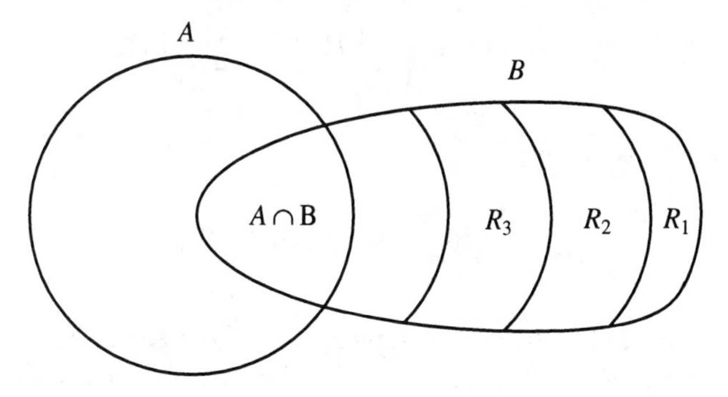

Figure 14.2.2:

Notice that the sets $B \cap A$ and $\bigcup_{j=1}^{n} R_j$ are separated (distance at least $1/n$), so we have

$$\mu\left((B \cap A) \cup \bigcup_{j=1}^{n} R_j\right) = \mu(B \cap A) + \mu\left(\bigcup_{j=1}^{n} R_j\right).$$

By monotonicity we have

$$\mu(B) \geq \mu\left((B \cap A) \cup \bigcup_{j=1}^{n} R_j\right),$$

so

$$\mu(B) \geq \mu(B \cap A) + \mu\left(\bigcup_{j=1}^{n} R_j\right).$$

To complete the proof we only need

$$\lim_{n \to \infty} \mu\left(\bigcup_{j=1}^{n} R_j\right) \geq \mu(B \backslash A)$$

since the reverse inequality

$$\mu(B) \leq \mu(B \cap A) + \mu(B \backslash A)$$

is automatic by subadditivity.

But by σ-subadditivity we know

$$\mu(B \backslash A) \leq \mu\left(\bigcup_{j=1}^{n} R_j\right) + \sum_{j=n+1}^{\infty} \mu(R_j),$$

so it suffices to show

$$\lim_{n\to\infty} \sum_{j=n+1}^{\infty} \mu(R_j) = 0,$$

which would follow from the convergence of the series $\sum_{j=1}^{\infty} \mu(R_j)$. We need one more trick to get this. Notice that the convergence would follow if we could show $\sum \mu(R_j)$ converges when summed separately over even and odd values of j. The point of this is that the sets $R_1, R_3, R_5, \ldots$ are all separated, as are the set $R_2, R_4, R_6, \ldots$. Thus $\mu(R_1) + \mu(R_3) + \cdots + \mu(R_{2j+1}) = \mu(R_1 \cup R_3 \cup \cdots \cup R_{2j+1})$ by the metric hypothesis on μ. Since $\mu(R_1 \cup R_3 \cup \cdots \cup R_{2j+1}) \leq \mu(B \backslash A)$ by monotonicity and we may assume $\mu(B \backslash A) < \infty$ (or else $\mu(B) = +\infty$ by monotonicity and the splitting is obvious), we have the convergence of the odd sums. A similar argument shows that the even sums converge. QED

14.2.3 Hausdorff Measures*

Lebesgue measure on the line has the simple *scaling property* that if we dilate a set by a factor t, then the measure is multiplied by t. In n-dimensions the corresponding factor is t^n (so if your height doubles, your weight should increase by about a factor of eight, assuming you live in three dimensions). Hausdorff had the brilliant insight that we can construct measures that scale with factor t^α for any positive α by making a simple change in the definition of Lebesgue measure. Hausdorff's construction works in any metric space. In recent years, these Hausdorff measures have played an important role in the development of fractal geometry.

For simplicity we begin with the definitions for subsets of the line. If A is a Borel set covered by a countable union of intervals $A \subseteq \bigcup_{j=1}^{\infty} I_j$, we may take $\sum_{j=1}^{\infty} |I_j|^\alpha$ as an upper approximation to the measure μ_α. However, the definition $\inf\{\sum_{j=1}^{\infty} |I_j|^\alpha\}$ over all such coverings does not work. The reason it fails is that if we split an interval I into pieces, say $I = \bigcup_{j=1}^{N} I_j$ (disjoint), then $|I| = \sum_{j=1}^{N} |I_j|$ but we do *not* have $|I|^\alpha = \sum_{j=1}^{N} |I_j|^\alpha$ for any $\alpha \neq 1$. In fact, for $0 \leq \alpha < 1$, which is the case in which we will be interested, $\sum_{j=1}^{N} |I_j|^\alpha$ is larger then $|I|^\alpha$. This means that if we have a covering $A \subseteq \bigcup_{j=1}^{\infty} I_j$ with $\sum_{j=1}^{\infty} |I_j|^\alpha$ small, we cannot automatically replace it with a covering where all the intervals are small simply by splitting the intervals into small pieces. Since the smallness of the covering intervals was a crucial fact that was used in proving the metric property of Lebesgue measure, we will have to build the corresponding fact into the definition of μ_α.

Definition 14.2.3 *For any fixed $\varepsilon > 0$, define $\mu_\alpha^{(\varepsilon)}(A) =$*

$$\inf \left\{ \sum_{j=1}^{\infty} |I_j|^\alpha : A \subseteq \bigcup_{j=1}^{\infty} I_j \text{ and the intervals } I_j \text{ all satisfy } |I_j| \leq \varepsilon \right\}.$$

Then define $\mu_\alpha(A) = \lim_{\varepsilon \to 0} \mu_\alpha^{(\varepsilon)}(A)$.

The limit defining $\mu_\alpha(A)$ always exists in the extended real numbers because $\mu_\alpha^{(\varepsilon)}(A)$ increases as $\varepsilon \to 0$ (it is the infimum over a smaller collection of coverings for smaller ε).

If A is a bounded set we will always have a finite value for $\mu_\alpha^{(\varepsilon)}(A)$, but we may nevertheless have $\mu_\alpha^{(\varepsilon)}(A) = +\infty$. In fact, if $\alpha < 1$ we will usually have $\mu_\alpha(A) = +\infty$ unless A is very thin. For example, if A is a nonempty open interval, then $\mu_\alpha(A) = +\infty$. The reason we require $\alpha \leq 1$ is that if $\alpha > 1$, then $\mu_\alpha(A) = 0$ for every set A. We leave the details to the exercises.

Theorem 14.2.3 *μ_α is a measure on the Borel sets.*

Proof: First we observe that μ_α is an outer measure. The proof of the required properties is routine and is left to the exercises. Then we observe that μ_α is a metric outer measure. The proof is almost the same as for Lebesgue measure. If A and B have distance apart δ, then for any $\varepsilon < \delta$ we have

$$\mu_\alpha^{(\varepsilon)}(A \cup B) = \mu_\alpha^{(\varepsilon)}(A) + \mu_\alpha^{(\varepsilon)}(B)$$

because any countable covering of $A \cup B$ by intervals of length at most ε splits into a disjoint union of coverings of A and B. Taking the limit as $\varepsilon \to 0$ we obtain the additivity for μ_α. Finally, we apply the Carathéodory theorem on metric outer measures. QED

It is easy to show (see exercises) that μ_1 is Lebesgue measure and μ_0 is counting measure and that μ_α has the desired scaling property, $\mu_\alpha(\delta_t A) = t^\alpha \mu_\alpha(A)$ for dilations $\delta_t A = \{tx : x \text{ is in } A\}$. To get a feeling for what μ_α is like for $0 < \alpha < 1$ consider the middle-third Cantor set C. Choose $\alpha = \log 2/\log 3$. We claim $\mu_\alpha(C) \le 1$. The idea is that we can cover C by 2^n intervals of length $1/3^n$ (the intervals that remain from the unit interval after we have removed middle thirds n times). Thus we obtain an upper bound of $2^n \cdot (1/3^n)^\alpha = (2/3^\alpha)^n = 1$ (because $\alpha = \log 2/\log 3$) for $\mu_\alpha^{(\varepsilon)}(C)$ for $\varepsilon = 1/3^n$, hence $\mu_\alpha(C) \le 1$ in the limit. It is trickier to show $\mu_\alpha(C) = 1$, and we leave the details to the exercises.

To extend the definition of μ_α to $\mathbb{R}^n$ or any metric space, we drop the requirement that the covering sets I_j be intervals and allow them to be arbitrary closed sets, interpreting $|I|$ to denote the *diameter* of the set (the sup of $d(x, y)$ as x and y vary over I). The definition is otherwise the same, and the proof that μ_α is a measure is essentially the same. (In $\mathbb{R}^1$ the two definitions coincide because every closed set is contained in a closed interval of the same diameter.) It is much harder to compute Hausdorff measures in $\mathbb{R}^n$ because we have to consider such general coverings. For example, it is true that μ_n is equal to a multiple of Lebesgue measure in $\mathbb{R}^n$, and one can even compute the constant ($\mu_n(B_r) = (2r)^n$ for a ball B_r of radius r), but this requires a deep geometric fact, the isodiametric theorem: the maximal volume of a set of fixed diameter is attained by a ball. It is usually easy to obtain upper bounds for Hausdorff measure since this only requires finding efficient coverings. Lower bounds are more problematic because they require proving estimates for all possible coverings by quite general sets. For example, if C is a rectifiable curve in $\mathbb{R}^n$, then $\mu_1(C) \le \text{length}(C)$. If

$\alpha(t)$ is an arc length parametrization for t in $[a, b]$, then the coverings by pieces of the curve corresponding to a partition of $\{a, b\}$ will give this estimate. However, it is necessary to make stronger assumptions on the curve in order to conclude that this is an equality.

Using Hausdorff measures with varying α, we can give the definition of Hausdorff dimension. It is based on the following simple lemma.

Lemma 14.2.1

a. *Suppose* $\mu_\alpha(A) < \infty$. *Then* $\mu_\beta(A) = 0$ *for* $\beta > \alpha$.

b. *Suppose* $\mu_\alpha(A) > 0$. *Then* $\mu_\beta(A) = +\infty$ *for* $\beta < \alpha$.

Proof: Consider any covering $A \subseteq \bigcup_{j=1}^{\infty} I_j$ with $|I_j| \le \varepsilon$. Then if $\beta > \alpha$ we have $\sum_{j=1}^{\infty} |I_j|^\beta = \sum_{j=1}^{\infty} |I_j|^{\beta-\alpha}|I_j|^\alpha \le \varepsilon^{\beta-\alpha} \sum_{j=1}^{\infty} |I_j|^\alpha$, while if $\beta < \alpha$ we have similarly $\sum_{j=1}^{\infty} |I_j|^\beta \ge \varepsilon^{\beta-\alpha} \sum |I_j|^\alpha$. Thus $\mu_\beta^{(\varepsilon)}(A) \le \varepsilon^{\beta-\alpha}\mu_\alpha^{(\varepsilon)}(A)$ if $\beta > \alpha$ and $\mu_\beta^{(\varepsilon)}(A) \ge \varepsilon^{\beta-\alpha}\mu_\alpha^{(\varepsilon)}(A)$ if $\beta < \alpha$. Taking the limit as $\varepsilon \to 0$ we obtain the desired result. QED

Thus if we graph $\mu_\alpha(A)$ as a function of α, there will be a unique value α_0 such that $\mu_\alpha(A) = +\infty$ for $\alpha < \alpha_0$ and $\mu_\alpha(A) = 0$ for $\alpha > \alpha_0$.

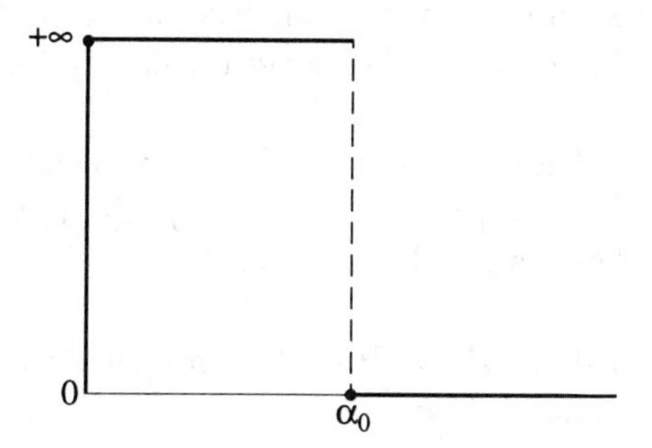

Figure 14.2.3:

The value $\mu_{\alpha_0}(A)$ may be anything in $[0, \infty]$. We define α_0 to be the *Hausdorff dimension* of A. Of course, if we can find α_0 such that $\mu_{\alpha_0}(A)$ is finite and positive, then α_0 is the Hausdorff dimension of A. Thus the Hausdorff dimension of the Cantor set is $\log 2/\log 3$.

14.2.4 Exercises

1. Verify that Lebesgue measure on the Borel sets in $\mathbb{R}^n$ is an outer measure, and in fact a metric outer measure, so it is indeed a measure by Carathéodory's theorem.

2. a. Prove that $\mu_\alpha(A) = 0$ for every Borel set A in $\mathbb{R}$ if $\alpha > 1$.

   b. Prove the same for every Borel set A in $\mathbb{R}^n$ if $\alpha > n$.

3. Verify that μ_α is an outer measure.

4. Show that μ_0 is counting measure on any metric space.

5. Show that μ_1 on $\mathbb{R}$ equals Lebesgue measure.

6. Show that μ_α on $\mathbb{R}^n$ is translation invariant, $\mu(A + y) = \mu(A)$ where $A + y = \{x + y : x \text{ is in } A\}$.

7. Show that $\mu_\alpha(\delta_t A) = t^\alpha \mu_\alpha(A)$ for Borel sets in $\mathbb{R}^n$.

8. *Show that $\mu_\alpha(C) = 1$ for the Cantor set C and $\alpha = \log 2 / \log 3$. (**Hint**: if I is an interval that has the form $I = I_1 \cup I_2 \cup I_3$ where I_2 is a deleted interval in the construction of C and I_1 and I_3 are adjacent to I_2 and of smaller length, then $|I_1|^\alpha + |I_3|^\alpha \leq |I|^\alpha$.)

9. Let C be the Cantor set and $\alpha = \log 2 / \log 3$. Let μ be the restriction of μ_α to C, so $\mu(A) = \mu_\alpha(A \cap C)$. Show that $\mu(A) \leq |A|^\alpha$ (where $|A|$ denotes the diameter of A).

10. *Let μ_α be n-dimensional Hausdorff measure on R^n, and let μ be Lebesgue measure on $\mathbb{R}^n$.

    a. Show that there exist constants c_1 and c_2 such that $\mu(A) \leq c_1 \mu_n(A)$ and $\mu_n(A) \leq c_2 \mu(A)$ for all Borel sets A.

    b. Show that there exists a constant c such that $\mu(R) = c\mu_n(R)$ for all rectangles R.

14.3 The Integral

14.3.1 Non-negative Measurable Functions

The Lebesgue integral is an absolutely convergent integral; $\int f(x)\,dx$ will be defined if and only if $\int |f(x)|\,dx$ is defined and finite. Therefore we will concentrate first on defining the integral for a non-negative function. Then for real-valued functions we will split the function into the difference of its positive and negative parts, $f = f^+ - f^-$, where $f^+ = \max(0, f)$ and $f^- = \max(0, -f)$, so that f^+ and f^- are non-negative and then define $\int f(x)\,dx = \int f^+(x)\,dx - \int f^-(x)\,dx$. Similarly for complex-valued functions we will integrate the real and imaginary parts separately.

Let X be a set on which we have defined a measure on a σ-field of subsets $\mathcal{F}$. We will let μ or $d\mu$ stand for the measure, as is conventional, so we write $\mu(A)$ for the measure of a set and $\int f d\mu$ for the integral with respect to the measure. It is standard terminology to refer to the pair $(X, \mathcal{F})$ as a *measurable space* and the triple $(X, \mathcal{F}, \mu)$ as a *measure space*, and we will refer to sets in $\mathcal{F}$ as *measurable sets.* It is usually safe to adopt the attitude that all sets that you will ever encounter are measurable. If you can write down a description of a set, then it will almost always be measurable, and usually it is a routine exercise to verify this. Of course these remarks apply primarily to Lebesgue measure on an interval of the line (or $\mathbb{R}^n$); there are situations in probability theory (as in the definition of conditional probability) where one deliberately takes a very small σ-field of sets so that not all reasonable sets are measurable.

Now suppose f is a non-negative function on X, in other words, $f(x) \geq 0$ for every x in X. The range of f is thus $[0, \infty)$, and for convenience we will even allow the possible value of $+\infty$. Now we partition the range $[0, \infty]$. For convenience let P_n denote the specific partition

$$\left[0, \frac{1}{2^n}\right), \left[\frac{1}{2^n}, \frac{2}{2^n}\right), \ldots, \left[2^n - \frac{1}{2^n}, 2^n\right), [2^n, \infty].$$

The reason for this particular choice is that it gives us a sequence of partitions, each being a refinement of the previous ones, such that in the limit every finite piece of the range is cut up into arbitrarily small pieces. Of course there are other ways to accomplish the same thing. Then we form the *Lebesgue approximate sum*

$$L(f, P_n) = \sum_{k=0}^{2^{2n}-1} \frac{k}{2^n}\mu\left\{x : \frac{k}{2^n} \le f(x) < \frac{k+1}{2^n}\right\} + 2^n\mu\{x : f(x) \ge 2^n\}.$$

Notice that for this to be well defined the sets $\{x : k/2^n \le f(x) < (k+1)/2^n\}$ must be measurable; this will put some sort of restriction on f but of a very weak nature since we believe most sets are measurable. Given that this condition is met, it is clear that $L(f, P_n)$ represents an underapproximation to the integral because we are multiplying the size $\mu(A)$ of each set A by the minimum value that f assumes on A, where $A = \{x : k/2^n \le f(x) < (k+1)/2^n\}$ or $\{x : f(x) > 2^n\}$. In other words, $A = f^{-1}(B)$ where B is one of the sets in the partition P_n. More generally, for any partition P of $[0, \infty]$ into a finite number of intervals we can define $L(f, P) = \sum_P (\inf B)\mu(f^{-1}(B))$ (here $\inf B$ means the lower endpoint of the interval B). Notice also that as we refine the partition we increase the Lebesgue approximate sum. If, say, a particular B_0 in P splits into $B_1' \cup \cdots \cup B_N'$ (disjoint) in a refined partition P', then $(\inf B_0)\mu(B_0) \le \sum_{k=1}^N (\inf B_k')\mu(B_k')$ because $\inf B_0 \le \inf B_k'$ for every k and $\mu(B_0) = \sum_{k=1}^N \mu(B_k')$ by the additivity of the measure. Since we have chosen the particular sequence of partitions P_n so that each one is a refinement of the previous ones, we have a monotone increasing sequence of Lebesgue approximate sums $\{L(f, P_n)\}$ and can define the integral as the limit, $\int f\,d\mu = \lim_{n\to\infty} L(f, P_n)$. This will be a nonnegative extended real number (the value $+\infty$ can occur either because $L(f, P_n)$ increases without bound or because $L(f, P_n) = +\infty$ for some n). We can be confident that this gives a reasonable definition because the maximum size of the intervals on any fixed bounded region goes to zero as $n \to \infty$ (since we are partitioning the unbounded range $[0, \infty]$ into a finite number of intervals, we must have one infinite interval).

Before going further with the definition of the integral we briefly pause to fill in the technical details of assuring that $\mu(f^{-1}(B))$ is defined. We say that a function $f : X \to \mathbb{R}$ is *measurable* if $f^{-1}(B)$ is a measurable subset of X whenever B is a Borel subset of $\mathbb{R}$ (recall that the Borel subsets are the σ-field generated by the intervals). We leave it as an exercise to verify that it suffices to show that $f^{-1}(B)$ is measurable for every *interval* B (or even every interval (a, ∞), or $[a, \infty)$, or $(-\infty, a)$, or $(-\infty, a])$ to conclude that f is measurable. The reason for this is that the intervals generate the Borel sets and f^{-1} preserves set-theoretic operations. It is of course simpler to verify that $f^{-1}(B)$ is measurable for every interval B (or every interval of the form (a, ∞), etc.) than for every Borel set B, so this remark is quite useful. We expect that every function we will encounter will be measurable, so it will be mostly a technical nuisance to have to verify it. Nevertheless, the Lebesgue theory of integration is restricted to measurable functions only. We can also define measurable functions taking values in the extended reals, by allowing the intervals B to contain $+\infty$ and $-\infty$. For technical reasons it is often convenient to do this, and we will not explicitly distinguish this minor variant.

There is a superficial resemblance between the definition of "measurable" function and one of the forms of the definition of "continuous" function: you need only interchange the words "measurable" and "open" (here we interpret "measurable" for subsets of the range $\mathbb{R}$ to mean "Borel set"). However, because there are so many more measurable sets than open sets—the axioms for a σ-field are quite generous—there are many more measurable functions than continuous functions. All the usual operations for generating functions preserve measurability; this explains why you won't "meet" a non-measurable function.

Theorem 14.3.1 *If f and g are measurable functions on X, then $af + bg, f \cdot g, f/g$ (if $g \neq 0$), $\max(f, g), \min(f, g)$, and $|f|$ are measurable functions. If $h : \mathbb{R} \to \mathbb{R}$ is measurable (with respect to the σ-field of Borel sets in $\mathbb{R}$), then $h \circ f$ is measurable. If f_n is a sequence of measurable functions on X, then $\sup_n f_n, \inf_n f_n \limsup f_n, \liminf f_n$, and $\lim f_n$ (if it exists pointwise) are measurable functions.*

Proof: We give the proof in some of the cases, leaving the others as exercises. Suppose $h = f + g$ and $f : X \to \mathbb{R}$ and $g : X \to \mathbb{R}$ are measurable. Let us define $F_a = \{x : f(x) > a\}$ and similarly for G_a and H_a, so $F_a = f^{-1}\{(a, \infty)\}$. We know that F_a and G_a are measurable, and we want to show that H_a is measurable. So we ask: how can $h(x)$ be greater than a? Clearly $h(x) > a$ if $f(x) > b$ while $g(x) > a - b$, and this must be true for some b if $h(x) > a$. Thus $H_a = \bigcup_b (F_b \cap G_{a-b})$. This shows how the sets H_a can be constructed out of the sets F_a and G_a. Unfortunately this construction involves an uncountable union that threatens to take us out of the σ-field of measurable sets. However a closer look at the argument suggests a trick to replace the uncountable union over all real b by a countable union. Since $f(x)$ is a real number, there must exist a *rational* number b such that $f(x) > b$. Thus we have $H_a = \bigcup_{b \text{ rational}}(F_b \cap G_{a-b})$ and this exhibits H_a as a measurable set. We have used tricks like this before, and we will have to use them again.

Next suppose $f_n : X \to \mathbb{R}$ are measurable functions, and let us show that $\sup_n f_n(x)$ is measurable. This is easy, since $\sup_n f_n(x) > a$ if and only if $f_n(x) > a$ for some n, so $\{x : \sup_n f_n(x) > a\} = \bigcup_n \{x : f_n(x) > a\}$ and a countable union of measurable sets is measurable. A similar argument works for $\inf_n f_n$. Then we obtain the result for $\limsup f_n$ and $\liminf f_n$ since $\limsup f_n = \inf_n(\sup_{k \geq n} f_k)$ and $\liminf f_n = \sup_n(\inf_{k \geq n} f_k)$. The same argument works for $\lim f_n$, if it exists pointwise, because then $\lim f_n = \limsup f_n = \liminf f_n$. QED

In order to use the theorem to show that all "usual" functions are measurable, we have to "get started" with some basic measurable functions. If X is $\mathbb{R}$ or an interval (or $\mathbb{R}^n$), then every continuous function is measurable, since $f^{-1}(B)$ for B open is open, and hence, measurable. Also, the characteristic function of a measurable set (the function that is one on the set and zero off it) is measurable. These functions enable us to get started constructing measurable functions. Finite linear combinations of characteristic functions of measurable sets are called *simple functions.* We write $f = \sum_{k=1}^{N} a_k \chi_{A_k}$ where χ_A denotes the characteristic functions of A. It is easy to see that we may assume that the measurable sets A_k are disjoint, even if they are not originally given disjoint, and *the simple functions are exactly the class of measurable functions that assume only a finite set of values.* Simple functions play a key role in our development of the integral, and we will need to establish some elementary facts about them along the way.

In terms of the simple functions, we can give a new interpretation to the Lebesgue approximate sums. Let f be a non-negative measurable function, and let P be a finite partition of $[0, \infty]$ into intervals. For each interval B in the partition, $f^{-1}(B)$ is a measurable set and so $\sum_P (\inf B)\chi_{f^{-1}(B)}$ is a simple function, obtained by replacing $f(x)$ by a possibly smaller value depending in which interval of the partition $f(x)$

lies. Then $L(f, P)$ is a kind of primitive integral of this simple function equal to the sum of the "areas" of the "rectangles" (the graphs of the functions $(\inf B)\chi_{f^{-1}(B)}$), which are the products of the height $\inf B$ with the measure $\mu(f^{-1}(B))$ of the "base" $f^{-1}(B)$. The situation is entirely analogous to the interpretation to the Cauchy approximate sums as integrals of step functions (more precisely the Riemann lower sums), but the simple functions are more versatile than the step functions, since the characteristic function of a measurable set is more general than the characteristic function of an interval. In fact, I dare not draw a picture of a simple function for fear of lulling you into thinking it is "simpler" than it might be—the measurable sets on which the function assumes its values might be Cantor sets, or worse.

Now suppose we consider the particular sequence P_n of partitions of $[0, \infty]$ described above, so P_n consists of $[2^n, \infty]$ together with $[0, 2^n)$ chopped up into intervals of length $1/2^n$, as indicated in Figure 14.3.1.

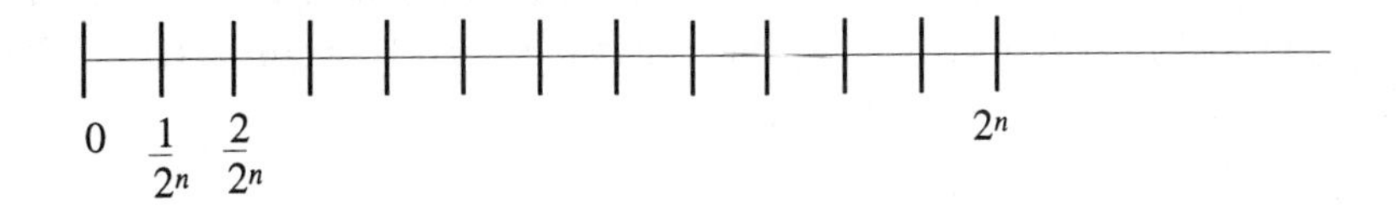

Figure 14.3.1:

Let $f_n = \sum_{P_n} (\inf B)\chi_{f^{-1}(B)}$ be the associated simple function. Then $\{f_n\}$ is monotone increasing and has limit equal to f pointwise. This is clear if we consider an individual point x and the value $f(x)$. For each n we locate the value $f(x)$ in one of the intervals of P_n, and then $f_n(x)$ is the inf of that interval. This value clearly increases as the partition is refined and approaches $f(x)$ since the size of the intervals is $1/2^n$ (once $f(x) \le 2^n$). In fact it is even true if f is allowed to assume the value $+\infty$, for then $f_n(x) = 2^n$ where $f(x) = +\infty$.

We have thus shown that every non-negative measurable function is obtainable as the pointwise limit of simple functions (it is then straightforward to obtain the same result for real-valued measurable functions). This justifies thinking of simple functions as the basic building blocks out of which general measurable functions are constructed. Actually we have proved a slightly stronger statement: *every non-negative measurable function is the pointwise limit of a monotone increasing sequence of non-negative simple functions.* Having the sequence monotone increasing may seem like only a minor improvement, but it turns out to be crucial for the theory of the integral. To understand its significance we need to look at some disturbing examples:

1. Let $X = [0,1]$ with Lebesgue measure and set $f_n = n\chi_{(0,1/n)}$. Then $\lim f_n(x) = 0$ pointwise, but the integral $n\mu(0,1/n) = 1$ does not approach zero. If instead we take $f_n = n^2\chi_{(0,1/n)}$, then still $\lim f_n(x) = 0$ but now the integral $n^2\mu(0,1/n) = n$ doesn't even have a limit.

2. Let $X = [0,\infty)$ with Lebesgue measure and set $f_n = (1/n)\chi_{(0,n)}$. Again $\lim f_n(x) = 0$, but the integrals $(1/n)\mu(0,n) = 1$ do not tend to zero.

Of course the limits are not monotone in these examples. The point is that without monotonicity, or at least some such restriction, there need be no relationship between the integral of f and the limit of the integrals of f_n, where f_n are simple functions such that $\lim f_n(x) = f(x)$ pointwise.

14.3.2 The Monotone Convergence Theorem

We start with the *definition of the integral:* If $f = \sum_{k=1}^N a_k\chi_{A_k}$ is a non-negative simple function we define $\int f d\mu = \sum_{k=1}^N a_k\mu(A_k)$. It is a simple exercise to verify that this definition is independent of the representation: if $\sum a_k\chi_{A_k} = \sum b_j\chi_{B_j}$, then $\sum a_k\mu(A_k) = \sum b_j\mu(B_j)$. Next if f is a non-negative measurable function we define $\int f d\mu = \lim_{n\to\infty} L(f,P_n) = \lim_{n\to\infty}\int f_n d\mu$ where $f_n = \sum_{P_n}(\inf B)\chi_{f^{-1}(B)}$. It is again easy to verify that this is consistent with the special case definition: if f is simple, say $f = \sum_{k=1}^N a_k\chi_{A_k}$, then $\lim_{n\to\infty} L(f,P_n) = \sum_{k=1}^N a_k\mu(A_k)$. To see this assume, as we may, that the sets A_k are disjoint and the values a_k distinct. Then for n large enough the values a_k will all be less than 2^n and fall into distinct intervals of length $1/2^n$ of P_n, so $L(f,P_n) = \sum_{k=1}^N b_k\mu(A_k)$ where $|b_k - a_k| \le 1/2^n$ and so $\lim_{n\to\infty} L(f,P_n) = \sum_{k=1}^N a_k\mu(A_k)$ (note that both sides are $+\infty$ if $\mu(A_k) = +\infty$ for some k).

This definition is extremely simple, but it has the defect that it appears to depend on the particular choice of the partitions P_n. If this were really the case, it would not be worth very much. Thus our next goal is to show that the same value for the integral would be obtained from other sequences of partitions. Recall that the sequence $f_n = \sum_{P_n}(\inf B)\chi_{f^{-1}(B)}$ of simple functions associated with the partitions P_n had the property that it was monotone increasing and had limit f. Clearly the same will be true if we take any other sequence of partitions such that each is a refinement of the previous one (to get monotonicity) and the maximum length of the subintervals on any bounded region goes to zero (to get the limit). Thus it suffices to show that $\lim_{n\to\infty}\int f_n d\mu = \int f d\mu$ if f_n is any monotone increasing sequence of non-negative simple functions converging to f. It turns out that the same is true even if the f_n are not assumed to be simple. This is the famous Lebesgue monotone convergence theorem. Since the proof is not much harder, we will go directly to the general case.

Theorem 14.3.2 (*Monotone Convergence Theorem*) *Let $0 \le f_1 \le f_2 \le \cdots$ be a monotone increasing sequence of non-negative measurable functions, and let $f = \lim_{k\to\infty} f_k$. Then $\int f d\mu = \lim_{k\to\infty}\int f_k d\mu$ (both sides may be equal to $+\infty$).*

Proof: Since we have defined $\int f d\mu = \lim_{n\to\infty} L(f, P_n)$ and $\int f_k d\mu = \lim_{n\to\infty} L(f_k, P_n)$, we need to show $\lim_{k\to\infty}\lim_{n\to\infty} L(f_k, P_n) = \lim_{n\to\infty} L(\lim_{k\to\infty} f_k, P_n)$. Now we claim that the inequality $\lim_{k\to\infty}\int f_k d\mu \le \int f d\mu$ is easy to obtain from the monotonicity. To prove it we observe that since $f_k \le f$, we have $L(f_k, P_n) \le L(f, P_n)$. This follows from the fact that the simple functions $f_k^{(n)} = \sum_{P_n}(\inf B)\chi_{f_k^{-1}(B)}$ and $f^{(n)} = \sum_{P_n}(\inf B)\chi_{f^{-1}(B)}$ (whose integrals give $L(f_k, P_n)$ and $L(f, P_n)$) satisfy $f_k^{(n)} \le f^{(n)}$ and the integral for simple functions is clearly monotone: $f \le g$ implies $\int f d\mu \le \int g d\mu$ (see exercises). From $L(f_k, P_n) \le L(f, P_n)$ we obtain $\int f_k d\mu \le \int f d\mu$ by letting $n \to \infty$, and then by letting $k \to \infty$ we obtain $\lim_{k\to\infty}\int f_k d\mu \le \int f d\mu$.

Now we work on getting the reverse inequality. Note that it suffices to show $\lim_{k\to\infty}\int f_k d\mu \ge L(f, P_n)$ for all n. Now $L(f, P_n) = \int f^{(n)} d\mu$ where $f^{(n)}$ is a simple function, and $\lim_{k\to\infty} f_k = f \ge f^{(n)}$. To simplify notation set $f^{(n)} = g$. Thus to complete the proof we need to show that if g is any simple function such that $\lim_{k\to\infty} f_k \ge g$, then $\lim_{k\to\infty}\int f_k d\mu \ge \int g d\mu$.

Write $g = \sum_{j=1}^{N} b_j \chi_{B_j}$ where the sets B_j are disjoint. We then restrict the functions f_k to the sets B_j by multiplying by χ_{B_j}, and observe that $f_k \geq \sum_{j=1}^{N} f_k \chi_{B_j}$ by disjointness, so $\int f_k d\mu \geq \sum_{j=1}^{N} \int f_k \chi_{B_j} d\mu$. If we can prove $\lim_{k\to\infty} \int f_k \chi_{B_j} d\mu \geq b_j \mu(B_j)$ then we will have $\lim_{k\to\infty} \int f_k d\mu \geq \int g d\mu$ by adding over j. In other words, without loss of generality we can assume $N = 1$ and so can drop subscripts and write $g = b\chi_B$. We need to show that $\lim_{k\to\infty} f_k = b\chi_B$ implies $\lim_{k\to\infty} \int f_k d\mu \geq b\mu(B)$.

Finally we can complete the proof of the theorem by appealing to the continuity from below for the measure. Suppose b and $\mu(B)$ are finite (we leave as an exercise the simple modifications necessary if either is $+\infty$). Fix an error ε and look at the set E_k where $f_k(x) \geq b - \varepsilon$. Since $\lim_{k\to\infty} f_k(x) \geq b$ on B, it follows that $\bigcup_{k=1}^{\infty} E_k \supseteq B$. Since f_k is increasing, so are the sets E_k; hence $\lim_{k\to\infty} \mu(E_k) \geq \mu(B)$ by continuity from below. But since $f_k(x) \geq b - \varepsilon$ on E_k, we have $\int f_k d\mu \geq (b-\varepsilon)\mu(E_k)$, so $\lim_{k\to\infty} f_k d\mu \geq (b-\varepsilon) \lim \mu(E_k) \geq (b-\varepsilon)\mu(B)$. Since this is true for every error ε, we have $\lim_{k\to\infty} \int f_k d\mu \geq b\mu(B)$ as desired. QED

With the aid of the monotone convergence theorem we can deduce the elementary properties of the integral of non-negative measurable functions quite easily from the corresponding properties of the integral of non-negative simple functions. For example, $\int (f+g) d\mu = \int f d\mu + \int g d\mu$ if f and g are non-negative simple functions. Then if f and g are merely non-negative measurable functions, let $\{f_n\}$ and $\{g_n\}$ be monotone increasing sequences of simple functions approximating f and g. Then $\{f_n + g_n\}$ is a monotone increasing sequence of simple functions approximating $f+g$, so from $\int (f_n+g_n) d\mu = \int f_n d\mu + \int g_n d\mu$ we obtain $\int (f+g) d\mu = \int f d\mu + \int g d\mu$ by passing to the limit. In a similar way we can prove the following theorem.

Theorem 14.3.3 *The integral of non-negative measurable functions is*

a. *linear,* $\int (af + bg) d\mu = a \int f d\mu + b \int g d\mu$ *if* a *and* b *are non-negative reals;*

b. *monotone,* $\int f d\mu \geq \int g d\mu$ *if* $f \geq g$;

c. *additive,* $\int_{A \cup B} f d\mu = \int_A f d\mu + \int_B f d\mu$ *for* A *and* B *disjoint measurable sets, where* $\int_A f d\mu$ *denotes* $\int f \chi_A d\mu$.

Note that if μ is Lebesgue measure and A is an interval (a,b), then $\int_A f d\mu$ plays the role of $\int_a^b f(x)\,dx$ in the Riemann theory of integration. In fact it is easy to see that if f is Riemann integrable, then the two must be equal. Indeed if f is Riemann integrable the Riemann upper and lower sums for any partition of (a,b) are integrals of step functions g and h such that $g \leq f \leq h$ on (a,b) and the step functions are just simple functions constant on intervals. Thus $\int g d\mu \leq \int f\chi_{(a,b)} d\mu \leq \int h d\mu$ by the monotonicity of the Lebesgue integral; hence, $\int f\chi_{(a,b)} d\mu = \int_a^b f(x)\,dx$ (of course a separate, technical argument must first be given to show that if f is Riemann integrable then f is measurable so $\int f\chi_{(a,b)} d\mu$ is defined). Since the Riemann and Lebesgue integrals agree on their common domain of definition, we will not insist on separate notation and will write $\int_a^b f(x)\,dx$ for the Lebesgue integral as well.

The monotone convergence theorem gives us a criterion for interchanging limits and integrals for non-negative functions, and we have seen some examples in which the interchange is not valid. One feature of these examples is that they both result in a "loss of mass" in passing to the limit—the limiting function has a smaller integral than the limit of the integrals. It turns out that this is always the case—it is impossible to gain mass by passing to a limit. This is a special case of a famous theorem of Fatou.

Theorem 14.3.4 (*Fatou's Theorem*) *Let* $f = \lim_{n\to\infty} f_n$ *where* f_n *are non-negative measurable functions. If* $\lim_{n\to\infty} \int f_n d\mu$ *exists, then* $\int f d\mu \leq \lim_{n\to\infty} \int f_n d\mu$.

Proof: The idea of the proof is to replace the sequence $\{f_n\}$ by a monotone increasing sequence $\{g_n\}$ with the same limit. A little thought shows that $g_n = \inf_{k\geq n} f_k$ will do the job. Note that $g_n \leq f_n$, so $\int g_n d\mu \leq \int f_n d\mu$ and passing to the limit $\lim_{n\to\infty} \int g_n d\mu \leq \lim_{n\to\infty} \int f_n d\mu$. But $\lim_{n\to\infty} \int g_n d\mu = \int f d\mu$ by the monotone convergence theorem. QED

Actually the full Fatou's Theorem says $\int \liminf_{n\to\infty} f_n d\mu \leq \liminf_{n\to\infty} \int f_n d\mu$ even when $\lim f_n$ and $\lim \int f_n d\mu$ are not assumed to exist, as long as f_n are assumed non-negative. We leave it as an exercise to verfiy that essentially the same argument works in this case.

14.3.3 Integrable Functions

We have now discussed all the important properties of the integral of non-negative measurable functions, and we would like to pass to the general case of measurable real-valued functions. We can always write $f = f^+ - f^-$ and so we will define $\int f d\mu = \int f^+ d\mu - \int f^- d\mu$ *provided both terms are finite.* For non-negative functions we could afford the luxury of allowing a value of $+\infty$ for the integral—this allowed us to define the integral for every measurable non-negative function and actually simplified the statements of the theorems. For real-valued functions, however, we have to avoid dealing with expressions like $\infty - \infty$ and so make the definition that a measurable function is *integrable* if f^+ and f^- both have finite integrals. We then define the integral $\int f d\mu = \int f^+ d\mu - \int f^- d\mu$ for integrable functions only. The terminology is slightly confusing in that we have previously defined an integral for all non-negative measurable functions, including those that are not integrable, and we do not wish to recant. It does mean that you will have to adjust to the possibility that a function that isn't integrable may have an integral (but the integral won't be finite).

Notice that $|f| = f^+ + f^-$, so a measurable function f is integrable if and only if $|f|$ is integrable and we have Minkowski's inequality $|\int f d\mu| \leq \int |f| d\mu$. It is in this sense that we say the Lebesgue integral is an absolutely convergent integral. In particular, certain improper Riemann integrals whose convergence depends on cancellation (such as $\int_0^\infty (\sin x/x)\, dx$) are not subsumed in the Lebesgue theory. If μ denotes counting measure on $\{1, 2, 3, \ldots\}$, then the integrable functions are the absolutely convergent series, with the integral equal to the sum of the series (see exercise set 14.3.5, number 9).

From the properties of the integral of non-negative functions it is simple to deduce the corresponding properties of the integral of real-valued integrable functions. In particular, it is linear (with real coefficients), monotone, and additive. For example, let's show $\int (f + g) d\mu = \int f d\mu + \int g d\mu$. This would follow immediately if we had $(f + g)^+ = f^+ + g^+$, but this need not be the case since the positive and negative parts of f and g can overlap. Nevertheless, from $f = f^+ - f^-$ and $g = g^+ - g^-$ we obtain $f + g = f^+ + g^+ - (f^- + g^-)$. We also have $f + g = (f + g)^+ - (f + g)^-$, so by equating the two we obtain $f^+ + g^+ - (f^- + g^-) = (f + g)^+ - (f + g)^-$, which we can rewrite as an equality between sums of non-negative functions, $f^+ + g^+ + (f+g)^- = f^- + g^- + (f+g)^+$. We integrate this equality and use the linearity of the integral of non-negative functions to interchange the finite sums and the integral:

$$\int f^+ d\mu + \int g^+ d\mu + \int (f+g)^- d\mu = \int f^- d\mu + \int g^- d\mu + \int (f+g)^+ d\mu.$$

Finally we rewrite this as

$$\begin{aligned} &\int (f+g)^+ d\mu - \int (f+g)^- d\mu \\ &\quad = \left(\int f^+ d\mu - \int f^- d\mu\right) + \left(\int g^+ d\mu - \int g^- d\mu\right), \end{aligned}$$

which says $\int (f+g) d\mu = \int f d\mu + \int g\, d\mu$. (A separate argument to show that all these integrals are finite must be given to justify the rearrangement of terms.) The derivation of the other properties is similar, and we leave it to the exercises. We can also define an integral for complex-valued measurable functions by splitting into real and imaginary parts. Again we leave the details as an exercise.

The monotone convergence theorem, on the other hand, does not extend to real-valued functions. Now some sort of interchange of limit and integral is vital to a good theory; essentially the only way to compute the integral of "new" functions f is to write $f = \lim_{k\to\infty} f_k$ where f_k are functions whose integrals we can compute by elementary means and then to compute $\int f d\mu = \lim_{k\to\infty} \int f_k\, d\mu$ by quoting the appropriate theorem. Thus we need to find a good substitute for the monotone convergence theorem. Lebesgue's dominated convergence theorem will serve this purpose very well. This theorem is motivated by the following elementary observation: if g is a non-negative integrable function and f is a measurable real-valued function such that $|f| \le g$, then f is integrable. This follows immediately from the two facts $|f| \le g$ implies $\int |f| d\mu \le \int g d\mu$ and $\int |f| d\mu < \infty$ implies f is integrable. If $|f| \le g$ we say that g *dominates* f. It implies that the graph of f lies between the graphs of $+g$ and $-g$, a region (shown in Figure 14.3.2) that has finite "area" since g is integrable. Now if $\{f_n\}$ is a sequence of measurable functions, each of which is dominated by g, then the graphs of f_n are all contained in this region of finite area and so it is plausible that there is no room for any mass to leak out when we pass to the limit.

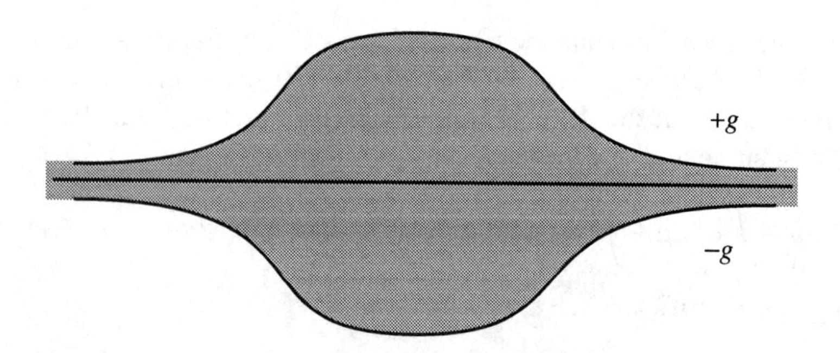

Figure 14.3.2:

If we look back at the examples where the limit and the integral cannot be interchanged, we see that the smallest function that dominates all the f_n is essentially $1/x$, and this function just fails to be integrable.

Theorem 14.3.5 (*Dominated Convergence Theorem*) *Let $\{f_n\}$ be a sequence of measurable functions converging pointwise to f. If there exists an integrable function g such that $|f_n(x)| \leq g(x)$ for all n and x, then f is integrable and $\int f d\mu = \lim_{n\to\infty} \int f_n d\mu$.*

Proof: Assume first $\lim_{n\to\infty} \int f_n d\mu$ exists. Then $\{g - f_n\}$ and $\{g+f_n\}$ are sequences of non-negative measurable functions converging to $g - f$ and $g + f$. Since g dominates f, f is integrable. By Fatou's theorem

$$\begin{aligned}\int (g \pm f)d\mu &\leq \lim_{n\to\infty} \int (g \pm f_n)d\mu \\ &= \lim_{n\to\infty} \left(\int g d\mu \pm \int f_n d\mu \right) \\ &= \int g d\mu \pm \lim_{n\to\infty} \int f_n d\mu\end{aligned}$$

and by subtracting the finite term $\int g d\mu$ from both sides the inequality we obtain $\pm \int f d\mu \leq \pm \lim_{n\to\infty} \int f_n d\mu$, which implies the equality.

Without the assumption that $\int f_n d\mu$ converges we first pass to a convergent subsequence (the sequence $\{\int f_n d\mu\}$ is bounded since $|\int f_n d\mu| \leq \int |f_n| d\mu \leq \int g d\mu$) and then conclude $\lim \int f_{n'} d\mu = \int f d\mu$ along the subsequence. Since this is true for any convergent subsequence, we conclude that the whole sequence converges to $\int f d\mu$. QED

One important special case of this theorem is the following: if the measure of the whole space is finite and the sequence f_n is uniformly bounded, then $\lim_{n\to\infty} \int f_n d\mu = \int \lim_{n\to\infty} f_n d\mu$. In this case we can take $g \equiv M$ where $M = \sup_{n,x} |f_n(x)|$.

14.3.4 Almost Everywhere

Suppose f and g are two measurable functions that are equal except on a set of measure zero, $f(x) = g(x)$ if x is not in E with $\mu(E) = 0$. Then we say $f = g$ *almost everywhere*, abbreviated a.e. More generally we say of any property that it holds a.e. if it is true for all x except for x in a set of measure zero. Because sets of measure zero are preserved under finite and even countable unions, the a.e. concept is suitably flexible. For instance, if $f_1 = g_1$ a.e. and $f_2 = g_2$ a.e., then $f_1 + f_2 = g_1 + g_2$ a.e.; or if $f_n = g_n$ a.e. for all n, then $\lim_{n\to\infty} f_n = \lim_{n\to\infty} g_n$ a.e. Since sets of measure zero cannot contribute to the integral, we expect that functions equal almost everywhere should be more or less interchangable when integrated.

Theorem 14.3.6 *Let f and g be integrable functions. Then $f = g$ a.e. if and only if $\int_A f d\mu = \int_A g d\mu$ for every measurable set A.*

Proof: Suppose $f = g$ a.e. Then we can write $f = g+h$ where $h = 0$ a.e. To show $\int f d\mu = \int g d\mu$ it suffices to show $\int h d\mu = 0$. Now $h = 0$ a.e. implies $h^+ = 0$ a.e. and $h^- = 0$ a.e., so it suffices to show $\int h d\mu = 0$ if $h = 0$ a.e. and h is non-negative. But $\int h d\mu = \lim_{n\to\infty} \int h_n d\mu$ for a sequence of simple non-negative functions increasing monotonically to h. Since $0 \le h_n \le h$, we have $h_n = 0$ a.e., so $h_n = \sum_{k=1}^N a_k \chi_{A_k}$ with $\mu(A_k) = 0$. It follows that $\int h_k d\mu = \sum_{k=1}^N a_k \mu(A_k) = 0$, hence $\int h d\mu = 0$.

More generally, if A is any measurable set, then $f = g$ a.e. implies $f\chi_A = g\chi_A$ a.e., so the above argument shows $\int_A f d\mu = \int_A d\mu$.

Conversely, suppose $\int_A f d\mu = \int_A g d\mu$ for every measurable set A. Apply this to the measurable set $A_n = \{x : f(x) \le g(x) - 1/n\}$. Then $\int_{A_n} f d\mu \le \int_{A_n} (g - 1/n) d\mu$ by monotonicity; hence,

$$\int_{A_n} f d\mu \le \int_{A_n} g d\mu - \frac{1}{n}\mu(A_n),$$

which contradicts $\int_{A_n} f d\mu = \int_{A_n} g d\mu$ unless $\mu(A_n) = 0$. Now $\mu\{\bigcup_{n=1}^{\infty} A_n\} = 0$ by σ-additivity, so $f(x) \geq g(x)$ a.e. Interchanging f and g gives $g(x) \geq f(x)$ a.e., hence $f(x) = g(x)$ a.e. QED

This result means that whenever we have a theorem about integrals, we can replace any condition that is supposed to hold for every x by the same condition a.e. For example, in the dominated convergence theorem, we need only assume $\lim_{n\to\infty} f_n(x) = f(x)$ a.e. and $|f_n(x)| \leq g(x)$. a.e.

The almost everywhere concept has some rather paradoxical consequences. For example, with respect to Lebesgue measure, an individual point, or even a countable set of points, has measure zero. Thus as far as integration theory is concerned, the value of a function $f(x)$ at an individual point $\tilde{x}$ is immaterial. We can always find $g = f$ a.e. that takes a different value at $\tilde{x}$. If we think of all the functions equal to f a.e. as forming an equivalence class, then there is no meaning within this equivalence class of the value at an individual point. Nevertheless, how can we know the function if we don't know its value at any point? There is a resolution to this paradox, but it is beyond the scope of this book. The Lebesgue Differentiation of the Integral Theorem says that the "average value" $\lim_{r\to 0}(1/2r)\int_{x-r}^{x+r} f(y)\,dy$ exists for almost every x, and this provides a "canonical" choice of value for $f(x)$ for such points. The "average values" are unchanged if we change f on a set of measure zero.

14.3.5 Exercises

1. Let $(X, \mathcal{F})$ be a measurable space. Prove that $f : X \to \mathbb{R}$ is a measurable function if and only if $f^{-1}(B)$ is in $\mathcal{F}$ for every set B in a collection $\mathcal{B}$ of sets with the property that the smallest σ-field containing $\mathcal{B}$ is the σ-field of Borel sets. Verify that this is the case if $\mathcal{B}$ consists of all intervals (a, ∞).

2. Prove that if $f : X \to \mathbb{R}$ and $g : X \to \mathbb{R}$ are measurable, then $\max(f, g)$ is measurable.

3. Prove that if $f : X \to \mathbb{R}$ is measurable and $h : \mathbb{R} \to \mathbb{R}$ is measurable (with respect to the σ-field of Borel sets in $\mathbb{R}$), then $h \circ f$ is measurable.

4. Prove that if $\sum_{k=1}^{N} a_k \chi_{A_k} = \sum_{j=1}^{M} b_j \chi_{B_j}$ for every x, then $\sum_{k=1}^{N} a_k \mu(A_k) = \sum_{j=1}^{M} b_j \mu(B_j)$.

5. Prove directly that if $\sum_{k=1}^{N} a_k \chi_{A_k} \leq \sum_{j=1}^{M} b_j \chi_{B_j}$ for every x, then $\sum_{k=1}^{N} a_k \mu(A_k) \leq \sum_{j=1}^{M} b_j \mu(B_j)$.

6. Write out a complete proof that the integral is linear, monotone, and additive first for non-negative simple functions, then for non-negative measurable functions, and finally for integrable functions.

7. Prove that a Riemann integrable function on a bounded interval is measurable. (**Hint**: it is the pointwise limit a.e. of the step functions involved in the Riemann upper and lower approximate sums.)

8. Prove that if $\{f_n\}$ is any sequence of nonnegative measurable functions, then $\int \liminf_{n\to\infty} f_n d\mu \leq \liminf_{n\to\infty} \int f_n d\mu$.

9. Prove that a series $\sum_{n=1}^{\infty} a_n$ is absolutely convergent if and only if the function $n \to a_n$ on the positive integers is integrable with respect to counting measure.

10. Prove that $\sum_{n=1}^{\infty} \int f_n d\mu = \int (\sum_{n=1}^{\infty} f_n)\, d\mu$ if f_n are non-negative measurable functions.

11. Suppose f_n are measurable functions and $\sum_{n=1}^{\infty} |f_n|$ is integrable. Prove that $\sum_{n=1}^{\infty} f_n$ is integrable and $\int (\sum_{n=1}^{\infty} f_n) d\mu = \sum_{n=1}^{\infty} \int f_n d\mu$.

12. Restate the results of problems 10 and 11 when μ is counting measure on the positive integers in terms of doubly indexed infinite series.

13. Let $F = f + ig$ be a complex-valued function. Define it to be integrable if f and g are integrable, and define $\int (f + ig) d\mu = \int f d\mu + i \int g d\mu$. Prove that F is integrable if and only if f and g are measurable and $|F|$ is integrable, and prove Minkowski's inequality $|\int F d\mu| \leq \int |F| d\mu$.

14. Prove that a non-negative measurable function has integral equal to zero if and only if it is zero a.e.

15. Prove that if f is an integrable function, then $\nu(A) = \int_A f d\mu$ is a measure on the same σ-field on which μ is defined.

16. Prove that under the hypotheses of the Dominated Convergence Theorem one has $\lim_{n\to\infty} \int |f_n - f| d\mu = 0$.

17. Explain why Fatou's Theorem shows that the "hard" inequality in the proof of the Monotone Convergence Theorem is always valid, even without the assumption of monotonicity. Did the proof of this part of the theorem use the hypothesis of monotonicity? Why can't we use Fatou's theorem to simplify the proof of the Monotone Convergence Theorem?

14.4 The Lebesgue Spaces L^1 and L^2

14.4.1 L^1 as a Banach Space

We have already seen the utility of thinking of a set of functions as forming a "space" on which certain structures are defined. We particularly made use of the space $C(I)$ of continuous functions on a compact interval I, which is a vector space and has a natural metric space structure given by the sup-norm $||f||_{\text{sup}} = \sup_I |f(x)|$. One of the reasons the space $C(I)$ with this metric is so useful is that it is complete, so we can apply the contractive mapping principle, for example. We observed that there are also other norms on this space, $||f||_1 = \int_I |f(x)|\, dx$ and $||f||_2 = (\int_I |f(x)|^2 dx)^{1/2}$, the 2-norm even being associated with an inner product $\langle f, g\rangle = \int_I f(x)\overline{g(x)}dx$. But the space $C(I)$ is not complete with respect to these norms. In this section we will see how with the aid of Lebesgue integration we can construct complete spaces with these norms, called $L^1(I)$ and $L^2(I)$. These will be concrete realizations of the "completions" of $C(I)$ with respect to these norms (analogous to the real numbers as completion of the rationals). Recall that a complete normed vector space is called a *Banach space.* Thus we are constructing two more examples of a Banach space. The study of Banach spaces is one of the central topics in twentieth century analysis. We will also give some applications of the space $L^2([-\pi, \pi])$ to Fourier series. This is not surprising if you recall the role of the inner product in defining Fourier coefficients and in the derivation of Parseval's identity.

Let us consider the space of integrable functions on an interval or more generally on a measure space $(X, \mathcal{F}, \mu)$. Recall the definition: these are function $f : X \to \mathbb{R}$ that are measurable and for which $\int f^+ d\mu$ and $\int f^- d\mu$ are finite. It is also important sometimes to consider complex-valued functions, $F : X \to \mathbb{C}$ with $F = f + ig$ and f and g real-valued integrable functions. Unless we specify otherwise, everything we say pertains to either case, although we will usually discuss the real-valued case for simplicity. Now it is easy to see that the integrable functions form a vector space, and we can define $||f||_1 = \int |f| d\mu$. However, when we check the definition of the norm, we find one problem: the statement "$||f||_1 = 0$ implies $f = 0$" is not correct. In fact we have established that $\int |f| d\mu = 0$ if and only if $f = 0$ a.e., and except in certain special cases (such as counting measure), there are functions $f = 0$ a.e. that are not identically zero.

To overcome this difficulty we are forced to consider equivalence classes of functions that are equal a.e. In other words we say f is *equivalent* to g if $f = g$ a.e. It is easy to see that this is an equivalence relation, so the measurable functions divide into equivalence classes. We then define $L^1(\mu)$ to be the set of equivalence classes of integrable functions. It is easy to show that the vector space structure of functions respects equivalence classes: $af + bg$ is equivalent to $af_1 + bg_1$ if f is equivalent to f_1 and g is equivalent to g_2 ($af + bg = af_1 + bg_1$ except on the union of the sets where $f \neq f_1$ and $g \neq g_1$, and the union of two sets of measure zero has measure zero). We can then define the norm of an equivalence class to be $||f||_1 = \int |f| d\mu$ for any f in the equivalence class since the value of $||f||_1$ is constant on the equivalence class. *We will follow the usual convention of confounding the equivalence class and a representative function f in the equivalence class.* This is, when we say "let f be a function in $L^1(\mu)$", we really mean "let f be a representative function in an equivalence class in $L^1(\mu)$", with the understanding that nothing we do to f depends on the particular choice of representative. This convention spares us from overburdening the notation, but it does mean that one has to be careful interpreting statements.

Now it is a straightforward matter to verify that $||f||_1$ satisfies the axioms for a norm on $L^1(\mu)$. The triangle inequality comes from integrating the pointwise inequality $|f(x) + g(x)| \leq |f(x)| + |g(x)|$, and the homogeneity is obvious. The troublesome positivity condition, $||f||_1 = 0$ implies $f = 0$, is built into the definition because now $f = 0$

means f and 0 are in the same equivalence class, $f = 0$ a.e. We want to prove that $L^1(\mu)$ is complete in this norm. Recall that this means that every Cauchy sequence converges.

Lemma 14.4.1 (*Chebyshev's Inequality*) *Let f be in $L^1(\mu)$, and let $s > 0$. Then $\mu(E_s) \leq ||f||_1/s$ where $E_s = \{x : |f(x)| \geq s\}$.*

Proof: $\int_{E_s} |f|\, d\mu \leq \int |f|\, d\mu = ||f||_1$. On the other hand $s \leq |f|$ on E_s, so $\int_{E_s} s\, d\mu \leq \int_{E_s} |f|\, d\mu$ and $\int_{E_s} s\, d\mu = s\mu(E_s)$. QED

Theorem 14.4.1 *$L^1(\mu)$ is complete.*

Proof: Let $\{f_k\}$ denote a Cauchy sequence in $L^1(\mu)$. As usual it suffices to show that a subsequence converges, for the convergence of the subsequence and the Cauchy criterion will give the convergence of the original sequence. Now we pass to a subsequence that comes together rapidly, say $||f'_{k+1} - f'_k|| \leq 4^{-k}$ (we choose 4^{-k} for convenience only; actually we only need a sequence ε_k with $\sum \varepsilon_k < \infty$). It is a routine matter to obtain such a subsequence: Given k, we choose $j(k)$ depending on k so that $j(k) \geq j(k-1)$ and $||f_{j(k)} - f_m|| \leq 4^{-k}$ for all $m \geq j(k)$, such $j(k)$ existing by the Cauchy criterion, and then set $f'_k = f_{j(k)}$. Since $j(k+1) \geq j(k)$, we have $||f_{j(k)} - f_{j(k+1)}|| \leq 4^{-k}$ as desired. For simplicity of notation we denote the subsequence by $\{f_k\}$.

Now the remarkable fact is that for this subsequence we have convergence at almost every point. This fact will enable us to get a hold of the limit function. To verify it we define the sets

$$E_n = \{x : |f_{k+1}(x) - f_k(x)| \geq n2^{-k} \text{ for some } k\}$$

and estimate the measure of E_n by Chebyshev's inequality as

$$\begin{aligned} \mu(E_n) &\leq \sum_{k=1}^{\infty} \mu\{x : |f_{k+1}(x) - f_k(x)| \geq n2^{-k}\} \\ &\leq \sum_{k=1}^{\infty} \frac{1}{n} 2^k ||f_{k+1} - f_k||_1 \\ &\leq \sum_{k=1}^{\infty} \frac{1}{n} 2^k \cdot 4^{-k} = \frac{1}{n}. \end{aligned}$$

The measure of E_n is small, but on the complement of E_n the sequence $\{f_k\}$ clearly converges pointwise by comparison; write $g_k = f_k - f_{k-1}$ for $k \geq 2, g_1 = f_1$, so $f_k = \sum_{j=1}^{k} g_j$ and $|g_k(x)| \leq n2^{-k}$ for every k on the complement of E_n and $\sum_k n2^{-k}$ converges. Thus $\{f_k\}$ converges pointwise on the complement of $\bigcap_{n=1}^{\infty} E_n$, which has measure zero.

We may now set $f(x) = \lim_{k\to\infty} f_k(x) = \sum_{k=1}^{\infty} g_k(x)$ on the set where the limit exists and set $f(x) = 0$ on the set of measure zero where the limit fails to exist. By redefining $f_k(x)$ to be zero also on this set of measure zero (this just means choosing a different representative of the equivalence class) we have $f(x) = \lim_{x\to\infty} f_k(x)$ at every point, so f is a measurable function. We need to show that f is integrable and $\lim_{k\to\infty} ||f_k - f||_1 = 0$ to complete the proof. To do this we will need both the monotone and dominated convergence theorems. We first consider $g = \sum_{k=1}^{\infty} |g_k|$, which converges a.e. by the above argument. This will essentially be our dominator. Notice that since $||g_k||_1 \leq 4^{-k}$ for $k = 2, 3, \ldots$ and the $|g_k|$ are non-negative measurable functions, we can apply the monotone convergence theorem to conclude

$$\int g d\mu = \sum_{k=1}^{\infty} \int |g_k| d\mu \leq ||f_1|| + \sum_{k=2}^{\infty} 4^{-k} < \infty,$$

so g is integrable. This shows that f is integrable also since $|f| = |\sum_{k=1}^{\infty} g_k| \leq g$. Finally

$$|f_k - f| \leq |f_k| + |f| = \left|\sum_{j=1}^{k} g_j\right| + |f| \leq \left(\sum_{j=1}^{k} |g_j|\right) + |f| \leq 2|g|,$$

so we may apply the dominated convergence theorem to the sequence $\{|f_k - f|\}$. Since $\lim_{k\to\infty} |f_k - f| = 0$ a.e, we have $\lim_{k\to\infty} \int |f_k - f| d\mu = \int 0\, d\mu = 0$ as desired. QED

14.4.2 L^2 as a Hilbert Space

Althought the space $L^1(\mu)$ of integrable functions is very natural to consider, it turns out for technical reasons to be less useful than the space $L^2(\mu)$ of measurable functions for which $|f|^2$ is integrable. Our discussion of $L^2(\mu)$ follows the same outline as before. We will again be dealing with equivalence classes of functions, although we will not indicate this in the notation. In the case of $L^2(\mu)$, however, we encounter a new difficulty: it is by no means obvious that $L^2(\mu)$ forms a vector space. If $|f|^2$ and $|g|^2$ are integrable, why is $|f+g|^2$ integrable? Taking the real case for simplicity, we can write $(f+g)^2 = f^2+2fg+g^2$, so we need to show that $f \cdot g$ is integrable. If this were the case, then we could define an inner product $\langle f, g\rangle = \int fg d\mu$ (put $\bar{g}$ in the complex case), and the norm $||f||_2 = (\int |f|^2 d\mu)^{1/2}$ would be associated to this inner product. Furthermore, the Cauchy-Schwartz inequality would say

$$\left|\int fg d\mu\right| \leq ||f||_2 ||g||_2,$$

which would in turn imply that fg is integrable since $|f|$ and $|g|$ have the same norm as f and g. We seem to be caught in circular reasoning— $f \cdot g$ integrable implies that the Cauchy-Schwartz inequality implies $f \cdot g$ integrable. The way out of this circle is first to consider simple functions where the integrability of fg is obvious.

Theorem 14.4.2 (*Cauchy-Schwartz Inequality*) *If $|f|^2$ and $|g|^2$ are integrable, then fg is integrable and*

$$\int |fg| d\mu \leq \left(\int |f|^2 d\mu\right)^{1/2} \left(\int |g|^2 d\mu\right)^{1/2}.$$

Proof: Let V denote the vector space of (equivalence classes of) simple functions $\sum_{k=1}^N a_k \chi_{A_k}$ where $\mu(A_k) < \infty$ for each k. It is s-traightforward to verify that V is a vector space and $\langle f, g\rangle = \int fg d\mu$ defines an inner product on V. Thus we have the Cauchy-Schwartz inequality $|\langle f, g\rangle| \leq ||f||_2 ||g||_2$ for f and g in V, and replacing f and g by $|f|$ and $|g|$ we obtain $\int |fg| d\mu \leq ||f||_2 ||g||_2$ on V.

Now let f and g belong to $L^2(\mu)$. Then fg is measurable, so we need to show $\int |fg| d\mu$ is finite. But if $\{f_n\}$ and $\{g_n\}$ are sequences of non-negative simple functions increasing monotonically to $|f|$ and $|g|$, then $\{f_n g_n\}$ is a sequence of non-negative simple functions increasing monotonically to $|fg|$. Furthermore it is easy to see that f_n and g_n must be in V. If say $f_n = \sum_{k=1}^N a_k \chi_{A_k}$ with all $a_k \neq 0$, then $f_n^2 = \sum_{k=1}^N a_k^2 \chi_{A_k}$ and $f_n^2 \leq |f|^2$ implies $\int f_n^2 d\mu \leq \int |f|^2 d\mu < \infty$, so $\mu(A_k) < \infty$ for all k. Thus

$$\int |f_n g_n| d\mu \leq ||f_n||_2 ||g_n||_2 \leq ||f||_2 ||g||_2$$

and taking the limit as $n \to \infty$ gives $\int |fg| d\mu \leq ||f||_2 ||g||_2$, showing that fg is integrable. QED

It is now a straightforward matter to verify that $L^2(\mu)$ forms a vector space and $\langle f, g \rangle = \int fg d\mu$ defines an inner product on $L^2(\mu)$. What is the relationship between functions in $L^1(\mu)$ and $L^2(\mu)$? If the measure of the whole space $\mu(X)$ is finite, then we have the containment $L^2(\mu) \subseteq L^1(\mu)$. Indeed taking $g \equiv 1$ in the Cauchy-Schwartz inequality we obtain

$$||f||_1 = \int |f| d\mu \leq \left(\int |f|^2 d\mu \right)^{1/2} \left(\int 1^2 d\mu \right)^{1/2} \leq \mu(X)^{1/2} ||f||_2.$$

On the other hand, if $\mu(X) = +\infty$, then there are functions in $L^2(\mu)$ that are not in $L^1(\mu)$. Looking at Lebesgue measure on the line, we can say roughly speaking that to be in $L^2(\mu)$ is a more restrictive condition concerning local behavior of singularities but a less restrictive condition concerning decay at infinity. Thus if $f(x) = (1 + |x|)^{-a}$, then f is in L^2 if and only if $a > 1/2$ while f is in L^1 if and only if $a > 1$. On the other hand if $f(x) = |x|^{-a} \chi_{|x| \leq 1}$, then f is in L^2 if and only if $a < 1/2$ while f is in L^1 if and only if $a < 1$. Finally, if μ is counting measure, the containment is reversed: L^1 is contained in L^2 (the usual notation for counting measure is l^1 and l^2).

Theorem 14.4.3 *$L^2(\mu)$ is complete.*

Proof: The proof follows the same pattern as the proof of the completeness of $L^1(\mu)$. We take a Cauchy sequence $\{f_k\}$ in $L^2(\mu)$, pass to a subsequence such that $||g_k||_2 \leq 4^{-k}$ where $g_k = f_k - f_{k-1}$, and then show that $f_k = \sum_{j=1}^k g_j$ converges a.e. The only difference in the proof is that we use the L^2 Chebyshev inequality $\mu\{x : |f(x)| \leq s\} \leq (1/s^2)||f||_2^2$, which follows by the same reasoning.

We then set $f = \sum_{j=1}^\infty g_j$ and $g = \sum_{j=1}^\infty |g_j|$ so that $|f|^2 \leq |g|^2$, and we prove $|g|^2$ is integrable by the monotone convergence theorem (here we estimate $|| \sum_{j=1}^k |g_j| \,||_2 \leq \sum_{j=1}^k ||g_j||_2$ by the triangle inequality for the norm). This shows that f is in $L^2(\mu)$ and then $2|g|^2$ serves to dominate the sequence $|f - f_k|^2$, so $\lim_{k \to 0} \int |f - f_k|^2 d\mu = 0$ by the dominated convergence theorem; hence $f_k \to f$ in the metric. QED

A complete inner product space is called a *Hilbert space.* $L^2(\mu)$ is a typical example of a Hilbert space. This is the setting for quantum mechanics, where μ is Lebesgue measure on some Euclidean space (or more generally the classical configuration space of the system being studied).

14.4.3 Fourier Series for L^2 Functions

We now consider the case where $X = [-\pi, \pi]$ and μ is Lebesgue measure, and we study Fourier series for functions in L^2. We observe that the definition of the Fourier coefficients

$$c_n = \frac{1}{2\pi}\int_{-\pi}^{\pi} f(x)e^{-inx}\,dx$$

makes sense for any f in L^1 if we interpret the integral as a Lebesgue integral. Indeed if $f(x)$ is integrable, then $f(x)e^{-inx}$ will also be integrable since it is the product of measurable functions, hence measurable, and $|f(x)e^{-inx}| = |f(x)|$, so

$$\frac{1}{2\pi}\int_{-\pi}^{\pi} |f(x)e^{-inx}|\,dx = \frac{1}{2\pi}\int_{-\pi}^{\pi} |f(x)|\,dx = \frac{1}{2\pi}||f||_1 < \infty.$$

We can then at least write the formal Fourier series $\sum_{-\infty}^{\infty} c_n e^{inx}$ and ask if it converges to f in any sense. Unfortunately there is no good answer to this question in L^1, so we restrict attention to L^2 functions (recall that $L^2 \subseteq L^1$ here because $\mu(X) = 2\pi$ is finite). Now for continuous functions we have established Parseval's identity

$$\frac{1}{2\pi}\int_{-\pi}^{\pi} |f(x)|^2\,dx = \sum_{-\infty}^{\infty} |c_n|^2$$

and the mean convergence

$$\lim_{N\to\infty}\int_{-\pi}^{\pi} |S_N f(x) - f(x)|^2\,dx = 0.$$

It would seem plausible that this would continue to hold for f in L^2. We will show that this is in fact the case and, furthermore, that every choice of Fourier coefficients for which $\sum |c_n|^2$ is finite corresponds to some L^2 function. What this means is that the Fourier coefficients give a one-to-one correspondence between L^2 functions and l^2 sequences $\{c_n\}$ such that $\sum_{-\infty}^{\infty} |c_n|^2 < \infty$. Furthermore this correspondence is linear and isometric so that in fact it shows the Hilbert space structure of $L^2(-\pi, \pi)$ and l^2 are isomorphic.

We begin by studying the correspondence in the direction $\{c_n\} \to f$.

Theorem 14.4.4 (*Riesz-Fischer*) *Let complex coefficients c_n be given with $\sum_{-\infty}^{\infty} |c_n|^2 < \infty$. Then there exists complex f in $L^2(-\pi, \pi)$ with $c_n = (1/2\pi) \int_{-\pi}^{\pi} f(x)e^{-inx}\,dx$ such that $(1/2\pi) \int_{-\pi}^{\pi} |f(x)|^2\,dx = \sum_{-\infty}^{\infty} |c_n|^2$ and $\lim_{N\to\infty} ||f - \sum_{-N}^{N} c_n e^{inx}||_2 = 0$.*

Proof: Let $f_N = \sum_{-N}^{N} c_n e^{inx}$. Because of the orthogonality of the functions e^{inx} with respect to the inner product we have

$$||f_N - f_M||_2^2 = \left|\left| \sum_{N<|n|\leq M} c_n e^{inx} \right|\right|_2^2 = 2\pi \sum_{N<|n|\leq M} |c_n|^2$$

if $N < M$ (the factor 2π appears here and not in Chapter 12 because we have not divided by 2π in defining the inner product). From this and the fact that $\sum_{-\infty}^{\infty} |c_n|^2 < \infty$ it follows easily that f_N is a Cauchy sequence in L^2. By the completeness of L^2 it converges to some f in L^2 in the norm, $\lim_{N\to\infty} ||f - f_N||_2 = 0$. Since $f - f_N$ and f_N are orthogonal, we obtain

$$\begin{aligned}
\frac{1}{2\pi}||f||_2^2 &= \frac{1}{2\pi}(||f - f_N||_2^2 + ||f_N||_2^2) \\
&= \frac{1}{2\pi} \lim_{N\to\infty} (||f - f_N||_2^2 + ||f_N||_2^2) \\
&= \frac{1}{2\pi} \lim_{N\to\infty} ||f_N||_2^2 = \sum_{-\infty}^{\infty} |c_n|^2.
\end{aligned}$$

Finally

$$\begin{aligned}
\frac{1}{2\pi}\int_{-\pi}^{\pi} f(x)e^{-inx}\,dx &= \frac{1}{2\pi}\int_{-\pi}^{\pi} f_N(x)e^{-inx}\,dx \\
&\quad + \frac{1}{2\pi}\int_{-\pi}^{\pi} (f(x) - f_N(x))e^{-inx}\,dx \\
&= c_n + \frac{1}{2\pi}\int_{-\pi}^{\pi} (f(x) - f_N(x))e^{-inx}\,dx
\end{aligned}$$

for any $N \geq n$. But

$$\begin{aligned}
&\left|\frac{1}{2\pi}\int_{-\pi}^{\pi}(f(x)-f_N(x))e^{-inx}\,dx\right| \\
&\leq \frac{1}{2\pi}\left(\int_{-\pi}^{\pi}|f(x)-f_N(x)|^2dx\right)^{1/2}\left(\int_{-\pi}^{\pi}1^2\,dx\right)^{1/2} \\
&= \frac{1}{\sqrt{2\pi}}||f-f_N||_2 \to 0 \quad \text{as } N \to \infty,
\end{aligned}$$

so $c_n = (1/2\pi)\int_{-\pi}^{\pi} f(x)e^{-inx}\,dx$. QED

For the converse result we will have to work harder, because it essentially depends on the fact that we haven't left out any functions (note that the Riesz-Fischer theorem remains true if we omit some of the exponentials e^{inx}). Now we have already verified this completeness of the exponentials e^{inx} in the context of continuous functions, but there remains the possibility that by enlarging to the class L^2 we have created some new functions that are orthogonal to all the exponentials. The reason this is not the case is that the continuous functions are already dense in L^2. Therefore we first must prove this density result. Incidentally, the analogous statement is true for L^1, and the proof is essentially the same.

Theorem 14.4.5 (*Density of Continuous Functions*) *Let f be in $L^2(-\pi,\pi)$. Then there exists a sequence of continuous functions f_n (with $f_n(-\pi) = f_n(\pi)$, so f_n extends to a continuous periodic function) such that $||f_n - f||_2 \to 0$ as $n \to \infty$.*

Proof: It clearly suffices to do this for non-negative f (approximate f^+ and f^- separately in the real case), and from the definition of $\int f^2 d\mu$ it is clear that there exists a sequence of simple functions f_n such that $||f - f_n||_2 \to 0$. Thus it suffuces to show that a simple function can be approximated by continuous functions. Again it is clear that it suffices to do this for characteristic functions $f = \chi_A$ where A is a measurable set.

Now recall that $\mu(A)$ was defined to be the infimum of $\sum_{j=1}^{\infty}|I_j|$ where $\bigcup_{j=1}^{\infty} I_j$ covers A and I_j are intervals. Without loss of generality we may assume the intervals I_j are disjoint, and given any $\varepsilon > 0$ we can choose the intervals so that $\bigcup_{j=1}^{\infty} I_j = A \cup B$, with A and B disjoint, and

$\mu(B) \leq \varepsilon$. We can also choose N large enough so that $\sum_{j=N+1}^{\infty} |I_j| \leq \varepsilon$. Thus χ_A and $\sum_{j=1}^{N} \chi_{I_j}$ are equal except on a set of measure at most 2ε, so $||\chi_A - \sum_{j=1}^{N} \chi_{I_j}||_2 \leq \sqrt{2\varepsilon}$.

Finally χ_{I_j} can clearly be approximated by continous functions in the L^2-norm, as indicated in Figure 14.4.1, and we can always arrange these approximating functions to vanish at $\pm\pi$.

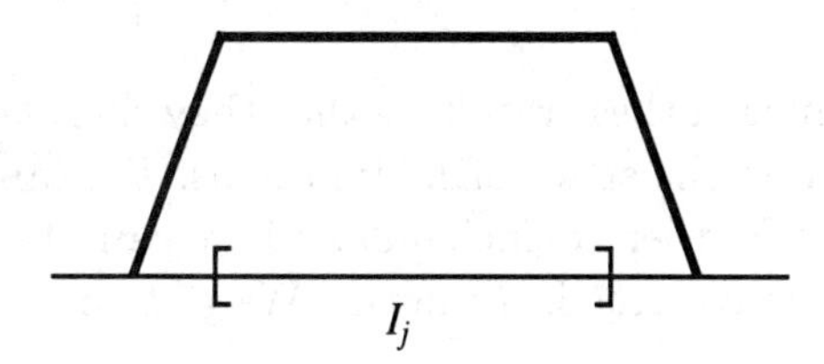

Figure 14.4.1:

Adding together all these approximations in the usual way we obtain the result. QED

Theorem 14.4.6 (*Parseval Identity*) *Let f be in $L^2(\pi, \pi)$. Then $||S_N f - f||_2 \to 0$ and $(1/2\pi)||f||_2^2 = \sum_{-\infty}^{\infty} |c_n|^2$.*

Proof: Let $\{f_k\}$ be a sequence of continuous functions such that $||f_k - f||_2 \to 0$ as $k \to \infty$. We have already proved the result for the functions f_k, so we want to get the result for f by passing to the limit. Given any error ε we can first find f_k so that $||f_k - f||_2 \leq \varepsilon/2$ and then choose N large enough so that $||S_N f_k - f_k||_2 \leq \varepsilon/2$, so $||S_N f_k - f||_2 \leq \varepsilon$. But we can apply the projection theorem to f to conclude that $S_N f$ minimizes the distance to f among all trigonometric polynomials $\sum_{-N}^{N} a_n e^{inx}$, so $||S_N f - f||_2 \leq ||S_N f_k - f||_2 \leq \varepsilon$ (recall that for the projection theorem we needed only to have an inner product space, so $L^2(-\pi, \pi)$ will do). This shows $\lim_{N\to\infty} ||S_N f - f||_2 = 0$ and so also $||f||_2^2 = \lim_{N\to\infty} ||S_N f||_2^2 = \lim_{N\to\infty} 2\pi \sum_{-N}^{N} |c_n|^2 = 2\pi \sum_{-\infty}^{\infty} |c_n|^2$. QED

The Parseval identity gives us the converse of the Riesz-Fischer theorem. If we start with $f \in L^2$ we produce a sequence $\{c_k\}$ in l^2, the Fourier coefficients of that return f as the limit of $\sum_{-N}^{N} c_n e^{inx}$

in the L^2 norm. The correspondence $f \leftrightarrow \{c_k\}$ is thus an isometric isomorphism, which is onto in both directions. We also obtain as a corollary the completeness of the orthogonal system e^{inx}.

Corollary 14.4.1 *If f in L^2 is orthogonal to all the functions e^{inx}, then f is zero.*

Proof: The orthogonality means $c_n = 0$ for all n, hence $||f||_2 = 0$ by the Parseval identity. This means $f = 0$ a.e.

The importance of these results is that they show that Fourier series are well behaved on the space of L^2 functions. We can recognize an L^2 function from its Fourier coefficients, and we can always interpret the Fourier series as converging in L^2 norm. We give one typical application to the heat equation.

Theorem 14.4.7 *Let $\{c_n\}$ by any sequence satisfying $\sum_{-\infty}^{\infty} |c_n|^2 < \infty$. Then $u(x,t) = \sum_{-\infty}^{\infty} e^{-n^2t} c_n e^{inx}$ for $t > 0$ is a solution of the heat equation $\partial u/\partial t = \partial^2 u/\partial x^2$ that is periodic of period 2π in x, and $\sup_{t>0} \int_{-\pi}^{\pi} |u(x,t)|^2\, dx$ is finite. Furthermore $\int_{-\pi}^{\pi} |u(x,t) - f(x)|^2\, dx \to 0$ as $t \to 0^+$ where $f = \int_{-\infty}^{\infty} c_n e^{inx}$. Conversely, every solution of the heat equation in $t > 0$ that is periodic of period 2π in x and such that $sup_{t>0} \int_{-\pi}^{\pi} |u(x,t)|^2\, dx$ is finite has the above form.*

Proof: For any fixed $t > 0$ the series $\sum_{-\infty}^{\infty} e^{-n^2t} c_n e^{inx}$ converges absolutely and uniformly in x by comparison with

$$\sum_{-\infty}^{\infty} |c_n| e^{-n^2t} \leq \left(\sum_{-\infty}^{\infty} |c_n|^2\right)^{1/2} \left(\sum_{-\infty}^{\infty} e^{-2n^2t}\right)^{1/2} < \infty$$

by the Cauchy-Schwartz inequality. Thus u is continuous. Furthermore if we formally differentiate with respect to t or x any number of times, the series remains uniformly convergent because the derivatives just produce polynomial factors in n, and we can repeat the previous argument since $\sum_{-\infty}^{\infty} n^{2k} e^{-2n^2t}$ is finite for any k. Thus u is C^∞ in $t > 0$ and the series may be differentiated term-by-term, so u satisfies the heat equation. By Parseval's identity we have

$$\int_{-\pi}^{\pi} |u(x,t)|^2\, dx = 2\pi \sum_{-\infty}^{\infty} |c_n|^2 e^{-2n^2t}$$

$$\leq \; 2\pi \sum_{-\infty}^{\infty} |c_n|^2,$$

so $\sup_{t>0} \int_{-\pi}^{\pi} |u(x,t)|^2 \, dx$ is finite; also by Parseval's identity we have

$$\int_{-\pi}^{\pi} |u(x,t) - f(x)|^2 \, dx = 2\pi \sum_{-\infty}^{\infty} |c_n|^2 (1 - e^{-n^2 t})^2,$$

and this goes to zero as $t \to 0^+$ by the dominated convergence theorem for counting measure ($|c_n|^2(1 - e^{-n^2t})^2$ goes to zero pointwise and is dominated by the integrable $|c_n|^2$).

For the converse, assume $u(x,t)$ is a solution to the heat equation in $t > 0$. Then we have already shown in Chapter 12 that $u(x,t) = \sum e^{-n^2 t} c_n e^{inx}$ for some coefficients c_n. Now we add the condition that $\int_{-\pi}^{\pi} |u(x,t)|^2 \, dx \leq M$ for all $t > 0$. By Parseval's identity this is $\sum_{-\pi}^{\pi} |c_n|^2 e^{-2n^2 t} \leq M/2\pi$. Finally we let $t \to 0$ and apply the monotone convergence theorem ($e^{-2n^2 t}|c_n|^2 \to |c_n|^2$ monotonically) to conclude that $\sum_{-\infty}^{\infty} |c_n|^2$ must be finite. QED

14.4.4 Exercises

1. Prove the L^p Chebyshev's inequality: if $|f|^p$ is integrable (for fixed $p > 0$), then $\mu\{x : |f(x)| \geq s\} \leq (1/s^p) \int |f|^p d\mu$.

2. Prove that $l^1 \subseteq l^2$ (recall l^p means L^p for counting measure).

3. Prove that if f is in L^1 and f is bounded, then f is in L^2.

4. Give an example of a function on the line that is in L^1 but not L^2 and one that is in L^2 but not L^1.

5. Prove that if f is any L^1 function such that $\int_{-\pi}^{\pi} f(x) e^{-inx} \, dx = 0$ for all n, then $f = 0$ a.e.

6. Prove the Riemann-Lebesgue Lemma: if f is in L^1, then $\lim_{n \to \pm\infty} c_n = 0$. (**Hint**: use the density of continuous functions in L^1 and Parseval's indentity.)

7. Let f be in $L^2(-\pi, \pi)$, and extend f to be periodic of period 2π. Prove the continuity of translation in L^2: if $h(t) = \int_{-\pi}^{\pi} |f(x+t) - f(x)|^2 \, dx$, then $\lim_{t \to 0} h(t) = 0$. (**Hint**: use density of continuous functions.) State and prove the analogous result for L^1.

8. Prove that if $u(x,t)$ is a solution to the heat equation in $t > 0$ that is periodic of period 2π and for which $\sup_{t>0} \int_{-\pi}^{\pi} |u(x,t)|^2\, dx$ is finite, then there exists a constant c such that $u(x,t) \to c$ uniformly as $t \to \infty$.

9. Consider the Laplace equation $(\partial^2/\partial x^2 + \partial^2/\partial y^2)u(x,y) = 0$ in $x^2 + y^2 < 1$ subject to the restriction

$$\int_{-\pi}^{\pi} |u(r\cos\theta, r\sin\theta)|^2 d\theta \leq M \text{ for } 0 < r < 1.$$

Prove that $u(r\cos\theta, r\sin\theta) = \sum_{-\infty}^{\infty} c_n r^{|n|} e^{in\theta}$ gives a solution for every choice of c_n with $\sum_{-\infty}^{\infty} |c_n|^2$ finite and conversely that all solutions have this form.

10. Give an example of a sequence of functions $f_n(x)$ on $[0,1]$ such that $\int_0^1 |f_n(x)|\, dx \to 0$ as $n \to \infty$ but $f_n(x)$ does not converge to zero for any x in $[0,1]$. Can you make the functions f_n continuous?

14.5 Summary

14.1 The Concept of Measure

Example *Dirichlet's function on* $[0,1]$

$$f(x) = \begin{cases} 0 & \text{if } x \text{ is irrational,} \\ 1 & \text{if } x \text{ is rational} \end{cases}$$

is not Riemann integrable, even though it is the pointwise limit of Riemann integrable functions.

Example *Let*

$$f_n(x) = \begin{cases} n & \text{if } 0 < x < 1/n, \\ 0 & \text{otherwise.} \end{cases}$$

Then $\int_0^1 (\lim_{n\to\infty} f_n(x))\, dx \neq \lim_{n\to\infty} \int_0^1 f_n(x)\, dx$.

Lemma 14.1.1 *Let $I_1, I_2, \ldots$ be disjoint intervals such that $I = I_1 \cup I_2 \cup \cdots$ is also an interval. Then $|I| = \sum_{j=1}^{\infty} |I_j|$ where $|I|$ denotes the length of the interval I.*

Corollary 14.1.1 *If I is any interval, then*

$$|I| = \inf \left\{ \sum_{j=1}^{\infty} |I_j| : I \subseteq \bigcup_{j=1}^{\infty} I_j \right\}.$$

Definition *A collection $\mathcal{F}$ of subsets of a fixed set X (the universe) is called a field of sets if*

1. *the empty set is in $\mathcal{F}$;*
2. *if A is in $\mathcal{F}$, then the complement of A is in $\mathcal{F}$; and*
3. *if A and B are in $\mathcal{F}$, then $A \cup B$ is in $\mathcal{F}$.*
   A field of sets is called a σ-field if in addition
4. *if $A_1, A_2, \ldots$ is a sequence of sets in $\mathcal{F}$, then $\bigcup_{j=1}^{\infty} A_j$ is in $\mathcal{F}$.*

Definition *The σ-field generated by the field $\mathcal{F}$, denoted $\mathcal{F}_\sigma$, is the smallest σ-field containing $\mathcal{F}$ or, equivalently, the intersection of all σ-fields containing $\mathcal{F}$. The σ-field generated by the field of finite unions of subintervals of a fixed interval X of the line is called the σ-field of Borel sets in X. The same definition applies to any metric space X.*

Definition *A measure on a σ-field $\mathcal{F}$ (referred to as the measurable sets) is a function $\mu : \mathcal{F} \to [0, \infty]$ satisfying $\mu(\emptyset) = 0$ and σ-additivity: if $A = \bigcup_{j=1}^{\infty} A_j$ with A_j disjoint sets in $\mathcal{F}$, then $\mu(A) = \sum_{j=1}^{\infty} \mu(A_j)$.*

Theorem *Any measure μ is*

1. *monotone: $A \subseteq B$ implies $\mu(A) \le \mu(B)$;*
2. *continuous from below: if $A_1 \subseteq A_2 \subseteq \cdots$ is an increasing sequence of measurable sets, then $\mu(\bigcup_{j=1}^{\infty} A_j) = \lim_{j\to\infty} \mu(A_j)$;*
3. *conditionally continuous from above: if $B_1 \supseteq B_2 \supseteq \cdots$ is a decreasing sequence of measurable sets and if the measures $\mu(B_j)$ are finite, then $\mu(\bigcap_{j=1}^{\infty} B_j) = \lim_{j\to\infty} \mu(B_j)$;*

4. *subadditive: If* $B \subseteq A_1 \cup A_2 \cup \cdots \cup A_n$, *then* $\mu(B) \leq \mu(A_1) + \cdots + \mu(A_n)$;

5. σ*-subadditive: if* $B \subseteq \bigcup_{j=1}^{\infty} A_j$, *then* $\mu(B) \leq \sum_{j=1}^{\infty} \mu(A_j)$.

Definition *The Lebesgue measure* $\mu(B)$ *of a Borel set in* $\mathbb{R}$ *is defined to be* $\mu(B) = \inf\{\sum_{j=1}^{\infty} |I_j| : B \subseteq \bigcup_{j=1}^{\infty} I_j\}$. *Without loss of generality we may assume the intervals* I_j *to be disjoint.*

Lemma *The Lebesgue measure of an interval is equal to its length.*

Theorem *Sets of Lebesgue measure zero are preserved under countable unions, and subsets of sets of measure zero also have measure zero.*

Theorem *Any Borel set* B *can be covered by a* G_δ *set* A *(a countable intersection of open sets) such that* $A \backslash B$ *has Lebesgue measure zero, and* $|A| = |B|$.

Definition *The Lebesgue measure* $\mu(B)$ *of a Borel set in* $\mathbb{R}^n$ *is defined to be* $\mu(B) = \inf\{\sum_{j=1}^{\infty} |R_j| : B \subseteq \bigcup_{j=1}^{\infty} R_j\}$ *where* R_j *denotes any rectangle* $I_1 \times I_2 \times \cdots \times I_n$ *with volume* $|R_j| = |I_1||I_2| \cdots |I_n|$.

Example *Counting measure is the measure on the* σ*-field of all subsets of* X *that assigns to each set its cardinality (the number of elements it contains).*

Definition *A probability measure is any measure such that* $|X| = 1$.

14.2 Proof of Existence of Measures

Definition *An outer measure on a* σ*-field* $\mathcal{F}$ *of sets is a function* $\mu(A)$ *satisfying:*

1. *(non-negativity)* $\mu : \mathcal{F} \to [0, \infty]$;

2. $\mu(\emptyset) = 0$;

3. *(*σ*-subadditivity) if* $A = \bigcup_{j=1}^{\infty} A_j$, *then* $\mu(A) \leq \sum_{j=1}^{\infty} \mu(A_j)$;

4. *(monotonicity) if* $A \subseteq B$, *then* $\mu(A) \leq \mu(B)$.

Theorem *Lebesgue measure is an outer measure.*

Definition 14.2.1 *For μ an outer measure on $\mathcal{F}$ and A in $\mathcal{F}$, we say A satisfies the splitting condition if $\mu(B) = \mu(B \cap A) + \mu(B \backslash A)$ for every B in $\mathcal{F}$.*

Theorem 14.2.1 *Let μ be an outer measure on $\mathcal{F}$. The sets satisfying the splitting condition (denoted $\mathcal{F}_0$) form a σ-field and μ restricted to $\mathcal{F}_0$ is a measure.*

Definition 14.2.2 *The distance between two sets A and B in a metric space is the infimum of $d(x, y)$ for x in A and y in B. If this distance is positive we say A and B are separated. An outer measure on the Borel sets of a metric space is said to be a metric outer measure if $\mu(A \cup B) = \mu(A) + \mu(B)$ whenever A and B are separated.*

Lemma *Lebesgue measure on $\mathbb{R}$ is a metric outer measure.*

Theorem 14.2.2 *(Carathéodory) A metric outer measure is a measure on the Borel sets.*

Definition 14.2.3 *Hausdorff measure μ_α of dimension α on the Borel sets in $\mathbb{R}$ is defined by $\mu_\alpha(A) = \lim_{\varepsilon \to 0} \mu_\alpha^{(\varepsilon)}(A)$ where $\mu_\alpha^{(\varepsilon)}(A) = \inf\{\sum_{j=1}^\infty |I_j|^\alpha : A \subseteq \bigcup_{j=1}^\infty I_j$ for intervals I_j satisfying $|I_j| \le \varepsilon\}$.*

Theorem 14.2.3 *μ_α is a measure on the Borel sets.*

Definition *On an arbitrary metric space, the Hausdorff measure of dimension α is defined as in $\mathbb{R}$ where I_j are allowed to be arbitrary closed sets and $|I_j|$ denotes the diameter of I_j (the supremum of $d(x, y)$ for x and y in I_j).*

Lemma 14.2.1

a. *$\mu_\alpha(A) < \infty$ implies $\mu_\beta(A) = 0$ for $\beta > \alpha$.*

b. *$\mu_\alpha(A) > 0$ implies $\mu_\beta(A) = +\infty$ for $\beta < \alpha$.*

Definition *The Hausdorff dimension of A is the unique value α_0 such that $\mu_\alpha(A) = +\infty$ for $\alpha < \alpha_0$ and $\mu_\alpha(A) = 0$ for $\alpha > \alpha_0$.*

Example *The Hausdorff dimension of the Cantor set is* $\log 2 / \log 3$.

14.3 The Integral

Definition *A measurable space is a pair $(X, \mathcal{F})$ where $\mathcal{F}$ is a σ-field of subsets of X. A measure space is a triple $(X, \mathcal{F}, \mu)$ where, in addition to the above, μ is a measure on $\mathcal{F}$. The sets of $\mathcal{F}$ are called measurable sets.*

Definition *A function $f : X \to \mathbb{R}$ is said to be measurable if $f^{-1}(B)$ is measurable for every Borel subset B of $\mathbb{R}$.*

Lemma *If $f : X \to \mathbb{R}$ and $f^{-1}(I)$ is measurable for every interval I (or even for every interval of the form (a, ∞), or $[a, \infty)$, or $(-\infty, a)$, or $(-\infty, a]$), then f is measurable.*

Theorem 14.3.1 *If f and g are measurable functions, then so are $af + bg$, $f \cdot g$, f/g (if $g \neq 0$), $\max(f, g)$, $\min(f, g)$, and $|f|$. If $h : \mathbb{R} \to \mathbb{R}$ is measurable (with respect to the Borel σ-field), then $h \circ f$ is measurable. If f_n is a sequence of measurable functions, then $\sup_n f_n$, $\inf_n f_n$, $\limsup f_n$, $\liminf f_n$, and $\lim f_n$ (if it exists pointwise) are measurable functions.*

Definition *A simple function is a finite linear combination of characteristic functions of measurable sets $f = \sum_{k=1}^{N} a_k \chi_{A_k}$ where A_k are measurable sets; and*

$$\chi_A = \begin{cases} 1 & \text{if } x \text{ is in } A, \\ 0 & \text{if } x \text{ isn't in } A. \end{cases}$$

Equivalently, a simple function is a measurable function that takes on only a finite set of values.

Lemma *Every simple function has a representation $f = \sum a_k \chi_{A_k}$ where the sets A_k are disjoint, and the representation is essentially* *unique.*

Theorem (*Approximation by Simple Functions*) *Every non-negative measurable function is the pointwise limit of a monotone increasing sequence of simple functions.*

Definition *The integral of a non-negative simple function* $f = \sum_{k=1}^{N} a_k \chi_{A_k}$ *is defined by* $\int f d\mu = \sum a_k \mu(A_k)$, *which does not depend on the representation. The integral of a non-negative measurable function* f *is defined by* $\int f d\mu = \lim_{n\to\infty} \int f_n d\mu$ *where* f_n *is a sequence of non-negative simple functions increasing monotonically to* f. *The integral does not depend on the sequence. A particular choice is* $f_n = \sum_{P_N} (\inf B) \chi_{f^{-1}(B)}$ *and* P_n *is the partition consisting of set* $[(k-1)/2^n, k/2^n)$ *for* $1 \leq k \leq 2^{2n}$ *and* $[2^n, \infty]$.

Theorem 14.3.2 (*Monotone Convergence Theorem*) *If* $0 \leq f_1 \leq f_2 \leq \cdots$ *is a monotone increasing sequence of non-negative measurable functions, then* $\int \lim_{n\to\infty} f_n d\mu = \lim_{n\to\infty} \int f_n d\mu$ *(both sides may be* $+\infty$*).*

Theorem 14.3.3 *The integral of non-negative measurable functions is*

1. *linear:* $\int (af + bg) d\mu = a \int f d\mu + b \int g d\mu$;
2. *monotone:* $\int f d\mu \leq \int g d\mu$ if $f \leq g$;
3. *additive:* $\int_{A\cup B} f d\mu = \int_A f d\mu + \int_B f d\mu$ *for* A *and* B *disjoint measurable sets, where* $\int_A f d\mu$ *denotes* $\int f \chi_A d\mu$.

Theorem *If* f *is Riemann integrable, then* f *is Lebesgue integrable and the two integrals are equal.*

Theorem 14.3.4 (*Fatou's Theorem*) *If* $f = \lim_{n\to\infty} f_n$ *where* f_n *are non-negative measurable functions and if* $\lim_{n\to\infty} \int f_n d\mu$ *exists, then* $\int f d\mu \leq \lim_{n\to\infty} \int f_n d\mu$. *More generally, if* $\{f_n\}$ *is any sequence of non-negative functions, then* $\int \liminf f_n d\mu \leq \liminf \int f_n d\mu$.

Definition *If* $f : X \to \mathbb{R}$, *then* $f^+ = \max(f, 0)$ *and* $f^- = \max(-f, 0)$, *so* $f = f^+ - f^-$ *and* $f^\pm$ *are nonnegative. If* f *is measurable we say* f

is integrable if $\int f^{\pm} d\mu$ *are both finite and then define the integral of* f *by* $\int f d\mu = \int f^{+} d\mu - \int f^{-} d\mu$.

Theorem *If* $f : X \to \mathbb{R}$ *is measurable, then* f *is integrable if and only if* $|f|$ *is integrable and Minkowski's inequality* $|\int f d\mu| \leq \int |f| d\mu$ *holds.*

Example *For counting measure on* $\{1, 2, 3, \ldots\}$ *the integrable functions are the absolutely convergent series, and the integral is the series sum.*

Theorem *The integral is linear, monotone, and additive on real-valued integrable functions.*

Theorem 14.3.5 (*Dominated Convergence Theorem*) *If* g *is a non-negative integrable function with* $|f_n(x)| \leq g(x)$ *for all* x *and* n *for a sequence* f_n *of measurable functions such that* $f_n(x) \to f(x)$ *as* $n \to \infty$ *for every* x, *then* f *is integrable and* $\int f d\mu = \lim_{n\to\infty} \int f_n d\mu$.

Corollary *If* $\mu(X)$ *if finite and* f_n *is a uniformly bounded sequence of functions converging pointwise to* f, *then* $\int f d\mu = \lim_{n\to\infty} \int f_n d\mu$.

Definition *Any statement about points* x *in* X *is said to hold almost everywhere (abbreviated a.e.) if it is true for all* x *not in a set* E *of measure zero.*

Theorem 14.3.6 *Let* f *and* g *be integrable functions. Then* $f = g$ *a.e. if and only if* $\int_A f d\mu = \int_A g d\mu$ *for every measurable set* A.

14.4 The Lebesgue Spaces L^1 and L^2

Definition *We say* f *is equivalent to* g *if* $f = g$ *a.e., where* f *and* g *are measurable functions on a measure space. We define* $L^1(\mu)$ *to be the vector space of equivalence classes of integrable functions, and the* L^1 *norm is defined to be* $||f||_1 = \int |f| d\mu$. *This is a norm on* $L^1(\mu)$.

Lemma 14.4.1 (*Chebyshev's Inequality*) *For any* f *in* $L^1(\mu)$ *and* $s > 0$ *we have* $\mu(\{x : |f(x)| > s\}) \leq (1/s)||f||_1$.

Theorem 14.4.1 *$L^1(\mu)$ is complete.*

Definition *$L^2(\mu)$ is the space of equivalence classes of measurable functions f such that $|f|^2$ is integrable, with L^2 norm $||f||_2 = (\int |f|^2 d\mu)^{1/2}$ and inner product $\langle f, g\rangle = \int fgd\mu$ (take $\bar{g}$ in the complex-valued case).*

Theorem 14.4.2 (*Cauchy-Schwartz Inequality*) *If f and g are in $L^2(\mu)$, then fg is integrable, so the inner product is well defined and the Cauchy-Schwartz inequality $\int |fg|d\mu \leq (\int |f|^2 d\mu)^{1/2}(\int |g|^2 d\mu)^{1/2}$ holds.*

Theorem *If $\mu(X)$ is finite, then $L^2(\mu) \subseteq L^1(\mu)$. For counting measure the spaces are denoted l^1 and l^2 and we have $l^1 \subseteq l^2$.*

Theorem 14.4.3 *$L^2(\mu)$ is complete.*

Definition *A complete normed space is called a Banach space, a complete inner product space is called a Hilbert space.*

Theorem 14.4.4 (*Riesz-Fischer*) *For any complex coefficients c_n with $\sum_{n=-\infty}^{\infty} |c_n|^2 < \infty$ there exists a complex-valued function f in $L^2(-\pi, \pi)$ with Fourier coefficients equal to c_n : $c_n = (1/2\pi) \int_{-\pi}^{\pi} f(x)e^{-inx}\, dx$. Furthermore, $(1/2\pi) \int_{-\pi}^{\pi} |f(x)|^2\, dx = \sum_{-\infty}^{\infty} |c_n|^2$ (Parseval's identity), and the partial sums $s_N f(x) = \sum_{n=-N}^{N} c_n e^{inx}$ converge to f in L^2 norm.*

Theorem 14.4.5 (*Density of Continuous Functions*) *The continuous functions on $[-\pi, \pi]$ with $f(-\pi) = f(\pi)$ are dense in $L^2(-\pi, \pi)$.*

Theorem 14.4.6 (*Parseval Identity*) *If f is in $L^2(-\pi, \pi)$, then the partial sums of the Fourier series converge to f in L^2 norm and Parseval's identity holds. In particular, the correspondence $f \leftrightarrow \{c_n\}$ between $L^2(-\pi, \pi)$ and l^2 is onto in both directions.*

Corollary 14.4.1 *If f is in L^2 and orthogonal to all the functions e^{inx}, then f is zero.*

Theorem 14.4.7 *If $\sum_{-\infty}^{\infty} |c_n|^2$ is finite, then $u(x,t) = \sum_{-\infty}^{\infty} e^{-n^2 t} c_n e^{inx}$ for $t > 0$ is a solution of the heat equation $\partial u/\partial t = \partial^2 u/\partial x^2$ that is periodic of period 2π in x, and $\sup_{t>0} \int_{-\pi}^{\pi} |u(x,t)|^2 \, dx$ is finite, with $\int_{-\pi}^{\pi} |u(x,t) - f(x)|^2 \, dx \to 0$ as $t \to 0^+$ where $f = \sum_{-\infty}^{\infty} c_n e^{inx}$. Conversely, every solution of the heat equation in $t > 0$ that is periodic of period 2π in x and such that $\sup_{t>0} \int_{-\pi}^{\pi} |u(x,t)|^2 \, dx$ is finite has the above form.*

Chapter 15

Multiple Integrals

15.1 Interchange of Integrals

15.1.1 Integrals of Continuous Functions

An important fact in the calculus of several variables is the equality of iterated integrals and multiple integrals, at least for continuous functions on well-behaved domains. For simplicity of notation we will state all results for $\mathbb{R}^2$, but the generalization to higher dimensions will be evident. Suppose $R = [a, b] \times [c, d]$ is a closed bounded rectangle and $f : R \to \mathbb{R}$ is a continuous function. The iterated integrals are

$$\int_c^d \left(\int_a^b f(x,y)\,dx \right) dy \quad \text{and} \quad \int_a^b \left(\int_c^d f(x,y)\,dy \right) dx.$$

The double integral $\int\int_R f(x,y)\,dx\,dy$ is defined as the limit of Cauchy sums

$$S(f,P) = \sum_{j=1}^{N} \sum_{k=1}^{M} f(\tilde{x}_j, \tilde{y}_k)(x_j - x_{j-1})(y_k - y_{k-1})$$

where P denotes a partition $a = x_0 < x_1 < \cdots < x_N = b, c = y_0 < y_1 < \cdots < y_M = d$, of the sides of the rectangle (hence determining a partition of the rectangle into MN little rectangles), and $(\tilde{x}_j, \tilde{y}_k)$ is any point in the rectangle $R_{jk} = [x_{j-1}, x_j] \times [y_{k-1}, y_k]$. The limit is taken as the maximum length of the subintervals $x_j - x_{j-1}$ and $y_k - y_{k-1}$ goes to zero. It is a straightforward matter to show that this limit exists, using the uniform continuity to show that the variation of f over any of the subrectangles R_{jk} goes to zero. Perhaps the only subtle point is that we must require that both sides of the rectangles R_{jk} get small; it is not enough that the areas of the rectangles go to zero, for we have no control of the variation of f over long skinny rectangles. We leave the details to the exercises.

We briefly sketch the proof of the equality of the double integral and one of the iterated integrals for a continuous function f on a rectangle R. Given any $\varepsilon > 0$ we first choose a partition P of the rectangle R into subrectangles R_{jk} such that $|f(x,y) - f(\tilde{x},\tilde{y})| \le \varepsilon$ if (x,y) and $(\tilde{x},\tilde{y})$ belong to the same R_{jk}. This is possible by the uniform continuity of f. Then $S(f,P)$ for this partition differs from the double integral $\int\int_R f(x,y)\,dx\,dy$ by at most $\varepsilon\cdot$ area(R). This follows by the same argument as in the one-dimensional case. Now consider the function g that is constant on each of the rectangles R_{jk} and equal to $f(\tilde{x}_j,\tilde{y}_k)$ there. g is discontinuous, but it has only a finite number of discontinuities, so the integrals involving g are covered by the theory of the Riemann integral. Notice that $|f(x,y) - g(x,y)| \le \varepsilon$ everywhere on R. Thus for any fixed y,

$$\left|\int_a^b f(x,y)\,dx - \int_a^b g(x,y)\,dx\right| \le \varepsilon(b-a),$$

so

$$\left|\int_c^d \left(\int_a^b f(x,y)\,dx\right) dy - \int_c^d \left(\int_a^b g(x,y)\,dx\right) dy\right| \le \varepsilon \text{ area}(R).$$

On the other hand a direct calculation of the iterated integral $\int_c^d (\int_a^b g(x,y)\,dx)\,dy$ shows it is equal to $S(f,P)$. Thus the iterated and double integral differ by at most 2ε area(R), and by letting $\varepsilon \to 0$ we obtain their equality.

For many applications it is desirable to relax the condition that the domain be a rectangle. Suppose the domain is a compact set K in $\mathbb{R}^2$. We can always set it inside some rectangle R and define the double integral $\int\int_K f(x,y)\,dx\,dy$ by partitioning R as before and taking Cauchy sums $S(f,P) = \sum\sum f(\tilde{x}_j,\tilde{y}_k)\text{area}(R_{jk})$ where the sum extends only over those rectangles R_{jk} that lie entirely in K, as indicated in Figure 15.1.1.

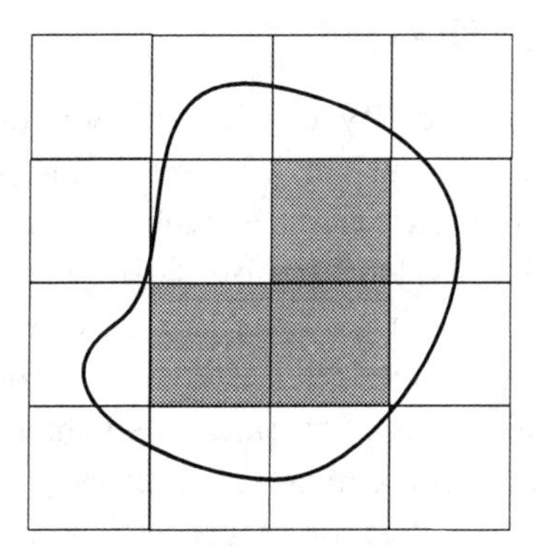

Figure 15.1.1:

In order for the double integral to exist and be equal to the iterated integrals for continuous functions on K we need to make some assumptions on K that will control the error that arises from omitting the rectangles that lie partially in K. It is clear that this error will be negligible if we assume that the sum of the areas of those rectangles that are partially in K goes to zero as the partitions are refined. It is not difficult to see this is the same as saying the *boundary of K has content zero*. Recall the boundary of K is defined to be all points of K that are not in the interior—the interior points being those lying in a neighborhood entirely contained in K. A subset of the plane is said to have *content zero* if given any $\varepsilon > 0$ it can be covered by a *finite* set of rectangles whose areas sum to at most ε. (Notice that this is a stronger condition than having measure zero because the covering must be finite—for measure zero the definition is the same except that countable coverings are allowed.) If the boundary of K is the union of a finite number of smooth curves, it is easy to show that it has content zero and this is the case in the usual calculus applications. We leave the details to the exercises.

Needless to say, the same theory of multiple integration is valid in $\mathbb{R}^n$. If $K \subseteq \mathbb{R}^n$ is a bounded set whose boundary has content zero and f is a continuous function on K, then we can define the multiple integral $\int_K f(x)\,dx$ and show that it is equal to the iterated integrals (n-fold) in any order. In particular, we can define the *volume* of K, written $\text{vol}(K)$, to be the integral of the function $f \equiv 1$ over K.

15.1.2 Fubini's Theorem

We discuss next the equality of iterated and double integrals in the context of Lebesgue integration. (We are deliberately skipping the case of Riemann integrable functions because the results are inconclusive and not very useful.) We let μ denote Lebesgue measure on $\mathbb{R}^2$. Recall that μ is the unique measure on the Borel subsets of $\mathbb{R}^2$ that assigns to each rectangle its usual area (it is convenient to adopt the convention $0 \cdot \infty = 0$ in defining areas of rectangles so that the area is defined for unbounded rectangles in a consistent manner). We would like to relate this measure on $\mathbb{R}^2$ to Lebesgue measure on $\mathbb{R}$ by summing the lengths of horizontal or vertical sections.

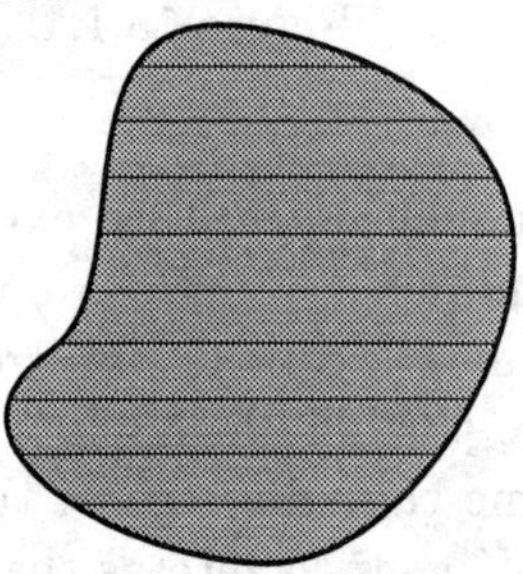

Figure 15.1.2:

Let A be a measurable subset of $\mathbb{R}^2$. Then for each y, the section $A_y = \{x \text{ in } \mathbb{R} : (x, y) \text{ is in } A\}$ is a subset of $\mathbb{R}$, as shown in Figure 15.1.2. If we take the Lebesgue measure of the section A_y and then integrate with respect to y we should get $\mu(A)$. Notice we use Lebesgue measure on $\mathbb{R}$ twice, once to measure A_y and once to integrate with respect to y. We can make this look like an iterated integral if we consider the characteristic function of A:

$$\chi_A(x, y) = \begin{cases} 1 & \text{if } (x, y) \text{ is in } A, \\ 0 & \text{if } (x, y) \text{ is not in } A. \end{cases}$$

Then the Lebesgue measure of A_y is $\int_{-\infty}^{\infty} \chi_A(x, y)\, dx$, so we are looking at the iterated itegral $\int_{-\infty}^{\infty}(\int_{-\infty}^{\infty} \chi_A(x, y)\, dx)\, dy$. If we were to section first in the y-direction (for fixed x) and then integrate in x, we would obtain the iterated integral in the other order. The double integral $\int \chi_A d\mu$ is just $\mu(A)$, so our claim that $\mu(A)$ is equal to the area of A

by sectioning is equivalent to the equality of the double integral and the iterated integrals for the particular function χ_A.

There are some technical difficulties with defining the interated integral for χ_A. First, in order to form $\int_{-\infty}^{\infty} \chi_A(x,y)\,dx$, we need to know that for each fixed y the function $\chi_A(x,y)$ as a function of x is measurable. Then we need to know that $\int_{-\infty}^{\infty} \chi_A(x,y)\,dx$ is measurable as a function of y in order to perform the next integration. The resolution of these difficulties is by no means trivial, and we postpone the details to section 15.1.3. Given that these functions are measurable, we indicate a simple proof that $\int_{-\infty}^{\infty}(\int_{-\infty}^{\infty} \chi_A(x,y)\,dx)\,dy = \mu(A)$. We note that the identity is trivially true when A is a rectangle. Since μ is characterized as being the unique measure on the Borel sets that gives the usual area for rectangles, it suffices to show that the iterated integral defines a measure. In other words, if we set $\nu(A) = \int_{-\infty}^{\infty}(\int_{-\infty}^{\infty} \chi_A(x,y)\,dx)\,dy$, then we have to verify the axioms for a measure for ν—the only non-trivial axiom being countable additivity. Thus we need to show $\nu(A) = \sum_{k=1}^{\infty} \nu(A_k)$ if $A = \bigcup_{k=1}^{\infty} A_k$ is a disjoint union of Borel sets. But we claim this follows easily from two applications of the monotone convergence theorem. Indeed the condition $A = \bigcup_{k=1}^{\infty} A_k$ (with A_k disjoint) translates into $\chi_A = \sum_{k=1}^{\infty} \chi_{A_k}$. Since these are non-negative functions, we have

$$\int \chi_A(x,y)\,dx = \sum_{k=1}^{\infty} \int \chi_{A_k}(x,y)\,dx$$

for each fixed y by the first application of the monotone convergence theorem. But we can then regard this equation as writing one non-negative function of y as an infinite series of non-negative functions, so by a second application of the theorem

$$\int \left(\int \chi_A(x,y)\,dx \right) dy = \sum_{k=1}^{\infty} \int \left(\int \chi_{A_k}(x,y)\,dx \right) dy,$$

which is the countable additivity of ν. Of course our technical assumptions about measurability were used implicitly in this argument.

This result admits an interesting interpretation. Suppose $f : \mathbb{R} \to \mathbb{R}$ is a non-negative measurable function. Then one interpretation of $\int f(x)\,dx$ is as the area under the graph of f. Thus let A denote this region under the graph of $f, A = \{(x,y) : 0 \leq y \leq f(x)\}$. If we section A first by varying y, as indicated in Figure 15.1.3, then the iterated integral $\int(\int \chi_A(x,y)\,dy)\,dx$ is equal to $\int f(x)\,dx$. Thus our result implies the equality of the integral of a non-negative measurable function and the area under its graph.

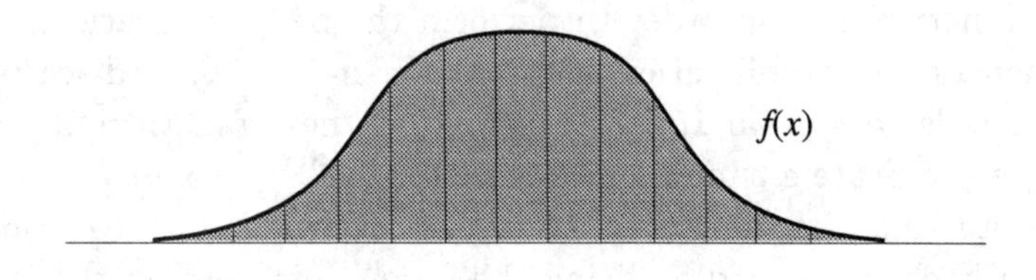

Figure 15.1.3:

We now turn to the problem of extending the equality of double and iterated integrals to more general functions than characteristic functions. To begin with we study non-negative functions.

Theorem 15.1.1 (*Fubini's Theorem, first version*) *Let f be a non-negative measurable function on $\mathbb{R}^2$. Then the double integral $\int f\,d\mu$ and the two iterated integrals $\int(\int f(x,y)\,dx)\,dy$ and $\int(\int f(x,y)\,dy)\,dx$ are all equal (they may be all $+\infty$).*

Proof: Let f_n be a montone increasing sequence of non-negative simple functions converging to f. Then $\int f d\mu = \lim_{n\to\infty} \int f_n d\mu$ by definition. But we have seen that the double and iterated integrals are equal for characteristic functions and, hence, for simple functions. Thus $\int f_n d\mu = \int(\int f_n(x,y)\,dx)\,dy$. The last step in the argument is to take the limit. Regarding $f_n(x,y)$ as a function of x (for fixed y), we know it is measurable (this will be proved in section 15.1.3); and as we vary n we have an increasing sequence of non-negative functions, so $f(x,y)$ is measurable and $\int f_n(x,y)\,dx \to \int f(x,y)\,dx$ by the monotone convergence theorem. Since $\int f_n(x,y)\,dx$ is measurable, this shows that $\int f(x,y)\,dx$ is measurable, and one more application of the monotone convergence theorem shows

$$\int\left(\int f_n\,(x,y)\,dx\right)dy \to \int\left(\int f(x,y)\,dx\right)dy,$$

completing the proof. QED

One word of caution in interpreting this theorem: even though the integral $\int f d\mu$ may be finite, some of the single integrals $\int f(x,y)\,dx$ may be infinite. This can only happen for values of y in a set of measure zero, of course. Thus the second integration in the iterated integral may involve a function taking values in the extended reals. For example, suppose $f(x,y) = (x^2+y^2)^{-1/2}\chi_{x^2+y^2\leq 1}$. Then

$$\int f(x,y)\,dx = \int_{-\sqrt{1-y^2}}^{\sqrt{1-y^2}} (x^2+y^2)^{-1/2}\,dx$$

is finite for $y \neq 0$ but $+\infty$ for $y = 0$ since $\int_{-1}^{1} |x|^{-1}\,dx = +\infty$. In this case the double integral is finite, but this is easiest to see using polar coordinates, which we discuss in section 15.2.

It is not necessary that f be defined on all of $\mathbb{R}^2$ to apply the theorem. If suffices to have f defined on a measurable set A, for then we can simply extend f to be zero outside of A. Notice that we do not need any assumptions about the boundary of A.

Next we consider functions that are not necessarily nonnegative.

Theorem 15.1.2 (*Fubini's Theorem, second version*) *Let f be a measurable function on $\mathbb{R}^2$, and assume that f is integrable with respect to μ (this means $\int f^+ d\mu$ and $\int f^- d\mu$ are finite). Then the double integral and iterated integrals are equal. In particular for almost every y, $f(x,y)$ as a function of x is integrable and $\int f(x,y)\,dx$ is integrable as a function of y and the same with x and y reversed.*

Proof: We need only apply the previous version to f^+ and f^-. Since $\int f^\pm d\mu$ is assumed finite, we have $\int (\int f^\pm(x,y)\,dx)\,dy = \int f^\pm d\mu$ and this implies $\int f^\pm(x,y)\,dx$ is finite for almost every y (or else the y integration would produce $+\infty$). Subtracting and using the linearity of the double integral and the iterated integrals we obtain $\int (\int f(x,y)\,dx)\,dy = \int f d\mu$. QED

This result says a lot less than appears since the hypothesis that f be integrable involves the double integral and normally we want to compute it by computing an iterated integral. The existence of the iterated integral, if it involves cancellation, need not imply the existence of the double integral. For example, consider

$$f(x,y) = \begin{cases} 1 & \text{if } 0 < x < 1, \\ -1 & \text{if } -1 < x < 0, \\ 0 & \text{otherwise.} \end{cases}$$

Then $\int f(x,y)\,dx = 0$ for every y, so $\int(\int f(x,y)\,dx)\,dy = 0$, but $\int f(x,y)\,dy = \pm\infty$ for x in $(-1,1)$, and of course f is not integrable. With a little more ingenuity one can construct examples with iterated integrals existing in both orders but not equal. All these examples exploit concealed cancellation of $\pm\infty$, as the final version of Fubini's Theorem makes clear.

Theorem 15.1.3 (*Fubini's Theorem, third version*) *Let f be a measurable function on $\mathbb{R}^2$, and suppose one of the iterated integrals for $|f|$ exists and is finite; say $\int |f(x,y)|\,dx$ is finite for almost every y and $\int(\int |f(x,y)|\,dx)\,dy$ is finite. Then f is integrable with respect to μ and both iterated integrals for f are equal to $\int f\,d\mu$.*

Proof: By the first version applied to $|f|$ we have $\int |f| d\mu = \int(\int |f(x,y)|\,dx)\,dy$, so $|f|$ is integrable. This means f is integrable, so by the second version the iterated integrals both equal the double integral of f. QED

This is the most useful of the three versions, since the hypothesis is one that we can hope to check by finding an upper bound for the iterated integral. It is significant that we have to deal with only one of the two iterated integrals, since it is often the case that one is easier to estimate.

The consequences of Fubini's theorem are often surprising. We give one application that is typical and an important theorem in its own right. Recall we defined the convolution of two functions $f * g(x) = \int_{-\infty}^{\infty} f(x-y)g(y)\,dy$. We originally defined this in terms of the Riemann integral, but now we can interpret it as a Lebesgue integral and can ask under which conditions on f and g it is defined.

Theorem 15.1.4 (L^1 *Convolution Theorem*) *Let f and g be integrable functions on the line. Then $f * g$ is defined as an integrable function in the following sense: for almost every x, the function $h(y) =$*

$f(x-y)g(y)$ is an integrable function and $f * g(x) = \int h(y)\,dy$ so defined for almost every x (it doesn't matter how it is defined on the remaining set of measure zero) is an integrable function. In addition we have the estimate $||f * g||_1 \le ||f||_1||g||_1$.

Proof: Since f and g are measurable, it follows easily that $h(x,y) = f(x-y)g(y)$ is a measurable function on $\mathbb{R}^2$. We want to apply the third version of Fubini's theorem to $h(x,y)$, and we will compute the iterated integral in the reverse order to what one might expect. For almost every $y, g(y)$ is finite and

$$\begin{aligned}\int |h(x,y)|\,dx &= \int |f(x-y)g(y)|\,dx \\ &= |g(y)| \int |f(x-y)|\,dx \\ &= |g(y)| \int |f(x)|\,dx = |g(y)|\,||f||_1.\end{aligned}$$

Here we have used the translation invariance of Lebesgue measure. Integrating this identity we obtain

$$\int \left(\int |h(x,y)|\,dx \right) dy = \int |g(y)|\,||f||_1\,dy = ||f||_1||g||_1,$$

which is finite. Thus by Fubini's theorem $\int h(x,y)\,dy = f * g(x)$ is finite for almost every x and

$$\int |f * g(x)|\,dx \le \int \left(\int |h(x,y)|\,dy \right) dx = ||f||_1||g|_1$$

by the equality of the iterated integrals. This shows $f * g$ is integrable and establishes the desired estimate. QED

Even the statement that $f * g$ is defined for almost every x is not obvious. It is easy to construct examples where $f * g$ fails to exist at a single point. Take $f = g = |x|^{-1/2}\chi_{|x|\le 1}$. Then f and g are integrable because the singularity at $x = 0$ is not too bad, but

$$f * g(0) = \int_{-1}^{1} |x|^{-1/2} \cdot |x|^{-1/2}\,dx = \int_{-1}^{1} |x|^{-1} dx = +\infty.$$

The idea here is simply that the ordinary product of two integrable functions need not be integrable. Since the convolution integrates the product of two functions, it would seem plausible that by placing the singularities f and g so that they reinforce each other in the product $f(x-y)g(y)$ it would be possible to construct integrable functions such that $f * g(x) = +\infty$ for every x. The theorem shows that this is not possible. There are many variants of this theorem: for example, if f is in L^1 and g is in L^2, then $f * g$ is in L^2; or if f and g are both in L^2, then $f * g$ is bounded and continuous.

Fubini's theorem generalizes not only to $\mathbb{R}^n$ but to any situation where there are functions of several variables. If $(X, \mathcal{F}, \mu)$ and $(Y, \mathcal{G}, \nu)$ are two measure spaces (a certain technical condition is required that X and Y be expressible as a countable union of sets of finite measure, but this condition is satisfied by most examples), then we can consider functions $f(x, y)$ of variables x in X and y in Y to be a function on the Cartesian product $X \times Y$. There is a σ-field of subsets of $X \times Y$ called $\mathcal{F} \times \mathcal{G}$, which is the σ-field generated by the "rectangles" $A \times B$ where A is in $\mathcal{F}$ and B is in $\mathcal{G}$ (of course in $\mathbb{R}^2$ these are not the conventional rectangles, since the "sides" are allowed to be arbitrary measurable subsets of $\mathbb{R}$). On this σ-field there is a product measure $\mu \times \nu$ that gives the correct area of a rectangle $\mu \times \nu(A \times B) = \mu(A)\nu(B)$. Then Fubini's theorem says the integral $\int f d\mu \times \nu$ is equal to the iterated integrals

$$\int \left(\int f(x, y) d\mu(x) \right) d\nu(y)$$

or

$$\int \left(\int f(x, y) d\nu(y) \right) d\mu(x)$$

provided f is non-negative, or integrable, or one iterated integral is finite for $|f|$. This is especially important in probability theory, where the choice of the product measure is interpreted as saying the events represented by x and y are independent. If μ and ν are counting measure on the positive integers, then Fubini's theorem is equivalent to the interchange of order of summation for absolutely convergent doubly indexed series.

15.1.3 The Monotone Class Lemma*

In this section we will explain the technicalities involved in proving measurability in Fubini's theorem. They key idea is a set-theoretic fact called the *Monotone Class Lemma.* The proof of this fact is very tricky although not very long. It serves as a device for finessing the difficulty of dealing with differences of sets. We define a *monotone class* $\mathcal{M}$ *to be a collection of sets (subsets of the universe* X*) that is closed under monotone increasing and decreasing sequences: if* $A_1 \subseteq A_2 \subseteq \cdots$ *with* A_j *in* $\mathcal{M}$*, then* $\bigcup_{j=1}^{\infty} A_j$ *is in* $\mathcal{M}$*; and if* $B_1 \supseteq B_2 \supseteq \cdots$ *with* B_j *in* $\mathcal{M}$*, then* $\bigcap_{j=1}^{\infty} B_j$ *is in* $\mathcal{M}$. These are conditions that are easy to check, especially using the monotone and dominated convergence theorems. It is easy to see that a σ-field is a monotone class, but a monotone class need not be a σ-field since it does not have to be closed under complements or differences. The Monotone Class Lemma gives a simple criterion for a monotone class to contain a σ-field.

Lemma 15.1.1 (*Monotone Class*) *Let* $\mathcal{M}$ *be a monotone class that contains a field* $\mathcal{F}$*. Then* $\mathcal{M}$ *contains the* σ*-field generated by* $\mathcal{F}$*.*

Proof: Let $\mathcal{F}_1$ denote the σ-field generated by $\mathcal{F}$, and let $\mathcal{M}_1$ be the monotone class generated by $\mathcal{F}$ (the smallest monotone class containing $\mathcal{F}$). Clearly $\mathcal{M}_1 \subseteq \mathcal{M}$ since $\mathcal{M}$ is a monotone class containing $\mathcal{F}$. We will show that $\mathcal{M}_1$ is a field, and this will complete the proof because it follows easily that $\mathcal{M}_1$ is a σ-field and, hence, contains $\mathcal{F}_1$ (actually $\mathcal{M}_1 = \mathcal{F}_1$ since $\mathcal{F}_1$ is also a monotone class).

For each set A in $\mathcal{M}_1$, define $\mathcal{M}_1(A)$ to be the collection of sets B in $\mathcal{M}_1$ such that $A \cap B, A \cup B, A \backslash B$, and $B \backslash A$ are also in $\mathcal{M}_1$. (This is the trick!) Notice the symmetry between A and B in the definition, so B is in $\mathcal{M}_1(A)$ if and only if A is in $\mathcal{M}_1(B)$. Of course there is no reason to believe a priori that $\mathcal{M}_1(A)$ contains any sets other than the empty set and A. But we will eventually show that $\mathcal{M}_1(A)$ is all of $\mathcal{M}_1$, which will imply that $\mathcal{M}_1$ is a field.

First we claim that $\mathcal{M}_1(A)$ is a monotone class. This is easy because each of the operations that defines $\mathcal{M}_1(A)$ preserves monotone sequences. For example, if $B_1 \subseteq B_2 \subseteq \cdots$, then $A \backslash B_1 \supseteq A \backslash B_2 \supseteq \cdots$ (notice the switch from increasing to decreasing, but this doesn't matter because the definition of monotone classes involves both). This means that if $\mathcal{M}_1(A)$ contains $\mathcal{F}$, then it must be all of $\mathcal{M}_1$. But how do we get started? We use that fact that $\mathcal{F}$ is a field; so if A and B are in $\mathcal{F}$, then B is in $\mathcal{M}_1(A)$. That means for any set A in $\mathcal{F}, \mathcal{M}_1(A)$ contains $\mathcal{F}$; hence, $\mathcal{M}_1(A) = \mathcal{M}_1$.

We are almost done, but first we need a clever use of the symmetry condition. For any B in $\mathcal{M}_1$ we know that B is in $\mathcal{M}_1(A)$ for A in $\mathcal{F}$, so A is in $\mathcal{M}_1(B)$. That means $\mathcal{M}_1(B)$ contains $\mathcal{F}$; hence, $\mathcal{M}_1(B) = \mathcal{M}_1$. Since B was any set in $\mathcal{M}_1$, we have succeeded in proving our main claim.

We can now see that $\mathcal{M}_1$ is a field, because A in $\mathcal{M}_1(B)$ for any A and B in $\mathcal{M}_1$ means $A \cap B, A \cup B, A \backslash B$, and $B \backslash A$ are also in $\mathcal{M}_1$. By using the field properties we can replace any countable union by a countable increasing union, so $\mathcal{M}_1$ is a σ-field. QED

Now we can fill in the missing arguments in the proof of Fubini's theorem.

Lemma 15.1.2 *Let A be a Borel subset of $\mathbb{R}^2$. Then every section A_y is a Borel subset of $\mathbb{R}$, and the function $|A_y| = \int_{-\infty}^{\infty} \chi_A(x, y)\, dx$ is a measurable function of y (taking values in the extended real numbers).*

Proof: First we prove the analogous result for Borel subsets of a finite rectangle $[a, b] \times [c, d]$. Let $\mathcal{M}$ denote the class of Borel sets A for which the conclusions of the lemma hold. Clearly $\mathcal{M}$ contains the field generated by the rectangles. We claim $\mathcal{M}$ is a monotone class. For this we simply apply the dominated convergence theorem to interchange the integral with limits. We can use a constant function as dominator since we are in a finite rectangle. For example, if $A_1 \subseteq A_2 \subseteq \cdots$, then the same is true for the sections, and

$$\int_a^b \chi_{\bigcup A_j}(x, y)\, dx = \lim_{j \to \infty} \int_a^b \chi_{A_j}(x, y)\, dx$$

is measurable, being the limit of measurable functions. Thus the monotone class lemma implies that $\mathcal{M}$ contains all Borel subsets of $[a, b] \times [c, d]$.

For a general Borel subset A of $\mathbb{R}^2$ we write it as the countable disjoint union of $A \cap [m_1, m_1 + 1) \times [m_2, m_2 + 1)$ as m_1 and m_2 vary over all integers. The conclusions are preserved under countable disjoint unions (the function $|A_y|$ may now assume the value $+\infty$). QED

This lemma also implies the measurability of $f(x, y)$ as a function of x for fixed y and $\int f(x, y)\, dx$ as a function of y for any simple function

f (just use the analogous facts for the sets A_k in the decomposition $f = \sum_{k=1}^{N} a_k \chi_{A_k}$), which was used in the proof of Fubini's theorem (first version).

The same argument can be used to justify the definition of the product measure $\mu \times \nu$ for any two finite measures μ and ν, and more generally for the case of *σ-finite measures*, which is defined as follows: *μ is σ-finite if the space X can be written as a countable union $X = \bigcup_{j=1}^{\infty} X_j$ with $\mu(X_j)$ finite.* For if μ and ν are σ-finite and we write $\mu = \sum \mu_j$ and $\nu = \sum \nu_k$ where μ_j and ν_k are the restrictions of μ to X_j and ν to Y_k (we can always take the decompositions $X = \bigcup X_j$ and $Y = \bigcup Y_k$ to be disjoint), then we can define $\mu \times \nu = \sum_{j=1}^{\infty} \sum_{k=1}^{\infty} \mu_j \times \nu_k$.

Another important application of the Monotone Class Lemma is the following uniqueness theorem of Hahn:

Theorem 15.1.5 (*Hahn Uniqueness Theorem*) *Let μ and ν be σ-finite measures on a σ-field $\mathcal{F}_1$ generated by a field $\mathcal{F}$. If μ and ν are equal on $\mathcal{F}$, then they are equal on $\mathcal{F}_1$.*

Proof: Suppose first that μ and ν are finite measures. Let $\mathcal{M}$ be the collection of sets A in $\mathcal{F}_1$ for which $\mu(A) = \nu(A)$. It is easy to see that $\mathcal{M}$ is a monotone class, so $\mathcal{M} = \mathcal{F}_1$ by the monotone class lemma. (The finiteness of the measures is needed in order to use the continuity from above for monotone decreasing sequences.) For the general case just split the space into a countable union of sets of finite measure for both μ and ν. A priori these sets may not belong to $\mathcal{F}$, but this can always be arranged (see exercise set 15.1.4 for details). Then apply the special case of finite measures to each of the pieces and sum, using σ-additivity. QED

In particular, this uniqueness theorem gives another proof of the uniqueness of Lebesgue measure.

It might seem possible to avoid using the Monotone Class Lemma in these applications and argue directly that the classes we called $\mathcal{M}$ are σ-fields. It isn't! Try it if you don't believe me.

15.1.4 Exercises

1. Write out the details of the proof of the existence of the double integral of a continuous function on a rectangle.

2. State and prove (by induction) the equality of all iterated integrals (how many are there?) for a continuous function on a rectangular parallelopiped in $\mathbb{R}^n$.

3. a. Prove that if K is a set in $\mathbb{R}^2$ whose boundary has content zero, then the double integral of a continuous function on K exists.

   b. Assume also that every section of K is a finite union of intervals. Prove that the iterated integrals are equal to the double integral.

4. Prove that a C^1 bounded curve has content zero.

5. Prove that a finite union of sets of content zero is also a set of content zero.

6. Prove that the graph of $f(x) = \sin 1/x$ on $[0, 1]$ has content zero.

7. Let $f : \mathbb{R} \to \mathbb{R}$ be measurable, and let $A \subseteq \mathbb{R}^2$ be the graph of f. Prove that A has measure zero. (**Hint**: use Fubini's theorem.)

8. Let A be a measurable subset of $\mathbb{R}^2$. Prove that A has measure zero if and only if almost every section A_y has measure zero.

9. Prove that if $f : \mathbb{R}^2 \to \mathbb{R}$ is bounded, measurable, and vanishes outside a bounded region, then the iterated integrals are equal.

10. Prove that the convolution of two L^2 functions is bounded.

11. Define a periodic convolution $f * g(x) = \int_{-\pi}^{\pi} f(x-y)g(y)\, dy$ where f and g are periodic of period 2π. Prove that if f and g restricted to $(-\pi, \pi)$ are integrable, then $f * g(x)$ is defined a.e. and is also integrable on $(-\pi, \pi)$.

12. Give an example of a function on $\mathbb{R}^2$ for which both iterated integrals exist and are finite but unequal. Can you do it with a continuous function?

13. State each of the three versions of Fubini's theorem for counting measure, in terms of doubly indexed series.

14. Give an example of a measure that is not σ-finite.

15. If μ is a σ-finite measure on a σ-field $\mathcal{F}_1$ generated by a field $\mathcal{F}$, show that in the decomposition $X = \bigcup_{j=1}^{\infty} X_j$ with $\mu(X_j) < \infty$ it is possible to take each X_j in $\mathcal{F}$.

15.2 Change of Variable in Multiple Integrals

15.2.1 Determinants and Volume

Our goal is to establish the change of variable formula

$$\int_D f(y)\,dy = \int_{D'} f(g(x))|\det dg(x)|\,dx$$

for suitable functions f and g, where $g : D' \to D$ is a one-to-one change of variable (D' and D are suitable regions in $\mathbb{R}^n$). This result is usually stated in calculus courses with an intuitive argument that $|\det dg(x)|$ is the correct magnification factor for relating the volumes in the x and $y = g(x)$ variables. This argument is based on the observation that $|\det dg(x)|$ is the correct magnification factor if g is a linear (or affine) transformation, and the general case follows by localization via the differential. We will follow the same outline, presenting the complete details. It is not an easy theorem to prove, even under strong assumptions on the domains D and D' and the functions f and g.

We begin with the case of a linear transformation $g(x) = Ax$, where A is an $n \times n$ matrix. Here $dg(x) = A$ is constant. The claim is that if R is any rectangle, then the volume of $g(R)$ is exactly $|\det A|$ times the volume of R.

Let us think of the $n \times n$ matrix A as being composed of n column vectors $A_1, \ldots, A_n$. These vectors are the image of the standard basis for $\mathbb{R}^n\,(e^{(1)}, \ldots, e^{(n)})$ under multiplication by A. Now the unit cube consists of all vectors $\sum_{j=1}^n b_j e^{(j)}$ where $0 \le b_j \le 1$, and this is mapped under multiplication by A into the parallelopiped of all vectors of the form $\sum_{j=1}^n b_j A_j$ for $0 \le b_j \le 1$. Let us call this the *parallelopiped generated* by $A_1, \ldots, A_n$. The unit cube is thus the parallelopiped generated by the standard basis vectors $e^{(1)}, \ldots, e^{(n)}$. The volume of the parallelopiped generated by $A_1, \ldots, A_n$ is clearly the magnification factor for which we are looking, because the unit cube has volume one and it is clear by linear algebra that the same factor relates the volume of $A(R)$ to the volume of R for any rectangle R (just decompose R into an approximate union of cubes).

We need to show that the volume of the parallelopiped generated by $A_1, \ldots, A_n$ is equal to $|\det A|$. The key to understanding why this is so is to get rid of the absolute value. Let us consider $\det A$ as a function of the n column vectors $A_1, \ldots, A_n$ that comprise A. With this understanding we will write $\det(A_1, \ldots, A_n)$. We now recall the basic properties of this function, properties that in fact will characterize it uniquely:

1. $\det\,(A_1, \ldots, A_n)$ is real-valued;

2. it is *multilinear*, meaning that for each $k = 1, \ldots, n$ we have

$$\begin{aligned}&\det(A_1, \ldots, aA_k + bB_k, \ldots, A_n)\\ &\quad = a\det(A_1, \ldots, A_k, \ldots, A_n) + b\det(A_1, \ldots, B_k, \ldots, A_n);\end{aligned}$$

3. it is *skew-symmetric* (or sometimes called *alternating*), meaning for any $j \neq k$

$$\begin{aligned}&\det(A_1, \ldots, A_j, \ldots, A_k, \ldots, A_n)\\ &\quad = -\det(A_1, \ldots, A_k, \ldots, A_j, \ldots, A_n);\end{aligned}$$

4. $\det\,(e^{(1)}, \ldots, e^{(n)}) = 1$.

Any function satisfying properties 1, 2, and 3 is called a *skew-symmetric multilinear form.* Property 4 is just a normalizing condition. A theorem of linear algebra says that the space of skew-symmetric multilinear forms in n vectors from $\mathbb{R}^n$ is one-dimensional, so together with the normalizing condition 4 this uniquely determines the determinant.

In view of this uniqueness result, in order to show that the volume of the parallelopiped generated by $A_1, \ldots, A_n$ is $|\det(A_1, \ldots, A_n)|$ we want to invent a signed volume for parallelopipeds that will have all the properties of the determinant. We give the argument first in the case $n = 2$ where we can draw pictures.

Let A_1 and A_2 be two vectors in $\mathbb{R}^2$. We define $m(A_1, A_2)$ to be $\pm$ the area of the parallelogram with sides A_1 and A_2, where we take the plus sign if the angle from A_1 to A_2 is less than $180°$ and the minus sign if it is more than $180°$, as indicated in Figure 15.2.1.

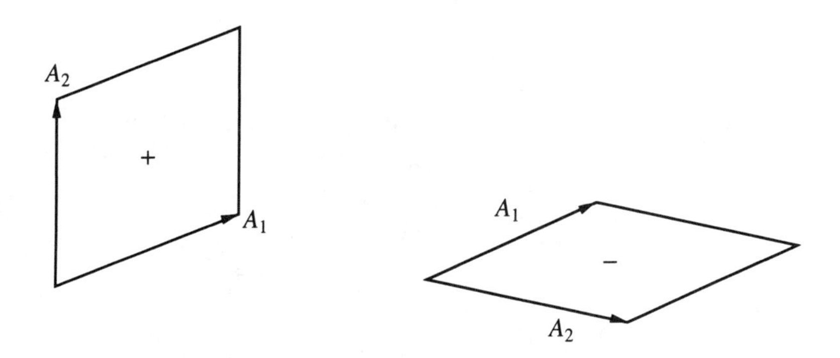

Figure 15.2.1:

If the angle is 0° or 180°, then the area is zero, so we need not determine the sign. Note that this definition automatically makes m skew-symmetric, $m(A_1, A_2) = -m(A_2, A_1)$, and it clearly satisfies the normalization condition $m(e^{(1)}, e^{(2)}) = 1$. Thus the only condition we need to check is the bilinearity; and by the skew-symmetry it suffices to establish it for one of the variables, say the second. Now fix $A_1 \neq 0$ (if $A_1 = 0$ there is nothing to prove), and let A_2 be any vector such that A_1 and A_2 are linearly independent. We claim $m(A_1, a_1A_1 + a_2A_2) = a_2m(A_1, A_2)$ for any real a_1 and a_2, and this will prove the linearity because every vector in $\mathbb{R}^2$ has the form $a_1A_1 + a_2A_2$ and $a_2m(A_1, A_2)$ is linear in (a_1, a_2). To prove this identity we compare the parallelograms with sides A_1, a_2A_2 and sides $A_1, a_1A_1 + a_2A_2$, as shown in Figure 15.2.2. Notice that they have the same base $(|A_1|)$ and the same altitude, hence the same area, and also the angle between the sides is always on the same side of 180°; so we have $m(A_1, a_1A_1 + a_2A_2) = m(A_1, a_2A_2)$. Next it is clear that the factor a_2 multiplying A_2 changes the area by $|a_2|$ since this is the change in the altitude. Also, the angle remains the same if $a_2 > 0$ and changes by 180° if $a_2 < 0$. Thus $m(A_1, a_2A_2) = a_2m(A_1, A_2)$.

By the uniqueness of the determinant we conclude that $m(A_1, A_2) = \det(A_1, A_2)$, which completes the proof that the area of the parallelogram generated by A_1, A_2 is $|\det(A_1, A_2)|$. In $\mathbb{R}^n$ the proof is similar. We define $m(A_1, \ldots, A_n)$ to be the volume of the parallelopiped generated by $A_1, \ldots, A_n$ multiplied by ± 1. The easiest way to specify the choice of the sign is to take the sign of $\det(A_1, \ldots, A_n)$. From

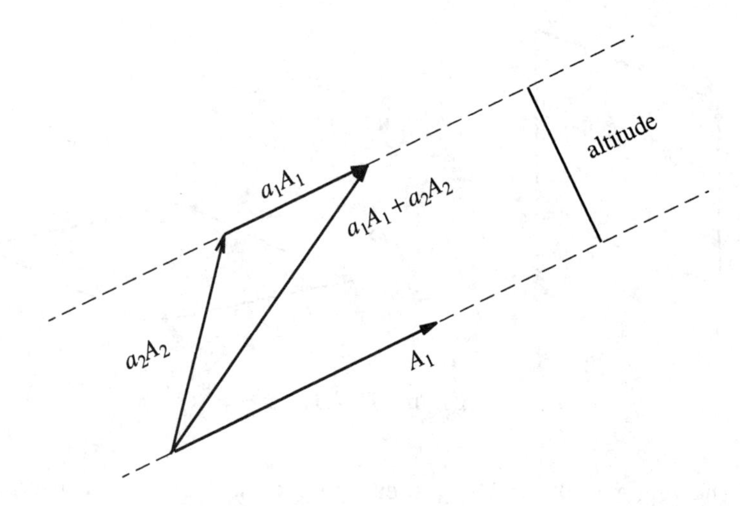

Figure 15.2.2:

the properties of det this guarantees the skew-symmetry and the correct normalization $m(e^{(1)}, \ldots, e^{(n)}) = 1$. To show the multilinearity it suffices to prove linearity in the last factor because of the skew-symmetry. Let $A_1, \ldots, A_{n-1}$ be fixed vectors that we may assume to be linearly independent, for otherwise $m(A_1, \ldots, A_{n-1}, A_n) = 0$ for any choice of A_n. We choose one more vector A_n, so $A_1, \ldots, A_n$ form a basis for $\mathbb{R}^n$. We will show $m(A_1, \ldots, A_{n-1}, a_1A_1 + \ldots + a_nA_n) = a_nm(A_1, \ldots, A_{n-1}, A_n)$, which implies the linearity. We can use the same argument as before to throw away the terms $a_1A_2 + \cdots + a_{n-1}A_{n-1}$ since they don't change the base (the parallelopiped of dimension $n-1$ generated by $A_1, \cdots, A_{n-1}$) and they don't change the altitude (the distance from $a_1A_1 + \cdots + a_nA_n$ to the base); hence they leave unchanged the volume = altitude $\times$ area of base and also the determinant (using properties of the determinant), so they don't change the sign. So $m(A_1, \ldots, A_{n-1}, a_1A_1 + \ldots + a_nA_n) = m(A_1, \ldots, A_{n-1}, a_nA_n)$. Finally the factor a_n changes the measure by $|a_n|$ since it doesn't change the base but multiplies the altitude by $|a_n|$; and the determinant is changed by a factor of a_n, so the sign changes appropriately. Thus we have established:

Theorem 15.2.1 *Let $g(x) = Ax$ where A is an $n \times n$ matrix. If R is any rectangle, then* $\mathrm{vol}(g(R)) = |\det A|\mathrm{vol}(R)$.

15.2.2 The Jacobian Factor*

Since we have shown that the absolute value of the determinant of a linear transformation gives the exact magnification factor for volumes under the transformation, the basic principle of differential calculus suggests that for a C^1 mapping $g : U \to \mathbb{R}^n$ (U an open set in $\mathbb{R}^n$) the absolute value of the determinant of the derivative $J(x) = |\det dg(x)|$ should give an approximate local magnification factor for volumes under the mapping near x. We call $J(x)$ the *Jacobian factor.* We then expect to get the exact volume of $g(U)$ by integrating $J(x)$ over U, if g is one-to-one. It turns out to be extremely difficult to prove this. Before beginning the proof we need to understand some of the difficulties we will encounter.

If we fix a point $\tilde{x}$ and use the differentiability of g at $\tilde{x}$ to write $g(x) = g(\tilde{x}) + dg(\tilde{x})(x - \tilde{x}) + o(|x - \tilde{x}|)$, then the mapping $g_1(x) = g(\tilde{x}) + dg(\tilde{x})(x - \tilde{x})$ satisfies $\text{vol}(g_1(R)) = J(\tilde{x})\text{vol}(R)$ for any rectangle R. We would like to say $\text{vol}(g(R)) \approx \text{vol}(g_1(R))$ with the approximation improving as the size of R decreases. But we will need to have an estimate that is uniform in $\tilde{x}$, and for that we will need to use the continuity of the derivative. We will use this idea to prove the upper bound $\text{vol}(g(U)) \leq \int_U J(x)\, dx$. But the lower bound is more delicate. Suppose $g_1(R)$ is a long skinny rectangle, say $[0, \varepsilon^{-1}] \times [0, \varepsilon]$ in $\mathbb{R}^2$, with volume one. Then by moving each point in $g_1(R)$ by a distance of only ε (send (x, y) to $(x, 0)$) we can squash the rectangle down to a segment of the x-axis, $[0, \varepsilon^{-1}] \times 0$, so the volume drops to zero. So even if we know that $g(R)$ is point-by-point close to $g_1(R)$, we cannot conclude that $\text{vol}(g(R))$ is close to $\text{vol}(g_1(R))$. Instead, we will use the inverse function theorem and the upper bound estimate for the inverse mapping, but for this we will have to avoid points where $J(x) = 0$, or even points where $J(x)$ is close to zero. Fortunately these points contribute very little to the integral, so we can omit them when we try to prove the lower bound $\text{vol}(g(U)) \geq \int_U J(x)\, dx$.

We begin with a simple geometric lemma that gives an upper bound for the volume of a neighborhood of set. We define the ε-neighborhood A_ε of A to be the set of points of distance at most ε from points in A. This definition makes sense for any set A; but we will only need to consider parallelopipeds, which are the images of rectangles under linear mappings.

Lemma 15.2.1 *Let A be a parallelopiped contained in a ball of radius r in $\mathbb{R}^n$. Then there exists a constant c depending only on n such that*

$$\text{vol}(A_\varepsilon) \leq \text{vol}(A) + c\varepsilon(r+\varepsilon)^{n-1} \text{ for all } \varepsilon \leq 1.$$

Proof: We give the proof for $n = 2$. Then A is a parallelogram with perimeter P. Clearly every point in A_ε that is not in A must be within distance of ε from P, so $A_\varepsilon \subseteq A \cup P_\varepsilon$. Thus $\text{vol}(A_\varepsilon) \leq \text{vol}(A) + \text{vol}(P_\varepsilon)$, and the result follows from the estimate $\text{vol}(P_\varepsilon) \leq c\varepsilon(r+\varepsilon)$, which is essentially trivial since P consists of four line segments each of length at most $2r$, and $\text{vol}(L_\varepsilon) \leq (l + 2\varepsilon)\varepsilon$ if L is a line segment of length l.

The argument for general n is similar, using induction. The key fact is again the uniform upper bound for the $(n-1)$-dimensional volume of the boundary of A that follows from the fact that A is contained in a ball of radius r. We leave the details to the exercises. QED

Now let C_δ denote a cube with center at $\tilde{x}$ and side length δ. If g is a C^1 mapping in a neighborhood of $\tilde{x}$ we want to compare $\text{vol}(g(C_\delta))$ with $J(\tilde{x})\delta^n$, which is the volume of $g_1(C_\delta)$, where $g_1(x) = g_1(\tilde{x}) + dg(\tilde{x})(x - \tilde{x})$ is the best affine approximation to g at $\tilde{x}$. We would like to have a good approximation for δ small, but it is important that we be able to control the size of δ independent of the point $\tilde{x}$. The next lemma tells us how to do this in terms of the variation of $dg(x)$ as x varies over C. Since $dg(x)$ is an $n \times n$ matrix, it is natural to measure its size by the matrix norm (recall that $||M||$ is the smallest constant c such that $|Mx| \leq c|x|$ for all x in $\mathbb{R}^n$).

Lemma 15.2.2 *Suppose $||dg(x)|| \leq M$ and $||dg(x) - dg(\tilde{x})|| \leq \varepsilon$ for all x in C_δ. Then $\text{vol}(g(C_\delta)) \leq (J(\tilde{x}) + c\varepsilon)\delta^n$ where c is a constant that depends only on n and M.*

Proof: The hypothesis $||dg(x)|| \leq M$ easily implies that $g_1(C_\delta)$ lies in a ball of radius $c\delta$ (here we use the letter c to denote a constant that depends only on n and M with the understanding that different occurences may stand for different constants). We will apply the previous lemma with $r = c\delta$. To do this we need an estimate for the difference $g(x) - g_1(x)$. This is provided by the fundamental theorem of calculus:

$$g(x) - g(\tilde{x}) = \int_0^1 dg(\tilde{x} + s(x - \tilde{x}))(x - \tilde{x})\, ds,$$

so

$$g(x) - g_1(x) = \int_0^1 (dg(\tilde{x} + s(x - \tilde{x}) - dg(\tilde{x}))(x - \tilde{x})\, ds.$$

Since $\tilde{x} + s(x - \tilde{x})$ lies in C_δ, we can use the hypothesis to obtain the estimate

$$|(dg(\tilde{x} + s(x - \tilde{x})) - dg(\tilde{x}))(x - \tilde{x})| \leq \varepsilon |x - \tilde{x}|.$$

Thus we have

$$|g(x) - g_1(x)| \leq c\varepsilon\delta,$$

hence $g(C_\delta)$ lies in the $c\varepsilon\delta$ neighborhood of $g_1(C_\delta)$. Thus the previous lemma gives the estimate

$$\begin{aligned} \mathrm{vol}(g(C_\delta)) &\leq \mathrm{vol}(g_1(C_\delta)) + c\varepsilon\delta(c\delta + c\varepsilon\delta)^{n-1} \\ &\leq (J(\tilde{x}) + c\varepsilon)\delta^n. \end{aligned}$$

QED

Lemma 15.2.3 *Let $g : U \to \mathbb{R}^n$ be a one-to-one map of a bounded open set U whose boundary has content zero, and assume g is C^1 on the closure of U. Then*

$$\mathrm{vol}(g(U)) \leq \int_U J(x)\, dx.$$

More generally, if f is any non-negative continuous function on the closure of $g(U)$, then $\int_{g(U)} f(x)\, dx \leq \int_U f(g(x)) J(x)\, dx$.

Proof: Since the closure of U is compact, we have a uniform upper bound $||dg(x)|| \leq M$. This enables us to use the previous lemma. It also allows us to conclude that the boundary of $g(U)$ has content zero (see exercises), so the integrals over $g(U)$ are well defined.

We subdivide U into cubes of side length δ, discarding those that do not lie entirely in U. Given any ε, we can choose δ so that

$$||dg(x) - dg(\tilde{x})|| \leq \varepsilon$$

for x and $\tilde{x}$ in the same cube, by the uniform continuity of dg on the compact closure of U. With these fixed ε and δ, write $\{C_j\}$ for the collection of cubes and $\{\tilde{x}_j\}$ for the center points of the cubes. By the previous lemma we have

$$(*) \qquad \mathrm{vol}(g(C_j)) \le (J(\tilde{x}_j) + c\varepsilon)\delta^n,$$

and summing over j we obtain

$$\mathrm{vol}\left(g\left(\bigcup_j C_j \right) \right) \le \sum_j J(\tilde{x}_j)\delta^n + c\varepsilon \sum_j \delta^n.$$

With ε fixed we can let $\delta \to 0$ (the condition relating δ to ε allows all values smaller than a fixed δ_0). The sum $\sum_j J(\tilde{x}_j)\delta^n$ converges to $\int_U J(x)\,dx$ and $c\varepsilon \sum_j \delta^n$ converges to $c\varepsilon\,\mathrm{vol}(U)$. Since the boundary of U has content zero, it is easy to show that $\mathrm{vol}(g(\bigcup_j C_j))$ converges to $\mathrm{vol}(g(U))$ (see exercises for the details). Thus we have the estimate

$$\mathrm{vol}(g(U)) \le \int_U J(x)\,dx + c\varepsilon \mathrm{vol}(U),$$

and $\mathrm{vol}(U)$ is finite. Since this holds for any $\varepsilon > 0$, we obtain the desired upper bound for $\mathrm{vol}(g(U))$.

To obtain the upper bound for the integral we simply multiply $(*)$ by $f(g(\tilde{x}_j))$ before summing to obtain

$$\sum_j f(g(\tilde{x}))\mathrm{vol}(g(C_j)) \le \sum_j f(g(\tilde{x}_j))J(\tilde{x}_j)\delta^n$$

$$+ c\varepsilon \sum_j f(g(\tilde{x}_j))\delta^n.$$

As $\delta \to 0$ the sum $\sum_j f(g(\tilde{x}_j))J(\tilde{x}_j)\delta^n$ converges to $\int_U f(g(x))J(x)\,dx$ and $c\varepsilon \sum_j f(g(\tilde{x}_j))\delta^n$ converges to $c\varepsilon \int_U f(g(x))\,dx$. It is also true that $\sum_j f(g(\tilde{x}_j))\mathrm{vol}(g(C_j))$ converges to $\int_{g(U)} f(x)\,dx$. We leave the details to the exercises. Thus

$$\int_{g(U)} f(x)\,dx \le \int_U f(g(x))J(x)\,dx + c\varepsilon \int_U f(g(x))\,dx,$$

and the desired upper bound for the integral follows by letting $\varepsilon \to 0$. QED

Theorem 15.2.2 (*Change of Variable Formula*) *Let $g : U \to \mathbb{R}^n$ be a one-to-one map of a bounded open set U whose boundary has content zero, and assume g is C^1 on the closure of U. Also let f be any continuous function on the closure of $g(U)$. Then*

$$\int_{g(U)} f(x)\,dx = \int_U f(g(x))J(x)\,dx.$$

Proof: It suffices to prove the result for non-negative f, and we already have the inequality in one direction. We start by proving the reverse inequality under the additional assumption that $J(x)$ never vanishes on the closure of U. Then the inverse function theorem implies that g^{-1} is a C^1 map from $g(U)$ to U (remember we are assuming g is one-to-one, which is not a consequence of the inverse function theorem). We may thus apply the inequality of the last lemma to the function $f(g(x))J(x) = F(x)$ on $U = g^{-1}(g(U))$ and the map $g^{-1} : g(U) \to U$. We have $\int_U F(x)\,dx \le \int_{g(U)} F(g^{-1}(x))J_1(x)\,dx$ where $J_1(x) = |\det dg^{-1}(x)|$ is the Jacobian factor for g^{-1}. But

$$F(g^{-1}(x))J_1(x) = f(x)J(g^{-1}(x))J_1(x) = f(x)$$

by the inverse function theorem (remember that the product of determinants is the determinant of the product, and $dg^{-1}(x)$ is the inverse matrix of $dg(g^{-1}(x))$. So we have the desired reverse inequality

$$\int_U f(g(x))J(x)\,dx \le \int_{g(U)} f(x)\,dx.$$

To deal with the general case we cut up U into a union of small cubes, except for a neighborhood of the boundary that does not contribute to the integral in the limit. For each cube C in the partition, either $J(x)$ does or does not vanish on the closure of C. If not, we are in the special case already completed, so

$$\int_C f(g(x))J(x)\,dx \le \int_{g(C)} f(x)\,dx.$$

When we sum over all such cubes we obtain

$$\sum_j \int_{C_j} f(g(x))J(x)\,dx \le \sum_j \int_{g(C_j)} f(x)\,dx \le \int_{g(U)} f(x)\,dx,$$

where the sum extends over the cubes for which $J(x)$ does not vanish. But if we form the lower Riemann sum to approximate the integral $\int_U f(g(x))J(x)\,dx$, the cubes where $J(x)$ vanishes contribute zero while the others contribute at most $\int_{C_j} f(g(x))J(x)\,dx$. Thus the lower Riemann sum is bounded above by $\int_{g(U)} f(x)\,dx$, hence so is the integral. This completes the proof of the reverse estimate in the general case. QED

15.2.3 Polar Coordinates

As an example we consider polar coordinates. The situation in $\mathbb{R}^2$ is familiar. We have

$$\begin{aligned} r &= (x^2+y^2)^{1/2}, \\ \theta &= \arctan \frac{y}{x}, \end{aligned}$$

relating the polar coordinates r and θ to the Cartesian x and y coordinates, and

$$\begin{aligned} x &= r\cos\theta, \\ y &= r\sin\theta \end{aligned}$$

in the reverse direction. To get a one-to-one correspondence we must omit the origin in $\mathbb{R}^2$ (a set of content zero that doesn't contribute to the integral) and suitably restrict θ, say $-\pi < \theta \le \pi$. Thus we let $D' = \{(r,\theta) : r > 0 \text{ and } -\pi < \theta \le \pi\}$ and $D = \{(x,y) : (x,y) \ne (0,0)\}$, with $g : D' \to D$ defined by $g(r,\theta) = (r\cos\theta, r\sin\theta)$. Note that g is one-to-one and onto. (The domain D' is not open because it contains the boundary strip $\theta = \pi$; but since this is a set of content zero, we can ignore it. Technically, to apply the theorems as stated, we should take D' to be given by $-\pi < \theta < \pi$ and D to be $\mathbb{R}^2$ with the negative real axis omitted.) We compute easily

$$dg = \begin{pmatrix} \cos\theta & \sin\theta \\ -r\sin\theta & r\cos\theta \end{pmatrix},$$

so $\det dg = r$ and, hence,

$$\int_{\mathbb{R}^2} f(x,y)\,dx\,dy = \int_{-\pi}^{\pi}\int_0^{\infty} f(r\cos\theta, r\sin\theta) r\,dr\,d\theta.$$

Next we consider the situation in $\mathbb{R}^n$. We have polar-spherical coordinates $r, \theta_1, \theta_2, \ldots, \theta_{n-1}$ related to Cartesian coordinates $x_1, \ldots, x_n$ by

$$\begin{aligned}
x_1 &= r\cos\theta_1 \\
x_2 &= r\sin\theta_1\cos\theta_2 \\
x_3 &= r\sin\theta_1\sin\theta_2\cos\theta_3 \\
&\vdots \\
x_{n-1} &= r\sin\theta_1\sin\theta_2\cdots\sin\theta_{n-2}\cos\theta_{n-1} \\
x_n &= r\sin\theta_1\sin\theta_2\cdots\sin\theta_{n-2}\sin\theta_{n-1},
\end{aligned}$$

which we write $x = g(r, \theta)$ and in the reverse direction

$$\begin{aligned}
r &= \sqrt{x_1^2 + \ldots + x_n^2} \\
\theta_1 &= \arccos\frac{x_1}{r} \\
\theta_2 &= \arccos\frac{x_2}{r\sin\theta_1} \\
&\vdots \\
\theta_{n-2} &= \arccos\frac{x_{n-1}}{r\sin\theta_1\cdots\sin\theta_{n-3}} \\
\theta_{n-1} &= \arctan\frac{x_n}{x_{n-1}}.
\end{aligned}$$

If we take $D'\{(r, \theta) : 0 < r < \infty, 0 < \theta_1 < \pi, \ldots, 0 < \theta_{n-2} < \pi, -\pi < \theta_{n-1} < \pi\}$, then g maps D' one-to-one onto $D = \{x^n : x_1 \neq 0, x_2 \neq 0, \ldots, x_n \neq 0\}$, which differs from $\mathbb{R}^n$ by a set of content zero. Next we compute $dg =$

$$\begin{pmatrix}
\cos\theta_1 & -r\sin\theta_1 & \cdots & 0 \\
\sin\theta_1\cos\theta_2 & r\cos\theta_1\cos\theta_2 & \cdots & 0 \\
\vdots & \vdots & & \vdots \\
\sin\theta_1\sin\theta_2\cdots\sin\theta_{n-2}\cos\theta_{n-1} & r\cos\theta_1\sin\theta_2\cdots\sin\theta_{n-2}\cos\theta_{n-1} & \cdots & -r\sin\theta_1\cdots\sin\theta_{n-1} \\
\sin\theta_1\sin\theta_2\cdots\sin\theta_{n-2}\sin\theta_{n-1} & r\cos\theta_1\sin\theta_2\cdots\sin\theta_{n-2}\sin\theta_{n-1} & \cdots & r\sin\theta_1\cdots\sin\theta_{n-2}\cos\theta_{n-1}
\end{pmatrix}$$

where the first column is $(\partial x_j/\partial r)$, the second column is $(\partial x_j/\partial\theta_1), \ldots$, and the last column is $(\partial x_j/\partial\theta_{n-1})$.

Lemma 15.2.4 $\det dg = r^{n-1}(\sin\theta_1)^{n-2}(\sin\theta_2)^{n-3}\cdots\sin\theta_{n-2}$.

Proof: We prove the result by induction, the case $n = 2$ having been previously computed. Suppose the result is true for $n-1$. Notice that the last column of dg (the θ_{n-1} derivative) has only two non-zero entries. We can eliminate one of these non-zero entries without changing the determinant by multiplying the last row by $\sin\theta_{n-1}/\cos\theta_{n-1}$ and adding it to the $(n-1)$th-row. This has the effect of reducing the upper left $(n-1)\times(n-1)$ submatrix of dg to that of the $n-1$ case except that the last row is multiplied by $\cos\theta_{n-1} + \sin^2\theta_{n-1}/\cos\theta_{n-1}$. Thus

$$\det dg = r\sin\theta_1\ldots\sin\theta_{n-2}\left[\cos\theta_{n-1}\left(\cos\theta_{n-1}+\frac{\sin^2\theta_{n-1}}{\cos\theta_{n-1}}\right)\right]\det dg_{n-1}$$

where g_{n-1} denotes the $(n-1)$-dimensional case. Note that the expression in brackets equals 1. Then by the induction hypothesis

$$\begin{aligned}\det dg &= r\sin\theta_1\cdots\sin\theta_{n-2}(r^{n-2}(\sin\theta_1)^{n-3}\cdots\sin\theta_{n-3})\\ &= r^{n-1}(\sin\theta_1)^{n-2}\cdots\sin\theta_{n-2}.\end{aligned}$$

QED

Thus we have the integration formula

$$\begin{aligned}&\int_{\mathbb{R}^n} f(x_1,\ldots,x_n)dx_1\cdots dx_n\\ &\quad=\int_{-\pi}^{\pi}\int_0^{\pi}\cdots\int_0^{\pi}\int_0^{\infty} f(r\cos\theta_1, r\sin\theta_1\cos\theta_2,\ldots,r\sin\theta_1\cdots\sin\theta_{n-1})\\ &\quad r^{n-1}(\sin\theta_1)^{n-2}(\sin\theta_2)^{n-3}\cdots\sin\theta_{n-2}dr\,d\theta_1 d\theta_2\cdots d\theta_{n-1}.\end{aligned}$$

This is especially useful if f is a radial function (a function of r alone). Then we can evaluate the polar-spherical integral as an iterated integral, and the θ-integrations just produce a constant c_n depending on the dimension

$$\int_{\mathbb{R}^n} f(|x|)\,dx = c_n\int_0^{\infty} f(r)r^{n-1}\,dr.$$

15.2.4 Change of Variable for Lebesgue Integrals*

In the context of the Lebesgue integral we can obtain a stronger change of variable theorem, allowing a general Lebesgue integrable function f and also weakening the assumption on the domain. Suppose U is an arbitrary open set in $\mathbb{R}^n$ and $g : U \to \mathbb{R}^n$ is a one-to-one C^1 mapping. Notice that we are *not* assuming anything about the behavior of g on the boundary of U. In particular, the Jacobian factor $J(x) = |\det dg(x)|$ does not have to be bounded. By the way, the boundary of a general open set can be quite bizarre; it can even have positive Legesgue measure, so the integral of a function over the closure of U might be different from the integral over U. In most applications of the change of variable formula the set U is rather tame, but frequently we need to deal with unbounded sets and mappings g that do not extend continuously to the boundary. Note that our hypotheses do imply that $g(U)$ is an open set (but this is a deep theorem, Brouwer's Invariance of Domain Theorem, which we will not prove here), but the Jacobian factor may have zeros in U (as is the case for $g(x) = x^3$ on $\mathbb{R}$).

To explain the change of variable formula from the Lebesgue point of view we need a simple construction that gives the image of a measure under a mapping. Suppose we have a measurable function $g : X \to Y$ for two measure spaces (so there are σ-fields $\mathcal{F}_1$ on X and $\mathcal{F}_2$ on Y such that $g^{-1}(A)$ is in $\mathcal{F}_1$ for every A in $\mathcal{F}_2$). In our case X will be U and Y will be $g(U)$, with the σ-fields of Borel sets. If ν is a measure on X, then define the *image measure* $\nu \circ g^{-1}$ on Y by

$$\nu \circ g^{-1}(A) = \nu(g^{-1}(A)).$$

This definition makes sense even if g is not one-to-one, and it is easy to show that $\nu \circ g^{-1}$ is a measure (see exercise set 15.2.5), because taking inverse images commutes with set-theoretic operations. Intuitively, if you think of ν as a mass distribution on X, then $\nu \circ g^{-1}$ moves the mass via g onto Y (if some regions of Y are hit more than once, then the mass transported by g is simply summed).

A basic property of the image measure is the integral identity $\int_Y f d(\nu \circ g^{-1}) = \int_X f \circ g \, d\nu$ for any non-negative measurable function f on Y (and more generally for any integrable f with respect to the measure $\nu \circ g^{-1}$). Indeed, if f is the characteristic function of a measurable set A in $\mathcal{F}_2, f = \chi_A$, then $f \circ g = \chi_{g^{-1}(A)}$ because x is in

$g^{-1}(A)$ if and only if $g(x)$ is in A. Thus the integral identity reduces to the definition for characteristic functions. It then follows easily that it holds for simple functions and then for non-negative measurable functions by the definition of the integral. The integral identity is in some ways more natural than the definition since it involves g rather than g^{-1}.

What does this have to do with the change of variable formula? If we denote by μ Lebesgue measure on $\mathbb{R}^n$ and $\mu|_U$ its restriction to U and let $\nu = J(x)d\mu|_U(x)$, then one side of the change of variable formula is $\int_U f \circ g\, d\nu$, which is just $\int_{g(U)} f\, d\nu \circ g^{-1}$. Thus the change of variable formula says that $\nu \circ g^{-1}$ is equal to Lebesgue measure restricted to $g(U)$.

Lemma 15.2.5 *If U is open in $\mathbb{R}^n$ and $g : U \to \mathbb{R}^n$ is one-to-one and C^1, then $\nu \circ g^{-1}$ is equal to Lebesgue measure on $g(U)$, where $\nu = J(x)d\mu|_U(x)$.*

Proof: Let R denote any closed rectangle contained in U. Then we can apply the change of variable theorem to the mapping $g : R \to g(R)$ to get

$$\int_{g(R)} f\, d\mu = \int_R f \circ g\, d\nu$$

for any continuous function f on $g(R)$. In particular, $\mu(g(R)) = \nu \circ g^{-1}(g(R))$. Similarly, if A is any subset of R whose boundary has content zero, then $\mu(g(A)) = \nu \circ g^{-1}(g(A))$.

Consider first the special case of a rectangle R such that $J(x) \neq 0$ on R. Then the inverse function theorem implies that g^{-1} is a C^1 mapping from $g(R)$ to R, so any rectangle B contained in $g(R)$ is of the form $B = g(A)$ where $A = g^{-1}(B)$ has boundary with content zero. Thus μ and $\nu \circ g^{-1}$ agree on rectangles contained in $g(R)$, hence by the Hahn uniqueness theorem they are equal as measures on $g(R)$.

What do we do with the set where $J(x) = 0$? This is called the *critical set* for the mapping g. Notice that it is a closed set, since $J(x)$ is continuous; and if we restrict attention to a bounded subset of U, say $U_r = U \cap \{|x| \leq r\}$, then it is compact. Given any ε, we can cover the set of points in U_r where $J(x) = 0$ by rectangles on which $J(x) \leq \varepsilon$ (because $J(x)$ is continuous), and by compactness can reduce to a finite subcover, say $R_1, \ldots R_N$. Also, by further decomposition if necessary, we can make the interiors of these rectangles disjoint. We can still apply the change of variable theorem on each rectangle R_k, so we have

$$\mu(g(R_k)) = \nu \circ g^{-1}(g(R_k)) = \int_{R_k} J(x) d\mu(x) \leq \varepsilon\mu(R_k).$$

This means

$$\begin{aligned} \mu\left(\bigcup_k g(R_k)\right) &= \nu \circ g^{-1}\left(\bigcup_k g(R_k)\right) \leq \varepsilon \sum_k \mu(R_k) = \varepsilon\mu\left(\bigcup_k R_k\right) \\ &\leq \varepsilon\mu(U_r) \leq cr^n\varepsilon \end{aligned}$$

since U_r is contained in the ball $\{|x| \leq r\}$.

Since $J(x) \neq 0$ on $U_r \setminus \bigcup_k R_k$ and we can write this set as a countable union of rectangles, we know that $\mu = \nu \circ g^{-1}$ on $g(U_r \setminus \bigcup_k R_k) = g(U_r) \setminus \bigcup_k g(R_k)$. But $\bigcup_k g(R_k)$ has measure at most $cr^n\varepsilon$ for both measures, so

$$|\mu(A) - \nu \circ g^{-1}(A)| \leq cr^n\varepsilon$$

for any measurable subset of $g(U_r)$. First let $\varepsilon \to 0$ to get $\mu = \nu \circ g^{-1}$ on $g(U_r)$, and then let $r \to \infty$ to get $\mu = \nu \circ g^{-1}$ on $g(U)$. QED

Theorem 15.2.3 *Let $U \subseteq \mathbb{R}^n$ be open and $g : U \to \mathbb{R}^n$ be one-to-one and C^1. Then for any non-negative measurable function f on $g(U)$,*

$$\int_{g(U)} f d\mu = \int_U f \circ g J d\mu.$$

More generally, if f is real- or complex-valued and measurable on $g(U)$, then f is integrable if and only if $f \circ gJ$ is integrable on U, in which case the change of variable formula holds.

Proof: This is an immediate consequence of the lemma and the integral identity

$$\int_{g(U)} f d(\nu \circ g^{-1}) = \int_U f \circ g d\nu.$$

QED

One consequence of the change of variable formula is the fact that the image of the critical set under g has Lebesgue measure zero,

$\mu(g(C)) = 0$ where $C = \{x : J(x) = 0\}$. Indeed, $\mu(g(C)) = \nu(C) = \int_C J d\mu = 0$ since J is zero on C and part of the proof of the lemma involved convering the set $g(C)$ by sets of small Lebesgue measure (after localizing to a bounded region). An examination of the proof shows that we really do not need the assumption that g is one-to-one, because all we used was the upper bound $\mu(g(R_k)) \leq \int_{R_k} J(x)d\mu(x)$ and the proof of this in the previous section did not use the one-to-one hypothesis. This result is of importance in the theory of differential topology, where it goes under the name of *Sard's Theorem.*

Theorem 15.2.4 (*Sard*) *If $g : U \to \mathbb{R}^n$ is C^1 and $C = \{x : \det g(x) = 0\}$, then $g(C)$ has Lebesgue measure zero.*

Returning to the example of the integration formula for polar coordinates worked out in section 15.2.3, we have the validity of this result for Lebesgue integrals. In particular, if $f(r)$ is a measurable function on $(0, \infty)$ and we consider the corresponding radial function $f(|x|)$ on $\mathbb{R}^n$, which is also measurable, we see that $f(|x|)$ is integrable on $\mathbb{R}^n$ if and only if $r^{n-1}f(r)$ is integrable on $(0, \infty)$. Recall that r^α is integrable near $r = 0$ if and only if $\alpha > -1$ and that r^α is integrable near $r = \infty$ if and only if $\alpha < -1$. That translates to the condition that $|x|^\alpha$ on $\mathbb{R}^n$ is integrable near the origin if and only if $\alpha > -n$ and integrable near infinity if and only if $\alpha < -n$. In particular, no power $|x|^\alpha$ is globally integrable. By cutting and pasting, we obtain the following useful criterion for integrability on $\mathbb{R}^n$:

Theorem 15.2.5 *Suppose f is continuous on $\mathbb{R}^n$ except for a finite set of isolated singularities $a_1, \ldots, a_n$. Suppose we have $|f(x)| \leq c_k|x - a_k|^{\alpha_k}$ for x near a_k and $|f(x)| \leq c|x|^\beta$ for all large x, where $\alpha_k > -n$ for all k and $\beta < -n$. Then f is integrable.*

15.2.5 Exercises

1. Evaluate $\int_{-\infty}^{\infty} e^{-x^2}\,dx$ by considering $\int_{\mathbb{R}^2} e^{-x^2-y^2}\,dx\,dy$ as an iterated integral and in polar coordinates.

2. Prove that a one-to-one C^1 mapping $g : U \to g(U)$ is measure-preserving ($\mu(g(A)) = \mu(A)$ for every measurable subset $A \subseteq U$) if and only if $\det dg(x) = \pm 1$ for every x in U.

3. Classify all continuous measure-preserving transformations $g : \mathbb{R} \to \mathbb{R}$. Give an example of a discontinuous measure-preserving transformation $g : \mathbb{R} \to \mathbb{R}$.

4. Prove that the set of non-invertible $n \times n$ matrices has measure zero in $\mathbb{R}^{n^2}$.

5. Let $\mathrm{GL}(n, \mathbb{R})$ denote the set of $n \times n$ invertible real matrices. Prove that for any y in $\mathrm{GL}(n, \mathbb{R})$,

$$\int_{GL(n,\mathbb{R})} f(xy)|\det x|^{-n}\, dx = \int_{GL(n,\mathbb{R})} f(x)|\det x|^{-n}\, dx$$

for any non-negative measurable function $f : \mathrm{GL}(n, \mathbb{R}) \to \mathbb{R}$, where xy denotes matrix multiplication. Is the same true if we replace xy by yx?

6. Define a multiplication on $\mathbb{R}^{2n+1}$ as $(x, y, t) \circ (x', y', t') = x+x', y+y', t+t'+x\cdot y' - x'\cdot y)$ for x in $\mathbb{R}^n$, y in $\mathbb{R}^n$, t in $\mathbb{R}^1$. Prove that $\int_{\mathbb{R}^{2n+1}} f((x, y, t)\circ(x', y', t'))\, dx\, dy\, dt = \int_{\mathbb{R}^{2n+1}} f(x, y, t) dx\, dy\, dt$ for any (x', y', t').

7. Prove $\int_{\mathbb{R}^n} f(tx)\, dx = t^{-n} \int_{\mathbb{R}} f(x)\, dx$ for any $t > 0$ where $tx = (tx_1, tx_2, \ldots, tx_n)$. Show also $\int_{\mathbb{R}^n} f(tx)dx/|x|^n = \int_{\mathbb{R}^n} f(x)dx/|x|^n$.

8. Prove $\int_{\mathbb{R}^n} f(x/|x|^2)dx/|x|^n = \int_{\mathbb{R}^n} f(x)dx/|x|^n$.

9. Complete the proof of the estimate $\mathrm{vol}(A_\varepsilon) \leq \mathrm{vol}(A)+c\varepsilon(r+\varepsilon)^{n-1}$ for $\varepsilon < 1$ if A is a parallelopiped in $\mathbb{R}^n$ contained in a ball of radius r, where c is a constant depending only on n.

10. Suppose U is a bounded open set in $\mathbb{R}^n$ whose boundary has content zero. If $g : U \to \mathbb{R}^n$ is C^1 on the closure of U, show that the boundary of $g(U)$ has content zero.

11. a. Suppose U is a bounded open set in $\mathbb{R}^n$ whose boundary has contant zero. For each fixed δ, decompose $\mathbb{R}^n$ by the standard tiling with cubes of side length δ; and let $\{C_k\}$ be the collection of cubes in the tiling that lie entirely in U. Show that $\sum_k \mathrm{vol}(C_k)$ converges to $\mathrm{vol}(U)$ as $\delta \to 0$.

b. Suppose that $g : U \to \mathbb{R}^n$ is one-to-one and C^1 on the closure of U. Show that $\sum_k \mathrm{vol}(g(C_k))$ converges to $\mathrm{vol}(g(U))$ as $\delta \to 0$.

12. Let U be a bounded open set in $\mathbb{R}^n$ whose boundary has content zero. Consider finite partitions $U = \bigcup_j A_j$ where the sets A_j have disjoint interiors and boundaries with contant zero. For f a continuous function on the closure of U, form the generalized Cauchy sums $\sum_j f(\tilde{x}_j)\mathrm{vol}(A_j)$ where $\tilde{x}_j$ is an arbitrary point in A_j. Prove that these sums converge to $\int_U f(x)\,dx$ as the maximum diameter of the sets A_j tends to zero.

15.3 Summary

Definition *If f is a continuous function defined on a rectangle $R = [a,b] \times [c,d]$ in $\mathbb{R}^2$, the Riemann double integral $\int\int_R f(x,y)\,dx\,dy$ is defined to be the limit of Cauchy sums*

$$S(f,P) = \sum_{j=1}^{N}\sum_{k=1}^{M} f(\tilde{x}_j, \tilde{y}_k)(x_j - x_{j-1})(y_k - y_{k-1})$$

where P denotes a partion $a = x_0 < x_1 < \cdots < x_N = b, c = y_0 < y_1 < \cdots < y_M = d$, of the sides of the rectangle, $x_{j-1} \le \tilde{x}_j \le x_j, y_{j-1} \le \tilde{y}_j \le y_j$, and the limit is taken as the maximum lengths of $x_j - x_{j-1}$ and $y_k - y_{k-1}$ both tend to zero.

Theorem *For a continuous function on a rectangle $R = [a,b] \times [c,d]$, the double integral is equal to each of the iterated integrals $\int_c^d(\int_a^b f(x,y)\,dx)\,dy$ and $\int_a^b(\int_c^b f(x,y)\,dy)\,dx$.*

Definition *A subset of the plane is said to have content zero if for every $\varepsilon > 0$ there exists a finite covering by rectangles whose areas sum to at most ε.*

Corollary *The double and iterated Riemann integrals are equal for a continuous function defined on a compact subset of the plane whose boundary has content zero.*

Lemma *If* μ *denotes Lebesgue measure on* $\mathbb{R}^2$, *then*

$$\mu(A) = \int \left(\int \chi_A(x,y)\, dx \right) dy = \int \left(\int \chi_A(x,y)\, dy \right) dx$$

for any measurable set A, *where the iterated Lebesgue integrals are well defined.*

Corollary *The Lebesgue integral of a non-negative function* f *on* $\mathbb{R}$ *is equal to the Lebesgue measure of the region under the graph.*

Theorem 15.1.1 (*Fubini's Theorem, first version*) *Let* f *be a non-negative measurable function on* $\mathbb{R}^2$. *Then the double and iterated Lebesgue integrals are equal.*

Theorem 15.1.2 (*Fubini's Theorem, second version*) *Let* f *be integrable on* $\mathbb{R}^2$. *Then the double and iterated Lebesgue integrals are equal; and for almost every* $y, f(x,y)$ *as a function of* x *is integrable and* $\int f(x,y)\, dx$ *is integrable as a function of* y, *and the same is true with* x *and* y *reversed.*

Example *The function*

$$f(x,y) = \begin{cases} 1 & \textit{if}\, 0 < x < 1, \\ -1 & \textit{if}\, -1 < x < 0, \\ 0 & \textit{otherwise} \end{cases}$$

is not integrable, but the iterated integral $\int(\int f(x,y)\, dx)\, dy = 0$.

Theorem 15.1.3 (*Fubini's Theorem, third version*) *Let* f *be measurable on* $\mathbb{R}^2$. *If one of the iterated integrals of* $|f|$ *is finite, then* f *is integrable and the double and iterated Lebesgue integrals (in both orders) of* f *are equal.*

Theorem 15.1.4 (L^1 *Convolution Theorem*) *The convolution* $f*g(x) = \int f(x-y)g(y)\, dy$ *is well defined for* f *and* g *in* L^1 *in the sense that for almost every* x, *the function* $h(y) = f(x-y)g(y)$ *is integrable. Furthermore,* $f*g$ *is in* L^1 *and* $||f*g||_1 \leq ||f||_1||g||_1$.

Definition *A monotone class $\mathcal{M}$ is a collection of subsets of X that is closed under monotone increasing and decreasing sequences.*

Lemma 15.1.1 (*Monotone Class*)*If a monotone class contains a field, it contains the σ-field generated by the field.*

Lemma 15.1.2 *If A is a Borel subset of $\mathbb{R}^2$, then every section A_y is a Borel set of $\mathbb{R}$ and the Lebesgue measure of A_y is a measurable function of y.*

Definition *A measure μ on X is σ-finite if $X = \bigcup_{j=1}^{\infty} X_j$ and $\mu(X_j)$ is finite for all j.*

Theorem 15.1.5 (*Hahn Uniqueness Theorem*)*If two σ-finite measures are equal on a field, they are equal on the σ-field generated by the field.*

15.2 Change of Variable in Multiple Integrals

Definition *Let $A_1, \ldots, A_n$ denote n vectors in $\mathbb{R}^n$. The parallelopiped generated is defined to be the set of vectors of the form $\sum_{j=1}^{n} b_j A_j$ with $0 \leq b_j \leq 1$.*

Theorem 15.2.1 *If $g(x) = Ax$ where A is an $n \times n$ matrix, then $\mathrm{vol}(g(R)) = |\det A| \mathrm{vol}(R)$ for any rectangle R.*

Definition *If $g : U \to \mathbb{R}^n$ is a C^1 function for U an open set in $\mathbb{R}^n$, then the Jacobian factor is $J(x) = |\det dg(x)|$.*

Definition *The ε-neighborhood A_ε of a set A is the set of points at distance at most ε from A.*

Lemma 15.2.1 *There exists c depending only on n such that $\mathrm{vol}(A_\varepsilon) \leq \mathrm{vol}(A) + c\varepsilon(r+\varepsilon)^{n-1}$ for $\varepsilon \leq 1$ for every parallelopiped A contained in a ball of radius r.*

Lemma 15.2.2 *Let C_δ denote a cube centered at $\tilde{x}$ of side length δ. Suppose g is C^1 on C_δ satisfying $||dg(x)|| = M$ and $||dy(x) - dg(\tilde{x})|| \leq \varepsilon$*

for all x in C_δ. Then $\mathrm{vol}(g(C_\delta)) \leq (J(\tilde{x}) + c\varepsilon)\delta^n$ *where c depends only on n and M.*

Lemma 15.2.3 *Let $g : U \to \mathbb{R}^n$ be one-to-one on a bounded open set U whose boundary has content zero, and suppose g is C^1 on the closure of U. Then* $\mathrm{vol}(g(U)) \leq \int_U J(x)\,dx$ *and*

$$\int_{g(U)} f(x)\,dx \leq \int_U f(g(x))J(x)\,dx$$

for any non-negative continuous function f on the closure of $g(U)$.

Theorem 15.2.2 (*Change of Variable Formula*) *Let $g : U \to \mathbb{R}^n$ be a one-to-one map of a bounded open set U whose boundary has content zero, and assume g is C^1 on the closure of U. Let f be continuous on the closure of $g(U)$. Then*

$$\int_{g(U)} f(x)\,dx = \int_U f(g(x))J(x)\,dx.$$

Example (*Polar Coordinates*)

$$\int_{\mathbb{R}^n} f(x,y)\,dx\,dy = \int_{-\pi}^{\pi} \int_0^{\infty} f(r\cos\theta, r\sin\theta) r\,dr\,d\theta.$$

Definition *Polar-spherical coordinates in $\mathbb{R}^n$ for $n \geq 3$ are $r, \theta_1, \theta_2, \ldots, \theta_{n-1}$ given by*

$$\begin{aligned}
x_1 &= r\cos\theta_1 \\
x_2 &= r\sin\theta_1\cos\theta_2 \\
x_3 &= r\sin\theta_1\sin\theta_2\cos\theta_3 \\
&\vdots \\
x_{n-1} &= r\sin\theta_1\sin\theta_2\cdots\sin\theta_{n-2}\cos\theta_{n-1} \\
x_n &= r\sin\theta_1\sin\theta_2\cdots\sin\theta_{n-2}\sin\theta_{n-1}.
\end{aligned}$$

Lemma 15.2.4 *If g denotes the mapping $(r, \theta) \to x$ given by polar-spherical coordinates, then g maps*

$$D' = \{(r, \theta) : 0 < r < \infty, 0 < \theta_j < \pi \text{ for } j \le n - 2 - \pi < \theta_{n-1} < \pi\}$$

one-to-one onto $D = \{x : x_j \ne 0 \text{ all } j\}$ and

$$\det dg = r^{n-1}(\sin \theta_1)^{n-2}(\sin \theta_2)^{n-3} \cdots \sin \theta_{n-2}.$$

Corollary *If g is radial (i.e., a function of r alone), then $\int_{\mathbb{R}^n} f(|x|)\, dx = c_n \int_0^\infty f(r) r^{n-1}\, dr$ where c_n is a constant depending on the dimension n alone.*

Definition *If $g : X \to Y$ is measurable function (with repect to σ-fields $\mathcal{F}_1$ on X and $\mathcal{F}_2$ on Y) and ν is a measure on X, then the image measure $\nu \circ g^{-1}$ on Y is given by $\nu \circ g^{-1}(A) = \nu(g^{-1}(A))$ for any A in $\mathcal{F}_2$.*

Lemma *If f is a non-negative measurable function on Y, then $\int_Y f d(\nu \circ g^{-1}) = \int_X f \circ g d\nu$.*

Lemma 15.2.5 *Let μ denote Lebesgue measure on $\mathbb{R}^n$ and $\mu|_U$ denote its restriction to U. Suppose U is open and $g : U \to \mathbb{R}^n$ is one-to-one and C^1. If $\nu = J(x)\, d\mu|_U$, then $\nu \circ g^{-1} = \mu|_{g(U)}$.*

Theorem 15.2.3 *Let U be open and $g : U \to \mathbb{R}^n$ be one-to-one and C^1. Then $\int_{g(U)} f\, d\mu = \int_U f \circ gJ\, d\mu$ for any nonnegative measurable function f on $g(U)$. Moreover, f is integrable if and only if $f \circ gJ$ is integrable on U, and the same identity holds.*

Theorem 15.2.4 *(Sard) If $g : U \to \mathbb{R}^n$ is C^1 and $C = \{x : \det dg(x) = 0\}$ is the critical set of g, then $\mu(g(C)) = 0$.*

Theorem 15.2.5 *Let f be continuous on $\mathbb{R}^n$ except for a finite set of singularities $a_1, \ldots, a_N$, and suppose $|f(x)| \le c_k |x - a_k|^{\alpha_k}$ for x near a_k and $|f(x)| \le c|x|^\beta$ for all large x, where $\alpha_k > -n$ for all k and $\beta < -n$. Then f is integrable.*

Index